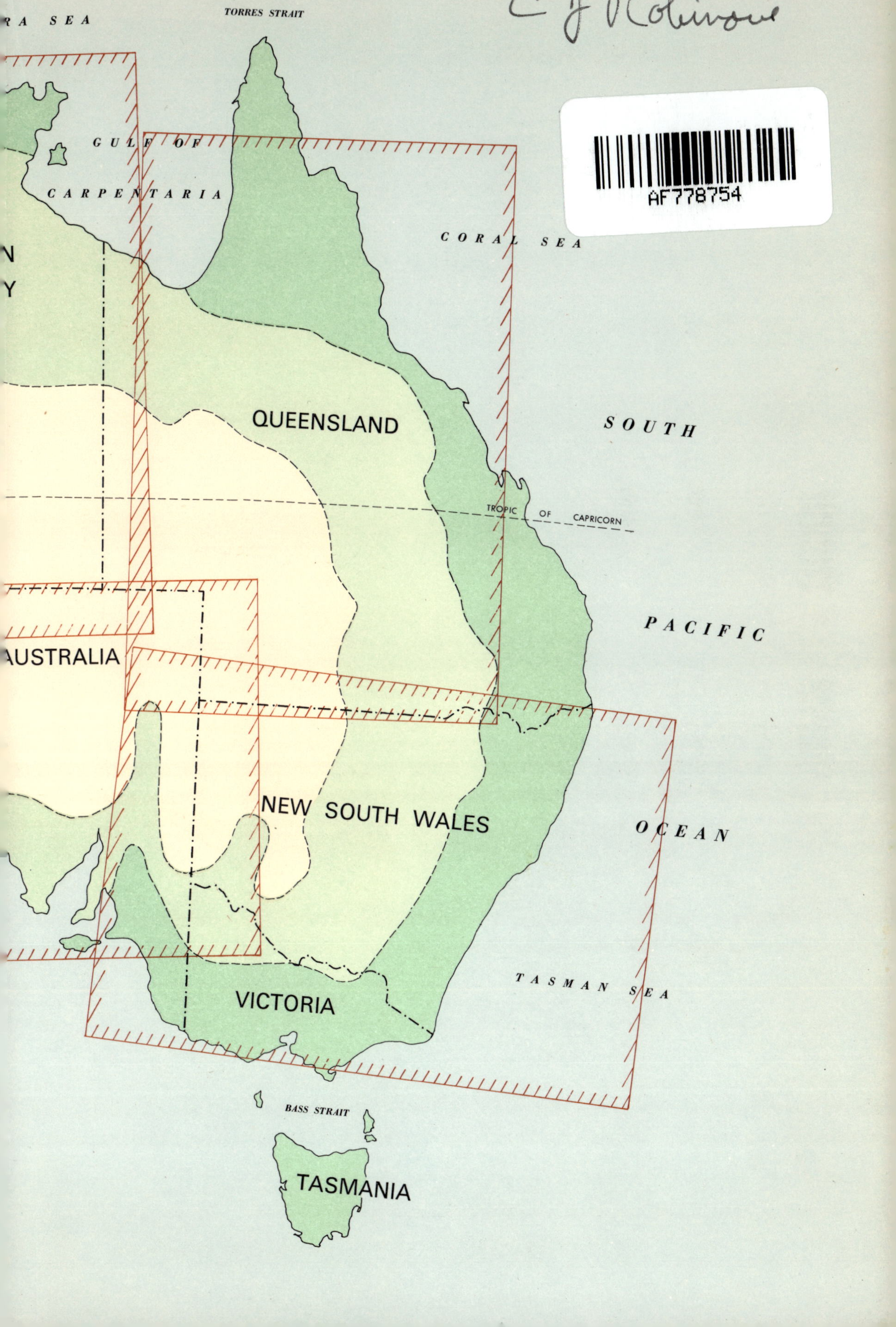
RA SEA
TORRES STRAIT
GULF OF
CARPENTARIA
CORAL SEA
N
Y
QUEENSLAND
SOUTH
TROPIC OF CAPRICORN
PACIFIC
AUSTRALIA
NEW SOUTH WALES
OCEAN
TASMAN SEA
VICTORIA
BASS STRAIT
TASMANIA

THE USE OF TREES AND SHRUBS IN THE DRY COUNTRY OF AUSTRALIA

A scene near Coolgardie, about 330 miles east of Perth, in the southern dry country of Western Australia. Climatically this is in the semi-arid zone but near the border with the arid zone; the average annual rainfall is about 11 inches. The small trees are mallee forms of *Eucalyptus*, whilst many of the low shrubs belong to the *Chenopodiaceae* (bluebushes, saltbushes and others). The photograph was taken at the end of a dry summer and there was very little herbaceous growth. The vegetation shown in the scene has only a small value for grazing, whilst it takes a long time to recover from fire or other forms of ill use.

DEPARTMENT OF NATIONAL DEVELOPMENT
FORESTRY AND TIMBER BUREAU

THE USE OF TREES AND SHRUBS IN THE DRY COUNTRY OF AUSTRALIA

AUSTRALIAN GOVERNMENT PUBLISHING SERVICE
CANBERRA 1972

Published for the Minister for National Development
by the Australian Government Publishing Service

Registered in Australia
for transmission by post as a book

ISBN 0 642 00021 2

O.D.C. 908.1:26

PRINTED BY JOHN SANDS PTY. LTD. HALSTEAD PRESS DIVISION

The Authors

The authors of this volume are

NORMAN HALL
Forest Officer
Forestry and Timber Bureau

and the following specialists

ROBERT BODEN
Executive Officer
Tourism and National Parks Section
Northern Territory Division
Department of the Interior

CLIFFORD STUART CHRISTIAN, C.M.G.
Member of the Executive of the
Commonwealth Scientific and Industrial Research Organisation

RICHARD CONDON
Assistant Commissioner
Western Lands Commission of New South Wales

FRANK DALE
Experimental Officer IV
Division of Forest Products
Commonwealth Scientific and Industrial Research Organisation

ALAN HART
Silviculturist (Nurseries)
Forests Department of Western Australia

JOHN LEIGH
Senior Research Scientist
Division of Plant Industry
Commonwealth Scientific and Industrial Research Organisation

JOHN MARSHALL
Senior Research Scientist
Division of Plant Industry
Commonwealth Scientific and Industrial Research Organisation

ALAN McARTHUR
Director
Forest Research Institute
Forestry and Timber Bureau

VANCE RUSSELL
Officer-in-Charge
Research Section
Parks and Gardens Branch
Department of the Interior

JOHN TURNBULL
Officer-in-Charge
Seed Section, Forest Research Institute
Forestry and Timber Bureau

Preface

THIS book provides, in one volume, information on many aspects of the use of trees and shrubs in the dry country, including an important section (Chapters 4-9) which deals with conservation, the principles of soil erosion, and its prevention and control in dryland Australia. Other chapters provide an outline of the environment in the arid and semi-arid zones, browse shrubs, factors to be considered in the selection of species to be used, as well as the raising of nursery stock and their planting and tending. Throughout the book emphasis is given to the amenity and conservation aspects of trees in the dry country, both on farms and in the new and old towns of the region. One chapter describes and illustrates what has been and is being done in several of the mining centres, where a large part of the population of the arid zone now lives.

The scope of this volume is so wide that it would not have been possible for one man to have written with authority on all aspects. About half of the text has been written by Norman Hall, a member of the staff, but I am deeply grateful to the distinguished authors from other organisations, both Commonwealth and State, who have contributed chapters on subjects about which they are acknowledged specialists.

This book has been written as a companion volume to *Growing Trees on Australian Farms* (1968), which was mainly concerned with the sub-humid and humid zones of coastal and tableland regions, and some of the chapters in the latter book are equally applicable in the dry country.

It is hoped that *The Use of Trees and Shrubs in the Dry Country* will prove valuable not only to landowners, but also to administrators and supervisors of towns and mining company operations concerned with amenity planting and conservation in the arid and semi-arid zones of the continent.

D. A. N. CROMER

Director-General
Forestry and Timber Bureau

Biographical notes on authors

HALL, Norman

Born 1906. Retired in 1967 from the position of Senior Lecturer (Engineering and Utilisation) at the Department of Forestry, Australian National University, and rejoined the Forestry and Timber Bureau, Department of National Development. Since then he has been mainly engaged in assisting with timber standards of the Standards Association of Australia and in the revision and enlargement of the third edition of *Forest Trees of Australia*, Forest Tree Series of leaflets and in the writing of, and photography for, twelve chapters of this book. He was one of the principal authors of *Growing Trees on Australian Farms* (1968). He commenced his career in private forestry in New Zealand but in 1937 transferred to the Woods and Forests Department of South Australia. During World War II he served in the 2/1 Aust. Forestry Company, Royal Australian Engineers, and later R.A.E., New Guinea Forests, 1940-45, and as a technical officer on the staff of the British Element of the Occupation in Austria, 1945-47. He then returned to South Australia but later joined the Forestry and Timber Bureau, Canberra, and became a senior lecturer at the Australian Forestry School in 1958. Published articles on botanical and forestry subjects have appeared in *Australian Journal of Botany*, *Unasylva*, *Australian Forestry* and other journals, and he has written or collaborated in writing leaflets, bulletins and books of the Forestry and Timber Bureau.

BODEN, R. W.

Born 1935. Formerly Research Arboriculturist, Parks and Gardens Branch, Department of the Interior, Canberra. Awarded Bachelor of Science (Forestry), Sydney, and Diploma of Forestry (Canberra) 1956. Worked in the Parks and Gardens Branch on tree selection and breeding with emphasis on *Eucalyptus* until 1960, when awarded a General Motors Holden Post-graduate Scholarship at the Australian National University. Submitted a thesis on 'Adaptation in the waterlogged environment in Eucalyptus' for Master of Science, Sydney (1963). Later that year he spent four months as Colombo Plan Adviser on eucalypt afforestation in Madras, South India and West Pakistan. He returned to Parks and Gardens until 1968 when he was granted three years leave to study rural land use planning, with emphasis on recreation and conservation management, at the Forestry Department of the Australian National University, which resulted in a Ph.D. (1972). At present Dr. Boden is Executive Officer, Tourism and National Parks Section, Northern Territory Division, Department of the Interior. He has written for *Australian Forestry, Australian Parks* and horticultural journals.

CHRISTIAN, C. S.

Born 1907. Mr Christian is a Member of the Executive of the Commonwealth Scientific and Industrial Research Organisation. He received his initial training at the University of Queensland in agricultural science and at the University of Minnesota and the Agricultural Research Station at Rothamsted in England. His early work with C.S.I.R.O., which he joined in 1930, was in the field of wheat genetics and wheat breeding. Subsequently he conducted a programme of pasture research in Queensland for many years. In 1946 he was appointed leader of the North Australian Regional Survey, a unit established by C.S.I.R.O., at the request of Commonwealth and State Governments to make scientific surveys of Northern Australia in order that the information necessary for the formulation of land development policies might be obtained. This unit has since become the Division of Land Research in C.S.I.R.O., and has been responsible for comprehensive natural resource surveys of about three-quarters of a million square miles of country on the Australian mainland and in the Territories of Papua and New Guinea. Mr Christian was responsible for the development of what is now widely referred to as the integrated approach to resource survey and assessment. This approach has been adopted by FAO and UNESCO and the Department of Technical Co-operation in the

United Kingdom. It has been practised in many countries and Mr Christian was recently appointed Vice-Chairman of the new international Advisory Committee for Natural Resource Research established by UNESCO. In 1958 Mr Christian was requested by UNESCO and the Government of India to visit that country and plan a national centre for arid zone research. He was responsible for planning what is now known as the Central Arid Zone Research Institute at Jodhpur in Rajasthan. Mr Christian was elected a Fellow of the Australian Institute of Agricultural Science in 1960, and he is also a recipient of the Institute's Medal. He was elected a Fellow of the World Academy of Art and Science in 1962 and in 1969 was awarded the Farrer Medal for Agriculture. In the New Year Honours he was appointed C.M.G.

CONDON, R. W.

Born 1924. Present position Assistant Commissioner, Western Lands Commission of New South Wales since October, 1968. After graduation in agricultural science from Sydney University in 1946 he joined the Soil Conservation Service of New South Wales as Research Officer and Botanist locally in charge of research and extension in Western New South Wales. In 1957 he was appointed Senior Soil Conservationist, Head Office, Sydney, as editor of the Soil Conservation Service journal and in charge of research stations. In 1962, appointed Special Soil Conservationist in charge of the Service's activities in catchment area investigations, coastal sand drift and arid zone research and extension in Western New South Wales. Mr Condon was member of a team of soil conservationists from the New South Wales Soil Conservation Service which carried out a survey of erosion problems in central Australia for the Northern Territory Administration from 1965 to 1967.

DALE, F. A.

Born 1920. Present position Experimental Officer, C.S.I.R.O., Division of Forest Products. After qualifying as a mechanical engineer he worked on aeronautical research until 1946. He joined C.S.I.R.O. again in 1949 to study railway sleeper breakdown; transferring to timber preservation soon after, he has specialised in methods and design of equipment for the preservative treatment of poles, posts and sawn timber. He developed the low-pressure process for treating hardwood posts and has been closely associated with commercial timber treatment since it started in 1957. In 1960 he visited Thailand to help establish a plant to treat transmission poles. Publications include papers and articles on structural timber design, pole and timber treatment and the applications of treated timber. Other interests include conservation, sculpture and bird watching.

HART, A. J.

Born 1928. Present position Silviculturist (Nurseries) with the Forests Department of Western Australia and is responsible for the plant propagation programmes of *Pinus* and *Eucalyptus* species for use in Departmental plantations, distribution to local authorities and the rural population. After initial undergraduate study at the University of Western Australia he joined the Forest Service of the Territory of Papua and New Guinea as a cadet in 1948, and completed the Diploma of Forestry course at the Australian Forestry School, Canberra, in 1951, and B.Sc. (For.) of the University of Western Australia in 1953. Following approximately five years of service in New Guinea he joined the Forests Department of Western Australia. His main interests are silvicultural research in karri (*Eucalyptus diversicolor*) and eucalypts in general, the native tree flora and tree establishment in the inland areas of the State. He has written for *Forest Notes* of his Forests Department and the *Newsletter* of the Institute of Foresters of Australia.

LEIGH, J. H.

Born 1936. Present position, Senior Research Scientist in the Ecology Section, C.S.I.R.O. Division of Plant Industry, Canberra. Graduated (Ph.D.) in 1960 from the University of the Witwatersrand, Johannesburg, South Africa, after working on certain aspects of the ecology, anatomy and physiology of the perennial grass, *Eragrostis curvula*. Dr Leigh later studied the significance of plants and grazing animals in the nitrogen cycle of sown pastures at the

Grasslands Research Institute at Hurley, England. In 1961 he joined the staff of the C.S.I.R.O. Riverina Laboratory, Deniliquin, N.S.W., where his main interests have been the ecology of a number of semi-arid woodland, shrub-steppe and grassland communities. Particular interest has been centred on diet selection by sheep and the effects of grazing on plant species. Between 1967-69 he was Officer-in-Charge of the Riverina Laboratory. He has published a number of papers in international journals, included amongst which are *Australian Journal of Agricultural Science*, *Australian Journal of Botany*, *Australian Journal of Experimental Agriculture and Animal Husbandry*, *Journal of Ecology*, *Empire Journal of Experimental Agriculture*, *Experimental Agriculture* (U.K.), *Nature*, *Journal of the British Grassland Society* and *South African Journal of Botany*. In 1965 he published, with W. E. Mulham, *Pastoral Plants of the Riverine Plain*, and in 1971 co-edited, with J. C. Noble, *Australian Plants for Sheep in Australia*.

MARSHALL, J. K.

Born 1938. Present position Senior Research Scientist with the C.S.I.R.O., Division of Plant Industry (now Rangelands Research Unit), Riverina Laboratory, Deniliquin, N.S.W., but 1972-73 with the Natural Resource Ecology Laboratory, Colorado State University, U.S.A. Graduated B.Sc. with first class honours in botany from St Andrew's University, Scotland in 1959. Proceeded to Cambridge University to carry out autecological research on the sand dune grass, *Corynephorus canescens*. Successfully submitted a doctoral thesis on that work in 1963. Since lectured in the Botany Department of Edinburgh University (1963-64) and carried out post-doctoral research on the effects of shelter on field crop and grassland productivity at the Department of Forestry and Natural Resources of the same University (1964-67). Joined the C.S.I.R.O., Division of Plant Industry dryland research programme at Deniliquin, New South Wales, in 1967 where he has been concerned with problems of environmental heterogeneity and their consequences, including soil erosion and pasture shelter. Dr Marshall has published papers in *Journal of Ecology*, *Oikos*, *Photosynthetica*, *Field Crop Abstracts*, *Agricultural Meteorology* and other journals, and in proceedings of conferences.

McARTHUR, A. G.

Born 1923. At present Director, Forest Research Institute, Canberra. Educated Yanco Agricultural High School, Sydney University and Australian Forestry School. Commenced career as a cadet forester with the Forestry Commission of N.S.W. in 1941. On graduation worked on *P. radiata* assessment and plantation establishment schemes in the Orange and Tumut Districts. In 1951 transferred to fire control duties as Senior Forester and became the first Fire Control Officer of the Hume-Snowy Fire District and was responsible for organising fire control services in the region. Transferred to the Forestry and Timber Bureau in 1953 as the first professional forester engaged full-time on fire research. Main avenues of research have been in the field of fire behaviour; has produced fire danger rating systems for eucalypt fuels and grassland; in later years mainly concerned with the development of advanced techniques for control burning in a variety of forest associations and the behaviour of mass fires and energy dynamics. In 1961 was Technical Adviser to the Royal Commission on Bushfires in Western Australia and a member of the Committee of Enquiry set up by the Tasmanian Government following the 1967 disaster fire. Overseas visits to India 1957, United States and Canada 1964, 1967 and 1971, New Zealand 1965 and 1969, and a member of the Trade Mission to Canada 1969, investigating the use of water bombing aircraft. Australian member of the Technical Co-operation Programme, Panel N_2, dealing with thermal nuclear effects; member of various I.U.F.R.O. Working Groups. Member of the Institute of Foresters of Australia, and foundation member of the Ecological Society of Australia; is the author of fifty-four papers on fire behaviour, fire ecology and watershed management which have been published in various journals, including *Australian Forestry*, proceedings of conferences, and publications of the Forestry and Timber Bureau.

RUSSELL, V.

Born 1928. Following graduation from the University of Adelaide and the Australian Forestry School in 1951, he joined the Forestry Commission of New South Wales and was employed in administration of pine plantations, native hardwood forests and rain forests. In 1955 he joined the Resources Division of the Commonwealth Forestry and Timber Bureau and later transferred to the Australian Capital Territory Forest Service where he was involved in the management of pine plantations. For the last five years he has been employed as an arboriculturist in the Parks and Gardens Branch of the Department of Interior, for three years as officer-in-charge of the Research Section. His main research interests are nursery techniques, irrigation systems and water pollution. He is currently working towards an M.Sc. degree on the subject of the effects of pollution of the Molonglo River and Lake Burley Griffin by mine wastes.

TURNBULL, J. W.

Born 1940. At present Officer-in-Charge of the Seeds Section, Forest Research Institute, Canberra. Educated University of Wales, Bangor, graduating B.Sc. (For.) with honours 1961. Joined Woods and Forests Department, South Australia 1962, where he was engaged on nursery management and plantation establishment. Joined Forest Research Institute 1965 and engaged on research into tree introductions and provenance testing; Officer-in-Charge Seeds Section 1969, and in the past two years has carried out an extensive programme of seed collection throughout Australia, some of which has been within the arid zone. Carried out overseas collection of provenances of various coniferous species in the western U.S.A. for the International Union of Forest Research Organisations, 1968; tropical pines in the Philippines 1969. Member of the Institute of Foresters of Australia.

Contents

Plates

2

Figures

Tables

Maps

Acknowledgments

THE preparation of this book was only made possible with the assistance of a large number of people and organisations. In addition to the help of many Commonwealth and State agencies a number of companies assisted with information on amenity aspects of their operations. The authors are deeply appreciative of the help which was so freely given and thanks are expressed to:

Dr D. A. N. Cromer, Director-General of the Forestry and Timber Bureau (1970-), for his interest in the project since it was first submitted to him. In the last two years his support has been instrumental in enabling the work to be completed.

Dr M. R. Jacobs, Director-General of the Forestry and Timber Bureau (1960-1970) who gave his full support to the early stages of the project and was largely responsible for the wide scope of the text.

The Secretary of the Department of Supply, for permission to visit Woomera and to Mr C. Duke, Area Administrator, and Messrs Pat Kelly and P. A. O'Donoghoe for assistance at Woomera.

The Director of Works, Department of Works, Adelaide, for permission to examine historical records of the initial planning and planting at Woomera and for the help of Mr R. W. Nash.

The Bureau of Meteorology, Department of Interior, and to the Director, Dr W. J. Gibbs, and his staff, notably Mr A. Garriock of the Statistics and Records Division in Melbourne, and Mr F. A. Daw, Regional Director for the Australian Capital Territory, for the supply of data and permission to use published maps.

Additional to the very considerable contributions made by officers of C.S.I.R.O. and which has been acknowledged in the preface, photographs have been made available by the Division of Radiophysics, Sydney, and the Pastoral Research Laboratories, Armidale. In addition the Division of Plant Industry supplied prints of dry country trees and shrubs from photography by Mr Lazarides, to whom thanks are made.

The Northern Territory Administration and the District Officer at Alice Springs (Mr D. T. Conway), who supplied information and checked the section on Alice Springs in Chapter 10. He also provided photographs which had been taken by the police photographer.

The Forestry Commission of New South Wales and members of the staff who provided information for that State or arranged for field visits by Mr Hall. Thanks are particularly due to Mr F. M. Bailey, Chief Commissioner, Messrs E. Regan and J. Dark of the Head Office and Mr T. O'Rourke, Extension Officer at Narrandera.

The Soil Conservation Service of the Department of Conservation of New South Wales and to Mr L. G. Kaleski, Commissioner, for permission to freely use published information and in the supply of photographs for illustrations, and the assistance of Messrs G. H. Knowles and J. C. Newman.

The Forestry Department of Queensland which was very helpful with information and arranging a field visit to the southern dry country. In particular it is desired to thank Mr J. Muir, Chief Silviculturist and Messrs J. Curtis and G. Shea of Head Office, and Mr E. J. Rider who guided Mr Hall in field work.

The Department of Primary Industries of Queensland and to Mr Rosser, Director of the Soil Conservation Branch, for discussion on soil conservation practice in that State, and for checking the section referring to Queensland in Chapter 12.

The Woods and Forests Department of South Australia which supplied historical information on the first Conservator, Mr J. Ednie Brown.

The Forests Department of Western Australia for help with information and field visits. It is desired to thank Mr W. R. Wallace, Conservator, Mr W. H. Eastman, Chief of the Division of Planting and Research and Messrs A. J. Hart and F. Batini, and to Mr D. H. Perry for his invaluable guidance and help on field work north of Perth. In addition Messrs Donovan and Brennan of the Kalgoorlie office were helpful in the Goldfields area.

The Department of North-West, Western Australia and especially to Mr J. C. Tozer, Administrator of North-West and to the previous Administrator, Mr H. L. McGuigan, for the very useful information supplied, as well as to Mr F. Lullfitz for a list of the tree and shrubs used for ornamental planting.

The office of the City Engineer, Broken Hill, for a copy of a plan showing existing and proposed extension of regeneration areas, and for permission to adapt the plan for use in the book.

The City Council of Kalgoorlie for access to a report on species useful for amenity planting.

The Broken Hill Proprietary Company Limited and the staff of the Public Relations Department for providing photos of Cockatoo and Koolan Islands at Yampi Sound, and for checking the section in Chapter 10.

The Mount Isa Mines Limited and particularly to the Resident Manager (Mr D. T. Buchanan) and his staff for the supply of a wide choice of photographs and for checking the section referring to Mount Isa in Chapter 10.

The Western Mining Corporation (Kambalda Nickel Operations) through the Resident Manager (Kambalda), for the provision of photographs and checking the section on Kambalda in Chapter 10.

The Zinc Corporation Limited at Broken Hill and to Mr J. L. Liebelt, General Manager—Broken Hill, for permission to visit the amenity plantings of the Company and for the supply of two historical photographs, and to Mr R. Gould whose personal knowledge of the early plantings and of the species used was invaluable.

Mr W. Bolton-Smith, Wilangee Station, Silverton near Broken Hill, for guidance and assistance when visiting his property.

Professor R. L. Specht, Professor of Botany, University of Queensland, for pre-publication information of vegetation classification from his contribution to the fourth (revised) edition of *The Australian Environment* (1971).

In the complex field of botanical nomenclature and the allied limitations of occurrence, thanks are due to the Heads of all herbaria and, in particular, to:

Dr N. Burbidge, Curator of the Herbarium Australiense, C.S.I.R.O., Canberra, for permission to examine herbarium material of dry country trees and shrubs, and also to Mr M. Gray for help.

Mr L. A. S. Johnson and Dr M. Tindale of the Royal Botanic Gardens and National Herbarium, Sydney.

Mr S. L. Everist, State Botanist, and Dr S. T. Blake of the Botanic Museum and Herbarium, Brisbane.

The half-tone and colour plates are an important part of this book and sincere appreciation is expressed to all the organisations and individuals who assisted. The following acknowledgements are made in order of first appearance in the book:

The News and Information Bureau, Department of the Interior, Canberra.

Mr G. M. Chippendale, Forest Research Institute.

Mr A. Clemson, Chief Apiarist, Department of Agriculture, New South Wales.

Dr J. H. Leigh, Division of Plant Industry, C.S.I.R.O., Canberra.

Dr J. K. Marshall, Riverina Laboratory, Division of Plant Industry, C.S.I.R.O.

The Deniliquin Historical Society for permission to use a photo of a dust storm.

Mr R. W. Jessup of the Division of Soils, C.S.I.R.O., and the Secretary of the Royal Society of South Australia for permission to use several photos.

The Soil Conservation Service of New South Wales.

The Pastoral Research Laboratory, C.S.I.R.O., Armidale (photo by Tamair Pty. Ltd.)

The Director, Division of National Mapping, Department of National Development for permission to use an air photo of the Peak Hill area of New South Wales.

Mr D. T. Conway, District Office, Alice Springs and the police photographer.

The Zinc Corporation Limited.

The Western Mining Corporation (Kambalda Nickel Operations).

Mount Isa Mines Limited.

Mr M. Lazarides, Division of Plant Industry, C.S.I.R.O., Canberra.

Professor L. D. Pryor, Professor of Botany, Australian National University, Canberra.

The Forestry Department of Queensland.

Dr R. D. Johnston, Forest Research Institute.

Mr G. M. Cunningham, District Conservation Officer, Condobolin, New South Wales.

Messrs V. Russell and P. Hahn, Parks and Gardens Branch, Department of the Interior, Canberra and Dr R. Boden, formerly of the Parks and Gardens Branch.

Mr A. G. Brown, Forest Research Institute.

The Division of Forest Products, C.S.I.R.O., Melbourne.

Plate 93 was reproduced from a photograph of POODINGA Photo Index Series DNM, produced by the Division of National Development.

Mr W. Hurditch, Forestry Commission of New South Wales, Sydney.

Acknowledgments of individual photos are shown on the plates. Photos for which no acknowledgments are shown on the plates were taken by the principal author. Mr A. G. Edward of the Forest Research Institute prepared many of the final prints for the plates, whilst a large number were done by Mrs Fisher.

Personally I am deeply appreciative of the comments and advice received from my colleagues in previous publications—Dr R. D. Johnston and Messrs A. G. Brown and G. M. Chippendale. Mr K. Lyons was responsible for the endpapers and most of the maps in the book, whilst Mr J. Curtis, the Publication Officer of the Bureau, was always ready to assist and advise on various matters connected with preparing the text and illustrations for printing.

A publication of this type cannot be produced without the assistance of a typist who is not only technically competent but also qualified in deciphering difficult drafts and for my part of the book I would like to express my appreciation of the work of Mrs F. Bagley, whilst Mr J. Kmieciak made a valuable contribution in communications.

NORMAN HALL

CHAPTER 1

The use of trees and shrubs for amenity and conservation; the resources and problems of the dry country

THE more severe the climate in any region the greater is the need for trees and shrubs to ameliorate the extremes of temperature or to reduce the effect of strong winds. Also, the more scattered the population and remote their homes the greater is the need to take full advantage of plants for amenity and ornamental purposes. The wise planting of trees and shrubs around a homestead or in a dry country township can make the difference between an environment in which it is a pleasure to live and one where life is simply an existence. The current striking increase in arid zone urban population as a result of the great development of mining operations provides an opportunity for Governments and the large companies to make provisions for the encouragement of tree planting through advice and the supply of plants either free or at cost price.

New planting is only one aspect of the wise use of trees and shrubs in the dry country. Of considerable and, sometimes, greater importance is the conservation of existing vegetation. The drier and hotter the climate of a district, the more important it is to give high priority to the retention of as much as possible of the natural vegetation; this is particularly important in belts half a mile wide or more around the edges of urban settlement. Trees and large shrubs should be kept for their value as shade, shelter and ornament, whilst all vegetation contributes to protection of the soil and, hence, reduction in the dust hazard. In much of arid Australia winds are usually light and many calm days occur, but when heat-wave conditions prevail these are often accompanied by strong winds which may attain gale force. These winds pick up very fine dust which may be lifted into the upper atmosphere and travel up to thousands of miles, for example to New Zealand. During the peak of the wind storms coarser dust may be carried shorter distances and material such as sand blown along the surface. Activity in the open becomes almost impossible during such conditions, whilst the life of the housewife becomes almost intolerable as the finer material penetrates into homes and covers everything with a layer of red or reddish-brown dust.

Nowadays there is a tendency to look only to the future and to ignore the lessons of the past, but the present writer believes that any administrator involved in living and industrial conditions in the arid zone should study the history of Broken Hill. During the fifty years from the time of the major mineral discoveries until 1936 the pressure of a large animal population on an ecologically delicately balanced habitat had almost denuded the land of vegetation, whilst the demand for wood fuel and mining timber caused a heavy drain on all trees and shrubs within a radius of 30 miles or more. Any subsequent reduction in the number of animals following the advent of motor vehicles was insufficient to enable re-establishment of the vegetation. The inevitable happened—strong winds caused movement of the soil, moving sand dunes were created and when dust storms prevailed in the district their load of fine soil particles was increased by the pick up of local material. On western and southern aspects of the city, fences were covered and roadways were kept open only by carting away the accumulating sands. In the early 1930s conditions were so bad that some homes were at times inaccessible to motor vehicles and within living memory wedding cars have been unable to collect the brides or hearses receive the bodies of the deceased as a result of the accumulation of sand. The Zinc Corporation initiated plantings in 1936 and the City Council followed soon after, but the most effective conservation practice was the fencing from all forms of grazing animals of wide belts around the city. These natural regeneration areas have been the principal factor in reducing the worst aspects of dust storms in the city.

Large scale industry operating in the arid zone of Australia has found that in order to retain a stable work force of the type which they wish to employ it is desirable, even essential, to initiate amenity planting in 'company' towns. In most other centres local authorities have the responsibility and many of them show a keen awareness of their responsibility and obligations in this type of planting. In addition to Broken Hill other centres where amenity planting has been done on a worthwhile scale include Mt Isa in Queensland, Kalgoorlie in Western Australia, Woomera in South Australia, Alice Springs in the southern part of the Northern Territory and, on a smaller scale, Cockatoo and Koolan Islands in Yampi Sound on the far north coast of Western Australia.

Kambalda, the first developed of the nickel mining centres, is an outstanding example of the wise retention of native vegetation. The site of the town is on low hills near a salt area, Lake Lefroy, some 35-40 miles south of Kalgoorlie and, fortunately, it carried an open stand of small trees 20-40 feet high. In the development of the residential and associated recreational areas a policy was adopted of leaving as many trees as possible, both on street verges and housing blocks. A visitor receives the impression of a long-established town in which successful tree planting has been conducted for two or three decades. The trees, nearly all eucalypts, being native to the area have the advantages of hardiness to dry conditions and of requiring minimum attention.

Brief descriptions and some illustrations of these operations are given in Chapter 10; there is also an outline of the North-West Tree Scheme in the north-west of Western Australia, where the State Government is assisting the needs of a rapid population expansion due to the spectacular development of the growing mining industry.

SHADE

The hotter the climate the greater is the necessity that man and both domesticated and native animals and birds should have access to suitable shade. In the dry country the mean maximum temperature in the shade of the hottest month

Top left: *Acacia pycnantha* in full bloom at Broken Hill, N.S.W.; it well merits the common name of golden wattle. This is a planted specimen and probably does not receive any additional watering.

Top right: There is beauty in the bark of this eucalypt planted at Woomera, S.A. It receives regular watering.

Bottom left: The bark of *Flindersia maculosa* growing naturally not far from St. George, southern Q'ld.; rainfall about 18 inches. The reason for the common name of leopardwood is obvious.

Bottom right: Naturally growing young *Acacia harpophylla* (brigalow) between Goondiwindi and Inglewood in southern sub-humid Q'ld.; rainfall about 23 inches.

Plate 1 THERE IS BEAUTY IN THE PLANTS OF THE DRY COUNTRY

Top left: Very lightly grazed natural vegetation in the southern arid zone near Woomera, S.A. The low shrubs are mainly saltbush (*Atriplex* spp.) and bluebush (*Kochia* spp.) of the family *Chenopodiaceae*. Rainfall about 7½ inches.

Top right: The result of drought and heavy grazing on a property in far western N.S.W. where the rainfall is about 8 inches.

Centre left: In some parts of the dry country there is only a shallow layer of sandy soil over the clay subsoil, so that there is not sufficient material for the creation of sand dunes. Widely spaced small shrubs grow in this type of habitat and in search of moisture their root systems extend for very long distances. If wind erosion is not too severe the plants survive, but if excessive sand is moved from the roots death may ensue. The surface, in general, is not sufficiently stable to allow the establishment of small plants.

Bottom left: Even in the most unlikely habitats the right selection will enable some trees and shrubs to be grown. In this case *Melaleuca lanceolata* (dry land teatree) is growing in salt areas at the foot of residue from adjacent mines. Planted at Broken Hill, N.S.W.; rainfall 9 inches.

Lower right: Over much of dry inland Australia there is no drainage to the sea; the result is that salts accumulate in depressions and strongly influence the type of vegetation which can grow. In this photograph, from near Southern Cross, W.A., there are patches where the salt content is so high that nothing will grow and over the rest only small bushes of salt-tolerant plants, mainly of the family *Chenopodiaceae*. Only on the far side of the depression on an elevation of a few feet are tree-size eucalypts able to grow.

Plate 2 NATURAL VEGETATION OF THE DRY COUNTRY

varies from about 85°F to over 105°F (Marble Bar 107.5°); the maxima on individual days may be 10-15 degrees F higher. Temperatures in the sun can be greater by 30-50 degrees, whilst research workers in Queensland have recorded values as high as 190°F at the tips of wool fibres on the back of sheep. On the evidence of these data there appears no further need to stress the necessity for shade in the arid and semi-arid zones of Australia.

In the case of homes and the sleeping quarters of institutions the effectiveness of insulation in the external vertical structures will be greater if the walls receive shade in the latter part of the afternoon, when maximum temperatures are attained. This is particularly important when the outer part of a wall is constructed of a heat retaining material such as brick, in which heat is stored during the day and only slowly released during the night. Lighter constructional materials without insulation will transfer heat more readily and this results in higher inside temperatures during the afternoon.

The benefit of shade to the physical well-being of man during periods of very high temperature is well known and does not need elaboration. In the case of animals suitable shade will reduce water loss and lower respiratory energy usage. It is no wonder that they eagerly seek any shade available during periods of above average temperatures. All stock owners should ensure that there is some shade in every paddock, however small, and in the case of large areas there should be scattered shade trees so that animals do not have to travel too far for protection. The bird lover will find that the provision of a few trees and water out of the reach of cats and dogs will be a powerful attraction to bird life.

SHELTER

The role of shelter in hot and dry regions is different from that in cool to cold areas, where a fairly high level of protection is needed for both animals and crops from cold and strong winds. In the arid and semi-arid zones of Australia only a limited amount of protection is required by stock from cool winds, and then mainly in southern parts. Shelter from the occasional very strong hot winds is of greater importance, not only for animals but also for fruit and garden crops. To the urban dweller, as well as to those living in isolated homesteads, the principal value of protection is through the lowering in wind velocity which reduces the dust and sand load which can be carried during periods of dust storms.

ORNAMENT

The current trend in public thought is increasingly to seek attractive surroundings in which to live, enjoy recreational time or among which to participate in sports. The wise use of trees and shrubs is a first step in any plans for beautification, and local authorities are giving greater attention to the landscaping of public areas. There are few householders who do not want to plant at least a few shrubs on their suburban block or around their country home. The appreciation of old and beautiful trees in Australia is growing, as evidenced by the public outcry which arises when there are proposals to remove such trees in order to make way for an expressway, or for buildings in a public park.

There are many material things which serve to provide an ornamental or beautiful background to the physical life; in the world of nature these include lakes, rivers or the ocean, mountain slopes and plants. Large bodies of water or mountains are only available to the fortunate few, but there are species of trees and shrubs which will grow in almost any locality. Some climates offer possibilities for a wide range of species, for example the higher rainfall regions of the temperate zones, but even in the dry country there are an adequate number of species from which to choose. With good selection and designing of layout they will provide that combination of shape, proportions, texture and colour which provides beauty in the eye of the beholder.

TREES AND SOIL CONSERVATION

The two most important agents causing soil erosion are wind and water. Trees and particularly forests are of considerable value in controlling loss of soil. Experiments in the United States of America have shown that the volumetric loss of top-soil under tree cover has been 1 per cent in 5 years; under cotton it has been 10 per cent and under cultivated vines 20 per cent in the same period. There are untold examples from the historic past as to what happens if the vegetation covering the soil is removed or badly managed or the soil exposed by cultivation, but man shows a marked reluctance to profit by the lessons of the past. During the dry seasons of the 1930s the people of the towns and country of the Mid-West of the United States of America frequently experienced dense dust storms which were sometimes so thick as to cause darkness at midday.

It was estimated in one dust storm in 1934 that about 500 million tons of good top-soil were blown away from the area forever. The devastation was so serious that hundreds of thousands of settlers were driven from their homes, in some cases the houses actually being covered with drifting soil. Remedial measures included one based on the fact that shelter belts of trees reduced the velocity of wind near the surface on level land for distances which may exceed thirty times the tree heights. Thus the great shelter belt planting scheme came into existence. In Australia the greatest wind erosion occurs in the arid and semi-arid areas and during dry years dust from dry grazing areas has been sufficient to colour the skies in New Zealand, whilst it is not uncommon for large amounts of dust to fall as fine mud on the tablelands and coastal lands of eastern Australia when the skies are filled with dust and rain occurs.

In the coastal-tablelands region and the higher rainfall parts of the semi-arid zone, water is more important than wind as an erosive agent and its activity is closely related to steepness of the topography. Control measures include good management of the vegetative cover, as well as various conservation techniques such as absorption banks and contour ploughing of cultivated areas. In the most vulnerable sites, for example steep valley heads, it may be desirable to plant woodlots following preliminary soil stabilisation measures. On steep slopes and with soils which are very susceptible to water erosion, for example fine loams, any form of ploughing or discing should be avoided and the area managed under permanent pasture.

CONSERVATION AND LAND USE

The subject of conservation is all important in a world where rising standards of living and a rapidly increasing population are putting an ever greater demand on the natural resources of all nations. Conservation is the study and practice of how man can satisfy his needs from the physical environment without reducing the capacity of the environment to continue to provide for these recreational aesthetic and material needs. Conservation means many things but especially wise land use and this may involve the retention of forest cover for recreation and wild life sanctuaries or, on certain types of catchments, for water harvesting, or the planting of other types of land with trees for timber production as the principal objective of management. Conservation does not mean leaving land in idleness and restricting all access. It means the management of a nation's natural resources in such a way that they continue to provide benefits for mankind. The only sound basis for conservation is adequate knowledge and the application of ecological principles to maintain or improve the stability of the various component parts of the material environment. Circumstances can arise in the community where sectional interests are in opposition, such as when the grazing of summer pasture in alpine regions conflicts with the maintenance of the soil and surface vegetation in the best condition for providing optimum water run-off for power and irrigation. In these cases it is always desirable, but not always seems necessary, to choose a policy which will combine the maintenance of the environment and the greatest public benefit.

Good land use has been described as using every acre according to its capability and protecting it according to its needs. The capability of any type of land depends on whether it can be cultivated or not, the fertility and type of soil, and the local climate. The protection it needs depends on the topography, the erodibility of the soil, the rainfall pattern and especially the incidence of high intensity falls, the nature of the local winds and whether there is any existing erosion. One classification used in Australia has seven classes which vary from I, which is fertile, level land upon which no special soil conservation practices are necessary to VII, which is land most suitable for protection forests because of steepness or shallowness, erodibility or infertility of the soil or the needs of water harvesting.

CLIMATE

It is perhaps an over-simplification to summarise the climate of the arid and semi-arid zones of Australia as being hot and dry, but it is correct to state that the area is one where potential evaporation greatly exceeds rainfall and that daytime temperatures are high for eight to twelve months of the year. The average annual rainfall of the arid zone is mainly in the range of 6-12 inches and shows a very high variability, with 10 percentile* values of 2-4 inches. Potential evaporation is high and exceeds 130 inches a year for a limited area in the far north of South Australia and an adjacent part of Western Australia; the lowest is about 60 inches for a narrow belt in the north. Temperatures are typically high; for example north of the Tropic of Capricorn many places have a mean maximum

* Probability of one year in ten.

of the hottest month exceeding 100°F and even in the south values are mainly 90-95°F. The semi-arid zone has a rainfall which varies from about 12 inches in the south to as high as 30 inches in the north; the 10 percentile values are mainly above 10 inches and, in the higher rainfall and less variable areas, may be as high as 15 inches. Potential evaporation shows a wide range from 50 to over 100 inches a year. Temperatures are also high and the mean maximum of the hottest month varies from 85°F in the southern parts to 95-100°F in the north. Mean maxima of the coldest month range from about 60°F in the south to 85°F in the north.

The temperatures which constitute heat-wave conditions are relative to those which prevail during the summer in a particular climatic region, but irrespective of the values for heat-wave temperatures it will generally be conceded that shade becomes particularly important both for man and beast when temperatures exceed 100°F in the shade. For a small area around Marble Bar, in the North-West of Western Australia, there are more than 150 days in the year when temperatures exceed 100°F, and for a larger belt which extends eastwards to central Northern Territory the number is over 100. For the remainder of the dry country north of about 28°S. latitude, but excluding eastern Queensland, there are still 40-100 such days a year; in the southern part of the dry country the number falls to 10-40 days.

Variations of climatic conditions are inevitable in an area of 2.4 million square miles in which the arid zone extends north and south for nearly 1,000 miles (circa 19¼-34°S. latitude) and the semi-arid for about 1,500 miles (circa 13½-36°S.). The greatest distance east to west is about 2,250 miles (circa 113-150½°E. longitude). Because climate is the most important factor in an appreciation of the conditions in and the management of the dry country a more detailed consideration of the various constituents are given in Chapter 3. Appendix I provides Bureau of Meteorology data for a number of representative stations.

CLIMATIC CLASSIFICATION

Like nearly all variable quantities, climate is capable of being classified, but there are greater difficulties with it than with the orderly grouping of plants and animals. In the plant world, for instance, there is usually agreement between botanists as to what constitutes a genus, though this concord may not apply in all cases to species. A genus in one family is usually quite distinct from one in another family, for example *Acacia* (Leguminosae) cannot be confused with *Eucalyptus* (Myrtaceae). The distinctions between genera within the one family, however, may not be so obvious. Meteorological phenomena, which constitute the elements of climate, can be measured with a high degree of precision, for example rainfall or temperature, for a particular spot on the surface of the earth. Meteorological data, however, are not constant from spot to spot and climatic zones essentially comprise at least moderately large areas; it therefore follows that the boundaries between adjacent zones are approximations based on various meteorological criteria. Table 1.1 is based on the major climatic zones as given by Meigs (1953), but area data for Australia is determined by the author. This classification follows the pattern of that made by Thornthwaite 1932-48, when he classified

TABLE 1.1

AREAS AND PERCENTAGES OF CLIMATIC ZONES BY STATES(*a*)

State	*Total area (sq miles)*	*Areas of climatic zones (sq miles)*			
		Humid and sub-humid		*Semi-arid*	*Arid*
		Temperate	*Tropical*		
Western Australia	975,920	53,333	3,251	250,834	668,502
Per cent of Australia	32.88				
Per cent of State	...	5.46	0.33	25.71	68.50
South Australia	380,070	26,913	...	63,017	290,140
Per cent of Australia	12.81				
Per cent of State	...	7.08	...	16.58	76.34
Victoria	87,884	67,504	...	20,380	...
Per cent of Australia	2.96	...	...	...	...
Per cent of State	...	76.81	...	23.19	...
New South Wales	310,372	95,936	...	137,078	77,358
Per cent of Australia	10.46		...	...	...
Per cent of State	...	30.90	...	44.17	24.92
Queensland	667,000	74,500	149,029	242,953	200,518
Per cent of Australia	22.47	...	...	...	...
Per cent of State	...	11.17	22.34	36.43	30.06
Northern Territory	520,280	...	64,827	213,315	242,138
Per cent of Australia	17.53	...	...	...	...
Per cent of Territory	...	...	12.46	41.00	46.53
Tasmania	26,383	26,383	...	...	...
Per cent of Australia	0.89				
Per cent of State	...	100.00	...	...	...
Totals(*b*)	2,967,909	344,569	217,107	927,577	1,478,656
Per cent of Australia	100.00	11.61	7.31	31.26	49.82
		561,676		2,406,233	
			18.92		81.08

(*a*) Based on the major climatic zones as given by Meigs, P. (1953). 'World distribution of arid and semi-arid homoclimates', in *Arid Zone Programme I—Reviews of research on arid zone hydrology.* UNESCO, Paris. Area data for Australian States determined by the author.

(*b*) Total area for Australia based on *Official Year Book of the Commonwealth of Australia* No. 51, 1965, which is a reduction of 3,172 square miles on previous estimates. It includes a decrease of 3,500 square miles for Queensland and a small increase for Tasmania.

the world into six zones on the basis of a moisture index derived from average potential evapotranspiration, water deficiency and water surplus. The zones were perhumid, humid, moist sub-humid, dry sub-humid, semi-arid and arid. For the purpose of this text the first two have been grouped to give the humid zone, and the second two grouped to make the sub-humid zone.

Earlier classifications include those of de Candolle and Koppen, in which temperatures were taken as the basis of the climatic limits. A brief account of these is given by Thornthwaite in the two references listed at the end of this chapter.

In the arid zone the higher rainfall/lower temperature parts grow sparse herbage and scattered shrubs which support a low stocking of sheep and cattle. The lower rainfall/higher temperature parts are mainly unused and are, to a large extent, deserts.

The semi-arid zone can be divided into an inner ring where native grasses and, to a less extent, shrubs support a large and more stable pastoral industry, with sheep predominating in the south and cattle in the north. The outer ring has still higher rainfall and somewhat lower maximum temperatures; where soil and climate are suitable, as in the southern half of the continent, it supports a productive farming community. It is the principal wheat-growing area of Australia and has a high sheep population.

The preceding climatic divisions may be described as a meteorologist's classification; ecologists and botanists, on the other hand, have a somewhat different classification in which their arid zone comprises the arid zone as shown on the maps in this text plus the inner ring of the semi-arid zone; the balance comprises a region which is known by various names, sometimes simply humid. The boundary used by the ecologists is soundly based on principles of plant growth. In southern Australia their arid zone is one in which the pastoral industry is based on native grasses and shrubs; in contrast with this is the non-arid zone where cropping and pasture improvement are possible to a varying degree.

DESERT V. DRY COUNTRY

There is no completely satisfactory definition of a desert, but the following factors are involved: aridity, sparseness of vegetation and extremes of temperature. Also, in popular conception, there is the idea of regions within the tropics where a high noonday sun desiccates both man and beast. The first three aspects are of critical importance and will be discussed further but, regarding the last point, the greater part of the world's deserts are in the 'temperate' regions and even the greatest desert of all, the Sahara, is nearly equally distributed on either side of the Tropic of Cancer. The North American, the Turkestan, Takla Makan and Gobi Deserts are in the 'temperate' regions of the northern hemisphere, whilst in the southern all the deserts, except the Patagonian which is wholly 'temperate', lie across the Tropic of Capricorn.

Aridity is, of course, the principal cause of the absence or sparseness of vegetation in desert regions. A distinction should be made, however, between sparse vegetation which will support a low stocking of domestic animals and an almost complete absence which leaves the soil exposed. In the latter case soil erosion is easily initiated and dune formation is normal where there is an adequate supply of sand. Sand dunes are a common feature of deserts but they occupy only relatively small proportions of many, sometimes as low as 10 per cent, and are not restricted to areas which climatically are deserts. Sparse vegetation is a characteristic of semi-arid zones, but not of the most arid parts of desert regions. Any vegetation present is, of course, useful in conservation but may require sympathetic management to maintain a satisfactory condition. Minimal rainfall is associated, obviously, with the very low humidity which is characteristic of deserts. Clear skies are a feature of most desert areas; the sun therefore heats up the land during the day, whilst at night radiation from the ground surface results in low minimum temperatures. Desert soils which have not the benefit of shading by vegetation may attain day temperatures 25-50°F higher than the maximum shade temperatures, which latter may reach 100-130°F. At night the high radiation is the principal factor in minimum temperatures being lower than in vegetated areas. Rapid inward and outward radiation results in very much greater ranges of temperatures in desert areas than prevail in more temperate regions with their greater cloud cover, rainfall and better cover of vegetation.

Desert soils may be fertile or barren, but are often highly impregnated with salts, especially those of sodium which may present problems in irrigation areas. This high concentration of salts, particularly on level land, is associated with the great excess of potential evaporation over rainfall resulting in the drawing up to the surface of any available moisture with its dissolved salts.

TREES AND MICROCLIMATE

The effect of trees on climate has been a subject of interest and observation from the times of earliest recorded history. In the historical past many writers have attributed the extremities of local climatic conditions in China, parts of India and in the lands bordering the Mediterranean to extensive deforestation. No scientific records of the nature and extent of the changes are available nor would it be possible to determine what was attributable to deforestation and what was due to changes in the general climate. Since the last half of the nineteenth century there has been a quickening of interest in the effect of forests on both climate and microclimate, that is climate close to the ground. A considerable amount of data has been accumulated on the difference in microclimate within and outside forests and on the effect of shelter belts. Unfortunately most of this work has been done in the northern hemisphere, especially between 40 and 57½°N. latitude, but there is an increasing amount of information available from some of the warmer and drier parts of the world, notably Israel. The Forestry and Timber Bureau of the Commonwealth has published data from measurements made inside eucalypt and pine forests, and outside.

Most of the supporters of the value of trees and forests in the latter part of the last century and in the earlier decades of the present century based their case on the claim that forests increased total rainfall. This, in general, is not correct and the position has been summarised on page 3 of *Growing Trees on Australian Farms*. Unfortunately the divisions between protagonists often attained acrimonious levels and had an impact on land use and forestry in Australia. The following account of the development in South Australia is taken from the Presidential Address by M. R. Jacobs (1956)* to Section K of ANZAAS in 1956. The two participants were George Woodroffe Goyder (1826-1898), who was Surveyor-General of South Australia (1861-1894) and John Ednie Brown (1848-1899), who was Conservator of Forests (1878-1899). The name of Goyder is best remembered nowadays in connection with 'Goyder's Line'. Not so well known was his interest in and very powerful influence on the general question of land use and in the preservation of an adequate proportion of wooded land both for purposes of timber production and the secondary benefits given by trees. He was

* Dr M. R. Jacobs was, at the time, Principal of the Australian Forestry School, but was later Director-General of the Forestry and Timber Bureau, Department of National Development. He retired in 1970.

instrumental in the formation of a Forestry Board and, later, a Forestry Department which was one of the first to be established in the British Commonwealth. In 1873 he recommended the introduction of *Pinus radiata* and was instrumental in initiating trials of the tree as a plantation species. The historical account of what happened in South Australia is quoted from a paper by Jacobs:

> The combination of the vision of Goyder and the energy of Ednie Brown should have provided a favourable beginning for forestry in Australia. Unfortunately, within two years of his arrival, Ednie Brown was at odds with Goyder. Looking back, the basic reason appears to be that Ednie Brown (1879) contended in a report to his government that the establishment of forests in the drier country in South Australia would improve the rainfall. Dr Schomburgh had made a similar suggestion in Parliamentary Paper No. 26 of 1873. This cut across Goyder's policy of restricting closer settlement to the better watered country. Ednie Brown was generous to Goyder in his writings, and at first rather mild in his contention that forests would improve the rainfall, but any suggestion along these lines was a stab in the back to Goyder. Goyder (1879) published a statement summarising the known effects of forests on climate, which is a reasonable summary 77 years later. He points out the beneficial effects of trees in ameliorating the microclimate generally, and he covered the available world evidence showing that forests do not influence the rainfall appreciably. *Goyder was right and Ednie Brown was wrong*, but in the resulting conflict Ednie Brown received most support. The political advantages of extending closer settlement in a boom period were too strong to combat sound reason. If South Australian reports only are examined to-day, the impression may be gained that Goyder took too much offence at the fairly mild statements made by Ednie Brown. On the other hand, when the report made by Ednie Brown (1891) to the Government of New South Wales is examined, another picture is obtained. He made strong claims that afforestation in far western New South Wales would improve the rainfall of the inland regions of that State. It may be, therefore, that he did more harm to Goyder than the South Australian papers suggest.

The most noticeable effect of a forest on microclimate occurs within the forest itself and, in contrast, the most important effect of narrow shelter belts is on the adjacent open areas which they protect. The most obvious effect of shelter belts is in the reduction of wind velocity on the lee side; this reduction on level land may be 50 per cent or more at a distance of fifteen times the height and still 25 per cent at thirty times the height. As a result of reduction in wind velocity there is also a lessened evaporation of moisture from crops and from dams. The former factor is of considerable importance in areas where desiccating winds are liable to occur during critical periods of crop growth. In the case of temperature, trees have little effect on the mean annual air temperature but there may be variations during the day, compared with those in exposed open situations; the daily extremes are less marked within a forest than in the open. Mean annual soil temperatures at depths up to 18 inches are slightly to significantly lower within a forest than in the open but in temperate regions the range of temperature between summer and winter will be considerably less under forest cover, especially in regions which experience extremes of air temperature. Where dense, evergreen shelter belts are orientated east-west, frost may remain much longer on the southern side than would otherwise be the case and where such belts are placed across mountain valleys they may impede air movement and cause frost hollows to form. This latter effect can be reduced by correct design of the shelter belts.

A suitable distribution of trees in the dry country will have several effects on the microclimate. The most obvious and, possibly, the most important is that of providing shade for both man and animals. Shade temperatures are often very high, 100-110°F being not uncommon, but temperatures in the sun may be 30-50 degrees F higher and the harmful effect on animal physiology of a temperature of 150°F or more compared with one of 100°F does not need elaboration.

WATER

Most people are aware that Australia is an arid continent but probably not everyone realises how dry it is in terms of average rainfall and run-off into the sea when compared with the rest of the world. It has been estimated that the average annual rainfall for the world is about 26 inches; South America ranks highest with 53, Asia has 25 and Europe 24, but in contrast with this Australia has only about 16 inches. When it comes to consideration of run-off the position is even less favourable; the average for the world is about 10 inches, that for the United States of America about 9, whilst that for Australia has been estimated at only one and one-third inches. It is pointed out in the section on Drainage in Chapter 2, that about one million square miles of the continent has either unco-ordinated or internal drainage which does not contribute to run-off.

The amount of run-off is of considerable importance, since it is a measure of the amount of water which could potentially be retained by additional dams and weirs, for increased domestic, industrial and agricultural use. The average annual flow into the sea of some individual rivers of the world in millions of acre feet,* is very high when compared with the total for Australia of about 280 million acre feet. The Amazon stands supreme with a run-off of about 1,780 million, the Mississippi has 474, the Danube 228, the Volga 148 and the Nile until recently, in spite of the dams on its course, had about 72 million acre feet. In Australia the run-off from the Murray is already very low and is likely to become less with the construction of further dams and weirs. Unfortunately, the greatest amount of run-off is in northern Australia where, in the dry country especially, there is limited demand for irrigation water and where dams' capacity would have to be comparatively large in view of the very seasonal nature of river flow. The opportunity in Australia for use and re-use of the water in any river system is limited due to the absence of large cities and industries along their courses; apart from the Murray there are few rivers which have more than one small city, and these are usually situated in the lower reaches. It is stated that the amount of water taken from the Mississippi is more than thirteen times its flow, that is, the water is used and re-used, but this degree of utilisation does not appear possible on any Australian river system.

The location of the large cities in Australia and nearly all the heavy industries near the sea precludes extensive reclamation of their water effluent and the re-use of the water. This is a disadvantage, but it avoids some of the pollution problems which occur in major rivers in other countries; it was only in 1969 that the dumping

* One acre foot represents an area of one acre covered by water to a depth of one foot, or approximately 0.273 million gallons.

of a small amount of a poison into the Rhine caused world-wide headlines as fish died and cities and countries on the lower reaches of the river were faced with problems of contaminated water for domestic consumption. Most people are aware that precautions need to be taken against the careless discharge of industrial waste into rivers, whilst periodically there are outcries against the outfall from sewerage plants. The discharge of effluent from correctly treated sewerage, however, is comparatively harmless compared with the more insidious danger from accumulation of many insecticides and herbicides, whilst some detergents do not readily break down and remain active in water for long periods.

Water is the most valuable asset in the dry country. In the arid zone and the drier parts of the adjacent semi-arid zone the water supply is at times insufficient to maintain the normally very sparse human and animal populations. There is little prospect that there is ever likely to be much change in the general position. The one town of any size in the centre of the continent is Alice Springs (population estimated at 10,000 in 1970), which draws its water from storage in old river gravels beneath the flat valley in which the town is situated. The water reserve appears sufficient for the size to which Alice Springs is likely to grow, but certainly would be completely inadequate for any large city or water-using industrial complex. In the fringing areas of the semi-arid zone the position is little better but, fortunately, much of the area in eastern Australia is within the Great Artesian Basin and water for animals can be secured from that source or from surface dams which are filled during the rainy season. Because of the high evaporation in these areas the dams should be as deep as practicable.

The northern section of the higher rainfall parts of the semi-arid zone has a well-marked summer maximum precipitation followed by a very dry winter and early spring; most of the streams seasonally dry up and pastoralists rely heavily on dams and artesian water where it is available. Agriculture is very limited. The eastern half of the southern semi-arid zone is comparatively well-watered; the Murray River and its principal tributaries, especially the Murrumbidgee, flow through the area and provide water for domestic purposes, the watering of animals and for irrigation. With the development of the dams of the Snowy Mountains Scheme and other projects on the headwaters of these rivers there are better prospects of continuity of flow, even during periods of prolonged drought. Away from these rivers and associated irrigation channels the rural population must rely on streams, some of which are not perennial, surface dams and bore water, possibly supplemented by limited tank rain water for human consumption.

MINERAL DEPOSITS

Even the briefest statement on the importance of the dry country of Australia would be incomplete without some reference to mineral deposits. With the notable exception of coal, most of the major mineral resources are in the arid and semi-arid zones. The importance of minerals as an export increased rapidly in the 1960s and in the year ending 30 June 1969 their value exceeded that of wool for the first time. During the last third of the nineteenth century many important deposits and fields were discovered and it was during this period that activity commenced at Broken Hill (1883) in silver-lead-zinc and at Kalgoorlie (1893) for gold. From 1893-1948 there were few significant discoveries in the dry country and still less in the humid zones; the notable exceptions were silver-lead-zinc and copper orebodies at Mt Isa and the gold fields at Tennant Creek. At the end of this period many people predicted that there was little more to find in Australia and that the continent did not have the quantities of minerals which occur overseas. From 1949, however, there have been many notable discoveries and the number and importance of them has tended to increase with time. Amongst the first was uranium, and somewhat later came iron ore on a very large scale, whilst in the late 1960s there were numerous discoveries of nickel, a metal in which formerly Australia was considered to be deficient. Also during this period copper, which had been previously discovered at Cobar, was found there in large quantities.

Most of the information in this account of mineral deposits has been drawn from the Department of National Development publication, *Mineral Deposits*.

Silver-lead-zinc

Australia is one of the world's principal producers of both lead and zinc. The two most important centres are Broken Hill and Mt Isa, whilst the third centre is on the humid west coast of Tasmania. Almost all the current production of silver is associated with the lead-zinc operations.

Copper

The largest known copper deposits are those at Mt Isa which were mined from 1943-46 but resumed on a much larger scale in 1953. There are also a number of smaller operations in Queensland. Prior to 1953 the most important source was on the coast of Tasmania and this is still an important producer. Copper had been first found at Cobar in 1869 and was mined intermittently until 1952; but following the discovery of more extensive orebodies large-scale mining commenced in 1965.

Iron

For a very long time iron ore, often of a high quality, that is 60-66 per cent iron, has been known to exist in Australia. In the period 1900-60 the two major areas were in the Middleback Ranges west of Whyalla, not far from Port Augusta in South Australia, and in the Yampi Sound area near Derby in north-western Western Australia. In the Middleback Ranges the most important orebody is at Iron Monarch, but there are also large quantities at Iron Knob where operations commenced in 1903 and at Iron Prince and Iron Baron. At Yampi Sound mining commenced on Cockatoo Island in 1951 and on Koolan Island in 1964. Then in the middle and late 1960s came the proving of immense quantities in the Hamersley and Ophthalmia Ranges situated east of Onslow and south of Port Hedland in Western Australia. Some of the best known names where large bodies have been proved are Mt Tom Price, Mt Newman and Mt Goldsworthy. These discoveries have given Australia some of the largest known reserves in the world.

Gold

Many of the early discoveries of gold were in the humid zones; the first officially recognised finds were near Bathurst (New South Wales) in 1851 and these were closely followed by others in the Ballarat-Bendigo districts (Victoria). Most of these were alluvial fields but the greatest orebody is that at Kalgoorlie where it was found in 1893 and since then there has been continuous mining, most of the output coming from the Golden Mile, an area of rich mineralisation only about two miles long and one mile wide, situated between Kalgoorlie and the adjacent town of Boulder. There are many other gold-producing centres both in Western Australia and elsewhere, for example, Mt Magnet (Western Australia), Mt Morgan (Queensland) and Tennant Creek (Northern Territory).

Nickel

Previous to the 1960s there had been only a small intermittent production of nickel; this was near Zeehan in Tasmania. Exploration in the early and middle 1960s revealed that nickel mineralisation was widespread in the arid and semi-arid centre and south of Western Australia. The main known occurrences up to 1969 are in the Kambalda area, south of Kalgoorlie, and near Mt Windarra to the north. These discoveries will make Australia a nickel exporter of world importance. Other prospecting has revealed lateritic nickel deposits at Wingellina in the same State and the adjoining arid areas of South Australia and the Northern Territory, with others at Greenvale and Marlborough in the higher rainfall areas of coastal Queensland, but none of these have yet been proved to be economic propositions.

The development of nickel operations in Western Australia has called for the construction of large new towns in which to house the labour force in areas where the average annual rainfall may be only 8-11 inches and other water resources very limited. Wise treatment of the natural environment in such regions normally calls for the retention for shade and ornament of as many as possible trees and shrubs in the urban area, and the conservation of ground cover on adjacent land for the purpose of minimising dust and drifting sand. In the first of the new nickel towns, Kambalda, this has been done very well; it is described briefly and illustrated in Chapter 10.

Uranium

Of the four deposits of uranium-bearing ore two (Rum Jungle and South Alligator River) are in the humid zone of the Northern Territory, one (Mary Kathleen) is near Mt Isa and one (Radium Hill) is in South Australia not far from Broken Hill. Some of the mineral deposits have been known for a very long time but were only positively identified after World War II when there was an increased interest in the metal.

Petroleum and natural gas

The search for oil and natural gas has proceeded intermittently since the beginning of the century, but it was not until the middle 1950s that intensive and systematic oil exploration commenced. In 1961 the first commercial field was discovered at Moonie, near Roma in Queensland; later there was Barrow Island off the coast of Western Australia. Still later in the 1960s came the discovery of the much larger oil and gas fields on the Gippsland Shelf off the coast of Victoria. In drilling for oil there have been discoveries of natural gas, several times in what have been considered to be very large quantities; the greatest of these have been Gidgealpa in South Australia near Innamincka in the far north-east, and off the coast of Victoria, with smaller but economic quantities in the Moonie field.

There appears little doubt that the story of mineral discoveries in Australia has still to be completed. What is of importance to the theme of improving living amenities by planting trees and shrubs is that most of the great mineral discoveries of the period 1950-70 have been in the arid zone where water supplies are limited. In some cases the water brought into the new townships for domestic and processing purposes may be sufficient, after suitable treatment, for watering plants if the salt content is not too high. Probably unique in the world is the source of water used when trees and shrubs were established for amenity purposes on Cockatoo Island, Yampi Sound, by Broken Hill Proprietary Ltd. Vessels which transport iron ore from Yampi Sound to the steel works at Newcastle returned with water in ballast for domestic and other purposes; a one-way journey of about six thousand miles. The results of some of the planting are shown in Plate 35.

TIMBER AND COMMERCIAL FORESTS

Although the humid zones of the continent are the main sources of commercial timber there is a small but significant production from the dry country, especially the semi-arid parts of eastern Australia. The most important are the cypress pine (*Callitris* spp.) and dry eucalypt forests of New South Wales and Queensland; of less importance are the riverain forests of the Murray River system and the very dry woodlands and mallee around Kalgoorlie in Western Australia.

1. *Cypress pine forests*

During the year ending 30 June 1968, cypress pine constituted about 3 per cent of total log removals from Australian forests, but in the important field of sawn timber it rose to 4.8 per cent, with production amounting to 65 million out of a total of 1,347 million super ft. The most important species is white cypress pine (*Callitris glauca*) but there is also some black cypress pine (*C. endlicheri*) in the south, whilst the closely allied northern cypress pine (*C. intratropica*) of northern Australia, especially the Northern Territory, is mainly found outside the boundary of the dry country.

The timber is renowned for its high resistance to decay and termite attack and its very small shrinkage on seasoning. The two major uses are as scantlings for building construction and as flooring, with lesser usage for linings and weatherboards; small diameter logs are utilised in the round as posts and light poles.

In New South Wales the forests fall wholely within the semi-arid zone; the most important managed area is that known as the Pilliga, which is located north of Coonabarabran and between Baradine and Narrabri, but there are smaller areas within a 100-mile radius of Condobolin. Formerly these forests covered very much larger areas which extended from the Queensland border southwards to Griffith, and between Bourke in the east and Coonabarabran and Wagga Wagga in the west. These areas, however, were on land which was suitable for pastoral and agricultural purposes and most of the tree cover was removed as settlement advanced. The Queensland occurrence of the more important commercial forests is on either side of the boundary between the semi-arid and sub-humid zones, especially north-west and south of Dalby. Scattered remnants of more open stands are found from these places southwards to the border with New South Wales.

2. *Dry sclerophyll forest of the semi-arid zone*

These are predominantly eucalypt forests in which the most common species belong to the box and ironbark groups. The timbers of these eucalypts are usually very heavy (60-70 lb cu ft air dry), very hard, tough, strong and most have an outstanding durability; unfortunately they are not fissile and most do not saw readily. The boxes are all outstanding species for firewood. The most important use for the ironbarks is as railway sleepers and, formerly, for hewn or sawn bridge timbers. Other uses include posts and strainers and, when of suitable form, heavy poles.

The principal ironbarks are red ironbark (*E. sideroxylon*) and narrow-leaved red ironbark (*E. crebra*), while silver-leaved ironbark (*E. melanophloia*) has an extensive occurrence. There are numerous boxes, especially overlapping the boundary with the sub-humid zone; species include grey box (*E. woollsiana*), narrow-leaved box (*E. pilligaensis*), poplar box (*E. populnea*) and fuzzy box (*E. conica*). Mainly on the more favourable eastern edges there are yellow box (*E. melliodora*) and white box (*E. albens*). Brown bloodwood (*E. trachyphloia*) also occurs over a large area.

3. *Riverain forest of the Murray River system*

This is a unique type of forest consisting of only one species, river red gum (*E. camaldulensis*), which has its centre on formerly seasonally-inundated land along the Murray River and its major tributaries. The construction of more weirs on the river and the regulation of flow through the dams of the Snowy Mountains Scheme has considerably altered the flow pattern so that the forests are of less importance than formerly. The timber of river red gum is hard, heavy and very durable but only of moderate strength because of wavy grain; it is still used for railway sleepers and for sawn timber, especially high quality flooring.

4. *Dry woodlands and mallee forests near Kalgoorlie*

These very open woodlands carry only a small amount of timber per acre, but are noteworthy in that trees 60-70 feet or more in height grow in a rainfall of only 9-10 inches a year. Prior to the greater availability of coal and fuel oil for raising steam the timber from these areas was used extensively for fuel at the gold mines of Kalgoorlie and Coolgardie. Amongst the larger tree species, all of which are eucalypts, are salmon gum (*E. salmonophloia*), gimlet (*E. salubris*) and red morrel or morrel (*E. longicornis*). There are a large number of mallee and small shrubby types of eucalypts and amongst the more common species are sand mallee (*E. eremophila*) and morrel (*E. gracilis*).

SHEEP

Australia is one of the great sheep raising countries of the world; in 1968 the numbers were 167 million, a drop from the peak of 171 million in 1965. Most of these are pastured in the semi-arid zone, especially south of about 23½°S. latitude, although there are large numbers in the coastal and tableland regions of Victoria and New South Wales.

Distribution by States is shown in Table 1.2.

TABLE 1.2

SHEEP POPULATION BY STATES AS AT 31 MARCH 1967 AND A COMPARISON WITH THE THREE YEAR AVERAGE TO 1939 (MILLIONS)

State	*Average of three years to 1939*	*Population as at 31 March 1967*
N.S.W.	51.202	63.848
Vic.	17.845	31.239
Qld	21.889	19.305
S.A.	8.916	17.864
W.A.	8.972	27.370
Tas.	2.460	4.321
N.T.	0.023	0.008
A.C.T.	0.251	0.281
Totals	111.558	164.237

The number in the arid zone is less than 10 per cent of the Australian total. There has been a general trend during the present century for sheep numbers in this zone to either remain static or to decrease, and the increase in population of from 71 to 171 million in the period 1900-65 was mainly in the better parts of the semi-arid zone or the adjoining sub-humid zone. The pattern of sheep population in the arid zone is cited by J. C. Newman and R. W. Condon (1969):

> In the Western Division of New South Wales sheep numbers rose from 6.5 million in 1879 to 15.4 million in 1891. In the period following, the reserve of perennial plants was depleted and soil erosion began. After the major drought of 1901-2, which was associated with severe economic depression, the sheep population fell to 3.5 million. Stock numbers have risen slowly since this time and reached 9-10 million before the drought in 1965, when numbers fell to approximately 6 million. Throughout this period stock numbers have fluctuated considerably, suggesting a dependence on annual rather than perennial species.

The Western Division is mainly in the arid zone.

Sheep numbers in a State or portion of a State may show considerable fluctuations with drought, but since it is unusual for all of Australia to be affected by drought at the one time, total population number does not fluctuate as widely. In the drought of 1965 the percentage change in sheep numbers 1965-66 (based on statistics for 31 March in each year) showed a drop of 15.2 per cent for New South Wales and 23.5 per cent for Queensland, although the national decrease was only 7.7 per cent. Again, in 1967 to 1968 when South Australia and Victoria experienced droughts the decreases were 8.2 and 10.7 per cent respectively although the total number for Australia increased by 1.6 per cent. It takes several years for sheep population to recover after a major drought; for example, after that of 1965 Queensland only had 20 million by 1968, whilst New South Wales

TABLE 1.3

REGIONAL FLUCTUATIONS IN WOOL PRODUCTION DUE TO DROUGHT; DATA FOR SELECTED DIVISIONS IN QUEENSLAND AND NEW SOUTH WALES, 1964-65 TO 1967-68

State and Division	*Numbers of bales of wool*			
	1964-65	*1965-66*	*1966-67*	*1967-68*
Queensland				
South-western	115,917	63,434	74,910	101,122
Roma	99,681	66,146	81,831	93,894
Downs	116,658	83,641	99,866	106,080
State totals	624,311	439,651	489,520	541,045
New South Wales				
Northern Tablelands	147,455	111,826	115,759	145,175
North West Slopes	128,855	92,856	90,416	115,797
North Central Plains	99,433	60,883	61,096	69,925
Western Division (north-eastern part)	114,905	63,570	61,522	96,809
Central Plains	186,293	122,861	106,861	143,019
State totals	1,823,455	1,486,387	1,543,568	1,667,866

had recovered to 68 million but was still under the pre-drought number of 72 million.

Drought not only affects sheep numbers but also results in a lower production of wool per sheep and a marked increase in the lower grades at the expense of the higher grades. The loss in meat production is not usually apparent in the drought year when many sheep are slaughtered for meat, but is in the one or two years immediately afterwards when pastoralists retain the maximum number for breeding. Fluctuations in wool production are best exemplified by data for regions where

TABLE 1.4

SHEEP AND CATTLE NUMBERS IN AUSTRALIA, 1860-1967 (MILLIONS)

Year	*Sheep*	*Cattle*
1860	20.1	4.0
1870	41.6	4.3
1880	62.2	7.5
1890	97.9	10.3
1900	70.6	8.6
1910	98.1	11.7
1920	81.8	13.5
1930	110.6	11.7
1940	119.3	13.1
1950	112.9	14.6
1960	155.2	16.5
1964		19.0(*a*)
1965	170.6(*b*)	
1967	164.2	18.3

(*a*) The highest number for cattle was in 1964 (19,055,000)
b) The highest number for sheep was in 1965 (170,622,000)

drought has been the most severe. Table 1.4 shows regional fluctuations for selected divisions in Queensland and New South Wales in the period 1964-68, based on data from a paper by G. D'A Chislett of the Wool Graziers' Association (1969).

The time required to recover from a severe drought is mainly a result of reduced sheep number but in the arid zone the recovery of deteriorated natural fodder may also be important. To a limited extent in parts of the higher rainfall semi-arid zone increased acreages of wheat may have had some impact.

The numbers of sheep and cattle during the past 107 years are shown in Table 1.4

Amongst other farm animals, horses attained their peak in 1920 with 2,527,000, whilst in 1967 the number had dropped to about 480,000. The peak for pigs was in 1967 with 1,804,000 head.

CATTLE

Australia's cattle population of 18.3 million in 1967, ranked only eleventh amongst countries of the world. The leading countries are India with 236 million (estimated), the United States of America with 109 million and the Union of Soviet Socialist Republics with 97 million, followed by Brazil, mainland China, Argentina, Pakistan, Mexico, Ethiopia and France.

About four-and-a-half million of the cattle are dairy stock and they are located mainly in the sub-humid and humid zones of the coastal and tablelands regions of temperate Australia. By contrast, the beef cattle are pastured chiefly in the dry country; this is particularly so for breeding stock, although fattening may be done on better pastures in higher rainfall areas more accessible to markets or freezing works. More than half of the 13.8 million beef cattle are located in northern Australia; Queensland, with just over six million is the leading State, whilst northern Western Australia and the Northern Territory each have about one million. In the southern half of the continent there are more than three million in New South Wales, only half a million in South Australia, but over one-and-a-half million in Victoria (mainly in the sub-humid zone), and about one-quarter of a million in Tasmania (solely in the sub-humid and humid zones). Most properties in the northern half of Australia which raise beef cattle do so to the exclusion of other pastoral activities; in the southern half the raising of beef cattle is usually mixed with sheep or some form of cropping.

Whilst cattle numbers do not fluctuate as markedly as those for sheep in times of drought, their numbers also do not return to normal as readily and the period of economic recovery by the pastoralist is longer. There is little doubt that the cattle industry in the arid zone, for example the southern half of the Northern Territory and the western half of Queensland, is particularly susceptible to the effect of long term droughts. In these areas, which are mainly suitable only for breeding and less for fattening purposes, any reduction in stock cannot but be at the expense of breeding animals and the natural recovery of numbers after a prolonged drought may take many years. The semi-arid zone, whilst still liable to drought, offers better possibilities for breeding and those pastoralists who own or control properties in higher rainfall coastal areas may be able to transfer stock at critical periods without serious reduction in numbers of breeding animals.

WHEAT

Wheat is the most important agricultural crop of Australia. It is grown mainly in the semi-arid zones of the southern half of the continent where rainfall is usually sufficient and the long hours of sunshine result in a quality of grain with a high protein content suitable for bread production and other uses. Production is usually near average when the rainfall is not less than about 85 per cent of average, provided that the distribution pattern is normal for the district concerned. The area of land under wheat production, together with associated fallow, was about 22.4 million acres in 1967-68, or over three-quarters of the total cultivated area. South-eastern Australia, especially New South Wales, is the main centre of production but during recent years there has been an increased amount grown in Western Australia. During the 1940s the average production was less than 150 million bushels a year but it increased to 300-400 million bushels during the 1960s, with a peak of over 500 million in 1968.

Periodically drought has taken a serious toll in production although, because of a wider geographical distribution of sowing and the introduction of strains of wheat which are more resistant to drought, the effect on total production has shown a tendency to decrease. In absolute quantities, however, the toll of drought has been and remains considerable and this is exemplified in information which has been extracted from a paper by A. R. Callaghan, Chairman of the Australian Wheat Board (1969):

TABLE 1.5

PRODUCTION LOSS OF WHEAT IN DROUGHT YEAR, BASED ON 5-YEAR AVERAGES WITH THE DROUGHT YEAR AS THE MID-POINT; FOR THE PERIODS 1939-42 TO 1965-70

Period	*Production*		*The impact of drought*			
			Year	*Loss*		*Per acre production in drought year*
	Total 10^6 bushels	*Per acre bushels*		*10^6 bushels*	*per cent*	*bushels*
1938-42	154	12.8	1940	72	46.8	6.5
1942-47	115	11.7	1945	63	45.2	6.3
1944-49	145	11.8	1947	27	18.6	8.0
1955-60	168	16.9	1957	71	42.3	11.0
1963-68	340	18.0	1965	81	23.8	14.8
1965-70	381(*a*)	17.4	1968	107	28.1	12.4

(*a*) Includes an estimate for 1969-70.

POPULATION

The estimated population of Australia at 31 June 1967 was 11,928,889 and by the end of 1970 it is likely to be about twelve-and-three-quarter million. Past and projected future populations are shown in Table 2.3 of *Growing Trees on Australian Farms*. The average population density of Australia as a whole is low. A comparison of the State and Commonwealth densities in 1901 with those in 1967 is given in Table 1.6. The figures are probably not very meaningful since, in accordance with general practice, allowance has not been made for exclusion of

TABLE 1.6

POPULATION DENSITY PER SQUARE MILE: AUSTRALIA, 1901 AND 1967

State	*1901*	*1967*
W.A.	0.20	0.91
S.A.	0.95	2.94
Vic.	13.77	37.59
Tas.	6.68	14.39
N.S.W.	4.45	14.05
Qld	0.76	2.58
N.T.	0.009(*a*)	0.12
Commonwealth	1.29	4.02

(*a*) Excludes aborigines prior to 1961 census.

the large area of the arid zone. This amounts to nearly 50 per cent of the continent. Even were this deduction made the average density per square mile is still very low by world standards.

Large portions of the arid zone, but only limited areas of the semi-arid, are virtually uninhabited on census criteria, or consist of a very large Aboriginal reserve on both sides of the Western Australian-South Australian border from approximately latitudes 20° to 28°S. The largest uninhabitated area comprises most of the eastern half of Western Australia and includes the Great Sandy Desert in the north, Gibson's Desert in the centre and the Great Victoria Desert in the south. In the Northern Territory there are several large areas classified as uninhabited, especially in the arid zone, whilst in the south-east the Simpson Desert extends into north-eastern South Australia and over the border into Queensland. Of the eastern States, Queensland has negligible areas so classified, although it has a very large area with a low density population. In New South Wales the lowest density area is on the mountains of the Great Australian Divide, mainly at altitudes over 4,000 feet. Victoria, with the highest overall density, has two locations uninhabited or with negligible population; these are in the high country of the Great Divide in the east of the State and two undeveloped low rainfall areas in the north-west. Tasmania, which is outside the dry country, has a very low population density in the south-western and western parts of the State.

The rural population not only shows a decreasing percentage but also a reduction in numbers. This pattern has become increasingly obvious since World War II and agrees with world-wide trends. Between the census of 1961 and that of 1966 when the total population increased from 10,502,842 to 11,550,462, the rural numbers decreased from 1,984,749 to 1,920,003; the corresponding percentages were 18.9 and 16.6 respectively. During the same period the change of population in the dry country showed a variable pattern, especially in the arid zone where there is only a very low density. Small influxes of people for mining and processing of ore may have resulted locally in small overall density increases. On the other hand, in many of the pastoral and extensive agricultural areas of the semi-arid zone there were small to moderately large decreases in population. This was common in all States but particularly in the more developed,

TABLE 1.7

THE CITIES AND LARGER TOWNS OF THE ARID AND SEMI-ARID ZONES OF AUSTRALIA: POPULATION AS AT 30 JUNE 1966

State and centre	*Climatic zone*	*Population (urban)*	*Industry*
Western Australia			
Kalgoorlie(*a*)	Arid	19,908(*c*)	Mining and processing
South Australia			
Port Augusta(*a*)	Semi-arid	10,103	Port (wool, minerals) salt, railway workshops, rural centre
Port Pirie	Semi-arid	15,536	Port (wool, wheat, minerals), very large silver-lead smelting works and by-products from same
Whyalla	Semi-arid	22,126	Port, mining, heavy industry (shipbuilding)
Victoria			
Mildura	Semi-arid	12,931	Dried (raisins, sultanas) and citrus fruits; pastoral and tourist centre
New South Wales			
Albury with Wodonga (Vic.)(*b*)	Semi-arid	32,032	Pastoral and wheat centre, light manufacture, communications and tourist centre
Broken Hill	Arid	30,014	Mining and processing, minor pastoral and tourist centre
Wagga Wagga(*b*)	Semi-arid	25,819	Pastoral and wheat centre, education and light manufacture
Queensland			
Mt Isa	Semi-arid	16,877(*d*)	Mining and processing

(*a*) On the boundary between the arid and semi-arid zones.
(*b*) On the boundary between the semi-arid and sub-humid zones.
(*c*) Includes urban area of Kalgoorlie and Boulder.
(*d*) By mid-1970 the population was estimated at 22,000.

relatively high-density rural areas of New South Wales and Victoria where decreases were as great as 4 per cent. Much of this decrease was probably due to increased size of pastoral holdings following amalgamation of farms, greater mechanisation of operations and, to some extent, by owners moving to the local towns and managing on a commuting basis. In general this was a period of increasing rural productivity, both in terms of total production and of wool per sheep and bushels of wheat per acre, so that there is no indication that the reduced population was prejudicial to the rural economy.

The dry country, based on the 1966 census figures, contained only nine towns or cities with populations greater than 10,000; see table 1.7 for a summary of these. It is noteworthy that in only three of these (Mildura in Victoria and Albury and Wagga Wagga in New South Wales) does grazing, cropping or fruit production comprise the principal industries. The remainder were based on mining, processing or associated heavy industry (for example, shipbuilding) although three were also ports of some importance which not only shipped ores and processed minerals and chemicals but also wheat and some other agricultural products. It is suggested that the establishment and growth of further large centres in the arid zone will only be those arising from mining and processing activities; in this regard greatest development is likely to be in Western Australia. In the more fertile semi-arid zone, especially in south-eastern Australia, there may be a number of towns which will exceed the 10,000 population mark due to the association of dried and citrus fruits and cotton growing, all under irrigation, with the local sheep, cattle-fattening and wheat industries.

LAND USE

The use of land for pastoral, agricultural and horticultural industries involves consideration of the environment, historical trends and the need for its products for local consumption and for cash exports. In much of Australia the limiting factor in the environment is rainfall; this and other aspects of climates are discussed elsewhere. Some of the economic aspects of land use are referred to briefly under the subjects of sheep, cattle and wheat. A distinction should be made between actual land use and wise land use, which may not necessarily be the same; when they are the same then that is a very satisfactory situation. Conservation is considered elsewhere.

Historically land use by Europeans spread out from coastal centres and, to a limited extent, along routes of natural transport such as the Murray River. In the case of Sydney, when drought and shortage of suitable land on the coastal lowlands in the early days forced an expansion towards the inland, a route was pushed through the Blue Mountains to the grazing lands of the Bathurst-Orange districts and beyond. In all this early settlement, whilst much of the land was suitable for the use to which it was put, the pressure of the later population resulted in some areas being used beyond their sustained capacity for production. As the young nation developed the frontiers of settlement were pushed still further into the semi-arid zone and, later, to the arid zone. The pioneer land use in the southern dry country was, in most cases, sheep. Their demands on the environment were not high but the periodic incidence of droughts took a serious toll of both man and animals, giving early evidence to the pioneers that Australia had a climatic environment which needed to be treated with caution.

From the expanding settlements and extensive land use pattern of the early days there has been a swing, especially in the middle twentieth century, to more intensive use of the better endowed regions. The introduction and breeding of more productive plants and animals, efficient pest and disease control by modern chemical and biological methods, mechanisation of production, and improved transport have all contributed to man's ability to make the most of his rural assets. Australia was fortunate in that the first sheep introduced were a hardy breed (merino) particularly suitable for a dry environment. The European and, especially, British breeds of cattle did well in southern Australia but were less suited for semi-arid tropical regions; the introduction of Santa Gertrudi and other breeds since World War II offers better prospects in those areas. New strains of wheat, mainly developed in Australia, are well adapted to the environment of the semi-arid zone and possess valuable resistance to pests, especially fungi. Their use, together with suitable fertilisers, has now almost doubled the yield per acre.

Another trend of recent decades has been the development, in districts of reasonable rainfall, usually by State governments or large insurance companies, of land which had been previously regarded as too poor for farming. Cultivation and the addition of suitable manures and minor elements have resulted in these areas becoming productive at a level sufficiently high to encourage further settlement.

Many crops have a large water requirement and some, for example sugar, are restricted to areas wholly outside the dry country, whilst dried fruits, rice and cotton, which are grown in the semi-arid zone, require irrigation. It appears unlikely that there will ever be more than very restricted cropping in the arid zone.

MULTIPLE USE OF LAND

The term *multiple use* in reference to land describes a concept in management, not a system or method of land use. It implies an idea whereby an area may be managed for more than one major objective at a time. The application of multiple use to the same area may be quite different for a private owner who, probably quite justifiably, requires maximum sustained profit, than for public lands. For government management the profit motive may be included, but could be subject to the overall public benefit. In the higher rainfall tablelands and coastal regions of Australia the concept of multiple use of large forested watersheds is an aspect of management policy which is yearly attaining greater importance, especially in those areas within reach of the large coastal cities. In many of these areas water harvesting, production of commercial timber and the provision of some recreational facilities can usually be readily incorporated into the management plan. The opposite of multiple use is *single use*; this is a concept which, in a world of growing population, higher standards of living, greater facilities for travel and the time for recreation, is hardly tenable. Even in wilderness areas two objectives can usually be readily met, for example the retention of the relatively undisturbed native flora and fauna, and water harvesting. In nearly all other areas more than two objectives of management can usually be incorporated in the management plan.

In the dry country, multiple use management is of less importance than in the sub-humid and humid zones where water, both in quantity and quality, is usually a matter of considerable public and Government concern. On steep topography in high rainfall regions there is ample evidence of the value of a well managed forest cover to regulate both infiltration and run-off rates, as well as to ensure the highest possible quality of the water. In marginal rainfall areas and, especially with gentle topography and sandy soils, the presence of a dense forest cover may considerably reduce the amount of water which will be available as run-off. There are considerable differences in the water demanded by different classes of forests; examples of extremes could be an open eucalypt woodland of dry country species and a healthy stand of *Pinus* species shortly after canopy closure. The desirable vegetation for maximum run-off may be only sufficient herbage to control wind and water erosion. There are many conflicting factors in deciding the best policy and it is beyond the scope of this book to discuss them in detail save to suggest that for any particular project all factors should be investigated before making a decision on management policy.

MULTIPLE USE OF WATER RESOURCES

Elsewhere in this chapter the *multiple use* of land has been discussed and the desirability, in many cases, of combining water harvesting with some aspects of food or timber production, as well as recreation. Here brief reference will be made to some aspects of the re-use of water combined where practicable with the preservation of an aquatic environment suitable for fish or other forms of animal and plant life.

Water is one of the critical factors in the long-term development of our nation and the importance of water is nowhere more appreciated than in the dry country. Our past practice of *single use* of water is something which Australia cannot afford in the future; wherever possible water should be re-used as many times as possible. In some industrial plants re-use up to 15 times or more is practicable, though it may require some form of filtration, aeration, chemical recovery and even chlorination between processes. Whilst there are fewer industrial plants in the dry country than in the humid and sub-humid coastal zones there are still opportunities for re-use; possibly the simplest case is that of water for domestic purposes being subsequently used for watering gardens. One important aspect of water in the dry country is its great demand for recreation. In some districts it may be stored in large dams or lakes suitable for boating, swimming and fishing, for example Lake Menindee near Broken Hill. Subsequently it may be pumped into reservoirs or dams adjacent to centres of population where picnicking may be permitted in the surrounds but other aspects of usage banned. In some places domestic supplies after use, including that from sewerage systems, may be treated and even chlorinated before use to water amenity plantings, for example at Woomera. In some cases the most lasting pollution arising during domestic use is the addition of detergents, especially those which do not break down readily.

The use and re-use of water is not only influenced by local demand but also by the extent to which there may have been pollution either at primary harvesting of the water or during usage. Pollution of fresh water before use may arise through three principal agencies; the first is that due to the careless or indiscriminate use of modern pesticides, especially DDT and allied chemicals, on the watershed or near streams and dams; the second is industrial discharge and the third is turbidity due to run-off from land subject to poor conservation practices, for example overgrazing or the poor siting of roads. In coastal areas, of course, there is the increasing risk of oil contamination due to either carelessness or marine shipping accidents.

The correct regulation of the quantity of water stored in large dams can be of immense value in ameliorating the effects of high run-off following heavy rain. The stage has now been reached at which stream flow data supplied to computers will indicate what discharge is permissible. In Australia many of the large dams on the major rivers and their tributaries are situated in the humid zone, but the benefits of flood mitigation may be felt in the dry country several hundred, or even a thousand, miles downstream.

Dams normally act as barriers to most forms of water life, particularly fish, and where free movement is desired the provision of water ladders will allow some species to move up or down a stream. The planting of attractive clumps of trees and shrubs adjacent to the foreshore of large bodies of water not only makes the area more attractive for recreation but provides an environment more suitable for many forms of bird life. Since nearly all large reservoirs are owned by the

State governments, there are usually few difficulties in declaring the foreshores national parks and so providing the authorities with greater power in maintaining sound conservation practice in such areas.

AMENITY ASPECTS OF THE RETENTION OF TREES AND SHRUBS ON COUNTRY ROADS

The conservation of natural vegetation along the sides of country roads is a subject in which the public is showing an increasing interest, and is voicing its opinions with greater strength through the journals of the various institutions associated with conservation, resources, parks and recreation,* as well as through the public press. Various State authorities, especially Departments of Conservation, Forestry and Works, are also showing a greater awareness of the desirability of retaining natural vegetation as far as this is practical in terms of road construction and maintenance.

The reserved width of many roads in the sheep and wheat country are greater than the conventional 66 or 100 feet; they are often three chains wide and some are up to five chains. With one chain road reservations and a road formation which may be up to 40 feet overall, there is little scope for shrubs, whilst the trunks of large trees may be prejudicial to safety; in the case of wider reservations there is a greater opportunity to retain trees and shrubs: see the examples shown in Plate 6. This retention is not only of benefit to the road user but also to the adjoining land owners, especially where the latter have devastated their properties of all natural shade and shelter. To the road user the retention of trees has an aesthetic value and this is of direct utility in reducing the dulling effects of a monotonous barren landscape. In addition it reduces the effect of very strong winds and provides shade in the late afternoon; the latter both tempers the heat and may reduce glare from a sun low in the sky. In the arid and semi-arid zones shelter also retards the drying out of unsealed roads and the development of dust (in humid areas drying out may be desirable). Local authorities and road engineers also benefit in that vegetation prevents or reduces soil erosion on the road verges and, in the case of dirt formation and gravel roads, diminishes the loss of road material.

The increased State interest in the planned conservation and regeneration of native vegetation along roads is well exemplified by the action of the Government of Western Australia which, in March 1970, set up the Committee for the Conservation of Road Verges. Very suitably the first chairman appointed was the Conservator of Forests, whilst the Committee included representatives of the Departments of Lands, Agriculture and Main Roads, as well as members from the State Electricity Commission, the Agricultural Protection Board, the Postmaster-General's Department and the Country Shire Councils Association.

In announcing the formation of the Committee the Premier of the State, Sir David Brand, said it was hoped to make an early start on planting native flora along road verges in selected localities. The programme was particularly directed to interstate and overseas visitors who go to Western Australia with the desire of seeing the striking displays of wild flowers for which the State is noted. He said

* For example, *Australian Parks*, Vol. 6 (3), Feb. 1970 which deals with recreation and motoring.

The attitude of man towards trees and shrubs varies with local climatic conditions and the time he has been settled in an area. These three photographs show widely differing aspects of the use of trees in Australia, both in time and geographically.

Top: Go-go Station near Fitzroy Crossing in the Kimberley Division of northern W.A. Here protection from stock, an auxiliary water supply and a tropical climate has enabled good use of native and exotic species. Natural tree growth is mainly along drainage channels, as seen in the background of this illustration. Mean annual rainfall about 20 inches, monsoonal pattern.*

Centre: Oodnadatta, S.A., is a communications centre in the north of the state, and has a surprising number of trees in the township for an average rainfall of just over 4 inches.*

Bottom: This is a photo of Hill End taken by Merlin in 1872. This is a pattern of early development in mining townships and only later attention is given to the amenities of life. Hill End is north-east of Orange in sub-humid zone of N.S.W. (This photo from the News and Information Bureau by courtesy of the Mitchell Library of N.S.W.).

(Photos News and Information Bureau.)*

Plate 3 TREES AND SETTLEMENTS

Top left shows lightly grazed low bushes, mainly species of saltbush (*Atriplex* spp.) and bluebush (*Kochia* spp.), with scattered eucalypts, whilst the adjacent fenced area in top right shows the effect of overgrazing. In the ecologically delicately balanced habitat of the dry country the damage shown in this second photograph may take many years to recover, even under conditions of full protection and above average rainfall. The next two photographs show a cemetery on the left and a commons on the right. The two areas are only separated by a road. *Second from bottom right*, goats on the outskirts of a country town; these animals will eat almost any kind of vegetation and have an unenviable reputation in many countries, for example, Cyprus, preventing regeneration of plant growth and adding the final touches to the destruction of dry country vegetation. Whilst small numbers of milking goats serve a useful role in many parts of dry country Australia it is imperative that there should be strict control of their numbers. *Bottom right* is a scene nearly 2,000 miles from the previous four, illustrating heavy usage on the left of the fence and the relatively protected area of a public highway on the right.

Plate 4 THE EFFECT OF GRAZING BY DOMESTICATED ANIMALS ON DRY COUNTRY VEGETATION—1

The top three illustrations were taken within a hundred yards of one another west of Euston in semi-arid N.S.W.; rainfall about 11 inches. The area was fairly heavily grazed by sheep.

Top left: Shows a mature shrub with a healthy crown beyond the reach of sheep. If a pastoralist wishes to lop such a tree during drought he should not cut off the complete top of the plant but thin the crown to about half. By this means the shrub will keep growing and, given one or two years of above average rainfall, recover most of the lost crown.

Top right: Shows a young shrub which, after heavy grazing, has managed to get its crown above the danger line.

Centre left: This young plant, of the same species, is on the road verge where it has been free from animals. The two bottom photos were taken about 40 miles north-west of Broken Hill, in the arid zone of N.S.W., where the rainfall is down to about 8 inches.

Lower left: Hopbush (*Dodonea attenuata*) is a shrub which some writers rate somewhat low as a fodder plant but, which, as can be seen in this illustration, is eaten readily in the very dry country.

Lower right: A comparison between an edible species in the background and a non-edible one in the foreground. Because of the low productivity of the arid country (the station upon which this photo was taken required about 20 acres to maintain one breeding ewe), the pastoralist cannot afford intensive control of natural regeneration and expansion of non-edible species in place of edible species may be a serious management problem.

Plate 5 THE EFFECTS OF GRAZING BY DOMESTICATED ANIMALS ON DRY COUNTRY VEGETATION—II

The retention of native trees and shrubs on road verges often presents design problems to the road engineer. Where the width of the original road reservation is sufficiently wide, for example five chains, his task is much simpler, as it also is when the route is through open country which belongs to the Crown or is of low value and presents few problems in resumption. Trees and shrubs along roads break the monotony when driving long distances, reduces the severity of strong winds and, in rural areas, affords shelter and shade to stock on adjacent farms which have been denuded of any natural cover. Trees should not, of course, be left in positions where they obscure vision or otherwise are a danger to road traffic.

Top left: The Great Eastern Highway near Perth, where the road engineers have made lanes for two-way traffic by leaving a wide central strip of native trees. The thick tree growth screens headlights of opposing traffic.

Centre left and lower left: Two scenes in the dry country where the natural vegetation provides both interest to the driver and shelter to stock in adjacent paddocks. The lower left illustration is a good example of a natural combination of eucalypts growing to about 35 feet and low shrubs 6-10 feet high, which together make an excellent shelter belt. These photos taken in W.A.

Lower right: A roadside scene near Chinchilla in southern Q'ld.

Plate 6 RETENTION OF TREES AND SHRUBS ALONG ROADS

that many of the one-chain road reserves were considered inadequate and that since 1961 many of the new reservations had been five chains wide. This, of course, not only makes easier the designing of modern roads but also provides space for native plants. Where farmers have heedlessly cleared every tree and shrub from large areas of their holdings the presence of these plants on the roadside are also of benefit in providing at least some shelter and shade to stock in adjacent areas.

The public announcement of the appointment of the Committee also drew to attention that in 1962 the Government introduced legislation prohibiting the picking of wildflowers on Crown land, State forests, land reserved for public purposes, *and along any road.*

TREES AND ROAD SAFETY

Trees and roads are complementary to one another and are not antagonistic as some authorities would appear to believe, if judged by their actions. Admittedly, trees must not be located immediately adjacent to the road pavement nor obscure vision on curves or at intersections; these and other aspects of the use of trees are discussed elsewhere in this book. One viewpoint of trees and roads from the safety angle was recently given in an overseas journal, *Road Safety* (1969):

> There was a time when roadside trees were rated fairly high among the causes of traffic accidents. Vehicles frequently crashed into them after skidding on wet roads, in many cases made more slippery by fallen leaves from overhanging branches. Cars do still collide with trees from time to time, of course, but vastly improved road surface and anti-skid tyres have considerably reduced this motoring hazard.
>
> Nowadays, trees are planted along highways not only to improve their appearance but to make them safer to use. The one is complementary to the other. Such are the motives behind the carefully planned landscaping programme for some of the new motorways and bypass roads in the U.K.
>
> Experience of motorway operations in Europe and the United States has proved, by reduced accident figures, that a well-landscaped road is safer than a neglected one. Trees planted at vantage points create interest on otherwise flat or monotonous stretches of road where drivers may, through boredom or fatigue, become less alert than they should be. Some of the new motorways in Britain, like the German Autobahns, have been specially routed and aligned to take advantage of the natural features of landscape through which they pass. Eventually, therefore, the roads will become part of the landscape itself.
>
> The trees are used to emphasise a good view or to screen industrial or other buildings along the way. New service areas require landscaping, of course, not only to screen the premises from the road, but to make them attractive to travellers who stop for rest and refreshments.
>
> One of the most useful contributions which trees can make to road and traffic safety is in the form of anti-dazzle screening. Clumps or rows of suitably-sized shrubs are planted along the centre strip which divides the carriageways and their opposing traffic streams. Such planting is also a safety measure at road junctions and roundabouts where excessive glare from headlights is a menace to drivers at night. The screens at junctions are carefully sited so that they do not obscure the driver's view in the daytime.

Highway planning is a sphere in which the civil engineer is the technical expert for road design, but in the overall task there is need for the assistance of many other professions—principally that of planner and landscape artist, but the help of agriculturist, arboriculturist and forester may also be of benefit. The present attitude of people in Australia towards trees is quite different from that of one to two hundred years ago; nowadays there is an increasing number who are sincerely attached to trees and deeply resent the unnecessary removal of attractive specimens which have been established for a long time. It is suggested that the civil engineer and town planner could benefit by looking at the possible effects of proposed plans and investigating whether alternative solutions were practical, before commencing the removal of existing trees and shrubs.

FURTHER READING

Australia. Bureau of Census and Statistics. *Official Yearbook*, 1966 (No. 55).

———. *Census of the Commonwealth of Australia, 30 June 1966.*

———. *Demography 1966*, Bull. 84.

Australian Institute of Parks and Recreation. Proceedings of the Sixth International and First Australasian Regional International Congress of Park Administration, Canberra, 18-24 October 1970.

———. *Report.* 41st Annual Conference, Perth, West Australia, 29 September-4 October 1968.

———. 'Man in the Park Environment', a Special Congress Issue of *Australian Parks.*

———. *Australian Parks.* (Official journal of A.I.P.R.—published quarterly).

———. Proceedings of Symposium on Street Trees conducted by A.I.P.R. Victorian Division, 8 June 1968.

———. 'Trees—How do We Meet the Problem?' Proceedings of Symposium conducted by A.I.P.R. South Australian Division, 26 February 1969.

———. Proceedings of Symposium on Turf and Synthetics in Recreation conducted by A.I.P.R. Victorian Division, 6 June 1970.

Australian Parks. (For example, 6 (3), Feb. 1970, which deals with recreation and motoring).

Callaghan, A. R. (1969). *The Impact of Drought on Australian Wheat Production and Marketing.* A paper presented at the Seminar on Drought, University of New England, May 1969.

Chippendale, G. M. & Jephcott, B. R. (1963). *Topfeed.* Extension Article 5, Animal Industry Branch, Northern Territory Administration.

Chislett, G. D'A. (1969). *Drought—its Effect on Wool Production and Marketing (1964-65 to 1967-68).* A paper presented at the Seminar on Drought, University of New England, May 1969.

Department of National Development. *Atlas of Australian Resources.* Commentary prepared by J. C. Foley and B. W. Newman to accompany map-sheet 'Climatic Regions' (1954).

———. Commentary prepared by A. J. Rose to accompany map-sheet 'Dominant Land Use' (1957).

———. Commentary prepared by W. C. Duggan to accompany map-sheet 'Distribution of Stock' (1955).

———. Commentary prepared by M. L. Benson to accompany map-sheet 'Forest Resources' (1967).

———. Commentary to accompany map-sheet 'Mineral Industry' (1959).

———. Commentary prepared by I. R. McLeod and Y. Miezitis to accompany map-sheet 'Population, Distribution and Growth' (1964).

———. Commentary prepared by E. S. Malikides and K. Liddy to accompany map-sheet 'Sheep and Wool' (1968).

Everist, S. L. (1969). *Use of fodder trees and shrubs.* Advisory Leaflet 1024, Division of Plant Industry, Queensland.

Commonwealth Scientific and Industrial Research Organisation (1971). *The Australian Environment*, 4th edition, revised.

Imperial Agricultural Bureaux (1947). *The use and misuse of shrubs and trees as fodder.* Joint publication 10.

Gray, S. G. 'The place of trees and shrubs as sources of forage in tropical and sub-tropical pastures'. *Tropical Grasslands*, 4 (I), 57-62.

Meigs, P. (1953). 'World distribution of arid and semi-arid homoclimates'. *Arid Zone Programme I—Reviews of research on arid zone hydrology*. UNESCO, Paris.

Newman, J. C. & Condon, R. W. (1969). *Land Use and Present Conditions.* A paper presented to the Symposium on Arid Lands, Canberra, May 1969.

Nicholas, H. T. (1953). 'Fish pond farming'. *Journ. Soil Conservation Service of New South Wales*, 9 : 2 (53).

———. (1955). 'Fish pond farming'. *Ibid.*, 11 : 3 (125).

———. (1958). 'Fish pond farming in New South Wales'. *Ibid.*, 14 : 3 (226).

Stannard, M. E., & Condon, R. W. (1968). 'Fodder trees of western New South Wales', *Journal of the Soil Conservation Service of N.S.W.*

Soil Conservation Service. *Journal of the Soil Conservation Service of New South Wales*, Vol. 1-25.

'Trees can make roads safer'. *Road Safety*, February-March (1969) p. 9.

Thornthwaite, C. W. (1943). 'Problems in the classification of climate'. *The Geographical Review*, 33 (2) : 233-255.

———. (1955). 'Climatic classification in forestry'. *UNASYLVA*, 9 (2) : 51-59.

Webb, L. J., Whitelock, D. & Le Gay Brereton, J. (1969). *The Last of Lands.* Jacaranda Press, South Melbourne.

CHAPTER 2

The environment

PART I: GEOGRAPHY AND VEGETATION

To have an adequate understanding of the use of trees and shrubs in the dry country and to make sound decisions when considering new plantings or the introduction of plants, either from elsewhere in Australia or from overseas, it is necessary to know something of the environment. This background information is presented in Chapters 2 and 3. Chapter 2 provides a brief outline of selected aspects of geography and vegetation, whilst Chapter 3 deals more fully with climate.

Most of the land surface of the dry country is at comparatively low altitudes; the Great Western Plateau, which comprises most of Western Australia, is 1,000-2,000 feet high and even the peaks of the sparsely-scattered ranges and hills only attain 4,000-5,000 feet. The absence of large, high mountain ranges and the prevailing dry nature of the air masses which pass over the continent, result in a region which is, in general, one of low rainfall. Associated with the gentle and often indeterminate topography are large areas of unco-ordinated or internal drainage with problems in salt accumulation and water storage. Collectively these two areas total about 1.5 million square miles, which represents 62.5 per cent of the dry country or half of the total area of Australia. Only in a small proportion of the area is there fresh water available from the few large rivers which flow through the dry country. Limited quantities of water, especially treated discharge from town and city sewerage plants, may be available for selected street and park trees but, in general, it is recommended that species planted should be able to thrive without watering. Because artesian water supply is important for stock it is mentioned briefly but, due to its salt content, it is rarely suitable for watering plants. Geologically most of Australia is very ancient and information on the geology is of greater interest to mineral and oil prospectors than to farmers, other than that it provides the latter with indications of the soil types which can be expected to develop under the prevailing climatic conditions. Because soils show a wide range in texture, salt content and fertility, land owners who are considering the introduction of tree and shrub species to their districts should compare local soil conditions with those which prevail where they grow naturally. A superficial examination of the soil is not sufficient, since in some areas a few inches of a sandy loam may overlie a tight, compact clay. A brief general account is given of the vegetation of the dry country and, especially, of generalised pasture types. The latter is, of course, very important as indicating the relative importance of areas dominated by shrub and tree species of value either as regular browse plants, for example saltbush and mulga, or for use during times of drought.

In the preparation of Chapters 2 and 3 considerable use was made of leaflets and maps issued by the Department of National Development in the *Atlas of Australian Resources* series, the Commonwealth Scientific and Industrial Research Organisation publication *The Australian Environment,* and the numerous publications of, as well as data supplied in personal communications from, the Bureau of Meteorology.

PHYSIOGRAPHY

For a geologically-long period of time the continent of Australia has been free from the major earth movements which have resulted in the fold mountains and areas of considerable uplift which now dominate the topography of the other continents. The Alps of Europe, the Rocky Mountains of North America, the Andes of South America, and the Himalayas of Asia rising in places to 20,000 to 30,000 feet in height, all have a major impact on the climate of those areas. There are no comparable mountain and alpine regions in Australia. The ancient highlands of this continent, arising from earth movements as remote as hundreds of millions of years ago, have long been weathered down to comparatively uniform lowlands and this is most marked in the arid and semi-arid zones. The modest size and heights of the Great Australian Divide, with its highest peak only 7,316 feet (Mt Kosciusko), is so near the eastern seaboard that, while locally important, it exerts negligible influence upon climate in the dry country. Because its western flanks form the main catchments of the Murray River and its principal tributaries it does, however, contribute substantially to the water which flows through the semi-arid and arid zones of south-eastern Australia.

The absence of large, high land masses in suitable positions, the latitudinal position (the mainland is chiefly between 12°S. and 39°S. latitude) and the generally dry nature of the upper air masses which pass over the continent in a west to east direction, result in a continent with a low rainfall and a sub-tropical to warm temperate climate. Most of the dry country falls within two major and one minor topographical divisions: the Great Western Plateau comprises nearly half of Australia, covering most of Western Australia, the southern half of the Northern Territory and the western half of South Australia; the Central Lowlands covers most of eastern Australia from the Great Western Plateau to the Great Australian Divide in the east and south-east and from the Gulf of Carpentaria in the north to the coast of parts of South Australia; the third and comparatively small division is the Western Coastal Plain of Western Australia which extends in a narrow belt from 33°S. to about 20°S. latitude, where it meets the Great Sandy Desert on the north-western coast.

The Great Western Plateau is an ancient plateau (shield area), mainly between 1,000 and 2,000 feet in altitude, but with smaller plateaus and ranges rising above the general level and some basins which fall below. The scattered ranges and hills have peaks which may attain 4,000-5,000 feet, but these are of comparatively minor importance in the overall structure compared with the vast distances without any significant variation in topography. For instance, from Broome on the Indian Ocean to Eucla on the Great Australian Bight, a distance of over 1,000 miles, there are no important hills or ranges. It is the same uniformity of plateau and plain which permitted the construction of the longest straight stretch on any railway system in the world, with 297 miles on the Nullarbor Plain section of the Transcontinental Railway between South Australia and Western Australia. Most of the central part of the Great Western Plateau is occupied by three major deserts, whilst the Nullarbor Plain is situated on the southern edge, bordering the Great Australian Bight for about 200 miles on either side of the Western Australian-South Australian border. The name Nullarbor is not Aboriginal as may be supposed, but was derived from the combination of the Latin words *nulla* and *arbor*, meaning 'no trees'. This is certainly quite suitable, since for long distances one can travel without seeing more than saltbushes (*Atriplex* spp.) and bluebushes (*Kochia* spp.) mainly less than two feet in height or, rarely, a few small shrubs, usually *Acacia* spp., which may grow to six feet. The Nullarbor Plain which was laid down beneath the sea, is noted for its limestone caves, many of the larger containing water. The three deserts are the Great Sandy Desert in the north, the Gibson Desert in the centre and the Great Victoria Desert in the south, the latter extending a considerable distance into South Australia. All this area is practically uninhabited except for railway maintenance staff at small settlements along the Transcontinental Railway. The deserts are noted for their long parallel sand dunes and areas of gibber plains.

The ranges are located in three principal areas: towards the coast near 22°S. latitude in Western Australia are the Hamersley and several smaller ranges and it is in this area that iron ore has been found in large quantities, for example at Mt Tom Price; north of Fitzroy Crossing and Hall's Creek, in the far north of the same State, are the lower King Leopold and Durack Ranges, whilst in the southern Northern Territory (formerly known as Central Australia) are the MacDonnell Ranges, which extend mainly west of Alice Springs, and in the south-west crossing the borders of Western Australia and South Australia are the Petermann, Musgrave and Blackstone Ranges, as well as the tourist attractions of Ayers Rock and Mt Olga. Amongst the greatest altitudes in the Great Western Plateau are 4,024 feet (Mt Bruce) in the Hamersley Ranges, 3,070 feet (Mt Ord) in the King Leopold Ranges, 4,955 feet (Mt Zeil) in the MacDonnell Ranges and 4,723 (Mt Woodroff) in the Musgrave Ranges.

The other major topographical division consists of lower land, mainly less than 1,000 feet, except for the upper slopes of the Murray Basin and parts of the Gulf of Carpentaria lowland. Some of the latter area is outside the boundaries of the dry country. Both the Murray Basin and the Lake Eyre Basin are described in the section on drainage.

HYDROLOGY AND DRAINAGE SYSTEMS

Much of arid and semi-arid Australia is a region where there is no surface flow of water, where it only flows in years of above average rainfall or when there are exceptionally heavy falls over a short period. The only large area where there are streams which are normally perennial is that of the Murray Basin in the south-east of the continent. In this latter area the headwaters of the Murray River and its major tributaries (the Darling and the Murrumbidgee) rise in the sub-humid and humid areas of the western side of the Great Divide. The higher rainfall and its greater reliability in the upper watersheds of these rivers contribute the greater part of their flow, but it should never be forgotten by Australians that all these major rivers have ceased to flow in the prolonged droughts of the past. During the succession of dry years in the period 1895-1903 the Murrumbidgee stopped flowing and it is recorded that horse races were held in its bed at Gundagai. Anyone who reads the history of early navigation on the Murray and Darling Rivers will be aware of the fact that river boats were sometimes stranded for long periods due to falling water levels.

No less than approximately one million square miles consists of unco-ordinated drainage; this comprises most of the dry country of Western Australia, a large portion of the Northern Territory and most of western and north-western South Australia. There is also a comparatively small area on either side of the South Australian-Victorian border, and while this does not have flowing streams there are a number of small swamps in the cooler southern portion.

The next largest area is the half a million square miles of internal drainage comprising the Lake Eyre Basin. It ranks as one of the world's largest basins of internal drainage. The Department of National Development's leaflet describes this area as follows:

> The focus of convergence of the component streams is a chain of playa lakes comprising Eyre, Gregory, Blanche, Callabonna and Frome. These lakes, often undistinguished on maps from normal lakes, give the false impression that the continental interior is well watered.
>
> The rivers of this system rise in areas of low and erratic rainfall averaging 10 to 15 inches annually, and are directed towards the most arid portion of the continent where average annual rainfall is less than 7 inches and the potential evaporation is 100 inches or more. It is believed that during Pleistocene times these rivers carried great quantities of water, but now the valleys are heavily alluviated and occupied by shrunken streams which flow only in times of heavy rain.
>
> A number of streams rise to the north of the western MacDonnell Ranges (in the southern Northern Territory) and after traversing them unite to form the Finke, which flows to the south-east. The Todd, Hale, Plenty, Hay and Field all disappear into the Simpson Desert and their lower courses are determined locally by sand ridges.
>
> In western Queensland the Cooper, Diamantina, Georgina and the Mulligan are active in summer. Though rainfall in their catchments is scanty, phenomenal falls occasionally occur, resulting in exceedingly erratic flow. In their lower reaches these rivers have an average fall of less than one foot to the mile and each diverge into a system of interlacing channels (the Channel Country) that traverse an aggregate of eight million acres, which they inundate during the occasional high floods.
>
> One noticeable feature of these seldom flowing streams is the presence of long, ribbon-like waterholes, sometimes in the main channel and sometimes on minor tributaries. Such permanent natural reservoirs vary from small holes only a few chains apart to large waterholes 20 miles or more in length.

It is only very occasionally that these rivers reach Lake Eyre but after rare exceptionally-heavy falls in the catchment they may carry large loads of silt. The waterholes left by the falling streams usually last through all but the worst droughts and are of critical importance in the management of cattle stations in this arid region.

The third large drainage area is that of the Murray Drainage System, which exceeds 0.4 million square miles or nearly one-seventh of the mainland area of Australia. It includes five-sixths of New South Wales, over one-half of Victoria and more than one-sixth of Queensland; there is also a small area of south-eastern South Australia. In New South Wales not all the watershed is in the dry country whilst a large part of Victoria is in sub-humid and humid areas. In the early days of settlement in southern Australia the Murray River and its major tributaries were the key to transportation in that area after the advent of the steam engine but prior to the development of a railway network. Today the waters of the river have assumed an even greater importance as they are essential to the several irrigation schemes developed in the southern portion of the basin, as well as supplying industry and several towns with domestic water; for example centres around the head of Spencer Gulf in South Australia. It appears highly probable that the discharge of this great river into the sea will become almost negligible as plans are developed for greater storage and use of the water in the system. South Australia is critically dependent upon storage and regulation of water from the Murray River. In the lower reaches of the river the presence of salt in the water presents some problems, especially for irrigation, but these can usually be overcome with barrages in critical positions and irrigation by overhead sprays in some locations. The desalination of sea water offers possibilities in the future for minimal quantities of domestic water but the prospects appear remote for production to be cheap enough for use in irrigation.

Elsewhere in the dry country water is drawn from the sub-humid and humid coastal zones; for example Kalgoorlie is supplied from Mundaring Weir near Perth by a pipeline which is about 350 miles long.

The fourth largest area is the Indian Ocean Drainage System of Western Australia between about 20°S. and 30°S. latitude. This is about 0.2 million square miles in extent and includes the DeGrey, Fortescue, Gascoyne and Murchison Rivers. The area is one of low rainfall and mainly exceptionally high temperatures, so that the rivers are intermittent and have many of the characteristics of desert streams. The Department of National Development leaflet states:

> Such rain as does fall comes frequently in violent downpours which cause the rivers to overflow their banks and inundate the country for many miles. These floods soon recede, however, and the water may never reach the ocean. In the dry season, flow ceases entirely, although some of the rivers contain permanent waterholes fed by springs.
>
> The DeGrey, Fortescue and the Ashburton have fairly well-drained channels, but the rivers further to the south traverse the plateau surface with little entrenchment and their valley floors are wide alluvial plains. Some of the rivers have been unable to maintain a passage to the sea against the encroachment of wind-blown sand; some discharge into large coastal salt lakes, whilst others dissipate upon alluvial plains far from the coast. The Gascoyne disappears in sand deposits and discharges to the sea as ground water.

Elsewhere the dry country includes portions, sometimes relatively small, of several coastal drainage areas, for example the Timor Sea and Gulf of Carpentaria drainage systems of Northern Australia. Only in north-eastern Queensland does the semi-arid zone cross the Great Dividing Range to the coastal side of the range. In this case it includes part of the North-Eastern Slopes Drainage System.

ARTESIAN WATER SUPPLY

Artesian and sub-artesian* water is known to exist in about one million square miles of Australia, mainly in the arid and semi-arid zones of the eastern half of the continent. Most of this is a region of limited surface water supplies except for some of the more permanent rivers in the south. In the latter part wool production is the main industry, but cattle for beef is of considerable importance in the northern half, especially in Queensland. All these artesian areas are of the open basin type with natural outlets at a lower level than the saturated intake-level, and the water is in a state of movement under hydraulic pressure.

With an area of nearly 0.7 million square miles the Great Artesian Basin not only dominates the Australian position but is by far the largest in the world. It comprises most of Queensland inland from the Great Dividing Range, a large part of north-eastern South Australia and portions of south-eastern Northern Territory, and northern New South Wales inland from the Great Divide. Other basins, all under 0.1 million square miles in extent, are the Murray located around the lower reaches of the Murray River and over much of which water is only obtainable in selected localities, the Eucla on the edge of the Great Australian Bight and stretching either side of the Western Australian-South Australian border, several smaller basins in coastal areas of Western Australia and one which extends eastwards from Wyndham in northern Western Australia into the Northern Territory (the Ord-Victoria River region).

The following has been adapted from the Department of National Development leaflet *Underground Water* (1953). It has been truly stated that the mineral most vital to Australian development is water—a statement which is no exaggeration when areas of the dry country are taken into consideration. Nearly 50 per cent falls into the arid zone and another 30 per cent falls into the semi-arid zone, whilst much of the remaining 20 per cent has an average annual rainfall which is only from 20 to 30 inches. Even in those parts of the sub-humid zone which receive an annual rainfall of 20-30 inches, underground water may be used to augment surface supplies. Pressure water is used extensively in the city of Adelaide for industry and, in times of drought, to increase domestic supply. In the dry country where artesian water is available, an even greater reliance is placed upon such sources with its advantages over surface storage from run-off, which is affected by a high evaporation and is dependent upon a low and erratic rainfall. Alice Springs, in the Northern Territory, for instance, draws its water from deposits of river gravel which underlie the town itself, whilst both the armed forces in time of war and mining operations in time of peace rely heavily on underground water.

The valuable role of the Great Artesian Basin is well known. Its waters support nearly all of the sheep population of Queensland and nearly one-quarter of

* Where pressure water is struck in a bore and rises to the surface, it is called artesian water, but if it rises above the level at which it is tapped in the bore, but not to the surface, it is known as sub-artesian water. The artesian bore is, therefore, a flowing bore but the sub-artesian bore is one where pumping is necessary in order to bring water to the surface.

its cattle, and make a material contribution in South Australia and New South Wales. Land development that is dependent upon artesian water is necessarily restricted to pastoral pursuits, since not only is the quantity available very limited in terms of irrigation but the quality, especially salt content, may limit the purposes for which it can be used. Some bores may have a low enough salt content to enable small-scale irrigation to be done on virgin soils but continued application in an arid climate with high evaporation inevitably leads to problems in salt accumulation in the soil.

GEOLOGY

The geology of inland Australia is dominated by two major features. In the west, a large part of the land surface consists of very ancient rocks, with little relief, but at a general altitude of 1,000-2,000 feet. East of a line joining the Gulf of Carpentaria to the Great Australian Bight, the rocks are much younger, and form an extensive sedimentary basin, limited in the east by the mountains of the Dividing Range. This sedimentary basin is generally below 1,000 feet elevation and at Lake Eyre the surface is below sea level.

The western shield consists of pre-Cambrian sediments, intensely folded and metamorphosed and with extensive injections of igneous material. Early intrusions were chiefly basic but there are large granitic bathyliths of middle pre-Cambrian age. The basic rocks carry a high content of minerals, such as gold in the Kalgoorlie district, also the lower and middle pre-Cambrian gneisses and schists of the Barrier Range and Mt Isa have high mineral content, mainly silver, lead, zinc and copper. It is the mineral wealth of these pre-Cambrian rocks which has been responsible for most of the non-pastoral development of the drier parts of Australia.

Upper pre-Cambrian lavas, interbedded with sediments, are found in the Hamersley Ranges of north-western Western Australia, which rise to over 4,000 feet, and metamorphosed sediments—slates, quartzites and limestones—of the same age occur in the Flinders Ranges of South Australia, together with Cambrian slates, shales and sandstones. Cambrian limestones form the Barkly Tableland in north-western Queensland and occur also in the Kimberleys of Western Australia. The latter region is a much-dissected plateau, rather complex geologically, with pre-Cambrian, Cambrian and Devonian formations represented. It is separated from the Hamersley Ranges by the Desert Basin, which is composed of Permian sediments covered by sand dunes. The remaining areas of relatively high elevation are the MacDonnell Ranges of Central Australia, which are pre-Cambrian schists and quartzites and Ordovician sandstones, shales and limestones, folded along an east-west axis, and the Musgrave Ranges, further south-west, which consist of pre-Cambrian granites.

Although large parts of the pre-Cambrian shield have been explored geologically because of the mineral potential, there are still large areas which are little known. This is partly because the basement rock is covered by a mantle of erosional detritus, but also partly because of the inhospitable nature of the country and the low population density.

In contrast, the eastern part of the dry country is relatively well known. The climate is less harsh and the presence of artesian water over much of it has enabled more pastoral development.

The major feature of this eastern part is the presence of the Great Artesian Basin, the aquifers of which are Cretaceous sandstones. This basin extends from the Gulf of Carpentaria into northern New South Wales and westward to the Lake Eyre district. It is limited in the south by Silurian and Devonian slates, quartzites and sandstones, but beyond this, the Murray Basin of Tertiary sediments provides another region of low relief with water-bearing limestones.

Much of the Great Artesian Basin was covered by Eocene sediments, chiefly conglomerates, sandstones and shales, but these are now greatly dissected to leave residual mesas over parts of the area.

Oil has been discovered in commercial quantities in the sedimentary formations of south-eastern Queensland, and strikes have been made on the western coast of Western Australia.

Late in the Tertiary there was extensive laterite formation over most of Australia with the creation of a resistant duricrust, relics of which are still to be seen as the protective capping of mesas. During the Tertiary, also, there was some volcanic activity, mainly confined to eastern Australia, and this has resulted in basaltic cappings in areas of central Queensland and New South Wales, bordering the semi-arid zone. A further feature of Tertiary age is the Nullarbor Plain, which was the site of an extension of the Southern Ocean and now consists of a large area of marine limestones, without surface water.

During recent times, erosion of the eastern highlands has supplied large quantities of material which have been deposited over the extensive flood plains of the eastern basin, and many of the soils in the area have been formed from this transported material rather than from the native rock. In contrast, much of the Western Shield has been subjected to a very long period of exposure and some of the features of soils from this area are the results of previous climates.

SOILS

Within any small local climatic area, soils and relative topography are the most important factors in controlling what species of trees and shrubs or, for that matter, any other vegetation will grow satisfactorily. On an Australia-wide scene, however, climate is the first factor to be considered when deciding what species can be planted with reasonable prospects of success. The soil requirements for trees differ somewhat from those for shallow-rooting pasture and cereal crops. To the prospective tree grower soil texture and depth are usually of greater importance than fertility as such, whilst in the arid zone and, to some extent, the semi-arid zone the presence or absence of a high salt content may be a controlling factor in the choice of species.

Most of the major soil groups have relatively low levels of fertility and many, especially the leached podzols of the sub-humid and humid zones, are very poor. There are exceptions, of course, for example the krasnozems, alluvials and recent volcanics which occur mainly in the humid zones, whilst black earths are notably fertile. Some of the terra rossa soils are moderately good but may be deficient in one or more minor elements. In the dry country many of the desert loams and arid red earths are at least moderately fertile; in the arid zone, in particular, low rainfall inhibits leaching and the comparatively sparse vegetation means a relatively low drain on nutriments. In addition, the root systems of most trees and shrubs spread for considerable distances from the plants, and so are able to draw moisture and nutriments from a large volume of soil. It is considered by the present writer that in areas of low rainfall, that is, below about 15 inches in southern Australia and 20-25 inches in the north, any introduced species should be capable of satisfactory growth in the soils without the addition of any fertiliser. Where the rainfall is higher or where watering is possible, the addition of manures such as superphosphate can be expected to give good results both during the year of establishment and later. In many cases it may be desirable to add appropriate minor elements. When a farmer or supervisor of parks and gardens wishes to induce optimum growth for developing a shelter belt or an area of ornamental planting as quickly as practicable, it is important to remember that, as with nearly all agricultural crops, trees respond to soil cultivation, especially deep ripping prior to planting in the case of shallow or very heavy soils, or those with a hardpan.

During the Tertiary era there was extensive peneplanation of the continental surface. This was followed by deep weathering and lateritisation, with the formation of duricrust. The peneplain was subsequently uplifted, probably in the late Tertiary, and this uplift initiated further erosion and dissection, with the formation, over much of the area, of a new plateau surface. This is generally from 50-200 feet below the old plateau.

Residuals of the old plateau surface remain as mesas which are characteristic features of the landscape, especially in Queensland, Central Australia, and Western Australia. There are also several regions where the harder basement rocks have been slow to erode and now project above the general level, to form the *tablelands and ranges* of Prescott. The most extensive of these regions are the Hamersley Range and the Kimberleys of Western Australia, the country west and south of the Gulf of Carpentaria, the Barrier Range of New South Wales, the Flinders Range of South Australia, and the MacDonnell Range of the Northern Territory.

The vegetation on these ranges is insufficient to prevent stripping of the weathered material, and the soils are skeletal.

A general decrease in the humidity of the climate during the late Tertiary uplift has resulted in alluviation, with the formation of numerous lakes, especially in Western Australia. Because of the inland drainage system many of these lakes are now saline, or gypsum clay pans have formed. At the same time, much of the alluvium has been cemented to form siliceous or calcareous hardpan.

Further increase in aridity, probably post-Tertiary, resulted in widespread wind-sorting of the erosional detritus and formation of sand plains and sand dunes.

Australia, with its great size and, notably, its range in latitude of about 33 degrees (approximately 2,300 miles) and differences in climate, shows a wide variety of soils types. This is particularly marked in the present arid and semi-arid zones. Table 2.1 shows the approximate areas of soil mapping units based on a soils map prepared by C. G. Stephens of C.S.I.R.O., with the allocation of climatic zones by the present writer. The purpose of the table is to show the relative importance of the various soil mapping units in the dry country and to provide a reference for the summary description of soils which follow. This outline is based on the work of Prescott (1947) and Stephens (1953, 1956 and 1963).

Podzolic soils (5)* occur on the wetter sides of the semi-arid zone in south-eastern and north-eastern Australia but as they are not typical of the dry country will not be described.

Krasnozems (6) have their principal occurrence in the humid zone of northern coastal Queensland, especially in the sugar-growing area around Cairns and Tully. Mention of them is made here because there is a limited distribution in the higher rainfall parts of the semi-arid zone, both in Queensland and New South Wales. They are red to brown in colour, deep friable clay soils with fine texture throughout the profile, and of high natural fertility.

There are appreciable areas of red-brown and brown earths (11) in the 15-25 inch rainfall area of southern Australia. They are the major wheat-growing soils, as well as supporting both native and improved pastures. The typical profile consists of a red-brown to brownish loamy surface horizon, which may be 12-24 inches deep, over clay. There may be considerable accumulations of limestone. When the surface layer is of light texture, wind erosion may present problems on cultivated areas. Prior to the advent of European man these soils supported a woodland formation with *Eucalyptus* as the principal genus.

Black earths (13) have their main occurrence in the sub-humid and humid zones of Queensland, but extend into the higher rainfall parts of the semi-arid zone. They are essentially deep profiles with dark-coloured surface and subsurface layers, and clay which is dark brownish-grey in the upper layers but yellow to brownish in the lower, with lime in the form of specks, pellets or concretions. These soils are characterised by marked swelling on wetting and shrinkage on drying. The vertical movement of blocks or strips of soils leads to an irregular or wavy surface, known as gilgai or crabhole and is not confined to black soils.

Black earths are used both for pasture and wheat production, and in sub-tropical areas their phosphate content is sometimes sufficiently high to preclude a response to the application of superphosphate. Arable land is subject to water erosion which may be difficult to control by conventional means. Few native species of trees and shrubs will grow satisfactorily on black earths and care needs to be taken in selection and establishment. The Forestry Department of Queensland has studied the problem of amenity planting on this type of soil and is in a position to give advice on the subject. Of interest, although outside the scope of this book, is the necessity for good design and care in construction of foundations for brick and stone buildings on black earths.

Weisenboden soils are a variation of black earths and are mapped with them.

Solonised brown soils (14) occur principally in the semi-arid zone of south-western Western Australia and in South Australia, north-western Victoria and southern New South Wales. The rainfall of the area is mainly 9-16 inches. The principal sub-group consists of grey-brown or red-brown, medium-textured soils up to 12-24 inches deep over deep accumulations of lime mixed with clay. They are used

* The numbers given in brackets after the first citing of soil type refers to Table 2.1, where areas in the various climatic zones are given for the type.

TABLE 2.1

SOIL AREAS IN AUSTRALIA

Approximate areas of soil mapping units and allocation to climatic zones(*a*)
(all areas in hundred square miles)

Soil mapping unit	*Total area*	*Per cent of total*	*Climatological zones*		
			Arid	*Semi-arid*	*Sub-humid, humid*
SOILS OF THE ALPINE AND PERHUMID ZONES					
1. High moor peats + alpine humus soils	110	0.5	—	—	110 100.0%
2. Peaty podzols	40	—	—	—	40 100.0%
SOILS OF THE HUMID ZONES					
Leached Soils—					
3. Acid swamp soils	10	—	—	—	10 100.0%
4. Podzols	60	—	—	—	60 100.0%
5. Grey-brown + brown + red + yellow + meadow podzolic soils	2300	7.5	—	223 9.7%	2077 90.3%
6. Krasnozems (and lateritic krasnozems)	270	1.0	—	69 25.6%	201 74.4%
7. Yellow earths	50	—	—	37 74.0%	13 26.0%
8. Prairie soils	10	—	—	—	10 100.0%
Soils on Calcareous Materials—					
9. Groundwater rendzinas (and rendzinas)	20	—	—	—	20 100.0%
10. Terra rossa soils	10	—	—	—	10 100.0%
SOILS OF THE SEASONALLY HUMID ZONES					
11. Red-brown earths (and brown earths)	650	2.0	—	487 74.9%	163 25.1%
12. Non-calcic brown soils	80	0.5	—	48 60.0%	32 40.0%
13. Black earths (and wiesenboden)	750	2.5	—	213 28.4%	537 71.6%
SOILS OF THE SEMI-ARID ZONES					
14. Solonised brown soils	1450	5.0	299 20.6%	1046 72.2%	105 7.2%
15. Grey + brown soils of heavy texture	2900	10.0	1307 45.1%	1492 51.4%	101 3.5%
16. Brown soils of light texture	900	3.0	215 23.9%	657 73.0%	28 3.1%
SOILS OF THE ARID ZONE					
17. Desert loams	210	0.5	155 73.8%	55 26.2%	—
18. Grey-brown to red calcareous desert soils	1400	4.5	1171 83.6%	229 16.4%	—

(*a*) The purpose of this table is to show the relative importance of various soil mapping units in the arid and semi-arid nes. The information on soil mapping unit and total area have been taken from p. 21 of the Soils Leaflet, 2nd ed. (1963) issued by the Department of National Development in the *Atlas of Australian Resources* series. The estimates for areas in the climatological zones have been made by the present writer, using climatological boundaries superimposed on the map-sheet *Soils*, prepared by C. G. Stephens, Head of the Soil Survey and Pedology Section, Division of Soils, C.S.I.R.O.

TABLE 2.1—*continued*

Soil mapping unit	*Total area*	*Per cent of total*	*Climatological zones*		
			Arid	*Semi-arid*	*Sub-humid, humid*
19. Arid red earths	2100	7.0	1837 87.5%	263 12.5%	—
20. Red and brown hardpan soils	1750	6.0	1667 95.3%	83 4.7%	—
21. Stony desert tableland soils	1500	5.0	1470 98.0%	30 2.0%	—
22. Desert sand-plain soils	1650	5.5	1266 76.7%	384 23.3%	—
23. Desert sandhills	4000	13.5	3879 97.0%	121 3.0%	—
SOILS WITH WEAK CLIMATIC ZONATION					
Lateritic Soils on Old Landscapes—					
24. Lateritic podzolic	1150	4.0	—	614 53.4%	536 46.6%
25. Lateritic red earths	1000	3.5	67 6.7%	633 63.3%	300 30.0%
26. Calcareous lateritic soils	10	—	—	10 100.0%	—
Saline and Solonised Soils—					
27. Solonchaks	130	0.5	20 15.4%	45 34.6%	65 50.0%
28. Solodised solonetz + solonetz	440	1.5	—	240 54.5%	200 45.5%
29. Solodic soils	650	2.0	—	200 30.8%	450 69.2%
SOILS WITH NO CLIMATIC ZONATION					
30. Alluvial soils	70	—	14 20.0%	3 4.3%	53 75.7%
31. Skeletal soils	3800	13.0	1257 33.1%	1835 48.3%	708 18.6%
32. Calcareous coastal sands	60	—	3 5.0%	35 58.3%	22 36.7%
Lakes and Salinas	220	0.5			

Total Area of Australia: 2,971,000 square miles.

for sheep grazing and, where rainfall is adequate, for wheat growing, but production is lower than on the red-brown earths. On arable areas wind erosion is a serious danger. With irrigation in the mid-Murray River area they are very suitable for horticultural crops.

Grey and brown soils of heavy texture (15) comprise large areas in both the semi-arid and arid zones of eastern and northern Australia; rainfall shows a wide range of about 10-25 inches. They occur mainly on the lower parts of extensive alluvial formations. These are soils of fine texture; clay dominates all horizons and, when lime or gypsum is present, gilgai formation may be present. They are characterised by grasslands which are frequently treeless and, where there is appreciable salt present, the typical vegetation may consist of saltbush and bluebush (*Chenopodiaceae*). These soils are largely used for grazing on the natural vegetation, since they do not respond well to pasture improvement but, under irrigation, rice is grown on them in the Riverina area of New South Wales.

Brown soils of light texture (16) are mainly found in the semi-arid zone around Cobar in New South Wales, between, but not at, Hughenden and Normanton in Queensland, and on several coastal areas of north-western Western Australia. Rainfall is mainly 11-14 inches in southern areas and parts of Western Australia, but up to 30 inches in the north. The surface horizon is usually a brown sand or sandy loam with little apparent organic matter; with increasing depth this changes to sandy clay and eventually to clay. Some lime may be present. They are used almost exclusively for grazing cattle and sheep on natural pasture.

Desert loams (17) are of limited extent and occur mainly in the southern arid zone; rainfall is 8-12 inches. The shallow upper horizon is brown to red in colour and varies in texture from loam to clay; frequently the surface carries a variable pavement of stones or a superficial layer of coarse sand only a fraction of an inch thick. The soils are usually alkaline. When rain does occur it is often of high intensity for short periods and soil erosion may result. The land is used for low intensity grazing on the mainly ephemeral natural herbage and the perennial edible shrubs.

Grey brown to red calcareous desert soils (18) are found on the Nullarbor Plain and in much of central northern South Australia with rainfall 6-10 inches. They consist of shallow powdery loams to clays with little or no differentiation in the profile. Lime is common. In places there is sparse grazing but much of the area is unused.

Arid red earths (19) occur mainly in the arid zone but also extend to the semi-arid zone in the Charleville area of Queensland with rainfall 7-18 inches. They are soils of fine to moderate texture and in the higher rainfall regions are used for grazing on natural pasture and edible shrubs.

Red brown hardpan soils (20) occupy a very large area in the arid zone of Western Australia with rainfall 7-10 inches. They carry mulga (*Acacia aneura*) and other shrubs, as well as ephemeral herbage which supports a very light grazing industry.

Stony desert tableland soils (21) occur almost exclusively in the arid zone with rainfall 5-12 inches. They are found on characteristic tableland relics of a former landscape and their surface is covered by a pavement of broken and polished siliceous stones. Beneath the stony surface is a shallow surface horizon of loam to clay texture. The driest areas of occurrence are mainly unused but some areas support a very sparse stocking of cattle.

The largest area of desert sand-plain soils (22) occur in the arid zone but they extend into the semi-arid zone of northern Australia with rainfall 7-12 inches. The surface soil is composed of brown to red loose sand over red brown to red sands and, at depths down to about 3 feet, ironstone. The spinifex and other natural vegetation is of limited value for cattle grazing.

Desert sandhills (23) are composed of deep, partly mobile, sands with a marked likelihood of wind erosion with rainfall 6-12 inches. Except on the fringes the type is mainly unused. This is an arid zone type.

Lateritic podzolic soils (24) are found in the semi-arid and humid south-west

Maps of Australia show a number of lakes, often of quite large size, in various parts of the southern arid and semi-arid zones. Early explorers often hoped to find usable water in these depressions and were nearly always disappointed, since they only hold water after exceptionally heavy rains. Even then the water is often so salty that it cannot be used either by men or animals. Likewise, most drainage channels shown on maps rarely contain water.

Top left: Lake Lefroy, near Kambalda, W.A. This is a dry salt area, some of which has been pegged for nickel prospecting.

Centre left: A drainage channel near the border of the semi-arid and sub-humid zones in W.A. shows the effect of the end of a cyclone a few weeks before the photo was taken. Rainfall for the area about 18 inches.

Lower left: Tree growth in the arid zone is often dependent upon access to underground water and if there is a change in the location of such water the trees may die. The species shown is river red gum (*Eucalyptus camaldulensis*) in arid western N.S.W. where the rainfall is about 8 inches.

Lower right: When water is available in the dry country many cool temperate species thrive; this illustration shows weeping willow (*Salix babylonica*) growing along the banks of an irrigation channel in semi-arid southern N.S.W.

Plate 7 WATER IN THE DRY COUNTRY—ON THE MAP AND ON THE GROUND

In many of the arid parts of the world the demand for water has resulted in major schemes to divert water from large rivers or high-rainfall coastal catchments to the dry country. The first major project in Australia was the Goldfields Water Supply to the mining centres of Kalgoorlie and Coolgardie with water from the coastal region of W.A.
The construction of the Mundaring Weir on the Helena River near Perth was started in April 1898 and water reached the Mt Charlotte reservoir in Kalgoorlie in January 1903. This plate shows scenes along the 348 mile route.

Top left: Mundaring Weir.

Upper centre left: The pipeline near the weir. Here the water is being pumped to higher levels.

Lower centre left: The pipeline goes through dry country vegetation,

Bottom left: and passes over railway lines.

Upper right: At higher elevations en route there are large storage tanks.

Lower right: Finally the water reaches Mt Charlotte reservoir. Part of the City of Kalgoorlie appears in the background.

Plate 8 WATER IN THE DRY COUNTRY—THE LONG PIPELINE

Water is of critical importance for stock raising in the dry country. Many property owners have a large part of their investment in the tapping of underground water and supplying it to stock.

Top: A travelling mob of cattle drinking at a Government bore on the main north-south stock route near Alice Springs, N.T.

Centre left: Stockmen and drought affected country in north-western N.S.W.

Lower left: Sheep in the dry country; old and new methods of droving.

Lower right: Due to the concentration of stock near watering points any feed present is eaten out; Alice Springs area of the southern N.T.

(*Photos News and Information Bureau.*)

Plate 9 WATER AND STOCK RAISING IN THE DRY COUNTRY

Top: Drought is a recurrent hazard to anyone who gets his living from the land and especially to the pastoralist in the arid and semi-arid zones, the risk of severe droughts is never from his mind.

Left: When rainfall has been markedly below average for some time a pastoralist must decide whether to dispose of a proportion of his stock or hold them in the hopes that the incipient drought will break. If he holds the animals they will lose condition if the drought persists, and are likely to die. Ravenous cattle will break down shrubs and small trees to get at any foliage available. This illustration was taken during a major drought and shows an overgrazed area about 75 miles north-east of Alice Springs, N.T. Many of the stumps left are those of supplejack (*Ventilago viminalis*), a small tree which is very palatable to animals.†

Bottom left: The making of better arterial roads and the introduction of *road trains* has helped the grazier in moving stock to markets without serious loss of condition.*

Bottom right: Even country which appears inhospitable can support animals in reasonable conditions if there is dry feed and water available. This photo was taken on Cordillo Downs in the far north-east of S.A., where the mean annual rainfall is under 6 inches.*

(† Photo G. Chippendale, Forest Research Institute, Forestry and Timber Bureau.)

(Photos to News and Information Bureau.)*

Plate 10 DROUGHT AND ANIMALS IN THE DRY COUNTRY

of Western Australia and in the same zones of northern Australia; rainfall has a wide variation from 10-50 inches. The surface layers are leached and usually shallow; beneath them is a profile of laterite over mottled or nearly white clay. Typically these soils are covered with woodland which, in some areas, may change to commercial forests. In suitable climatic areas, for example, the higher rainfall parts of south-western Western Australia, they can be developed to carry improved pasture.

Lateritic red earths (25) are found in the semi-arid and humid zones of northern Australia with rainfall mainly 15-50 inches. They support woodland and tall tropical grasses.

Calcareous lateritic soils (26) are of minor importance whilst solonchaks (27), solodised solenetz (28) and solodic soils (29) have limited importance, occurring mainly in the semi-arid zone. Solonchaks are saline soils; solonised solenetz soils are deficient in phosphate and many minor elements but when these are added satisfactory pastures can be produced. Solodic soils mainly occur in semi-arid and humid Queensland. They are mainly utilised for natural pasture.

Skeletal soils (31) occupy extensive areas in semi-arid and humid northern Australia, as well as some of arid Western Australia; rainfall mainly 20-45 inches. They are essentially stony or gravelly soils in which there is no profile development. Normally they are shallow and a large proportion of their volume is filled with rock fragments. They mainly occur on topography which has sufficient slope for natural erosion to remove surface soil. Skeletal soils are not favourable for utilisation, though sparse grazing is available for cattle. Open woodlands occur through most of this type of area but few of the trees are suitable for commercial timber production.

DETERMINATION OF SALINITY

It has been indicated elsewhere that the salinity of many soils in the dry country of Australia, especially in the arid zone, is so high that it is prejudicial to the growth of most species of trees and shrubs. The salt level may change over quite short distances if there is even the slightest variation in altitude, and the ready estimation of the salt content is of importance to all people concerned with the planning of amenity planting. The Division of Soils, Commonwealth Scientific and Industrial Research Organisation, has developed equipment which is simple to operate and by means of soil reaction gives adequate indications for the salt level, and it gives semi-quantitative values like those obtained with field pH kits. The C.S.I.R.O. saltmeter is described in the *Journal of the Australian Institute of Agricultural Science,* Vol. 24 (1), pp. 51-53, March 1958, whilst new scales for use with the saltmeter are published as *Technical Memorandum* 38/1969 of the Division of Soils. The saltmeter is manufactured by Fulham Instruments Manufacturers, 29 Coral Sea Road, Fulham, S.A. 5024.

VEGETATION

Natural vegetation tends to occur in types which are more or less homogeneous, that is with considerable similarity in growth habits of the various tiers or strata which compose a unit or stand. The primary classification is usually into formations on the basis of their physical structure, with further division into associations.

The *formation* is a type which tends to occur throughout the world under similar environmental conditions of temperature, rainfall and soil type, and is therefore an international criterion. *Rainforest*, for example, is found in regions of high rainfall, *broad-leaved temperate forests* occur in areas of moderate rainfall and temperatures, whilst the extensive, nearly pure, *coniferous forests* belong to the cool temperate regions. When the latter occur nearer the equator they are invariably at high altitudes. The first formation is represented in coastal eastern Australia, especially north-eastern Queensland, northern coastal New South Wales, parts of southern Victoria and much of Tasmania. Australian examples of the broad-leaved forests are the wet sclerophyll eucalypts of the southern part of the continent, whilst the typical cool temperate coniferous forests, which are a feature of the higher latitudes of the northern hemisphere, are only represented by fragmentary areas. *Woodlands*, that is areas with trees mainly 33-100 feet high and grasses or shrubs occupying much of the ground cover, are characteristic of areas which are warmer and have a lower rainfall than the closed forests. They occupy very large parts of the semi-arid and sub-humid zones in Australia. Grasses and other herbaceous formations are found in widely-differing climates and frequently on soil types which are unsuitable for tree growth.

Associations are usually classified on the basis of floristics and, hence, are of more local application than *formations*. Australian examples include such types as the *Eucalyptus populnea* (poplar box)—*Callitris glauca* (white cypress pine) association of eastern temperate regions, especially New South Wales and southern Queensland, the *Acacia aneura* (mulga) association which is widespread in southern arid and semi-arid areas, the association of saltbush (*Atriplex*) and allied small shrubs which are members of the Chenopodiaceae and the Mitchell grass (*Astrebla*) association which occurs in the warmer semi-arid parts.

The classification of vegetation presents many problems, but in recent years a number of Australian plant ecologists have endeavoured to prepare a simple and workable classification of the major structural forms found in this land. The result has been treated by R. L. Specht in Chapter 6 for the revised fourth edition of *The Australian Environment* (1971). The book is recommended for further reading.

The formations referred to earlier have been classified on the basis of growth form and height of the tallest layer, that is trees in three height categories, shrubs in two categories, hummock grasses and herbs, together with a density grouping of the foliage cover of the tallest stratum.* The height classes are greater than 30 metres, 10-30 m and 6-10 m, the shrub heights are 2-8 m and 0-2 m. A tree is defined as a woody plant more than 6-8 m tall, usually with a single stem and a shrub as a woody plant less than 6-8 m tall, frequently with many stems arising at or near the base. Foliage cover is divided into four categories on the basis of the percentage of the ground cover of the tallest stratum, as viewed from the air; these are dense (70-100 per cent), mid-dense (30-70 per cent), sparse (10-30 per cent) and very sparse (less than 10 per cent).

* Acknowledgment is made to Professor Specht and the publishers of *The Australian Environment*.

The broad pattern of distribution of vegetation in Australia is one of concentric zones, corresponding to the zones of increasing rainfall, but modified by the seasonal incidence of the rain, so that grasslands are characteristic of the semi-arid summer rainfall regions of the north but shrubs occupy the corresponding winter-rainfall regions of the south.

Particularly in the arid zone the presence of parallel sand dunes and low ridges, alternating with depressions, give rise to a characteristic banding of the vegetation. On the top of the dunes there may be a complete absence of plants or clumps of *Triodia or Zygochloa* grass, but part way down the slopes there is also sometimes a very sparse scattering of small shrubs, often species of *Acacia.* When the ridges are of a stony or lateritic nature there may be some of the tussock type of grasses and small shrubs may occur in clumps or along drainage lines; the principal genera of the latter are *Acacia, Cassia, Eremophila, Grevillea* and *Hakea.* The vegetation of the interdune areas depends largely upon the amount of sand present and the depth of the soil; where there is deep sand some of the tussock type grasses may dominate the cover but where it is shallow the herbaceous growth becomes sparse and widely scattered small shrubs are the principal form of vegetation.

This pattern of vegetation zoning is further modified by topography and soils through their effect on drainage, moisture supply and soil nutrients. In many parts of southern Australia, however, the occurrence of soils which are either relics of a past climate, or influenced by salt, may have overriding effects, so that the present climate is of secondary importance.

Some of the more important vegetation types which occur in the arid and semi-arid zones are described briefly.

Woodland is the term applied to vegetation in which the tallest stratum consists of trees 10-30 m in height and where the foliage cover is 10-30 per cent. When, as is often the case, the trees are associated with a grassy ground cover the type is qualified as *grassy woodland.* Where the density is very sparse the term *open woodland* is used by some ecologists. Woodland in some form or another is mainly found in the 15 to 30-inch rainfall belt and is a characteristic type of the semi-arid zone. The principal tree species are eucalypts (*Eucalyptus* spp.), but locally cypress pine (*Callitris* spp.), sheoaks (*Casuarina* spp.) and some of the larger wattles (*Acacia* spp.) may dominate the tallest stratum. Typical dominants are, for the Northern Territory, *Eucalyptus terminalis, E. dichromophloia* and *E. papuana*; for Queensland *E. crebra* and *E. tessellaris,* with *E. brownii* and *E. melanophloia* on coarse, stony soils; for New South Wales *E. woollsiana* and *E. albens* or *E. populnea* and *Callitris glauca*; for Victoria *E. microcarpa* and *Casuarina luehmannii*; for South Australia *E. odorata, E. leucoxylon* and *Casuarina stricta*; and for Western Australia *E. loxophleba* and *E. salmonophloia.*

When there is a well-developed lower layer of shrubs associated with the trees the name given is *shrubby woodland.*

Open-scrub formation (density 30-70 per cent) is found over large areas of southern Australia where the mean annual rainfall is 10-20 inches. The typical dominant growth is mallee eucalypts. When the density falls to the range of 10-30 per cent the type is then known as *tall shrubland* and mulga (*Acacia aneura*) communities are the chief examples of this type of vegetation. Mulga, however, also occurs in dense stands and could, accordingly, also be classified as *closed-shrub* (70-100 per cent) or *open-shrub* (30-70 per cent). The rainfall for the mulga areas is mainly 8-12 inches but on some soils it may be as high as 12-15 inches. Depending upon the density of the upper canopy of these two types there may be a lower shrub stratum of species of the genera *Cassia* and *Eremophila,* and a herbaceous layer of perennial grasses and seasonal herbs. Many of the shrub species are an important source of forage for stock.

Low shrubland formation, previously known as shrub steppe, is common in large areas of southern Australia. Typically it is dominated by low shrubs of the family *Chenopodiaceae,* especially saltbush (*Atriplex* spp.) and bluebush (*Kochia* spp.). Many of these are eaten by stock and in some areas form an important part of the food supply.

Hummock grasses consist of large clumps which may be up to several feet in diameter and with the flowering stems growing to five feet or more. The principal genera are *Triodia* and *Plectrachne*, the common name of which is spinifex. This type is typical of large areas of sandplains and sand dunes throughout the arid zone.

Tussock grassland is represented by several formations; the most important in the dry country being that dominated by Mitchell grass (*Astrebla* spp.) which extends across northern Australia from the Kimberley area of Western Australia, through the Barkly Tablelands of the Northern Territory and Queensland and as far south as northern New South Wales. This type occurs on very large areas of rolling plains with calcareous, cracking clay soils which are treeless with the exception of occasional lines which grow along the drainage channels where soil moisture relationships are more suitable for tree growth. Away from these channels tree establishment will always be very difficult due to the seasonal drying out and cracking of the soils.

PASTURE ZONES

Over much of the dry country livestock are carried on natural vegetative communities which may be little changed in species of grasses and shrubs since the advent of European settlers. The proportion or dominance of number of plants of any one species or group may, however, be considerably altered due to selective feeding by the introduced animals. In the arid zone the change in native species is the least; there is a greater removal of trees and shrubs and some introduction of exotic grasses and clovers in the semi-arid zone. The major artificial pastures due to soil cultivation, the sowing of introduced grasses and clovers and the addition of fertilisers only attains their maximum development in the sub-humid and humid zones of the continent and in Tasmania.

In view of the importance of 'topfeed'—accessible shrub and tree growth—in the dry country it has been considered desirable to give a brief summary of natural pastures in Australia; this has been based on Chap. 7 of *The Australian Environment* (1971), and due acknowledgment is made to the authors.

With the exception of moist temperate perennial grasses, mixed alpine and moist temperate perennial grasses and blue grasses, all types occur within the boundaries of the dry country; but tropical tall grass and south-western and

TABLE 2.2

AREAS OF GENERALISED PASTURE TYPES IN AUSTRALIA

Approximate areas of pasture zones and allocation to climatic zones(*a*)
(all areas in hundred square miles)

	Generalised pasture zone	*Total area*	*Per cent of total*	*Climatic zones*		
				Arid	*Semi-arid*	*Sub-humid, humid*
1.	Moist temperate perennial grass	1070	3.6	—	7 0.7%	1063 99.3%
2.	Winter annuals (south-western)	980	3.3	—	467 47.7%	513 52.3%
3.	Winter annuals (southern)	550	1.8	—	374 68.0%	176 32.0%
4.	Temperate short grass	1800	6.1	130 7.2%	935 52.0%	735 40.8%
5.	Mixed alpine and moist temperate perennial grass	160	0.5	—	—	160 100.0%
6.	Tropical tall grass	1940	6.5	—	698 36.0%	1242 64.0%
7.	Bunch spear grass	1180	4.0	—	184 15.6%	996 84.4%
8.	Brigalow	620	2.1	—	256 41.3%	364 58.7%
9.	Blue grass	120	0.4	—	56 46.7%	64 53.3%
10.	Mitchell-Flinders grass	2310	7.8	875 37.9%	1412 61.1%	23 1.0%
11.	Semi-arid open forest	690	2.3	—	690 100.0%	—
12.	South-western semi-arid pasture	790	2.7	199 25.2%	591 74.8%	—
13.	Arid scrub	750	2.5	410 54.7%	340 45.3%	—
14.	Saltbush steppe	740	2.5	663 89.6%	77 10.4%	—
15.	Mulga scrub	4810	16.2	4226 87.9%	584 12.1%	—
16.	Spinifex (western)	1120	3.8	967 86.3%	153 13.7%	—
17.	Spinifex (northern)	4610	15.5	1718 37.3%	2470 53.6%	422 9.1%
18.	Desert	5250	17.7	5112 97.4%	138 2.6%	—
	Lakes and Salinas	220	0.7			

Total Area of Australia: 2,971,000 square miles

(*a*) The purpose of this table is to show the relative importance of generalised pasture zones, especially those including important shrub species such as mulga (*Acacia aneura*) and saltbush (*Atriplex* and *Kochia* spp.), in the arid and semi-arid zones. The classification of pasture is that given in Fig. 24 in *The Australian Environment*, published by C.S.I.R.O. in association with the Melbourne Press, and was prepared by C. S. Christian, C. M. Donald and R. A. Perry. The estimates of areas of types and allocation of climatic zones were made by the present writer using climatological boundaries superimposed on a copy of Fig. 24.

southern winter annuals are of limited occurrence. Table 2.2 has been compiled from Fig. 24 (Pasture Map of Australia) in *The Australian Environment,* but with climatic boundaries added by the present author. Some of the characteristics of the principal pasture zones are summarised below.

Temperate short grass

This occupies most of semi-arid New South Wales and has a limited extension into the southern part of the arid zone; in addition it occurs in part of semi-arid Victoria. The original vegetation type was mainly savannah and savannah woodland, chiefly with eucalypts as the main species in the higher rainfall parts, but also cypress pine; in the drier sections there are sheoaks and many other small trees and shrubs. The principal grasses are *Danthonia* (wallaby grasses) and *Stipa* (spear grasses), the former being far the superior for grazing; in the northern part where rainfall tends to a summer maximum, genera include *Panicum, Digitaria* and *Eriochloa.* Carrying capacity is high.

Unlike most types of the dry country this is one where there has been considerable pasture improvement; subterranean clover and *Phalaris tuberosa* are amongst the more common introduced species, and where there is regular top-dressing the native perennials are suppressed to permit the introduction of annuals.

Tropical tall grass

Most of this type falls in the sub-humid and humid zones of tropical north Australia, with only a narrow fringe within the dry country. The pastures constitute the ground vegetation of woodlands or open forest, usually dominated by eucalypts. The grasses are mainly perennials of the tussock type four to eight feet high, with some tall annuals which may assume local dominance. Genera include *Sorghum, Chrysopogon, Themeda* and *Heteropogon*; on the more unfavourable sites there may be species of *Triodia* and *Aristida.* In general these pastures are of low nutritive value except for short periods after fire or rain.

Bunch spear grass

This mainly occurs in coastal areas of Queensland but there is a small extension into semi-arid country in the north-east of the State. Pasture genera includes some which occur in the adjoining tropical tall grass type. On the more favourable sites, mainly outside the dry country, introduced tropical grasses and annual legumes show promise for pasture improvement.

Brigalow

This type is mainly restricted to south-western Queensland on either side of the semi-arid-sub-humid boundary. It takes its name from brigalow (*Acacia harpophylla*), which is the dominant tree over much of the type; associated small trees and shrubs include belah (*Casuarina cristata*), wilga (*Geijera parviflora*), budda (*Eremophila mitchellii*), as well as teatree (*Melaleuca pubescens*), yellowwood (*Terminalia oblongata*) and gidgee (*A. cambagei*). Most of these provide useful topfeed and when they are thinned or cleared altogether pastures of mixed grasses develop; these include species of *Chloris, Eragrostis* and *Dichanthium.* Throughout most of this type climate and soil are suitable for pasture improvement and even cropping following clearing of the trees and shrubs.

Mitchell-Flinders grasses

This extensive type has its main centre in central and western Queensland, but there are small extensions into all other States. The type takes its name from the two major perennial pasture genera *Astrebla* (Mitchell grasses) and *Iseilema*

(Flinders grasses), but there are many others. Adequate rain at the right season results in rapid growth but variation in the rainfall pattern can have unfavourable results both on palatability and the keeping qualities of the dried grasses. Most of the grasslands are treeless or only carry a sparse cover of small trees and shrubs, some of which, such as *Acacia farnesiana*, have limited value as topfeed.

Within the broad boundaries of this type the Channel Country is an area which receives periodic flooding from the Georgina, Diamantina and Cooper River systems; after flooding there is a growth of numerous grass and dicotyledonous species which are excellent for cattle fattening.

Semi-arid open forest

This is a comparatively narrow belt which lies almost wholly within the semi-arid zone of Queensland. Several species of eucalypts dominate the open forest cover; these include ghost gum (*Eucalyptus papuana*) in the north, bimbil box (*E. populnea*) in the south and blue-leaved ironbark (*E. melanophloia*) throughout the type. Grass species include elements of both the Mitchell-Flinders and the brigalow types. Topfeed is an important element of the natural fodder.

South-western semi-arid pasture

This is a Western Australian type occurring mainly within the semi-arid zone of the south-west of the State. Parts are treeless heaths on sandy plains whilst there are very open forests of eucalypt species, a few of which may attain tree size, such as salmon gum (*Eucalyptus salmonoploia*), but most are only mallees. Some grasses occur and there is usually a growth of ephemeral herbage after rain, but the type has only a very low sheep-carrying capacity.

Arid shrub

Arid shrub is mainly located in the southern part of the dry country of South Australia, both in the arid and the semi-arid zones; there is a limited extension into Western Australia.

Saltbush (*Atriplex* spp.) and blue bush (*Kochia* spp.) are important browse perennials; whilst there is also a sparse cover of numerous southern perennial grasses. Topfeed species include inland rosewood (*Heterodendron oleifolium*) and several *Acacia* spp. Livestock carrying capacity is low.

Saltbush shrub steppe

This is located near the Great Australian Bight on either side of the Western Australian-South Australian border, mainly within the arid zone. Characteristically this type is treeless and the dominant vegetation of low shrub species, often only one to two feet high, consists mainly of saltbush (*Atriplex*) and bluebush (*Kochia*). These plants are very drought-resistant and most are readily grazed by sheep. As a result of the harsh climate, weak regeneration of the shrubs and overgrazing, the condition of the steppe has deteriorated in many places and soil erosion has developed as a result of reduced soil cover.

Mulga shrub

Mulga is *Acacia aneura* and whilst it is the dominant shrub species in this very large type, there are also a number of *Acacia* spp. and representatives of other genera. There is a very large area of the type in Western Australia but it is also an important type in south-western Queensland, north-western New South Wales, and between 30 and 32½°S. latitude in South Australia. Most of the climate is harsh; the rainfall being both very low and erratic. There are a large number of grass genera but non-grass ephemerals and topfeed species are of considerable importance. In times of drought the lopping of small trees and shrubs is one means of keeping stock alive but a combination of heavy lopping and prolonged droughts has reduced the amount of topfeed in some localities. This type is one which has a low carrying capacity even under favourable conditions. There is also a large area of mulga interspersed with spinifex in the southern half of the Northern Territory.

Spinifex (western)

This covers an area of moderate size bordering the coast of Western Australia north of about 23½°S. latitude. The dominant ground vegetation is spinifex (*Triodia*) but on included heavier soils there may be *Astrebla*. Amongst tree and shrub cover there are a number of species of *Acacia* and *Eucalyptus*. The climate is harsh, with very high temperatures in summer (Marble Bar falls within the type) and erratic rainfall. Stock-carrying capacity is low except on selected sites.

Introduced species, especially in the coastal region of Port Hedland and Roebourne, include Birdwood grass (*Cenchrus biflorus*) and buffel grass (*C. ciliaris*).

Spinifex (northern)

This covers a very large area and comprises much of the Northern Territory north of about 23°S. latitude and most of Western Australia north of about 19½°S. There is some extension into north-western Queensland. Spinifex, together with species of *Plectrachne,* dominate the ground flora but included in the type are a variety of sparse to moderately dense low shrub and woodland communities with many acacias and eucalypts, as well as representatives of other genera. These include a number of topfeed species. In the depressions there are softer grasses, sometimes including *Astrebla* spp.; whilst these are of limited extent they provide the greater part of the feed for grazing. The drought resistant spinifex and the associated topfeed species are mainly a reserve for times of drought. The carrying capacity of this type is low.

Desert

The desert type is one of the very large areas; it occupies most of the eastern half of Western Australia between 19° and 29°S. latitude, some of the southern half of the Northern Territory and most of South Australia north of about 30°S. Vegetation is sparse.

FURTHER READING

Australia. Bureau of Census and Statistics. *Official Yearbook.* 1969 (No. 55).

Department of National Development. *Atlas of Australian Resources.* Commentary prepared by Anne F. T. Bauer to accompany map-sheet *Drainage Systems.* (1955).

———. Commentary prepared by L. C. Noakes to accompany map-sheet *Geology.* (1966).

———. Commentary prepared by I. R. McLeod and Y. Miezitis to accompany map-sheet *Mineral Deposits.* 2nd edition. (1965).

Department of National Development. Commentary prepared by H. W. H. King *et al.* to accompany map-sheet *Physical Features.* (1965).

———. Commentary prepared by C. G. Stephens to accompany map-sheet *Soils.* (1963).

———. Commentary to accompany map-sheet *Surface Water Resources.* (1967).

———. Commentary to accompany map-sheet *Underground water.* (1953).

———. Commentary to accompany map-sheet *Vegetation Regions.* (1955).

Leeper, G. W. (1957). *Introduction to Soil Science.* 3rd (revised) edition. Melbourne University Press, Melbourne.

Northcote, K. H. in collaboration with various others (1960-). *Atlas of Australian Soils.* Sheets 1 and 2 *et seq.* C.S.I.R.O. and Melbourne University Press, Melbourne. This series of maps is to cover Australia in ten sheets, all but one of them at 1:2,000,000. Each sheet is accompanied by a booklet of 'explanatory data'.

Prescott, J. A. (1931). *The soils of Australia in relation to vegetation and climate.* C.S.I.R.O. Bulletin No. 52.

Stace, H. C. T. *et al.* (1968). *A Handbook of Australian Soils.* Rellim Technical Publications for the Commonwealth Scientific and Industrial Research Organisation and the International Society of Soil Science.

Stephens, C. G. (1962). *A Manual of Australian Soils.* 3rd edition. C.S.I.R.O., Melbourne. This book contains morphological descriptions of the forty-seven great soil groups recognised in Australia together with coloured photographs showing typical profiles.

Stephens, C. G. & Donald, C. M. (1958). 'Australian soils and their responses to fertilisers', in Norman A. G. (ed.) (1958). *Advances in Agronomy,* Volume 10. Academic Press, New York.

CHAPTER 3

The environment

PART II: CLIMATIC FACTORS

MOST people in the country are keenly aware of changes in weather conditions since the weather not only affects their daily occupation but, particularly in the rainfall pattern and total amount, has a very direct influence on the material prosperity of farmers and their dependants. Rainfall is particularly important in the arid and semi-arid zones where relatively small amounts may be sufficient to maintain development in a wheat crop or to keep grass and other fodder plants growing at critical times of the year. There is a difference in the response to rainfall which is shown by grasses and wheat on the one hand and small shrubs and trees on the other. The former are comparatively shallow-rooted and are very dependent upon adequate soil moisture at critical periods; for example, short-term droughts can have serious effects on production. By comparison, shrubs and trees have deeper and more extensive root systems and are not so readily affected by short-term droughts but, also, they need more rain to break a drought than does the shallow-rooted vegetation. Because rainfall, both total annual amount and variability, is so important in the arid and semi-arid areas of Australia a large part of this chapter is allocated to rainfall and the allied subjects of dew and evaporation. At least some knowledge of rainfall is necessary in the right selection of trees and shrubs for planting in the dry country and to an appreciation of the practice of conservation.

RAINFALL, EVAPORATION AND DROUGHT

Compared with the sub-humid and humid zones of Tasmania and coastal Australia, the rainfall of the arid and semi-arid zones is not only much lower but also shows an increasingly greater variability from the coast towards the centre of the continent. An exception is coastal Western Australia between 18 and 28°S. latitude, which is in the arid zone and has high variability. Associated with this is a striking difference in average annual potential evaporation from as low as 20-30 inches in most of Tasmania and the south-eastern highlands of the mainland and 30-50 inches in most of the remaining temperate sub-humid and humid zones, to over 130 inches in southern central Australia; even on the coast of Western Australia in the region of 17½-21°S. latitude, it may be as high as 90-100 inches. In the arid zone, especially, it is realistic to consider the mean annual rainfall as data of only restricted value, since the rainfall pattern tends to be one where the mean annual rainfall in any one year is unusual. Typical are fluctuations between a few to several years with markedly sub-normal values and scattered years with

TABLE 3.1

VARIATION IN ANNUAL RAINFALL DURING THE PERIOD 1911-50 FOR SELECTED LOCATIONS IN THE ARID AND SEMI-ARID CLIMATIC ZONES OF AUSTRALIA

	Hall's Creek Northern semi-arid	*Port Lincoln Southern semi-arid*	*Griffith South-eastern semi-arid*	*Charleville Eastern semi-arid*	*Clermont North-eastern semi-arid*	*Alice Springs Central arid*
	altitudes in feet					
	1,321	13	412	963	875	1,795
Year	*rainfall in points (0.01 inch)*					
1911	2,170	1,854		1,975	2,509	709
12	1,190	2,001		1,393	2,378	892
13	1,559	1,323		2,681	3,184	844
14	3,622	1,190		1,752	2,046	1,052
1915	1,677	1,967	1,218	1,133	1,514	433
1916	2,292	2,418	2,426	2,376	4,870	1,361
17	2,220	2,999	2,378	3,133	4,462	905
18	2,019	1,752	1,725	1,484	3,513	420
19	1,718	1,278	873	1,167	1,089	1,168
1920	1,369	2,239	1,545	2,729	2,581	2,857
1921	1,975	1,815	1,799	2,914	3,114	2,117
22	3,336	1,617	1,259	1,178	3,225	1,280
23	1,994	2,272	1,607	1,413	2,102	1,456
24	869	1,468	1,983	3,067	2,989	538
1925	1,873	1,868	1,505	1,936	2,267	915
1926	3,997	1,913	1,527	1,550	768	844
27	2,396	1,814	945	1,261	3,411	785
28	1,229	1,473	1,504	1,230	2,200	237
29	987	1,615	1,232	985	2,463	560
1930	2,738	1,419	1,768	2,013	2,551	1,106
1931	1,184	1,763	2,087	2,020	1,679	629
32	1,634	2,557	1,486	1,034	1,237	932
33	1,207	1,825	1,455	2,520	4,213	1,011
34	1,577	1,594	2,059	2,287	1,763	597
1935	1,724	1,808	1,300	1,042	1,334	612
1936	880	1,418	1,686	1,915	2,375	1,040
37	1,612	1,841	994	1,319	1,818	908
38	1,780	1,743	894	1,258	3,079	851
39	2,028	2,114	2,624	2,153	2,197	1,648
1940	1,299	1,772	734	1,001	3,067	1,068
1941	1,723	1,838	1,099	2,814	4,325	1,038
42	2,241	2,348	1,571	1,923	2,938	966
43	1,782	2,059	1,222	1,564	1,736	1,036
44	1,991	1,635	1,256	1,251	2,129	1,198
1945	1,370	1,642	1,338	1,344	1,791	1,062
1946	1,192	2,062	1,333	814	1,712	1,177
47	2,263	2,203	1,645	2,170	2,187	1,917
48	1,317	1,706	1,376	1,422	1,021	1,061
49	2,431	1,715	1,578	3,333	2,291	1,301
1950	1,948	1,617	2,397	4,134	4,917	1,506
	Mean annual rainfall for periods shown above					
	1,859	1,839	1,535	1,866	2,527	1,049

relatively high rainfall. This pattern is illustrated to different degrees in Table 3.1 which gives data for the 40-year period 1911-50, for several locations: Hall's Creek (northern semi-arid); Port Lincoln (southern semi-arid); Griffith (south-eastern semi-arid); Charleville (eastern semi-arid); Clermont (north-eastern semi-arid); and Alice Springs (central arid).

This unreliability of rainfall and the fluctuations from below average, possibly for up to several years at a time, to above average when rivers which have not had a surface flow for years run full for short periods is, then, a characteristic of the arid zone. In the semi-arid zone there is still wide variation in annual rainfall and whilst droughts can be severe and sometimes last for several years, they tend to occur only at longer intervals. Because the rainfall pattern has a profound effect on vegetation, especially that of an annual nature, and consequently on the amount of fodder available and the production from crops, there is a close correlation between it and management practices in the pastoral industry. This also applies to wheat raising in the semi-arid zone. It is a natural corollary that the farming community in the dry areas of the continent is much more dependent upon seasonal conditions and farmers are keener observers of weather than people who live in an urban environment.

As mean annual rainfall becomes lower and variability higher it is better conservation practice to retain native trees and shrubs where required for shade, shelter or soil protection than to make large scale clearances and then subsequently be forced into planting for protection against wind and sun. When planting does become necessary it is desirable to examine the possibility of using native species which grow naturally in the zone before considering species either from other zones or exotics. This is discussed more fully in later chapters.

RAINFALL

The area of Australia with which this book is concerned has an average annual rainfall which may be less than 6 inches in the southern central part of the continent, mostly between 23½-30½°S. latitude; this is located mainly in northern South Australia, but there are extensions into Western Australia, the Northern Territory and Queensland. The highest annual rainfall of the semi-arid zone is in the north where it may attain about 35 inches in the Kimberley region of Western Australia and the north of the Northern Territory; from the Gulf of Carpentaria region of Queensland where the rainfall is about 30 inches there is a gradual drop to about 20 inches in the south of the State and along most of the boundary in New South Wales. In parts of Victoria, South Australia and Western Australia the boundary is near the 14 inch isohyet. The isohyets between the arid and semi-arid zones vary from about 10 inches in most of the southern half of the continent to 12-16 inches in the northern half.

The most commonly used measure of annual rainfall is the arithmetical mean. For areas which have a relatively low variability this is a useful figure and is commonly used in descriptions of climate. This criterion applies reasonably well for locations in the sub-humid and humid areas of the continent, but is of less significance in the dry country with its lower annual values and much higher variability. The Commonwealth Bureau of Meteorology favours presentation of

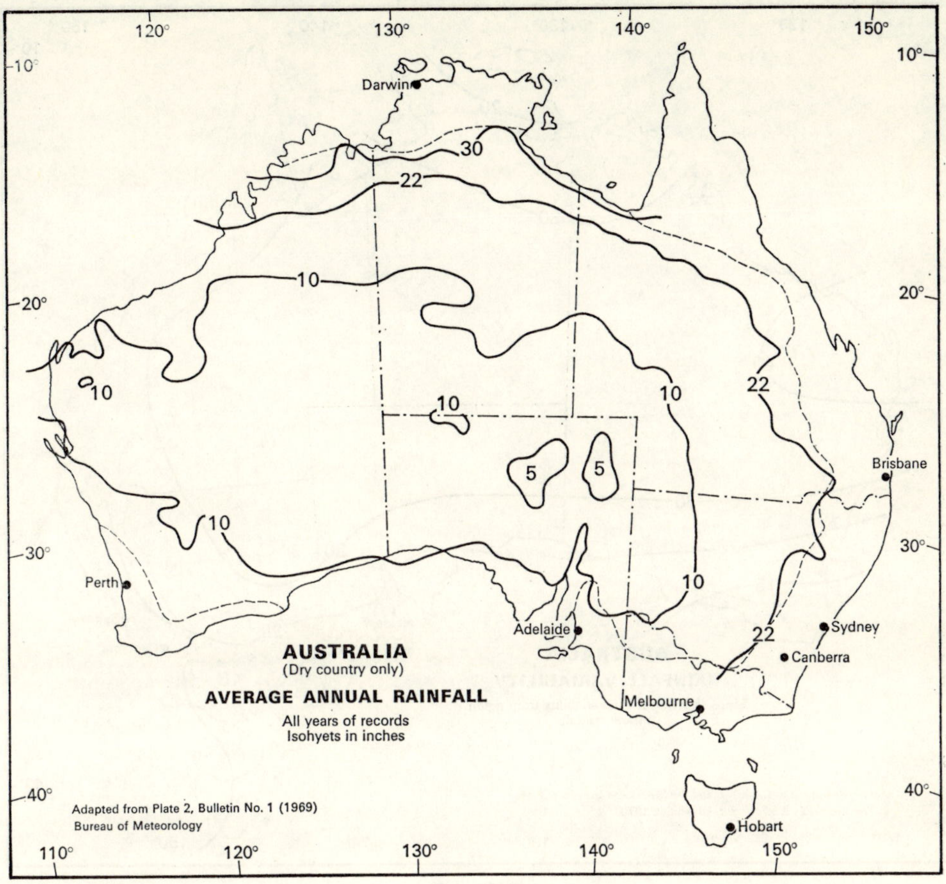

FIG. 3.1

data giving the lowest and highest on record and various percentile values, e.g., the 10, 50 and 90, or even the 10, 30, 50, 70 and 90 percentile amounts. The 10 percentile value simply means that on an average of one year in ten the rainfall will be less than the amount shown, the 50 percentile value is the midway amount and means in half the number of years rainfall will be higher and half the number of years it will be lower, and the 90 percentile is that value below which the rainfall will be 90 per cent of the time or, in different terms, the amount which will only be exceeded on an average one year in ten. The 50 percentile value may be very close to the arithmetical mean but, by definition, is not necessarily the same and only by chance will the two values coincide.

Data published by the Commonwealth Bureau of Meteorology for Coober Pedy in northern South Australia (18°30′S., 124°50′E.), give the 50 percentile value of 4.81 inches. This is one of the lowest annual rainfall values for any recording station in Australia. The lowest value for a year was 1.30 inches, one year in ten it was likely to fall below 2.12 inches, one year in ten it was likely to be above 10.26 inches and the highest recorded was 12.94 inches. This type of presentation is of much greater significance than an average value, even with the percentage

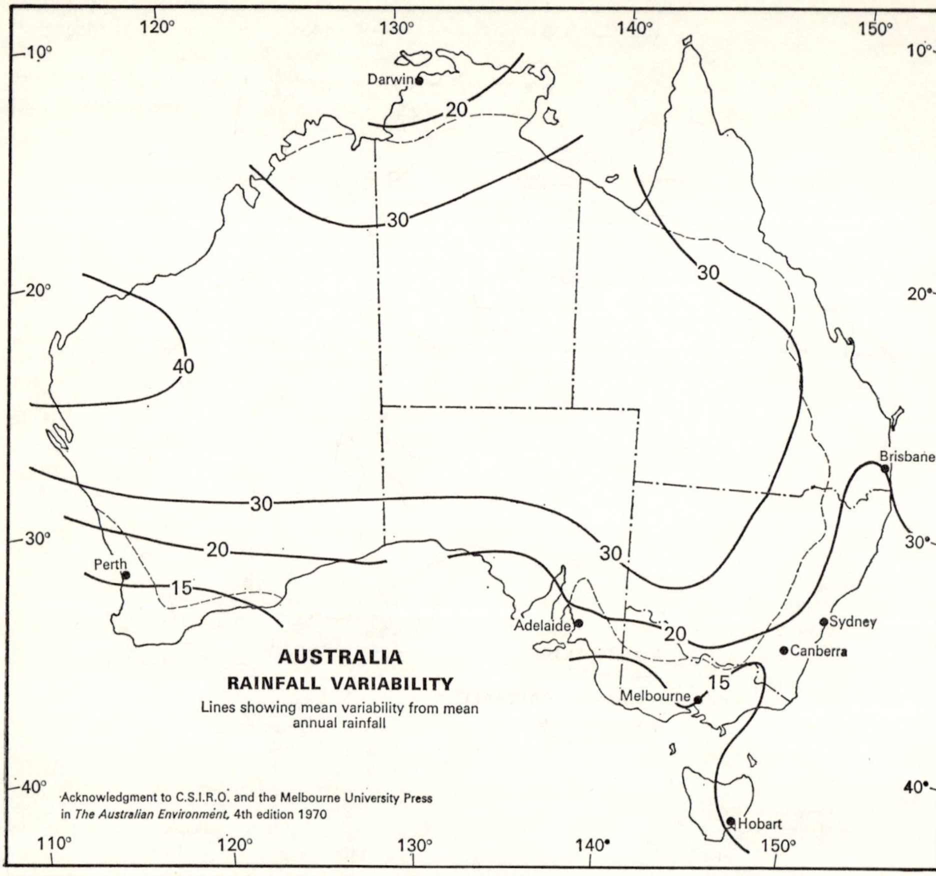

FIG. 3.2

probability given, and is of considerable use to pastoralists and anyone wishing to introduce trees, shrubs or other plants into the dry country. In other fields of activity the information is very useful to engineers and conservationists planning large water storage schemes.

Where the data is available from the Bureau of Meteorology the lowest, the 10, 50 and 90 percentile and the highest annual fall on record are given in Appendix I for most of the locations shown on the maps, pp. 526-531.

SEASONAL DISTRIBUTION

It is an over-simplification to summarise the rainfall pattern of the continent in recording that northern Australia experiences a summer maximum and southern parts a winter maximum. It is more accurate to state, however, that the average annual rainfalls of the sub-humid and humid zones show a well-defined summer maximum north of the Tropic of Capricorn in Western Australia and north of the New South Wales-Queensland border in eastern Australia, whilst the coast of New South Wales north of Sydney exhibits a summer-autumn maximum, and that southern Australia shows a winter maximum which is most marked in the south-west of Western Australia and in South Australia. Across the north of Western Australia, the Northern Territory and Queensland summer monsoonal conditions develop, with a very wet period from about December to March, followed by several months which may be almost rainless, even where the mean annual rainfall may be over 50 inches.

In the semi-arid and, especially, in the arid zone average monthly rainfalls for long periods do not give a realistic pattern of rainfall distribution. This is because of the low average values and the high variability. In the driest areas with an average of 6-8 inches there may be years of markedly below average rainfall, followed by the equivalent of a year's fall in a few weeks and a monthly value of, say 0.50 inches, may be averaged from a few years with 3-6 inches fall in the month followed by many years which were either rainless or received only a few points. It is for this reason that the presentation of data for the three driest and the three wettest months as given in Appendix II of *Growing Trees on Australian Farms* was not considered applicable to the arid and semi-arid zones.

In spite of all the limitations of using average monthly or annual values for rainfall in the dry country it was considered that the only practical way of presenting seasonal distribution was by showing average data for all months of the year. This is given in Appendix I. It will be noted that whilst northern Western Australia and the semi-arid zone of the Northern Territory show a well-developed summer monsoonal pattern, much of Queensland, whilst retaining a summer maximum may also have a secondary maximum; compare Charters Towers with 137 and Hughenden with 98 points, both in June, whilst further south Charleville has a well-developed double maximum when peaks are attained in November to March and again in June and July, and Cunnamulla follows the same general pattern. This double maximum also occurs in adjacent areas of New South Wales; compare Bourke where the three wettest months are December (164 points), January (121) and June (127); this pattern is also obvious at Cobar but less apparent in the lower rainfall area of Broken Hill.

The effect of seasonal distribution of rainfall upon plant growth is of considerable importance. Except for the southern fringe of the dry country, rainfall may be more important than temperature in influencing growth, but there is a difference between various plant forms under dry conditions. Following adequate rainfall in the arid zone annual herbs will grow, flower and seed within a few weeks and, as the surface soil dries they wither and die, leaving seed which will remain dormant until suitable conditions occur for development. Trees and shrubs which are native to the dry country, on the other hand, have a different pattern of behaviour in that they usually have extensive root systems which not only extend deeper than those of the herbs and grasses but spread laterally for long distances. The length of roots may be up to 10 times plant height, and they draw all possible moisture from a greater volume of soil. These plants are also adapted to going into a semi-dormant stage for very long periods, when conditions for growth are very unfavourable. Some species can stand several years of drought and revive and show new growth when good rains come. Species selection, spacing and soil protection are very important factors in planning the establishment of shade and shelter in the dry country; they are considered in some detail in a subsequent chapter.

TABLE 3.2

A COMPARISON OF AVERAGE ANNUAL RAINFALL AND EVAPORATION FOR SELECTED LOCATIONS IN THE ARID AND SEMI-ARID ZONES OF AUSTRALIA

Place	*Altitude (ft)*	*Rainfall (in)*	*Evaporation (in)(a) approx*	*Excess or deficit of rainfall over evaporation (in)*	
				Excess	*Deficit*
Arid zone					
Western Australia					
Carnarvon	15	9	85	—	76
Giles	1,900	9	123	—	114
Kalgoorlie	1,247	9	95	—	86
Laverton	1,530	9	114	—	105
Marble Bar	595	12	104	—	92
Meekatharra	1,676	9	100	—	91
Port Hedland	25	11	96	—	85
Rawlinna	607	7	90	—	83
South Australia					
Marree	160	6	110	—	104
Tarcoola	395	6	103	—	97
New South Wales					
Broken Hill	1,000	9	82	—	73
Ivanhoe	288	11	70	—	59
Tibooburra	400	8	110	—	102
Queensland					
Birdsville	150	6	125	—	119
Boulia	478	10	115	—	105
Thargomindah	402	9	110	—	101
Windorah	390	10	110	—	100
Northern Territory					
Alice Springs	1,901	10	99	—	89
New Crown	645	5	120	—	115
Semi-arid zone					
Western Australia					
Derby	53	24	94	—	70
Fitzroy Crossing	450	20	100	—	80
Geraldton	13	19	70	—	51
Hall's Creek	1,225	19	103	—	84
Merriden	1,046	12	80	—	68
Norseman	927	11	70	—	59
South Australia					
Ceduna	43	10	60	—	50
Port Augusta	18	9	80	—	71
Port Lincoln	13	18	45	—	27
Renmark	80	10	62	—	48
Victoria					
Mildura	177	10	60	—	50
Ouyen	165	12	56	—	44
New South Wales					
Bourke	361	12	85	—	73
Coonabarabran	1,673	25	60	—	35
Deniliquin	311	15	60	—	45
Moree	686	21	60	—	39
Wagga Wagga	612	21	52	—	31
Queensland					
Charleville	965	18	90	—	72
Clermont	870	25	70	—	45
Cloncurry	633	17	107	—	90
Hughenden	1,074	18	90	—	72
Roma	1,010	20	72	—	52
Northern Territory					
Daly Waters	691	25	95	—	70
Katherine	355	35	89	—	54
Tennant Creek	1,075	14	111	—	97

(*a*) Estimated from locations on Bureau of Meteorology map of annual evaporation in inches.

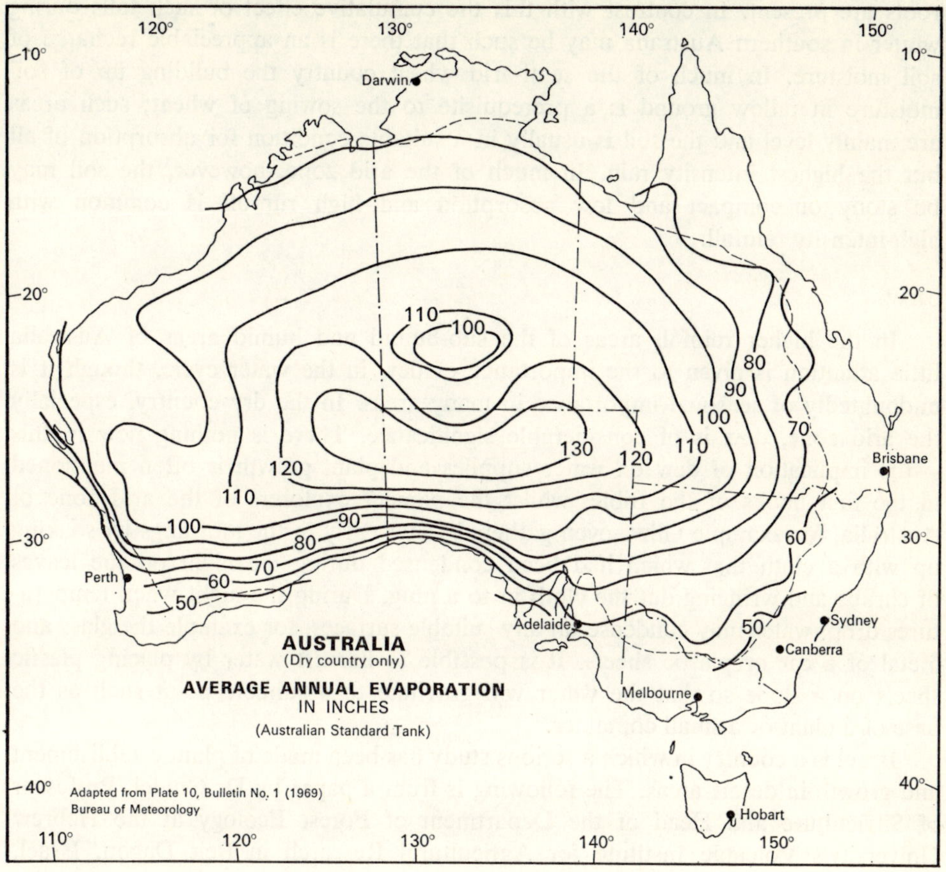

FIG. 3.3

EVAPORATION

The amount of evaporation from a free-water surface is influenced by a number of factors, of which the most important are temperature, wind and relative humidity. A consideration of evaporation in conjunction with rainfall gives a general guide to the aridity or humidity of an area, although it is not as useful a guide to water availability for plant growth as a study of evapotranspiration. In the arid zone, also, some consideration should be given to the dew pattern. The contrasts which can be found between rainfall and evaporation are shown in Table 3.2 whilst data for a small number of stations in the sub-humid and humid zones will be found in Table 3.2 of *Growing Trees on Australian Farms*.

THE EFFECTIVENESS OF RAINFALL

Two major factors which influence the effectiveness of rainfall are seasons of the year and intensity. These may assume an even greater importance in the dry country than they do elsewhere. Light falls during the peak of summer may do no more than dampen the leaves of plants and possibly temporarily lay the dust, but they certainly do not penetrate the soil to the depths at which absorption roots are present. In contrast with this the cumulative effect of such falls during winter in southern Australia may be such that there is an appreciable recharge of soil moisture. In much of the semi-arid wheat country the building up of soil moisture in fallow ground is a prerequisite to the sowing of wheat; such areas are mainly level and the soil is usually in a suitable condition for absorption of all but the highest intensity rain. In much of the arid zone, however, the soil may be stony or compact and low absorption and high run-off is common with high-intensity rainfall.

DEW

In the higher rainfall areas of the sub-humid and humid areas of Australia little attention is given to the importance of dew in the water cycle, though it is undoubtedly of seasonal importance in many areas. In the dry country, especially the arid zone, dew is of considerable significance. There is nothing new in this —the implication of dew on water supplies and plant growth is often mentioned in the first books of the Bible, whilst the pioneer explorers of the arid zone of Australia, for example Giles, even gathered water early in the mornings by soaking up with a cloth that which had been condensed during the night on the leaves of shrubs and wringing out the cloth in to a mug. During the night when temperatures drop, water may condense on any suitable surfaces, for example the glass and metal of a car or plastic sheets. It is possible to harvest water by placing plastic sheets on a slope so that the water will gravitate to a suitable point such as the base of a plant or a small container.

Israel is a country in which a serious study has been made of plant establishment and growth in desert areas. The following is from a paper by Dr Gindel, Professor of Silviculture and Head of the Department of Forest Ecology at the Hebrew University's Volcanic Institute for Agricultural Research in Beit Dagon, Israel.

> It is widely believed that dew and mist are composed of distilled water, but this is not true for the desert and sub-tropical climates of Israel. Analyses have shown that the concentration of positive and negative ions in such water is ten times higher than it is in rainwater, and includes ions of bicarbonate, nitrate, sulphate, chloride, sodium potassium, magnesium and calcium. There are approximately 200 mg/1 of soluble minerals in condensed dew and 400 mg/1 in mist; the quantity varies with the seasons and districts. These ions originate in dust particles suspended in the air, which form nuclei for dew drops. During the dry season there are many sources of ions, mainly evaporation of sea water (the Mediterranean, Red and Dead Seas), of the Sea of Galilee and the Jordan River, and of water used for irrigation in adjacent districts, and the weathering of rock in the mountains.
>
> The fact that atmospheric moisture has 40-70 mg/1 of chloride and its pH is 6.4-6.9 is very important for plant nutrition in a desert environment, where the salinity of both soil and air is high. For purposes of comparison—the salinity of water coming from northern Israel via the national conduit is 300 mg/1.

DROUGHT

From the first years of the settlement of Port Jackson in the latter part of the eighteenth century droughts have been known as a feature of the Australian environment which have been inescapable as grazing and cropping spread over the continent. Since then there has been a record of a series of minor and major droughts in the settled areas which have had variable, sometimes disastrous, effects upon the national economy. As the economy becomes less dependent upon the primary products from animals and crops there is a tendency by urban dwellers to overlook the fact that serious droughts can still have a major influence on our welfare. Certainly most city people are familiar with periodic water restrictions but this is all too often blamed on insufficient dam capacity. It is probably more correct to say that with increasing population even larger and more numerous dams are unlikely to permit complete relaxation of restrictions because of the limited number of major rivers and the necessity for authorities to retain large reserves against a sequence of dry years. We must recognise that Australia is a dry country. The effect of drought on animal numbers and wheat production is discussed elsewhere but it can be noted that since World War II there have been occasions when Queensland has had to import butter and New South Wales wheat.

Because of the large size of Australia it is unusual for major droughts to prevail throughout the country at the one time, but on an average of about one year in four there is a major drought affecting one or more States. Sometimes there have been a succession of years with below average rainfall, culminating with one or two exceptionally dry years; the resultant droughts may result in widespread loss of stock, serious reduction in crop yields and sometimes abandonment of rural holdings.

Various writers have referred to the record of droughts in Australia and one of the most comprehensive is that by J. C. Foley (1957). Undoubtedly one of the most widespread and severe droughts was that which was spread over the years 1895-1903 and covered most of Australia, but was particularly persistent on the coast of Queensland, inland New South Wales and in South Australia and Central Australia (now the southern part of the Northern Territory). Another very widespread occurrence was in the years 1939-45 when most parts of the continent were affected for at least part of the time. More recently, 1966-68, parts of New South Wales, Queensland and Victoria have been seriously affected. Any detailed study of drought since settlement will show that very few years pass without a drought in at least one or more States; some of these may be locally serious but they do not attract the national attention as much as more widespread occurrences.

Strange as it may seem there is by no means general agreement as to definitions of drought. The meteorologist thinks mainly in terms of conditions when rainfall is appreciably below average; a mild drought could be when annual rainfall is only 85 per cent of normal. Some climatologists represent drought in terms of numbers of consecutive days without rain, experienced in various periods of the year. Another definition is when rainfall is less than 50 per cent of average for three months or more. All these definitions have originated overseas and highlight the point that adequate definition of drought in one country may not apply in another. For instance in the British Isles an 'absolute drought' is defined as a period of at least 15 consecutive days without one point (0.01 inch) of rain; this could have some relevance in parts of Tasmania but is a normal condition during the dry season in most parts of the arid and semi-arid zones of Australia. Any satisfactory definition of drought must take into consideration many factors, such as total deficit of rainfall not only in the current season and year but also over the immediately preceding years, variation from normal distribution and the effect on plant growth and water supplies. Also, the effect of hot and dry winds on vegetation will accentuate any deficit in rainfall.

It is fairly obvious that no one index can satisfactorily take all factors into consideration and in recent decades there has been a trend to define droughts in terms of their ecological effect. This necessarily takes local conditions into account and makes allowance for the fact that crops in different parts of the world or even within one country are more or less adjusted to the normal moisture conditions prevailing there, whether natural or adjusted to irrigation. Studies in Australia have adopted definitions based on *effective* and *influential* rainfall; the former is rainfall which promotes germination and the latter one which sustains growth over a period. It has been pointed out by Everist: (1969).

> Serious departures from the normal pattern of rainfall distribution may cause disturbance of pasture growth whether the amount is normal or subnormal.
>
> In regions where rain normally falls during the summer months and the winter and spring months are dry, pastures are dominated by perennial grasses which are incapable of making growth in cold weather, whether soil moisture is available or not. In regions where the winter is usually wet and the summer dry, the pastures generally are dominated by winter growing plants, many of them annual, which dry off in summer.
>
> In either situation, reversal or serious disturbance of the rainfall pattern can result in serious shortage of feed since the dominant species do not make their normal growth and there are few or no plants available in the pasture which are capable of utilising the soil moisture when it is available. These off-season rains give temporary relief but they do not maintain normal production. In some regions and with some plants, e.g. Flinders grass (*Iseilema* spp.), off-season rains cause positive harm by rendering them unpalatable or causing them to rot, whereas if the pasture had remained dry it would have continued to support some animals for months.

It is probable that the measure of soil moisture is the best indicator whether drought conditions are present; certainly soil moisture in relation to the time of year is a good measure of the ability of that soil to maintain plant growth of different types. Unfortunately soil moisture varies with soil type and topography and also because it is not so readily measured has not the more general application of rainfall data. In specialised fields, for example wheat growing, experienced farmers can tell by inspection of the soil whether the moisture conditions are suitable for sowing the grain.

TEMPERATURE

A general indication of the temperature pattern in arid and semi-arid Australia is provided by Fig. 3.4, which shows the isohyets for the normal daily maximum in January and the normal daily minimum in July. In addition, Table 3 of Appendix I gives the altitude of selected meteorological stations, together with the mean maxima and the mean minima for January and July, except that where the highest mean maxima occur in other months than January, data is shown for the hottest month. In the north of the continent, which is subject to summer

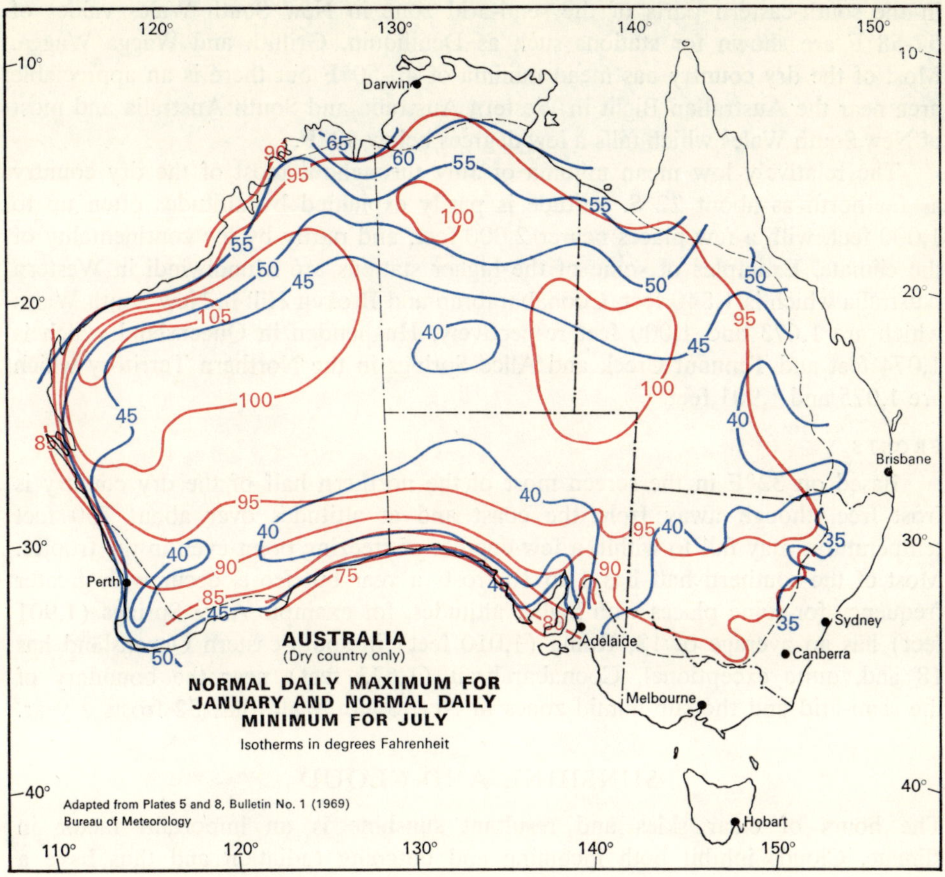

FIG. 3.4

monsoonal rains, the highest monthly maximum is often November or December just before the monsoon breaks. Most of the dry country has mean maxima for the hottest month in the range of 95-105°F, whilst there is a small area in the north-west of Western Australia where mean temperatures may exceed 105°F. Marble Bar, claimed to be the hottest settled place in Australia, has a mean maximum of 107.5°F for December, whilst Winning Pool has 104.5°F for January. Summer temperatures are lowest along the Australian Bight region of Western Australia and South Australia, where the mean maxima falls to about 80°F. Absolute maxima in the hottest month are often up to 10 degrees higher than the mean and may be as much as 15 degrees in some areas. Corresponding mean minima for the same month are 65-75°F over most of the area, but exceed 75°F over much of Western Australia and the Northern Territory north of about 22°S. latitude and the Gulf of Carpentaria region of Queensland.

Throughout most of the region the mean maxima of the coldest month, which is uniformly July, are in the very pleasant range of 65-75°F, but in the northern part of the continent they may attain to 85°F, whilst in the south there is a wide belt which has values of 60-65°F. Only in Victoria does it fall below 60°F and in the south-eastern parts of the semi-arid zone in New South Wales values of 57-58°F are shown for stations such as Deniliquin, Griffith and Wagga Wagga. Most of the dry country has mean minima of 40-50°F but there is an appreciable area near the Australian Bight in Western Australia and South Australia and most of New South Wales which falls a few degrees below 40°F.

The relatively low mean minima of July throughout most of the dry country as far north as about 22°S. latitude is partly explained by altitudes often up to 1,000 feet, with a few places nearer 2,000 feet, and partly by the continentality of the climate. Examples of some of the higher stations are Mundiwindi in Western Australia which is 1,840 feet, Coonabarabran and Broken Hill in New South Wales which are 1,673 and 1,000 feet respectively, Hughenden in Queensland which is 1,074 feet and Tennant Creek and Alice Springs in the Northern Territory which are 1,075 and 1,901 feet.

FROSTS

Based on 32°F in the screen most of the northern half of the dry country is frost free, though away from the coast and at altitudes over about 750 feet temperatures may fall to within a few degrees of freezing point even in the tropics. Most of the southern half has 1 to 10 frosts a year but frosts occur with greater frequency for some places with higher altitudes, for example Alice Springs (1,901 feet) has an average of 12, Roma (1,010 feet) in south-eastern Queensland has 18 and, quite exceptional, Coonabarabran (1,673 feet) near the boundary of the semi-arid and the sub-humid zones in New South Wales has 52 frosts a year.

SUNSHINE AND CLOUD

The hours of clear skies and resultant sunshine is an important factor in climate. Clouds inhibit both incoming and outgoing radiation and thus have a profound effect on temperature and other aspects of climate; a high level of sunshine, especially when associated with clear skies at night, leads to high day and low night temperatures, humidity of the air is usually low and evaporation from water surfaces and evapotranspiration from plants are high. One reason for selecting the period of maximum cloud cover as the season for planting trees and shrubs is not only because increased cloud means reduced evapotranspiration but because it is usually associated with the principal rainy season. Hence the planting season in southern Australia usually starts as soon as adequate rain falls in winter, whilst in the north of the continent planting may be done in early summer when increasing cloud and the break of the monsoon provides suitable conditions. Two extremes of the season of maximum cloudiness and rain are shown by data for Darwin (12°28′S. latitude) and Adelaide (34°56′S. latitude); the former has about 150 hours of sunshine in January but nearly 300 in July; Adelaide, on the other hand, has over 300 hours in January and only 135 in July.

The greater part of the dry country receives more than 3,250 hours of sunshine a year, whilst there are two areas, one in the Carnarvon to Port Hedland coastal region of Western Australia, and a larger one north and south-east of Alice Springs and extending into south-western Queensland and north-eastern South Australia, where the value exceeds 3,500 hours a year. Only in fringe areas of southern and northern Australia do hours of sunshine fall into the range of 3,000-3,250 hours and, in the south only, still smaller areas where it is 2,750-3,000 hours a year.

The seasonal pattern of sunshine is influenced both by latitude and the incidence of the rainy season. Throughout much of the dry country, especially the arid zone, with its very low rainfall, clear skies prevail most of the year. There is a belt, approximately between 20-25°S. latitude and extending for nearly 1,500 miles from Western Australia through the southern half of the Northern Territory to western Queensland, where the monthly hours of sunshine throughout the year vary only between 275 and 300 hours (or nearly 3,500 hours a year). In January most of the dry country south of Tennant Creek (19°34′S. latitude) receives more than 275 hours whilst south of about 28°S. this increases to over 325 hours. There are two areas, one in Western Australia and one in New South Wales, where the monthly sunshine figure is over 350 hours, or the very high daily average value of nearly 11½ hours. Winter sees a change, mainly in the north of the continent where monthly values may rise to over 300 hours and in the southern fringe where July figures may fall to 200 hours in southern Western Australia and adjoining parts of South Australia; further eastwards in Victoria and New South Wales there is a further fall to 175 hours.

WIND

Wind direction and intensity at the various seasons of the year are necessary background information for farmers and those concerned with town beautification and shelter planting schemes, when deciding the location and design of shelter belts for the protection of homes, animals and crops, or in ornamental plantings and shade for towns. Those who have lived for a long time in one locality in the dry country will have gained this background requirement through experience, but for those who have only recently moved into new areas the information can be secured from various sources such as observations of local vegetation, information from older residents and, where available, data from the Bureau of Meteorology.

The most readily-appreciated form of presentation of wind data is by means of wind roses. These can be drawn for whatever hours wind observations are made at meteorological stations, which are usually 9.00 a.m. and 3.00 p.m. for minor stations and at three-hourly intervals from 3.00 a.m. onwards for the capital cities and important aerodromes. The usual form of wind rose is that given by the Bureau of Meteorology (1967), in which lines, the length of which represent wind frequencies, for the eight cardinal and semi-cardinal directions, for each month are drawn from the inner of two concentric octagons. The distance of the outer octagon from the inner one is made equivalent to 12½ per cent of the distance for a 100 per cent frequency; that is, if one-eighth of the total observations are from the north then the line from the inner octagon will stop at the outer octagon. The conventional method is to draw lines on each side of the octagon for each month of the year. Calms are shown by figures within the inner octagon. The wind roses so far discussed only show wind direction; those which include a graphical representation of wind velocity are more involved and not considered necessary for the purpose of this text. The wind roses shown in Figure 3.5 have been simplified by using only four values on each side of the octagon, that is, January to represent summer, April for autumn, July for winter, and October for spring. These roses have been drawn in a computer programme from data supplied by the Bureau of Meteorology. There are 23 pairs of roses and the stations were selected to give as extensive a coverage as practical of wind conditions in dry Australia; there are eight locations for Western Australia, three for South Australia, one for Victoria, three for New South Wales, five for Queensland and three for the Northern Territory.

For comparative purposes Figure 3.6 shows four pairs of wind roses with frequencies for the twelve months of the year. These have been taken from Bulletin No. I of the Bureau of Meteorology and the stations are Geraldton and Kalgoorlie in Western Australia and Charleville and Cloncurry in Queensland; there are substantial differences between the patterns of the four stations.

It is emphasised here and elsewhere that the wind roses only show an analysis of wind direction for the meteorological station named and that the local wind pattern in hilly country, possibly only 10-20 miles away, may be different. The farmer, and parks and gardens overseer will find the wind roses of help in designing tree plantings for amenity purposes but they must also use local knowledge on the strength of very hot or cold winds, and variations in direction due to local topography.

There are some areas of the dry country, for example the west coast of Western Australia north of Geraldton, which are periodically subject to cyclonic disturbances in which exceptionally strong winds are usually accompanied by drenching rains. The influence of these cyclones on tree selection and planting is discussed in Chapter 18.

FIG. 3.5

WIND ROSES (9.00 A.M. AND 3.00 P.M.) FOR TWENTY-THREE LOCATIONS IN AUSTRALIA, SHOWING FREQUENCIES FROM EIGHT CARDINAL AND SEMI-CARDINAL DIRECTIONS FOR THE FOUR SEASONS OF THE YEAR, AS REPRESENTED BY JANUARY (SUMMER), APRIL (AUTUMN), JULY (WINTER) AND OCTOBER (SPRING)(*a*)

Meteorological station *latitude* °*S.* *longitude* °*E.* *altitude (feet)*	*Wind roses* *9.00 a.m.*	 *3.00 p.m.*
WESTERN AUSTRALIA		
Corrigin 32.3° 117.9° 968 ft	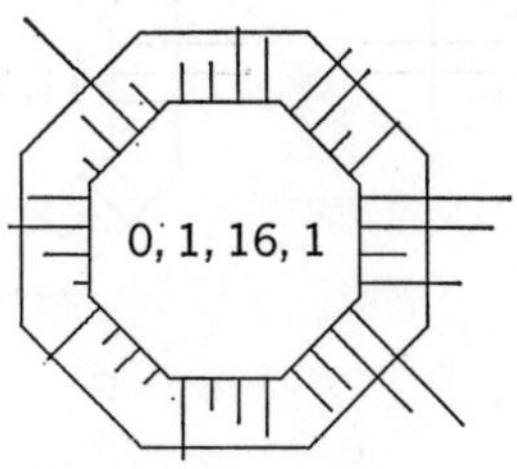	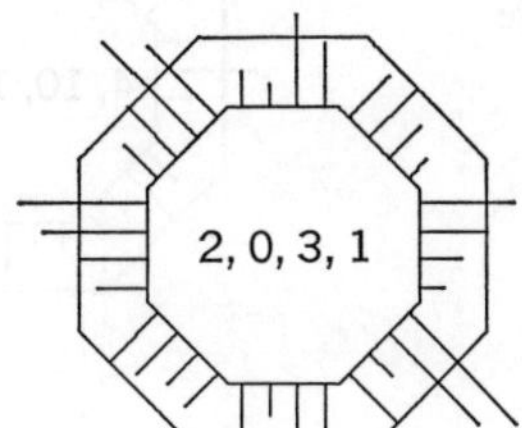
Eucla 31.7° 129.0° 15 ft	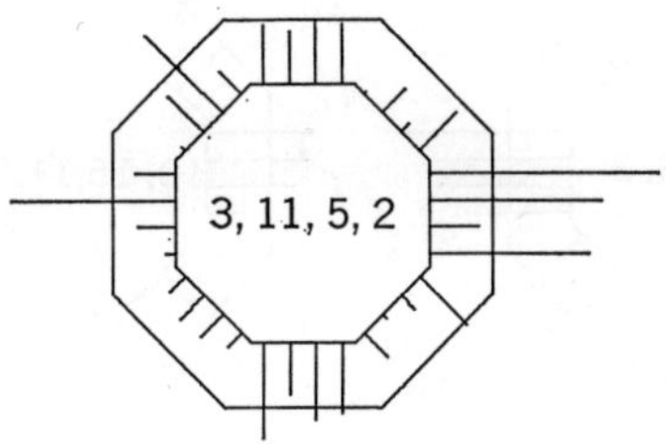	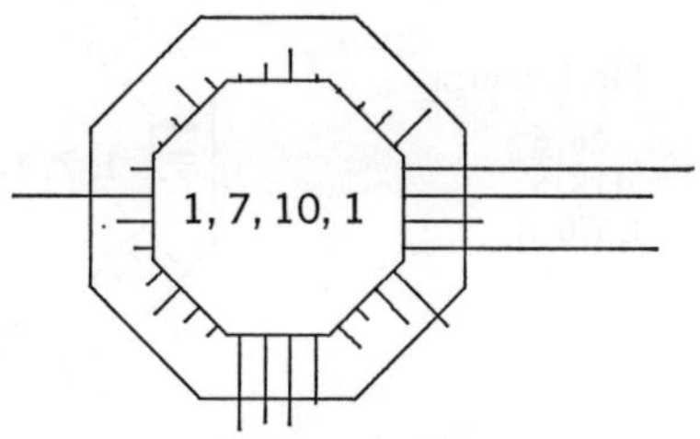
Fitzroy Crossing 18.2° 125.6° 450 ft	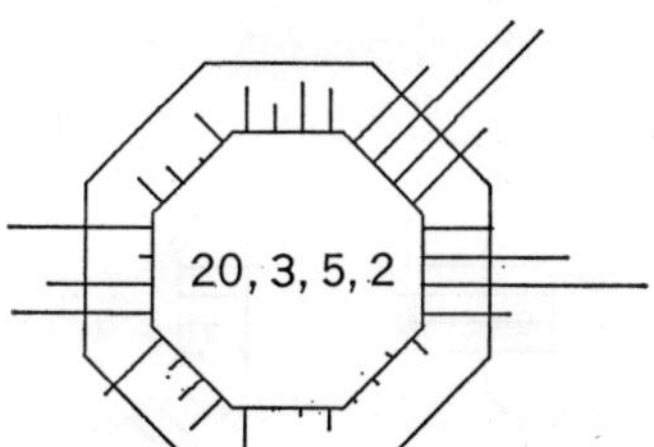	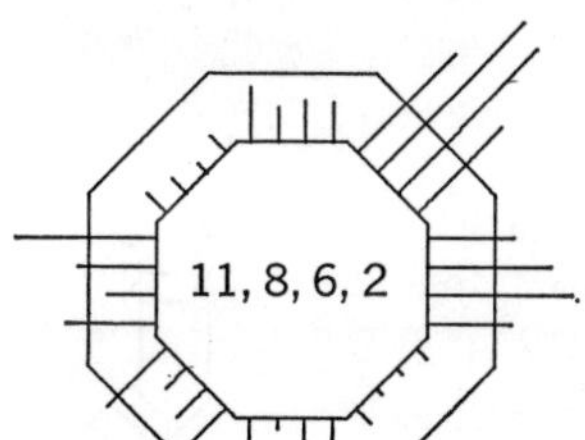
Geraldton 28.7° 114.6° 13 ft	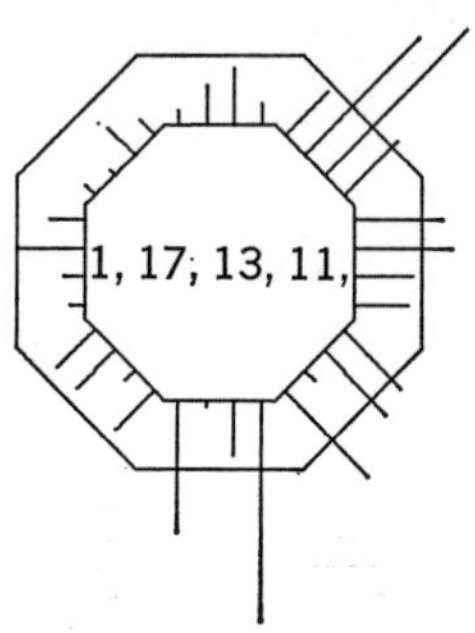	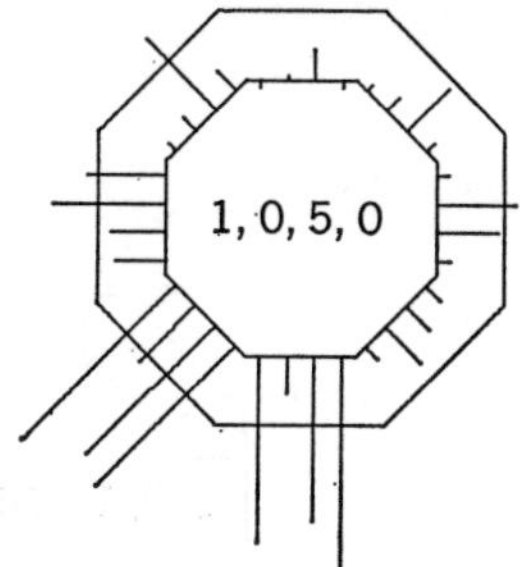

(*a*) The basic information for these wind roses was supplied by the Bureau of Meteorology, Department of the Interior, Melbourne.

The Fortran computer programme was written by Mr G. A. V. Bary of the Forest Research Institute, Forestry and Timber Bureau, and the machine work done at the Division of Computing Research, C.S.I.R.O., Canberra, using a CDC 3600 computer and a Calcomp plotter.

FIG. 3.5—*continued*

Meteorological station *latitude* °*S.* *longitude* °*E.* *altitude* (*feet*)	*Wind roses*	
	9.00 a.m.	*3.00 p.m.*
WESTERN AUSTRALIA—*continued*		
Kalgoorlie 30.7° 121.5° 1,247 ft	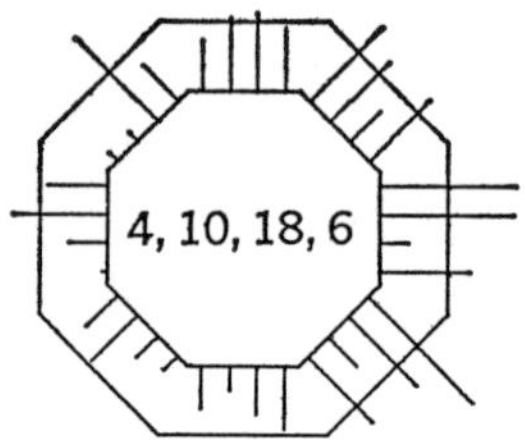	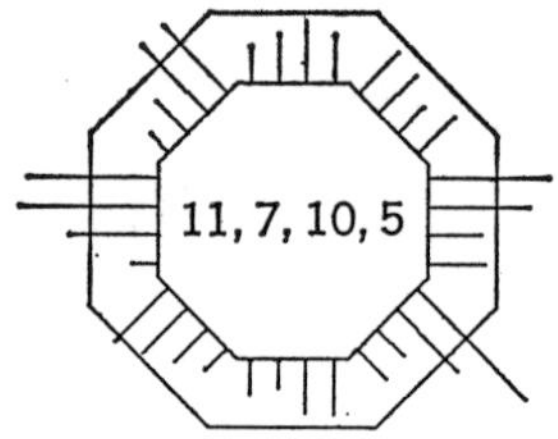
Meekatharra 26.6° 118.5° 1,676 ft	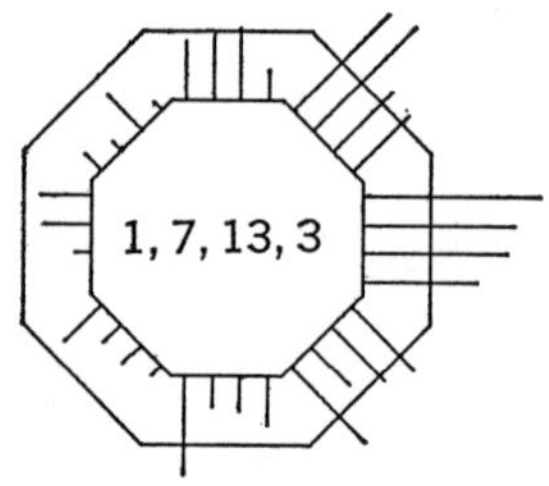	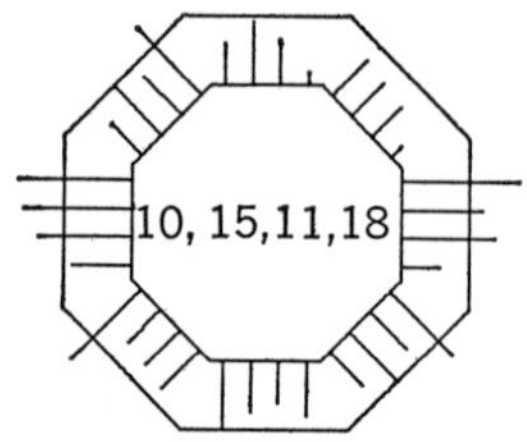
Mundiwindi 23.9° 120.2° 1,840 ft	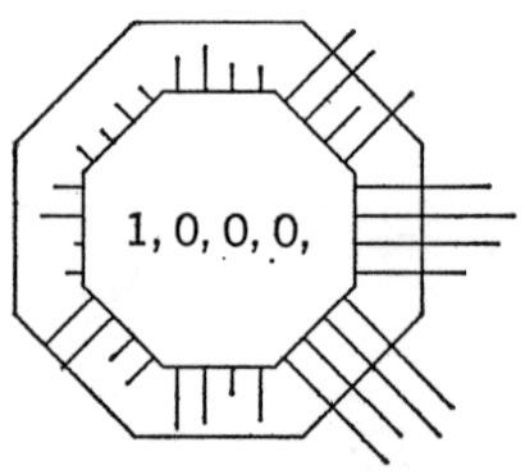	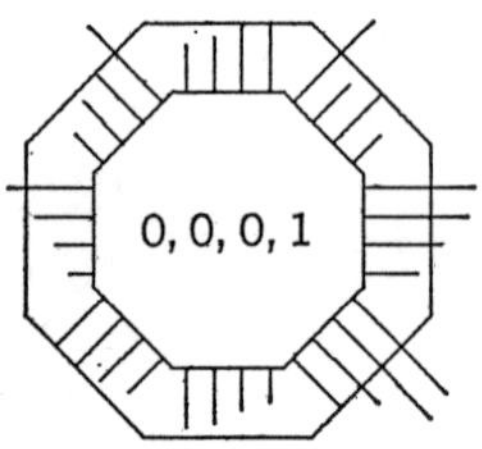
Port Hedland 20.3° 118.4° 25 ft	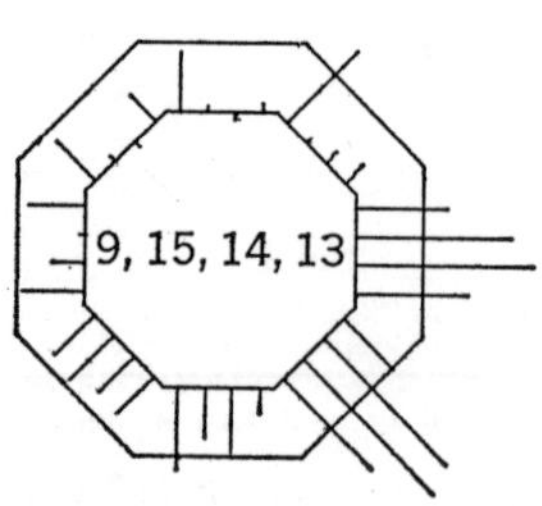	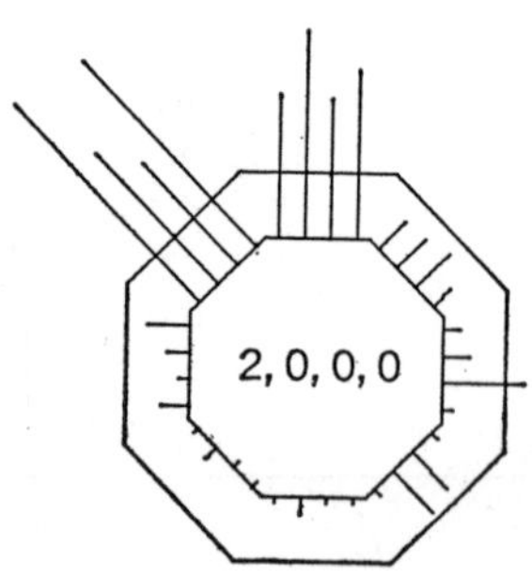

FIG. 3.5—*continued*

Meteorological station *latitude* °S. *longitude* °E. *altitude* (*feet*)	*Wind roses* 9.00 a.m.	3.00 p.m.

SOUTH AUSTRALIA

Marree
29.6°
138.1°
160 ft

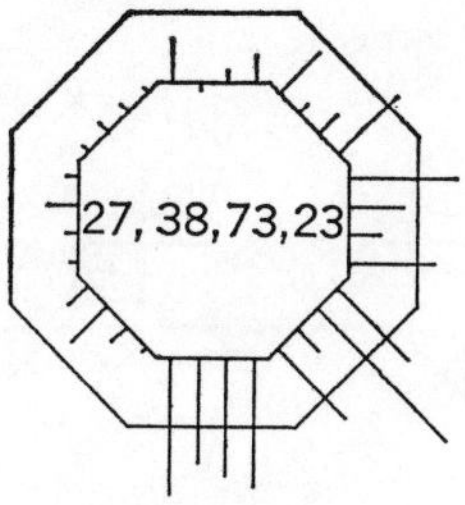

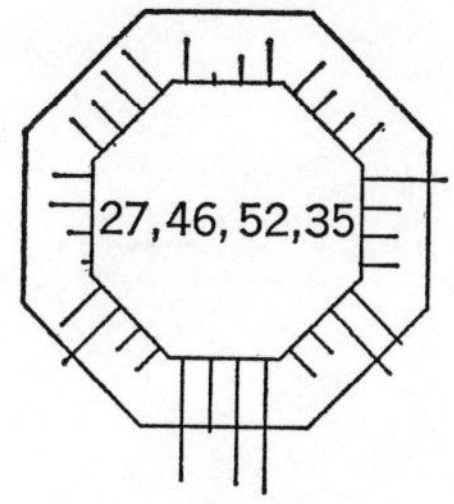

Port Lincoln
34.7°
135.9°
13 ft

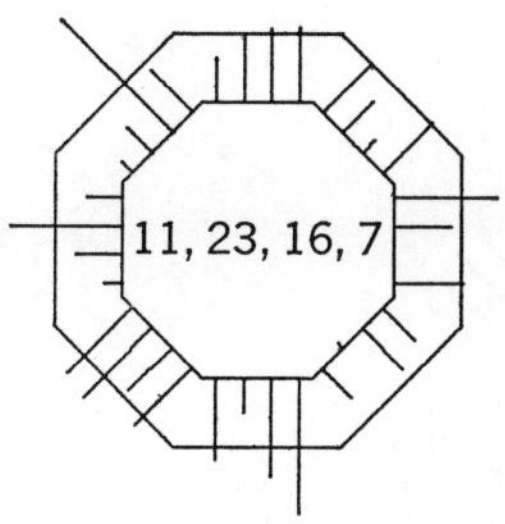

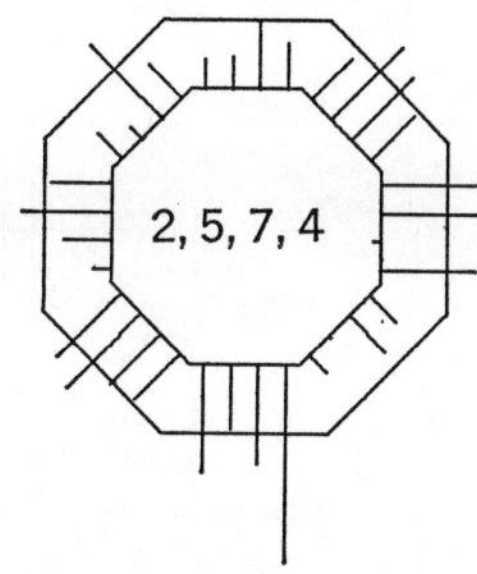

Port Pirie
33.2°
138.0°
16 ft

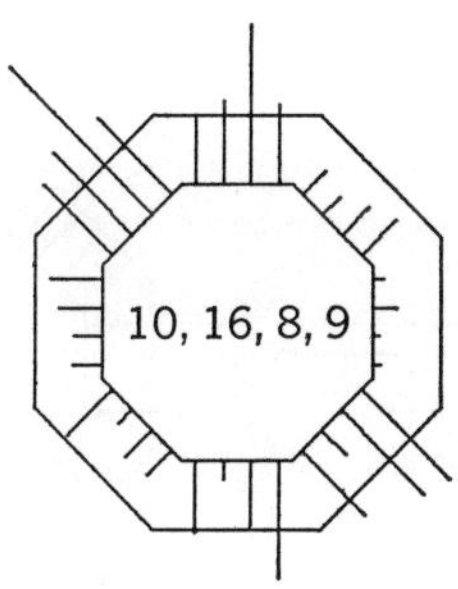

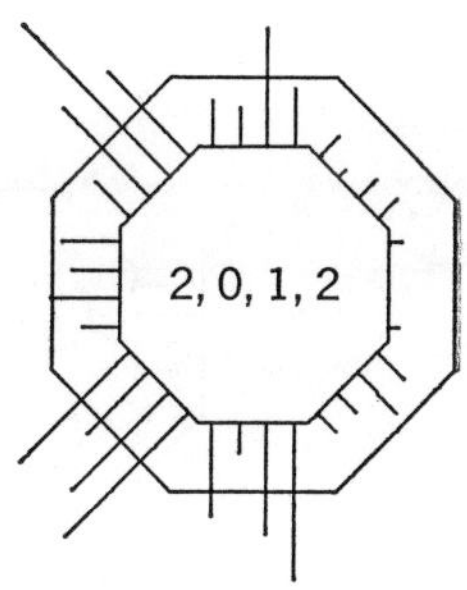

VICTORIA

Mildura
34.2°
142.2°
177 ft

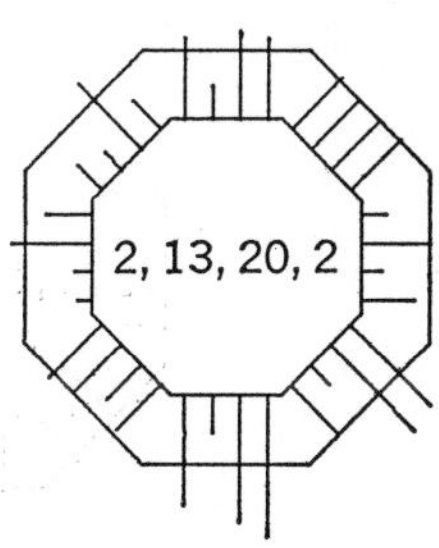

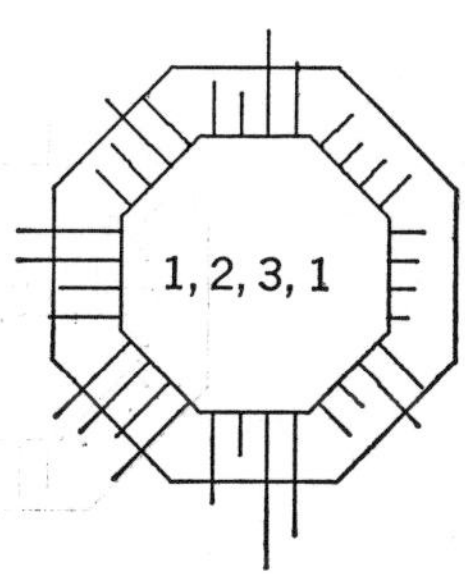

FIG. 3.5—*continued*

Meteorological station latitude °S. longitude °E. altitude (feet)	*Wind roses* 9.00 a.m.	3.00 p.m.

NEW SOUTH WALES

Broken Hill
31.9°
141.5°
1,000 ft

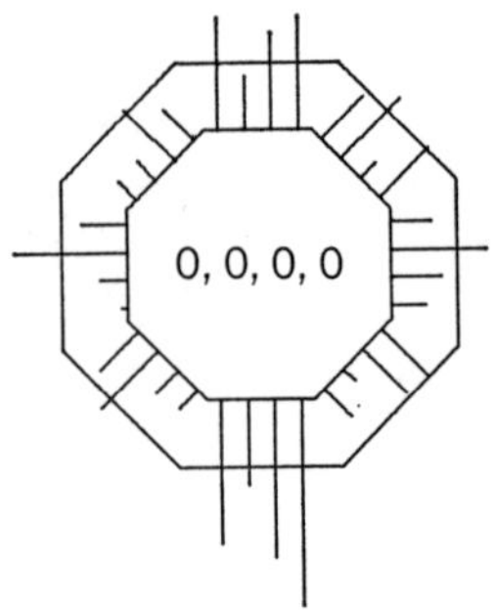

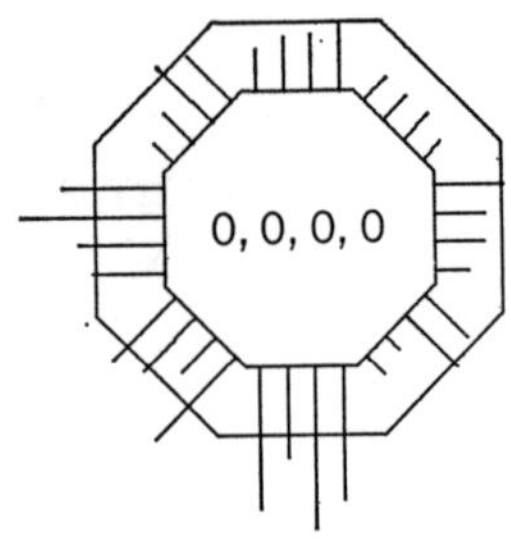

Moree
29.5°
149.9°
680 ft

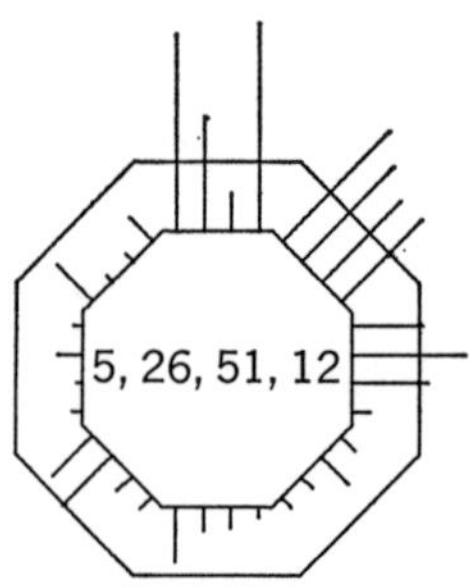

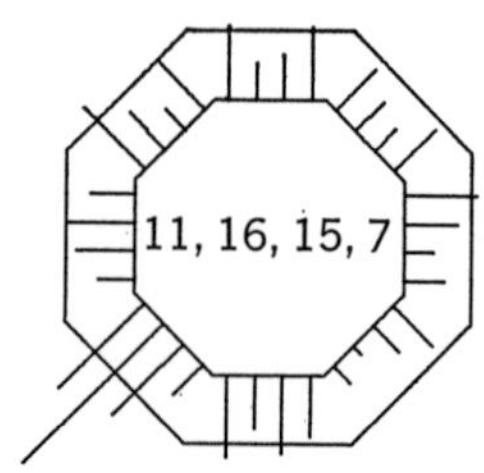

Wagga Wagga
35.1°
147.4°
612 ft

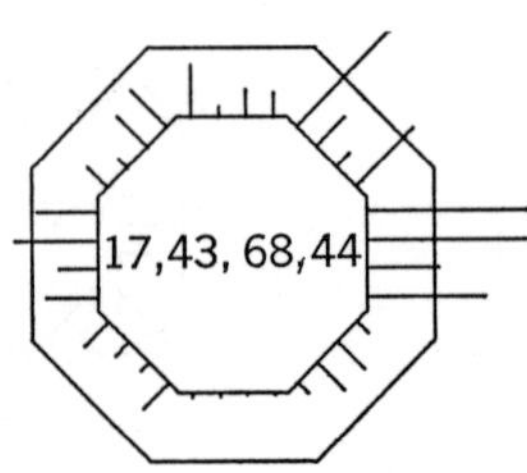

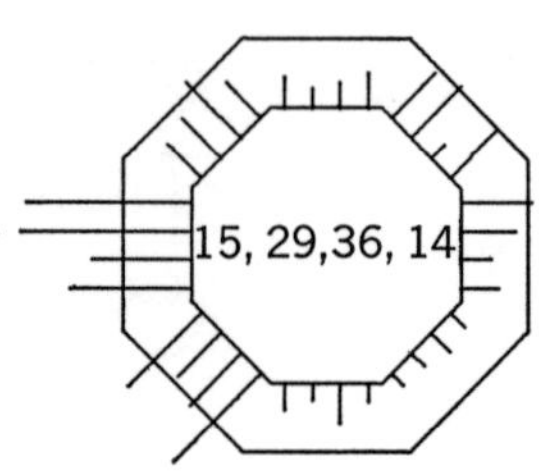

QUEENSLAND

Charleville
26.4°
146.2°
965 ft

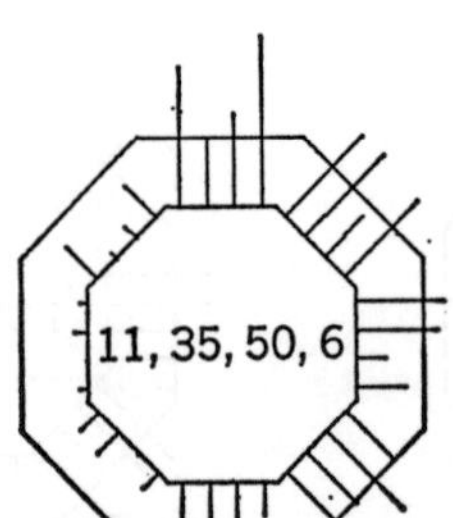

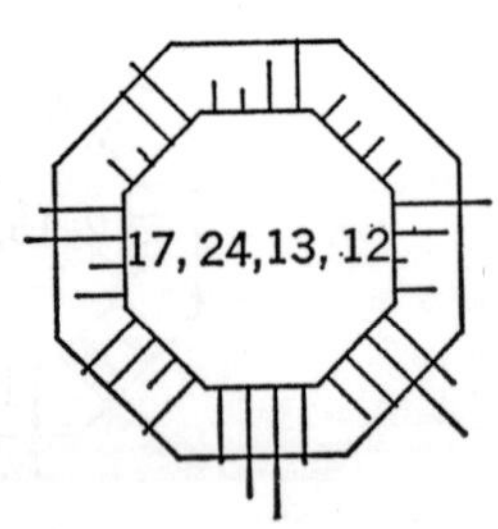

FIG. 3.5—*continued*

Meteorological station *latitude °S.* *longitude °E.* *altitude (feet)*	*Wind roses* 9.00 a.m.	*Wind roses* 3.00 p.m.

QUEENSLAND—*continued*

Clermont
22.8°
147.6°
870 ft

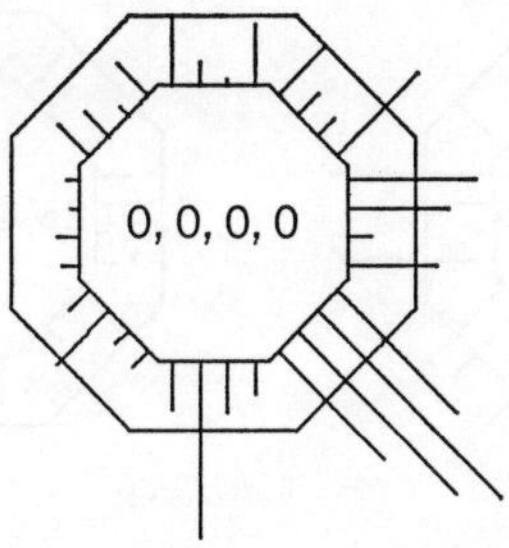

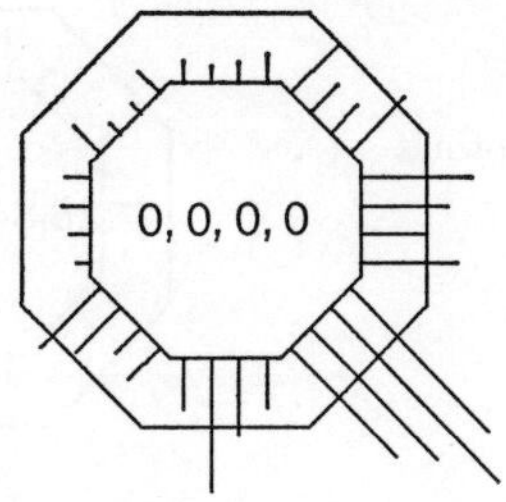

Cloncurry
20.7°
140.5°
633 ft

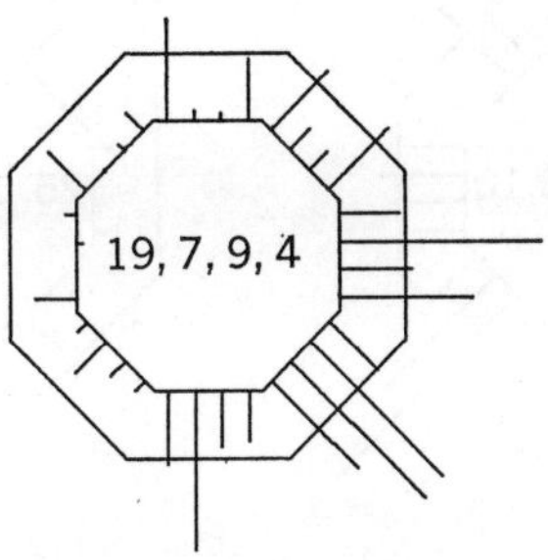

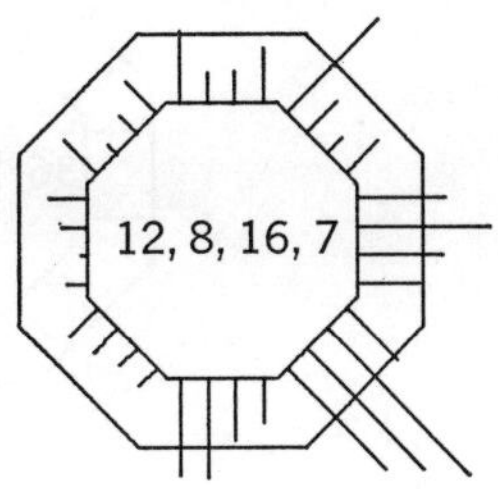

Hughenden
20.8°
144.2°
1,074 ft

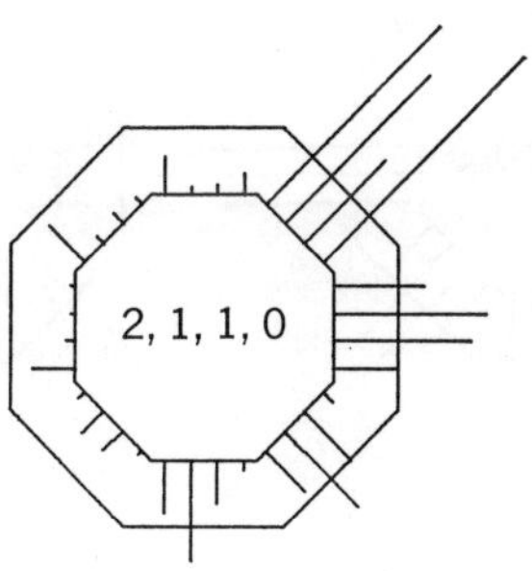

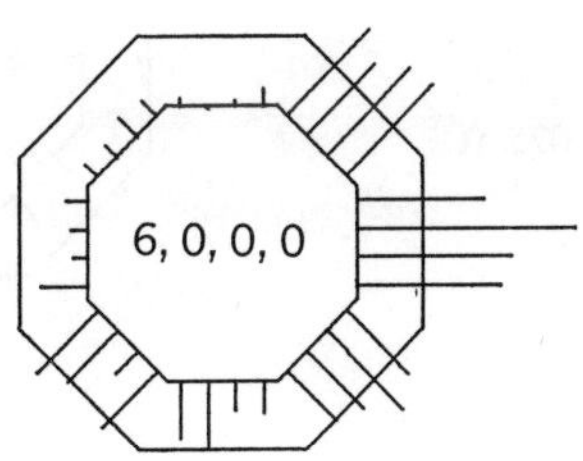

Windorah
25.4°
142.6°
390 ft

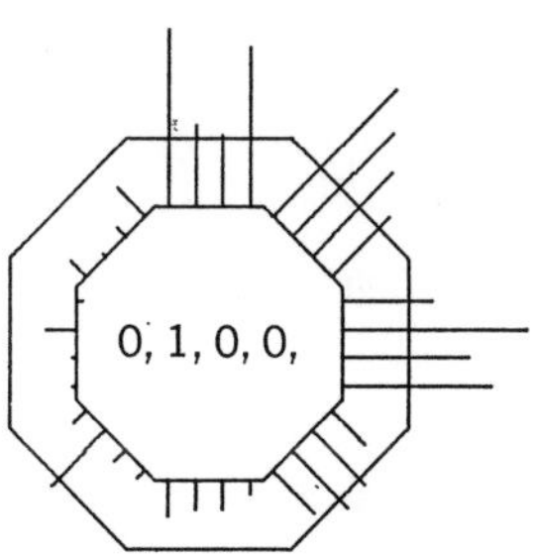

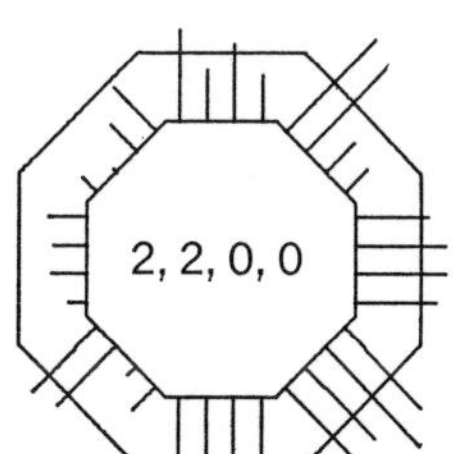

FIG. 3.5—*continued*

Meteorological station *latitude °S.* *longitude °E.* *altitude (feet)*	*Wind roses* *9.00 a.m.*	 *3.00 p.m.*
NORTHERN TERRITORY		
Alice Springs 23.6° 133.6° 1,901 ft	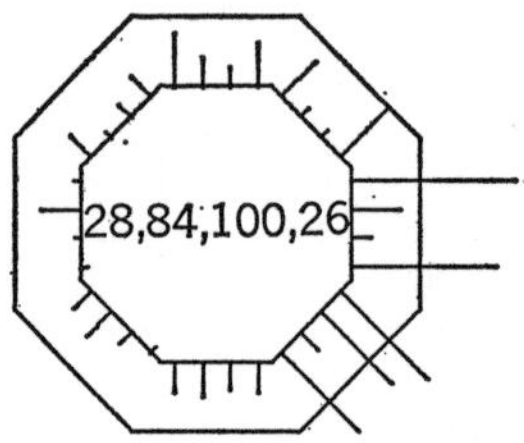	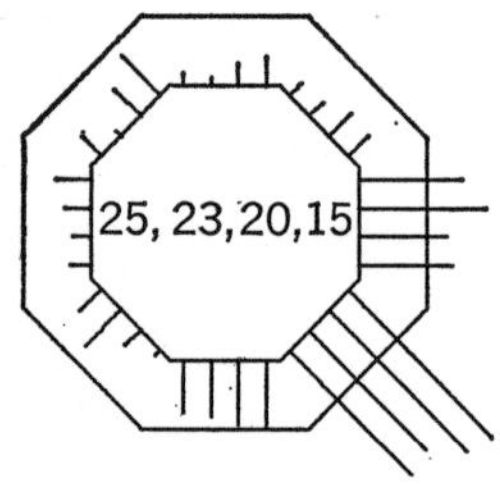
Daly Waters 16.3° 133.4° 691 ft	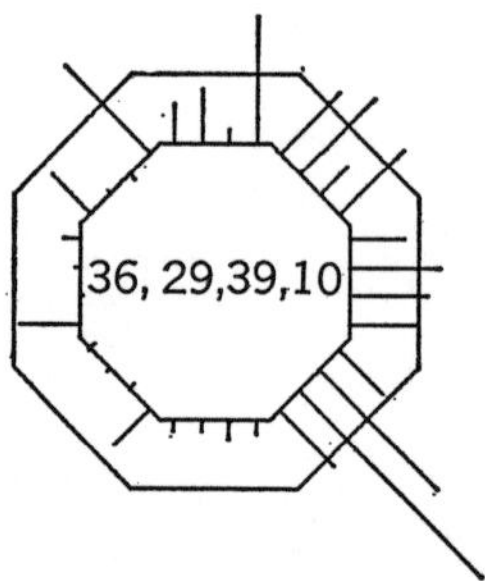	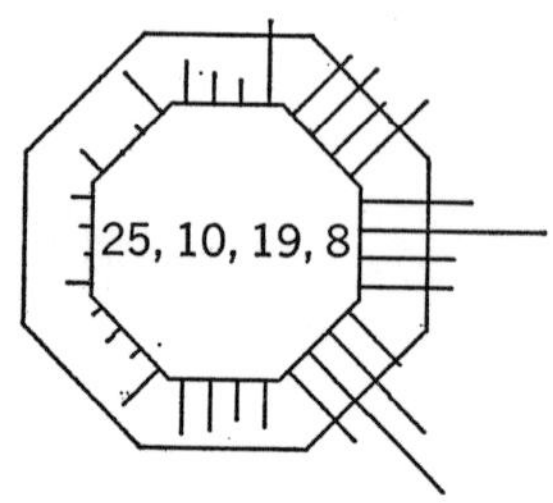
Tennant Creek 19.6° 134.2° 1,075 ft	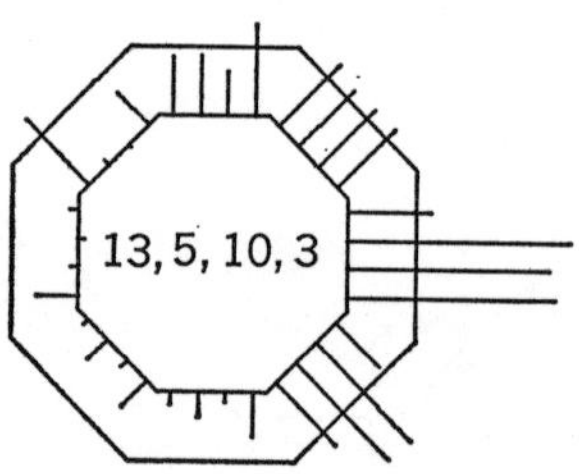	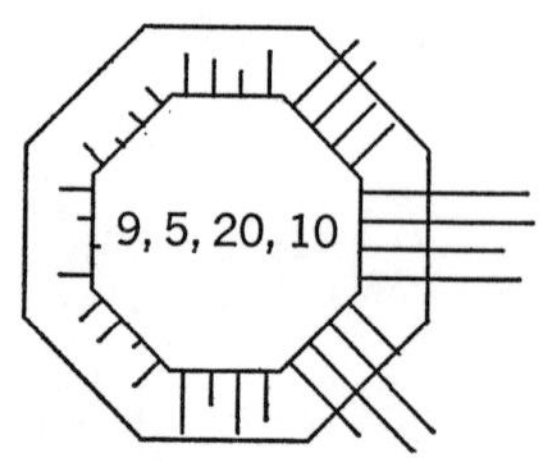

FIG. 3.6

WIND ROSES (9.00 A.M. AND 3.00 P.M.) FOR FOUR LOCATIONS IN AUSTRALIA, SHOWING FREQUENCIES FOR EIGHT CARDINAL AND SEMI-CARDINAL DIRECTIONS FOR EACH MONTH OF THE YEAR(*a*)

Meteorological station *latitude °S.* *longitude °E.* *altitude (feet)*	*Wind roses* *9.00 a.m.*	 *3.00 p.m.*

WESTERN AUSTRALIA

Geraldton
28.7°
114.6°
13 ft

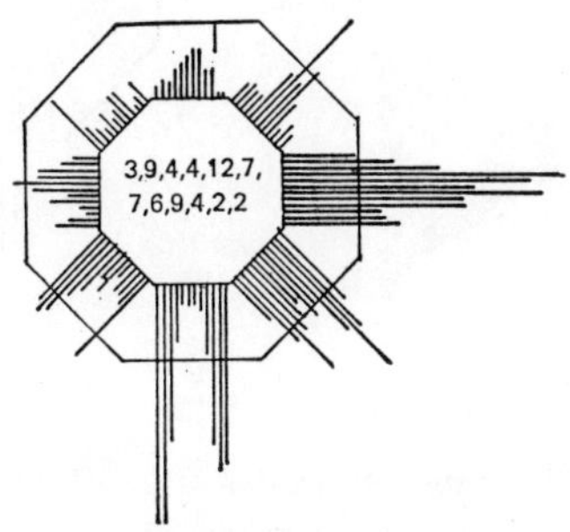

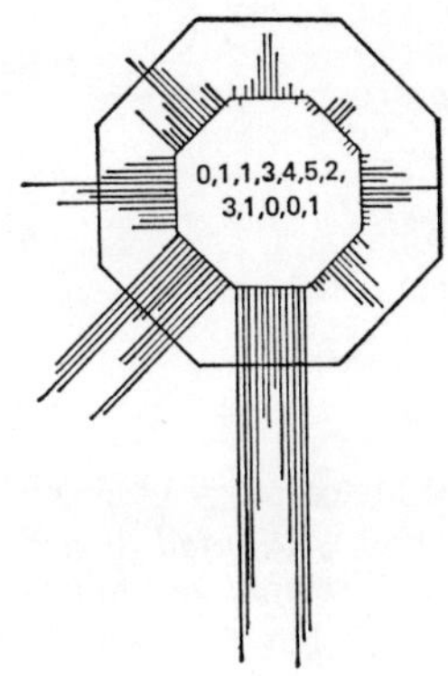

Kalgoorlie
30.7°
121.5°
1,247 ft

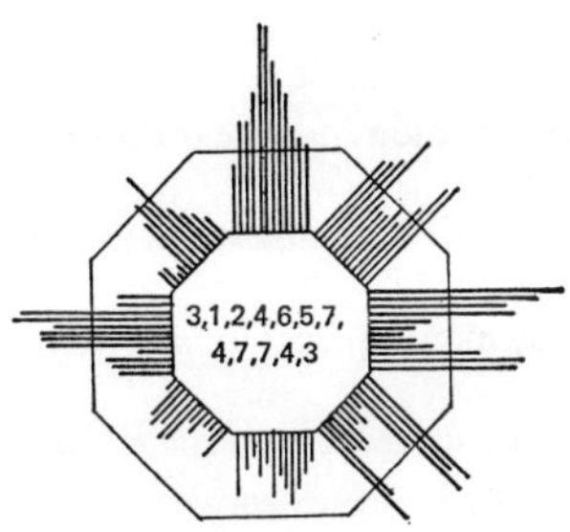

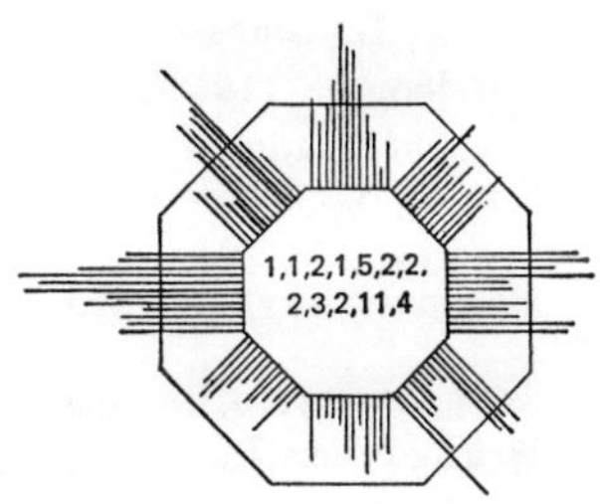

QUEENSLAND

Charleville
26.4°
126.2°
965 ft

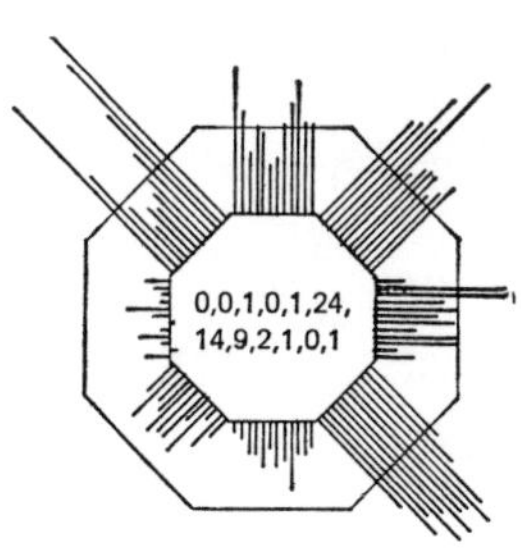

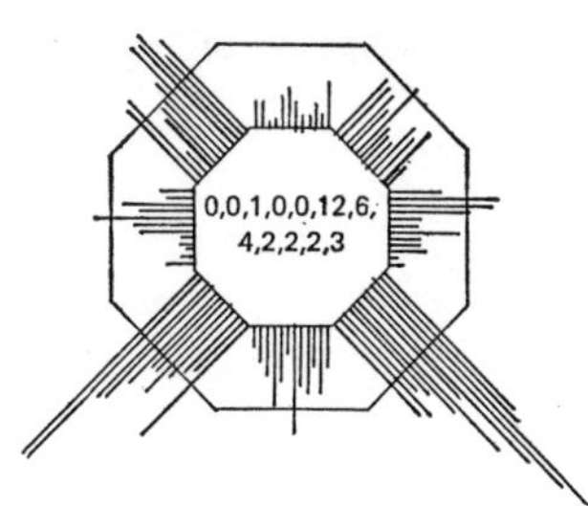

(*a*) These wind roses have been taken from the Bureau of Meteorology publication, Bulletin No. I, revised edition (1967), *The climate and meteorology of Australia.*

FIG. 3.6—*continued*

Meteorological station *latitude °S.* *longitude °E.* *altitude (feet)*	*Wind roses* 9.00 a.m.	3.00 p.m.

QUEENSLAND—*continued*

Cloncurry
20.7°
140.5°
633 ft

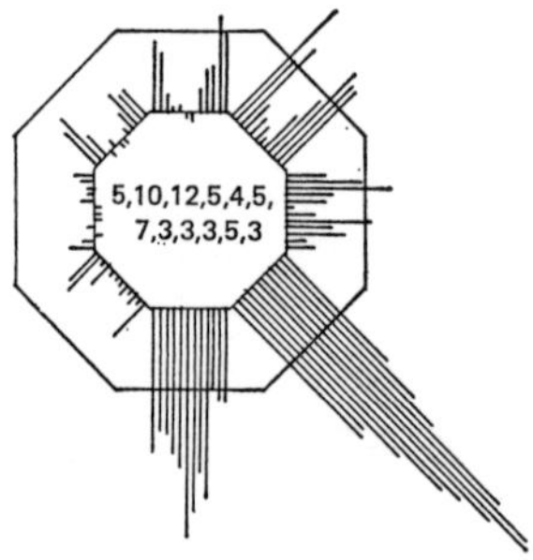

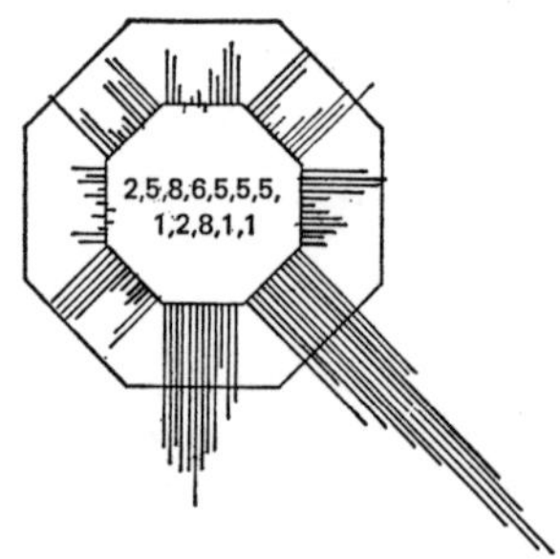

FURTHER READING

Bureau of Meteorology (1956). *Climatic Average Australia.*

——— (1966). *Rainfall Statistics Australia—Long term averages (all years), standard period normals of rainfall and raindays (1931-1960).*

——— (1967 but frequently revised). *The Climate and Meteorology of Australia.* Bull. I of the Bureau of Meteorology, Commonwealth Government Printer, Canberra.

——— (1968). *Review of Australia's Water Resources—monthly rainfall and evaporation,* parts I and II. Bureau of Meteorology for the Australian Water Resources Council.

———, Melbourne (1969-70). Private communications to the Forestry and Timber Bureau.

Department of National Development. *Atlas of Australian Resources.* Commentary prepared by J. C. Foley and B. W. Newman to accompany map-sheet *Climatic Regions.* (1954).

———. Commentary prepared by B. W. Newman to accompany map-sheet *Temperatures.* (1953).

Everist, S. L. (1969). Some thoughts on drought. (A paper presented at the Drought Seminar, University of New England, May 1969).

Foley, J. C. (1957). *Drought in Australia: review of records from years of settlement to 1955.* Bull. 45 of the Bureau of Meteorology.

Gibbs, W. J. and Maher, J. V. (1967). *Rainfall Deciles as Drought Indicators.* Bull. 48 of the Bureau of Meteorology.

Marshall, T. J. (1969). *Soil-Water in the Driest Continent* (Presidential address to the Australian Society of Soil Science, 5 Aug. 1968).

CHAPTER 4

Conservation

by C. S. Christian

THE CONCEPT

In exploiting the environment on which they depend all living species modify it to some degree and are in turn influenced by the physical changes that result and by the responses and interactions of other species. In a natural mixed community these responses and interactions are complex and lead to many fluctuations and adjustments in the populations of individual species. However, in a physical environment free of violent changes mixed communities establish an apparent stability even though longer term trends may be taking place. Primitive man was no exception. Even at the evolutionary stage reached by the time of the European invasion the Aboriginal of dryland Australia demanded little more from the environment and his associated species than the necessities for his own survival. With little in the way of tools or weapons, and with some measure of self-imposed population control, he existed as one of the components of a natural, seemingly stable ecosystem.

Modern man demands much more. He differs from primitive man in a number of important respects. His greater intellect has enabled him to establish a strong competitive role with respect to other species. By reducing his death rate and extending his longevity he has enabled the human population to expand rapidly. His desire to accumulate material wealth and to achieve new standards of eating, clothing and protection, beyond mere needs for survival, has increased his demands upon his environment. His intellect has enabled him to develop tools, weapons, and technologies which permit him to exploit the resources of his environment at a level of activity far beyond the powers of any other species. He has achieved a mobility that enables him to apply these over a much wider territory than that necessary for his existence. His technologies and mobility have extended his demands while the availability of leisure time has developed a strong desire for recreation opportunities and for means of satisfying his personal aesthetic and cultural interests.

All this has happened over a short period of time relative to the life span of the major terrestrial ecosystems with which he exists. The result is that he is in the process of making rapid changes in, and is even destroying, many of these ecosystems. In some respects he is no longer a part of them, but an opponent to them. In the exploitation of natural resources to meet his own magnified demands, at a rate which could lead to the exhaustion of some, he is modifying his total environment in a way which not only threatens its potential to supply his needs but also imposes dangers to his physical welfare, and tends to destroy those facets of the environment which are important to his expanding cultural and aesthetic aspirations. But man's intellect, which has made this situation possible, also enables him to recognise it and its implications. The forward-looking members of the species see the dangers to man's own existence and happiness of allowing past and current trends to continue unchecked and uncontrolled. It is from this that the concept of conservation emerges, a concept as essential in practice as any other concept relating to the physical, social or cultural development of the human species.

THE MULTIPLE ROLE OF CONSERVATION

The dryland areas of Australia form one continuous tract of country of low biological productivity, nearly 2,000 miles long and 1,000 miles wide, extending over a range of climatic zones and land types occupying virtually the inner two-thirds of the continent. With the exception of the Murray River Basin, and the ephemeral surface drainage of the Queensland Channel Country and a few other smaller inland drainage systems, its rainfall is not supplemented from outside the region as are many other dry regions of the world. Nor does it have highland areas within it receiving substantially higher rainfall. Its continuity and extent, relative to the area of the whole continent, and its virtual dependence upon its low and erratic rainfall are factors of considerable importance in planning its long-term use and its conservation.

The nature of the lands and climate of this region makes the pastoral industries the major form of land use in terms of area occupied and, because their impact is so widespread, a major concern of conservation. For this reason this aspect will be given more attention than others in this chapter. Mining and the search for minerals are rapidly expanding and having their own particular kind of impact at an increasing number of sites. These are small compared with the areas occupied by the grazing industries, but because of their concentrated effects on restricted localities they also demand attention. Tourism is a growing industry in dryland areas and this use of the resource requires, amongst other things, good transportation facilities and suitable man-made amenities consistent with the preservation of those interests which attract tourists as well as an environment kept as free as possible of unnecessary man-made inconveniences and distractions. Finally, the dryland areas, like all natural regions, inherit features of cultural and scientific interest, unique geological features, vegetation types, plant and animal species, scenic and other natural phenomena which, once lost, cannot be artificially replaced by man, and which, therefore, require protection and preservation for the benefit of this and future generations of mankind.

The objectives of conservation must be to recognise the multiple uses and values of these resources and to encourage the selective or multiple use of them in a rational way which will permit just return to land users compatible with the perpetuation of the resource for future generations. A more deliberate allocation of some areas in accordance with their suitability or unsuitability for different purposes will become more important if present trends in land resources are allowed to continue.

ECOLOGICAL LIMITATIONS OF DRYLAND AREAS

As with other dry regions of the world the sparse perennial vegetation, supported by and adapted to limited and erratic rainfall, provides a degree of stability to the land surface, but this is insufficient on many landscape types to prevent major losses of surface soil material by wind erosion in drought periods, and to create surface flow conditions which lead to water erosion during the early falls of rain when droughts break. Climatic crises of this kind, with contrasting and severe climatic stresses at different times, are characteristic of most dry regions. The vegetation and land surface maintain a fluctuating equilibrium with irregular wide departures from the 'norm' involving long periods of recovery. Ephemeral species quickly establish when suitable rains fall, and vary in kind and density with the season, but after severe drought crises, even these pioneering species may fail to cover satisfactorily those parts of the land surface which have been most severely degraded by natural processes. During drought periods most native perennial species persist, with perhaps some losses of individuals, but survivors may be reduced in size and vigour so that the completeness of the community as a whole may take some time to be re-established. Suitable moisture and temperature conditions for the germination and establishment of perennial species occur infrequently and erratically so that many years may elapse before regeneration takes place in a depleted community even under natural conditions. It is this natural fluctuation and time lag in recovery which makes dryland areas so susceptible to use and so dependent upon the application of sound conservation attitudes and practices if their usefulness is to be maintained.

DRYLAND AREAS AS A PASTORAL RESOURCE

Sparse dryland plant communities, restricted in total growth by rainfall and in many areas by infertile soils, and in seasonal growth by the erratic occurrence of rain, are characteristically low-producing communities when productivity is expressed in terms of animal products per unit of area and time. The depasturing of stock is, therefore, practised on an extensive scale, allowing large areas of land per animal. This allows both for some feed selection by the animal as well as some provision for the irregular distribution of rain. The low return per unit area places limits on the amount of money that can be spent economically on attempts to improve productivity, or to exercise control of the grazing animal. Thus the pastoral industry of dryland areas exists mainly by the direct exploitation of the vegetation that exists naturally, a vegetation inherently in a somewhat precarious balance with natural crises. The maintenance of the vegetation in a healthy, protective and productive state is the major objective of conservation. The deliberate introduction of domestic stock to these areas, mainly sheep, cattle and horses, but to a less extent camels, goats and donkeys, and the unfortunate invasion of many southern areas by rabbits, add additional pressures to these susceptible plant communities whenever a climatic crisis occurs. Australia has been fortunate in the absence of large indigenous human populations dependent upon locally produced agricultural and pastoral products and of intensive nomadic grazing over large territories. This, and the practice of land tenure involving the sedentary occupation of large tracts of land as stations under one control, have provided some measure of protection to the land resource. Furthermore, in the absence of permanent surface water over much of the region the necessity to establish watering points for stock at a relatively high cost, and the sparseness of sources of water or difficulty in locating them, have led to a low density of watering sites over much of the area. The consequence has been concentrated grazing near these points, but a much reduced grazing rate extending outwards from them, although the actual pattern of grazing varies with the type of animal and the circumstances.

The impact of grazing stock on dryland areas varies with the type of land and vegetation, and the controls exercised by the property manager on stock numbers and the time and periods when they are concentrated or dispersed on the different land types under his management, but in all cases some change in the vegetation takes place. Examples are the reduction in the stands or vigour of the dominant edible trees or shrubs of many communities such as mulga and saltbush; the depletion of the perennial grasses in many vegetation types and their replacement by annual grasses, forbs, inedible woody shrubs, or by bare scalded areas; and the reduction in the proportion of edible annual species such as in the perennial Mitchell grass grasslands and their invasion by unpalatable annual grasses such as *Aristida* species. In some circumstances one vegetation type may be replaced by another not entirely to the disadvantage of the grazing stock or the stability of the land surface as for example the replacement of saltbush by grass in some of the higher rainfall parts of the Riverine Plain. Changes induced by grazing sometimes produce conditions more suitable for wild herbivores leading to a build-up in their numbers, thus adding to the total grazing pressure and the ultimate degradation of the economic resource. On many land surfaces, depletion of the natural vegetation following grazing leads to an accelerated loss of surface soil and soil nutrients by wind and water erosion and the production of surface conditions much less satisfactory for new plant establishment and plant growth. This can proceed to a stage where natural reclamation is virtually impossible. It is apparent that most dryland communities in Australia have deteriorated in one way or another since domestic stock were introduced. The table on p. 79 prepared by Newman and Condon* gives estimates of the extent of degeneration which has occurred on different rangeland types.

The occurrence of widespread deterioration resulting in lower productivity and lower durability was confirmed at a conference on the arid and semi-arid zones sponsored by the Australian Standing Committee on Agriculture in 1970. Further, it was considered that in most of the degraded areas a downward trend in the condition of the vegetation and the land resource as a whole was continuing.

In view of the fact that the period of domestic grazing in inland areas has been limited to a period of about 100 years these assessments reflect a very serious situation which must be controlled unless more disastrous consequences are to result.

* J. C. Newman & R. W. Condon, 'Land use and present condition', in R. O. Slatyer R. A. Perry (eds), *Arid Lands of Australia.* (Proceedings of a symposium held in the Academy of Science, Canberra, May 1969.)

TABLE 4.1

PERCENTAGE OF RANGELANDS TYPES IN PARTICULAR DEGENERATION CLASSES

	The present pastoral value in relation to expected pristine condition			
	Little or no degeneration (80-100%)	*Minor degeneration (60-80%)*	*Moderate degeneration (40-60%)*	*Severe degeneration (<40%)*
1. Grasslands	30	55	15	negligible
2. Shrublands	10	25	40	25
3. Low woodlands	40	30	20	10
4. Flood plains and alluvial pans	40	30	20	10
5. Spinifex grassland	70	20	5	5
6. Mountains and hills	50	20	20	10

It is true that at times the assessments of the research worker, the layman, or the grazier, may differ. This is because of the wide variation in appearance presented by native vegetation from season to season and year to year. In drought times most communities appear ominously degraded, but the flush of green growth that follows rain, especially of annual ground storey species, encourages the belief that the vegetation has recovered and is in good condition. The presence of trees and shrubs can be deluding in this respect also, as a stand of either can be near senility, with few replacement seedlings, yet create an impression of well-grown vegetation. Unfortunately, precise criteria for assessing the actual status of Australian dryland plant communities are not available and one of the urgent tasks for research workers is to produce these. When they are available a continuing monitoring system could be initiated to give guidance on the need or not to modify the prevailing grazing practices in any area.

Important, too, is the fact that knowledge of the precise responses of different plant communities to different systems of grazing management is limited and a study of management systems of different types of communities in different stages of degradation in the various climate-land type situations requires further research. Only when it is possible to specify such management systems, or at least the basic principles on which they must be based, will it be possible to manage the dryland regions of Australia for optimum production in a manner that also gives confidence in their long-term conservation.

The first and obvious consequences of degradation are a lower and more widely fluctuating plant and animal productivity and, therefore, reduced and more erratic returns to the grazier. The extent to which productivity will recover will vary greatly with the vegetation and land types and the prevailing climatic conditions as well as the degree of degradation that has occurred. A reduced vigour of an otherwise healthy vegetation complex is not regarded as degradation and may require little more than temporary protection. Degradation that involves a substantial reduction in the numbers of, or elimination of, edible perennial components is far more serious and may call for a drastic change in stock management. The erratic and infrequent occurrence of suitable combinations of soil moisture and temperature essential for the germination and establishment of these perennial species means that re-establishment may occur by chance at any time from one to many years after stock have been excluded or stocking rates reduced. Such chance variations have led to many conflicting opinions concerning vegetation responses to changes in management and this is one of the important aspects on which research needs to give more guidance.

Where degradation has proceeded to the point that soil erosion has been significantly accelerated, the time for recovery may be greatly extended. In fact, where the surface horizons of a well-differentiated soil profile have actually been removed natural recovery may be virtually impossible and for all practicable purposes the change may be regarded as irreversible. Under such conditions the economics of reclamation by artificial means need careful attention and in many instances such degraded land must, for all practical purposes, be regarded as a lost resource, at least for this generation.

Except at the more favourable margins with higher rainfall, or on specially-favourable parts of the topography, the replacement of a native vegetation by re-seeding is not widely practicable.

There are exceptions where reclamation is essential. A major example is the Ord River catchment where the need to regenerate vegetation to retard silting of the Ord River Dam added a special incentive. In parts of this catchment continuous uncontrolled grazing over many years had led to extensive deterioration in the vegetation and to serious soil erosion. Although confined largely to one group of soils in the region, the degraded area occupied about 8 per cent of the total catchment and provided a major threat of siltation for the Ord River Dam. Many of the perennial grass species of the natural savannah woodlands were replaced by annuals or, as the surface soil eroded, by bare ground. The lower permeability of the exposed soil horizons, and the higher temperatures of the bare surfaces were followed by the death of many of the trees of the upper storey. Fencing and control of stock were necessary before re-establishment of vegetation was possible. As supplies of seed of the native perennials had been virtually eliminated it was necessary to re-seed with introduced species. Those used were buffel and birdwood (*Cenchrus* spp.) grasses and kapok bush (*Aerva javanica*). To provide a seed bed, improve water penetration and reduce wind activity, cultivation on the contours with banks at intervals was practised. The work has now been in progress about 10 years and large areas have been re-vegetated. From these, seed is now spreading to re-vegetate some of the non-treated parts. In addition seedlings of native trees and shrubs, as well as ground storey species, are reappearing. Grazing of the area will eventually again be feasible, but because of the susceptible soils of the region it will need to be carefully controlled. The area in question is one which probably should never have been grazed on a continuous basis. The natural fodder on the calcareous soils was no doubt regarded highly, but, with a limitation of 17-18 inches of rainfall per annum, the natural productivity was low. The most severe damage was done nearest the watering points along the natural streams where grazing was heaviest. The fact that the area was a source of large quantities of silt to the Ord River was the main inducement for initiating control and treatment. Had this not been done and had serious erosion continued a natural land resource of a million acres could well have become a virtual useless waste.

In some localities, especially in southern areas, it has been proved economically possible to achieve the re-vegetation of localised eroded claypan areas by the use of furrows, pits, or banks on the soil surface in order to restrain the surface flow of water, and to provide suitable resting places for seed. The Soil Conservation Service of New South Wales has made extensive studies of these practices.

In general terms, however, the main weapon available to the land user is the management of his grazing stock. Most property managers consciously use this in the way their experience dictates, but long periods of climatic stress, economic conditions, lack of proven practices for particular land types, property quarantine restrictions, and even restraints applied by bankers and other financial institutions can dictate the action taken. It is clear that, in spite of the best intentions, the practices adopted in the past have not been adequate to contend with the combined pressures of climate and grazing in many of our pastoral dryland regions. Degradation has occurred and is continuing. Current low wool prices could induce some landholders in sheep areas to extract the greatest possible return from the resource and, hence, accelerate degradation. This aspect needs to be watched very closely in the coming decade.

The consequences of inaction in face of the evident downward trend of so many areas could be disastrous. The immediate results could be further diminished returns from the capital resources already invested in the area. As degradation proceeds the rate of degradation would almost certainly accelerate and it is at such a stage that the situation could conceivably get out of hand. Interruptions to transport and other communications and increased costs of maintenance of public utilities would follow from any local mass movement of surface soil from degraded areas. Hazards to human life could be increased. The hydrology of degraded areas would be changed resulting in an even less efficient utilisation of the limited rainfall of these areas and, therefore, reduced production. Pollution of some of the lower parts of the topography by salt moving through soil profiles could follow. Silt would be carried more abundantly in surface drainage systems. Dust in the area would increase, affecting both livestock and human beings. Dust storms could become intolerable in certain parts of the region and even far beyond it. The affected parts of the region would become far less attractive for tourism and this alternative use of the resource could be endangered. While this is admittedly an unduly pessimistic total picture, any of these could happen. The objective of conservation is to prevent them; to ensure that the various parts of this geographically large resource remain stable for whatever continuous use they are most suitable.

Reference has been made to the need for research to obtain more information for future management and for the development of techniques for assessing vegetation and land condition and trend. This should be encouraged and supported, but the problem also requires more immediate decisions. While research is busy seeking the long-term answers much could be achieved by mobilising the knowledge that already exists from past research work and from the experiences of informed land users and other observers in order to formulate the principles on which the best possible advice available at the present time could be given for each area. Such advice will not be perfect and will be improved in time by the results of further research, but if formulated with the guidance of sound ecological principles and managerial feasibilities, it should serve to meet the requirements of the immediate future by retarding the downward trend, and give time for research to produce the long-term answers.

CULTIVATED AREAS

At the edge of the inland pastoral zones are areas which are marginal for crop production. Early efforts to extend clearing and cultivation into and beyond these limits had unfortunate results. An example was the extension of wheat-growing in South Australia to the north of Goyder's line. The problems of sand drift in this area became so important that C.S.I.R. appointed the late Dr F. N. Ratcliffe to investigate the problem in 1935. The early experience in the mallee of north-western Victoria is another example of a similar nature, but fortunately the subsequent development of better farming practices adapted to the light-textured soils and erratic rainfall of this region has enabled a more stable agriculture to be achieved. The measures necessary have included an increase in farm size and greater mechanisation, the introduction of the clover ley and the grazing animal together with a crop-pasture rotation system in place of continuous cereal cropping and bare fallows, and also the sowing of a deep-rooted legume, lucerne, on erodible sandy rises.

While the major problems of sand drifts intruding on the public utilities, such as irrigation channels, roads, railways, and interfering with other forms of communication can be avoided by proper land use measures, once they have occurred costly remedial measures are necessary until such time as prevention practices have had their effect. It is obviously wise planning to avoid the destruction of the vegetation cover of such marginal land for cultivation purposes until there is some assurance that an appropriate technology for safe and economic land use is available. One approach adopted by the Western Lands Commission of New South Wales is to limit the area of land cleared in single blocks to 300 acres and to require a 5-chain strip of uncleared and uncultivated land to separate these blocks. This not only restricts the potential areas of erosion should hazardous conditions occur, but also provides seed sources from which recolonisation of native species may commence.

TREES AND SHRUBS

Many of Australia's dryland areas are fortunate in having a variety of trees and shrubs which provide not only nutritive grazing for stock in good seasons, but also a reserve feed supply in dry periods.

They often give the only sparse shade available for stock and, of course, are important components of the habitats of wildlife. As many trees and shrubs can be expected to tap water and nutrients from different parts of the soil profile than the ground flora, they contribute to the greater exploitation of the natural environment. They play a significant role in reducing wind activity, in protecting the land surface and in maintaining the hydrological status of the land type. They assist in reducing the local dust hazards around townships. It is important that this resource should be conserved. Some species have been seriously depleted around centres of human population as a result of harvesting for mine timbers or for firewood.

In many grazed areas there has been a reduction in the density or vigour of trees and shrubs. However, even in natural stands, there may be a proportion of dead trees and their presence should not always be regarded as induced. From the conservation point of view the important need is to ensure recruitment of seedlings to the population. It is where recruitment is not taking place that there is need for concern. Seedling establishment or growth can be inhibited by changes in the micro-environment of the habitat. These environmental changes can be induced by grazing and subsequent erosion, by harvesting of timber, by abnormal climatic stresses, and even at popular sites by the traffic of tourists. Grazing at inappropriate times can be damaging to the recruitment seedlings. Rabbits are particularly important in this respect. Each region and vegetation type needs to be assessed individually. For example, in some parts of western Queensland, the density of mulga has increased so greatly, possibly following a very favourable recruitment period, that the dense tree cover has reduced the grazing productivity of the land. Its destruction and replacement by introduced grass species is attempted to increase economic return. Similarly, periods of high rainfall and the reduced use of fire are factors put forward to help explain the dense regrowth of timber and shrub species in the Cobar-Byrock district of New South Wales. Periods of regrowth of this kind have occurred on several occasions in the last century or so and are regarded by some as natural phenomena in the history of the vegetation.

Some 'pruning' of dryland trees, such as is provided by grazing or even judicious fodder harvesting can sometimes be beneficial and even lead to greater productive growth. As long ago as 1932 it was recorded at Koonamore that moderately heavy grazing by sheep produced an apparently healthier stand of *Atriplex vesicarium* than light grazing or even protection from grazing. In grazed woodland communities some edible trees are eaten to head height and effectively grow above the range of the grazing animal. The presence of these taller trees can inhibit the growth of younger members of the species which may accelerate their growth if the older trees are cut. It is not possible to generalise as responses vary so much with species and conditions, but it is worth stressing that protectionism is not the only approach to conservation. Conservation can be combined with practices aimed at achieving a higher economic return and such a combination may in the long run be in the interests of the best and most prolonged economic return.

WILDLIFE

When settlement began in inland Australia there were many species of marsupials and rodents. Some of these were very numerous. They included small wallabies, bandicoots, marsupial mice and native rats. Most are now very rare and several are thought to be extinct.

For instance, in the boree woodlands in the eastern Riverina there were at least 30 species of mammal; today only two kangaroos, a marsupial mouse, a possum and a few species of bat remain. The situation in the mulga of central Australia is similar.

Of the 119 species of marsupial in Australia five are considered extinct and 34 are endangered. The rodent situation is similar. Nearly all of the endangered animals are inland species. By contrast the mammal fauna of the coastal and near-coastal forests and woodlands that have not been so destructively grazed is still practically entire and numerous.

Several birds characteristic of open woodlands and adapted to feed on seed of native grasses have almost disappeared following the reduction in these grasses. These include a quail (*Pedionomus*) several species of parrot of the genus *Neophema* and some ground living pigeons (*Geophaps*).

The probable reason for the decline of mammals and birds is the destruction of food plants and the tall dry herbage cover needed for shelter, nesting etc. This has undoubtedly been influenced by the intrusion of exotic grazing animals into these areas.

On the other hand a few mammals have prospered because of domestic stock. Grazing has produced a disclimax community that is more suitable for these animals than was the climax. These animals are red kangaroo (*Megaleia rufa*), euro (*Macropus robustus*) and probably also the grey kangaroos (*Macropus giganteus* and *M. fuliginosus*). It is often thought that the proliferation of water points for stock has also helped these animals increase in numbers and range, but there is no decisive proof of this.

A few birds also are extending their ranges. These include the galah (*Cacatua roseicapilla*) and the wood duck (*Chenonetta jubata*). The former is a seed-eating bird favoured by cereal crop production; the latter is a grazing bird favoured in much the same way as were the kangaroos. In this case there is little doubt that the proliferation of water points is strongly involved in the increase in numbers of both birds.

The impact of land use changes on wildlife may occur long before those changes are sufficient to reduce economic return of livestock properties. It follows that adequate conservation of many wildlife species can be achieved only if special areas are set aside for their conservation.

SALINITY

A characteristic of dryland areas of some importance is the occurrence of salt-pans and saline drainage lines. These are particularly prevalent in the southern part of the region and especially in South Australia and Western Australia. Salt in the soil profile is a common feature in Australian dryland areas and interference with normal infiltration and movement of water can lead to concentrations of salt at the surface either at the site involved or at lower places in the topography. A reduced productivity of the affected parts of the land resource is usually the result. Mismanagement of irrigation water can lead to this, but it has also happened following clearing of land for cultivation in Western Australia. It can also result from changes in vegetation and surface condition following uncontrolled grazing. Efforts are being made to find salt-tolerant fodder plants adapted to such areas, but the problem is another example of a deleterious change in a natural resource resulting from land use which, in terms of common conservation outlooks, is better avoided rather than ameliorated after the damage has been done. On the other hand a rational use of the total resource may suggest that mankind is better served by accepting some degradations of parts of the land surface as a consequence of achieving greater usefulness in associated parts. Much will depend upon the permanency and magnitude of both effects and the ultimate repercussions of degrading parts of the landscape.

NATIONAL PARKS AND RESERVES

Irrespective of how land is used for production purposes some changes will take place. It is important, therefore, that in all regions there should be set aside areas of land for other purposes. These include recreation and tourist areas, scenic reserves, historic and archaeological sites, nature reserves, and special features of scientific or educational interest.

As at June 1970 there were some 50 separate national parks and reserves in the dryland areas of Australia with a total acreage of about 30 million. Seventeen of these exceeding 100,000 acres each are listed below:

		(Area: acres)
QLD	Simpson Desert National Park	1,248,000
N.S.W.	Kinchega National Park	106,700
S.A.	Simpson Desert National Park	1,708,800
	Elliot Price Wilderness National Park	160,000
	Yumbarra National Park (in part)	262,000
	Oraparrina National Park	c. 150,000
W.A.	Hamersley Range National Park	1,458,430
	Chichester Range National Park	372,483
	Barlee Range Reserve	257,995
	Northern Nullarbor Reserve	6,167,200
	Queen Victoria Springs Reserve	618,750
	Kalbarri National Park	358,000
	Cape Arid National Park	843,056
	Nullarbor Cliffs Reserve	1,535,500
N.T.	Tanami Desert Sanctuary	9,273,600
	Ayers Rock-Mt Olga National Park	311,680
	Palm Valley Flora and Fauna Reserve	113,280

Several additional ones of this order are contemplated. One hundred thousand acres may be regarded as the minimum size necessary to preserve representative samples of land and vegetation, but in some landscape types this area could be quite inadequate to sample the landscape pattern. A figure quoted in international circles for the desirable percentage of a country to be set aside as reserve is 5 per cent. The total reserves in Australia's dryland areas are of the order of 2 per cent and half of this area occurs in two large parks. Apart from those listed there are numerous small scenic and nature reserves and historic sites. Although the present extent of reserves indicates a big advance in recent years, many land and vegetation types, and wild life habitats, are still not represented. Australia is predominantly a dry country and, hence, much of its characteristic biological inheritance occurs in the dry regions.

In the absence of a comprehensive biological survey of Australia, which could take decades to complete, a practicable approach to conservation of fauna and flora is to ensure that samples of habitats are preserved in the hope that this will automatically protect most biological species. The contents of the reserved habitats can be explored later. Habitats are represented by combinations of topography, geology, soils and vegetation with their associated climatic, microenvironmental and biological characteristics. The land system approach to landscape analysis is one method of identifying patterns of landscape, which are in effect patterns of associated habitats. It has proved valuable in wildlife studies and provides a systematic method for identifying distinctive types of landscape worthy of consideration for conservation purposes for both fauna and flora protection.

Apart from the objective of setting aside parks and reserves for recreation, conservation, or scientific and educational purposes, the dryland areas of Australia include numerous historic places which could commemorate the activities of the early explorers and settlers, and many Aboriginal grounds and antiquities. There is a special importance in recognising and protecting these before time and civilisation erase them completely.

CONSERVATION ACTION

The recognition of the need for conservation measures in the dryland areas of Australia is not new. In fact, in the latter part of the last century there was a good deal of public concern about the effects of grazing and of drought in dryland areas. A consequence of this was that Lands Departments brought in various land reforms aimed at protecting these resources. Since then there have been many significant developments collectively contributing towards better recognition of the problems and their treatment. For example, in 1901 the Western Lands Commission of New South Wales was established following a Royal Commission which examined land settlement problems in the West. These had become prominent because of the combination of drought, rabbit infestation, over-stocking, erosion and consequential sand storms. One of the major actions taken was to subdivide large properties into home maintenance areas, but it has been found by experience that this concept must be treated with some flexibility. In more recent years land holders with home maintenance areas have been permitted to increase the size of their holdings and areas up to two home maintenance areas are now allowed. Leases contain many requirements and restrictions aimed at better land use and many activities are dependent upon written consent. The Commission has combined with the Soil Conservation Service and the Forestry Commission to establish standard conditions for various classes of country according to their erosion hazard.

The Soil Conservation Service of New South Wales was established in 1938 and subsequently all mainland States established similar activities in some form or other. There have been many surveys of resources for particular purposes. In Western Australia extensive surveys were made of the saline lands in the wheat areas and later special studies were made of some pastoral areas which had deteriorated following sheep grazing and the accompanying increase in density of marsupials. Numerous authorities have made vegetation surveys. New South Wales and Victoria have made detailed studies of the lands of their dry areas, the Western Districts and the North-western Lands respectively, with particular attention to soil erosion and soil erosion potential. Queensland has initiated in association with the C.S.I.R.O. and the Bureau of Agricultural Economics a study of the pastoral lands of that State receiving less than 15 inches rainfall annually. The first trial survey is proceeding in the south-west of the State. In 1946 the C.S.I.R.O. commenced its comprehensive and integrated survey and assessment of the land resources of northern Australia. This included large parts of the northern and central dryland areas and was extended to some southern regions. The Soils Division of C.S.I.R.O. completed a soil map of Australia in 1968. More recently the Organisation has established its Rangelands Research Unit and has converted its research station at Deniliquin from an irrigation research station into one dealing mainly with arid zone problems. New South Wales, Victoria, Queensland and Western Australia have all established research stations within their dryland areas. A number of universities has special interests in the dryland areas, some examples being the long-standing work of the Botany Departments of the University of Adelaide and of the University of New England, the Fowler's Gap Research Station of the University of New South Wales, and the Arid Zone Research Institute in the Zoology Department of the University of Western Australia. There has been a number of conferences on arid zone problems organised under the sponsorship of the Australian Agricultural Council.

In the last few years there has been an accelerating interest in conservation throughout the country, and most States have introduced or have considered legislation to set up authorities with responsibilities in the broad field of environment protection or conservation. Some of these have prominent interests in land use. In this regard two significant recent developments have been the establishment of the Land Conservation Council in Victoria and the National Parks and Wildlife Act in Tasmania. Many departments and authorities have responsibilities related to conservation such as the departments of lands, agriculture, and forestry, and soil and water conservation authorities, etc. Collectively the activities which have been referred to and others which have not been enumerated in detail have increased our knowledge of the dryland areas of Australia and of their conservation problems very considerably. However, these areas are not only very extensive geographically; they are also extremely variable, and the total effort is small relative to the needs.

As the dryland regions cover a large part of all mainland States, and the continuation of present trends could have repercussions far beyond State boundaries, there would seem to be a strong case for a combined national effort to ensure that these environmentally sensitive areas are conserved for future generations by the further development and rigid application of sound land use and management policies.

FURTHER READING

Beadle, N. C. W. (1948). *The vegetation and pastures of western New South Wales with special reference to soil erosion.* Government Printer, Sydney.

Bettenay, E. & Hingston, F. J. (1961). *The soils and land use of the Merredin area, Western Australia.* C.S.I.R.O., Division of Soils. (Soils and Land Use Series, No. 41).

Boyland, D. E. (1970). 'Ecological and floristic studies in the Simpson Desert National Park, South Western Queensland'. *Proc. Roy. Soc. Queensland*, Vol. 82, No. 1.

Campbell, Keith O. (1948). 'The development of soil conservation programmes in Australia'. *Land Economics*, 24 : 63-78.

Carder, D. J. (1969). 'Land use in the Kulin-Kondinin-Hyden district'. *J. Dept. Agri. W. Aust.*, Vol. 10 (Fourth series), No. 10.

Christian, C. S. (1970). 'Land Resources and Land Research' (Farrer Memorial Oration, September 1969). *J. Aust. Inst. Ag. Sc.*, June 1970, p. 18.

Christian, C. S. *et al.* (1954). *Survey of the Barkly Region, 1947-48.* C.S.I.R.O. Aust. Land Research Series No. 3.

Christian, C. S. & Slatyer, R. O. (1958). 'Some observations on vegetation changes and water relationships in arid areas'. *Proc. Canberra Symposium, Climatology and Microclimatology, UNESCO* pp. 156-8.

Condon, R. W. (1969). The assessment of grazing capacity on arid grazing lands in Australia. International Arid Lands Conference, Tucson, Arizona.

Condon, R. W., Newman, J. C. & Cunningham, G. M. (1969). 'Soil erosion and pasture degeneration in central Australia'. *J. Soil Cons. Serv. N.S.W.* Part I: 'Soil erosion and degeneration of pastures and topfeeds', Vol. 25, No. 1. Part II: Vol. 25, No. 2. Part III: 'The assessment of grazing capacity', Vol. 25, No. 3. Part IV: Vol. 25, No. 4.

Downes, R. G. (1970). *Soil conservation law in Australia.* Soil Conservation Authority, Victoria.

Downes, R. G. (1961). 'Soil salinity in non-irrigated arable and pastoral land as the result of unbalance of the hydrologic cycle in salinity problems in the arid regions'. *Arid Zone Research*, XIV, pp. 105-110.

Everist, S. L. (1969). *Use of fodder trees and shrubs.* Queensland Department of Primary Industries. (Division of Plant Industry Advisory Leaflet No. 1024).

Fitzgerald, K. (1968). 'The Ord River catchment regeneration project'. *J. Dept. Agri. W. Aust. Bull.*, No. 3599, November 1967, March 1968 and September 1968.

Hall, Elizabeth A. A., Specht, R. L. & Eardley, Constance M. (1964). 'Regeneration of the vegetation on Koonamore Vegetation Reserve, 1926-1962'. *Aust. J. Bot.*, Vol. 12, No. 2.

Jones, R. (Ed.) (1970). *The biology of the* Atriplex. (Studies of the Australian Arid Zone). Division of Plant Industry, C.S.I.R.O.

Kaleski, L. G. (1969). 'Address to the Western Division Pasture Protection Boards and Graziers Conference at Brewarrina, October 1968'. *J. Soil Cons. Serv. N.S.W.*, Vol. 25, No. 1.

Kaleski, L. G. (1945 and 1963). 'The erosion survey of N.S.W.' *J. Soil Cons. Serv. N.S.W.*, Vol. 1, No. 1 and Vol. 19, No. 4.

Mabbutt, J. A. *et al.* (1963). *General Report on lands of the Wiluna-Meekatharra Area, Western Australia, 1958.* C.S.I.R.O. Aust. Land Research Series No. 7.

Malcolm, C. V. (1969). 'Use of halophytes for forage production on saline wastelands'. *J. Aust. Inst. Ag. Sci.*, Vol. 35, No. 1.

Moore, C. W. E. (1969). 'Application of ecology to the management of pastoral leases in northwestern New South Wales'. *Proc. Ecol. Soc. Aust.*, 4 : 39-54.

Moore, R. Milton, Condon, R. W. & Leigh, J. H. (1970). 'Semi-arid woodlands', in *Australian Grasslands*, Ch. 16. A.N.U. Press.

National Parks and Reserves in Australia. (Report of the Australian Academy of Science, 1968).

Newman, J. C. & Condon, R. W. (1969). 'Land use and present condition in arid lands of Australia'. (Eds. Slatyer and Perry). A.N.U. Press, Canberra.

Osborn, T. G. B., Wood, J. G. & Paltridge, T. B. (1932). 'On the growth and reaction to grazing of the perennial saltbush *Atriplex vesicarium*—an ecological study of the biotic factor'. *Proc. Linn. Soc. N.S.W.*

Perry, R. A. (comp.) *et al. General report on lands of the Alice Springs Area, Northern Territory, 1956-57.* C.S.I.R.O. Aust. Land Research Series No. 6.

Ratcliffe, F. N. (1938, Repr. 1947). *Flying fox and drifting sand.* Angus and Robertson, Sydney.

Ratcliffe, F. N. (1936). *Soil drift in the arid pastoral areas of South Australia* (Pamphlet 64). Melbourne.

Report of the Inter-departmental Committee on Scrub and Timber Regrowth in the Cobar-Byrock district and other areas of the Western Division of N.S.W., February 1969.

Rowan, J. N. & Downes, R. G. (1963). *A study of the land in north-western Victoria.* Soil Conservation Authority, Victoria.

Speck, N. H. *et al.* (1964). *General report on lands of the West Kimberley area, W.A.* C.S.I.R.O. Aust. Land Research Series No. 9.

Stewart, G. A. *et al.* (1970). *Lands of the Ord-Victoria area, Western Australia and Northern Territory.* C.S.I.R.O. Aust. Land Research Series No. 28.

Suijdendorp, H. (1955). 'Changes in pastoral vegetation can provide a guide to management'. *J. Dept. Agri. W. Aust.,* (Third series) Vol. 4, pp. 683-7.

Suijdendorp, H. (1953). 'Problems of the spinifex areas—the decline in carrying capacity'. *J. Dept. Agri. W. Aust.,* Vol. 2, pp. 489-91.

Teakle, L. J. H. (1968). 'Soil salinity in Western Australia'. *J. Dept. Agri. W. Aust.,* (Second series) 15 : 434-452.

Teakle, L. J. H., Southern, B. L. & Stokes, S. J. (1940). 'A soil survey of the Lakes District, Western Australia'. *J. Dept. Agri. W. Aust.,* (Second series) 17 : 251-294.

Wilcox, D. G. (1960). 'Studies in the mulga pastoral zone'. *J. Dept. Agri. W. Aust.,* (Fourth series). Part I: 'Grazing of Wondarrie grass associations', Vol. 1, pp. 475-9. Part II: 'Some aspects of the value of the mulga scrub', Vol. 1, pp. 581-6.

Wood, M. S. (1964). *Understanding the Mallee.* (Victoria's Resources).

For articles on scald reclamation, pasture and tree establishment, the use of contour furrowing and water ponding, refer to the *Journal of Soil Conservation Service of New South Wales.*

CHAPTER 5

Principles of soil erosion and its prevention

By J. K. Marshall

EROSION is an inevitable consequence of the dryland environment largely because that environment is unable to support a permanently adequate cover of vegetation. Soil erosion by wind and by water is therefore a natural feature of the dryland. Any use of the dryland which further reduces the cover of vegetation tends to lead to accelerated soil erosion, that is, erosion exceeding that which is a natural consequence of the environment. It is this accelerated erosion which is to be discussed here with a view to indicating how it can be minimised and, thereby, how the renewable biological resources of the dryland may be conserved.

PASTORAL USE OF THE DRYLAND

Of the uses of the dryland by man, the most important in accelerating soil erosion is the extensive pastoral use. Effects of free-ranging, grazing animals which are of particular importance are the reduction of the vegetation cover, and the disturbance and compaction of the soil surface.

The extent of the dryland which is subject to pastoral use is shown in Figure 5.1. The region of used dryland is bounded externally by the transition from dryland to semi-humid country and internally by those arid and remote regions which are unused by the pastoral industry and in which the extensive influence of man is negligible. Both boundaries fluctuate as climate fluctuates with time and so those shown in Figure 5.1 provide only an indication of the position of what is, in fact, a broad region of change. However, given this delineation of boundaries it can be seen that the area used by man occupies 62 per cent of the total dryland, and 44 per cent of the entire country. In terms of animal numbers, the used dryland supports about 23 per cent of the Australian sheep and about 30 per cent of the Australian beef cattle populations.

THE PROBLEM OF EROSION

In the used dryland both natural and accelerated erosion are taking place. The rate of accelerated erosion can exceed that of natural erosion many times, probably by as much as 10 to 100 times in places. Thus, while natural erosion has persisted in balance with soil-forming processes and vegetation development to present a harmonious landscape, accelerated erosion often does not. Severe accelerated erosion frequently gives rise to marked soil truncation and semi-permanent changes in vegetation including complete denudation. Reduction or avoidance of the

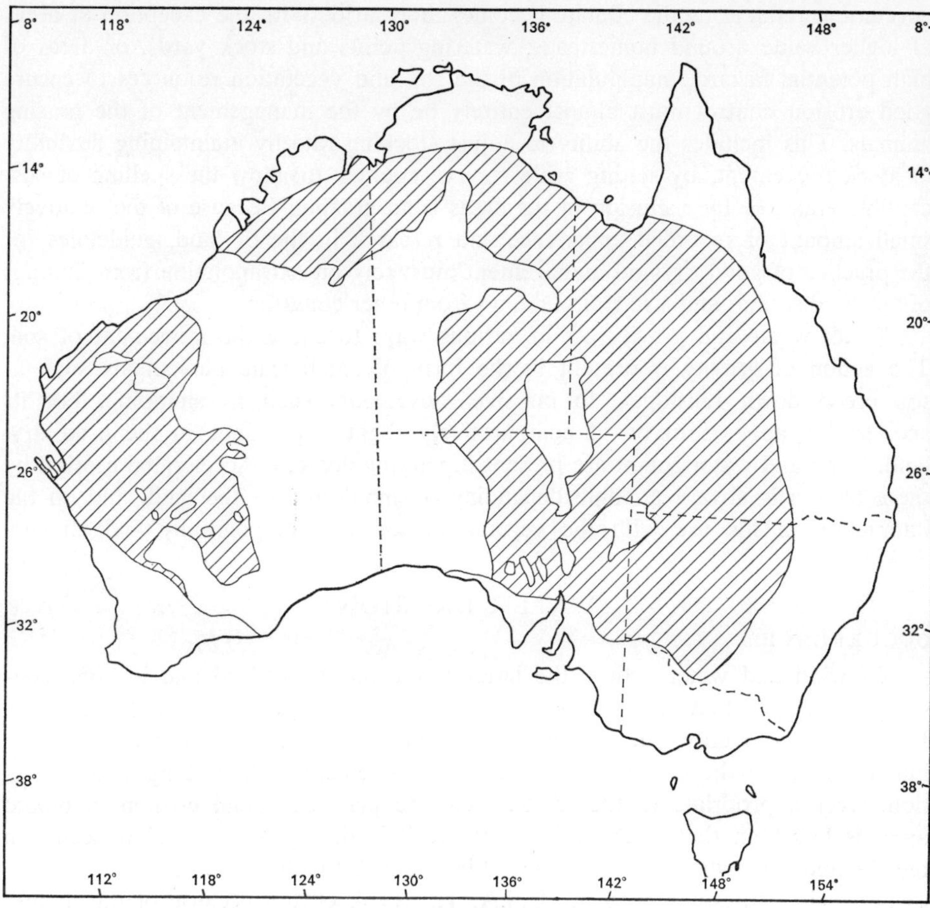

FIG. 5.1

Map of Australia showing the area of the dryland subject to pastoral use (cross-hatched), the outer boundary with the sub-humid country and the internal boundary beyond which extensive influence by man is negligible. The used dryland occupies about 62% of the total dryland (used plus unused) area and about 44% of the entire country (incl. Tasmania). (Sources used in compilation include, for outer boundary, Perry, R. A. (1967) *Proc. ecol. Soc. Aust.* 2, 1-14, and several maps in Slatyer, R. O. and Perry, R. A. (Eds) (1969) *Arid Lands of Australia,* A.N.U. Press, Canberra; for internal boundary, Department of National Development (1957) *Atlas of Australian Resources. Dominant Land Use,* Government Printer, Canberra.)

undesirable consequences of accelerated erosion requires an examination of the underlying causes of erosion, identification of actions which accelerate the process, and of practices which reduce its occurrence.

It should be mentioned that while our understanding of erosion processes in general is reasonably good, our understanding in detail is not yet complete enough for a definitive account of preventive control to be given. In particular, the lessons learnt and experiments carried out in more humid regions cannot be expected to apply directly to the dryland, where practices which will result in the prevention of accelerated erosion are often quite different because the areas involved are more extensive, the economics of control measures are more precarious, and the scope for manipulation of the situation, particularly by means of vegetation, reduces as the climate becomes more arid. With the exception of areas of higher value around homesteads, watering points and stock yards, or areas of high potential hazard, manipulation of the soil and vegetation resources to ensure good erosion control must almost entirely be by the management of the grazing animals. This includes the ability to adjust stocking rate by maintaining flexibility of stock movement, by setting aside reserve country to allow the spelling of susceptible areas, or by exclusion of the areas from grazing. Because of the relatively small amount of recorded experience and research in the dryland, guidelines for the practice of good dryland management must rely on extrapolation from findings of localised studies and also from findings from other countries.

Wind, water and gravity act in various ways to cause the movement of soil. The action of gravity is present in the form of small-scale movements such as soil creep down hillslopes, to massive movements such as slumping, but its acceleration appears to be of comparatively slight importance in the used dryland. Wind and water action are present in varying degrees. Accelerated erosion by these factors is always possible depending on man's use of the land, and on the interaction of his use with the vegetation, soil type, topography and climate.

WATER EROSION

OCCURRENCE

Of wind and water action, the latter is the more localised and is often considered less important in the dryland in comparison with the higher rainfall country. However, this is only true of the potential of the landscape to erode in a more humid climate if the natural vegetation cover were removed. As long as a vegetation cover appropriate to the regional climate persists, actual erosion in humid areas is less than that in the drier country. Indeed, a peak in erosion occurs in the dryland somewhere between the 10-inch and 20-inch rainfall limits according to Langbein and Schumm in the U.S.A. This peak occurs because of the natural decrease in protective vegetation cover due to the lower rainfall which itself is still high enough, with storms of adequately-high intensity, to erode the poorly-protected soil. In a still lower rainfall region, while the vegetation is generally quite inadequate for soil protection, the rainfall is so infrequent that water action contributes progressively less as an erosion agent. In spite of the comparatively poor natural protection of dryland soils by vegetation in dry seasons and at times of drought, there is evidence to suggest that erosion of these soils by water can be increased as a consequence of the use of the land by stock. Compaction of the soil by trampling and destruction of the vegetation cover by grazing may, together, result in an increase in the rate of water erosion by as much as three to four times.

CONSEQUENCES

Water erosion has two important consequences which, combined as they inevitably are, can lead to very rapid deterioration of the country so affected. The more obvious of these is the removal of soil material, including plant nutrients, by run-off water. This results in gullies, tunnels and sheet-eroded surfaces, and silted dams, creeks, and outwash fans where the eroded material is deposited. The second consequence, though less obvious, precedes the first and is the beginning of a sequence of interrelated events. This is the loss of water by run-off which would otherwise be stored in the soil, and some of which would subsequently be available for plant growth.

The water balance equation

The reduction in stored soil water due to increased run-off in dryland country, where water for plant growth is by definition in short supply, obviously has serious repercussions. An examination of the equation expressing what happens to rain once it has fallen allows a ready appreciation of the source of water available for erosion in relation to other possible pathways of water use.

Thus:

$$P = S + \Delta D + \Delta M + \Delta U + \int Edt$$

where P is rainfall; S is net surface run-off, that part of the rainfall available for water erosion; ΔD is the change in surface water storage provided by dead and living plant material, and by surface hollows; ΔM is the change in soil water storage in the plant rooting zone; ΔU is the change in storage below the plant rooting zone and $\int$ Edt is the total evaporation over the period (t) under consideration.

The net surface run-off (S) is that water which has not entered the area of ground under consideration and which, therefore, is available to move soil away from the area. It can also be expressed in terms of the rate of infiltration of water into the soil (I) thus:

$$S = P - (\int Idt + \Delta D).$$

Surface run-off then depends on the rate at which the falling rainwater can enter the soil, and the amount of water which can be retained in the soil surface store, (D). The capacity for increased surface storage is limited. Infiltration rate is related to several environmental factors and it is the variation of one or more of these which usually determines the amount of run-off.

Infiltration

The rate at which rainwater will enter the soil depends on several factors. Of particular importance is the presence of fairly large pores opening to the soil

TABLE 5.1

INFILTRATION RATES OF SOILS OF VARYING TEXTURE(*a*)

Soil texture	*Maximum infiltration rate (in/hr)*	*Soil texture*	*Minimum infiltration rate (in/hr)*
Clay loam	0.1–0.2	Heavy clays, highly swelling soils and some saline soils	0–0.05
Silt loam	0.3–0.6	Clay rich soils, many clay loams, soils poor in organic matter, shallow sandy loams	0.05–0.15
Loam	0.5–1.0	Sandy loams and shallow loessic soils	0.15–0.3
Loamy sand	1–2	Deep sands, deep loess and aggregated silts	0.3 –0.45

(*a*) Sources used: H. Kohnke & A. Bertrand, *Soil conservation* (McGraw-Hill, New York, 1959); C. W. Musgrave, 'How much rain enters the soil', in *U.S.D.A. Yearbook of Agriculture* (U.S. Govt Printing Office, Washington, 1955), pp. 151-159.

surface and the sub-surface distribution and size of pore spaces. Because of this, there is an approximate correlation of soil texture with infiltration rate, light, sandy soils having higher rates than heavy clay soils with loamy soils intermediate as shown in Table 5.1.

The correlation of infiltration rate with soil texture is also illustrated by the range of accumulated water infiltrated into soils of varying texture in Figure 5.2. The characteristic fall-off in infiltration rate with time is also apparent in that

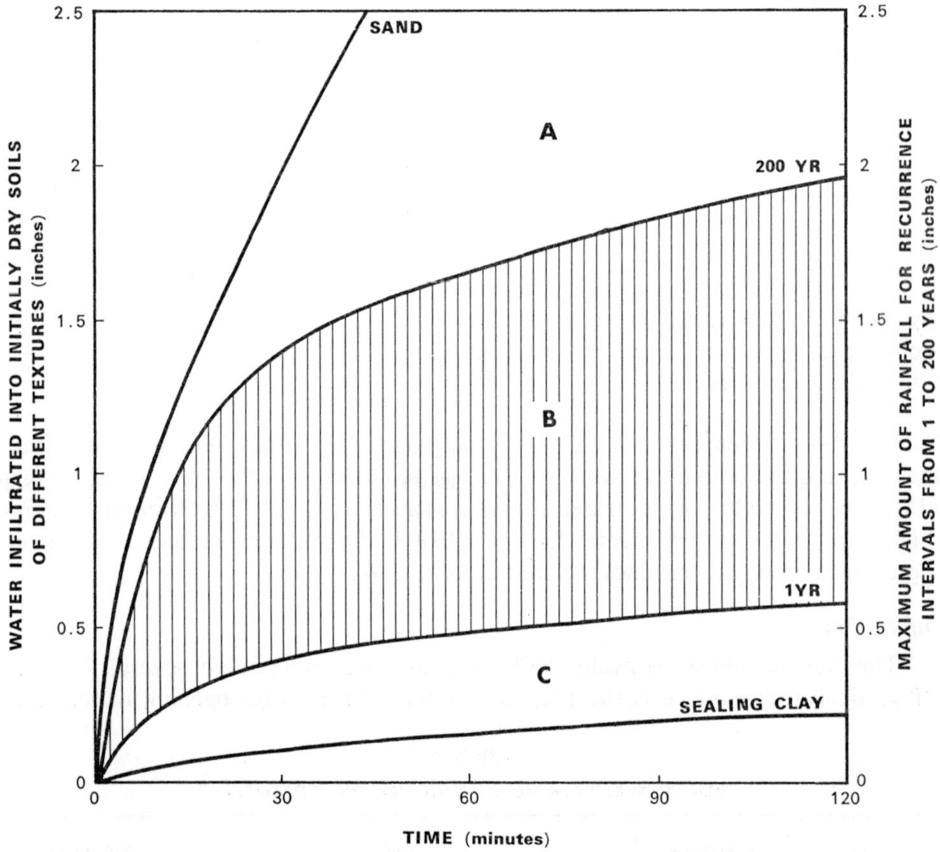

FIG. 5.2

The range of infiltration rate of water into dryland soils of different textures and the range of maximum expected amounts of rainfall for recurrence intervals of 1 to 200 years. The range of infiltration rates refers to initially dry soils. The rates are approximately related to soil texture with the lighter, sandy soils having high infiltration rates and the heavy, swelling clay soils having extremely low infiltration rates. Loamy soils tend to lie within these extremes. The range of rainfall intensities refers to the average maximum intensities expected in the Western Division of New South Wales. From the superimposition of the infiltration and rainfall ranges it can be seen that, in the absence of a vegetation covering, there are some soils (A) which are capable, when initially dry, of absorbing all of even the heaviest rains; others (C) will shed water from all rains in the range plotted, and many (B) will absorb or shed water depending on the intensity of the rainstorm. For initially wet soils (data not plotted) most would, irrespective of texture, shed water from rainfall of intensities in the range plotted. (Sources used in compilation include, for infiltration rates, Jackson, E. A. (1958) *Soils Ld Use Ser.* C.S.I.R.O. Aust. 24, 66 pp., Stannard, M. E. (1961) *J. Soil Conserv. Serv. N.S.W.* 17, 253-263; for rainfall intensities, Stewart, J. (1960) *J. Soil Conserv. Serv. N.S.W.* 16, 231-244, Wiltshire, G. R. (1960) *J. Soil Conserv. Serv. N.S.W.* 16, 54-70.)

figure where it can be seen that water enters the soil most rapidly in the initial stages of infiltration and more slowly as time goes on.

Infiltration rate also depends on the condition of the soil water storages, M and U in the water balance equation. The rate is highest when the soil is initially dry, except when that condition is associated with a hydrophobic soil surface crust. The wetter the soil prior to infiltration, the lower is the rate of infiltration as is shown by the contrast in Table 5.1, between maximum and minimum infiltration rates. For a given soil type, the shallower the soil the smaller is the capacity of the soil water storages, M and U. Infiltration rate consequently falls off earlier the shallower the soil, an effect which is aggravated when successive soil layers have poorer infiltration properties.

Roughening of a soil surface can assist infiltration in two ways. First, it allows more rapid water entry into the soil. Second, by providing more surface hollows in which water can be held (i.e. by increasing the capacity of the soil surface water store), greater infiltration can be achieved by extending the time over which the opportunity for infiltration exists. This second effect is made use of in water-ponding and other techniques.

The adverse effects of direct raindrop action and of the concentrated trampling of animals on infiltration rate are related to the importance of soil pore distribution, and particularly the presence of fairly large surface pores. The first effect of the direct impact of raindrops and of trampling by animals is the breaking down of soil aggregates into smaller components and even into discrete soil particles. Following this break-down, there occurs a redistribution of the finer soil material by water into the surface pores. When blocked by finer material these pores lose their ability to transmit water and a cycle is set up in which the infiltration capacity of the soil decreases and run-off increases.

The presence of vegetation can affect infiltration of water into the soil in several ways. One of the more important is the reduction of direct impact of raindrops on the soil surface. Rainwater which has dissipated its energy in colliding with plants or plant residue characteristically emerges clear at the soil surface. This contrasts with rainwater which has shattered soil aggregates by direct impact. It is usually turbid as it redistributes the soil material. Another effect of vegetation is the improvement of soil texture with the increase in pore space due to old root channels and the addition of organic matter to the soil. In some circumstances, the formation of a soil surface crust is reduced by the presence of vegetation and in this way soil infiltration capacity is improved. Occasionally, however, the presence of organic matter will adversely affect infiltration if it is present as dry, small particles covering the surface, because as such it is extremely difficult to wet.

Run-off

Rainwater which fails to enter the soil or be detained in hollows and by vegetation runs off the soil surface and is responsible for erosion. Once the available storage for detention of water above the soil has been filled, run-off will occur whenever the rate of addition of water to the surface (rainfall intensity or run-on from an adjacent area) exceeds the infiltration rate. The relationship of infiltration rate and soil texture has already been mentioned and the range of cumulative infiltrated water for dryland soils is shown in Figure 5.2. Superimposed on the infiltration data in that figure is a range of expected rainfall intensities typical of the Western District of New South Wales. The range of intensities plotted is from the maximum intensity for rainstorms of up to 120 minutes duration expected to recur yearly, to the maximum intensity expected to occur, on average, once every 200 years. From the combination of the rainfall intensity and soil infiltration data in Figure 5.2 it is clear that, under similar rainfall intensities, some soils will yield run-off water while others will adequately absorb all the rain falling on them. Many of the soils are intermediate in character in that water-shedding from them depends greatly on the rainfall intensity. It would be expected that any reduction in the infiltration capacity of these soils as a consequence of land use such as trampling or the reduction of vegetation cover would lead to an increase in run-off, and hence, erosion.

The relation of rainfall intensity and infiltration rate, together with a frequently inadequate cover of vegetation result in run-off from most dryland soils, sometimes quite frequently. This conclusion underlies the earlier stated finding that water erosion tends to be highest in the dryland environment.

Given that run-off is frequently inevitable it is necessary to consider how it is related to erosion of the soil. The relationship is fairly complicated. In any particular situation it depends on the soil material available for transport by run-off water, the slope of the ground and length of slope, the quantity of water available and its velocity, and the roughness of the surface. A similar relationship of threshold velocities to that shown in Figure 5.5 for wind holds for the movement of particles by water, except that the range of highly-susceptible particles is wider, from about 0.004 to 0.024 in. (0.1 to 0.6 mm) in diameter. Erosion increases with length of slope, angle of slope and the quantity of water available, but the most spectacular relationship is that of erosive capacity of the run-off water and its velocity. The erosive capacity is proportional to the square of the velocity so that if the velocity is doubled, the amount of material which can be eroded is increased fourfold. The erosion of soil material is less the rougher the surface over which the run-off water flows, partly because of the reduced velocity and partly because of increased infiltration.

The most effective way of reducing the erosive powers of run-off water is to reduce its velocity. This can be done by roughening the soil surface and is one of the ways by which vegetation reduces erosion. Herbaceous vegetation is particularly effective in reducing the velocity of surface run-off. While the effect of vegetation is, in general, beneficial it is also possible that by locally obstructing run-off the water is diverted and concentrated into more, not less, erosive channels. Reduction in the velocity of run-off on sloping ground is particularly important.

PREVENTION

The preceding outline of water action indicates several ways in which accelerated erosion by water can be prevented. Erosion-susceptible situations include sloping ground, particularly hills with shallow soils, soils with inherently low infiltration rates, and sites where denudation of vegetation is likely. Rainfall of high intensity can occur throughout the dryland but there is a tendency for falls to be of higher intensity further north where summer rainfall becomes increasingly predominant. Dense vegetation is the best means of reducing the water erosion hazard, but the occurrence of an adequate cover cannot be relied upon. Instead, the importance of reducing the velocity of run-off water suggests preventive measures which can be applied in the absence of a vegetative cover or supplemented by vegetative cover when it occurs. Preventive control of accelerated water erosion follows the recognition of these various facts when making management decisions which could affect the water balance of a particular area.

The consequences of adversely affecting the water balance of an area can be well illustrated using the example of a soil in which the successive layers, when exposed, present surfaces of decreasing infiltration rates. This holds for the so-called texture contrast soils common in New South Wales, and which occur frequently throughout the dryland country. In these soils the top few centimetres of the original soil is a fairly light, sandy loam overlying layers progressively richer in clay. Erosion of the top layer can result in a 10-fold reduction in infiltration rate. Prevention of loss of the original surface of this soil by protection against wind and water action will avoid a progressively deteriorating situation, with a self-perpetuating water erosion problem as successive soil layers are exposed. Where preventive measures have not been taken the end-point of this erosion is the familiar denuded scald. The treatment of scalds is discussed in Chapter 8 but it is costly in relation to pastoral land use.

There are several situations in which preventive control can be exercised and no attempt will be made here to give an exhaustive account of these. The guiding rule is that the land user should carefully consider the principles of water action in relation to each management decision he must make, whether this concerns animal numbers, fencing and subdivision of land, installation of watering points or of stock yards. The procedure in some situations is, however, fairly obvious from what has already been said. Thus, clearing of hill tops and slopes should be avoided and grazing of these should be carried out with careful attention to its effect on the covering vegetation. If possible, grazing should be concentrated on the plains country, valley floors and lower hillslopes. The building of earthworks and laying of access tracks should be planned with local topography and soil type in mind, giving particular consideration to siting in relation to the destination of run-off water and eroded material. Access tracks can channel water which would otherwise be more evenly spread over the country, with undesirable consequences for the track and the vegetation of the former run-on country (Figure 5.3). Similarly, the siting of bores, waterholes and stock yards is of importance in that denudation of the vegetation surrounding these facilities is largely inevitable. Where the likely consequence of the provision of these various facilities is the increase in volume and velocity of run-off, preventive works such as contour furrowing, damming or other means of encouraging water spreading, and strip ploughing should be carried out in anticipation of the consequences rather than in retrospect. That is, the potential consequences of the actions of the land user should be countered by means which reduce their erosion effects and so lead to a balanced, planned development which will have the effect of maintaining the resource rather than degrading it.

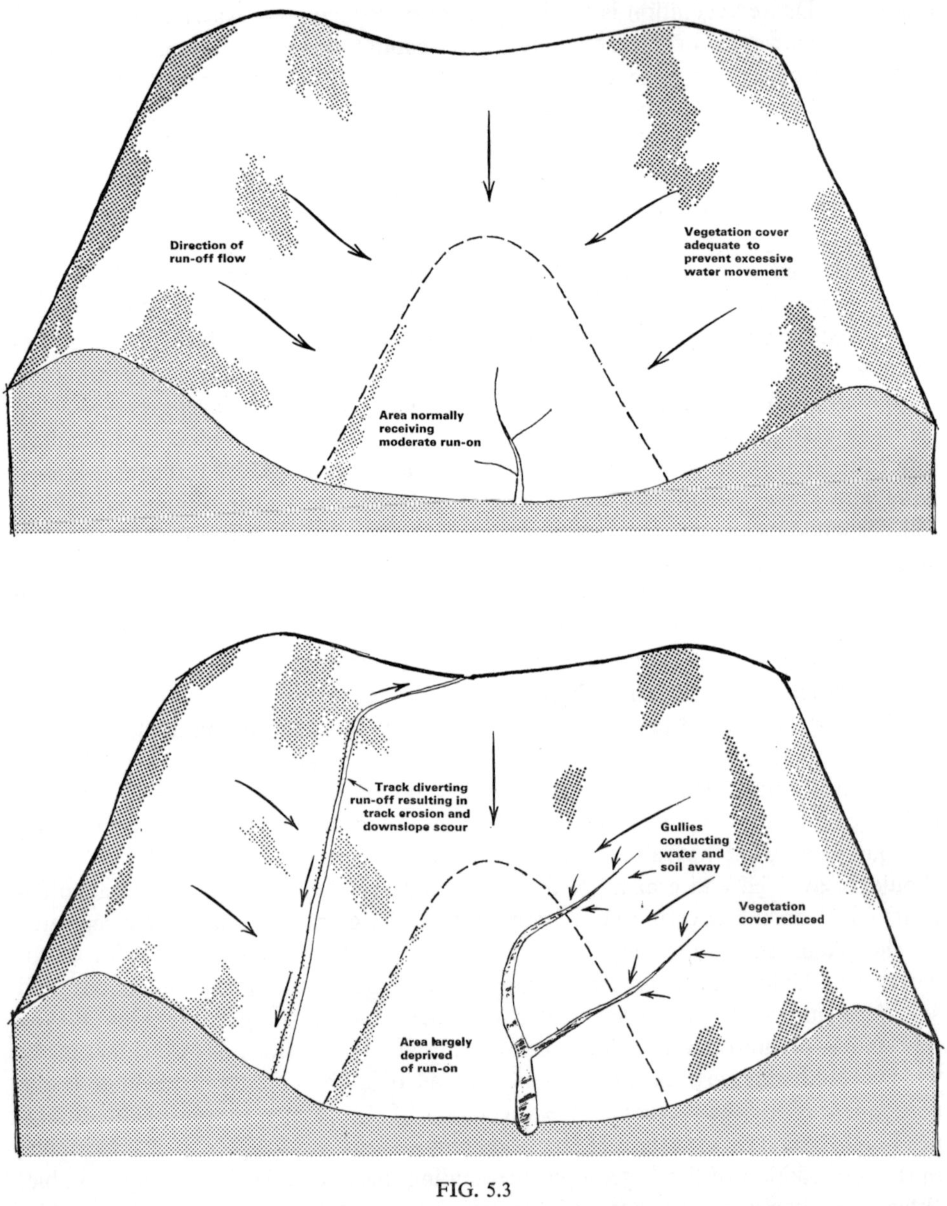

FIG. 5.3

Diagrammatic sketch of hillslopes to illustrate the consequences of track positioning and of heavy grazing of the slope vegetation. In the former, a natural run-on area is deprived of some of its water due to diversion of slope run-off by the track. Local scouring of soil occurs immediately downslope of the track as well as erosion of the track itself. In the latter, reduction of the vegetation cover by stock leads to greater downslope movement of water and the tendency for gullies to form by headward gully erosion. A formerly even spread of water over the hillslope is thus concentrated into the gullies which also serve as channels to conduct eroded soil material away.

WIND EROSION

TYPES

In considering the results of wind action on soils, it is convenient to define two sorts of erosion depending on the relative sizes of the donor and recipient areas. Regional erosion is the net effective loss of soil over a period of time (say a year or more) from a substantial area (say of many acres to square miles in extent). Net effective loss refers both to the complete loss from a region of small and light material carried away as dust in the atmosphere, as well as to large-scale redistribution which results in the concentration of material from the original surface into effectively useless deposits along fence lines, around homesteads, in dunes and in rivers. Local erosion, in contrast to regional erosion, results in the redistribution of surface material by wind over a variable time scale, from seconds to years, to and from small areas of a scale similar to the average distance between prominent vegetation elements, such as trees and shrubs. The areas of surface here are usually of the order of square yards in extent. One result of redistribution of material due to local erosion is the frequently encountered mound around bladder saltbush (*Atriplex vesicaria*).

Local erosion is inevitable in an arid environment where a sparse vegetation is the sole protector of the soil at times of the year when it is susceptible to erosion. Accelerated regional erosion should, however, be controllable where serious deterioration has not already set in and where the climate is favourable enough for regeneration to be fairly rapid both from seed and by regrowth of the perennial species. This latter condition usually applies to the higher rainfall part of the dryland, conditions in the more arid areas frequently being too extreme to allow rapid recovery.

CONSEQUENCES

Our understanding of the implications of these two types of erosion is still incomplete. In land under cultivation for crops the consequences of the loss of top soil in terms of fertility and microbiological activity are well known, but our knowledge of the way in which dryland plant communities function is not yet complete enough to allow similar conclusions to be drawn. It is not certain that gradual removal of soil will have profound effects on the productivity of dryland vegetation. However, with more rapid removal both regeneration of plant communities and their productivity can certainly be affected. This is particularly true on the lighter soils where the accumulated material associated with surface obstructions normally provides seed beds essential for regeneration of species within the community, material which is also readily lost in the absence of these obstructions.

While it is difficult to know the biological cost of erosion in its less extreme forms, a rough indication of the economic cost is possible. Thus, the cost of the nitrogen and phosphorus lost per inch of soil depth eroded per acre of dryland community would be about \$16.70 for nitrogen and \$6.30 for phosphorus—this being the bulk purchasing cost of the appropriate fertilisers at 1969 prices without contemplating the cost of application or the loss of other plant nutrients. Put in another way, erosion of the top 5 inches of soil from an area of 100 sq miles would represent a loss in terms of nitrogen and phosphorus of $7,340,600 calculated as stated. The costs, both to the primary producer and to the nation, of other aspects of the dust-storm part of wind erosion are less easily estimated. They include, however, additional wear on machinery operating in dust-laden conditions; the time required by citizens of townships and cities to rid themselves of the wind-borne material; areas lost from native pasture through excessive erosion and even the eventual abandonment of settled areas.

To summarise, regional erosion is responsible for the large-scale export of soil material, either as atmospheric dust or as accumulated surface material. It is likely that, as accelerated regional erosion proceeds, the productivity of many dryland plant communities will decline. Certainly, if the soil is shallow, it is only a matter of time before bedrock is reached and the land can no longer be used for grazing. If the soil is deep, much will depend on the textural and permeability properties of the newly exposed layers. The cost of this export of material is high relative to the value of the land. By contrast, local erosion plays an essential part in providing material for seed beds for community regeneration. It is inevitable in an arid environment in the presence of scattered vegetation, but the redistribution of material depends on the continued presence of perennials otherwise it will become source material for regional erosion.

FUNDAMENTALS OF WIND ACTION

Prevention of soil erosion by wind is best understood by first examining the response of individual soil particles to wind action. These respond in four ways which are, in fact, intergraded rather than distinctly separate. Large, heavy particles, such as quartz grains of diameter exceeding 0.035 in. (0.9 mm) tend, under natural wind conditions, to be non-erodible. Many grains in the range of about 0.004 to 0.035 in. (0.1 to 0.9 mm) move by rolling or creeping over the surface. Particles at the smaller, lighter end of this range, and down to about 0.0008 in. (0.02 mm), receive enough energy relative to their weight to bounce or saltate, gaining greater momentum as they rise into the higher wind velocities above the ground. However, they eventually fall back to the ground surface and there displace other particles or themselves bounce upwards again. Saltating particles rarely reach heights greater than a few feet with their greatest concentration occurring within an inch or two of the ground surface (Plate 11, top). Finally, light or very small particles (e.g. quartz particles smaller than about 0.0008 in. [0.02 mm]) can be carried in suspension. That is, they remain aloft in the atmosphere because up-currents in the air are great enough to counteract the force of gravity on them. But unusually high wind velocities are required to actually raise these smaller particles off the ground surface in the absence of disturbance provided by saltating particles, by animals or by vehicles. This is mainly due to the fact that on their own these smaller particles form too smooth a surface for turbulent wind action to affect them. The usual mechanism for lifting these particles into the atmosphere is the energy provided by the return of the saltating particles to the surface. When saltating particles land amongst these smaller particles they are 'splashed' into the air and may be carried up to heights of a mile or so to form the familiar dust storm (Plate 11, bottom). It follows that prevention of saltation will result in the prevention of regional wind erosion.

CONDITIONS FOR WIND EROSION

Wind erosion of soil is fairly unusual in that, to take place, it requires the coincidence of a special set of conditions of climate, soil and vegetation. Unless all are simultaneously in a condition conducive to erosion, no erosion will occur. The conditions favouring wind erosion are shown in Figure 5.4. Pastoral use of the land tends to increase site susceptibility to wind erosion through the effects of the animals on both the soil and the vegetation.

Climate

For wind erosion, fairly strong winds must coincide with prolonged hot weather to dry out the soil surface because a moist soil surface is scarcely erodible. Such conditions are typical of the dryland country.

Soils

The susceptibility of various soils to erosion depends, as outlined above, on the availability of particles which are capable of saltating. But the question of what

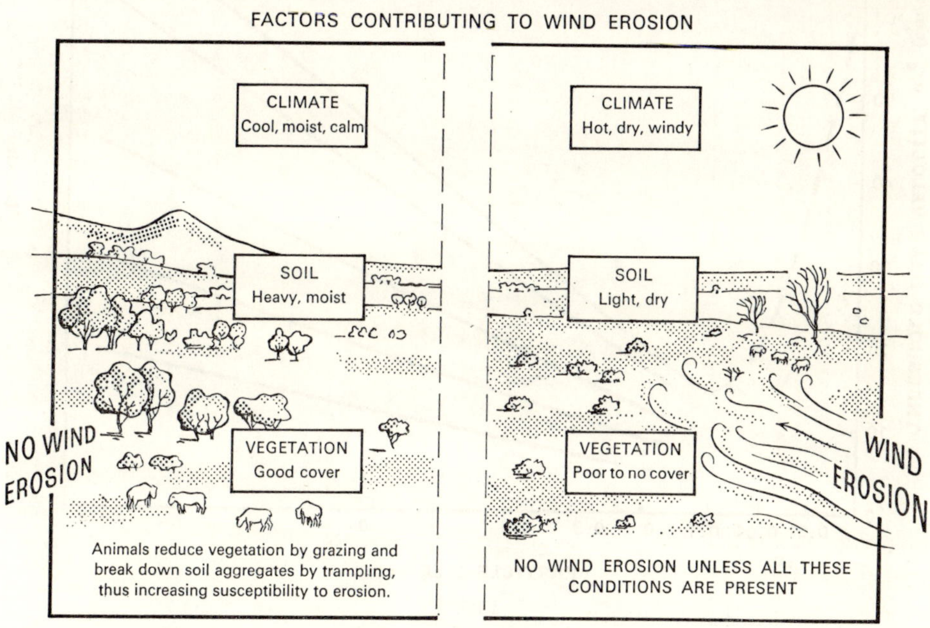

FIG. 5.4

Diagrammatic representation of the conditions of climate, soil and vegetation required for soil erosion by wind to take place. Erosion will only occur if all three factors favour its occurrence; the existence of only one of the factors in a favourable condition (e.g. light, dry soil) is not adequate on its own for erosion to occur. Grazing by stock increases the susceptibility of a site to erosion by reducing the vegetation cover and breaking-down the soil surface aggregates. From the diagram it is clear that the same rate of stocking on any particular site may be conservative or excessive relative to the erosion hazard depending on the condition of the other three factors at the site. For this reason, the practice of varying stock numbers with conditions has arisen and is of great importance for the long-term maintenance of the soil and vegetation resources.

particles it is important to retain in the soil must also be considered. Susceptibility is often judged in terms of the mineral particles only, but should there be a reason for retaining the lighter, organic matter particles for which the wind velocity threshold for movement is much lower, then the definition of susceptibility will be wider. The threshold velocity for particle movement, given as a drag velocity (u_{*t}) can be calculated for particles of different diameters (d) and densities (σ) by an expression due to Bagnold:

$$u_{*t} = A_B \, (gd \, (\sigma - \rho)/\rho)$$

where g is the acceleration due to gravity, ρ is air density and A_B is a coefficient due to Bagnold which, in air, takes the value of about 0.1 for particles of diameter down to about 0.008 in. (0.2 mm) and which increases after that. Curves from this expression are shown in Figure 5.5 for particles of different size and density. There it can be seen that substrates with a predominance of particles in the fine

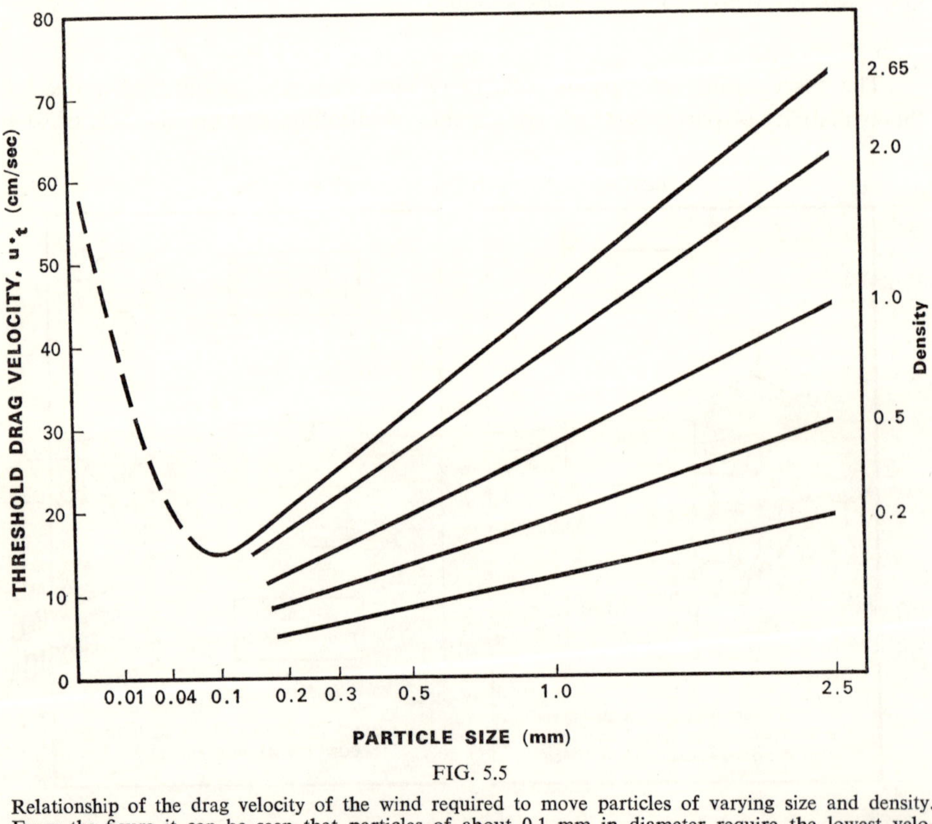

FIG. 5.5

Relationship of the drag velocity of the wind required to move particles of varying size and density. From the figure it can be seen that particles of about 0.1 mm in diameter require the lowest velocities of all to be set in motion; particles smaller and larger than this require higher velocities to move them. Consequently, soil with a predominance of particles of the most susceptible size are the most easily erodible. Also, particles of different density vary greatly in the velocities required to set them in motion. Quartz (sand) particles, of density 2.65 are relatively much less susceptible to movement than the lightest organic matter particles (density 0.2). Reduction of wind velocity must therefore be much greater if it is of importance to retain organic material at a site than if it is only necessary to retain quartz particles. Conversions: 1 mm = 0.3937 in; 10 cm/sec = 0.22 mi/hr; drag velocity is very approximately 1/10 of actual mean wind speed. (Source used in compilation: Bagnold, R. A. (1941) *The physics of blown sand and desert dunes.* Methuen, London, 265 pp.)

and coarse sand category (0.0008 in. [0.02 mm] to 0.08 in. [2 mm]) are especially prone to particle movement, saltation, and hence to erosion, while a preponderance of larger or smaller particles forms a more resistant substrate. The presence of unhumified, lighter organic matter particles, especially if they remain discrete and only weakly incorporated into the soil, makes the substrate more susceptible to erosion. However, the tendency of both humified, incorporated organic matter, and of clay and silt particles, to promote aggregation in a soil can lead to a more resistant substrate.

In practice, the soils most susceptible to wind erosion are those which have weak aggregate stability (i.e. with discrete soil particles or with aggregates which are easily broken down), a preponderance of coarse and fine sand particles, and a substantial quantity of organic matter on their surface. Sandy soils, and soil with surface litter often associated with the base of perennial shrubs, for example, occur frequently in the dryland country.

Accelerated wind erosion results from two effects of trampling by grazing animals. The first effect of trampling is to break down the soil aggregates into smaller components which are more susceptible to movement by wind. The second occurs on those tracts of dryland country which are naturally protected from wind erosion by a resistant soil-surface crust formed by algae, or by a continuous covering of lichens. Both the crust and the lichen mat are highly susceptible to damage by the trampling of stock and the underlying soil particles are thus exposed to wind action.

A classification of dryland soils into categories of highly erodible, moderately erodible, or effectively non-erodible by wind allows their proportions in that part of the dryland country used by man to be calculated. These proportions are presented in Table 5.2.

Vegetation

An inadequate cover of vegetation is also required before erosion can occur. Vegetation, and any other surface obstructions, such as stones and persistent plant remains, decrease the force of the wind near the erodible surface so that the average force available for particle movement is reduced. While herbaceous plant species often occur so densely that they provide complete protection to the soil surface, they are often seasonal in their appearance. For much of the dryland, the seasonal herbaceous growth dies off at the time of year which is most favourable,

TABLE 5.2

DRYLAND SOILS SUSCEPTIBLE TO WIND EROSION AS A PERCENTAGE OF THE USED DRYLAND COUNTRY(*a*)

Category	*Soil clay and silt content*	*Percentage of used dryland*
Highly erodible	< 20%	14.8
Moderately erodible	20–40%	58.7
Non-erodible	> 40%	26.5

(*a*) Calculated using as sources: Department of National Development. *Atlas of Australian Resources*, Second Series. Soils, Second Edition. Government Printer, Canberra, 1963. Stace, H. C. T. *et al.*, *A Handbook of Australian Soils*. Rellim, Glenside, South Australia, 1968, 435 pp.

climatologically, for erosion. The turbulent winds and long hot dry spells of summer favour erosion most strongly. The role of the perennial elements in the vegetation is now obvious. The trees and shrubs, and to a smaller extent the perennial grasses and forbs which persist throughout the year are the major obstructions available in summer to reduce the wind force at the soil surface.

Accelerated erosion occurs when the natural vegetation cover is reduced by grazing animals. The reduction may be either in the length of time each year that the soil is adequately protected by the closely-growing annual herbaceous species, or in the protection afforded by the more sparsely-growing perennial species.

To summarise, erosion susceptibility requires a dry, windy climate, light soils lacking particle cohesion, and the absence of an adequate vegetation or stony covering. Animals can increase this susceptibility by trampling the soil, and by grazing and trampling the vegetation.

PREVENTION

The ways in which the factors contributing to wind erosion can be manipulated to reduce excessive erosion can now be considered. Because of the extensive grazing use of the dryland, and its relatively low value soil modification using capital input is only a slight possibility with the exception, perhaps, of areas surrounding the homestead and capital intensive improvements. The climate has to be accepted except to the extent that it can be modified near the ground surface where the site of erosion initiation lies. The interrelating factors shown in Fig. 5.4 include animals, and there can be little doubt that it is largely by the management of those animals husbanded by man that erosion control and prevention must be effected. It is here that an apparent conflict of interests arises because the only means of climate modification and hence of erosion prevention is by the vegetation elements which persist throughout the year, especially the shrubs and trees, but these are also frequently looked upon as a drought reserve for the animals (Chapter 15).

Clearly, the two functions of the perennials must be reconciled and managed according to the principle of maintaining the vegetation and soil resources in the long-term. Thus, perennial species which are capable of ready re-establishment and regrowth may be used fairly heavily as a drought reserve. Some trees, such as mulga (*Acacia aneura*) in south-west Queensland, fall into this category. Other perennials, however, are highly susceptible to heavy grazing and may fail to survive in an area following injudicious grazing in drought periods. The saltbushes (*Atriplex vesicaria, A. nummularia*) and bluebushes (*Kochia sedifolia, K. astrotricha*) fall into this category as does mulga in much of dryland New South Wales. When the shrub species mentioned occur on erosion-susceptible soils they should be retained as erosion preventers in the long-term rather than used in the short-term as drought feed.

With the recognition of the protective role of woody perennials the question that follows is, how much perennial vegetation is required to protect the soil surface from erosion? Shrubs, for example, occupy ground which could otherwise, in the short-term at least, produce seasonal, herbaceous feed more acceptable to sheep than the shrubs. A balance between the proportions of ground occupied by seasonal herbaceous growth and woody perennials must be sought. The results of two studies can be used to indicate the advisable, minimum number of shrubs required just to prevent regional erosion of soil, and consequently to provide a safe near-maximum area for seasonal pasture growth.

The results of a field study by R. W. Jessup are shown in Plates 12 and 13 for bluebush (*K. sedifolia*) and saltbush (*A. vesicaria*) country respectively. After detailed studies in the field, he concluded that both bluebush and saltbush country could be found, sometimes for reasons of past management, sometimes because of local soil and climate conditions, carrying different numbers of shrubs per ground area and showing varying degrees of erosion. These observations led him to the conclusion that at certain sparse densities the ground was inadequately protected against wind erosion, while at certain very high densities the ground was not only adequately protected but was probably overprotected from the point of view of grazing use. The series of photographs in Plates 12 and 13 illustrate the sequence of decreasing ground cover in five stages where Stage 3 represents an adequate balance between shrub presence, grazing use and erosion protection.

The results of wind-tunnel experiments in conjunction with field observations by the author point to a similar conclusion and also provide a means for calculating the minimum number of perennials expected to be necessary for the control of regional erosion. The question asked in the wind-tunnel experiments, in which blocks of wood of varying shapes were used to represent shrubs, was, at what density (i.e. number per unit area) of wooden blocks did the average force at the wind-tunnel floor in the direction of the wind become negligible? That is, at what density would saltation be expected to be reduced to a practical minimum? Not surprisingly, the answer was seen to depend on the shape and size of the roughness elements. When the results are applied to the field situation in which both the irregular distribution of shrubs and trees as well as their degree of openness is taken into account they agree substantially with Jessup's findings. In Table 5.3, the results are expressed in terms of the average distance separating shrubs of bluebush (*K. sedifolia*), saltbush (*A. vesicaria*) and cottonbush (*K. aphylla*) predicted to be the distance (termed 'critical' distance) which, if exceeded, would leave the intervening surface erosion-susceptible.

An approximate indication can also be given for this critical distance between perennials in terms of their height. This indication is necessarily approximate

TABLE 5.3

CRITICAL AVERAGE SPACING OF SOME DRYLAND SHRUB SPECIES FOR EROSION CONTROL

Species	*Average shrub height in inches (cm)*	*Average shrub diameter in inches (cm)*	*Critical average separation distance in inches (cm)*
Kochia sedifolia	31.5 (80)	31.5 (80)	117 (297)
Atriplex vesicaria	14.6 (37)	14.6 (37)	53.5 (136)
Kochia aphylla	15.7 (40)	31.5 (80)	67.3 (171)

because the calculation assumes a standard openness, shape and diameter to height relationship of the perennial. For highly efficient obstructing elements such as solid wooden blocks, the protective effect can begin to diminish when their distance apart exceeds 7 times their height. For natural vegetation, where there is a degree of openness in the elements and where their height is variable, the critical distance is more nearly given by multiplying the average of the maximum heights of the trees, shrubs or grasses by 3.5. A spacing exceeding this value is likely to be inadequate for erosion protection of a susceptible soil. To obtain an indication of the adequacy of cover of perennials, the simplest way is to count their number (N) over an area (A) large enough to be representative, and calculate the average area occupied per perennial (A/N). Their average distance apart (D) is *approximately* the square root of this value ($\sqrt{(A/N)}$), less the average diameter (d) of the perennials. From this, and knowing the average maximum height (H) of the perennials, the distance apart in terms of 'heights' can be calculated (D/H) and compared with the above value of 3.5. If D/H exceeds this value then a potential erosion hazard exists. Alternatively, if bluebush or saltbush country is being assessed then comparison of the condition of the country with Plates 12 and 13 or of some simple measurements with the data in Table 5.3 should allow a rapid initial assessment to be made.

These relationships of element height, spacing and openness can be used to indicate ways in which erosion can be reduced or prevented. The problem in an erosion-susceptible situation is that the perennial elements in the vegetation are too far apart, too small, too open or a combination of these things. Increase in height and width by growth of established perennials is the most effective means because of the increase in frontal area. The effect of an increase in height and width is to reduce the ratio D/H because D decreases and H increases. Additional control can be gained if the openness of the perennials can be reduced, but this is less effective and is unlikely to be of major importance. Finally, an increase in the number of perennials per unit area of ground is another way in which control can be gained. This is a less attractive means than extra height and width growth of already established plants because it is more long-term. Nevertheless, in situations where rehabilitation of a denuded area of ground necessitates planting, an indication of plant number to gain erosion control is of considerable value.

SUMMARY OF PREVENTIVE ACTION

WATER EROSION

1. Avoid land use practices leading to reduced soil-infiltration rate.
2. Encourage grass and herbaceous cover of the soil for as long as possible every year.
3. Do not clear hilltops or steep slopes or graze these areas heavily.
4. Take care in the layout of tracks, especially where these will canalise run-off water at storm times.
5. Take care in siting bores, waterholes and stock yards in relation to the effect and destination of run-off and siltation.
6. Remember that the more water that goes into the soil, the better is the chance

Top: Sand particles saltating across a sand dune by wind action. Note the varying abrasive power of the particles with height (greater nearer the surface) as illustrated by the cut-back edges of the more resistant cores of original dune surface.
(*Photograph by Dr. Marshall.*)

Bottom: Dust storm at Deniliquin aerodrome, New South Wales, in the summer of 1944 during a period of severe drought throughout the State in which the vegetation cover in the dryland was negligible.
(*Photograph by courtesy of the Deniliquin Historical Society.*)

Plate 11 SAND PARTICLES SALTATING AND A DUST STORM

Five bush density stages (1-5) on country dominated by pearl bluebush (*Kochia sedifolia*) with occasional myall (*Acacia sowdenii*). The following comments refer only to the bush components of the photographs (i.e. as if the trees and tall shrubs were not there). The number of bushes per acre for the five stages is (1) exceeding 750, (2) 570-750, (3) 240-569, (4) 75-239 and (5) 15-74. Country with a bush density equal to middle to upper stage (3) is considered by Jessup to constitute an ideally balanced pasture. That is, one with sufficient bush present to maintain soil stability and to provide a drought reserve of feed but at the same time not so dense a bush cover as to eliminate the growth of herbaceous species in between the bushes. At densities lower than stage (3), soil erosion by wind can be expected, given a susceptible soil, when the ground between the shrubs is unprotected by herbaceous vegetation.

(Photographs from Jessup, R. W. (1951) Trans. Roy. Soc. S. Aust. 74, 189-273, reproduced by permission of the author and the Secretary of the Royal Society of South Australia.)

Plate 12 *(above and facing)* FIVE DENSITY STAGES OF DRY COUNTRY VEGETATION—I

Five bush density stages (1-5) on country dominated by bladder saltbush (*Atriplex vesicaria*) alone (1), and by bladder saltbush and low bluebush (*Kochia astrotricha*) (2-5). The number of bushes per acre for the equivalent five stages as in Plate 12 is (1) exceeding 2630, (2) 2220-2629, (3) 900-2219, (4) 240-899 and (5) 15-239. The bush numbers here are markedly greater than for pearl bluebush country because of the smaller size of the bushes and correspondingly lower protective capacity per bush. As before, middle to upper stage (3) country is considered by Jessup to constitute an ideally balanced pasture and at densities lower than this stage, wind erosion of the soil is a potential hazard. (*Photograph acknowledgements for 2-5 as for Plate 12.*)

Plate 13 *(above and facing)* FIVE DENSITY STAGES OF DRY COUNTRY VEGETATION—II

of sustaining plant growth. The water which runs off is mostly lost underground or to creeks and ponds where much of it evaporates and to which it has transported valuable soil material.

WIND EROSION

1. Avoid management strategies which will lead to the elimination, over large areas, of perennial plant species, particularly woody perennials.
2. Carefully consider soil type when siting water holes and bores and avoid the more erosion-susceptible soils if possible. If on erodible soils, site in or near tree cover if possible.
3. Examine the status of the woody perennials on each paddock, assessing their number in relation to the advisable spacing to prevent wind erosion. Ideally each paddock should contain only one major vegetation type (e.g. saltbush, bluebush, grassland, mallee scrub) because the soils associated with different vegetation types vary in their susceptibility to erosion.
4. Manage the grazing of paddocks in accordance with the findings of 3 above to maintain a good balance between herbaceous feed for the grazing animals and perennial plants for erosion prevention.
5. Make use of your knowledge of both the vegetation and soils on your property, moving animals to more resistant soil types or paddocks more adequately protected by vegetation when paddocks on lighter soil types are approaching the advisable limit of cover of perennial species for erosion prevention.
6. Control erosion near the homestead by plantings of hedges and shelter belts (see Chapters 9, 10 and 11).

FURTHER READING

Bagnold, R. A. (1941). *The physics of blown sand and desert dunes.* Methuen, London. 265 pp.

Bennett, H. H. (1939). *Soil conservation.* McGraw-Hill, New York and London. 993 pp.

Chepil, W. S. & Woodruff, N. P. (1963). 'The physics of wind erosion and its control'. *Adv. Agron.*, 15 : 211-302.

Kohnke, H. & Bertrand, A. R. (1959). *Soil conservation.* McGraw-Hill, New York. 298 pp.

Parr, J. F. & Bertrand, A. R. (1960). 'Water infiltration into soils'. *Adv. Agron.*, 12 : 311-363.

Pick, J. H. & Alldis, V. R. (1944). *Australia's dying heart: soil erosion and station management in the inland.* 2nd edition. Melbourne University Press, Melbourne.

TEXT REFERENCES

Jessup, R. W. (1951). 'The soils, geology and vegetation of north-western South Australia'. *Trans. Roy. Soc. S. Austr.*, 74 : 189-273.

Langbein, W. B. & Schumm, S. A. (1958). 'Yield of sediment in relation to mean annual precipitation'. *Am. Geophysical Union Trans.*, 39 : 1076-1084.

Marshall, J. K. (1971). 'Drag measurements in roughness arrays of varying density and distribution'. *Agric. Meteorol.,* 8 : 269-292.

CHAPTER 6

Land use classification in dryland Australia

by R. W. Condon

THE term *land use classification* implies classification of land for various forms or degrees of land use. The object of land use classification should be to classify land so that the best use can be determined. The scheme of classification will depend on the purpose for which it is required.

Most of Australian farming lands have been settled and developed on a trial and error basis, in which the successes and mistakes of the early settlers have built up a fund of experience which later settlers have used as a guide. This experience was also used by the early land surveyors, who categorised land into various classes of agricultural and grazing land, and described the soils and vegetation so that others could also recognise these categories.

Special problems in settlement led to soil surveys, initially by the State departments of agriculture and generally on a small scale. Irrigation areas raised special problems which could be traced back to soils. In the 1930s, the Council for Scientific and Industrial Research (later C.S.I.R.O.) undertook detailed soil surveys of a number of irrigation areas as a basis for understanding the causes of the problems and for planning the best forms of horticultural land use. These soil surveys also included consideration of vegetation, climate and other important aspects of the environment. Later regional surveys of soils were undertaken by C.S.I.R.O. in various states. The first Australia-wide soils map was published by Prescott (1944), and showed the distribution of the major soil groups within Australia. The latest is the *Atlas of Australian Soils,* presently being compiled in several sheets, (Northcote, 1960), and culminating in the preparation of the *Handbook of Australian Soils* by Stace *et al.* (1968).

Vegetation surveys, generally on a broader scale, were also undertaken by some State departments of agriculture, forestry authorities and universities. These mapped the distribution of ecological units and related their occurrence to soils, climate and other aspects.

Geology, too, is important to the understanding of the natural environment and although geological surveys were generally undertaken for purely geological purposes, people concerned with planning or managing land resources could use them to further their understanding of the natural environment. Farmers quickly learnt the differences between granite, basalt, trap, alluvial and other geological formations, and applied this knowledge to the development of their lands.

In 1943 an erosion survey was carried out by the Soil Conservation Service of New South Wales, in which various classes of erosion were recognised and mapped over the Eastern and Central Divisions of the State (Kaleski, 1945) and the areas affected in various regions summarised to provide basic information for a State-wide attack on erosion. At the same time, maps were compiled, on a county basis at 2 miles per inch, showing the distribution of 8 major land use classes. These have been used mostly to facilitate local planning, but some have been used to compile the land use map of the Hunter Valley in New South Wales (Anon., 1960).

The initial broad-scale survey of erosion was followed up by more detailed surveys in areas warranting special attention, such as the catchments to major water storages (Taylor, 1957-58 and Morland, 1958-59). These were based primarily on plant ecology, the natural vegetation serving to reflect the other features of the environment, and most certainly to show up symptoms of abuse or misuse.

It will be seen from this summary that surveys are carried out for a definite purpose, generally to map the distribution of a particular resource, such as soils, vegetation or geology, and that in the process much information is obtained about other resources because of the relationship which exists between the various factors of the environment. These developments lead in turn to the broad-scale survey of land resources such as those carried out by the C.S.I.R.O. Division of Land Research from the early 1950s. These were done by teams of specialists in various fields such as soils, vegetation, climate, geomorphology, agronomy and others.

This integrated approach gave a survey of all resources likely to be important in developing and using land for primary production. The mapping units were land systems which might be defined as combinations of land units such that the various land units comprising a land system were present in a recurring pattern which gave a distinctive appearance to the land surface as viewed from above, as in an aerial photograph (Christian, 1952). For example, a large belt of level red earth soils may comprise one land system. Elsewhere such soils may be mixed with undulating stony soils to form a different land system. In another area it may be also mixed with hills of a particular geological form to comprise a further land system.

The map showing distribution of land systems is accompanied by a report which tabulates the land units making up each land system, and lists the particular characteristics of each land unit, especially those of importance to pastoral or agricultural development. From the information gained from the integrated survey it would be possible also to compile maps of soils, geology and vegetation, as well as other features such as erosion, land use and, most important, potential land use.

The integrated land survey has been used primarily by C.S.I.R.O. to survey land resources in undeveloped areas, particularly in northern Australia and the Territory of Papua and New Guinea (Haantjens *et al.*, 1964). It has also been used in already-settled areas in the Hunter Valley (Story *et al.*, 1963) and the Southern Tablelands (Gunn *et al.*, 1969) of New South Wales. A similar technique has been used by the Soil Conservation Authority of Victoria (Rowan and Downes, 1963) in surveys of land resources in various regions of that State.

These surveys present a complete inventory of all the resources and factors important to land use.

In a similar category, but different context, are the surveys which have been carried out over parts of Australia by engineers (Grant, 1968) to map terrain types in relation to their suitability for engineering purposes (roads, airports, ability to support heavy structures, etc.).

The types of surveys described above are concerned with broad regions. Rarely are they mapped on a scale that the landholder can apply to his particular holding. It is also necessary to have them interpreted in a fashion which the landholder can use.

Land use classification at the farm level needs to show those features which can help the farmer make decisions on the best form of land use. The accumulation of data from the land use and erosion surveys discussed earlier provided the basis for survey of individual farms to facilitate planning of land use. Several States have instituted farm planning schemes involving survey of the farm and mapping of land use classes. Such a farm plan sets out the potential of various classes of country, establishes guide lines for land use to ensure permanent productivity, and points up the need for soil conservation measures to control existing erosion or prevent erosion. Its principal value is that it enables the landholder to establish priorities for development.

The most important reason for classifying agricultural and grazing land and planning its use is to preserve the land in a high state of productivity, by ensuring that it is not used beyond its capability and that it is sufficiently protected against erosion.

The greater part of the capital invested in a farming or grazing property is in the purchase of the land or, more correctly, the soil. To gain the maximum return from this investment in the soil its use must be carefully planned so that it is protected against loss by erosion (which means loss of capital) and, if possible, its value increased by making it more productive.

Management of the land requires recognition that its productive capacity, whether in terms of crops or grazing animals, or other form of production, depends upon many factors such as climate, soil conditions, slope, susceptibility to erosion and other physical characteristics. Management should aim at using each parcel of land towards gaining the greatest return without injuring the physical capacity of the soil to produce.

The occupation and development of Australia's farming and grazing lands in the various States has been somewhat haphazard, this being a natural result of the differences in the ideas, knowledge and purposes of individual landowners, the entirely different conditions in a new country and the limitations of labour and finance. The background of knowledge of land use which the early settlers brought with them from Europe had to be modified to suit the conditions of soil and climate as encountered in the various parts of Australia. Subsequently rapid advances have taken place in both the practical and scientific aspects of primary production. New types of machinery, new farming methods, advances in pasture improvement, soil and water conservation, use of fertilisers, control of pests and diseases, and breeding of plants and animals have transformed the rural industries.

Subdivision of the land by the land administration authorities has been based on the use of straight-line boundaries, often running north-south or east-west. The landholder has tended to follow this practice by placing his subdivision fencing parallel to the boundaries. This has frequently made land use difficult in country of irregular topography with natural boundaries lying in every direction.

So that the rural landholder can make maximum use of the knowledge now available he needs to look very closely at his principal resource—the land—and decide whether it is all being used to its maximum capability consistent with efficiency of use and the need for preserving productivity. Much country in Australia is being used for purposes for which it is best suited. However, there is a great deal of land capable of producing more, while other land has been worked beyond its capability or without sufficient protection against erosion.

Land use classification is the tool for determining land use capability. Many schemes have been developed for application in the humid and sub-humid areas in various parts of the world. These schemes are generally similar in nature, with modifications to suit the particular environments or special forms of land use.

The system of classification serving as the basis for farm planning for soil conservation in New South Wales (Anon., 1968) sets out eight land use capability classes, which are listed in Table 6.1. Similar schemes, with slight modifications to suit local conditions, are used by other States in Australia. These capability classes are applicable to any land use situation involving a mixture of land suitable for cropping and grazing, and hence can be used in areas with rainfall as low as 10-12 inches.

The broad principle applied here is to grade the classes from the best to the worst. While some exceptions may occur where grazing or forest are more profit-

TABLE 6.1

LAND CAPABILITY CLASSES USED FOR SOIL CONSERVATION PURPOSES IN NEW SOUTH WALES

Land suitable for cultivation

1. No special soil conservation practices necessary.
2. Simple soil conservation practices such as contour cultivation and adequate crop rotations.
3. Intensive soil conservation practices such as diversion banks, graded banks and grassed waterways, in addition to the practices for Class 2.

Land not suitable for cultivation

Grazing land

4. No special soil conservation practices necessary.
5. Intensive soil conservation measures necessary, such as pasture furrows, absorption and diversion banks, silt traps and pasture improvements.
6. Intensive soil conservation measures necessary on very steep lands, such as broadcasting seed and fertiliser, destruction of vermin, prevention of fire and limitation of stock.

Timbered land

7. Land best suited for green timber for any of the following reasons: erosion hazard, steepness, shallowness of the soil, infertility or farm requirements.

Unusable land

8. Cliffs, lakes, swamps and other unusable land.

able than cultivation, it is generally the case that the productivity of the land follows the order used.

This system of land use in classification, as with similar systems in other States and countries, depends firstly on the recognition of the capability of various types of country for the maximum type of production, having in mind the need for protection against erosion. Both require consideration of the landscape, and characteristics such as degree of slope, soil fertility and erodibility, and other features of the environment, especially those governed by climate. The objective is maximum productivity in accordance with needs for protection from erosion and degeneration.

Urban and recreational land use planning

The system of land classification described above is used mainly as a tool for determining land use capability for primary production. There are other forms of land use. Urban land use planning is becoming increasingly important and another form of land use, to which belated recognition is being given, is for recreation purposes. Increasing population, the concentration of population in large cities, increasing leisure time and increasing facility of travel is resulting in ever-increasing demands for recreation areas where people can use and enjoy the natural environment. The great majority of the land of any value for agriculture, grazing or timber production was taken up many years ago. This means that recreation areas of relatively unspoilt natural environment are almost wholly confined to classes 7 and 8.

However, for the purpose of conserving and preserving the natural environment for recreation or similar uses, there is plenty of scope for further land classification. Land set aside for recreation needs to be carefully planned. The same objectives apply, that is maximum productivity (in terms of visitor enjoyment or some similar measure) in accordance with the needs for protection from erosion and degeneration. Over-use of natural environments can cause despoliation of those features which attract people to it. Authorities responsible for the preservation and conservation of the natural environment for outdoor recreation are moving increasingly towards the complete planning of these areas. This planning involves the same broad approach—surveys to determine the nature of the resources available, and classification of the land in accordance with its capability for types of recreation use to which it is best suited, and recognition of the need to protect against over-use. This also implies that once the recreational use of the land is decided, planning should provide for the need to control use so that the risk of over-use or abuse is minimised.

Because developments such as pasture improvement, ringbarking and clearing, and the installation of structural soil conservation practices are rarely economically feasible in arid areas, it might be thought that there is no need to consider land use classification in arid areas. However, dryland Australia is composed of many types of country with many different degrees of erodibility and natural protection. Some soils are highly fertile with little risk of degradation by erosion. Other soils of high fertility are highly susceptible to erosion and need special management if their productivity is to be preserved. Some soils may be of low productivity and high susceptibility to erosion, but have a vegetative cover largely unpalatable to the grazing animal, and this affords sufficient protection.

Between these extremes there is a wide variety of soil types upon which are superimposed varying forms of vegetation and differing types of climate. The differences in soils are due to variation in origin or geology, which causes differences in physical and chemical nature of the soil, and the shape of the land surface. Different types of topography create differences in moisture relationships with resultant differences in vegetation and consequently productivity.

Arid and semi-arid grazing land is in a delicate state of balance. For this reason the differences in various land classes and soil types need to be recognised by the land manager and the characteristics of different landscapes appreciated so that they can be managed according to their particular requirements. This means that there is also scope for land use classification in the arid zone, with the adoption of the concept of maximum production consistent with sustained productivity.

An early example of this kind of approach was that by Condon and Stannard (1956-57) following research into the erosional behaviour of soils in western New South Wales. Nine major erosion classes were recognised, these being classified on the basis of type of erosion surface produced in the case of wind erosion, and the pattern of sheet and gully erosion in the case of water erosion. Special classes were established for those erosion types which depended on a special set of circumstances. It was found that the soil characteristics which influenced the form and degree of erosion were also responsible for differences in vegetative cover, and the nine erosion classes were used as a basis for erosion survey mapping over much of western New South Wales (Stannard, 1957-58; James, 1959-60).

An example of the application of integrated survey to the practical problems of land management in arid areas is given in central Australia where the C.S.I.R.O. Division of Land Research, in conducting a survey of land resources in that region, mapped 88 land systems (Perry *et al.*, 1962). Land systems are various arrangements of one or more land units which form a typical recurring pattern. In a subsequent survey the Soil Conservation Service of New South Wales, in a report dealing with various problems in this area (Condon *et al.*, 1969) grouped the many different types of country into five broad land classes. These were classes which warranted different kinds of grazing management because of the pastoral resources available in terms of pastures or topfeeds, and susceptibility or otherwise to erosion.

As an example of the type of approach used, the broad land classes are described briefly below in the order of intensity of management needed to make optimum use of their particular characteristics.

1. *Run-on country*

Country receiving run-on from nearby hilly country or by spreading of floodwaters over land adjacent to the middle and lower reaches of watercourses. This is, in most cases, highly-productive country with little risk of erosion, although often associated with highly erodible country.

2. *Country highly susceptible to erosion and pasture degeneration*

Generally highly-productive country of which maximum use should be made while conditions are good but from which stock should be removed completely during prolonged or severe droughts. Care is needed in the selection of sites for watering points and holding yards to avoid damage to pastures and soils and to the improvements themselves.

3. *Medium-textured (loams and sandy loams) country with plentiful top-feed and low to moderate susceptibility to erosion*

Country of moderate productivity. The availability of top-feed as drought forage and the low erosion risk make this ideal country for use during drought conditions.

4. *Coarse-textured sandy soils with sparse top-feeds and low susceptibility to erosion*

Country of moderate productivity with sparse resources of drought forage.

5. *Dune fields and spinifex sand plains with low susceptibility to erosion*

The dominance of the coarse spinifex of low palatability limits productivity but protects against erosion. Top-feeds are variable but where plentiful this country can be used in drought with little risk of damage.

6. *Mountains and hills*

Although many hilly areas are inaccessible, or support spinifex of low palatability, the narrow valley areas often provide a good flush of feed because of local run-on and the tendency to attract storms.

The scheme of classification can be refined or broadened to suit the particular purpose in mind. Much depends on the type of information available from broad scale surveys.

Such a scheme of land classification is applicable to the individual holding. This is being done in the Western Division of New South Wales where the Soil Conservation Service is providing plans of pastoral leases on which land classes are mapped in some detail on a scale of one mile per inch, and recommendations made as to appropriate methods of management for each land class.

With the view to encouraging recognition of the various types of country, and the associated management problems, landscape types in dryland Australia are described in Chapter 9.

FURTHER READING

Christian, C. S. (1952). 'The concept of land units and land systems'. *9th Pacific Science Congress Proceedings.*

Condon, R. W. & Stannard, M. E. (1956-57). 'Erosion in western N.S.W.' *J. Soil Conservation N.S.W.*, 12 : 4 (192); 13 : 1 (17); 13 : 2 (68); 13 : 3 (122).

Condon, R. W., Newman, J. C. & Cunningham, G. M. (1969). 'Soil erosion and pasture degeneration in Central Australia. Part II: Prevention and control of soil erosion and degeneration'. *J. Soil Conservation N.S.W.*, 25 : 2 (161).

Grant, K. (1968). *Terrain classification for engineering purposes of the Rolling Downs Province, Queensland.* C.S.I.R.O. Division of Soil Mechanics. (Tech. paper No. 3).

Gunn, R. H. *et al.* (1969). *Lands of the Queanbeyan—Shoalhaven Area, A.C.T. and N.S.W.* C.S.I.R.O. Land Research Series No. 24.

Haantjens, H. A. *et al.* (1964). *General report on lands of the Buna-Kokoda Area, Territory of Papua and New Guinea.* C.S.I.R.O. Land Research Series No. 10.

James, J. W., (1959-60). 'Erosion Survey of the Paroo-Upper Darling Region'. Parts I-IV, *J. Soil Conservation N.S.W.*, 15 : 4 (315); 16 : 1 (41); 16 : 3 (185); 16 : 4 (253).

Kaleski, L. G. (1945). 'Erosion Survey of N.S.W. (Eastern and Central Divisions)'. Reprinted in *J. Soil Conservation N.S.W.*, 19 : 4 (171).

Morland, R. T. (1958-59). 'Erosion Survey of the Hume Catchment Area'. Parts I-V, *J. Soil Conservation N.S.W.*, 14 : 3 (191); 14 : 4 (293); 15 : 1 (66); 15 : 2 (172); 15 : 3 (208).

Northcote, K. H. (1960). *Atlas of Australian Soils.* C.S.I.R.O., Melbourne University Press.

Perry, R. A. *et al.* (1962). *General Report on the Lands of the Alice Springs Area, Northern Territory, 1956-57.* C.S.I.R.O. Land Research Series No. 6.

Prescott, J. A. (1944). *A Soil Map of Australia.* C.S.I.R.O. Bull. 177.

Rowan, J. N. & Downes, R. G. (1963). *A study of the land in north-western Victoria.* Soil Conservation Authority of Victoria. TC2.

Soil Conservation Service of N.S.W. (1960). *Land use and topography of the Hunter Valley Catchment, N.S.W.* (Map). Government Printer, Sydney.

——— (1968). *Planning land use.* (Extension Handbook No. 3).

Stace, H. C. T. *et al.* (1968). *Handbook of Australian Soils.* Rellim Tech. Publns, Glenside, South Australia.

Stannard, M. E. (1957-58). 'Erosion Survey of the South West Cobar peneplain'. Parts I-III, *J. Soil Conservation N.S.W.*, 13 : 4 (176): 14 : 1 (30) and 14 : 2 (137).

Story, R. *et al.* (1963). *General report on the lands of the Hunter Valley.* C.S.I.R.O. Land Research Series No. 8.

Taylor, A. C. (1957-58). 'A soil conservation survey of the Snowy River Catchment'. *J. Soil Conservation N.S.W.*, 13 : 4 (159); 14 : 1 (5); 14 : 2 (104).

CHAPTER 7

Soil erosion in dryland Australia

by R. W. Condon

SOIL erosion is a serious problem affecting every member of the community. Its significance to the farmer and the rural population can readily be appreciated but its importance to the community may not be so obvious to the man in the street. The detrimental effects of erosion, however, are not confined to eroded farming and grazing lands and to the consequent loss of productivity. They are seen also in the siltation of streams, reservoirs and harbours, in more frequent and more devastating floods, and in a lowered standard of living for the community.

Soil erosion is by no means a new problem peculiar to the twentieth century. It is a problem to which fresh attention has been drawn by the extensive damage suffered by lands in those continents which have more recently felt the impact of European settlement. An account of the historical aspects of soil erosion and conservation is given in Chapter 8.

Soil erosion is not a process for which man is entirely responsible. In its broadest sense, erosion may be divided into two categories:

(a) Geologic erosion
(b) Accelerated erosion

Geologic erosion is a natural process by means of which the land surface is cut down to a form which is in equilibrium with the existing climate and other factors of the environment. Once this equilibrium is reached the combination of conditions of climate, slope and vegetation produces a balance in which the rate of soil removal is constant. However, the rate of weathering or soil formation in the particular environment is also constant, so that a balance is reached which maintains a constant cover of soil.

Any changes in the above three factors produce a new environment and the rate of erosion changes. Most such changes affecting geological erosion are brought about by a marked change in slope due to geological uplift or climate.

Wind and water are the principal agencies acting on the land surface to remove soil material. Other agencies also become important in localised areas—alternate freezing and thawing and the production of needle ice are agencies operating in very cold climates. While the land surface has a protective covering of vegetation, these agencies are unable to operate. It is only when the cover of vegetation is removed or depleted, and the delicate balance of the environment thus destroyed, that erosion is able to take its toll of the soil.

Accelerated erosion occurs when this delicate balance has been destroyed by man's influence. The soil is a body of material which has been built up over a period of many thousands of years, and which develops certain physical and chemical characteristics under the interaction of the principal features of the environment—climate and topography—and with which it is closely integrated. Accelerated erosion may remove the complete soil mantle from large areas within the space of a few decades, or may remove sufficient of it to seriously impair its ability to support sufficient plant growth to supply man's needs.

Man's use of the land for the production of foodstuffs and material for clothing has been the principal factor leading to accelerated erosion. The destruction of vegetation for the growth of crops and for the pasturage of livestock has created considerable change in the natural environment. Forest cover has been removed to encourage pasture growth for livestock and to make room for the growth of crops. Many extensive areas of natural grassland have also been put to the plough for the growth of crops. Roads, railways and fences by interfering with the natural drainage (an important part of the environment) have caused widespread and serious erosion.

Soil erosion results from land use not in harmony with the environment. In the higher rainfall areas it is easier to keep land use in harmony with the environ ment because the growth of protective vegetation is not limited by climate.

In drier climates it is difficult to maintain land use in harmony with the environment. One of the most important aspects of the environment in this context is the soil. Soils vary in their susceptibility to erosion with the result that a form of land use which may be stable on one soil type may result in serious erosion on another. This becomes an increasingly important factor as we pass through the range of climates from sub-humid to semi-arid and arid.

THE EROSION PROCESS IN AN ARID CLIMATE

Contrary to what might be expected, water is the most destructive agent of erosion in desert environments. Wind and water are frequently complementary in their role of removing soil, and it is frequently difficult to determine whether wind or water is the dominant agent in the erosion process on a particular site.

The desert environment typically alternates between extreme dryness and varying degrees of wetness. When the period of dryness is a long one the desiccated pasture cover is gradually broken down or blown away and the exposed soil becomes subject to soil removal by wind action. The process is aggravated where grazing livestock assist in the removal of plant cover and disturbance and powdering of the desiccated soil surface. When drought, especially in combination with grazing, leaves extensive areas of soil unprotected by plant cover, heavy rains will cause serious erosion damage.

The period during which the soil may be subject to wind erosion is generally a long one, and the process of removal gradual. On the other hand the period of subjection to water erosion is generally short, and the removal process is often violent.

Warren (1965) has described the prominent part played by water in the removal of soil from scalds, for long considered a result primarily of wind erosion.

The complementary roles of wind and water erosion can be seen in many forms and patterns of erosion. As Warren (*loc. cit.*) has shown, the extension of a scald can be attributed as much by attrition of the sides by water erosion as by the process of destruction of adjacent vegetation by sand-blasting and smothering and the exposure of the soil to wind action.

Warren also showed water to have a considerable down-cutting influence, even on almost flat surfaces. Water frequently removes soil particles from the place where they have lain for centuries, carrying them a short distance in run-off and spreading them in a shallow sheet, as a layer of silt. As the skin of silt dries out, it forms small polygonal plates which curl up at the edges, and are readily broken up and blown away by the first strong gusty wind.

The converse process can also be seen in action. Sloping sandy soils subjected to overgrazing may develop gullies as a result of storm rains on extensive areas of drought-bared soils. Subsequent drought may see sand and silt blown across the surface by wind action and trapped in the gullies from where it will be washed downstream and will become part of the debris of water erosion.

In some cases the debris left in the gullies by wind erosion may encourage an extension of existing water erosion. Shallow rills often support clumps of desert plants which cause the sand and silt-laden wind to drop its load to completely block the rills. In the next rain the flow is diverted to a new path where it cuts a new rill. In the almost flat, very long slopes typical of much of central Australia, the almost imperceptible flow lines often become dissected by a series of parallel rills. This causes large amounts of sand and silt to be moved downslope, where they may be spread in a broad delta, colonised by weeds, easily activated by wind, and consequently a focal point for wind erosion of nearby areas.

From the above description, one of the insidious characteristics of erosion will have become evident—once erosion becomes serious, it becomes a cumulative or self-regenerating process.

The by-products of erosion from one site are likely to initiate an erosion cycle in another site. The aridity caused by the rapid shedding of water from eroded surfaces hastens the demise of the plant life struggling to hold the soil together, and the area affected by erosion extends. The sand and silt blown from an eroding area may kill the vegetation around its fringes by sand-blasting and smothering, leaving more soil bared to the action of the wind.

Once an area becomes bare due to soil removal by wind during drought, the heavy rains which break the drought are likely to cause further erosion by sheeting and gullying.

In spite of the complementary roles of wind and water action in causing erosion, one or the other will generally show features of dominance due primarily to the characteristics of the site. The presence of rills, or the accumulation of sand in drifts, indicates clearly the predominance of water and wind respectively. However, when extensive areas become eroded with little or no signs of drift or rilling, it becomes difficult to label the dominant agent as wind or water. Traditionally, sloping areas are regarded as subject to water-sheeting, it being assumed that more definite symptoms of water erosion will show up in due course if erosion is allowed to continue.

Eroded areas which are level, or mainly so, are regarded as being subject to wind erosion. Where there is little scope for surface flow to take place, wind can be regarded as the principal removing agent, the role of water being confined to local rearrangement of soil material during rain and while sheet flow is taking place, and to conditioning (or more strictly de-conditioning) of the eroded surface as drying takes place.

Although the following text discusses forms of erosion caused by wind and water, it will be appreciated that the division is often arbitrary.

WIND EROSION

Wind is an agent of erosion by reason of its ability to lift soil particles into the atmosphere, or move them along the surface. Depending on the size of the soil particles and the strength of the wind, the particles may be carried thousands of miles, or only a few inches.

In this section the process by which wind removes and transports soil are described, together with the influence of soil characteristics. The wind erosion process and the soil characteristics are jointly responsible for the forms of wind erosion described at the end of this section.

WIND EROSION PROCESSES

Wind may remove soil particles from the surface by either lifting them up into the atmosphere and carrying them away, or by making the particles jump along the ground. In the latter case the jumping particles also apply energy to the soil surface each time they hit the ground, and in doing so dislodge and break up other particles so that they can be moved by wind. This process is called *saltation.*

The ability of the wind to lift large amounts of soil into the atmosphere may be gauged by a simple exercise in arithmetic. If one particle is plucked from the soil surface, and once every second dislodges another particle from the soil surface, and all particles so dislodged continue to dislodge another particle every second—within the space of 21 seconds new particles of soil are being set in motion at the rate of over one million every second.

For wind to be able to lift and move soil particles several conditions must apply.

(a) Soil aggregates must be small enough to be lifted by wind.
(b) Soil must be dry and thus lack cohesion.
(c) Soil must be poorly protected by vegetation.
(d) The wind must have a velocity of not less than 11 m.p.h. at the soil surface—anything tending to reduce this velocity at the surface will reduce the effect of wind in removing soil particles.

The process of wind erosion is aided if the soil surface is powdery as a result of disturbance.

Characteristics of Wind Flow

Gentle breezes have streamlined or laminated flow. The ground surface creates friction which drags on the undersurface of the wind flow and reduces its velocity in relation to upper parts of the flow. This friction drag depends upon the roughness of the surface, and the height and density of the roughness have a direct effect on the velocity of wind flow to a considerable height.

Strong winds have turbulent flow, having velocity in upward and downward, as well as horizontal directions. This produces gusts of strong wind alternating with short periods of lighter winds. By their great velocity and upward movement gusty winds are able to dislodge soil particles and carry them upward.

The flow of wind over a soil surface can be likened to dragging a piece of hessian over the soil—the hessian would produce a drag which dislodges soil particles.

The principal factor in minimising the effect of wind is roughness—this roughness may be composed of grass and herbage (as on grazing land), trees (as in windbreaks), a growing crop (as is often used in sand-drift control), stubble ploughed into fallow cultivation to leave a rough, cloddy surface, or ridging to protect bays between the ridges.

The largest soil particle which can be moved by wind is about 1 mm. Well-aggregated clay soils are therefore resistant to erosion unless trampled to a powdery condition by livestock. Sandy soils are very susceptible to wind erosion when dry, because they lack cohesiveness, and have a large proportion of particles capable of being moved by wind.

Where the size of the soil particles extends over a wide range wind has a sorting effect on the soil. Very fine clay and silt particles (<0.02 mm) are lifted into the atmosphere and carried away. Fine sand particles are blown along the ground surface till they reach an obstruction, where they pile up into dunes. Coarse sand particles (>1 mm) are left on the eroded surface and if they eventually form a thick layer they will effectively prevent further erosion. The tendency to form dunes depends on the depth of sandy soil down to resistant horizon—such as clay or gravel.

SOIL CHARACTERISTICS

Wind erosion may take several forms, ranging from hard and soft denuded surfaces free of loose sand, to surfaces with varying degrees of loose sand moving around a denuded surface, and to deep sand drifts and dunes. The nature of the surface left by wind erosion depends on a number of factors which fall under the heading of soil characteristics, namely:

The structure and texture of the surface horizon.
The depth of any surface horizon of erodible material.
The presence of a sub-surface horizon of different character to the surface.
The character of the surface.

The effect of these four factors in determining the form of wind erosion is discussed below.

(a) *The structure and texture of the surface horizon.* Self-mulching clay soils are virtually resistant to wind erosion by reason of the aggregation of the surface soil into coarse particles when dry. On the other hand, compact clay soils are prone to degeneration of the surface to a bare scalded condition on which regeneration of plant cover is difficult, although there may be relatively little loss of soil.

Fine sandy soils are highly susceptible to wind erosion, drying out quickly after rain and being easily disturbed. Sandy soils with a sufficient proportion of coarse particles are reasonably resistant to erosion (Crocker, 1936). When the surface is disturbed, movement of soil particles will take place by wind action until sufficient coarse particles are exposed to provide a protective covering. The proportion of coarse sand needs to be only about 10 per cent of the total sand to have this protective effect. Loamy surfaces of friable structures are lightly susceptible to wind erosion, while those with a compact surface structure are moderately susceptible.

(b) *The depth of any surface horizon of erodible material.* In sandy soils of high susceptibility to erosion the depth of sand able to be moved by the wind determines the amount of material available to form drifts. The proportion of sandy material is also a factor. When an area of dry soil is subject to wind action, being unprotected by vegetation or other cover, little happens until the wind becomes strong enough to pluck particles of sand off the top of the soil and start them bouncing across the surface. Each bounce dislodges other particles which are in turn set in motion. The fine clay particles are lifted into the atmosphere to form a dust haze which may take several days to settle and may travel hundreds of miles. The silt particles remain airborne while wind conditions are turbulent. Silt and clay comprise the majority of the material, along with clay and some of the finer sand fractions, which is carried by the typical dust storm which sometimes accompanies a fast-moving cold front. As velocity and turbulence drop, the airborne silt and fine sand particles settle quickly.

During the soil-removal process most of the airborne fine sand is carried within a few feet of the surface and accumulates against and around any obstacle sufficient to cause wind velocity to drop in the near vicinity. Fences, yards, buildings and surface tanks in the lee of the large supply of sandy material are prone to sanding-up in dust storms. This is the process behind the events of the 1901-02 drought described on page 141.

If the depth of sandy material is considerable (2-3 feet or more) continued erosion will lead to the formation of dunes.

Accumulations of fine sandy material on the surface, in the form of shallow drifts, bring forth a wealth of weedy species when rains come. However, these dry off quickly and the drifts are easily re-activated in short dry spells. The erosion process is easily set in motion where drifty conditions prevail.

Loose sand, of 1-2 mm or more in diameter, is not readily moved by wind and is generally arranged on the surface in shallow ripples and waves which effectively prevent any further erosion while the wind is at right angles to their alignment.

(c) *The presence of a sub-surface horizon of different character to the surface.* Most soils with clay, clay loam and loam at the surface show little difference in texture with increasing depth, although there may be a change in structure. In these soils any change in profile characteristics normally takes place as a gradual transition. Susceptibility to erosion is only slight, and if erosion does occur the exposed surface is little different to the original surface. The plant species normal to the environment are readily able to colonise and ensure early regeneration as soon as the cause of erosion is removed.

Sandy loams and loamy sand soils are more easily removed. In most of such soils there is a gradual increase in clay content with depth. When erosion takes place, the same conditions as described in the preceding paragraph apply, and recolonisation and regeneration take place readily with good seasonal conditions. Wind erosion on these soils has been termed wind-sheeting to distinguish it from the situation where there is a sharp transition in texture and structure between the surface horizon and the sub-soil.

Wind-sheeted surfaces regenerate fairly readily, the exposed loamy surface being reasonably permeable to water. Colonisation will take place by annuals with a reasonable sprinkling of perennial species normal to the environment. In 2-4 years the slightly wind-sheeted surface may have completely recovered. This explains why many areas which appear bare and windswept during drought conditions regenerate fairly readily.

The rate of regeneration depends on the depth to which erosion has taken place and the size of the eroding areas. If the erosion exposes a horizon of high clay content the process of erosion is slowed down because of the lower susceptibility of clay soil to erosion. However, the rate of regeneration is much slower, for two reasons—the low permeability of eroded clay surfaces, and the need for plants other than those native to the original surface to effect colonisation on the denuded surface where soil characteristics differ considerably from the original.

Large bare areas are slow to regenerate because there is no obstruction to wind velocity. Consequently seeds which might lodge in the surface are quickly blown away. In such circumstances, colonisation commences around the edges on sites where conditions are more favourable. The larger the eroded area, and the greater the depth of soil loss, the smaller are the chances of successful regeneration and the greater the likelihood of the erosion extending. This is why it is important to recognise erosion in the early stages, so that protective measures can be taken while the chances for natural regeneration are good.

The process of erosion is slowed down and the rate of regeneration speeded up when there is a reservoir of coarse sand moving about on the surface, generally in the form of small well-spaced ripples. The coarse sand not only protects the soil surface from erosion, but also seems to carry a large reservoir of seed ready to germinate when conditions are right. The ripples of coarse sand also form a useful mulch which ensures good soil moisture conditions for germination.

Soils which are characterised by a sharp transition between a sandy loam or sandy surface horizon and a heavy clay sub-soil are the most prone to serious erosion damage. The sandy surface is easily activated by wind and erosion proceeds rapidly until the clay sub-surface is exposed. When this stage is reached the rate of erosion probably slows down, but the chances of regeneration are remote.

The clay is usually solonetic in character, having a high proportion of sodium ions on the clay molecule. This imparts certain characteristics which make colonisation very difficult. The most notable property is high dispersibility, which makes the soil unstable to water so that it slakes or 'melts' when wetted. The exposed surface is slaked with a thin surface crust of sand, silt and clay, with fine cracks causing a pattern of polygons 2-4 inches in diameter. Below the crust the clayey soil is generally loose and well-aggregated into small clods $\frac{1}{8}$-$\frac{1}{2}$ inch in diameter.

Top: Typical profile of a texture contrast soil showing the sharp contrast between the heavy clay B horizon and the sandy or loamy A horizon. The light-textured surface is easily removed by wind and water erosion to produce a scalded surface.

Centre: A typical scald surface produced after the removal of the surface horizon and the exposure of the clayey B horizon. The vegetation in the background is on small sandy "islands" which still retain top-soil.

Bottom: A stony scald produced on bluebush country in central Australia. As soil loss continues, the layer of stones and pebbles becomes denser until it is able to protect the soil against further soil loss. However, there is little chance of regeneration unless moisture relationships can be improved.

(Photos courtesy Soil Conservation Service of N.S.W.)

Plate 14 SOIL EROSION—I

Top: Soil subject to serious drift during drought conditions in central Australia. Although the surface is protected to some extent by coarse sand, there is a loss of the fine clay particles carrying the plant foods. This is shown by the dust haze obscuring the mountains in the background.

Centre: The same scene as in figure 1 above after good rains. Although a remarkable transformation, soil loss has resulted in some pasture degeneration as shown by the predominance of the inferior buckbush (*Salsola kali*).

Bottom: Wind and water erosion often operate together in dry land areas. In this scene, wind erosion of the surrounding areas has swept fine sand into a gully cut by water erosion. The bed of sand will prevent further downward cutting but will encourage sideways cutting of the gully walls.

(*Photos courtesy Soil Conservation Service of N.S.W.*)

Plate 15 SOIL EROSION—II

Top: Water is equally as important as wind as an agent of erosion in dry land areas. An example of serious sheet erosion on gravelly red earth soils in western New South Wales.

Centre: As sheet eroded areas increase in size, the high run-off leads to rilling in upper-slope situations.

Bottom: Serious erosion invariably has a cumulative effect, causing damage to surrounding areas. Frequent gullying of texture contrast soil in a lower slope position has resulted from large volumes of run-off from extensively sheet eroded areas up-slope. It could be expected that the drainage line down slope would also be seriously affected by erosion.

(Photos courtesy Soil Conservation Service of N.S.W.)

Plate 16 SOIL EROSION—III

Top: In many of the marginal wheat growing areas, very sandy loams carrying mallee vegetation have been cleared of timber and sown to crops, resulting in serious drift and dune activation on deep sands.

Centre: Sand dunes which adjoin more productive country are often subjected to excessive grazing and trampling and become activated by wind erosion.

Bottom: Deep sandy loams and clayey sands are subject to wind-sheeting and moderate drift when exposed to heavy grazing and trampling during drought conditions.

(*Photos courtesy Soil Conservation Service of N.S.W.*)

Plate 17 SOIL EROSION—IV

The exposed surface may vary in 'hardness' from soft and spongy to quite compact. This type of surface has become known as a 'scald'. It is frequently referred to by local people as 'claypan' but this latter term is better reserved for the small shallow 'lakes' or area of local run-on where clay-laden water leaves behind a scald-like surface after it evaporates. Claypans and the process involved in their formation will be described separately later. The principal feature of the scald surface which militates against regeneration is impermeability to water. Although the initial infiltration is rapid, because of the loose fine cloddy structure, the slaking of the clods and the swelling of the clay forms a seal through which further penetration of water is extremely slow.

This results in high surface run-off in heavy rains and rapid evaporation of the small amount of moisture which may penetrate the surface. Conditions for colonisation by plants are therefore very difficult. Spasmodic colonisation occurs in small 'islands' but is quickly swept away by sand blast when dry windy conditions return. The run-off from the scalded surface may drain to adjacent non-eroded areas where it may provide favourable moisture conditions over a very limited area.

Often a proportion of the run-off may accumulate in low-lying areas on the scald surface. If this occurs frequently, as may happen in an exceptionally good year, an initial colonisation may be boosted by repeated shallow floodings, and all areas affected by flooding will begin to regenerate. If good seasonal conditions continue the regenerating area will continue to extend. This has happened over many large scalded areas in western New South Wales as a result of the prolonged run of favourable seasons, commencing in 1949 and extending, with a few short drought periods, until 1964 (Jones, 1966). Prolonged drought conditions are likely to cause a reversion of all but the more favoured sites.

Claypans are of frequent occurrence in dryland Australia and generally give the impression of being an erosion surface. They are formed in areas where water draining from the surrounding country forms a shallow sheet until it is evaporated, leaving its deposit of clay behind.

Claypan surfaces are often similar to scalds, in appearance and nature. The main difference is that scalds normally shed water (although there will often be small areas where water lies for some time before evaporating), whereas a claypan accumulates water, generally as a very shallow sheet. Because of this intermittent flooding they invariably support a stand of canegrass (*Eragrostis australasica*) and are frequently known as 'canegrass swamps', although they may be dry for 90 per cent of the time. They frequently have much sand and silt brought in by wind and water and this moves about as drift, and adds to the general impression of being an eroded surface. Although active erosion may take place on claypan surface because of the difficult environment, it is more correctly a depositional surface rather than an erosional surface.

In some areas where large amounts of salt are brought in from the surrounding country, the claypan surface is often very saline, being loose and friable beneath a weak surface crust. The presence of samphires and saltworts (*Arthrocnemum* and *Salicornia*) indicates high salinity. The next step in this process is the saltpan which may store several feet of water after heavy rains. The evaporation of large quantities of water over thousands of years has left a considerable depth of salt which may be found as coarse salt under an inch or so of surface sand, silt and clay. The typical saltpan normally has a soft crumbly surface and is completely bare of vegetation except for samphires and saltworts around the fringes.

(d) *The character of the surface.* Some surfaces have characteristics such as a scattering of stone or gravel which helps to protect them against wind erosion. However, stone or gravel cover has to be very heavy for effective protection against sheet erosion by water.

Other soils have a surface contorted by gilgais, which are puffs and depressions caused by differential expansion of the clay. The usual arrangement is a small depression 3-12 inches deep and perhaps 3-4 yards in diameter, of heavy, deeply cracking clay interspersed among compact clays or clays which are often solonetzic or saline in nature. The inter-depression areas are generally poorly covered and shed water into the depressions in heavy rains. The plants in the depression are thus irrigated. Sometimes the depression may be linked to form figure-8 depressions or they may be much elongated to form linear gilgais.

Although the wavy nature of the soil surface should produce a sufficient degree of roughness to prevent wind erosion, this form of protection is not sufficient in arid areas subject to over-grazing. In the more solonetzic clays, removal of soil from the inter-depression areas by both wind and water erosion has often proceeded to the point where the depressions are filled with sandy material. In some cases, the accumulation of wind-blown silt amongst the vegetation in the former depression causes the area to be raised above the general level of the ground surface. This produces a surface of small islands of struggling vegetation separated by ever-increasing spaces of eroding soil.

FORMS OF WIND EROSION

Having discussed the various factors responsible for different forms of wind erosion it is proposed to list the predominant forms and describe them briefly. More complete descriptions are given by Condon and Stannard (1957) in respect of western New South Wales, and by Condon *et al.* (1969) in respect of central Australia.

SCALDING

Scalds are formed when a resistant clay horizon, solonetzic in character, is exposed by the removal of overlying sandy material. The surface is relatively soft and strongly slaked to form a shallow crust over a loose, very fine cloddy structure. The clay swells when wet to form a seal which prevents moisture from penetrating deep into the soil. The results of much research into scald behaviour and characteristics is given by Jones (1966-67).

Scalding is the typical erosion form of texture-contrast soils. These soils are characterised by a sandy or loamy surface horizon with sharp transition to a solonetzic clay which has been strongly affected by sodium. Wherever these soils occur they have been seriously damaged because of their high susceptibility to erosion and because of the attractiveness of the forage to livestock.

Texture-contrast soils are often found in river flood plain situations where they are mixed either with areas of heavy clay soils or coarse alluvial sands. They are also widely associated with the saltbush-bluebush landscape type, a large proportion of which has suffered severe damage in the past.

WIND-SHEETING

This term is used to describe the eroded surface exposed by wind erosion of soils in which there is only a gradual change in texture down the profile, or no change at all. The denuded surface is firm to hard, with little or no structure. The surface is normally loamy in texture and reasonably permeable to rainfall moisture and natural regeneration is relatively easy.

Wind-sheeting is the typical erosion form of the medium-textured soils: the red earths, red clayey sands and red earths of central Australia, the red and brown sandy loams and loams throughout much of New South Wales, Queensland and the drier parts of Western Australia.

The susceptibility to erosion of soils liable to wind-sheeting is generally low to moderate. These soils generally support light to moderate, and sometimes heavy, stands of timber and scrub which also give some degree of protection.

Because of the relative ease of natural regeneration, the presence of wind-sheeting is often of a temporary nature, appearing in drought times and recovering during good seasons. Although an area may appear to recover after being bare, there will have been loss of soil and plant nutrients, and loss of physical condition. This loss is often accompanied by a deterioration in the quality and quantity of pasture. If loss of soil continues the time will come, because of unfavourable moisture characteristics with increasing depth of soil loss, when there will be no recovery.

In arid regions a certain amount of soil movement takes place as a natural process. Loose material on the surface is collected around the base of plants. If a plant dies the loose material is moved across the surface to accumulate against the base of adjoining plants. Thus during a drought there is a considerable amount of soil movement but relatively little soil loss. If the surface is unduly disturbed by excessive stocking, the movement of soil over the surface will be accelerated and soil loss will be considerable. The result of this is the appearance of small and large bare spaces where normally they should be absent. This is an indication of accelerated erosion and is also regarded as wind-sheeting, the exposed surface being little different to the original in texture and nature. Regeneration should take place naturally, but once the affected areas become large, or are able to drain run-off away, regeneration is only likely in exceptional seasons.

DRIFT

This term refers to the presence of loose sand on the denuded surface whether it be scalded or wind-sheeted. It results from the erosion of soils with a foot or more of sandy soil in an erodible surface horizon. In this case the sandy material, being too coarse to be moved far by the wind, is retained on the eroded surface where it forms shallow drifts separated by expanses of scalded or wind-sheeted surface.

In addition to the inhospitable conditions of the exposed surface, plants attempting to colonise such surfaces also have to withstand the effects of sandblasting.

Attempts at reclamation are also overwhelmed by sand. The sandy drifts grow short-lived annual grasses and herbage while conditions are good but these dry out quickly, and the sand is easily activated.

Areas suffering from severe drift over extensive areas, invariably carry a degenerate pasture of short-lived annuals. In some areas the deterioration in soil conditions often leads to invasion of woody shrubs which serve to stabilise the area.

HUMMOCKING

Hummocking is frequently a by-product of serious wind erosion and may take two forms. One form is characterised by small residuals of uneroded soil left behind after the surface soil from surrounding areas has been completely removed. This form of hummock is generally 9-18 inches in height with a flat top representing the original surface, perhaps with some addition of windblown soil, and sloping shoulders. It occurs in the deeper texture-contrast soils where the resistant clay horizon is one foot deep or more.

The second form of hummocking is caused by the accumulation of drift sand and other material into large hemispherical mounds around bushes of *Nitraria schoberi* (nitre bush). These mounds may be up to 4-5 feet high with a base of 12-18 feet, and are generally scattered thickly around the leeward fringes of scalds on saline soil types. The nitre bushes apparently become established during favourable conditions. The bushy and scrambling habit encourages accumulation of drift sand and clay. As the bushes grow so the hummocks of drift material increase in size.

This form of erosion is typical of scalds with a highly saline surface. The hummocks are normally composed of a mixture of clay and sand. The accumulation of clay as drift is unusual but can be explained by the breakdown of the clay aggregates by high salinity into grains small enough to behave like sand.

Hummocking of this kind is similar to the mesquite-dunes of some degenerate North American desert areas. Mesquite forms a low scrambling bush on sandy soils. When the spacing gets to about 10 yards, competition for moisture removes all the ground cover. When this happens, the sandy material from the bare spaces between the bushes is swept into the bushes by the wind to form large hummocks separated by bare wind-sheeted or scalded surfaces.

DRIFT AND DUNE ACTIVATION

This form of wind erosion occurs on sandy soils where the depth to a resistant horizon is considerable (3 feet or more). It may occur as drift and small dune formation on level sands, or it may occur as a reactivation of dunes. The extensive areas of parallel dunes which occupy much of the centre of the Australian continent invariably have an unstable crest of bare drift sand. These develop small slip faces and wind ripples, and constantly change shape, but do not move for any distance across the trend of the dunes. The nett direction of sand movement is along the dune crest in the direction of the predominant wind.

Drift is initiated by sand being blown from disturbed areas and covering vegetation on adjacent undisturbed areas. When the smothered vegetation dies this provides a larger area from which loose sand may be moved. The drift soon takes the typical form of a 'blow out' (being the basin-like hollow from which sand is removed) with an apron of moving drift on the leeward side. Changes in wind direction are able to spread sand from the apron in all directions.

Reactivation of dunes over small areas of a few square chains is common in arid areas, but it rarely affects large areas. Unless valuable improvements or property are affected or threatened it is of little consequence. There is frequently some dune activation where dune sands adjoin river flood plains, particularly where there has been a long history of grazing. This is a reflection of over-use of the dune areas because of the much higher carrying capacity of the adjacent soils. Dune activation is a common problem in coastal areas and is more serious because of the threat to property and services.

WATER EROSION

In an arid region it would be expected that wind would be the major agent in soil removal. Wind erosion does account for a considerable amount of soil movement and loss, but water erosion, by sheeting and rilling, is the principal agent in the degeneration of the majority of seriously eroded country in arid areas.

WATER EROSION PROCESSES

As in humid regions, the three processes of water erosion in dryland areas are, in order of their occurrence in the erosion cycle:

—Raindrop splash
—Surface or sheet flow
—Channelised flow

A fourth process, resulting from the action of the above three processes, is very important in desert areas, namely deposition of debris.

These four processes will be described briefly so that the forms of water erosion may be more readily understood.

RAINDROP SPLASH

In the absence of vegetative cover on the ground surface, the softening of the surface by wetting, and the pounding by falling raindrops causes a breakdown in the structure at the soil surface, the soil crumbs being broken down so that the fine silt and clay particles fill up the spaces between the larger sand particles and tend to form a surface seal which significantly reduces infiltration. The greater the proportion of the silt and clay fraction in the exposed surface the worse the effect of puddling. This accounts for the difference in surface characteristics between scalds and wind-sheeted surfaces. The greater the proportion of sand and the coarser the sand, the less likelihood there is of forming a surface seal.

It is the puddling action of raindrop splash which causes the surface sealing. This action has an immediate adverse influence on infiltration while rain is falling, and probably more so after the soil has dried out after rain. Through the action of raindrop splash, the soil may develop an eroded surface or 'erosion pavement' without any soil having been lost or removed from the site. Once this erosion pavement is formed, the reduced infiltration, leading to greater run-off and greater difficulty of establishment of plant cover, causes the soil to become very prone to erosion.

SURFACE FLOW

When the rate of rainfall exceeds the entry of water into the soil, water accumulates on the surface and, under the influence of slope, surface flow takes place, carrying away the soil particles loosened by raindrop splash. As the velocity of surface flow increases, further soil material is removed from the surface. This process takes place much quicker where infiltration has been reduced by raindrop splash, and where there is no vegetative cover or plant litter on the surface to impede flow. This is typical of the process involved in sheet erosion, and although the actual amount of soil lost may be small, the greatest damage is done by the breakdown of the structure of the soil surface which seriously limits the rate of penetration of water into the soil, making conditions difficult for the establishment of plants.

The exposure of a large proportion of the land surface to the action of raindrop splash and surface flow causes very large volumes of run-off from storm rains. In central Australia it is a common experience to see sheet flow 1-2 inches deep over distances up to half a mile wide through country which appears almost flat. In other areas, run-off from low stony ridges during and after storm rains runs water 1-2 feet deep and 2-5 chains wide in the narrow flats between the ridges.

CHANNELISED FLOW

Increasing run-off from increasing bare surfaces leads to the next stage in water erosion. Flowing water picked up by cattle or sheep pads, or vehicle tracks, starts to abrade, cutting shallow rills, generally in lower slope situations. These rills fret up-slope and laterally, fanning out to command a greater catchment and concentrating the flow into the main channel.

The presence of rills, or small channels of the order of 3-12 inches deep and similar width, on slopes away from drainage lines is an indication that sheeting has proceeded much further than nature intended, and that both run-off and soil loss are excessive. Occasional rilling can be expected under most situations where there is any slope. Where the areas concerned are large and frequent, and rilling is associated with shallow gullying of flats and adjacent slopes, the degree of degeneration is serious.

DEPOSITION OF DEBRIS

The effect of the preceding three processes is to loosen soil material from the soil surface and transport it down-slope, either in sheet flow or channelised flow. The greater the volume and velocity of flow the greater is the amount of solid material which can be transported. In the process of transportation water also has a sorting action in much the same way as wind.

Soil material carried in sheet flow is generally the finer silt and clay fractions. Often the distance of transport is relatively short, the run-off water lying in a shallow sheet on a small nearby run-on area. When the water evaporates, the silt and clay left behind form a thin veneer which cracks as it dries and curls up at the edges to form small polygonal wafers. These are very easily broken up and blown away by wind. When confined to a narrow gully running down a steep grade, water is able to transport a high proportion of soil material, including sand, which helps in the abrasive process of removing further soil material.

When some factor, such as a change in grade of the gully floor, or the presence of an obstruction, causes the velocity to decrease, the heavier sand particles drop out of suspension. The result is a delta of coarse and fine sand. Because the sand is poor in nutrients, plants colonising such sites are ephemeral, and the sand is easily activated by wind to form drifts which will spread over adjoining areas, smothering plant life, and becoming focal points for wind erosion.

The finer silt fraction is carried on further, generally falling out of suspension when the flowing water is brought almost to a standstill. This happens if the flowing water finds its way to a dam or surface tank. Surface tanks fed from eroding catchments have a very short life, or need frequent de-silting. The fine clay fraction may remain in suspension for several days after the water has been stilled, and in some types of soils the clay in the water in surface tanks may remain in suspension almost indefinitely.

The usual endpoint for the clay fraction of transported material may be a local run-on area where it serves to enrich the soil, a local claypan (see under 'Forms of wind erosion') or a saltpan. In the latter two cases the plant nutrients tied to the clay are lost from the point of view of effective plant growth.

FORMS OF WATER EROSION

The two principal forms of water erosion are:

—Sheet erosion where the loss of soil is in a relatively shallow layer over a relatively large area.

—Gully erosion where the loss of soil is from a deep, relatively small area, such as in an eroded channel.

Whether the pattern of erosion is determined by sheet erosion or gullying depends to a large extent on the nature of the landscape, particularly the topography and soil characteristics.

Some landscapes are prone to gullying, and may show evidence of gullying of drainage lines before sheet erosion becomes obvious on the slopes. In most cases, however, gullying only takes place as a result of serious degeneration of up-slope areas by sheet erosion. In many cases there is an intermediate stage where shallow rilling may also be a strong feature in the pattern of erosion.

There are other types of erosion which can be regarded as water erosion, such as the pulverisation of bare soil by alternate freezing and thawing in alpine climates. In dryland areas, the forms of water erosion are limited to sheet erosion and gullying. These two forms of erosion are described below.

SHEET EROSION BY WATER

To determine whether sheet erosion has taken place, it is necessary to keep in mind the natural or pristine condition which would apply under the particular soil characteristics and slope. Desert environments vary considerably in the nature of the ground cover. Sandy soils normally support a uniform cover of grasses and forbs, although the individual plants may be well spaced. As the surface texture moves towards loam or clay loam, and especially if there is some degree of slope, the natural condition is a pattern of 'islands' of vegetation separated by bare soil. The bare spaces are subject to wind and water erosion but the soil material dislodged is moved on to the vegetated areas. These are 'enriched' by the addition of soil material washed and blown on to the islands and 'irrigated' by the water draining from the bare spaces. Land of this type, in good conditions, could have as much as 50 per cent bare space.

Prolonged over-grazing, particularly if associated with drought conditions, will cause a weakening of the plant cover on the islands with the result that the proportion of bare space becomes greater. Because of the greater area from which soil can be readily removed, the rate of build-up on the islands becomes sufficient to cause them to be elevated above the adjacent bare surface. The tendency then is for the run-off water from the bare spaces to run around the islands of vegetation instead of on to them. The increased volume and velocity of flow causes greater soil loss so that the bare spaces become 'depressed' in relation to the vegetated islands.

When this happens, the islands become starved of their normal irrigation, the environment becomes more arid and the vegetation is further weakened—creating more bare space and releasing the accumulated soil material to further the erosion process. The final result is a scattering of small vegetated islands in an ocean of bare soil.

Some landscapes are much more susceptible to sheet erosion than others, and where these have been subjected to heavy grazing use in the past, serious damage has occurred. In arid and semi-arid regions it would seem that land with slopes more than about two per cent, and surface soil textures ranging from loam to clay loam, are moderately susceptible to sheet erosion. Sandy loams and sands are generally able to absorb a sufficient proportion of rainfall to prevent sheet erosion becoming a serious problem, although it will occur in small localised areas. The self-mulching granular clays with strongly cracking features are also highly absorptive and rarely suffer from serious or even mild erosion.

Soils with a compacted clay surface, or easily-slaked clay soils, are very susceptible to sheet erosion by water where any slope exists, particularly if vegetative cover has been depleted by overgrazing or drought. Texture-contrast soils with solonetzic clay in the subsoil are very highly susceptible to sheet erosion and deterioration sets in quickly where use has been too heavy.

Extensive areas of country in the arid and semi-arid regions of Australia have been seriously damaged by sheet erosion. In many cases productivity has been lowered to less than half that of similar country in good condition.

The fine-textured calcareous soils of the East Kimberley region have been seriously eroded over very extensive areas, necessitating a large scale regeneration project to rehabilitate much of the Ord River catchment (Fitzgerald, 1968).

Gently sloping texture-contrast soils in central Australia have been seriously damaged over extensive areas (Condon *et al.*, 1969). These texture-contrast soils carrying saltbush-bluebush vegetation extend south-easterly into South Australia in a broad belt around Oodnadatta and Marree, and have also been seriously damaged by widespread erosion. Similar country in the far west of New South Wales, occurring as undulating stony and gravelly clays, has also been seriously damaged. In the central-west of New South Wales, undulating hard, red loamy soils have suffered severely from sheet erosion.

As indicated previously, wind and water erosion proceed concurrently in the dryland environment, and wind erosion could be equally as important in removing soil from the areas described above as suffering from extensive sheet erosion. It is assumed that where slopes are in excess of 2-3 per cent, water erosion would be the dominant agent in removing soil.

GULLY EROSION

When the proportion of bare space becomes very large the proportion of rainfall lost as run-off also becomes very large. Advanced sheet erosion is invariably accompanied by shallow rilling of lower slopes. As the areas affected by sheet erosion, and associated rilling, become larger so does the pressure on the drainage system increase.

Should the vegetative cover of a drainage line be weakened by over-grazing or drought, increasing volumes of run-off from increasingly bare sheet-eroded areas will lead to gullying of drainage lines.

The cross-sectional shape of the gullies will depend on soil characteristics. Deep clayey soils will erode vertically by down-cutting and horizontally by under-cutting of walls and subsequent collapse. The result is a deep channel of square or rectangular cross-section. Deep loamy soils or friable clays form U-shape or V-shape gullies.

The depth in both cases is generally controlled by the exposure of a more resistant layer, often of coarse stone. In stony or gravelly soils the removal of soil leaves stone and gravel behind and this eventually becomes sufficient to form a stable bed. Further erosion is then in a horizontal direction.

Soils with a high proportion of sand form a broad shallow gully. As the finer fractions are carried away in the run-off, the coarse fractions are left behind to form a stable bed at a depth of about $1\frac{1}{2}$-2 feet. This process is accentuated by the gully trapping windblown sand during drought conditions. In smaller gullies in shallow drainage lines, the gully is often diverted by windblown sand piled up against the scattering of plants encouraged by the additional moisture which finds its way to the drainage line. This might be expected to stabilise the drainage line by allowing the water to spread, but the usual result is a second gully alongside the first.

As sheet erosion and gullying expand in the up-slope areas or minor drainage lines, greater pressure is placed on the large watercourses and flats by increased run-off. As a result these also become prone to erosion and deterioration.

In most gently undulating country in dryland areas there are normally broad flat valleys between slight ridges. These 'flats' may vary from a few chains to half a mile or more wide. They are flat in cross-section with perhaps a slight channel an inch or two deep and half a chain or so wide. In the normal situation in heavy rains, the run-off from the surrounding areas floods out over the slight channel and spreads across the whole width of the flat. This flooding causes a flush of growth which remains green for much longer than that on the surrounding areas from where the water comes. The flats or run-on areas, which normally occupy a small proportion of the landscape, may provide the greater proportion of the available forage.

The first effect of the flats having to handle increased volume and velocity of run-off, and especially if the ground cover has been weakened by over-grazing, is to sweep the floor clean of dead plant material and seed. Lack of regenerative ability leaves the swept area prone to gullying in subsequent flows, or gullying may be initiated by a track or cattle pad.

Once the floor of a flat is gullied, all flow is directed to the gully so that it enlarges with each flow. Flooding of the floor takes place only in exceptional run-offs which, because of the weakening condition of ground cover due to increasing aridity, are likely to cause more damage. Branch gullies start to work up tributary drainage lines. Small rills and gullies on the lower slopes of adjoining ridges link up with the main gully. Once the link-up is accomplished deepening of the tributary by head-cutting takes place rapidly.

When the flats become dissected by gullies their ability to provide forage is very drastically reduced and productivity may fall to one-fifth to one-tenth of its former level. Not only is there a reduction in total ground cover, but a deterioration in pasture quality, as only weedy and ephemeral species are able to survive in the increasingly arid site.

Most people tend to regard gullying, particularly in arid and semi-arid areas, as of little consequence. The area occupied by the actual gullied area is generally very small in relation to the total area available.

Because of the large amounts of surface flow which result from heavy rains, especially when drought has caused most of the landscape to be bare, some degree of gullying must be expected as a natural condition of the dryland environment. However, when gullies are the result of widespread sheet erosion, they become a symptom of advanced accelerated erosion.

Gullying is often initiated by vehicle tracks or cattle pads running diagonally across the slope of the flat or broad shallow flow line. Once a gully is initiated in this way, water which previously spread as a shallow sheet to irrigate a broad area is confined to a narrow channel. The environment of the formerly irrigated area becomes more arid. The change-over to a plant community capable of providing stability to the drier site is not likely to be a smooth one, especially if the situation is complicated by the presence of grazing livestock. Because of the adverse change in the moisture regime, such areas are likely to become seriously deteriorated, adding further to the erosion problem.

Gullying is normally a reflection of extensive sheet erosion in areas up-slope. In some types of country it may become pronounced with little or no associated sheet erosion. Generally this occurs in steeper country with narrow U-shaped drainage lines, on relatively steep grades.

CAUSES OF ACCELERATED EROSION

Having discussed the processes by which wind and water act as agents of erosion, it is pertinent to consider the various influences that can be labelled as causes of accelerated erosion.

If a land surface is completely untouched by man, some movement of soil will take place. This erosion, whether wind or water, is responsible for shaping the landscape, and is known as geologic erosion. The rate of erosion is in balance with other aspects of the environment, being controlled largely by the effectiveness of the natural vegetative cover in protecting the land surface against the influences of wind and water.

When man uses land for the growth of crops and for pasturing his livestock, this may upset the natural balance which existed between the various aspects of the environment. When erosion takes place as a result of man's influence on the land surface, this is known as 'accelerated erosion'. If allowed to proceed unchecked the damage to the environment can be considerable.

When land is used for the grazing of livestock the trees are killed, and generally removed, to encourage a greater yield of forage. This is drastic treatment of mother nature, and the system of land use which replaces the natural forest or woodland must be in harmony with the various aspects of the environment. If it is not, accelerated erosion will result. Even more drastic is the preparation of the land for the growth of crops. Not only are the trees removed, but the pasture is ploughed up and cultivated frequently to bring about conditions suitable for the establishment and growth of crops. While the soil is in a cultivated condition it is unprotected by plant growth, and is very vulnerable to the action of wind and water and is easily removed.

Accelerated erosion may take one or more of several forms—sheet erosion and gullying by water; or scalding, wind-sheeting, and drift and dune activation by wind. Accelerated erosion results from man's use of the land in a manner which is not in harmony with the environment, or from activities which leave the soil prone to the forces of erosion. As much of dryland Australia is used for cropping, it is pertinent to discuss the causes of erosion from this form of land use as well as from dryland grazing. The causes of erosion are therefore discussed under two headings—arable land and grazing land.

ARABLE LAND

Cultivated land may be subject to wind or water erosion, or both. The drier the climate, the slower the process of recovery of natural or sown pastures after cropping. Drought periods are also more frequent and generally more prolonged in drier areas so that the land surface is left in a vulnerable condition for a greater proportion of the time.

Much of the serious erosion in Australia's crop lands was brought about by over-use of cultivated land in the period leading up to World War II. Rest periods were short and served only as rest periods, rather than as pasture phases designed to restore physical condition and chemical fertility. Sown pastures were largely non-existent and the recovery of natural pastures was very slow. The only time that the land surface was reasonably protected by vegetation was in the late crop stage and before the stubble was burnt. Run-off from bare or nearly bare surfaces was shed quickly to the drainage lines—also bare, and unaccustomed to such volumes of flow. The flow lines became dissected by gullies. The first stage was normally the removal of all the loose soil to plough depth from the centre of the flow line—resulting in a shallow channel 3-30 feet wide and 3-6 inches deep. Subsequent cultivation across these flow lines merely provided more loose soil to be washed away by the next flow.

The pattern of cultivation—up and down the slope, and diagonally across the slope—hastened the delivery of water to the flow lines, ensuring bigger and better flows with greater scouring capacity. Eventually the gullies became too deep to plough across and the farmer was forced to plough his paddocks in small blocks between the gullies. The pattern of gully erosion in the drainage lines was often reinforced by tributary gullies caused by the diversion of overland flow and its concentration in stock pads or plough-out furrows, fences and tracks. Many cultivation paddocks became a mosaic of browns and yellows as sheet erosion stripped the top soil from the ridges and slopes and exposed the paler sub-soil.

Sheet erosion is brought about by the pulverisation of the surface soil by frequent cultivation and its exposure to the puddling action of raindrop splash, the removal of the loosened soil particles in the overland flow, leading to the gradual loss of the friable and fertile top-soil. In addition to the loss of top-soil, with its valuable store of available plant nutrients, there is also a deterioration in the physical condition of the soil so that permeability to water is also drastically reduced.

The widespread application of superphosphate failed to halt the continuing decline in yield which took place in spite of improvements in wheat varieties, and machinery for cultivating and harvesting. The common term appled to the ailing soil at the time was 'wheat-sick', a recognition by the farmer that the land was over-used.

The cure to wheat-sickness required rest and recuperation of physical and chemical fertility under a pasture. The introduction of sown pastures such as sub-clover and Wimmera rye grass in the pre-World War II period provided the medicine for the cure. The medicine was widely used in the post-war period, enabling land use in harmony with the environment. Long pasture phases between periods of cropping restored the physical condition of soil. The legume component of the pasture, restored the nitrogen depleted during the cropping phase. The pasture revolution was the most important factor in the increase of the average wheat yields from about 13 bushels per acre in the pre-war period to over 20 bushels per acre in the 1960s—an increase of 50 per cent.

Where the land was dissected by gullies, or stripped by sheet erosion, or where slopes were steep, or the soil highly erodible, additional measures were necessary to effect the cure. The process of erosion had to be halted, run-off had to be controlled. Banks and waterways and small water storages created new patterns on the land surface. Contour cultivation held the rainfall on the surface where it fell. Excess water was taken away slowly across the slope by gently graded banks, delivering into the grassed waterways which carried it safely down slope, dams being used to store the surplus for stock water supplies and to slow the velocity of flow.

Without effective soil-conservation measures much of our wheatland country will continue to erode. A recent re-survey of erosion in New South Wales (Stewart, 1968) showed that over a period of 25 years, there had been a considerable increase in the area of land suffering from moderate gully erosion. This was attributed to deterioration of lands formerly sheet-eroded (and not protected by soil conservation measures), and to the expansion of cultivation for wheat without adequate soil conservation measures.

The many soil conservation measures used to control run-off and water erosion in crop lands are discussed later, with emphasis on the special techniques developed for the drier areas.

Arable lands on very sandy soils in the drier areas are rarely prone to water erosion. Slopes are slight, the sandy soil is absorptive, and the lack of clay makes it unlikely to deteriorate in physical condition. Lack of clay, however, means a loose single-grained soil, drying out quickly on the surface and prone to erosion by wind action.

In New South Wales, Victoria, South Australia and Western Australia large areas of sandy country carrying a dwarf many-stemmed eucalypt known generally as 'mallee' were developed for cropping on the drier fringes of the wheat belt. The small trees were easily removed by rolling and later burning. These lands were also subject to excessive cropping with frequent fallow and short periods of 'rest' with little chance of establishing a reasonable pasture cover before being again put under the plough. The stubble was burnt or grazed by sheep, as were the weeds and natural pastures which managed to grow during the rest period.

Almost constant exposure, and the disturbance of soil by cultivation after each shower, left the soil prone to wind action which sifted out the small fraction of finer clay and silt (with plant nutrients attached), leaving behind infertile sand to be blown about in drifts and pile up into dunes. Fences were sanded-up, tanks filled with drifting sand, and homesteads became a hell of choking dust in windy weather. Yards, out-buildings and homesteads were often in danger of being drifted up. Large sums of public money were spent in keeping water supply channels, roads and railway lines clear of drift.

Much of this land was settled in the post-World War I period when wheat was in world-wide demand. By the time the farmers were established, the world wheat market was over-supplied, and prolonged drought was on the land. Farmers, with no other enterprise available to them, over-cropped their land with disastrous results. The depression of the early 1930s was followed by reconstruction of the farms to a size which would allow a prolonged period under pasture between cropping phases. As in the case of wheat lands affected by water erosion, special measures were necessary to control erosion and restore eroded lands to productivity. These measures are also discussed later.

GRAZING LAND

Land used for grazing may be subject to both wind and water erosion. Generally speaking the higher the rainfall the better the pasture cover and consequent protection against erosion. As average rainfall decreases, dry periods become more frequent and longer in duration so that it becomes more difficult to maintain a balance between the amount of feed and the number of stock to utilise it.

From about 1880 to about 1950 the problem of balancing feed and stock numbers was complicated by the presence of the rabbit. Under these conditions pastures were constantly overgrazed. Bald patches of sheet erosion appeared on the slopes, and a greater proportion of the run-off was shed to the drainage lines, causing gullying of the drainage lines and larger watercourses.

This period of over-use coincided closely in time with the widespread over-use of cultivation lands. The increasing use of sown pasture in the postwar period since 1946, the application of superphosphate and the virtual elimination of the rabbit in the 1950s have changed the picture entirely in areas where rainfall is sufficient to support improved pastures. These developments have enabled land use to be more in harmony with the environment. As a consequence the rate of new erosion should have slowed down, but there remains a large legacy of erosion which needs to be brought under control.

In the semi-arid and arid regions the story was similar. Rabbits kept pastures grazed close. Feed disappeared quickly and the term 'rabbit drought' was a common expression. Dust storms were frequent and the evidence of erosion can still be seen in old fences buried in sand and silt.

Undulating country became badly sheet-eroded, the drainage lines being dissected by rills and gullies. The rapid removal of run-off increased the aridity of the environment, tending to greater disparity between the forage available and the animals using the country.

The disappearance of the rabbit following the introduction of myxomatosis in the early 1950s has brought about a very considerable improvement in the condition of the pastures in the semi-arid zone. In south-eastern Australia this has been assisted by a prolonged period of higher rainfall. However, where erosion damage was serious there could be little or no recovery. Large areas of the more fragile soil types have remained seriously eroded as a reminder of past misuse.

The causes of accelerated erosion on grazing land come back to the underlying principle of land use not in harmony with the environment. The drier the climate the more difficult it is to match livestock to the vagaries of seasonal production. Stock numbers need to be at a level which will ensure an adequate carrying-over of feed into the dry periods which are normal to the environment.

The land manager in the arid or semi-arid regions needs to appreciate that serious deterioration is virtually an irreversible process. Perry (1968) has pointed out that the land manager in the high rainfall regions can carry out many kinds of manipulations to improve the productivity of his land. If his pasture is destroyed by overgrazing or other forms of mis-management, he can re-sow it. The opportunities of the pastoralist in the arid zone to manage the ecosystem are very limited. A single period of mismanagement, perhaps forced by circumstances outside his control, may cause serious deterioration from which recovery may be very slow or perhaps impossible.

Research has devised ways of restoring seriously eroded country to productivity. Methods which are applicable to the dryland situation are discussed in the chapter on soil conservation. However, the important lesson to be learnt from this is the need to prevent deterioration of soils or pastures. This requires recognition that there are many different types of country which respond in different ways to pastoral use. Some types of country are fragile and need careful management at all times. Other types of country are stable and will stand limited heavy use without deterioration.

These matters are discussed in the chapter on land use classification. The many types of landscape which may be found in dryland Australia, and their susceptibility or otherwise to erosion, are described in Chapter 9.

FURTHER READING

Condon, R. W., Newman, J. C. & Cunningham, G. M. (1969). 'Soil erosion and pasture degeneration in central Australia. Part I—Soil erosion and degeneration of pastures and topfeeds'. *J. Soil Conservation N.S.W.*, 25 : 1 (47).

Condon, R. W. & Stannard, M. E. (1956-57). 'Erosion in Western N.S.W. Parts I-IV.' *J. Soil Conservation N.S.W.*, 12 : 4 (192); 13 : 1 (17); 13 : 2 (68) and 13 : 3 (122).

Crocker, R. L. (1946). 'The soils and vegetation of the Simpson Desert and its borders'. *Trans. Roy. Soc. South Australia,* 70 : (235).

Fitzgerald, K. (1968). 'The Ord River regeneration project. Part II—Dealing with the problem'. *J. Agric. West Australia,* 9 : 90-5.

Jones, R. M. (1966a). 'Scald reclamation studies in the Hay district. Part I—Natural reclamation of scalds'. *J. Soil Conservation N.S.W.*, 22 : 3 (147).

——— (1966b). *Ibid.* 'Part II—Reclamation by ploughing'. *J. Soil Conservation N.S.W.*, 22 : 4 (213).

——— (1967). *Ibid.* 'Part III—Reclamation by ponding banks'. *J. Soil Conservation N.S.W.*, 23 : 1 (57).

Perry, R. A. (1969). 'Rangelands research and extension', in *Arid Lands of Australia.* (Proc. Symp. Australian Acad. Science, Canberra). Aust. National Univ. Press, Canberra, 1969.

Warren, J. G. (1965). 'Scalds of Western N.S.W.—a form of water erosion'. *Aust. Geographer*, 9 : 5 (282).

CHAPTER 8

Soil conservation in dryland Australia

by R. W. Condon

SOIL conservation has two aspects—prevention and control. Any consideration of soil conservation in dryland areas must be based on the principles and practices applicable to higher rainfall areas, but the practice may often be different because of very different environmental conditions. Of particular interest is the influence of historical factors on soil erosion and the development of techniques especially suited to Australian conditions of climate and land use.

This chapter will be introduced by an outline of the historical aspects of soil erosion and control. This is followed by a brief discussion of the principles of prevention of erosion. The main part of the chapter is devoted to the control of soil erosion by both water and wind, dealing firstly with the design and use of soil conservation structures for the control of water erosion, with special reference to the adaptation of conventional soil conservation techniques to the dryland situation. There is also a description of wind erosion control techniques, again with particular reference to those developed for dryland Australia.

HISTORICAL ASPECTS

Soil erosion is by no means a phenomenon of recent historical times. History provides us with an almost continuous record of the consequences of erosion. Earliest historical records and results of archaeological excavations provide evidence of the changes that have taken place in the lands around the eastern Mediterranean Sea. The early organisation of people into tribes required the cultivation of plants. Existing vegetation was removed to allow of cultivation, and trees were killed to improve the growth of grass as well as for use as firewood and for shelter. Little damage was done to the soils as the tribes were nomadic, but once the techniques of agriculture improved, the tribes settled in fixed places and established cities. It was at this stage that accelerated erosion began.

Biblical records refer to the extensive forests of Lebanon, and to a large population of men and animals which could have only been supported by fertile soils. The fertile soils of the uplands have been removed through exposure to excessive cultivation and overgrazing, and only vestiges of a former healthy civilisation remain.

In later times, the huge increase in population of China and India have resulted in the destruction of forests and barren wastes in the uplands. The rivers draining these countries carry huge loads of silt, progressively burying evidence of human occupation.

In the last century the same process has begun in the new countries of North and South America, South Africa and Australia. The vegetation has been removed or reduced and the balance of the environment upset. The advent of modern agriculture has rushed the process forward at an unprecedented rate by large-scale clearing and cultivation coupled with great numbers of sheep and cattle.

Erosion was recognised and steps taken to prevent it in prehistoric times. The terrace systems of the Incas in Peru are probably the most ancient and extensive scheme to prevent water flowing rapidly downhill. Other terrace systems have been developed, notably in China. The ancient Greek and Roman authors refer to the value of fallow and pasture in improving the next crop.

The alarming spread of erosion in the new countries during the last century has resulted from applying farming methods suited to the environment of Europe to a totally different set of conditions. The arable lands of Europe which gave rise to its agricultural tradition were gently sloping or flat; the rainfall was reliable and of low intensity. Under these conditions there was adequate vegetative protection, the amount of run-off was small and erosion was not apparent.

Other countries with unreliable rainfall and with storms of high intensity presented very different conditions. The vegetation was eaten out during periods of drought or the arable land worked in cultivated fallow to conserve moisture, leaving the soil in a much more vulnerable condition than ever occurred in Europe. Heavy storms on the exposed soil caused serious erosion that was previously unknown. The greatest areas of cultivated land and the largest numbers of grazing animals were concentrated in North America and, as that country had been settled longer than the other new countries, erosion was most serious there. Although various methods were used to check erosion, it was not until the second quarter of this century that any great advance was made.

The history of erosion in Australia has two phases, before and after European settlement. Prior to European settlement the vegetation of Australia ranged from forest in the humid coastal and mountain region, through woodland in the subhumid winterland, into the grasslands and shrublands of the inland deserts. The vegetated areas were lightly grazed by a restricted number of marsupials which were kept in check by Aboriginal hunting, so that the protective cover of vegetation was maintained. The Aborigines were not cultivators—they used only naturally growing plants for food. Their greatest influence on the environment resulted from the use of hunting fires to drive the game, or to attract it to the sweet feed which follows burning. In this environment the pattern of erosion reached stability between climate, slopes and vegetation. This pattern set the shape of the topography.

The introduction of European agriculture, domestic animals and rabbits produced a new pattern through removal of vegetation by cultivation and grazing. Forests were cleared and perennial grasslands ploughed without regard for the result, because the farming techniques and traditions of Britain had been established where no erosion problem had been encountered. The most extensive areas of cultivation are those where wheat is grown and it is there that erosion is most serious. The pattern of erosion damage has been very similar in all States, being governed by two major factors—topography and climate.

The coast and nearby mountain and tableland areas enjoying high rainfall are relatively free of erosion. Much of this country is unsuitable for agricultural development and the native forest is largely undisturbed, providing protection to many important catchment areas. Cleared country is used mainly for grazing and the reliable rainfall maintains a good sward of native grass or permits the establishment of sown pastures. Cultivation is generally confined to alluvial flats along the rivers. Serious erosion has resulted in small areas where cultivation has been extended on to steep slopes for the production of horticultural or vegetable crops.

Further inland, the extensive cultivation of wheat, with its practice of bare fallow and stubble burning, has caused extensive water erosion. Where wheat grazing has been pushed on to the sandy soils in the drier marginal country, there has been serious damage by wind erosion.

Serious damage has also occurred on vulnerable soils in the semi-arid areas of Queensland, New South Wales and South Australia. The gold rushes of the 1850s were followed by a move of pastoral settlement which spread quickly over the western plains of New South Wales and southern Queensland and into the inland areas of South Australia.

The early use of Australia's arid zone rangelands was characterised by the occupation of areas served by natural waters persisting in deep waterholes along ephemeral streams, or from wells in the floodplain of these streams. In good seasons stock could move out to the back country but as temporary waterholes dried up the stock returned to more permanent watering points. The permanent waterholes were also an important link in the drought strategy of the early pastoral barons who moved large mobs of sheep and cattle into the desert areas in good seasons, and returned back along the chain of watering points as the normal condition of 'drought' returned to the desert.

In much of Australia's arid zone the deterioration of river-frontage country has been almost catastrophic in nature, particularly in northern Australia. On many ephemeral streams a consequence of this deterioration has been the silting-up of the old waterholes with sand and silt washed down from seriously eroding catchments. Degeneration of river-frontage country is not so obvious in western New South Wales where the heavy grey clays of the flood plains associated with the major river systems are largely resistant to erosion.

Later use of arid zone pastures saw the development of fencing, which permitted control of the movements of livestock, and of the excavated tank (where catchment conditions and soil-holding qualities were suitable) and artesian and sub-artesian bores where underground water of suitable quality was available. The level topography of much of Australia's outback has also permitted the use of the bore drain as a means of getting water from a single bore into paddocks quite remote from the bore.

The occupation of the dryland pastoral zone was so rapid that by 1887, less than 40 years after the first flocks started moving in, there were 15 million sheep depastured on the 80 million acres of the Western Division of New South Wales. Rabbits increased even more rapidly than sheep and rabbit plagues were a common occurrence in the years after 1884.

The occurrence of severe drought conditions during this expanding phase of pastoral settlement, when stock population was much higher than the country could be expected to carry, caused serious degeneration of the native pastures and in many cases serious erosion. The severe drought of 1864 was said to have destroyed much of the perennial saltbush in the saltbush country of South Australia.

In western New South Wales, the run of five dry years ending in two years of intensive drought in 1901-02 devastated pastures and soils. The effect of the combination of excessive stock numbers, very serious rabbit infestation and prolonged drought is best described in the words of the Royal Commission (1902) appointed at that time to enquire into the serious depression which had decimated the pastoral industry in the Western Division.

> The effect of the operations of the rabbits and 'overstocking' during recent years has been the destruction of almost all of the vegetation on the face of large areas of the drought-stricken country. The prevailing westerly winds, under these conditions, have produced sand-storms the effects of which it is no exaggeration to describe as calamitous . . . The manager of Teryawynia Station, in the Wilcannia district, states that, out of the leasehold area of 460,000 acres, an area of 100,000 acres, is as bare as a floor in spite of a great rain which they have had . . . On Tarella Station, Wilcannia district, in some instances soil has been carried away to a depth of from 4 and 5 inches to a foot . . . Witnesses also stated that large quantities of feed had been destroyed by being buried under the sand deposits . . . In many instances rabbit-proof fences have been completely buried. In some cases 'a second storey' has been added to the buried fence, and one witness mentioned an instance of the second fence being almost covered . . . Sheep yards have fared the same as fences, being in numerous instances completely buried in sand . . . The most serious form of the evil at present, perhaps, is the filling-up of tanks and drains. A 4,000-yard tank on a plain, one witness stated might fill up in three months. Twelve feet of sand had been deposited in one of his tanks in that period. Another witness stated that upon his holding in the Mossgiel district he left one morning some weak sheep watering at a 400 or 500-yard tank. A duststorm occurred during the day, and by night-time all traces of the tank, except the embankment, had disappeared, a number of sheep having been buried alive.

Time and favourable seasons, especially the lush period from 1950 to 1956, have healed many of the scars of the 1902 drought, and much country described by witnesses as 'windswept and barren' is now producing again, albeit with much less vigour. The great majority of the erosion and pasture degeneration evident today can be traced back to the period of abuse towards the end of the nineteenth century.

The serious erosion which accompanied the agricultural and pastoral settlement, especially in the wheat belt and pastoral zones of New South Wales, South Australia and Victoria, was largely the result of a lack of understanding of the nature of the environment. Spasmodic attempts to control erosion have been made in the 1920s and 1930s and so little was known of the principles of erosion control that these attempts often created additional erosion problems.

Soil conservation made little headway until the New South Wales Government sent Mr E. S. Clayton to the United States of America in the late 1930s to study the administrative and practical aspects of soil erosion control in that country. It is interesting to note that the U.S.A. had gone through a similar pattern of settlement of the drier agricultural and pastoral regions.

New South Wales established a Soil Conservation Service in 1939 and began the task of adapting the soil conservation techniques imported from the United States of America to the Australian conditions of climate, land use and farming methods. Other States followed suit so that most Australian States now have a soil conservation organisation able to offer landholders advice and assistance on soil conservation and to carry out research. It is again stressed that landholders proposing to undertake soil conservation work on their properties should seek assistance from their appropriate State authority.

PREVENTION OF EROSION

Prevention of erosion requires an understanding of the environment and the adoption of land use practices in harmony with the environment. As the environment is constantly changing, both in time and space, there also needs to be recognition of erosion at an early stage, so that preventive measures may be taken or control measures installed.

An understanding of the environment requires a knowledge of soils and their likely behaviour under various forms and levels of land use. It requires a knowledge of the natural vegetation cover, and what it can tell in regard to soil characteristics and local and broad climatic conditions. It also requires a knowledge of the climate and how it may affect soils and vegetation under adverse conditions, and the limitations imposed by climate on cropping.

Understanding of the environment therefore requires a survey of the resources. At the farm level this is best accomplished by the implementation of a farm plan based on a survey of the land resources, and mapping according to the capability of the land for various forms of land use and its needs for protection against erosion. This method of land use planning has been described in the chapter on land use classification.

Given the inventory of resources provided by the farm plan, the landholder is able to establish priority in development, use and protection of the whole farm, and plan these in accordance with his financial resources. The plan may show that the first priority is the protection against erosion of land already developed. This may require structural soil conservation measures or it may require an adjustment of cultural practices. Sometimes it may mean re-design of subdivision to conform closely to natural boundaries.

On the other hand the plan may show that certain undeveloped land could be safely developed for cropping or improved pasture, or that land currently being cropped should be retired to improved pastures or natural grazing. Agriculture and soil conservation authorities in all Australian States are able to provide farm plans of this type as part of their advisory services.

The adoption of wise land use practices also depends on the understanding of the environment. It means the adoption of practices in accordance with the capabilities of the land and its need for protection against erosion. On arable land this means adequate pasture periods in cropping rotations to ensure maintenance of chemical fertility and physical condition. It also means contour cultivation on sloping lands, and the protection of steeper lands, or lands more susceptible to erosion, by structural soil conservation measures. On grazing lands it means conservative levels of stocking designed to leave vegetative cover on the soil and have feed available at all times. This implies conservation of fodder and use at times of shortage to avoid over-grazing of pastures. Wise land use on grazing land also requires careful observing of the condition of the pastures, especially on more vulnerable areas. Poor condition of pastures will be reflected in poor condition of livestock. If the stock owner waits until his animals are showing poor condition, the damage has been done—both to the pastures and to the value of the animals.

Recognition of erosion requires that the landholder is able to recognise the early symptoms and do something about it before the loss of productivity or the cost of control becomes too great. The best time to look for early symptoms on crop lands is after heavy rains, when the scars from sheeting or rilling are still fresh. The scars may be hidden by subsequent growth but will reappear next time the land is stressed by drought or heavy rain. The most vulnerable period for crop lands is during the fallow or early crop stage, especially if this coincides with the period when heavy storms may be expected.

On grazing lands the vulnerable period is when growth has stopped due to drought and pasture cover has been depleted by grazing. If the drought-breaking rains are heavy, the exposed soil will be battered by raindrop splash and swept by surface flow. Patches of bare soil, or patches on which the response to rain is poor, are early signs of sheet erosion. Shallow rills are early signs of gullying.

When the signs are recognised they should not be disregarded in the hope that nature will effect a cure. Nature can only effect a cure if everything is on her side. Good seasonal conditions and light stocking may effect a cure in the early stages, but it is best to seek professional advice immediately from a soil conservationist. His experience in a wide range of situations will help him decide whether the situation is mild or serious, and what type of measures are necessary. There are ample reminders around the Australian countryside that soil conservation is not a 'do it yourself' job.

APPLICATION TO DRYLAND AUSTRALIA

The broad principles behind the prevention of erosion in higher rainfall areas are also applicable to dryland areas with very little difference in application.

Sloping croplands in semi-arid areas are just as susceptible to water erosion and the same techniques are required for wise land use. Croplands on sandy soils subject to wind erosion need an adequate pasture phase or period of rest to restore soil nitrogen and re-build physical condition. Special cultural practices may be necessary such as ridging or leaving in the rough ploughed state during the windy period, or the use of stubble mulching so that the residue from last year's crop is used to protect the soil during the preparation for the next crop.

On arid and semi-arid grazing lands the early symptoms of erosion are the same—the appearance of bare spaces with no response to rains. In country where a lot of bare space is a natural feature, the presence of erosion is indicated by an increase in the proportion of bare space. Details are given in the chapter on erosion processes.

The increase in unpalatable weed species is invariably a sign that the land is being over-grazed and productivity will be reduced, either by erosion, or by invasion of useless species. Such invasion may serve a most useful purpose in protecting the soil against erosion.

Action to prevent erosion when the early symptoms are recognised require either resting from grazing until nature has a chance to effect a cure, or reducing grazing pressure. Much can be done by the recognition of different types of country, especially those which are highly susceptible to erosion, and management according to its capabilities (in terms of forage production) and needs (in terms of protection). Recommendations on the management of dryland grazing areas to prevent erosion are outlined in the chapter on land use classification.

CONTROL OF EROSION

Under this heading the various structural measures used for the control of erosion are discussed. There are a wide number of techniques available and the choice of which to use depends on circumstances at the site—the type of land use, the nature of the problem, the contribution of surroundings to the problem, climate, soil type and others. The cure will require an understanding of hydrology, which encompasses such things as rainfall intensity, the time taken for water to flow from one point to another, and the effect of channel size, grade, and degree of roughness of the channel floor on the volume of flow which can be carried by a channel. This is why it is important to seek professional advice in handling erosion problems, whether they be caused by wind or water. Only with the knowledge available to the professional soil conservationist can the problem be properly evaluated, and control measures designed to give maximum effect for minimum cost.

Soil conservation techniques may be divided into two broad groups—those used for control of water erosion and those used for control of wind erosion.

CONTROL OF WATER EROSION

In dealing with the various soil conservation techniques for the control of water erosion it is proposed firstly to describe briefly the techniques applicable to higher rainfall areas, most of which are applicable to some degree in arid areas, and later deal with the special techniques which have been developed for use in arid areas.

When a soil conservationist inspects a property to evaluate an erosion problem and design control measures, his principal thoughts are of run-off:

—Where does run-off come from?
—How much run-off and how often?
—Can run-off be reduced in any way before it reaches the problem area?
—Can land use be improved?
—Can structural soil conservation measures be used?
—Does the land upslope belong to someone else?
—How is run-off to be handled within the problem area?
—How is run-off to be disposed of once it has passed through or around the problem area?

These thoughts on run-off have to be tempered by consideration of costs of installing structures or changing land use—the prevention of damage which may be reducing productivity or increasing costs—or improvement in productivity which may result from control and better use of run-off or changes in land use.

DESIGN OF SOIL CONSERVATION MEASURES

The principal erosion control measures are designed to prevent water running directly down a slope, where it would increase in volume and velocity and thus in erosive power. There are various types of structures suited to different conditions, but they have a number of features in common.

The primary object of all types of banks or furrows is to divide a slope into small sections to prevent run-off increasing to sufficient volume and velocity to abrade the soil and to carry soil particles. The spacing of banks is determined by the distance water may travel before it causes erosion. On soils resistant to erosion the safe spacing is wider than on susceptible soils. Arable land requires a closer spacing than pasture because the loose bare soil erodes more readily than compact soil protected by grass. Steep slopes require closer spacing because run-off is greater and erosive velocity is reached in a shorter distance due to the slope; also the channel above the bank holds less water on steeper slopes. The second object of banks is to hold water so that it has time to be absorbed into the soil. A third object is to reduce the velocity of the water thus reducing both its erosive and transporting power so that soil in suspension settles. In the case of banks designed to divert water a further object is to carry the run-off at low velocity to an outlet where it may run down the slope without causing erosion.

In designing pasture furrows the main considerations are the capacity of the individual furrows and their spacing in relation to the steepness of slope and the amount of run-off. The amount of run-off to be expected varies according to the amount of rainfall and the condition of the surface.

The design of size and spacing of absorption banks depends solely on the total amount of run-off from the rain storm for which provision is to be made.

Dams are also impounding structures and their size will depend upon the amount of run-off to be expected from the catchment. However, it is also necessary to provide a spillway to carry flood flows which may be experienced when the dam is full. The spillway therefore must be designed in accordance with the requirements for structures required to carry flowing water.

For these structures it is necessary to calculate the amount of flowing water they will have to carry. This will govern the size and spacing of graded and diversion banks, it also being necessary to ensure that the velocity of flow will not be great enough to cause scouring of the channel. In the case of waterways and spillways it will be necessary to calculate the width required to carry the flow at a safe velocity, this being governed by the degree of slope and the amount of cover, the cover acting to reduce velocity as well as protect the surface from scouring.

A number of formulae have been developed to assess the amount of flow that a structure will be required to handle. The simplest of these is

$$Q = CIA$$

in which the discharge or amount of flow (Q), in cubic feet per second, is the product of three factors—the coefficient of run-off (C) expressed as a decimal representing the proportion of rain not absorbed by the soil; the rainfall intensity (I), being the amount of rainfall (in inches per hour) which can be expected during the time that it takes water to flow from the furthest point of the catchment to the site under consideration; and the area (A) of the catchment in acres.

Values for C differ for different types of land use and vegetative cover, being low for forest country with a deep bed of litter and high for undulating arable land. The value for I also differs in accordance with the importance of the structure. For a bank which is relatively cheap to construct and repair the rainfall intensity factor is for a storm which might be expected once in 5 years. For a bridge the rainfall intensity factor might range from a 20-year storm to a 50-year storm, depending on such things as cost of construction, cost of replacing or repair, anticipated life, public safety and inconvenience in the event of failure.

STRUCTURAL SOIL CONSERVATION MEASURES

Structural measures, or earthworks, are designed to control run-off by one or more of the following ways:

(a) by holding it on the land—contour cultivation, pasture furrows;
(b) by leading it slowly across the slope so that it cannot cause erosion—graded banks, diversion banks;
(c) by storing it in special structures—level absorption banks, storage dams;
(d) by leading it down the slope in specially-prepared waterways.

The various types of erosion control measures in common use in Australia, and the circumstances in which they are used are described below.

1. *Contour cultivation*

In this method, which is used on cultivation land, contour lines are surveyed at intervals down the slope, and all cultivation is carried out on the contour between these lines. Each tyne mark or plough furrow then serves to hold excess run-off, giving it more time to soak in, and providing more moisture for crop growth.

This method is used only on very gently sloping lands (less than 5 per cent slope) which are not suffering from erosion. The contour guide lines are commonly marked out by constructing a small bank by using a disc plough throwing soil inwards. However, as the banks so constructed are close to the contour, they are likely to be overtopped in heavy rains. If this happens serious rilling could be initiated downslope from the breach.

A technique developed for cropland on the fringes of the wheat belt in central-western New South Wales used contour ditches as guide lines. Instead of having soil thrown in to form a bank, the disc plough was used to throw the soil outwards so that a small ditch was formed. Any tendency to form a bank can be prevented by cultivating close up to each side of the ditch. Grades in the ditch can be varied so that run-off tending to concentrate in valley areas or drainage lines can be directed out into the adjoining slopes, where it would be spilled over an unlimited length of level ditch.

Apart from conserving soil this technique has the advantage of making maximum use of extraneous run-off to the benefit of the crops or pastures. It also ensures uniform moistening of the whole paddock and avoids the situation where part of the paddock may be too dry to cultivate while machinery is getting bogged in other parts.

2. *Contour (or pasture) furrows*

These are furrows surveyed along the contour and constructed with a plough, or with a ripper followed by a grader to clean out the furrow. They are used only on grazing land, and are designed for the purpose of holding water which cannot be absorbed by the soil and which would otherwise run off.

In addition to run-off, contour furrows hold valuable seed, fine soil, organic matter and soluble plant nutrients which would otherwise be carried away by run-off. Consequently they promote a vigorous growth of pasture along the furrows, hence the name 'pasture furrow', and are particularly valuable for reclaiming sheet-eroded land or improving the vegetative cover and grazing capacity of land with poor surface cover.

In the higher rainfall areas pasture furrowing has been largely replaced as a soil conservation technique by improved pastures. Aerial sowing and fertilising have made it possible to establish improved pasture on almost any terrain, and provide a double benefit of run-off control with improved productivity. Contour furrows still have a place where sown pastures cannot be established and their use in semi-arid regions will be discussed later.

3. *Banks*

These are earthworks constructed with a bulldozer or grader, usually with the borrow pit on the top side serving as a channel, with the spoil thrown on the downhill side to form a bank.

Banks are of three main types according to the purpose for which they are used.

(a) *Graded banks.* These are banks constructed on a grade of 3 to 6 inches per 100 feet (0.25-0.50 per cent) and serve to carry excess run-off from the land at a low velocity so that it cannot cause erosion. They are generally arranged in series down the slope, serving to break up the slope into short sections so that run-off flowing straight down the slope cannot reach an erosive velocity.

The size of the bank will depend upon the area of the contributing catchment, but as it is generally easier to maintain a standard size of bank, the area of catchment is varied to suit the bank by varying the interval between banks. This interval may range from 1½ chains on steep slopes (8-10 per cent) to 3 or 4 chains on gentle slopes (2-4 per cent). Because the banks carry water from one point to another it is generally necessary to construct specially-prepared waterways to carry the water discharged from the banks safely down the slope.

(b) *Diversion banks.* These banks are used in situations where it is necessary to carry large volumes of water from one place to another, e.g. into or out of farm dams, to or from road culverts, or any other place where water may be concentrated.

They are most frequently used to protect arable or grazing lands lying below steep hills which are too steep, rocky or thickly-timbered to treat with graded or absorption banks. Large diversion banks are built to carry the run-off from such areas across the slope to a point where they can be safely discharged into a waterway, dam or natural watercourse. Such banks may have a cross-sectional area ranging from 20-40 square feet, and are constructed on grades similar to graded banks, that is 3-6 inches per 100 feet. Their carrying capacity is determined by the amount of run-off expected.

(c) *Absorption banks.* This type of bank is used where it is possible to absorb all run-off except from very big rains, or where a suitable safe discharge point is not available to enable use of diversion or graded banks.

They are level banks, located on the true contour. One or both ends of the bank are normally partly open so that if their capacity is exceeded, they may overflow at a pre-determined point where it will do no damage. The size and spacing of absorption banks depends solely on the total amount of run-off to be expected from the catchment area. This type of bank is widely used on grazing land, and is often used in series down a slope, trapping all run-off except that from the occasional exceptional storm.

Special types of banks have been developed for use in semi-arid areas. These will be discussed later.

4. *Grassed waterways*

These are specially-prepared structures designed to carry large volumes of water straight down the slope without causing erosion. They are prepared by throwing up a small levee bank on each side of a suitably grassed area.

If a good naturally-grassed area exists, this is left undisturbed and the cover improved by sowing special grasses and fertilising. Good naturally-grassed areas are rare on cultivation land, and it is generally necessary to cultivate the area and sow down to special grasses. Where the waterway is to be fed by banks, as is usually the case, it may be one to two years before the cover is sufficient to allow the banks to be built and discharge into the waterway.

Once established the waterway has to be carefully managed to maintain a good cover. Waterways are thus fenced off so that stocking can be carefully controlled. As such areas receive frequent irrigations they yield large quantities of feed.

Waterways may be used to carry the discharge from series of banks on either cultivation or grazing land. Natural watercourses are used where the run-off can be discharged safely into them without causing erosion.

The most suitable grasses for sowing down a waterway are phalaris, paspalum, couch, Rhodes and kikuyu grass, preferably sown with a legume to maintain vigour in the sward.

5. *Dams*

Dams may be used in erosion control for many purposes.

(a) *Silt dams.* Are used frequently in the catchments of large storage reservoirs to trap silt and prevent it reaching the storage. Small silt traps are also used in front of farm dams with badly-gullied catchments. All the silt carried by the run-off is deposited in the silt trap when it slows the velocity of flow. As the trap fills, silt-free water flows into the main dam. Such silt traps are built shallow so that they can be easily cleaned out from time to time.

(b) *Storage dams.* Are frequently used as part of an erosion control scheme to reduce the amount of run-off discharging from a catchment area, or on to a waterway system. They are also used to control gullying by placing at the head of a gully, and so preventing its extension by drowning the head. They may also be located above the head of a gully and the spillway located so that the overflow will not fall back directly into the gully. Such dams may also be used as stock water supplies or homestead supplies and thus serve a dual purpose. Care must be taken to ensure that the spillway and any structures below the dam are able to carry the full flow expected from the catchment, as the dams may be full when storm rains are received.

(c) *Flood control or cushion dams.* Flood control dams, whether they be large or small, must be empty at the time the flood rains begin. They are generally used below steep catchments delivering a large volume of run-off. They may be fitted with siphons or pipes through the wall fitted with valves so that the dam can be emptied as soon as the flood flow passes. Others are fitted with small pipes without valves so that they begin to empty as soon as the water level rises above the inlet of the pipe. The small flow from the pipe, often called a 'dribble-pipe', contributes little to the general flood flow and the dam empties automatically.

The peak flow of the flood is reduced by the amount of water held by the dam. Such dams must also be fitted with normal spillways so that flood flows in excess of the dam's capacity may be by-passed safely. Flood control dams of this type are rarely used on farm lands as farm dams are almost invariably used for water storage. However, the same principle is often used for irrigation of land below the dam, the irrigation water becoming available by turning on the valve or siphon.

6. *Gully filling*

Gullies may range in size from rills less than a foot deep and wide to near-canyons 30 feet deep and 100 feet wide. Small gullies may be filled in with a plough or grader but rippers and bulldozers are necessary for large gullies.

On crop lands gullies are usually filled to enable cultivation and access. The decision to fill any gully depends on the relationship between costs of filling and the increased value, which includes return from production, convenience, saving in time or reconstruction of roads or fences, and other factors.

To fill a gully, and bring it to a level with the surrounding soil, it must be remembered that top-soil has to be scraped from a very large area to provide material for filling. Ideally sufficient top-soil should be stock-piled so that the exposed subsoil can be covered thinly when filling and shaping is completed. All run-off entering the gully must be diverted to another area where it may be discharged safely, otherwise the loose soil in the freshly-filled gully will be washed out again.

If a gully is too large to warrant filling it should be fenced off, stock excluded, and vegetation encouraged by seeding, fertilising or planting trees so that it remains stable.

7. *Gully control structures*

Structures used for the control of gullies may be divided into two types:

(a) *Structures which carry water from the head or side of the gully down into the gully*. These structures may be built of concrete or stone to protect the soil where the water flows at an erosive velocity, and must have a stilling pool at the base to reduce the velocity again. They are very expensive, and trouble is often experienced in obtaining a close bond with the underlying soil, so that they may be easily undermined. More recently, vegetated flumes or chutes have been used and are generally more satisfactory. These are gently graded at a slope of 1:6 or less, and stabilised by planting sods of kikuyu grass to protect the soil.

(b) *Structures in the floor of the gully*. These may be built of timber, brush, wire netting, loose rock or concrete. They are generally intended as temporary structures to trap silt on the gully floor and build it up to the level of the structure. The structures are placed at intervals along the gully so that the grade from the bottom of each to the top of the next downstream is stable for the soil type concerned.

These structures are also difficult to bond into the soil on the sides of the gully, and many failures have resulted from water undermining the sides of the structure. They are also very expensive and make no attempt to control the problem at its source.

They have been used commonly in the control of roadside gullies, but because of the many failures and the high cost of the structures, their use has been discontinued in favour of treatment of the catchment with conventional soil conservation techniques. Even though this may involve treating a very large area, it is generally cheaper, is much more effective and of much greater general benefit.

8. *Special soil conservation measures*

There are many situations in which soil conservation practices normal to farming and grazing lands are not practical or need adaption to meet special needs. One of these is the dryland situation, and the adaptations and special techniques applicable to such areas will be discussed in detail later. There are many other situations in which special techniques for special problems warrant brief mention. Most of these are concerned in some way with engineering.

(a) *Roadside erosion*. Roadside gullies add very greatly to the cost of maintaining roads in trafficable and safe condition. The extension of gullies into adjoining paddocks, both upslope and downslope contributes to serious and costly erosion on private lands.

Where damage is caused by run-off from adjoining lands, these can be treated by normal soil conservation techniques, to the advantage of the road authorities and the landholder. Roadside areas should be reconstructed, sown to stabilising vegetation and fenced to exclude stock. Where grades are too steep to enable run-off to be handled safely, diversion banks can be used to disperse run-off safely over adjoining lands, or absorption banks to hold all expected run-off.

Often it is necessary to handle the flow from a large catchment in a confined space. Special engineering structures may be used, or nature may be aided by using straw mulch or other temporary surface protectors such as bitumen emulsion or fibreglass matting until vegetation is safely established.

These latter techniques are used where high road batters or cuts and fills need to be revegetated to prevent them eroding, but may also be used for flumes and chutes designed to carry flow down steep slopes for short distances.

(b) *Construction projects.* The removal and replacement of large volumes of earth in major engineering projects leaves extensive areas of denuded soil exposed to the elements. Such projects would include borrow pits for dam construction and road maintenance, opencut mining operations, expressway interchanges, batters associated with highway cuts and fills, and suburban real estate developments.

The techniques appropriate to these situations were introduced into Australia to facilitate stabilisation of engineering projects in the Snowy Mountains Scheme. Extreme conditions arising from frost, snow, high winds and short growing season in the sub-alpine environment made stabilisation measures both very necessary and very difficult. The many different techniques have been described by Durham (1956) and later developments by Condon and Clothier (1968a and 1968b).

These techniques usually embody some form of run-off control to minimise surface flow over denuded surfaces, and the establishment of vegetation under a protective cover of straw mulch. The latter protects the soil against raindrop splash, and ensures ideal conditions for growth and early establishment of plant cover. In sub-alpine and alpine regions the mulch also insulates the soil against the pulverising effect of frost-heave. In situations exposed to high winds it is necessary to hold the straw down with netting.

To enable treatment of large areas at reasonable cost, special machinery has been developed which is able to blow a mixture of short straw and bitumen emulsion to provide a mulch over exposed surfaces. This might precede or follow the application of seed or fertiliser. A still later development has been the use of a hydro-mulcher, in which seed and fertiliser are added to a large tank of water and mixed with cellulose fibre or similar material. The whole mix is then sprayed on to the area to be treated.

(c) *Streambank erosion.* This form of erosion becomes a serious problem on the alluvial reaches of streams, particularly on New South Wales coastal streams and especially where they follow a meandering course. In time of high flow the banks are undercut on the bends and large sections of earth fall into the stream. The condition may affect large strong-flowing rivers, or it may affect small tributary creeks which may flow only occasionally.

The collapse and disappearance of areas of rich alluvial land from some reaches may be accompanied by the deposition of infertile silt and sand over rich alluvial flats in other reaches. Structures and buildings may be undermined as soil undercut from banks leaves shoals of sand and gravel which may direct damaging flows on to previously stable areas. The deposition of silt, sand and gravel in the stream bed also reduces the capacity of the channel to carry flood flows.

There are two methods of approaching the problem of protecting the banks of streams. One involves the covering of the bank with protective material to hold it in place, while the other depends on deflecting the main current away from exposed banks so that the velocity of the water in contact with the bank is reduced and its erosive power minimised. In both cases the bank should be battered to a stable slope to prevent collapse of the steep section.

The protective materials used are either vegetation or stone. The banks of small streams may require no more than a cover of sward-forming grass on the re-formed slope. Fibrous rooted trees such as willows or poplars are effective in binding the soil.

Stone covering is usually laid loose and uniform or it may be grouted with concrete. Where large rocks are readily available they may be dumped in irregular fashion for best effect as they give a wider zone of low velocity out from the bank and on broad streams give better protection from wave action.

Deflection of the current is usually done by building groynes pointing downstream. These may be permeable or impermeable, and may be constructed of concrete, rock, piles or mesh. The permeable types are safer, as the pressure of water against them is reduced, there is less turbulence around their extremities and they collect silt, which reforms the bank and permits vegetation to establish.

SPECIAL MEASURES ADAPTED FOR WATER EROSION CONTROL IN DRYLAND AREAS

1. *Contour furrows and similar techniques*

Since pasture improvement by aerial seeding and fertilising became general on higher rainfall country it has superseded contour furrowing as a means of controlling run-off and restoring depleted land. Where pasture improvement is not possible for climatic reasons, contour furrowing still has a place as a means of reclaiming sheet-eroded country.

Experimental work in severely sheet-eroded red loams at Cobar in western New South Wales, showed contour furrowing to be very successful in restoring productivity (Cunningham, 1967). It was found that a horizontal interval of 5 feet between furrows gave a similar response to furrows at 2 and 3 times this distance. This means that closer spacing would give the best total forage yield.

To minimise the cost of taking levels, contour lines are run at 2-3 chain intervals, the area between being filled in by eye, each 'bay' being progressively split in half. The furrower is lifted from the ground for a few feet every chain, and these 'breaks' are staggered so that if a length of furrow fills and overflows around the ends, the overflow is caught by a length of furrow immediately below.

Short lengths of furrow overcome the problem inherent in continuous furrows ploughed by eye. A long length of furrow slightly off contour will drain water towards the low spots and probably break over and drain most of the water out of the furrow. This overtaxes the next furrows downslope and causes them to break down the whole slope.

Small run-off plots at Cobar showed that run-off to range up to 80 per cent for individual rains with an average of 20 per cent over a period of six months. A year-long average of 20 per cent means that the run-off from the 5 feet of inter-furrow space would have the effect of increasing the amount of water received by the furrows to double the annual rainfall.

The greatest benefit comes from storm rains. A fall of 100 points with 50 per cent run-off is equivalent to 350 points in furrows at a spacing of 5 feet. This explains why the vegetation in the furrows remained green throughout the 1965-67 drought in the area.

Contour furrowing has been used on an extensive scale with good results by the Government of Western Australia in a scheme to restore the seriously-eroded uplands of the Ord River catchment (Fitzgerald, 1968).

Furrowing gives a reasonable response only on eroded soils with high run-off. On slopes of less than 1 per cent, the result would be disappointing. Trials on sloping friable clays gave a very poor response because of lack of run-off.

There are other techniques similar in nature to furrowing which may be successful on different types of country. Elongated shallow pits can be made by a disc with two large scallops removed from opposite sides of the circumference. These scoop out a shallow basin 3-4 feet long, 3-4 inches deep and 6-9 inches wide. Several discs in a gang give several lines of staggered basins which fill with run-off and overflow safely around the ends when filled. This technique requires no levelling. Arid zone graziers who have used the scalloped discs report good results on some soils, poor results on others. As with contour furrows it is probable that a high run-off situation is necessary for success.

Deep ripping is successful in some soils and not in others. Soils which responded well at Cobar to furrowing slaked over too quickly after ripping so that the beneficial result lasted less than 12 months.

2. *Spreader banks*

The management of run-off for erosion control in croplands in higher rainfall areas depends either on the storage of water in large absorption banks and in dams, or the use of graded banks to carry run-off safely from the land. In the latter case grassed waterways are necessary to carry flow discharged from the banks directly down the slope.

To be successful, waterways need to be thickly grassed. As the climate becomes drier, the chances of establishing a well-grassed waterway are lessened until it is no longer feasible to use waterways in the design of erosion control works. In drier areas it is also desirable to make every drop of water work for crop and pasture growth rather than store it behind banks until it evaporates or soaks away.

Another requirement of soil conservation structures in dryland areas is that the benefits must exceed the costs. This is often difficult where the level of productivity per acre is low. Assuming a grassed waterway could be successfully established, the cost of installing conventional soil conservation structures would be too great in relation to the value of the land and its level of productivity.

Contour cultivation is a practice that can be undertaken with little or no extra cost. The contour or level bank was an early development in dryland cropping areas, often constructed with a disc plough by the farmer against the advice of the soil conservationist who may have intended small banks as guide lines for contour cultivation. This led to the development of the contour ditch (see page 146) as a guide line and also a small water spreader which could not cause damage if its capacity was exceeded. A technique developed from this idea was the spreader bank (Newman, 1963), in which the spoil is pushed uphill to leave a channel on the downhill side. By using normal downhill push construction, banks could be used to pick up flow concentrated in drainage lines or in eroded gullies, and transport it on a safe grade to a point where it could be safely spread. At this point the construction is changed to uphill push so that the channel continues in the same line. The grade of the channel is gradually reduced to level, or zero grade, and continued at zero grade as far as required.

In operation, run-off water which would normally flow down the drainage line or gully is picked up by the channel and flows along the graded section and then along the level section to the end of the channel. When the level section is filled, it overflows uniformly along its whole length. In this way, run-off water creating erosion problems in valley areas can be carried away and used on adjacent slopes for crop and pasture growth. Cultivation on the contour, parallel to the banks, aids the infiltration of this additional water into the soil. Further banks may be located downslope, leaving gaps in the uphill push banks to allow water through to be redistributed by the level channel. Contour ditches may also be used so that banks can be widely spaced. The spreader bank has been applied widely in undulating wheat lands on the fringes of the wheat belt in central western New South Wales. It provides an effective method of protecting crop land where slopes are gentle but very long. It promotes maximum use of water for crop and pasture growth.

There is a limit to the degree of slope to which the technique can be applied, as the steeper the slope the smaller the capacity of the channel in an uphill push bank. It has been used quite successfully on slopes as steep as 8 per cent where there is no extraneous catchment. On slight slopes of 2 per cent or less the banks have a very considerable capacity. In this situation, spreader banks are able to handle large flows of several hundred cusecs and spread them safely over valuable land which would otherwise be seriously dissected by gullies if cultivated without the protection of spreader banks.

In more arid areas where it is not normally possible to grow crops, spreader banks can be used to harness the occasional prodigious run-off and use it for the production of fodder crops or grain crops or improved pastures (N.S.W. Interdepartmental Committee, 1969). It is necessary to issue a warning that water-spreading cannot be accomplished by merely pushing a bank across a flattish area. To avoid damage, and to get maximum effect with a minimum of construction, careful attention has to be given to design and construction.

CONTROL OF WIND EROSION

Control of wind erosion is based on the same broad principles as applies to control of water erosion—structures of some kind are used to provide some degree of protection, while stabilizing vegetation is being established. Wind erosion has several forms ranging from freely mobile drift and dune sands, often with no 'base' or firm surface, to cement-like surfaces with little or no moving sand or silt. Structural methods used in the moving sand situation are concerned with the control of wind, while those used on the smooth exposed surface with relatively

Top: Checkerboard furrowing is used as a means of reclaiming scalded areas. Water held on the bays and in the furrows promotes regeneration of native or sown species. To be successful under most conditions, the initial furrowing must be followed up by further furrowing treatments (see fig. 2).

Centre: This area has been furrowed on a half-chain grid, and produced excellent early results. However the furrows have become silted up with windblown and water washed silt and the furrow-ridges slaked down by water erosion. Plants are dying as moisture relations become worse and the area will revert to scald.

Bottom: A comparison of results with a single checkerboard furrowing in the foreground on a half-chain grid, and a progressive checker-board furrowing in the background. In the latter treatment the bays are subdivided by further furrowing each year to provide freshly culti-vated soil for colonisation by plants and to give more effective control of water.

(Photos courtesy Soil Conservation Service of N.S.W.)

Plate 18 SOIL CONSERVATION—1

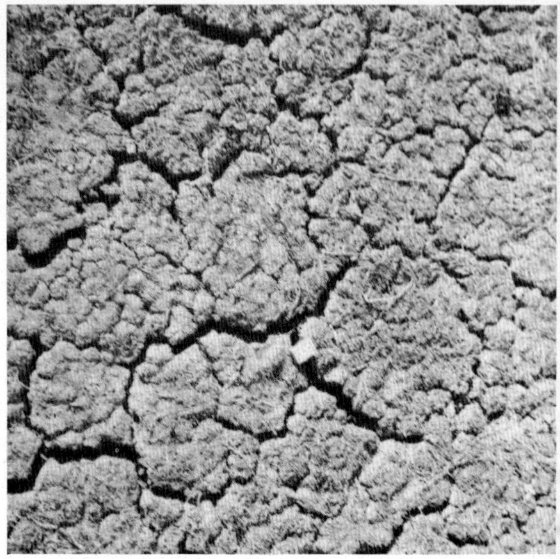

Top left: A close-up of a typical scalded surface. There is a firm crust on the surface composed of silt sand and clay, broken by fine cracks. Beneath the crust the soil is of a fine blocky structure, and generally easily disturbed. The surface seals on wetting and most of the rainfall runs off. Seed falling on the scalded surface is easily blown off by wind or floated off by water.

Top right: The main requirement in reclaiming a scalded surface is to retain water and trap seed on the eroded surface. This photo shows a close-up of a formerly scalded surface after repeated flooding as a result of water ponding. Alternate wetting and drying has caused deep cracking and the rough surface is able to trap seed and water penetrates deeply. This surface was covered with thick vegetation which was removed by fire prior to the photograph being taken.

Centre: Water ponding has been successful in reclaiming scalds and returning them to productivity at reasonable cost. A small bank built with a road grader is used to pond 3-6 inches water over the scald surface. The shallow flooding causes the clayey soil to crack after drying so that it can effectively trap seed. The frequent flooding causes rapid colonisation by native and sown species.

Bottom: A further example of water ponding showing ponded water backed up for a considerable distance above the bank. In this instance the bank has been constructed after an initial colonisation by checkerboard furrowing.

(Photos courtesy Soil Conservation Service of N.S.W.)

Plate 19—SOIL CONSERVATION—II

Top left: Staggered contour furrowing has been successful in reclaiming severely sheet eroded red earths in western New South Wales. This photo shows the initial colonisation of furrows by thistles and weeds which serve to provide protection and improve soil conditions for the normal pasture plants of the locality. The effect of furrowing is to trap the run-off from the inter-furrow spaces. At a spacing of 5-6 feet an inch of rain yielding 30 points of run-off from the inter-furrow spaces would put the equivalent of a further 1½ inches of water into the space occupied by the furrows. The total rainfall over the area influenced by the furrow would then be approximately 2½ inches.

Top right: The same area as top left taken from a slightly different angle some 4 years later. Grasses dominate in the furrows, although thistles are still prominant, and serving a very useful purpose in protecting the more palatable species from grazing. There has also been a good development of cover on the inter-furrow spaces contrasting with the bare condition apparent in figure 1. Because the furrows trapped run-off from small showers, the vegetation along these furrows was fresh and green throughout the 1965-67 drought.

Bottom: An excellent result some 4 years after contour furrowing in 1947 on severely sheet eroded red earths in western New South Wales. A broad swathe of native grasses and herbage occurs along each furrow, with annuals colonising the inter-furrow spaces. Furrows may need replacing after 10-12 years.

(Photos courtesy Soil Conservation Service of N.S.W.)

Top: Special techniques have been developed to control water erosion and at the same time to make effective use of excess water on marginal wheat lands in western New South Wales. Graded ditches, constructed with a disc plough, serve as guide lines for contour cultivation and also direct excess water away from valley areas out towards ridges.

Centre: A diversion spreader-bank picks up the overflow from a farm dam and carries it out on to the lower slopes away from a gullied drainage line. At the change-over point shown in the picture, bank construction has changed from down-hill push to up-hill push but the channel is continued through in the same line. Run-off water flowing along this channel spills in a shallow sheet over a length of several hundred yards, serving to irrigate the gentle slope below, but avoiding the concentration of flow which might cause erosion.

Bottom: Another water-spreading device is the gap-spreader bank shown in operation during a storm. Water overflowing from the dam in the background and from the slopes above, is picked up by the bank and directed through gaps into the channel on the down-hill side of the bank. The channel then spreads the water on a broad front. Spreader-banks serve the dual purpose of controlling erosion and making use of run-off for crop and pasture growth.

(Photos courtesy Soil Conservation Service of N.S.W.)

Plate 21 SOIL CONSERVATION—IV

little wind-moved material are more likely to be directed towards controlling or manipulating water movement on the surface.

The control of sand drift and dune activation is very expensive per unit area. It is frequently necessary in the coastal dune situation to protect land, or developments of high value. Even with the aid of high rainfall of 30-60 inches, complete stabilisation is likely to cost from $200 to $400 per acre. Stabilisation of active dunes in the inland, with rainfall ranging from 5-20 inches per annum, is extremely difficult in the absence of irrigation.

The reclamation of bare exposed surfaces is much easier and in the semi-arid areas can be accomplished at a cost which will be covered by returns from increase in productivity. It is proposed to discuss the various types of structural measures used in wind erosion control, dealing firstly with those involved in sand drift control.

SAND DRIFT CONTROL

Drift and dune activation occur in sandy soils where the depth to a resistant horizon is considerable, 3 feet or more. It is common in coastal areas where it may be caused by some factor which is not in harmony with the rather delicate balance of the environment. The destruction of the protective cover on the more exposed situations by over-grazing, fire, traffic, construction works or wave action is sufficient to allow the wind to act on the sandy surface.

Once initiated, drift spreads rapidly by smothering and killing vegetation on adjacent areas, and may cover up to two acres within a couple of years. The larger the area, the more difficult and more costly the stabilisation treatment, so that early treatment is essential. Often the cost of treating the drifted land will be more than its value. However, the cost of treatment must be balanced against the value of the land or property which may be destroyed if the drifted area is not treated. Small areas of the order of one to two square chains are best treated by laying brush over the surface and planting sand-binding plants which will normally be growing in the locality. If the eroded area consists of a blow-out and a downwind apron of moving sand, the drift sand should preferably be first pushed back into the blow-out to restore a uniform contour.

The method chosen to treat large areas of coastal drift will depend on the situation. If the frontal dune is breached or badly damaged, the first attack must be to stabilise it. If it has completely disappeared, as is often the case, stabilisation of other areas will be virtually impossible until a frontal dune is re-established and stabilised.

Re-establishment of the frontal dune may be accomplished by pushing sand up with earth-moving equipment (which is quick but expensive) or by building a dune-forming fence of suitably permeable material. This may be of brush, or of an open weave hessian or similar material (Barr, 1966) or of slat-fencing.

The effect of the wind-permeable fence is to slow down the velocity of the wind carrying sand off the beach and cause it to drop its load on each side of the fence. The small dune so formed is planted with seashore grasses, and protected by brush matting if necessary. This assists the accumulation process and the dune continues to grow in size. If necessary another fence may be built on top to gain additional height.

When the frontal dune is sufficiently stabilised it serves to protect the area behind it from the force of the wind and cuts off the supply of sand that would normally overwhelm any attempt at stabilisation without the protection of the frontal dune. Stabilisation in this zone consists of planting of seashore grasses and shrubs, protected by brush matting. In less-exposed situations it is possible to effect stabilisation by sowing cover crops such as cereal, rye or oats with shrubby species such as lupins.

Beach mining for minerals such as rutile and zircon is normally conducted by a dredge and processing plant which floats in a large pool and chews its way through the landscape. The raw sand which has passed through the separation process is returned to fill the area through which the plant has operated. To ensure efficient stabilisation of this raw sand it is necessary to follow well-established procedures. With variations due to local circumstances, these are generally as follows:

1. Remove the cover of scrub or brush, generally by bulldozing.
2. Remove the surface 1-2 feet of organic sand and stockpile it along the edges of the area to be mined, ready for later use.
3. Dredge and return the processed sand to the mined area.
4. Use a bulldozer to spread the sand, shaping it to gentle contours.
5. Spread the stockpiled surface sand over the whole area.
6. Carry out stabilisation techniques appropriate to the situation.

The stabilisation process is aided considerably if step 5 can be carried out within 5-6 weeks of the original disturbance. Much of the plant material will be still living and will regenerate vegetatively. Although it will still need to be supported by artificial stabilisation, the return to a stable community, similar to the original, will be hastened. This is especially important in heath lands where low fertility creates problems for normal stabilisation techniques.

The need to stabilise extensive areas of unstable sand following beach mining operations has led to much trial and experiment into improved methods of stabilisation. Most of the operations involved in stabilisation require hand labour. When large areas need to be stabilised quickly, it is very difficult to get adequate labour, especially in remote areas. Mechanical planters have been used successfully in many places to plant vegetative clones of marram grass, sand spinifex and other seashore plants, as well as tree and shrub species.

As an extension of the need for mechanisation, many types of surface-sealing agents have been tested with the view to replacing brush for temporary protection while permanent vegetation is being established. Cutting and laying brush requires large amounts of labour and in many areas it may be in short supply.

Many types of surface-sealing agents have been tested. These include bitumen emulsion, resins and oils. Some have been successful in certain situations but the very wide range of conditions of exposure, even in a single dunal area, make it difficult for a chemical sealant to be effective in all cases. If brush is readily available it should be used. For large-scale work, straw can be spread mechanically and is quite effective.

In stabilising coastal dunes it is important to consider the type of land use to which the stabilised surface will eventually be exposed. A pasture-type cover needs continual fertilisation to maintain a vigorous sward, and unless this can be assured in the future management of the area, the stabilisation programme should provide for the establishment of a shrub or scrub cover which will give permanent protection. Species natural to the environment are preferable, but if not available there are other plants which can be established from seed or by planting. This is best done as soon as the initial stabilisation by grass is becoming effective.

Another important aspect on coastal dunes is the construction of buildings or other forms of development on the frontal dune. The frontal dune is a natural barrier which provides a buffer zone against the action of high seas. It provides a reservoir of sand which circulates between the dune, the surf-zone sea bed and the beach according to sea and wind conditions. Consequently any construction on the frontal dune is likely to be undermined by the natural process of wave erosion. Condon and Barr (1968) have set down guide lines for building and other developments near coastal dunes and beach areas.

The scope of this chapter does not permit a detailed description of the stabilisation techniques available. Techniques appropriate to various situations have been described in detail by Barr (1965). However, the local circumstances are extremely important in drift sand control, and as control measures are relatively expensive it is always wise to seek expert advice from the State soil conservation authority.

Drift and dune activation is common in some dryland grazing areas, but rarely affects large areas. It has caused considerable damage by silting up and undermining fences and by silting up homesteads, yards and even stock watering tanks.

On the fringes of the wheat belt, particularly in the Victorian and South Australian mallee, drift and dune activation is prevalent on sandy lands used for wheat growing in a sub-humid and semi-arid climate. Strong winds during the late summer-autumn period, while the land is being prepared for sowing, causes sand movement, often resulting in serious drift and dune formation. Young crops may also be buried.

Unstable sand in these areas has caused considerable problems by the filling in of water channels and the covering of railway lines and roads. Authorities responsible for these public utilities have established schemes of financial assistance to enable adjoining landholders to undertake control measures with the view to controlling the problem at its source.

Control can be accomplished by growing cereal rye, which is a tough-strawed crop which can withstand sand-blasting in its early stages and thus is able to establish protective cover. Leguminous crops such as vetches and lupins are used with the cereal rye to improve the physical condition of the soil and to restore fertility.

Drift on these lands can be prevented by using safe cultivation methods such as stubble mulching which retains trash cover on the surface after the land has been ploughed.

Where drift and dune activation occurs in areas of insufficient rainfall to support crop growth, stabilisation by vegetative means is virtually impossible unless irrigation is available. Because of this it is wise to avoid locating improvements such as homestead and other buildings, yards and watering points in places liable to inundation by drift sand. Ideally, improvements of this nature should be located on firm loamy soils in clumps of open timber, with a belt of fairly thick timber on the side from which the sand-carrying winds will come.

If sand drift does become a problem, the only recourse is to transport it away, taking care to dump the sand on the downwind side. In some cases it may be possible to trap a proportion of the sand by placing dune-forming fences on the upwind side.

Trees and shrubs provide a means of preventing the wind from getting at the surface and of causing it to drop its load of sand. In the drier parts this requires some form of irrigation or hand-watering in order to establish the young plants, which is also expensive.

RECLAMATION OF SCALDS AND SIMILAR SURFACES

In the chapter on soil erosion, the various forms of wind erosion were described. These were based on the type of surface exposed by erosion and the amount of drift associated with it.

The exposed surface may be a wind-sheeted surface of fairly good permeability differing little from the character of the original surface. The texture is normally a firm or hard loam or clay loam. Colonisation takes place readily in good seasonal conditions with plants native to the locality, and succession of moderate or good seasons will assist wind-sheeted areas to regenerate naturally. A shallow veneer of coarse sand on the surface appears to help the process by trapping and holding seed, and providing good moisture conditions when the sand is arranged in shallow ripples on the surface. If natural regeneration is slow the techniques described for reclamation of scalded areas can be used with success under much lower rainfall conditions.

Natural regeneration of the scalded surface is much more difficult. The slaked clay swells rapidly when wet so that after a few points of rain the surface is sealed against further infiltration unless it should be flooded. The exposed clay is normally very different in texture and other characteristics from the original surface, so that the plant species native to the original surface are normally not suited to the eroded surface. Plants which are able to germinate on the limited amount of soil moisture after rain can survive only as long as the shallow zone of moisture lasts. Often they are subject to sand-blasting as sand grains are whipped across the exposed surface by moderate breezes. The smooth condition of the scald surface makes it very difficult for seed to lodge.

To reclaim scalded country it is necessary to improve the absorptive capacity of the scalded surface, and to ensure a seed supply of suitable plant species to effect colonisation of the scalded area. Ploughing of the scalded surface improves its water-absorbing capacity and the rough surface is also instrumental in reducing wind velocity at the surface and trapping windblown seed. However, as the scalded surface is a heavy clay and of totally different characteristics to the surrounding non-eroded soils it is normally necessary to introduce seed of plants especially suited to the habitat. Saltbushes and related drought-resistant species have been found most suitable for this role.

In the higher rainfall parts, reclamation of scalded areas can be successfully accomplished by complete ploughing, and sowing of saltbushes, preferably with a crop of cereal rye or oats to protect the ploughed surface from the slaking action of rainfall, and to assist in the even dispersal of seed from established saltbush plants. In the drier areas, 11-14 inches rainfall a year, complete ploughing has not been successful, because light falls of rain will often slake the ploughed surface back to scald before sufficient rain is received to germinate the seed.

Initial attempts at scald reclamation were by furrowing, either as parallel furrows or in the form of a checkerboard. The effect of the furrows is to trap the large volume of water shed from the inter-furrow space and also windblown seed. The result is generally a good colonisation by species suited to the saline clay soils, such as saltwort and annual saltbushes.

Parallel furrows are of little benefit because unless they are on the true contour they are liable to drain water away, one end being flooded and other left dry. Checkerboard furrows overcome this problem to a certain extent, serving to trap the water, and hold it in the bays and along the furrows. However, furrows fill with wind-blown and water-washed silt to such an extent that the furrow ridge eventually serves to drain the water away from the lines of vegetation established along the furrows. The plants die and the ridge is gradually eroded back to the level of the scald. Often the only indication that an area has been furrowed are the slightly discoloured lines on the scald surface.

This problem was overcome by a technique which became known as progressive checkerboard furrowing (James, 1954). In this technique the initial bays are subdivided by a further set of furrows at a stage when the initial colonisation is in a vigorous condition and producing plentiful seed. This has the effect of quartering the bays, providing freshly roughened soil in which wind-blown seed is able to lodge and germinate, and effecting much greater control over the run-off. A third subdivision would reduce the size of the bays to 5-6 feet squares and encourage much more colonisation. This technique was successful in most applications down to an average annual rainfall of about 11 inches. However, the rough ride for the tractor driver involved in effecting the second and third subdivisions has made it an unpopular method of reclaiming scalded country.

A variation of furrowing technique is the spiral or concentric furrowing in which the operator starts furrowing in a tight circle, gradually increasing the diameter to several chains.

Furrowing is a useful device for protecting homesteads from dust during drought conditions. The surface roughness caused by the furrows at about half chain intervals is sufficient to prevent the wind from picking up soil particles. Expanses of bare ground which raise a pall of dust into the air with each moderate gust of wind can be made dust-free by furrowing.

Water ponding has superseded furrowing as a means of reclaiming scalded areas. In this technique, small banks are constructed on the contour across the scalded surface, and turned up at the ends so that the banks store about 3 inches of water. The ponded water causes the clay soil to open up with frequent cracks as it dries out. Colonising plants are able to get established as the water dries back and the cracking surface is able to collect and hold seed. The next flooding normally results in a heavy growth of vegetation which goes through a typical ecological succession, commencing with plants adapted for colonising, and finishing with perennial grasses native to the heavy soils in the locality. Once the soil surface becomes protected by vegetation the formerly slaked surface becomes crumbly and absorptive due to the leaching of sodium from the surface and its retention in the sub-soil (Jones, 1966).

For water ponding to be successful the banks should leave sufficient bare scald above them to shed water to the ponding area. On scalded surfaces with a high potential for run-off, a ratio of about two of 'catchment' to one of ponding area is reasonable.

The depth of ponding will vary with the type of eroded surface. On impermeable scald surfaces a ponding depth of no more than 3 inches should be used. Greater depth will cause drowning of the regenerating vegetation. On more permeable surfaces where the water soaks away in a few days, a ponding depth of 6 inches may be allowed. On a gradient of 1 in 200, provision for a depth of 6 inches will back water horizontally 100 ft.

Banks need to be built to settle to 9-12 inches for a 3-inch ponding, a road grader being the ideal implement. If possible the bank should be pushed upslope to keep the ditch or borrow area on the lower side. A ditch on the upper side of the bank will accept a large volume of water before any is available to flood bank over the normal surface.

The results of trials with water ponding on several sites in the Hay district of New South Wales have been described by Jones (1967). Subsequent work in which the increase in forage production has been measured has indicated that the cost of treatment is more than covered by the improvement in productivity.

Pitting is another technique which can be used to reclaim wind-eroded surfaces, being well suited to the more permeable wind-sheeted surfaces, but meeting with little success in the strongly slaked scald surface. In this technique, a series of scalloped discs are used to construct shallow basins about 2-4 feet apart, 3-4 inches deep, 3-4 feet long and 6-9 inches wide with a small ridge of earth adjacent.

The ridges create sufficient roughness on the eroded surface to trap wind-blown seed, while the basins collect run-off from the intervening spaces. Each pit thus forms a small area of improved moisture conditions. Pitting can also be used on sloping surfaces subject to sheet erosion by water, and have the advantage that no levelling is required as in laying out contour furrows.

FURTHER READING

Barr, D. A. (1965). 'Restoration of coastal dunes after beach mining'. *J. Soil Conservation N.S.W.*, 21 : 4 (199).

Barr, D. A. (1966). 'Jute-mesh dune-forming fences'. *J. Soil Conservation N.S.W.*, 22 : 3 (123).

Condon, R. W. & Barr, D. A. (1968). 'Guidelines for real estate development near coastal dunes and beach areas'. *J. Soil Conservation N.S.W.*, 24 : 4 (237).

Condon, R. W. & Clothier, D. P. (1968a). 'Soil conservation in alpine catchments'. *J. Soil Conservation N.S.W.*, 24 : 2 (96).

Condon, R. W. & Clothier, D. P. (1968b). 'Bitumen-straw mulching for better stabilisation on the Kosciusko Road'. *J. Soil Conservation N.S.W.*, 24 : 4 (218).

Cunningham, G. M. (1967). 'Furrowing aids revegetation at Cobar despite worst drought on record'. *J. Soil Conservation N.S.W.*, 23 : 3 (192).

Durham, L. J. (1956). 'Soil erosion problems in the Snowy Mountains area', Parts I and II. *J. Soil Conservation N.S.W.*, 12 : 3 (121); 12 : 4 (156).

Fitzgerald, K. (1968). 'The Ord River regeneration project'. Part II: 'Dealing with the problem'. *J. Agric. West. Aust.*, 9 : 90-5.

James, J. W. (1954). 'Scald reclamation experiments in the Bourke district'. *J. Soil Conservation N.S.W.*, 12 : 1 (44).

Jones, R. M. (1966). 'Scald reclamation studies in the Hay district'. Part I: 'Natural reclamation'. *J. Soil Conservation N.S.W.*, 22 : 3 (147).

Jones, R. M. (1967). *Ibid.* Part III: 'Reclamation by ponding banks'. *J. Soil Conservation N.S.W.*, 23 : 1 (57).

Newman, J. C. (1963). 'Waterspreading on marginal arable areas'. *J. Soil Conservation N.S.W.*, 19 : 1 (49).

N.S.W. Inter-departmental Committee report on scrub and timber regrowth in the Cobar-Byrock district and other areas of the Western Division of N.S.W. (1969). Govt. Printer, Sydney.

Royal Commission to inquire into the condition of the crown tenants: Western Division of New South Wales (1901). Government Printer, Sydney.

CHAPTER 9

Landscape types in dryland Australia

by R. W. Condon

THE differential response of the various types of country to grazing use, and particularly the serious deterioration which has taken place on certain soil types, points to the need for recognition of those factors which predispose country to deterioration. It is proposed to make some contribution towards this need by describing the principal landscape types in dryland Australia, with special reference to the factors which make for stability or instability in their environments.

The landscape types chosen for description are based on a combination of topography, soils and vegetation; the former two factors often being interdependent and the third generally dependent on the other two. Vegetation had been included because it invariably gives the landscape type its distinctive appearance.

MOUNTAINS AND HILLS

Much of central Australia, parts of Western Australia, and western New South Wales and Queensland, are occupied by stony, rocky and steep hills and mountains. Many of these have little or no grazing value, supporting only harsh spinifex and stunted mulga, or being composed mostly of rock and stone with little soil, or being too steep for livestock to utilise. The hilly areas are broken by narrow valleys with gentle slopes and good pastures accessible to livestock.

The soils in the valley areas range from stony and gravelly skeletal soils to red and brown loams or sandy loams, with coarse alluvial sands along the watercourses.

Timber cover may range from tall river red gums (*Eucalyptus camaldulensis*) and coolibah (*E. microtheca*) along the watercourses, to sparse and scattered smaller trees such as whitewood (*Atalaya hemiglauca*), beefwood (*Grevillea striata*), and ironwood (*Acacia estrophiolata*), with belts of mulga and other acacias on the firmer soils.

Pastures in the valley areas are normally short grasses and forbs of reasonable palatability. Hilly and mountainous areas tend to attract storms which run the creeks and watercourses. As a result these areas often carry a flush of feed not available elsewhere in the locality.

In sheep-raising country the hilly areas are little used in normal seasonal conditions but provide a valuable reserve of dry feed in times of drought. Because of the stony and gravelly nature of the soil on the steeper hills and the reduced accessibility to livestock, there is little danger of erosion in this landscape type. The condition of the lowlands mixed with the hilly areas is generally good, but some areas, where texture-contrast soils predominate, have been badly damaged by wind and water sheeting and rilling and gullying.

MANAGEMENT CONSIDERATIONS

Because of the greater frequency of storms and the run-on from adjacent hills and mountains, foothill and valley areas have a much higher productivity per unit area than nearby country with similar soils and pastures. Management should aim at preserving this high productivity by avoiding overgrazing. There is a tendency to over-estimate the acreage of usable country, which would tend to lead to overgrazing. Because of their value as drought reserves, management should aim at light use during average and good seasonal conditions, so that there will be plenty of forage available when it is needed in drought times.

TREE PLANTING

Fodder and shade trees present are generally adequate and the arid climate of the majority of this landscape type militates against successful tree planting without irrigation. There could be scope for planting trees to protect homestead areas from dust and to provide shade and shelter. The presence of sloping country nearby means that there will be run-off during storms and this can be utilised to promote growth of planted trees. If plantings are undertaken, they should be planted to take maximum advantage of run-on. This can be done by using crescents and short drains on slopes in excess of 2-3 per cent and by using spreader banks to bring flows from shallow gullies and watercourses to flood planted areas on gentler slopes.

SPINIFEX SAND PLAIN

This landscape type occurs widely in the Northern Territory, extending northwards from the MacDonnell Ranges to near the Victoria River, and westwards to Western Australia. It is mixed with extensive occurrences of the spinifex dune fields.

The sand plain areas are broad expanses of nearly level red, deep sandy soils and sandy red earths, supporting a characteristic growth of spinifex (*Triodia* and *Pletrachne* spp.) with a scattering of stunted trees such as beefwood (*Grevillea striata*), corkwood (*Hakea* spp.), horseradish (*Codonocarpus cotonifolius*) and others, with small belts of mulga (*Acacia aneura*). The spinifex is coarse and tough and of low palatability, and it leaves little room for other species and, consequently, grazing capacity is very low. Further north and where the loamier soil types support the more palatable species of spinifex and associated species, the grazing capacity is somewhat higher.

In the higher rainfall areas of the Northern Territory and north-western Western Australia, burning of the spinifex before summer rains is practised with the view to improving palatability by encouraging fresh green shoot. This has led to serious deterioration in limited areas. In the north-west of Western Australia, productivity has been improved on some areas by sowing kapokbush (*Aerva tomentosa*) and buffel grass amongst the spinifex.

Much of the spinifex grassland is in pristine condition, and most of the remainder has been only lightly used, so that it is still in good condition, although of low productivity. The spinifex provides effective protective cover. Where it has been removed in small areas, drift is only of a minor nature, there being sufficient coarse sand present to protect the soil against wind erosion.

MANAGEMENT CONSIDERATIONS

The spinifex sand plains are used for cattle grazing, and although of low grazing capacity, the spinifex can carry stock through prolonged drought without undue risk of erosion. Where topfeeds are plentiful, it is ideal drought refuge country. The main consideration is whether the returns from grazing will cover the cost of providing water. There is normally little chance of finding good catchments and soil of good holding quality for surface tanks in spinifex sand plain. In the lower rainfall areas receiving less than 10-12 inches per annum, year round grazing capacity may be only 2-3 beasts per square mile. Assuming a five-mile radius as reasonable for cattle to utilise around a watering point, the area commanded by a watering point would be approximately 75 square miles with a carrying capacity of 225 beasts. For this reason, underground water needs to be reasonably shallow to warrant locating bores in country of such low productivity.

Although there is little risk of erosion under normal grazing pressures, the continual disturbance which takes place around watering points, holding yards etc., is likely to encourage drift and the risk of damage to improvements. For this reason, watering points and similar areas of stock concentration should be located in belts of timber as far as possible. Apart from the protection afforded to the soil by timber cover, the presence of moderate to thick timber in this landscape would normally be an indication of firmer loamy soils and favourable moisture conditions due to run-on.

TREE PLANTING

Native trees on this landscape are generally stunted. The soils are so absorptive and generally so level that there is little prospect of run-on. Plantings would need to be irrigated and probably fertilised to make up for the nutrient deficiency in the soil.

SPINIFEX DUNE FIELDS

Throughout much of the desert areas of inland Australia, extensive areas of parallel sand ridges occur at spacings of 10-20 chains. There are three principal areas

- —The Simpson Desert, generally north and north-east of Lake Eyre, with the dunes aligned in a south-north direction.
- —The Great Victoria Desert, extending westward across the northern end of South Australia into Western Australia, with the dunes trending generally west-east.
- —The Great Sandy Desert, across the northern end of Western Australia, with dunes trending generally east-west.

This pattern of trends appears to form a giant anti-clockwise arc, closely, but not exactly, related to the resultant dominant wind (Brookfield, in press). The dunes range from 30-100 feet in height, with long sloping flanks leading up to a steep crest of bare sand. The soil type in the swales between the ridges may range from sands to clay loams. Sometimes a reticulate pattern is formed with small clay pans between the dunes. The dune fields generally carry a light stand of mulga in the swales, with scattered patches of desert oak (*Casuarina decaisneana*) on the flanks. The variety and frequency of tree species increases with increasing rainfall.

Pastures contain varying proportions of spinifex, generally on the upper slopes of the broad flanks of the dunes, but also include a much wider and more plentiful range of short grasses and forbs of value for grazing in the swales and adjoining lower slopes of the dune flanks. A large proportion of this country is remote from settlement, and has not been used pastorally. Parts that have been used are generally in good condition. Although the crests of the dunes may be bare drift sand, there is relatively little encroachment on to the dune flanks. The long sloping flanks of the dunes are composed of red clay sands, with a good proportion of coarse material which protects the smaller particles from wind removal. The light to moderate cover of spinifex also serves to protect the flanks from erosion and to contain any sand which may move off the crests on to the flanks. The swales may sometimes suffer from scalding or windsheeting. Serious erosion may occur where dunes occasionally overlap harder and more productive country. Sand activated by excessive grazing and trampling is carried across the plain, stripping vegetation by sand blast, or burying it in shallow or deep drifts.

MANAGEMENT CONSIDERATIONS

This landscape type is used primarily for extensive cattle grazing. Because of the greater quantity of short grasses and forbs and in many places good stands of mulga, it is more productive than the spinifex sand plain, but the same management considerations apply. There is little risk of erosion and where there is plentiful topfeeds it makes useful drought refuge country.

TREE PLANTING

In spite of the coarse nature of the soils, there is sufficient clay present to cause run-off in heavy storms. Any tree-planting undertaken in this landscape will normally need supplementary irrigation and watering but advantage should be taken of run-on water from the sloping flanks to minimise this need.

SALTBUSH PLAINS AND TABLELANDS

The vegetation on this landscape type is dominated by the perennial shrubs of the saltbush (*Atriplex*) and bluebush (*Kochia*) genera, invariably with a wide range of other species of the family Chenopodiaceae, such as copper burrs (*Bassia* spp.) and smaller herbaceous species of *Atriplex* and *Kochia.*

The dominant feature of the landscape is the almost total absence of trees of any kind, although these may occur in small clumps or belts in situations favoured by run-on or more favourable soil conditions. In some cases, saltbush and bluebush may occur as an understorey to mulga (*Acacia aneura*) or belah (*Casuarina cristata*) scrub.

There are several different sub-types due to differences in soils or topography.

(a) The heavy grey and brown clay soils of the riverine plain in south-western New South Wales. This area supported a climax community of perennial saltbush (*Atriplex vesicaria*), although over much of the area it has been altered by grazing to grasses and herbage.

(b) Slightly undulating sandy texture-contrast and solonised brown soils carrying bluebush (*Kochia pyramidata* in New South Wales, *K. sedifolia* and *K. astrotricha* in South Australia and central Australia).

(c) Undulating stony brown clays often with gilgais (or small depressions) carrying good saltbush, copper burrs and grasses and herbage in the depressions. This sub-type occurs over large areas in north-western New South Wales as foothills to large hills and ranges, and as a broad swathe north-westerly across South Australia and into southern central Australia.

(d) Broad level plains of brown clay soils mixed with sandy texture-contrast soils extending from the Broken Hill region in western New South Wales westerly into South Australia. Perennial grasses are often co-dominant, forming a mixture of grass land and shrub land. Scrub species such as mulga and myall (*Acacia sowdenii*) occur on the sandier soil types.

(e) Broad stony clay plains such as the Nullarbor Plain where pearl bush (*K. sedifolia*) is dominant on strongly calcareous soils, and the Sturt's Stony Desert area in north-eastern South Australia, which is very sparsely vegetated in the rare good seasons.

The saltbush-bluebush shrub lands are largely confined to the winter rainfall zone with average annual rainfall ranging from about 5 to 15 inches. Although the dominant perennial bushes are of relatively low palatability, and are little eaten while there is plenty of green feed available from other sources, they provide high-quality green forage when grasses and herbage have dried off. There is normally a wide range of associated species—perennial and annual grasses and herbs—which in most areas make up an attractive pasture.

The saltbush-bluebush landscape type is predominantly wool-growing country. The texture-contrast soils are very prone to wind erosion. Because of the attractiveness of the pastures on these soils, and the rapid response to rains, they have been subject to serious overstocking since early occupation. Under these conditions recurrent droughts have taken heavy toll of the pastures and the perennial bush has disappeared from extensive areas. Serious erosion is also widespread in many areas.

The undulating stony brown clays have also been heavily used, and the perennial saltbush and bluebushes which were native to this landscape sub-type have disappeared over very extensive areas in northern South Australia, southern central Australia and north-western New South Wales. This has been accompanied by a gradual filling in of the gilgais or depressions with wind-blown sand and water-washed silt. Where rainfall was once trapped in these small depressions to promote growth, it now runs off directly into the drainage lines and is lost to the immediate environment.

Deterioration of saltbush-bluebush country is common in all States in which it occurs. Ratcliffe (1937) has described the conditions which led to the deterioration of South Australia's arid pastoral lands, in particular the saltbush country. Beadle (1948) has described the changes in the saltbush community which resulted from overgrazing in various areas of western New South Wales, and has listed and described a number of degenerate communities which were apparently once dominated by perennial saltbush or bluebush. Most of the deterioration in New South Wales dates back to the 1901-02 drought period.

Wilcox (1963) reported that the saltbush type shrub lands in the Murchison area of Western Australia had been overgrazed early in the development of the area, and were susceptible to irreversible damage. The shrub lands of central Australia were also heavily utilised in the early phases of pastoral settlement, and are now in a seriously deteriorated condition. The heavy grey clay soils of the riverine plain in south-western New South Wales are mostly resistant to erosion, although the associated slightly elevated areas of texture-contrast soils are invariably severely scalded.

The saltbush-bluebush country has suffered more than any other of the dryland landscape types from the effects of the severe overgrazing which took place in the early phases of pastoral settlement in the outback.

MANAGEMENT CONSIDERATIONS

In the face of such extensive deterioration of both soils and pastures, and the wide range of communities resulting from deterioration, it is difficult to be specific in recommending management practices for this landscape type. In good condition this landscape type is highly productive. This, and its susceptibility to deterioration, should be recognised. Maximum use of this type of country should be made during good seasons when forage is plentiful but adjusting grazing so that there is adequate feed in reserve. Past experience suggests that heavy prolonged grazing during drought should be avoided.

Many graziers in the past have grazed saltbush country heavily with the object of eliminating the saltbush and encouraging a pasture of short grasses and annuals. This gives high yields of forage in good seasons, but very low yields in average and poor seasons. Studies by Leigh and Mulham (1967) showed that the less palatable saltbush made up a very high proportion of the diet during dry conditions. A heavy stand of saltbush is probably less productive than a grass-dominated pasture, but the ideal would be to maintain an open stand of shrubs and gain the benefit of both the grasses and shrubs.

TREE PLANTING

Saltbush country is naturally treeless, because trees find it difficult to grow on heavy clay soils unless moisture conditions are very favourable. Consequently plantings would need much supplementary watering for successful growth. The protection and shelter afforded many outback homesteads by groves of planted trees in an otherwise treeless plain, show that trees can be grown successfully in this landscape.

MITCHELL GRASS PLAINS

Several species of Mitchell grass (*Astrebla* spp.) occur as dominant tall grasses on red and brown cracking self-mulching clay soils throughout central and south-western Queensland, and extending north-westwards into the Northern Territory on the Barkly Tablelands. Smaller areas extend southward into northern New South Wales, generally on heavy grey cracking clays on river flood plain landscapes.

The Mitchell grass landscape is largely treeless, although sometimes broken by belts and clumps of trees in favourable moisture situations or on lighter-textured soils. It may be slightly undulating or rolling, sometimes strewn with stone and gravel, and generally with some small depressions or gilgais. In other areas it occurs on the broad flood plains of the almost ephemeral streams flowing through south-western Queensland into South Australia or northern New South Wales. These soils often show strong gilgai features. Although Mitchell grass is dominant over most of the area, there are sub-dominant grasses which occasionally become locally dominant over small areas. These include brown top (*Eulalia fulva*), neverfail (*Eragrostis setifolia*), Queensland bluegrass (*Dicanthium sericeum*) and panics. There is normally a wealth of annual herbage and grasses which respond well to good rains, and a scattering of chenopodiaceous plants such as copper burrs (*Bassia* spp.) and saltbushes.

The various forms of Mitchell grass pasture have an ability to withstand heavy grazing, and the heavy cracking soils are not susceptible to deterioration by soil erosion. However, there has been some degeneration, generally along heavily used stock routes and around watering points. Continued heavy stocking will lead to elimination of Mitchell grass and other perennial grasses, and lead to its replacement by annual forbs and grasses of low value. On the Ord River catchment in northern Western Australia, much of the Mitchell grass and Queensland bluegrass have disappeared from heavy soils because of uncontrolled grazing.

In northern New South Wales some graziers control stock to maintain their pastures in the neverfail dominant stage because it forms a more palatable pasture than Mitchell grass. The condition of the Mitchell grass grasslands is generally good, although degeneration will take place with heavy grazing.

MANAGEMENT CONSIDERATIONS

Pastures on this landscape type are fairly stable, although they can suffer deterioration by constant overgrazing. The soil is resistant to wind erosion by reason of the aggregated nature of the clay soil creating particles too large to be moved by wind. The level topography also militates against water erosion, although severe gullying can occur where flowing water drops from the natural land surface.

For these reasons it is reasonably safe to make heavy use of Mitchell grass pastures while seasonal conditions are favourable in order to make full use of the plentiful ephemeral herbage. Stocking should be reduced before use of the Mitchell grass or other perennial grasses becomes too heavy. If the grasses should show signs of opening up it is time to look carefully at the grazing management policy and adjust this to allow return to normal density. There is a considerable difference in productivity between a stable grass-based pasture and a pasture of ephemerals which yields nothing in poor seasons and too much to be used in good seasons.

The heavy yield of annual type forage in good seasons is of little value when there is so much feed around that the stock cannot use it. Its main advantage is that its presence takes the grazing pressure off the perennial grasses and allows them to make plenty of growth before they begin to be utilised. However, good seasonal conditions will normally bring a high yield of annual herbage in spite of the competition from the perennial grasses.

TREE PLANTING CONSIDERATIONS

Mitchell grass country is naturally treeless and much the same considerations apply as have been outlined for the saltbush country.

MALLEE LANDS

Mallee is the name given to the dwarf species of eucalypt which have several stems arising from a swollen stock or base. Individual 'trees' are invariably closely spaced (from 5-15 yards apart) and of uniform height, often 6-20 feet. Ground cover is generally sparse and there is often a sprinkling of shrubs. Spinifex (*Triodia* spp.) may occur on very sandy soil types, either as a continuous understorey or on parallel sandy rises or well-developed sand hills. Because of the dearth of ground forage the productivity of the mallee country for grazing by sheep and cattle is very low, being from one-third to one-sixth that of neighbouring landscape types. Some mallee country is relatively open, or trees are arranged in clumps separated by open areas.

Large belts of mallee landscape occur in central New South Wales on red loams and sandy loams. Extensive areas occur on brown calcareous sands with sand hills or large sandy rises in the south-west of the State, continuing south into north-western Victoria and west into South Australia, forming an extensive area around the lower reaches of the Darling and Murray Rivers. Further examples occur in South Australia around the Great Australian Bight from St. Vincent's Gulf to the Nullarbor Plain. In Western Australia, a further extensive belt of mallee country extends from the Bight westward towards the Indian Ocean near Geraldton.

Mallee country has been used extensively for wheat growing in Victoria and South Australia. The clearing of the protective vegetation from the sand hills, and the disturbance of cultivation, led to much drift and dune activation, and deterioration was very serious in the 1920s and 1930s. Unstable dunes are unproductive and because they threaten productive adjacent lands they need to be stabilised. This can be done using cereal rye as a cover crop with lupins or evening primrose.

Wheat cropping has extended into the leasehold country in the Western Division of New South Wales, but following widespread erosion resulting from earlier essays in cropping, restrictions have been placed on clearing which prevent the removal of protective mallee on the sand hills, and restrictions on the frequency of cropping.

MANAGEMENT CONSIDERATIONS

It is important to recognise the susceptibility of the sand hills and sandier soils to drift and dune activation if cleared of the protective mallee cover. Although the sandy soil produces good crops for a few years, there will come a time when crop failure leaves the soil unprotected and the area becomes drifted, declining in productivity, threatening adjacent productive lands and requiring expensive control measures. In the drier parts, cropping is undertaken with the view to developing and improving the country for grazing. Because the mallee regenerates from seed and root suckers, it is necessary to undertake cropping for several years to ensure that the mallee is killed. Erosion can be prevented by keeping the natural timber cover on the sand hills and sandy rises, particularly those in areas which have spinifex as an understorey. The loamier soils between the sandy rises can be cultivated, the risk of erosion being very much reduced because of the lower susceptibility of the soil to erosion.

Restrictions on clearing of mallee country which apply to Western Lands leases in the Western Division of New South Wales, require that the mallee timber cover on sand hills and sandy rises shall not be destroyed and that cultivation shall be confined to blocks not exceeding 300 acres, separated from neighbouring cleared areas by uncleared strips 5 chains wide. This ensures that susceptible areas are not exposed to erosion and that if serious erosion does occur in the cleared areas it would be contained within the 300-acre block and therefore unable to overwhelm roads, railways or other utilities, or the land of neighbouring landholders.

The five-chain wide uncleared strips are not intended to serve as windbreaks. It is well known that the effectiveness of a windbreak is limited to a horizontal distance of 15-25 times the height of the trees in the windbreak. However, there is apparently some influence on the wind. This was shown during the drought of 1967 in western New South Wales. Dust palls from country which had been over-cleared were very evident on windy days, whereas on neighbouring country, cleared in accordance with the required conditions, there was negligible dust being lifted into the atmosphere by strong winds.

The effectiveness of these conditions covering the development of mallee land in the Western Division of New South Wales is shown by the fact that the only places where any erosion problem has developed or threatened, is on areas where the conditions were not strictly observed. If over-clearing does take place it is possible to correct the situation merely by requiring that areas wrongly cleared remain uncultivated. Mallee very quickly regenerates to provide the necessary protection.

It is important in clearing mallee not to leave isolated trees or small clumps of trees for shade and shelter in a large cleared paddock. The surface-feeding roots keep the ground under the trees bare. Stock camping in the shade help to keep the soil in a loosened condition. The effect of a single tree (or small clump) with a V-shaped habit of growth is to increase the velocity of the wind at the soil surface as the wind is deflected around the tree. For this reason, isolated trees or small clumps are likely to start erosion on sandy soils, and the ensuing drift may affect quite a large area downwind.

If it is desired to provide shelter by leaving some trees, it should be done by leaving clumps of not less than half an acre in area and keeping the plough at least half a chain away from the edge, that is beyond the zone where the surface roots compete for moisture.

Top: Two landscapes typical of arid Australia. In the foreground, spinifex sandplain with stunted trees and a protective cover of coarse unpalatable spinifex. In the background, rugged mountains rise several thousand feet above the level of the plain.

Centre: Spinifex dune fields in central Australia showing the typical sequence of well grassed swales with useful grasses and forbs, long sloping sandy flanks with coarse unpalatable spinifex, and a steep dune in the background also protected by spinifex, but often with a crest of bare drifting sand. The trees on the sloping flanks are desert oak (*Casuarina decaisneana*). Mulga may occur in the swales where moisture relationships are good.

Bottom: Mountainous and hilly areas are often flanked by sandy or clayey alluvial outwash plains with open to sparse timber cover and excellent grass and herbage. Storm run-off draining from the hills frequently floods over the plains, adding to the yield of forage.

(*Photos courtesy Soil Conservation Service of N.S.W.*)

Plate 22 LANDSCAPE TYPES—I

Top: A typical saltbush plain on heavy clay soils. During good seasons there is an abundance of grasses and herbage. The saltbush provides a valuable reserve of green forage in short dry spells and droughts. Note the absence of trees except on an island of sandy soil in the background, or along shallow water courses.

Centre: Much of the saltbush country of highly erodible texture contrast soils in central Australia have been seriously degenerated by past grazing use. Vegetation in this scene is limited to remnants of the original bluebush stand, with grass on shallow accumulations of sand and silt.

Bottom: Mitchell grass is the dominant species on broad clay plains across western Queensland and parts of central Australia. This scene shows part of a Mitchell grass plain at the foot of mountains in central Australia.

(*Photos courtesy Soil Conservation Service of N.S.W.*)

Plate 23 LANDSCAPE TYPES—II

Top: A typical landscape of the flood plain alluvium. Scattered trees of coolibah (*E. microtheca*) or other species with a flush growth of grass and herbage following heavy rains and flooding from nearby streams.

Top centre: Acacia shrubland on red earth soils in central Australia. This photo shows the typical pattern of relatively bare spaces (foreground) alternating with groves of mulga with good grass. Water running off the bare intergrove areas during rains serves to irrigate the groves.

Bottom centre: Acacia shrubland. Mulga recently killed by drought but still providing adequate protection to the soil. The soil is in good condition and should respond well to rains.

Bottom: Overuse of the acacia shrublands, normally reasonably stable, can lead to serious degeneration. There has been no regeneration of mulga following the death of this stand and the soil drifts badly during drought and supports useless ephemerals in good seasons.

(*Photos courtesy Soil Conservation Service of N.S.W.*)

Plate 24 LANDSCAPE TYPES—III

Top: Mallee (*Eucalypt* spp.) vegetation occupies large areas of dry land Australia in the lower Murray-Darling area in New South Wales, Victoria and South Australia, and across South Australia into Western Australia. In the eastern areas of its occurrence, mallee forms a thick stand of many-stemmed small trees with sparse ground forage. Sandy soils often support a stand of coarse spinifex (*Triodia* spp.).

Centre: Shrub woodland covers large areas on the eastern fringes of dry land Australia. Tall eucalypts interspersed with small trees and shrubs, many edible, and with plenty of room for ground forage.

Bottom: In some areas of western New South Wales, on red earth soils, the shrub component of the shrub woodland landscape has increased alarmingly, following a long period of above average rainfall in the 1950's. The invasion of shrub species has caused a serious reduction in grazing capacity of these lands.

(*Photos courtesy Soil Conservation Service of N.S.W.*)

Plate 25 LANDSCAPE TYPES—IV

Another important aspect to consider in preventing an erosion problem from developing is the location of watering points. The greatest factor pre-disposing the soil to erosion is to have bare soil unprotected by vegetative cover and kept in a pulverised state by trampling or cultivation. Watering points on soil types susceptible to erosion are invariably surrounded by a sacrifice area extending for several chains out from the watering point. This is a zone on which little vegetation grows because of the intense over-grazing and trampling which takes place. On sandy mallee soils this is likely to lead to drift which may overwhelm the watering point and be a focal point for the spread of erosion into stable areas.

To avoid erosion in this situation, it is wise to locate watering points within belts of uncleared mallee, or leave the actual mallee cover uncleared around watering points for a radius of at least five chains. The natural tree cover will not only protect the soil from wind erosion, but also provide shade and shelter close to the watering point where stock spend much of their time during the summer.

TREE PLANTING CONSIDERATIONS

There is no shortage of trees in mallee country in its natural state. The provision of trees for shade and shelter and other purposes merely requires careful planning of the removal of the timber cover so that the required trees can be left. Requirements for the protection of the soil against erosion have been outlined under 'Management considerations'.

If a homestead is to be located in mallee country, consideration should be given to selecting a site within a belt of taller timber which can then be judiciously cleared to provide space for buildings, roads and other developments, but leave plenty of shelter from taller trees. Mallee country frequently has small patches of firmer, more fertile soils with taller and more open timber cover better suited aesthetically and practically for a homestead complex than the mallee country itself. There are many examples of a homestead complex stuck out on a broad plain from which the trees have been effectively removed. It takes many years for planted trees to grow sufficiently to provide the necessary shelter and aesthetically-pleasing situation which should surround any homestead complex.

If it is necessary to plant trees in mallee country, this presents no great difficulty. Physical soil conditions are ideal and it is normally only necessary to make provision for regular cultivation during the first 2-3 years of growth, with limited artificial watering to get young trees through a dry time. Mallee country can be expected to be chemically poor and growth of planted trees may be stunted or poorly shaped for this reason. Soil testing may provide the answer to what may be needed for best growth, but in most situations a dressing of manure from the wool shed when the trees are 2-3 years old, should overcome the deficiency.

FLOOD PLAINS AND ALLUVIUMS

This landscape type includes the outwash plains adjacent to the mountain ranges and flood-out plains alongside ephemeral streams in central Australia and Western Australia, the so-called 'channel country' of south-western Queensland, and the broad plains of the fine-textured alluviums associated with the large river systems in New South Wales and Queensland. The soil types are very variable and consequently changes in vegetation may be most marked. The dominant feature of the landscape is flatness, with a propensity for occasional flooding and a generally sparse cover of small trees in extensive grasslands or shrublands.

Included in this landscape type are many vegetation communities which are strictly woodlands or low woodlands, being composed of small trees with no or very few shrub species. This is generally the case on the heavier soil types where bimble box (*E. populnea*), coolibah (*E. microtheca*), black box (*E. largiflorens*), gidgee (*Acacia cambagei*), brigalow (*A. harpophylla*), myall (*A. pendula*) or whitewood (*Atalaya hemiglauca*) may form mono-specific communities with no or few shrub species.

In New South Wales this landscape type occupies the broad plain extending westward from the foot of the North West Slopes of the Great Dividing Range. The plains form a broad fan extending from the Bogan River in central New South Wales in an arc northward into southern Queensland, with the apex at the point where the Barwon River becomes the Darling upstream from Bourke. They also extend as broad and narrow plains along the Darling and Lachlan Rivers.

Soil types may range from heavy grey and brown clays of the river flood plains, through loams and sandy loams to limited areas of deep sands and sandhills. Because of the wide range of soils there is considerable variation in the composition and structure of the communities. Many areas of heavy grey clays are treeless, carrying either a saltbush shrubland or Mitchell grass grassland. These soils are largely resistant to wind and water erosion, although the pastures may be severely depleted by constant over-grazing. Other areas of heavy grey clay soils carry an open woodland of coolibah or black box.

Associated with the heavy grey clays of the river flood plains are texture-contrast soils occurring as islands or belts of slightly elevated country. The shallower texture-contrast soils are generally treeless with saltbush shrublands, or grasslands of windmill grass (*Chloris truncata*) and wallaby grass (*Danthonia caespitosa*). Deeper texture-contrast soils carry bimble box as the dominant species with moderate to heavy shrubs and smaller trees such as budda or sandalwood (*Eremophila mitchellii*), wilga (*Geijera parviflora*) and other species. White cypress pine (*Callitris glauca*) appears on texture-contrast with 12 inches or more of loam or sandy loam over the sub-soil clay, the proportion of pine increasing with the depth of the light-textured surface horizon. The vegetation on the deeper texture-contrast soils is a shrub woodland and strictly belongs to that landscape type.

The texture-contrast soils are very susceptible to wind erosion, removal of the light-textured surface horizon exposing a highly-dispersible clay which slakes down rapidly under the action of rainfall to form a scalded surface.

Included in this landscape type is the 'channel country' of south-western Queensland, forming the broad flood plains of the Diamantina and Georgina Rivers, with myriad distributaries forming a net-like pattern of channels in grey clays, with small belts of red sandy and silty soils scattered throughout. Alongside the channelled zone are broad clay plains with silty islands and sand dunes. The channel country supports scattered coolibah with lignum (*Muehlenbeckia cunninghamii*) and canegrass (*Eragrostis australasica*) with annual saltbushes (*Atriplex* spp.) and copper burrs (*Bassia* spp.), whilst buttongrass (*Dactyloctenium radulans*) and other annuals are produced in an abundance after floods.

In central Australia and Western Australia this landscape includes the outwash plains adjacent to mountain ranges, and the flood-out plains alongside the beds of ephemeral streams. The soils in these areas are variable with coarse and fine-textured sandy soils, texture-contrast soils and small areas of heavy clay soil. These soils support a sparse woodland of small trees such as ironwood (*Acacia estrophiolata*), whitewood and others, with belts of mulga and gidgee with good short grass and forb pastures.

MANAGEMENT CONSIDERATIONS

Because of the considerable variation in soils and pastures over relatively short distances, there is plenty of scope for management of country in accordance with the soil characteristics and the nature of the pastures.

The cracking grey and brown clays are highly productive, although they may require a fairly heavy rainfall in order to get a reasonable response. The more self-mulching the structure of the surface the greater the amount of rain needed to produce a response from the pasture.

These soils are resistant to erosion and can be stocked heavily to make use of the flush of ephemeral forage which is usual in good seasons. However, continuous over-stocking will lead to deterioration by removal of perennial grasses, often leaving only an annual type pasture of value only in good seasons. Because of the low susceptibility to erosion it is normally quite safe to destroy timber cover by ringbarking or poisoning in order to improve the productivity of pastures.

Texture-contrast soils are highly productive but highly erodible. The pastures are of high quality and high palatability, and because of their particular characteristics, respond very quickly to light and moderate rains. Consequently they are favoured by stock while feed exists and pastures on neighbouring clay soils may be relatively unstocked. Being slightly higher and well drained, the texture-contrast soils are also favoured by stock as camping places in wet weather and especially during flood periods. As a result, such soils, which in a non-eroded state are probably the most highly productive, quickly degenerate to a state of low or nil productivity. The fact that the texture-contrast soils are generally treeless, and invariably slightly elevated above the surrounding country, also makes them more susceptible to erosion by wind action.

Because of these special characteristics of soils and pastures, it is very necessary to be able to manage such soils as a unit in accordance with their particular needs. The most effective way of doing this is to fence them off from adjoining heavy-textured soils. This should also apply in those instances where these soils occasionally adjoin light-textured loamy soils, the differences in erodibility, and pasture composition and quality, warranting the management of the texture-contrast soils as a separate unit.

Separation will invariably be difficult due to the complexity of occurrence of the texture-contrast soils within a general area of heavy-textured clays. However, effective separation can often be accomplished on the basis of the proportion of texture-contrast soils, areas with a high proportion being separated from areas with a low proportion of such soils.

Most areas of texture-contrast soils have already been seriously deteriorated. Scalded areas respond very well to waterponding as described in Chapter 8, and their results in terms of increased pasture growth would seem to indicate that this gives a reasonable return on the investment in areas with rainfall greater than 12-14 inches per annum.

Although there is no research in Australia to prove it, it would seem that depleted pastures on the texture-contrast soils should respond well to deferred grazing aimed at encouraging the establishment, growth and seeding and, if necessary, further establishment of desirable species. This can be done by excluding stock on a paddock for a suitable period after good rains to permit the more desirable species to grow unchecked and to produce mature seed. This should be followed by a short period of heavy stocking to scatter seed and trample it into the ground. Depending on the condition of the pasture it may be desirable to carry out this treatment for two or more successive seasons.

Becausc of the high erodibility, care should be taken in the removal of natural timber cover to improve productivity on the texture-contrast soils. Pulling in a face and burning would create a serious erosion risk. It is preferable to ringbark or poison, leaving dead timber standing and retaining green trees at a spacing of 20-40 yards to provide reasonable protection against wind action. Careful consideration should always be given to whether killing-off timber species is likely to encourage an invasion of shrubby species which would in turn leave the land less productive than under timber. Past experience of similar lands in the district will provide the most valuable guide to this likelihood.

TREE PLANTING CONSIDERATIONS

With such a wide range of soils and climate, it is difficult to make recommendations on this aspect. Trees will be generally easier to grow on the light-textured sandy soils, but as the majority of the landscape type is in the arid zone, supplementary watering is essential, whether on light or heavy soils.

ACACIA SHRUBLANDS

This landscape type is composed of level to slightly undulating red earth soils, mostly earthy loams and sandy loams, and sometimes stony and gravelly soils, which support a variety of low trees of the genus *Acacia*.

There are four main species of *Acacia* comprising the shrub cover on this landscape type. Mulga (*A. aneura*) is the predominant species and is very widespread. Gidgee (*A. cambagei* and *A. georgiana*) extends from western New South Wales through southern Queensland into central Australia, but is generally confined to relatively small areas. It also occurs on a wide range of soil types from calcareous loamy soils to texture-contrast soils and sometimes brown clays. In the latter two cases it belongs to another landscape type. Witchetty bush (*A. kempeana*) also occurs on relatively small areas of calcareous earths in central

Australia, while myall (*A. sowdenii*) is the dominant species on calcareous soils over an extensive area across South Australia.

In south-western New South Wales there is an area of level solonised brown soils, also calcareous, which support a shrubland dominated by belah (*Casuarina cristata*) and inland rosewood (*Heterodendron oleifolium*), generally as scattered clumps, sometimes as extensive stands; Although not strictly an *Acacia* shrubland, the structure of the community is similar to other examples of this landscape type.

Mulga is a very widespread species throughout dryland Australia. It occurs as the dominant species on light-textured red-brown soils from the Darling River in New South Wales, westward across South Australia and Western Australia to the Indian Ocean, and northward into Queensland. Extensive areas also occur in central Australia.

The large belt of mulga throughout the West Darling area of New South Wales occurs on red sandy loams and loams on the calcareous red earths, extending into southern Queensland on red earth soils. In the drier areas the mulga occurs as sparse or scattered trees with an understorey of perennial grasses and herbage. On firmer soils, in higher rainfall areas, it forms extensive stands of fairly dense scrub at 5-15 yards spacing, with open well-grassed areas scattered throughout. As average rainfall increases, the mulga becomes increasingly dominated by taller species such as the bimble box or poplar box (*E. populnea*).

In central Australia, mulga occurs on level and slightly undulating red earthy sands. Here it is typically arranged in groves or contour-aligned clumps of trees at 5-10 yards spacing over an area of one to several square chains, separated by inter-grove areas of mostly bare soil. Run-off from the bare inter-groves accumulates in the groves, providing a good growth of grass and herbage amongst the trees. Mulga also occurs frequently in the swales between the sand hills of the spinifex dune fields, and in small areas favoured by better moisture relationships in the spinifex sand plain.

In Western Australia, the mulga communities occur on a variety of soils, including shallow stony and gravelly skeletal soils, red earths, red-brown earths and the shallow loams of the red and brown hard-pan soils. Consequently the structure and composition of the mulga communities is also variable and many communities have a strong component of associated smaller shrubs. Mulga is also the dominant tree species on the stony hills and ranges throughout practically all dryland Australia. These areas have been described under the landscape type 'Mountains and hills'.

The acacia shrublands carry a good pasture of perennial grasses (*Aristida* spp. *Eragrostis* spp. *Stipa* spp. and others), with a fair range of ephemeral grasses and herbage when seasonal conditions are good. Copper burrs (*Bassia* spp.) occur in most pastures, sometimes being the principal herbaceous component.

Many areas, particularly throughout South Australia, carry an understorey of saltbush and bluebush, together with good grasses. The Western Australian mulga shrublands also carry a greater component of small shrubs which sometimes includes *Atriplex* and *Kochia*.

MANAGEMENT CONSIDERATIONS

This landscape is normally stable with only slight to moderate susceptibility to erosion, although serious erosion will occur with over-use. Most examples of severe degeneration of mulga country, as shown by serious wind or water erosion and widespread death of mulga, occur on areas where mulga is growing out of its normal habitat. This is often the case where texture-contrast soils are mixed up with red earths and the mulga has established a community on the texture-contrast soils.

On sandy loams in western New South Wales, over-use occurs along stock routes and around watering points and, as often happens adjacent to more productive country, is often reflected by an increase of useless shrubby species such as turpentine (*Eremophila sturtii*), hopbush (*Dodonea* spp.) and puntybush (*Cassia* spp.).

The mulga and other tree and shrub species provide valuable drought forage. Because of this and the low susceptibility to degeneration, this landscape type is a most useful one for drought refuge. Landholders who have a mixture of country composed of these stable acacia shrublands and/or highly erodible types should make maximum use of the latter while seasonal conditions are good, and stock the acacia shrubland lightly so that it will contain a good reserve of ground forage, in addition to the topfeed, for use during drought periods.

The same principle can be applied to open and heavily-timbered country on this same landscape type. Open well-grassed country should be used to the maximum while seasonal conditions are good, saving the well-timbered country for use during drought periods. Maximum use in this context does not mean grazing the ground bare of vegetation. It means making maximum use of the flush growth of ephemeral short-lived grasses and herbage while they are available, but maintaining the stand of perennials in good condition.

A large proportion of the mature mulga on sandy soils in western New South Wales has died out over extensive areas, leaving only occasional trees alive. Most noticeable is the lack of young mulga on these soils, although on the hard red loamy soils regeneration is plentiful, so much so that it often competes seriously against ground forage in many areas. It is probable that the deaths are due to the trees approaching the end of their normal life span and being killed by drought or disease. Drought appears to have been responsible for widespread death of mulga in a large belt of country north of Alice Springs in the early 1960s.

Extensive stands of mulga were also killed in the early developmental phase of the arid zone pastoral industry by lopping for stock feed during drought. By taking only part of the tree and leaving a strong leader, mulga can be used for drought feeding in a way which will ensure the continued existence of the tree. Whether mulga should be lopped in this way to preserve it, or destroy it by pushing or cutting, depends on the nature of its occurrence. On loose sandy soils mulga is fairly sparse, and regeneration limited to small areas favoured by run-on. In these circumstances care should be taken to take only part of the tree and allow the remainder to continue growing.

On firm and hard red loams, mulga often occurs as thick and tall forests, and there is adequate regeneration. In the circumstances the productivity of the country can be improved without destroying the resource of drought forage, by pulling or cutting the mulga so that the tree dies. However, this should be done so that a reasonable number of trees is retained to provide seed for future regeneration, to give protection against erosion during drought periods when all ground forage is removed, and to preserve for future drought feeding.

If mulga is being cut or pushed individually, sufficient well-grown young trees should be left to provide an average spacing of 15-25 yards. At this spacing there is little competition against ground forage and an adequate resource of topfeed is left for the future, as well as adequate protection to the soil.

If mulga is being pulled by chain or cable, it is difficult to leave trees evenly spaced. However, to ensure adequate topfeed for the future, and protection of the soil, each run with the cable or chain should be separated by at least one chain distance of untouched mulga. The elongated strips of fallen mulga left by this treatment should be broken at intervals of half a mile or so by a chain or so of untreated mulga. Clearing contractors will be loath to work in this fashion, and the cost to the landholder per acre treated will be a little higher, but the benefits in future productivity will greatly outweigh the slightly increased cost.

The thinning or removal of mulga on hard red loams in the higher rainfall areas should always be approached with caution in areas where it is not feasible to follow timber treatment with cultivation. Experience has shown that in many areas the killing of a species in a face frequently results in an invasion by another species which may be of no value as a browse species, and which may compete for pasture growth much more vigorously than the species removed.

SHRUB WOODLANDS

This landscape type occupies the semi-arid zone in New South Wales and Queensland, occurring on undulating and level red earth soils, with some areas in central and northern Queensland on black earths and yellow earths. It extends from the Murray River north-eastward through New South Wales to the Queensland border, thence generally northward to near the Tropic of Capricorn. Average annual rainfall ranges from 10-16 inches in the winter rainfall zone at the southern extremity to 20-25 inches, with a marked summer predominance at the northern extremity.

Shrub woodlands are composed of an open to moderate stand of taller trees with one or more layers of smaller trees and shrubs. Bimble box is the most widespread of the taller trees, while species such as ironbarks (*Eucalyptus melanophloia* and *E. crebra*), belah (*Casuarina cristata*), brigalow (*Acacia harpophylla*), gidgee (*A. cambagei*), ironwood (*A. excelsa*) white cypress pine (*Callitris glauca*) and others may be predominant species over small or large areas. The range of shrub species is extremely wide. In the far north, this landscape type is bounded on the west by the Mitchell grass plains on brown and red self-mulching clays, and in the far south by the saltbush plains on the Western Riverina. However, over a long distance from central western New South Wales into southern Queensland, the shrub woodland grades into the acacia shrublands, the former occupying the better-favoured sites as rainfall decreases from east to west.

In New South Wales the soils are level to very gently undulating red and brown loams and gravelly loams of the red earth soils, with some areas of sandy loams of the calcareous red earths. Belts of country with deeper texture-contrast soils in the broad expanse of river flood plain country in north-western New South Wales also carry a typical shrub woodland type of vegetation.

Red earth soils composed of hard loams and gravelly loams occur in a belt of country in western New South Wales lying between the Lachlan and Darling Rivers. This country supports a shrub woodland dominated by bimble box (*E. populnea*) and white cypress pine, with several smaller trees and shrubs strongly represented. On harder loams in the Cobar district, the white pine becomes increasingly confined to run-on situations such as the flats between ridges, as rainfall decreases. Its place is taken by a number of tall shrub species such as mulga, ironwood, yarran (*Acacia homalophylla*) budda (*Eremophila mitchellii*) wilga (*Geijera parviflora*) and others. In recent years the smaller shrubs, hopbush (*Dodonea* spp.), turpentine (*Eremophila sturtii*) and puntybush (*Cassia nemophila* etc.) have increased very considerably and have taken over large areas of formerly open woodland.

Pastures are composed principally of the perennial grasses; spear, wire, windmill and wallaby grasses, with Queensland blue grass in some favoured areas. There is a fair range of herbage species such as crowfoot (*Erodium* spp.), tah-vine (*Boerhavia diffusa*), various copper burrs (*Bassia* spp.) and bindi-eye (*Calotis* spp.).

Much of this country in the higher rainfall areas above about 15 inches per annum has been cleared in the early development of the lands and now supports a reasonably stable grassland of spear grass (*Stipa variablis*), wire grass (*Aristida* spp.), wallaby grass (*Danthonia caespitosa*) and windmill grass (*Chloris truncata*). Much of this landscape type has also been developed for wheat cropping. The trend in the marginal areas has been to improve the land for grazing by clearing and burning and sowing down to about three crops. This initial cropping phase has the dual purpose of providing a cash crop to pay for the heavy clearing costs and to prevent regeneration of the timber species.

In the case of leasehold land in the Western Division of New South Wales, this form of land treatment is subject to conditions similar in nature to those described for the mallee landscape type with the object of preventing over-cultivation and erosion. In Queensland this landscape type is very variable, being mostly gently undulating country with red-brown earths and red earths, with some yellow earths in the far north. This is broken by areas of hills and ranges in the central area with shallow sandy soils. Extensive belts of grey-brown self-mulching clays occur with some belts of black earths.

The poplar or bimble box is also dominant over a wide area throughout the length and breadth of this landscape type in Queensland, occurring on a wide variety of soils, from heavy grey cracking clays to deep sands, but most typically on red earths and the deeper texture-contrast soils. The silver-leaved ironbark (*E. melanophloia*) forms extensive woodlands in the northern section of the zone, extending southward as far as the New South Wales border. In southern and central Queensland, brigalow and belah are frequently the dominant species on more fertile soils with higher clay contents.

Because of the wide variation in soil characteristics within this landscape type, and also the wide range of climate, there is a considerable variation in the character of the pastures. In the sub-tropical areas in Queensland, the light-textured soils support pastures dominated by species of *Bothriochloa*, with *Eriachne* and *Aristida*, with some areas of the harsh *Triodia* generally associated with the silver-leaved ironbark. Further south in southern Queensland and northern New South Wales, the *Bothriochloa* becomes less frequent, with *Chloris*, *Neurachne*, *Eragrostis*, *Enneapogon*, and *Aristida* and *Stipa* all being represented.

In recent years there has been considerable activity in the development of the brigalow, belah and gidgee country on the heavy clays and clay loams in Queensland. These soils have a high fertility status and develop very productive pastures after clearing of the heavy timber cover.

MANAGEMENT CONSIDERATIONS

One of the most important management problems on this landscape type is that of timber treatment to improve the productivity of the natural pastures.

On the firmer red earths and deeper red-brown earths bimble box may occur in dense stands at 5-10 yards spacing, or budda (*Eremophila mitchellii*) and other species may occur with it at similar spacings, and effectively choking out pasture growth. Ringbarking of bimble box needs to be followed up by sucker-bashing one or more times to give an effective kill. If this is not done, the stand resulting from the regrowth of suckers will be worse than the original stand. Budda may be killed by ringbarking if it has a single well-developed stem. Poisoning has had variable results from good to poor, and in any case increases the cost of treatment.

Experience has shown that opening up of the natural timber stand is likely to lead to an invasion of native shrubs such as budda, turpentine (*Eremophila sturtii*), hopbush (*Dodonea* spp.) or puntybush (*Cassia* spp.) which will reduce productivity within the space of a few years to less than it was prior to treatment. A prolonged run of above average seasons in the early 1950s has led to a serious increase in the above shrub species in the Cobar district of western New South Wales. As the growing shrubs take up more and more space normally occupied by pasture, the landholder continues to run stock at the same level as usual, with the result that the pastures, depleted by competition from shrubs, are subjected to serious over-grazing. Under these conditions the ground surface is nearly always bare and subject to erosion from sheet flow as huge volumes of run-off are drained from extensive areas of bare country.

In addition to the reduction in grazing capacity brought about by increasing shrub growth and consequent soil erosion, the landholder also has problems in mustering country in which it is difficult to see more than 20-30 yards. Mating under such conditions produces a very low conception rate, while the poor quality of the pastures makes it difficult for ewes to carry a healthy lamb. Consequently lambing percentages are very low. Attempts at breeding meet with success only in exceptionally good years.

The problems posed by the increase of scrub and timber, and proposals put forward to meet these problems, have been outlined (N.S.W. Inter-departmental Committee, 1969). Contour furrowing of sheet-eroded slopes will add considerably to the quantity and quality of pasturage on undulating country sufficiently open to permit of contour furrowing. In some areas the very heavy run-offs can be harnessed by spreader banks (see Chapter 8) to provide conditions suitable for growing fodder crops and improved pastures such as lucerne. This will allow improved nutrition of the ewe at mating and lambing and provide better quality forage for growing lambs and weaners.

From past experience it would seem that the only way to prevent timber and scrub from regenerating and taking over country which has been treated, is to cultivate for 2-3 years after clearing, and to be prepared to cultivate again in the future should there be a further invasion of tree and scrub species.

Large areas of this landscape type in the southern portion of its occurrence in New South Wales was successfully cleared by ringbarking or pulling. This was accomplished in the early days of settlement when labour was cheap and plentiful. However, with the exception of some areas invaded by heavy seedling growth of white cypress pine (*Callitris glauca*) in the 1950s the scrub regrowth problem is not as serious there as it has been on the harder loams further north.

FURTHER READING

Beadle, N. C. W. (1948). *The vegetation and pastures of Western New South Wales with special reference to soil erosion.* Govt. Printer, Sydney.

Brookfield (in press). On wind and the trend of desert dunes—a case study in Central Australia.

Costin, A. B. & Mosley, J. G. (1969). 'Conservation and recreation in arid Australia'. In Slatyer, R. O. & Perry, R. A. (eds.) *Arid Lands of Australia.* A.N.U. Press, Canberra.

Leigh, J. H. & Mulham, W. E. (1966). 'Selection of diet by sheep grazing semi-arid pastures of the Riverina Plain. Bladder saltbush (*Atriplex vesicaria*)—cottonbush (*Kochia aphylla*) community'. *Aust. Journ. Exp. Agric. & Animal Husb.*, 6 : 460.

Leigh, J. H. & Noble, J. C. (1969). 'Vegetation resources of arid Australia'. In Slatyer, R. O. & Perry, R. A. (eds.) *Arid Lands of Australia.* A.N.U. Press, Canberra.

Mabbutt, J. A. (1969). 'Land forms of Arid Australia'. In Slatyer, R. O. & Perry, R. A. (eds.) *Arid Lands of Australia.* A.N.U. Press, Canberra.

Moore, R. M. (ed.) (1970). *Australia's Grasslands.* A.N.U. Press, Canberra.

N.S.W. Inter-departmental Committee. (1969). *Report on scrub and timber regrowth in the Cobar-Byrock district and other areas of the Western Division of N.S.W.* Govt. Printer, Sydney.

Newman, J. C. & Condon, R. W. (1969). 'Land Use and Present Condition'. In Slatyer, R. O. & Perry, R. A. (eds.) *Arid Lands of Australia.* A.N.U. Press, Canberra.

Perry, R. A. (1969). 'Rangelands research and extension'. In Slatyer, R. O. & Perry, R. A. (eds.) *Arid Lands of Australia.* A.N.U. Press, Canberra.

Ratcliffe, F. N. (1936). *Soil drift in arid pastoral areas of South Australia.* C.S.I.R.O. Pamphlet No. 64.

Wilcox, D. G. (1963). 'The pastoral industry of the Wiluna-Meekatharra area'. In *Lands of the Wiluna-Meekatharra area, Western Australia.* C.S.I.R.O. Aust. Land Research Series No. 7.

See also reports on various regions of dryland Australia by C.S.I.R.O. Division of Land Research in the Land Research Series.

CHAPTER 10

Amenity planting

THE most important role of trees and shrubs in the dry country is in amelioration of the local microclimate and in the beautification of homesteads and towns. Also of value in the arid zone and during droughts elsewhere, are those species which make a contribution as fodder reserves for stock. Only of local and incidental importance is the production of timber or other useful products. Trees and shrubs influence soil conditions and are, of course, of considerable value in erosion control; through the soil they also have an effect on the ground vegetation and the fauna of the area. The first has an effect on vegetation and, therefore, the growth and condition of pastures, and the vigour and health of both domesticated and native animals. There is a marked feeling of relaxation experienced by man when passing from sunshine into shade on days of heat wave temperatures or from an exposed site into a sheltered one during periods of either very cold winds, especially if accompanied by rain squalls or snow, or very hot winds of high intensity. The effect of shade and shelter on vegetation and animals can be measured in terms of productivity and resultant financial return; on man the effect is less susceptible to measurement but is nonetheless real. The benefits of the beautification of homesteads and towns also cannot be measured in terms of dollars and cents but are important as they have a profound effect on man's psychological attitude towards life. Well-designed planting of trees and shrubs changes houses without trees into homes where the family delights to gather for rest and relaxation; they alter bare towns which the traveller wishes to pass through as quickly as possible to places where he and his family, especially if on holiday, wish to linger. There is little need to stress the different reaction of the family group to a picnic or camping ground with pleasant open areas surrounded by trees and shrubs which have an aesthetic appeal, as well as providing a screen from winds and high density heavy traffic, from a site where a treeless area with minimum toilet facilities is all that is offered.

Amenity planting will be discussed in connection with the beautification and protection of homesteads, the use of trees and shrubs in urban areas and roadside planting. The provision of a suitable level of shade and shelter for animals and crops is considered in the following chapter.

There are a number of striking examples of the successful use of trees and shrubs in the larger towns and cities of the dry country and descriptions and illustrations are given for Alice Springs, Broken Hill, Cockatoo and Koolan Islands, Kalgoorlie, Kambalda, Mt Isa, the North West Scheme of the Government of Western Australia, and Woomera.

BEAUTIFICATION AND PROTECTION OF HOMESTEADS

The homestead is the focal point of family living. It is the centre from which work and recreation, rest and social activities, community interests and private family life are developed. This focal point aspect is more marked in the sparsely-populated rural areas than it is in large towns or the more densely-populated cities where many of the family, especially the younger members, go away from the home for recreational and social activities. A house located in a comparatively barren landscape without trees and shrubs lacks appeal and there is certainly a different psychological attitude towards it, compared with the position of a home which has been beautified and protected by wisely-selected trees and shrubs.

Hamilton (1969) has very appropriately written:

> It is vital that an appropriate atmosphere is developed around a homestead so that living can be fully enjoyed and work become less tedious in the pleasant surroundings. The appearance of the home should proclaim its haven-like qualities, being happily positioned within the landscape so that, even from a distance, there is a suggestion of the peace and enjoyment waiting within. Trees can do much to provide this kind of external setting for family living. . . . . . .
> Homestead planting is usually regarded as a beautification, but it has far greater significance, exercised through its deep psychological impact on man and by physical modification of his environment. Open and unprotected spaces are too large and too frequent in most Australian rural areas–unmodified they form a poor location for a home, even though the contours of the land may approach the ideal. It is necessary to soften outlines, to unite buildings to the land by means of an appropriate arrangement of vegetation which extends into the surrounding fields, thereby avoiding the formation of unpleasantly abrupt boundaries. This involves ornamental planting at some distance from the home as well as in its immediate vicinity, since by this means the homestead is integrated with the rural scene.

Compared with urban dwellers on a small block, often less than a quarter of an acre in extent, people on the land are fortunate in having more scope for tree planting. In the case of large pastoral holdings there is no practical limit to the area which can be used for planting or for the conservation of natural vegetation. In these areas the location of the homestead is of critical importance in controlling the extent and nature of planting and conservation; where it has been sited near a drainage channel or, in fortunate cases, accessible to a large pond or perennial stream, there is the greatest scope for modification of the landscape. Where the homestead is on a well-drained elevated rocky rise there is a severe restriction on the species which can be planted and considerable reliance may need to be placed on the limited number of hardy native species which grow there. In the case of small rural holdings, say only one or two hundred acres in extent, there is still scope for growing large trees around the homestead but consideration must be given to proximity to telephone and power lines, and to the possibility of tall trees casting unwanted shade on the homes or paddocks of neighbours.

Factors in the selection and use of tree and shrubs species which are of general application are discussed in Chapter 18. Other factors which should be taken into consideration when planting around homesteads are considered here:

1. Do not commence a planting scheme haphazardly. Any long-term project of this nature should not only be carefully considered in advance of commencing

planting but should be committed to writing and plans which show the proposed location of all the trees and larger shrubs, having regard to the position of telephone and power lines and also sewers. In very dry areas tree roots will travel a long distance in order to obtain water from a sewer. In very hot areas shade for buildings in the mornings and afternoons may be more desirable, and also more practicable, than at midday.

2. Probably the first point to be considered should be the aspect of the home in relation to strong and cold or very hot winds, or to views which it is desired to maintain. Decisions made at this stage will control where trees and large shrubs can be located.
3. Decide on the purpose for which planting is to be done, that is shelter, shade, ornament or as a source of food for birds. Some plants will fill only one or two of these objectives whilst a few may fill all. The landholder who is fortunate enough to be able to plant even a few trees and shrubs, especially if they bear attractive fruit, will be well rewarded by the increase in birdlife. Tall trees may also help by providing safe nesting sites.
4. At this stage read up descriptions of likely trees, not only with regard to filling the requirements of item 3, but also as to suitability for the local environment. If possible have a look at neighbouring homesteads or the local towns which have planted the trees and shrubs in which you are interested, taking care to note whether the specimens are mature or only partly grown. Many people who have been attracted by a beautiful young tree when it was only a quarter to half full height, have later been burdened by a tall tree dropping leaves into gutters, undesirably shading the house or even endangering the fabric. At this stage the cost and difficulty of removal without damage to the house may be considerable.
5. The next stage is to consider the planting programme, desirably planned over a number of years. When quick early shelter is wanted for a newly-built house it may be practical to plant a temporary shelter belt of a fast-growing but probably short-lived shrub and further away establish a slower growing but more permanent species—often a long-lived tree.

Further suggestions on tree planning around a homestead is given by Hamilton (1969) and details on planting and watering are given in other chapters of this book and in *Growing Trees on Australian Farms*. Descriptions of species are given in Chapter 17 and in the several books listed at the end of this chapter and of Chapter 17.

AMENITY PLANTING FOR COMMUNITY PURPOSES

The present attitude of the Australian public towards the role of trees and shrubs in the suburban and city landscape has changed considerably from the early pioneering days of the late eighteenth and early nineteenth centuries. The earliest settlements by Europeans were mainly on the forested coastal areas where trees were an impediment to initial development and, following the pattern in most newly-settled lands in the temperate zones of the world, the pioneers proceeded to fell and burn all tree growth. Unfortunately they often went beyond the stage necessary for optimum cropping and grazing conditions and proceeded almost to the point of devastation in many centres. Even in those days, however, there were a thoughtful few who remembered the lands of their birth and retained at least some specimen trees or small pockets of native forest; this was particularly the case among those who had received large grants of land. In those days township sites were, all too often, completely cleared of trees and shrubs and the result was barren-looking settlements which offered few attractions as residential areas. As the initial difficulties of carving a living in a new land were overcome people were able to give attention to the amenities of life. Local authorities commenced to plant trees and shrubs in order to provide shade and shelter, as well as beautify their towns, and many were fortunate in that the early surveyors for the State governments often made provision for future parks or designed very wide principal thoroughfares where trees could either be retained or planted. Today there are few local government authorities who do not allocate at least a small part of their budget to amenity planting.

There was a contrast between conditions in the relatively high rainfall coastal areas of southern Australia and the drier inland regions. In the former there grew naturally many fine trees which could be retained along wide thoroughfares or in parks and other reservations and, in addition, the climate was usually suitable for the introduction of some of the good shade and ornamental trees which the early settlers had known in their native countries. The drier country, on the other hand, had fewer large trees, and a more difficult environment into which European broad-leaved trees, such as elms (*Ulmus* spp.), could be introduced. In the course of time there was an increase in the knowledge of native species suitable for town and park use and also of exotic species which could be introduced; many of the most successful of the latter have come from the drier and hotter parts of the world.

For any form of amenity planting it is recommended that at least outline plans should be prepared at the initiation of a scheme; these should not only record existing trees and shrubs but also show the long-term proposals. Somewhere in the records, possibly on the sides of the plans, should be noted the expected maximum heights and spread of trees at maturity for that locality. This latter information is of considerable importance and not only should reliable reference books be consulted but measurements made of trees growing in the local town or adjacent areas. Care should also be exercised in using data from a book which has been written for an entirely different locality to the one in which it is proposed to grow the species concerned. It is suggested that copies of the plans should be available for inspection by the public as well as being readily available in the engineers' office, so that there can be co-ordination with all other aspects of town planning, that is with the position of telephone and power lines, sewerage and water pipes and proposed alterations to the principal thoroughfares. Plans and other records of this type are of critical importance for street planting and are useful for other types of tree establishment.

Only some of the factors which should be taken into consideration in selecting trees for amenity planting are considered here and the subject is considered more comprehensively in Chapter 18. Other chapters deal with methods of establishment and tending.

STREET PLANTING

For use in street planting trees should:

(a) Be easy to establish, preferably with the ability to be transplanted as advanced nursery stock, grow relatively quickly to the stage that they provide some amenity value and are beyond the stage of greatest risk from vandalism.

(b) Be healthy in the environment, relatively long-lived and not subject to wind-throw or the breakage of large limbs.

(c) Be as maintenance-free as possible; in this regard many of the species suitable for the dry country compare more than favourably with deciduous species used in wetter areas, which may need regular pruning and removal of fallen leaves.

(d) The form and height of the species must be suitable for the width of the street in which they are to be planted and for distance from sewers and so on.

(e) Where a thoroughfare crosses different types of topography, for example from somewhat dry ridges with shallow soils to moister soils in valleys, it is desirable and possibly essential to change the species used.

It has been pointed out elsewhere (Chapter 18) that the selection of the best species for use in streets presents many difficulties; it should not be done without due consideration of all information which is readily available and this includes visits to neighbouring towns in the same climatic zone.

SMALL SUBURBAN AND CITY PARKS

The selection of species for this purpose, in general, presents fewer problems than choosing trees for street planting, since growing conditions are less restricted and telephone lines and sewers can usually be avoided. In addition, the mature size of the trees and shrubs can be readily related to the size of the park. In the case of a small area, perhaps only two or three acres in extent, possibly one or two moderate-size trees will provide all the shade desirable, whilst small belts of shrubs will give any shelter needed from wind or act as screens to traffic noise. In the selection of species allowance should be made as to whether regular watering of the lawns in summer is included in the maintenance plan or not. In the arid country where watering is not practised some of the denser-crowned Western Australian eucalypt species, such as coral-flowered gum (*E. torquata*) merit first consideration. On the other hand, in the semi-arid zone where some watering can be done there is a wider choice of species possible, including exotics such as elm (*Ulmus* spp.). In any park planting it is undesirable to concentrate on only one species or on two closely-allied species of the same genus, since in the event of serious insect attack or outbreak of a fungal disease the value of the park could be seriously reduced; a mixture of species, especially of widely different genera, is a good form of insurance against catastrophes of this type. Most of the general factors which have been mentioned for street trees also apply here.

One aspect of amenity planting which receives less attention than it merits is the provision of shade trees in car parks, especially in small towns where space is not restricted and at tourist centres which are mainly used in the summer. The provision of some shade need not have any real effect in restricting the number of cars which can be parked in a given area; the trees can be spaced 30-50 feet apart in lines between double rows of vehicles. Each tree needs a concrete or stone wall or other type of guard around it, to afford protection from vehicles. If the car park is paved the protective wall should be a few feet, that is two to four, from the tree trunk in order to allow rain to soak into the soil. If necessary the trees can be watered in summer and also entrance holes through any solid protective wall should be provided, so that some of the run-off from a paved surface will find its way into the depression around each tree.

LARGE PARKS AND RESERVES

Large parks, which may be located either within urban or rural areas, are often considered as public sporting and recreational areas and, as such, the role of trees and shrubs is to provide the degree of shade, shelter and beautification desirable for the type of activities for which the park is planned. Sporting ovals, for instance, need to be free of trees but a judicious spacing of shade trees or a shelter belt on the windward side can ameliorate extremes of temperature and wind to spectators. The provision of shade is more desirable in the hotter inland districts of the dry country than it is in the comparatively temperate coastal areas. When species are being introduced which are not native to the area, it is desirable to consider only trees which are hardy to the locality, will not need watering and only need the minimum of other maintenance. In most cases formal patterns of layout of trees, for instance in long straight rows, do not fit happily into the landscape pattern and planting in small clumps or as individual specimens is usually more suitable.

Reserve is the term commonly used when referring to areas which have plant, animal or other features of sufficient importance to warrant preservation in as near the natural state as possible. Other than protection from fire and possibly animals such as rabbits, the natural vegetation is usually not touched and planting of either native or exotic species is not considered.

ROADSIDE PLANTING

The thoughtful use of trees and shrubs can not only improve the aesthetic appeal of a public thoroughfare but can add to the safety of driving by reducing the monotony of long stretches of bare, uninteresting highway. Trees ameliorate highway conditions by providing shelter from wind or snow on exposed sites, shade in summer and attractive vistas at all times. Roadside planting, and retention of native species where present, is a concept which includes not only planting along country roads and major highways but the provision of amenities at roadside picnic areas, stopping points for visitors and rest areas for heavy transport. The need for such amenities will vary with the local topography and natural vegetation, but planners are only too well aware that tree cover on private property, especially on the edges of growing cities, can disappear and it is their responsibility to make provision within the highway area for the amenities considered desirable. It is probably only alongside roads through national parks or dedicated forests that the permanence of tree cover can be guaranteed.

Amongst the important factors which should be considered in planning the use of trees along thoroughfares are:

(a) Selection of the species for hardiness, longevity, freedom from windthrow and breakage, attractive appearance and minimal maintenance. This is a formidable list of requirements but there are many trees and shrubs which substantially fill all of them. In the semi-arid zone there are a large number of native small trees and large shrubs which can be used, as well as some exotic species. Where the environment is suitable consideration should be given to small patches of deciduous species with colourful autumnal foliage.

(b) Suitability of the species to the climate, topography and soil. Even on plain country where there is little change in topography there can be variation in soils which call for different species—for instance, in the semi-arid zone of eastern and south-eastern Australia most of the box group of eucalypts, such as bimble box (*E. populnea*), are suitable for planting on heavy soils which may sometimes be waterlogged for short periods during winter, but on adjacent well-drained sandy rises white cypress pine (*Callitris glauca*) is one of the better trees to use. In general, on well-drained sandy or rocky ridges it is desirable to select from local hardy species, but on fertile, moist river flats it is possible to introduce a much wider range of species, including some of the northern hemisphere deciduous trees, for example Lombardy poplar (*Populus nigra* var. *italica*) and English elm (*Ulmus procera*). Where a road runs alongside a perennial stream very fine effects can be obtained by the combination of Lombardy poplar and weeping willow (*Salix babylonica*).

(c) Location of the trees in relation to road formation. Firstly consideration should be given to the existing road formation so that trees are not planted close to the inside of curves or near road junctions where they could obscure vision and so create a driving hazard. In any case, trees should not be planted near the edge of the formation so that there could be serious risk of a vehicle out of control hitting the trunks and so endangering the lives of occupants; this factor is particularly critical on the outside of curves and at the approaches to bridges. On more or less level terrain and on straight stretches of road, 20 feet might possibly be considered a reasonable distance; on the lower side of steep embankments the distance could be reduced, say to about 10 feet, as the trees would form a safety barrier which could prevent a vehicle falling a much greater distance. Secondly, consideration should be given to the possibility of the future widening of road formation, including the development of double traffic lanes. Whilst no one can completely foresee the future, the engineers and planners of most local authorities can make a sound forecast of the possibility of such developments and whether tree planting could be done within defined distances from the existing formation. Intelligent anticipation of the future could provide well-grown trees in places where they would not need to be removed when later road development takes place. Many of the three and five chain wide road reservations of the sheep and cattle country offer considerable scope for tree planting where there are no natural species present, but the one chain wide roads present much greater problems in tree location if there is any chance at all of a road being widened.

In areas where moderately heavy frosts occur it is not desirable to have a dense shelter belt of trees on the eastern and northern sides of a road; this is to allow the morning sun to melt the layer of ice on the sealed surface as early as possible.

The use of trees on expressways is outside the scope of this book, but such thoroughfares are not likely to be numerous in the arid and semi-arid zones. This is a field of planning for which the landscape designer is responsible; here it will only be noted that large trees should be placed well back from the formation, including the tops of cuttings, that wide spacing (one to two hundred feet) is usually desirable and that scattered clumps may increase the aesthetic value.

The establishment of 'remembrance driveways' have periodically had a strong popular appeal, especially in the decade after a conflict such as World War II. When expertly designed, well executed and satisfactorily maintained they do much to enhance the roadside environment, besides fulfilling their original purpose as a memorial to the lives of men who served their country faithfully in time of national crisis. Experience from existing plantings emphasises the importance of the selection of species suitable for the variations in soil and topography which occurs along most roads. In hilly or undulating country it is rare that one species will be satisfactory for continuous planting along a mile or more of highway, and nothing detracts more from the overall aesthetic appeal than patches of stunted, diseased or dying trees. Where funds are limited or where the best effect is desired for a given effort it is suggested that small belts, say 10-20 chains in length and 1-2 chains wide, should be established on reasonably uniform sites. The soil should neither be too wet or stony if good tree growth is required. Another form of remembrance planting is the planting of 'memorial groves'; either one species can be used or a small number which will grow satisfactorily together. There is nothing new in this suggestion of memorial groves since they have been planted by mankind since the earliest historical times, notably to the gods of the pre-Christian era. Nowadays these groves instead of being the site of sacrifices, sometimes human, to unknown gods, should be suitably located for picnic places, sites where scenic views are obtainable or rest areas for heavy vehicles. Because of the relatively small size of such groves it is usually possible to select areas of relatively uniform soil quality, with the result that even tree growth can be expected. The trees can be spaced close together so that vehicles cannot enter the inside of the grove or wide apart so that each tree will have room for crown expansion and holiday makers can park and picnic throughout the area; it is suggested that allowance of vehicle entry is more suitable for large groves, which tend to resemble a small park, than it is for small groves.

AMENITY PLANTING IN INLAND CENTRES

Australia has many interesting examples of the use of trees and shrubs in the towns and cities of the dry country. Conservation of the native vegetation has been practised by some centres, notably the city of Broken Hill which initiated natural regeneration areas in 1936 and planned extension in 1969, and the new nickel town of Kambalda which retained native trees and shrubs when the residential and recreational areas were established in 1967-71. Descriptions and illustrations are given for eight places of the dry country; geographically these range from the centre of the continent (Alice Springs) to the coast (Cockatoo and Koolan Islands), and from southern Western Australia (Kalgoorlie and Kambalda) to the north of Queensland (Mount Isa). Whilst most of the towns and cities have originated from the mining and processing of minerals, others have a different background; for instance Alice Springs goes back to 1871 when a telegraph station was established there, whilst Woomera is a residential area for Commonwealth Government research work.

It is considered that these and other places should be of interest to all Australians concerned with the initiation and administration of amenity planting elsewhere in the dry country. It is also suggested that other countries with large areas of arid and semi-arid land may find much to interest them in many of the centres which have been described and illustrated in this chapter.

ALICE SPRINGS

Alice Springs is situated on the Todd River in the southern half of the Northern Territory and not far from the centre of the continent; it is 1,028 road miles from Adelaide and 954 miles from Darwin. The name Alice Springs was originally given to a useful waterhole by the builders of the overland telegraph, when the local telegraph station was established in 1871, and was not officially applied to the town until 1933. Previously it had been known as Stuart, that name having been given to the site in 1889 in honour of the early explorer John McDouall Stuart. The name Alice commemorates Alice Todd, wife of a Postmaster-General of South Australia. The establishment of the telegraph station, the availability of good water in the gravels underlying the plains of the Todd River (which only flows at irregular intervals) and the presence of arid pasture suitable for cattle resulted in Alice Springs developing as the most important cattle and administrative centre in the area including the northern half of South Australia and the southern part of the Northern Territory. Rail connection from Adelaide, but with break of gauge at Marree, was made in 1929. The importance of the town greatly increased in World War II with the construction of a bitumen road to Darwin and its development as an important staging centre. More recently, especially in the 1960s it has shown striking growth as a tourist centre, both for overseas and Australian visitors. Amongst leading tourist attractions are Ayer's Rock and Standley's Chasm, but there are many others, including the school of Aboriginal painting made famous by Albert Namatjira (1902-59).

The town is managed through the District Officer, who is the representative of the Administrator of the Northern Territory.* He receives advice from the Alice

* Some change was made in 1971 and a mayor was elected in June.

Springs Town Management Board, which comprises five elected citizens and four officials appointed by the Northern Territory Administration. The estimated population at 30 June 1970 was 10,000, which represented a growth rate of 11 per cent per annum during the three preceding years.

The summers are hot to very hot, with a mean maximum for the hottest month of 95.3°F and many days with temperatures between 100 and 110°F, but the mean minimum for the same month is not unduly high at 69.8°F. Visitors from the cooler southern coastal areas find the day temperatures in winter very pleasant, for example the mean maximum of the coldest month is 66.9°F; night temperatures are comparatively cool, with a July mean minimum of 38.9°F. Although Alice Springs, at 23° 38′ S. latitude, is almost on the Tropic of Capricorn there are an average of 12 frosts a year—partly a reflection of the altitude of 1,901 feet. Rainfall is low and highly variable—the 50 percentile value is 10.09 inches, whilst that for the 10 percentile is 5.41 inches and the lowest on record over a period of 92 years is 2.37 inches. It is not unusual for there to be several consecutive years with rainfall below the 50 percentile value, and this is prejudicial to the profitability of cattle raising.

There are several species of trees which grow naturally in the area and are retained in suitable locations or used for new public and private plantings. Amongst the eucalypts ghost gum (*Eucalyptus papuana*) and river red gum (*E. camaldulensis*) are the two main species. Ghost gum grows both on alluvial flats and on

TABLE 10.1

SOME OF THE MORE IMPORTANT TREES AND SHRUBS USED FOR PUBLIC AND PRIVATE PLANTING AT ALICE SPRINGS

Botanical names	*Common name*	*Growth form*
NATIVE SPECIES (*a*)		
Acacia aneura	mulga	Shrub
A. pycnantha	golden wattle	,,
Brachychiton gregorii	desert kurrajong	,,
Callistemon macropunctatus syn. *C. coccineus*	bottlebrush	,,
Cassia nemophila	cassia, yellow cassia	,,
Casuarina cunninghamiana	river sheoak	Tree
Eucalyptus salmonophloia	salmon gum	,,
E. torquata	coral-flowered gum, coral gum	Small tree
Gossypium sturtianum syn. *Cienfugosia gossypioides*	Sturt desert rose	Shrub
Melaleuca incana	grey honey myrtle	Shrub
EXOTIC SPECIES		
Dalbergia sissoo	Himalaya raintree	Tree
Schinus molle	pepper tree, pepperberry tree	Small tree
Tamarix articulata syn. *T. aphylla*	athel tree	Small tree
Thevetia peruviana syn. *T. nereifolia*	yellow oleander	Shrub

(*a*) *Native* is used here to indicate species which are indigenous to Australia but do not necessarily occur naturally in the Alice Springs area.

plateaux and cliff faces and has often been the subject of paintings by Aboriginal artists. River red gum grows to moderately large size on the relatively low-lying country which receives run-off from adjacent slopes or where there are water-bearing sands and gravels within reach of its root systems. Table 10.1 lists some of the more important trees and shrubs which are used either in public planting or in private gardens.

The District Officer of the Northern Territory Administration at Alice Springs reports that trees and shrubs which are native to Queensland or Western Australia are, in general, more successful than species from southern States. This agrees well with principles of plant introduction, especially similarity of the general climatic pattern.

Following planting young trees are watered three times a week but, as they become well established, this is reduced to once a week and eventually gradually tapered off until at 5 to 6 years of age they can survive without watering. The Administration has developed a useful device whereby the driver of a water tanker can control the flow of water in a pipe and hose so that he does not need to leave the vehicle when watering trees. The operation of the equipment is shown in Plate 32.

In Alice Springs, as elsewhere, there are limitations on the species of trees which may be planted within specified distances of sewers and other water-carrying pipelines. The athel tree (*Tamarix articulata*) is not allowed to be planted on the land of Government house sites because of its tendency to block sewer lines, whilst white cedar (*Melia azedarach* var. *australasica*) is not favoured for street planting due to its ability to disturb kerbing and because of the annual leaf shed.

BROKEN HILL

Broken Hill, in the western area of New South Wales, has several claims to fame; one is that it is the site of the richest silver-lead-zinc lode in the world and another is the wise use made by the city authorities, mining companies and citizens, of trees and shrubs for the amelioration of the local climate. There were many factors, mainly associated with the establishment and growth of a large mining community in an arid climate, which contributed towards deterioration of the local environment. At times, especially during strong, hot and dry winds, living conditions could become very unpleasant. This city is situated in the arid zone of Australia. The mean annual rainfall 1889-1964 was 9.33 inches, but this only tells part of the story, since there is a one in ten probability that it will be less than 5.15 inches, as well as a one in ten probability that it will exceed 14.47 inches. The lowest on record is 2.24 inches. For most of the vegetation it is the lower one in ten probability which is critical to survival and introduced plants which are expected to survive without artificial watering must be able to survive such low annual falls. Site, of course, is important, since many species will grow satisfactorily on extensive flats but will not thrive on rocky hills during periods of below average rainfall.

Broken Hill has a hot summer, but maximum temperatures are not as high as in many other parts of inland Australia; the mean maximum of the hottest month is only 90.5°F, whilst the mean minimum of the same month is comparatively moderate at 64.5°F. In winter, days are usually sunny and warm but nights cool.

TABLE 10.2

MONTHLY RAINFALL FOR BROKEN HILL, N.S.W.
FROM JANUARY 1965 TO DECEMBER 1969
(in points; 100 points = 1 inch)

Month	*Years*				
	1965	*1966*	*1967*	*1968*	*1969*
Jan.	0	104	86	457	58
Feb.	0	105	98	0	81
Mar.	10	51	49	57	228
Apr.	0	2	0	155	10
May	60	49	9	332	64
June	101	14	31	72	70
July	25	59	5	89	76
Aug.	146	54	12	55	5
Sept.	49	11	1	5	45
Oct.	38	52	30	20	55
Nov.	5	132	7	49	7
Dec.	66	176	0	75	25
Totals	500	809	328	1366	724

This table illustrates the low and variable rainfall of the dry country; in this case Broken Hill. Many eucalypts and paper-barks (teatrees) grow successfully under these conditions without watering. The photographs of trees and shrubs at Broken Hill in this and Chapter 17 were taken in August 1969.

with temperatures sometimes falling below freezing point. Like most inland areas there are many days of very light winds from all directions of the compass; the worst local winds associated with dust storms and sand movement are westerlies to southerlies. The monthly rainfall from January 1965 to December 1969 are shown in Table 10.2. Other climatic data is given in Appendix I.

The Barrier Range, on the southern end of which Broken Hill is located, was discovered by Charles Sturt in 1844 and it was from the broken appearance of the hills that the name of the city was derived. The first discovery of minerals was silver-lead deposits at Thackeringa, some 20 miles from Broken Hill, in 1875. The realisation of the richness of the find was delayed because the first consignment of 20 tons sent to England was jettisoned from the boat during a storm at sea. The results of the assay of the second consignment were so good that there was a rush for mining rights in the Thackeringa and Umberumberka districts. Umberumberka was later renamed Silverton and for a short period showed faster early development than Broken Hill. These finds were, however, of minor importance compared with the discovery of the Broken Hill lode by Charles Rasp and his colleagues in 1883. At the time of the discovery they were under the impression that the contents were mainly tin and it was only on assay that silver chloride was found to be the most valuable mineral present. Assays of some of the earlier samples exceeded 10,000 ounces of silver to the ton of ore; the richest part of the lode is of comparatively restricted extent, being only about three miles long, up to 500 feet wide but of variable depth. Rasp and his six associates formed the Broken Hill Proprietary Ltd (B.H.P.) in 1884; this giant of Australian heavy industry subsequently expanded into steel manufacture, shipbuilding and mining elsewhere and eventually withdrew from active operation at Broken Hill in 1939. At the present time there are only four companies engaged in mining—the Zinc Corporation Limited and the associated New Broken Hill Consolidated Ltd, North Broken Hill Limited and Broken Hill South Limited.

The mineral discoveries resulted in the usual rapid growth of population and because of the richness and nature of the finds the population has remained comparatively stable at a high level since the late 1880s. By 1914 Broken Hill had a population of 35,000. To meet the demand for food, fuel and transport local resources were used to a considerable extent. Sheep and cattle for meat were pastured adjacent to the city prior to slaughtering, whilst the large numbers of horses, bullocks and even camels needed for local transport made a heavy toll on ground and shrub vegetation. In addition there were appreciable numbers of goats, mainly kept for milk, and these animals maintained their reputation for eating almost any type of plant growth. Trees and the larger shrubs were felled for timber for domestic cooking and heating, use in the mines and for steam raising, and horses and bullocks were used for transporting it for distances up to 30-40 miles from the city. The rail connection from Adelaide in 1887 did not affect the use of animals for local transport and it was only after World War I that motor vehicles commenced to replace animals. For communications to outlying homesteads camels persisted until 1940. The first half a century of heavy animal population on an ecologically delicately-balanced habitat had already denuded the land of vegetation around Broken Hill by the early 1930s. Any subsequent reduction in the number of animals was insufficient to enable re-establishment of the vegetation. The inevitable happened: strong winds caused movement of the soil, moving sand dunes were created and when dust storms prevailed in the district their load of fine soil particles was increased by pick-up of local material. On western and southern aspects of the city fences were covered and roadways were only kept open to traffic by carting away the accumulating sands. In the early 1930s conditions were so bad that some homes were at times inaccessible to motor vehicles and within living memory wedding cars have been unable to collect the brides or hearses receive the bodies of the deceased, due to the accumulation of sand.

During the late 1920s and early 1930s Albert Morris, chief metallurgist of the Central Mine of the Sulphide Corporation (later incorporated in the Zinc Corporation) had put forward the proposition that protection from grazing animals of land adjacent to the city and the planting of suitable trees and shrubs would control sand drift and reduce the local severity of dust storms. Fortunately the general manager of the Zinc Corporation at the time was A. J. Keast, a man who not only realised the deterioration in the local environment but had the vision to appreciate the proposals of Albert Morris. With the approval of the Managing Director, W. S. Robinson, small plantings at the Zinc Corporation were initiated in 1936. These trees received adequate watering and showed such a favourable response that, in co-operation with the City Council and various local associations, more extensive plantings were started in 1937. Other mining companies have also contributed to operations. There have been two major aspects in the rehabilitation of the locality. The largest scheme consisted of encircling the city with belts half a mile or more in width, fenced against grazing and burrowing animals. These areas have been allowed to regenerate with the shrubs and herbs natural to the area—especially various species of acacia, such as mulga (*Acacia aneura*), eremophilas (*Eremophila* spp.) cassias (*Cassia* spp.) and hopbush (*Dodonea* spp.). Some of the areas on the southern side of the city have had small trees and shrubs planted in them, particularly several species of eucalypts (*Eucalyptus* spp.) as well as some paperbarks (*Melaleuca* spp.). The present condition of these regeneration areas is now probably better than at the time of the discovery of the lode at Broken Hill, since the fencing has protected them from native as well as domesticated animals. Plate 34 illustrates the condition of land surrounding Broken Hill prior to 1937 and at the present time. Given continuation of the present degree of protection from grazing these areas should remain in a stable state. The City Council, comprised largely of men who have lived in Broken Hill for most of their lives,

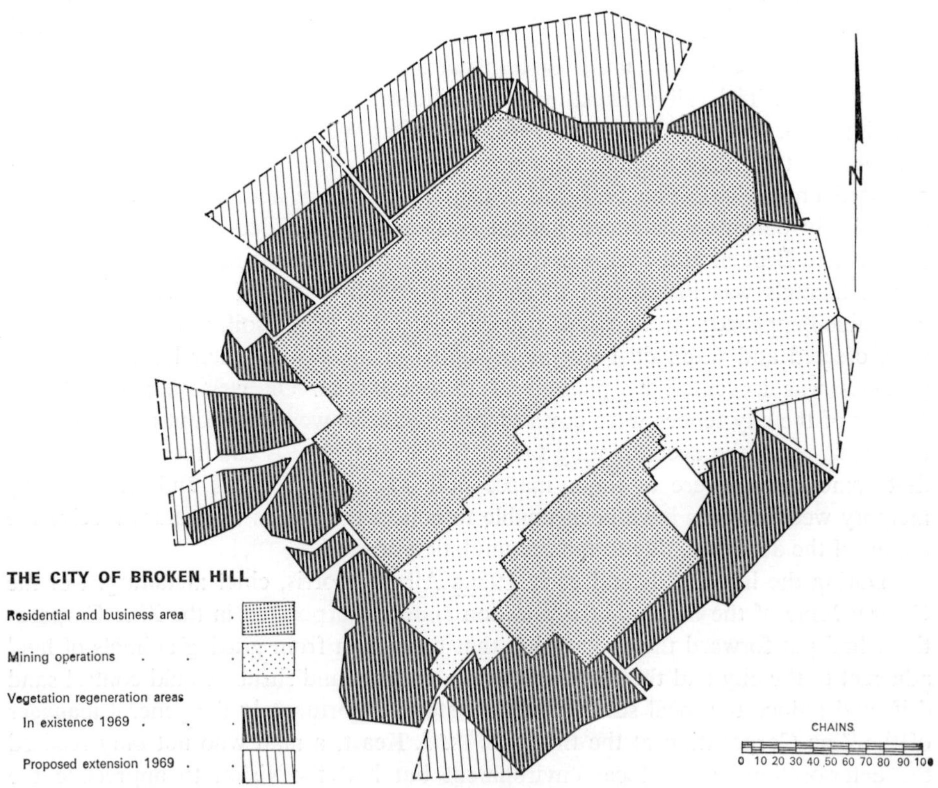

FIG. 10.1

Plan of the City of Broken Hill, showing existing and proposed extension of the regeneration areas (1969).

The removal of trees for fuel and the grazing of shrubs and other ground cover had almost denuded the country around the arid zone city by the early 1930's. The inevitable happened—strong winds caused movement of the soil, moving sand dunes were created and when dust storms prevailed in the district their load of fine particles was increased by pick-up of local material. Albert Morris promoted the idea that protection from grazing animals of the land adjacent to the city was the first step in abating the sand and dust nuisance. In the late 1930's the City Council fenced wide belts around the city and in 1969 it was planned to extend the area. This project is described and illustrated elsewhere in Chapter 10.

There is no doubt of the demand by people for shade in the dry country of Australia. Temperatures in the shade often attain 95-110°F., whilst those in the sun may exceed 150°F. Good shade does not come by chance but calls for a consideration of many factors—especially selection of species, soil and moisture conditions and the positioning of the trees or shrubs so as to give maximum shade during the periods of highest temperatures.

Top left: Moreton Bay fig (*Ficus macrophylla*) on the outskirts of Dongara, W.A.; annual rainfall 18 inches. This is a magnificent specimen but could present problems in road designing were the highway to be widened.

Second from top left: A new house near Cunderdin, W.A.; rainfall 14 inches. This house was designed to leave the tree which is showing above the roof, whilst the young trees and shrubs were only planted after the house was constructed. They are planned to give shade to the living rooms, in the afternoon, (the lounge has no large windows to the west but has its main openings to the eastern facing courtyard).

Second from bottom left: These are Athel trees (*Tamarix articulata*) almost completely shading a house at Woomera, S.A. Some people may consider this excessive shade, since there is also considerable reduction in wind movement.

Lower left: Another Moreton Bay fig, this time in a town of semi-arid N.S.W. When planting any tree there should be an awareness of its potential size.

Upper right: This house is in Kalgoorlie, W.A., rainfall 9½ inches. Trees provide a large amount of shade but are so located as to permit free air movement.

Lower right: In very hot areas shade for bicycles and cars is a valuable amenity. This illustration shows bicycles in a school ground at Boulder, adjacent to Kalgoorlie, W.A.

Plate 26 SHADE NEAR HOMES AND IN PARKS

Throughout most of the southern dry country for at least six months of the year the traveller seeks shady locations in which to park his caravan. Maximum shade temperatures are usually reached in the late afternoon and shade at this time of day gives the greatest ameliorating benefits to the traveller. Most of these illustrations were taken in W.A., but they could have been in almost any state. Some caravan parks even have roadside signs advertising the fact that they have tree-planted areas.

Top left: Part of the caravan park of the shire authority at Cunderdin, W.A.; rainfall 14 inches. Small size, strong timbered eucalypts are excellent trees for this use.

Centre left: Caravans parking on a long term basis at Southern Cross, W.A.; rainfall 11 inches.

Bottom left: A caravan park with very little shade but where the proprietors are planting trees.

Upper right: Shows excellent shade for this caravan at a private home in semi-arid W.A.

Lower right: Another view of a caravan park where trees have been planted for the future; this is in an area of about 10½ inch rainfall. When watering is possible fast early growth can be expected so that in 3-5 years worthwhile shade is available.

Plate 27 SHADE AND CARAVAN PARKS

The selection of the most suitable species of trees and shrubs requires a knowledge of many factors. The presence of power or telephone lines above ground and sewers and drainage pipes underground are important, as well as the maintenance of visibility for traffic.

Top left: Excellent street shade provided by large eucalypts in a street of Kalgoorlie, W.A.; rainfall 9½ inches. These are planted trees.

Top right: Outside the post office of Roma, western southern Q'ld.; rainfall 20½ inches; mean maximum temperature of the hottest month 94·4°F. This bottletree (*Brachychiton rupestre*) provides welcome shade.

Centre left: Planted river red gum (*Eucalyptus camaldulensis*) and sugar gum (*E. cladocalyx*) at Wongan Hills, W.A.; rainfall 15 inches.

Centre right: The centre placing of poles and the planting of small trees or shrubs in the centre of a main road leading out of Merredin, W.A.

Bottom left: Healthy young trees in the parks and sports area at Cunderdin, W.A.; rainfall 14 inches.

Bottom right: This shows trees planted at Cunderdin, but they do not obscure the vision of motorists.

Plate 28 TREES IN STREETS AND PARKS

This low level aerial photograph of the Pastoral Research Laboratory, C.S.I.R.O., Armidale, was taken specially for the use of the architects responsible for planning extensions to the establishment. They were interested in the most suitable location of new buildings with the minimum removal of trees and the retention of the greatest benefit from the shade and shelter already present. The large-crowned trees in the centre and right are eucalypts native to the area, but most of the other trees and shrubs have been planted. The work of the laboratory is the investigation of fertilisers, plant or animal practices on annual production from grazing stock. Low level aerial photographs are an invaluable aid to architects, landscape planners and road engineers.

(The scale of the photograph can be judged from the length of the long building near the top centre of the photo, which is about 240 feet. The top of the photo is north.)

(Photograph taken by Tamair Pty. Ltd. in April, 1970 and reproduced by permission of the Pastoral Research Laboratory, C.S.I.R.O.)

Plate 29 THE USE OF TREES AND SHRUBS FOR A LARGE RESEARCH ESTABLISHMENT

This air photo shows a pattern of rural development which is typical of much of the higher rainfall semi-arid country of Australia. There has been adequate conservation of the original tree cover in much of the district, but there are also large areas where there is inadequate shade for stock, whilst protection from cold or very hot winds is almost non-existent. The excellent shelter belts showing in the north-western sector of the photo can be mainly attributed to the retention of the trees along the road verges—most of which are eucalypts of the box group (grey or bimble boxes), but there are some white cypress pine in the north-west corner. There is an adequate number of trees remaining on the undulating country along the north-south road (part of the Newell Highway), but their distribution is not everywhere the optimum for farm amenity purposes combined with the maximum crop or grass production. The moderately steep slopes of the small hill to the east of the town are covered with a small red gum and some ironbark. Within the town there is a small park, mainly with exotic species; some more recent plantings of eucalypts have been made in the wider streets.

(The air photo was taken in 1958 and the scale is approximately 54 chains to 1 inch; the north point is at the top of the photo. Mean annual rainfall for 74 years is 21·34 inches, with a moderately uniform distribution throughout the year. The mean maximum temperature of the hottest month is over 90°F. and the mean minimum of the coldest month below 40°F.)

(Photo reproduced by permission of the Director of National Mapping, Department of National Development.)

Plate 30 AN AERIAL VIEW OF TREES AND SHRUBS IN A RURAL AREA OF SEMI-ARID NEW SOUTH WALES

This is the town shown in the air photograph of the preceding plate and prior to settlement most of the land was covered with open forest. The extent to which trees are retained in any area depends upon the value of the land for farming or other purposes, the presence of Crown Lands over which felling is restricted and the attitude of the land owners towards conservation.

Top left shows one of the inland grey boxes (*Eucalyptus microcarpa* or *E. woollsiana*) and bimble box (*E. populnea*) left on a road about a mile from the town. *Top right* shows river red gum (*E. camaldulensis*) on a tributary of the Bogan River about two miles from the town and is a site of high potential for additional recreation areas in the locality.

Upper centre: This Memorial Park is well located on the highway which passes through the town. Most of the species in the park are exotics but there are a number of native trees and shrubs which could be used to advantage.

Lower centre: One of the older and larger homes has an adequate area of land for the planting of shade trees. Some shade is highly desirable where the mean maximum temperatu of the hottest month exceeds 90°F.

Bottom left: A new house where possible amenity planting is limited by the narrow width of the block and the proximity of the building to the street.

Bottom centre: This hedge is oldman saltbush (*Atriplex nummularia*) which does not need watering and can tolerate both drought and very high temperatures.

Bottom right: Shows recent planting of eucalypts in the centre of one of the wide streets.

Plate 31 TREES AND SHRUBS ADD TO THE AMENITIES OF A SMALL TOWN IN THE DRY COUNTRY OF NEW SOUTH WALES

This rapidly growing town and tourist area near the centre of Australia has a mean annual rainfall of 10 inches, but it is highly variable and the lowest on record is only 2·37 inches. The town was managed by a representative of the Administrator of the Northern Territory, with advice from a Town Management Board. Tree planting has been done in the streets and parks for a very long period. The illustrations were taken in 1970.

Top left: White cedar (*Melia azedarach* var. *australasica*) in Hartley Street (20 years of age).

Second from top left: River red gum (*Eucalyptus camaldulensis*) along the Gap Road (42 years of age).

Second from bottom left: Forrest's mallee (*Eucalyptus forrestiana*) and yellow-flowering blackbutt (*E. stricklandii*) in Giles Street (5 years of age).

Lower left: Coral-flowered gum (*E. torquata*) in Raggatt Street (11 years of age).

Lower right: A young eucalypt, staked for protection, showing depression in the soil for use in watering.

Centre bottom: A tanker used in watering, showing swinging arm which is controlled by the driver of the vehicle. Immediately after planting watering is done three time a week, which is gradually reduced to once a week as the trees become established; the trees are usually able to manage without watering by the time they are 5-6 years of age.

Plate 32 ALICE SPRINGS—AMENITY PLANTING

The first proposals came from Albert Morris (1887-1939), who was Chief Metallurgist at the Central Mine of the Sulphide Corporation (later incorporated in the Zinc Corporation) when he put forward the proposition that protection from grazing animals of land adjacent to the city would control sand drift and reduce the local severity of dust storms. Fortunately the General Manager of the Corporation at that time was A. J. Keast, a man who not only realised the deterioration in the local environment but had the vision to appreciate the proposals of Albert Morris. Much of the early work in the Corporation area was carried out by J. Scougall.

Top and centre left: The simple grave of Albert Morris in the cemetery of Broken Hill.

Lower left: Memorial gates to Albert Morris in the grounds of the Zinc Corporation.

Top right: Memorial gates to A. J. Keast in a small city park.

Lower right: The J. Scougall gates at the entrance to private recreational area of the Zinc Corporation.

Plate 33 BROKEN HILL—PIONEERS OF TREE PLANTING AND CONSERVATION

Top left: A view of the southern part of the city.

Top right: A small park which receives regular watering.

Upper centre left: A rabbit and stock proof fence on a side of a natural regeneration area.

Upper centre right: Views of a grazed area and one where there is protection from animals. (*Two photos*)

Lower centre left: Some species of native plants will grow on even the most unfavourable sites, in this case *Myoporum* spp.

Bottom left: The top of an area of industrial waste (skimp dump) showing native plants slowly revegetating the area.

Bottom right: The side of a skimp dump; salts are washed out of the waste material and shrubs growing at the base of the dump must be very tolerant of salt.

Plate 34 BROKEN HILL—THE CITY AND THE ENVIRONMENT

The photos show the change which has been wrought by tree planting, conservation and the use of water in an arid climate.
Top: The Zinc Corporation area in 1936. Several of the buildings in this photo also appear in the central photo.*
Centre: The same area taken in the late 1960s.*
The lower photos: Within the private recreational area of the Corporation. The left hand one shows river red gum (*E. camaldulensis*) growing on the edge of watered lawns.
The right hand one shows the lake, which also serves as a reservoir.
(* *Photos Zinc Corporation.*)

Plate 35 BROKEN HILL—TREE PLANTING BY AN INDUSTRIAL ORGANISATION

Yampi Sound was named in 1838 during the voyage of H.M.S. Beagle. The islands are adjacent to the northern coast of Western Australia; the latitude is 17°S. and the mean annual rainfall about 26 inches, but with a strong monsoonal pattern and nearly half of the year is almost rainless. In spite of this, and a very shallow soil, the Broken Hill Proprietary has carried out tree and shrub planting which has helped to improve the surrounds of homes and amenity centres of this iron ore operation.

Top left: The swimming pool on Koolan Island, which is a great gathering centre for the children. At the right background can be seen the screen of an open air theatre.

Top right: Young children enjoy themselves on a merry-go-round on Cockatoo Island. The surround had been planted with attractive shrubs.

Centre left: One of the assistant gardeners on Cockatoo Island admires a fine *Bougainvillaea* in the gardens near the administration offices.

Bottom left: Ornamental planting in front of the Koolan Island Recreational Club.

Bottom right: The supermarket, bank, post office and newsagency on Koolan Island.

(*All photos Broken Hill Proprietary.*)

Plate 36 COCKATOO AND KOOLAN ISLANDS, YAMPI SOUND

Kalgoorlie is situated on the border of the arid zone in W.A. and, in spite of an annual rainfall of only 9 inches, has a fine display of trees. Most of the species planted in the streets are native to the area and thrive without watering, but many of the small parks are watered regularly and support trees from more humid areas.

Top left: A scene of the city from Mt Charlotte.

Centre left: Part of the main thoroughfare, Hannan Street.

Bottom left: Shade for cars outside the railway station. The larger tree is a species of kurrajong (*Brachychiton*) and the one to its right is a carob bean (*Ceratonia siliqua*).

Centre right: Water is valued so highly in Kalgoorlie that run-off from the streets is stored in this large earthen tank and used for watering playing fields and other amenities. This photo was taken in March when the tank was empty.

Bottom right: A delightful shady area in a small park; the large tree is Moreton Bay fig (*Ficus macrophylla*). The trees and palms benefit from the watering of the lawns.

Plate 37 KALGOORLIE

This town came into existence as a result of the finding of commercial quantities of nickel ore in 1966. A few years later it was a thriving town of several thousand people. This centre is in the Goldfields area of W.A., a region famous for the many hardy and attractive small eucalypts which thrive naturally in a rainfall of 9-11 inches. Kambalda is fortunate in a combination of a site where there was a good cover of eucalypts with mature heights of 20-50 feet, a company which had an enlightened attitude towards conservation and town planning, and a State Government which encourages retention of native vegetation.

Top left: A scene in Kambalda West where construction was in progress at the time of photography.

Centre left: This sign speaks for itself.

Bottom left: The playing area of the Kambalda public school.

Top right: Part of the shopping centre of Kambalda.*

Bottom right: Another view showing mature small eucalypts about 20-25 feet high, which form an excellent screen between the street on the right and the almost obscured houses on the left.

(* *Photo Western Mining Corporation Limited* (**Kambalda** *Nickel Operations*).)

Plate 38 KAMBALDA

Mount Isa, located in tropical northern Q'ld., has an annual rainfall of about 16 inches and high temperatures throughout the year. With watering many species of both native and exotic trees grow very fast. Photographs taken in early 1970.

Top left: A local kurrajong (*Brachychiton* spp.).

Centre left: Beach sheoak (*Casuarina equisetifolia*).

Bottom left: Drooping bottle brush (*Callistemon viminalis*).

Top right: White cedar (*Melia azedarach* var. *australasica*) only two years from planting; it receives water only when the lawn is watered.

Bottom centre: The larger tree is satinash silky oak (*Grevillea robusta*) and the smaller a species of *Bauhinia*.

Bottom right: A really fine specimen of a fig (*Ficus hillii*). This is one of the outstanding trees for shade in dry northern Australia.

(*All photos Mount Isa Mines Limited.*)

Plate 39 MOUNT ISA—NATIVE TREES AND SHRUBS

Many of the exotics planted at Mount Isa are species often used for ornamental planting in the warm temperate and tropical regions of the world. Some of them show remarkably good early growth.

Top left: Jacaranda (*Jacaranda mimosaefolia*) and *Plumeria rubra*. Very fine specimens about 15 years old.

Centre left: The bushy tree is mango (*Mangifera indica*), whilst the small shrub near the house is yellow oleander or thevetia (*Thevetia peruviana*).

Bottom left: A flame tree (*Schotia brachypetala*) about 15 years of age.

Bottom centre: Leopard-tree (*Caesalpinea ferrea*) and spathodea (*Spathodea campanulata*).

Lower centre right: A very fine albizzia or siris (*Albizzia lebbek*).

Bottom right: A camphor laurel (*Cinnamomum camphora*).

(*All photos Mount Isa Mines Limited.*)

Plate 40 MOUNT ISA—EXOTIC TREES AND SHRUBS

The north west of this State has seen unparalleled expansion in the 1960s as a result of the development of mining operations, mainly iron ore for export, and it appears probable that a high rate of growth will continue in the 1970s. Both the Government of Western Australia and the large mining companies are giving attention to planting trees and shrubs for amenity purposes but this presents many problems in a region which is mainly arid or semi-arid, with very high temperatures and, in much of the coastal area, the risk of winds of cyclonic intensity. These illustrations show a few aspects of planting in Port Hedland, which is destined to be one of the largest centres in the North West.

Top left: The start of planting in front of a new home; note the halves of 44-gallon drums protecting young trees planted by the Shire.

Top right: Further words are not necessary.

Centre left: A fine example of river red gum (*Eucalyptus camaldulensis*) near the foreshore in the older part of the city.

Bottom left: An exotic, *Sesbania grandiflora*, also in the older part of the city.

Bottom centre: Young trees in a caravan park.

Bottom right: Two native species (*E. camaldulensis* and *E. grandifolia*) and an exotic (*Sesbania grandiflora*) in a caravan park.

(Photos—F. Batini, Forests Department of Western Australia.)

Plate 41 THE NORTH WEST OF WESTERN AUSTRALIA

rightly considers that not only should the present regeneration areas be maintained but that it is desirable to expand them. Fig. 10.1 shows both the present areas and proposals for extension.

Albert Morris was born at Bridgwater, South Australia, in 1887 and came to Broken Hill in 1900, and from his early years took a keen interest in the natural vegetation of the district and introduced plants. Unfortunately he died at the early age of 52 and although he saw his recommendations put into practice, he did not live long enough to see them come to fruition. Whilst an area of the early plantings at the Zinc Corporation has been named the Albert Morris Park and entry gates named in his honour, his principal memorial is in the regeneration areas and planting which have changed Broken Hill from a barren locality to a place where many people not only spend their working lives but remain for their years of retirement. Albert Morris preached that it was very important to select only plants which are naturally suitable to the climate and in that environment adequate protection from grazing animals was essential for natural regeneration. These are fundamental principles in plant ecology and Albert Morris well merits the credit he has received not only in Broken Hill but also overseas for his part in initiation of a sound planting and regeneration programme. The extremes of environments—the arid zones on the one hand and alpine regions on the other—need greater protection and take longer to recuperate after misuse than areas in more temperate climates and, fortunately, Australians are becoming increasingly aware of the need for greater care in the management of such environments.

The most important lesson to be learnt from Broken Hill planting is not the examples of fine trees growing in regularly watered parks, because given sufficient

TABLE 10.3

A SELECTED LIST OF TREES AND SHRUBS WHICH HAVE PROVED SUITABLE FOR GROWING AT BROKEN HILL WITHOUT REGULAR WATERING

(Information from Mr. R. Gould, Zinc Corporation)

	Very good	*Good*
NATIVE SPECIES		
Eucalypts (*Eucalyptus* spp.)	*E. dundasii* (S) *E. stricklandii* (S) *E. torquata* (S) *E. sargentii* (S) *E. oleosa* var. *angustifolia*	*E. brockwayi* (S) *E. salubris* (S)
Acacias (wattles = *Acacia* spp.)	*A. salicina* *A. gladiformis*	*A. notabalis* *A. pycnantha* *A. saligna* *A. sowdeni* *A. victoriae*
Saltbush (*Atriplex* spp.)	*A. nummularia*	
Paperbarks (*Melaleuca* spp.)	*M. pubescens* *M. halmatorum*	*M. armillaris*
EXOTIC SPECIES	*Schinus molle*	*Tamarix articulata* *Olea europaea*

(S) = tolerant of salt in the soil

15

water many species can be grown satisfactorily in such a climate, but in the species which will grow at least moderately well without watering other than in the first few months after planting. There is a limit to the number of trees which a local authority can water throughout the year, even if supplies are available, but large areas can be planted with trees which will grow satisfactorily without attention. Mr. R. Gould, a gardener of the Zinc Corporation who has been associated with the planting since the early years, considers the trees in Table 10.3 amongst the most suitable for use without regular watering.

COCKATOO AND KOOLAN ISLANDS: YAMPI SOUND

Yampi Sound was named by J. L. Stokes when he and others entered it in H.M.S. Beagle in 1838; the name coming from 'yampee', the local Aboriginal word for fresh water. These two islands are amongst many located between the northern end of King Sound and Collier Bay on the north-western coast of Western Australia, and form part of the Buccaneer Archipelago. Major factors in the climate are the latitude of close to 16°S. latitude and the location within a dry monsoonal belt. Rainfall for Beagle Bay (16° 59′ S. and 122° 40′ E.) gives a 50 percentile value of 26.45 inches, with 7.65 inches for the lowest on record in a period of 61 years, a 10 percentile value of 18.08 inches, a 90 percentile value of 39.60 inches and the highest on record of 58.70 inches. Some 85-90 per cent of the rain occurs in the four months of December to March, with five to six months completely dry. Temperatures are high at all seasons of the year and, based on long term records for Derby (17° 18′ S.) and Wyndham (15° 27′ S.), the mean maximum of the hottest month (November) is in the range of 95-100° F, whilst that of the coolest month is still high at about 85°F.

Although the first discovery of iron ore deposits was in the 1880s the initial steps for the development of the resources were not made until 1938. Cockatoo Island, the first developed, was leased in 1927 by the Hoskins Iron and Steel Co. Ltd, and in 1935 the company, now Australian Iron & Steel Ltd, became a subsidiary of the Broken Hill Proprietary Co. Ltd. Proposals that the Yampi iron ore should be exported, in this case to Japan, were vetoed by the Commonwealth Government which imposed a ban on such exports which was not lifted until 1959. Initial work by the B.H.P., through its sudsidiary the Australian Iron and Steel Co. Pty, was stopped in 1942 as a result of events in World War II but resumed in 1945, and in 1951 the first shipments were made to the Port Kembla and Newcastle works of the company.

The development of large scale open-cut mining involved the establishment of a prefabricated town which now has a population of about 350, and a jetty with a luffing head because of the exceptionally great tidal range. Once wives and children joined the company employees the question of living amenities arose as a factor in stability of the work force. Amongst out-of-door amenities a high priority was given to the establishment of small trees, shrubs and other plants for shade, shelter and ornament. The soils are shallow light sands or sandy loams, reddish in colour due to the presence of iron oxides (over sandstones, shales, slates and schists). The sites where trees are planted need to be excavated to a suitable depth, often 3 feet or more, and top-soil added. In the case of children's playgrounds, for instance, where lawns were desired it was necessary to level the sites and import sufficient top-soil. The native vegetation is sparse and usually of poor development though better in depressions where the natural soil is deeper and there is the benefit of drainage. Amongst the small trees and larger shrubs are several species of *Eucalyptus, Acacia, Grevillea, Terminalia* and a few representatives of *Brachychiton* and *Ficus*. The dominant tree is Darwin woollybutt (*E. miniata*).

A decision to start amenity planting on Cockatoo Island was made by the then chairman of B.H.P., Mr Essington Lewis, C.H. (1881-1961), late in 1951 and Commonwealth advice was sought on species to be used and techniques of establishment. The result was that by 1953 plans had been prepared and the first plants established. This work has continued and was extended to Koolan Island when ore extraction commenced there in 1964. The growing of trees and shrubs in this area involves a number of difficulties, not all of which can be overcome. The method of dealing with shallow soils has been described and there is also the need for additional watering on islands without any worthwhile local supply. The townships on Cockatoo and Koolan Islands are probably unique in the world in that water is transported by sea from the opposite side of the continent—a distance of about 6,000 miles. This is done by ore boats returning from Kwinana, Port Kembla and Newcastle. Some control of the very damaging termite (*Mastotermes darwiniensis*) is only possible within selected areas and a careful watch needs to be kept for incipient attack on the larger woody plants. In the case of the occasional cyclone wise selection of species can reduce the amount of damage but nothing can prevent all damage. One means of ensuring quick recovery is to have a suitable proportion of fast-growing, though possibly short-lived, plants which can either be replaced or cut back to shoot again. Plants such as those of the genus *Musa* (bananas) are particularly susceptible to wind damage but, on the other hand, make a quick recovery. A very large number of trees and shrubs have been planted and amongst those which have proved satisfactory or promise success are:

Adenanthera pavonina (red sandalwood-tree or bead tree),
Albizia lebbeck (albizzia, siris tree or East Indian walnut),
**Calophyllum inophyllum* (beach calophyllum or Indian oil nut),
**Callitris intratropica* (northern cypress pine),
Cassia fistula (golden shower or purging cascara),
Kigelia pinnata (sausage-tree),
Lagerstroemia speciosa syn. L. *flos-reginae* (tabek or Pride of India),
**Melia azedarach* var. *australasica* (white cedar),
Peltophorum pterocarpum syn. *P. ferruginea* (golden flame of Malaya or golden shower),
Poinciana regia (flame tree or flamboyante),
Samanea saman syn. *Pithecellobium saman* (raintree),
Tamarindus indica (tamarind).

Amongst the smaller shrubs and other garden plants are aralias, acalyphas, crotons, frangipani, gardenias, hibiscus, oleanders, poinsettias, and bananas as well as kalamansi introduced by the then managing director of B.H.P., Sir Ian McLennan.

* Trees so marked are indigenous.

Most of the species listed grow well in Darwin, which has similar temperatures but a much higher rainfall. They are mainly exotics and it is considered that there is greater scope for the use of indigenous species, especially those which are native to northern Australia.

In the early days a number of dry country southern Western Australian eucalypts were planted but these are reported to have been destroyed by fire. More recently *E. phoenicea* (scarlet gum), native to northern Australia, and *E. torquata* (coral-flowered gum), from southern Western Australia, have been recommended for planting on Koolan Island. There is little doubt that other attractive Australian species could be used which would have a smaller demand for water than many of the exotic species which are at present used.

KALGOORLIE

The discovery of gold on the site of what is now the city of Kalgoorlie in Western Australia, led to the rapid growth of a centre of population which soon reached 20,000. Originally the name given was Hannans, in honour of Patrick Hannan who, with two mates, found gold at the foot of Maritana Hill on 15 June 1893; but two years later the name was changed and the Municipality of Kalgoorlie proclaimed. Earlier there had been prospecting in the nearby Coolgardie area, where gold had been discovered in September of the previous year. Before the first located alluvial deposits were exhausted rich gold-bearing lodes were discovered beneath an area which has become famous as the Golden Mile. From the time of this discovery there was no doubt on the future development of Kalgoorlie as one of the major gold-mining centres of the world. After World War II reduced recovery of gold and its comparatively static price pattern led to a serious lowering in the profitability of operations and threatened the survival of the city. Fortunately, in the late 1960s nickel was found in several localities near Kalgoorlie, so that the future of the city is assured. The construction of the town of Kambalda, about 36 miles south of Kalgoorlie, was commenced late in 1967, whilst there will probably be another new town near Mt Windarra, the site of the Poseidon strike some 140 miles north of Kalgoorlie.

With an altitude of 1,247 feet, Kalgoorlie is located on the border of the arid and semi-arid zones, about 360 miles east of Perth and 220 miles from the southern coast of the State. The summers are distinctly hot but maximum temperatures are not as high as in many parts of inland Australia; the mean maximum of the hottest month is 93.2°F; whilst the mean minimum of the coldest month is 42.9°F. On an average only one frost occurs a year. Rainfall is low, with a 50 percentile value of 9.05 inches and there is a moderately marked winter maximum. Variability is not quite as high as in some parts of the dry country; for example the 10 percentile value is 5.74 inches and the lowest on record for 69 years is 4.75 inches.

In spite of the aridity of the climate and that the water for the city is pumped from Mundaring Weir, near Perth—a distance of 348 miles—Kalgoorlie is one of the better examples of successful tree planting in dry Australia. Some street and park tree establishment was done in the first half of the century, but an extended programme was initiated in 1946. Based on earlier experience the City decided to concentrate on eucalypts native to the dry country, especially species from the arid

TABLE 10.4

TREES AND SHRUBS USED FOR STREET AND PARK PLANTING IN KALGOORLIE

Botanical name	*Common name*	*Comments*(*a*)
Eucalypts		
The more important species		
Eucalyptus brockwayi	Dundas mahogany	Suitable for rocky sites and sandy loams on flats near hills. A fast grower in early life.
E. camaldulensis	river red gum	Very good on deep retentive soils. A fast grower in early life. One of the best eucalypts for the area.
E. cladocalyx	sugar gum	This is an excellent shade tree in parks or other places where it receives some extra water, but where unwatered has a life span which may not exceed 20 years.
E. dundasii	Dundas blackbutt	Considered the best tree in the State for hot, dry inland areas. Will grow on loams of varying texture, including gravels and alkaline soils.
E. gardneri	blue mallet	A useful species on sandy to moderate textured loams.
E. gracilis	yorrell	Considered one of the best which has been tried in Kalgoorlie; a small tree with attractive shiny foliage. Adaptable as to soil.
E. longicornis	morrell	An excellent shade tree; adaptable as to soils and tolerant of salt.
E. salmonophloia	salmon gum	The largest and best of the inland trees, but requires good soils for best results. Somewhat slow early growth but hardy and long lived. Very attractive glossy foliage.
E. salubris	gimlet	A small tree with dense, glossy foliage; does best on heavy soils.
E. torquata	coral-flowered gum	A very ornamental small tree which will grow on rocky sites. A first class shade tree for streets and parks.
Other species		
E. astringens	brown mallet	An excellent shade tree on good, deep loams.
E. campaspe	silver-topped gimlet	Hardy to drought, tolerant of soil conditions. In Kalgoorlie a hybrid form is superior to the species.
E. erythronema	red-flowered mallee	One of the most attractive of the smaller growing eucalypts, with blood-red flowers.
E. flocktoniae	merrit	An attractive small tree; does best on sandy to moderate-textured loams.
E. kondininensis	Kondininen blackbutt	A useful shade species on good quality loams.
E. occidentalis var. *stenantha*		A small tree with attractive foliage and large greenish-yellow flowers.
E. stricklandii	yellow-flowered blackbutt	Considered one of the best eucalypts in Kalgoorlie; has large yellow flowers; adaptable as to soil.
E. wandoo (inland form)	inland wandoo	A good street species; grown on lateritic soils.
E. woodwardii	lemon-flowered gum	A hardy, small tree which does well on light loams. A hybrid, *E. woodwardii* x *E. torquata*, is extensively used for street planting and, as would be expected, shows a variety of colour in its flowers.

(*a*) Based on report of the town of Kalgoorlie and observations by the writer.

and semi-arid zones in the south-west of the State. The Forestry Department commenced the development of an arboretum in 1954, and this is providing useful information on a number of species.

Of the native trees and shrubs other than eucalypts which have been tried there are only a few worthy of note. Pittosporum (*Pittosporum phylliraeoides*) grows satisfactorily on light soils and develops a dense crown; desert kurrajong (*Brachychiton gregorii*) is ranked as an excellent shade species but is a slow grower. Species such as kurrajong (*B. populneum*) and some of the cypress pines, for example black cypress pine (*Callitris endlicheri*), find the climate too dry and the soils, in general, either not fertile enough or too heavy.

The gardening staff of the city of Kalgoorlie has had good results with the raising and planting of street trees and an outline of their methods is given. About the first week in September seed is sown direct into four-gallon containers, kerosene tins being used when available. The soil is selected as close as possible to the type occurring where it is intended to plant the trees; and after sowing, the seed is covered with lawn clippings to keep the surface soil moist (there would be less risk of damping-off if one of the plastic-based shade materials were used). Germination of the seeds takes 10-21 days and as the young seedlings become established they are progressively thinned until only one plant remains. By the end of the following June, a period of about nine months, the young trees are a few feet in height. The planting sites are prepared by digging holes about four feet in diameter and two feet deep, with the soil in the bottom of the hole broken up to allow still deeper penetration of roots. Unless rock has been removed from the hole no imported soil is added. On placing the young plant in its permanent position the tin or drum is cut away to leave the block of soil intact; by this method the success is 99-100 per cent. Guards are placed around the young trees to protect them from dogs and vandals, and watering at the rate of 30 gallons a week is done during the first summer and autumn until there is adequate rain. It has been found that most trees only need watering for the first year; rarely does it need to be continued into the second year. The guards are retained for two years.

KAMBALDA

Major finds of nickel ore were made at Kambalda in 1966 and commercial operations commenced the following year. As the need arose for a large work force there was the necessity to establish a town for a population of several thousand inhabitants; a number which is likely to increase to about 6,000 within a few years as mining operations are expanded. Kalgoorlie, which is less than 40 miles distant, is likely to remain the major centre for transport, secondary education and various forms of shopping and entertainment. Kambalda and Kambalda West are, however, self-contained with regard to primary education, local shopping, clubs and a hotel.

Kambalda is located within semi-arid country but near the border with the arid zone; the latitude is about 31¼°S. Climatic conditions are very close to those of Kalgoorlie and will not be further described other than to note that the annual rainfall is probably about one inch greater and the mean maximum temperature of the hottest month slightly lower.

The Western Mining Corporation, which operates the nickel mine and has built the town, has shown a sensitive awareness of the desirability of conservation of as many as possible of the numerous mallee and other small eucalypts which occur plentifully on the low hills of the Kambalda area. The result is a pleasing retention of plants in streets and on building sites and gives an impression of a town which has had careful tree planting for a generation. The attitude of the company is shown by a notice which is displayed in a prominent position in an area under construction at Kambalda West, located on flat country some two miles away from the first town.

KAMBALDA WEST TOWNSHIP SITE
IMPORTANT

Extra special rules apply on this site for the maximum possible preservation of natural trees and bushes including saltbush. No vehicle is to be driven across or soil removed from or materials dumped on any uncleared area.

Resident Manager.

This protection of native vegetation, is, of course, required on many forms of leases where the land is held from the Government of Western Australia. The excellent effects of the town planning by the company are shown in Plate 38.

MOUNT ISA

The town of Mount Isa is still another centre in the dry country of Australia which has come into existence as a result of the discoveries of significant quantities of important minerals. In this case the initial find of silver-lead deposits was made by John Campbell Miles in 1923. The origin of the name Isa in Mount Isa is not known, since the long-held belief that Miles named it after his sister, Isabelle, was denied by him shortly before his death. Substantial reserves of copper were located in 1942 and diamond drilling in 1963 disclosed previously unknown ore bodies. Following several changes in the ownership of the principal lease the firm of Mount Isa Mines Limited was incorporated in Queensland during 1931. The deep mining of sulphide ores was commenced in 1935 and the milling of copper ore in 1942, but the latter was discontinued 1946-1953. Since the middle 1950s there has been an increasing tempo of development—a copper refinery was opened in 1959 at Townsville, a port some 603 miles distance by rail, a new shaft (K57) for the extraction of copper-bearing ore, additions to the copper treatment plant at Mount Isa and reconstruction of the State Government railway line to Townsville. There appears little doubt of the growth and future stability of the town. In addition to the mining activities in Mount Isa it also serves as a railhead serving the Mary Kathleen uranium field and the cattle industry.

The City of Mount Isa was officially proclaimed in the late 1960s when it was separated from the Cloncurry Shire. The municipality is claimed to be the largest in the world, and extends from Mount Isa north-west 117 miles to Camooweal, only a few miles from the border of Queensland with the Northern Territory. The population of permanent residents at June 1970 was estimated at 22,000. Local government administration is under the control of the Mount Isa City Council, with elections every two years.

TABLE 10.5

SOME OF THE TREES AND SHRUBS PLANTED AT MOUNT ISA

Botanical name	*Common name*	*Notes on growth*
NATIVE SPECIES		
Trees		
Brachychiton acerifolium	flame kurrajong or flametree	Very good
Brachychiton spp.	kurrajong	Good
Bauhinia hookeri	Hooker's bauhinia	Satisfactory
Callistemon viminalis(*a*)	drooping bottlebrush	Fair
Callitris columellaris	cypress pine	Satisfactory
Casuarina equisetifolia	beach sheoak	Only fair
Eugenia smithii syn. *Acmena smithii*	lillipilli satinash	Very good
Ficus hillii	fig	Very good
F. macrophylla	Moreton Bay fig	Satisfactory
Grevillea robusta	satinash silky oak	Satisfactory
Melia azedarach var. *australasica*	white cedar	Fast, healthy
Pleiogynium timorense syn. *P. solandri*	tulip plum or Burdekin plum	Promising
EXOTIC		
Trees		
Albizia lebbek	albizzia or siris	
Caesalpinea ferrea	leopard-tree	Satisfactory
Cinnamomum camphora	camphor	Satisfactory
Cupressus arizonica	Arizona cypress	Fast growth but only fair form
Delonix regia	royal poinciana	Promising
Ficus benjamina	fig	Good growth, healthy
Jacaranda mimosifolia	jacaranda	Very good
Mangifera indica	mango	Very good
Parkinsonia aculeata	parkinsonia	Fair growth without watering
Populus sp.	evergreen poplar	Very fast early growth
Plumeria rubra		Very good
Salix babylonica	weeping willow	Fast, form fair
Schinus molle	pepper tree	Fair only
S. terebinthifolius	Brazilian pepper tree	Moderately good
Schotia brachypetala	flame tree	Very good
Spathodea campanulata	Santo Domingo mahogany or spathodea	Very good
Tamarindus indica	tamarind	Promising
Tamarix articulata syn. *T. aphylla*	athel tree	Very good
Shrubs		
Manihot esculenta	bitter cascara or manioc	
Thevetia peruviana	yellow oleander or thevetia	Promising

(*a*) A shrub

Mount Isa is located in north-western Queensland on the Barkly highway leading to the Northern Territory; it is about 76 road miles west of Cloncurry and over 200 miles south of the Gulf of Carpentaria. The altitude is 1,157 feet but the hills of the nearby Selwyn Ranges rise to 2,000 feet. The town is situated within the tropics (20° 42′ S. latitude) near the border of the arid and semi-arid climatic

This Commonwealth Government-established town has an annual rainfall of about 7½ inches, but in spite of this, it has been possible to establish sufficient trees and shrubs to make a pleasant setting for houses and offices. All the planted trees receive regular watering from treated effluent.

Top left: The natural vegetation of most of the plains. The small bushes are mainly species of saltbush (*Atriplex* spp.) or bluebush (*Kochia* spp.) and few are more than two feet in height.

Top right: A few miles from the town is this small natural pond with a few shrubs up to 8 feet in height. This country is an area of internal drainage and ponds such as this only occur in a few localities.

Centre left: Tuart (*Eucalyptus gomphocephala*) at the arboretum and recreational area. This fine Western Australian tree is growing very satisfactorily but is, of course, heavily dependent upon the regular watering which the lawns receive.

Bottom left: An exceptionally fine specimen of coral-flowered gum (*Eucalyptus torquata*) near Government offices at Woomera West. Although this tree receives some watering the species is one of the hardiest and most promising for the southern dry country.

Bottom right: A house where there has been extensive planting, which enhances the appearance of the home but too dense a shrubbery will also reduce the beneficial effect of evening breezes and may not be acceptable to all people.

Plate 42 WOOMERA

zones. Data cited will be based on that for nearby Camooweal (738 feet) and Cloncurry (633 feet); the mean maximum for the hottest month (December) for both places is over 100°F and the mean minimum is comparatively high at about 75°F, which is the same as the mean maximum of the coldest month, the mean minimum of which is around 45°F. Only very rarely do temperatures fall to freezing point. The mean annual rainfall for Mount Isa in the period 1926-1964 was 15.77 inches. There is a well-marked summer maximum precipitation but only December to March have an average over one inch; the total for the three months July to September is only 59 points. Based on the two adjacent towns cited it can be expected that the 10 percentile value will be about 8½ inches and the lowest on record down to 5½ inches. A substantial water supply for the town was completed in 1958 with the construction of a dam on the Leichhardt River, with a storage capacity of 17,000 million gallons. The body of water is known as Lake Mondarra.

The northerly latitude of Mount Isa and the high temperatures which prevail throughout the year make it possible, with the use of water, to grow many tropical trees and shrubs which would not thrive in southern locations. Many of the exotic plants are species which are commonly planted for ornamental purposes in tropical and sub-tropical areas throughout the world. In this sense they are well tried, though not necessarily under the conditions which exist at Mount Isa. A few species from temperate regions have shown very fast early growth but the writer has some doubts as to whether they will maintain good form and health to old age. A few species have shown satisfactory or even very good height growth in the first few years but they do not exhibit the typical form and vigour which is seen in more suitable climate. The information in Table 10.5 is based on data supplied with photographs from Mount Isa Mines Limited.

THE NORTH-WEST TREE SCHEME OF WESTERN AUSTRALIA

The North-West Tree Scheme of northern Western Australia came into operation in 1964 when the Hon. C. W. M. Court, Minister for the North-West, approved the appointment of a tree adviser. The project was established with the specific task of encouraging local authorities and individual residents to plant suitable trees and shrubs in streets, reserves and home gardens. The object was to provide shade and shelter, to improve the aesthetic appeal of residential and recreational areas, and to assist in the control of wind-blown dust. The State Government planned the scheme both as an advisory service and as a supplier of young plants to a wide range of interests. These are provided on a free distribution basis to other government departments and local authorities, as well as direct to householders and local organisations; in the case of mining companies special arrangements are made.

The area involved in the scheme is practically all the northern half of Western Australia or about half a million square miles. This compares with a total area of 397,277 square miles for New South Wales and Victoria or 561,313 square miles for the combined areas of the United Kingdom, France, Germany (both West and East), and Italy. There are, however, two parts of this vast region which are of principal importance. The first and much the more important at the present time is the rapidly developing iron ore mining district south of Port Hedland and east of Onslow; this is located between 20 and 23½°S. latitude. The second part consists of the Kimberleys in the north at latitudes 14-16½°S., and includes both mining and agricultural operations. Except on the coastal and river plains much of the topography is rugged, with some peaks attaining 3,000-4,000 feet in height. Soils are often skeletal and natural vegetation on the hills is usually sparse.

The most spectacular development has been in the southern area, where continuing growth since the early 1960s is resulting in iron ore mining operations which are large by world standards. One result has been that small towns or settlements which previously had populations only numbered by a few score or hundreds have grown to towns of several thousand people. In addition, a few new centres are being established. These towns are of two types: a number of 'company' centres created by the mining companies to house their employees at or near the source of the iron ore; the others are sometimes referred to as 'open' towns, that is centres which are developing under normal government guidance and administered by a local authority. These latter towns are mainly at or near ports on the coast where the iron ore is shipped.

The more important of the 'open' centres is the existing town of Port Hedland and the new town of Karratha. A few years ago Port Hedland was a small port with a resident population of only a few hundred people. Now, as the port for the ore from both the Goldsworthy and Newman operations it has grown to between six and seven thousand, and this is expected to double in each of the five year periods to give about 25,000 people by 1980. Karratha is near Roebourne and the erection of the first houses was started late in 1969. It is estimated that the ultimate population is likely to be about 25,000.

There are a number of 'mining' towns, some already established and some only at the initial stages of development. Goldsworthy (about 1,000*) serves the mine at Goldsworthy, an operation of Goldsworthy Mining Ltd; Tom Price (about 2,500) serves the Mt Tom Price mines of Hamersley Iron Ltd; Dampier (about 3,000) is the port for Hamersley Iron ore; Newman (about 1,500) serves the mine at Mt Whaleback on Mt Newman Station, operated by Mt Newman Mining Co. Pty Ltd, and the size of the town is likely to treble by the mid 1970s. Amongst new centres at the advanced planning stage is Paraburdoo, which will be nearly as big as Tom Price; this will be developed by the Hamersley Iron Company. Another new town, but smaller (probably 500-1,000 people), is one to be built at Shay Gap by the Goldsworthy organisation. Many other centres are subject to negotiations and is likely that towns will be built at Mt Enid on the Robe River, and near Cape Lambert for export operations.

The Kimberley area in the north of the State has two types of development. First there is that associated with the mining industry, which includes iron ore and bauxite and alumina operations. The extraction of iron ore from Cockatoo and Koolan Islands in Yampi Sound not far from Derby has been carried on by Broken Hill Proprietary Ltd, (B.H.P.) or its subsidiary, Dampier Mining Pty Ltd, since shortly after World War II. There are two well laid out settlements with a total population of about 1,000. The amenity planting work on these two islands is outlined in a separate section. Further north in the Admiralty Gulf area, American Metal Climax is establishing a town which is likely to have a population of about

* Estimated population at the beginning of 1970. This also applies to other estimates.

3,000. The second type of development is that of the cultivation of tropical crops, notably cotton and rice, on irrigated areas of the Fitzroy and Ord Rivers.

With the current growth of population there has developed a need for amenity benefits which only trees and shrubs can supply. Since a wise selection of species is so important it is most appropriate that a suitable government agency is available to provide advice to local bodies and householders on all aspects of tree planting. All the large mining companies have planning officers and they are, of course, in a position to call in specialist consultants, but where required the service of the Government Tree Adviser is available.

Climatically the area south of Port Hedland is in the arid zone except for a small area at the highest latitudes; Yampi Sound is in the semi-arid zone and the Admiralty Gulf area is on the border with the sub-humid zone. The climatic conditions of the southern area are quite different from those in the two northern areas and will be described separately. In the area south of Port Hedland the annual rainfall is low (8-11 inches) and variability is the highest in Australia; for instance in the belt from Exmouth Gulf to Onslow it is over 60 per cent and none of the area is below 40 per cent. Table 10.6 gives rainfall data for several meteorological stations.

Most of the rain is in the first half of the year, especially February to April, whilst the second half is dry, notably August to October, but varying with the locality.

Potential evaporation from a free water surface is 90-105 inches a year, indicating a theoretical deficiency of 80-95 inches. Combined with low rainfall and high evaporation are very high temperatures. The coastal towns have mean maxima of the hottest month (which may be as late as March) of about 95°F, whilst

TABLE 10.6

RAINFALL DATA FOR SELECTED STATIONS IN NORTHERN WESTERN AUSTRALIA(*a*)

Region and station	*Rainfall in inches*					
	Lowest on record	*Percentile values*(*b*)			*Highest on record*	*Years of records*
		10	50	90		
Port Hedland region						
Onslow (coastal)	0.57	2.35	9.18	21.94	32.97	80
Port Hedland (coastal)	1.25	4.82	10.73	22.74	40.13	68
Roy Hill (inland)	0.88	4.86	9.02	17.49	25.15	66
The Kimberleys						
Derby (coastal)	4.39	14.00	23.61	36.18	57.01	80
Fitzroy Crossing (inland)	7.41	11.53	20.48	28.67	35.19	72
Wyndham (tidal waters about 50 miles from the open sea)	14.38	17.46	25.68	38.24	56.34	79

(*a*) Data published by the Bureau of Meteorology, Department of the Interior.

(*b*) A description of percentile values is given in Chapter 3; briefly, the 10 percentile value is that below which the rainfall is likely to be one year in ten; the 50 percentile value is the mid value but it is only coincidental if it agrees exactly with the arithmetical mean rainfall; the 90 percentile value is that above which the rain is likely to be one year in ten.

inland temperatures are usually much higher; for example Marble Bar on the eastern edge of the area has a mean maximum of 107.5°F. The mean maximum for even the coolest month is around 80°F.

The Kimberley area, by contrast, has a rainfall which is at least double and a variability which is in the range of 20-35 per cent. The monsoonal pattern is better developed and rains start in late November or December and continue to March or early April; the months June to September are almost rainless, but the pattern differs between the three stations listed in Table 10.5. Temperatures are high throughout the year, with the mean maximum of the hottest month for Wyndham of 98.5°F (November) and the corresponding value for the coolest month (July) of 85.0°F.

Water is one of the critical factors in amenity planting in this area. Where fresh water is available it is rarely in unlimited quantities or cheap enough for the generous watering of extensive areas of trees, shrubs and lawns. Where bore water is available it is, unfortunately, often so saline (up to 3,500 p.p.m.) as to be unsuitable for use.

The selection of trees and shrubs for the difficult climate of these two areas presents problems to planning officials, especially if they are not thoroughly familiar with both the local environment and the plant species which grow there. The present writer has discussed factors in the selection of species in Chapter 18. In this area it appears that it is desirable:

1. That the adviser on this work should thoroughly know the conditions under which trees have to grow.
2. Give preference to native species which grow naturally in the area.
3. In the case of introduced plants, both from elsewhere in Australia and from overseas:
   (a) Give preference to those which have already been tried in the area.
   (b) Proceed with caution in the case of all new species, but not hesitate to establish a few trial areas.

This has been the policy of the Department of the North-West in the North-West Tree Scheme and is one which will undoubtedly give the best long-term results. With the permission of the Administrator of the North-West, the species recommended by his tree adviser are given in Table 10.7. In addition to these trees and shrubs there are a number of other plants, mainly vines and climbers, which are also recommended.

In addition to the trees and shrubs listed in Table 10.7 the tree adviser recommends the exotic plants *Cocos nucifera* (coconut palm), *Phoenix dactylifera* (date palm), *Asimina triloba* (pawpaw) and *Musa sapientum* (banana). Amongst vines, climbers and scrambling plants there are *Bougainvillea* sp. (bougainvillea), *Ipomoea* spp. (mainly exotic but *I. pes-capri* is native), *Passiflora flanicarpa* (passion flower) and *Vitis* sp. (grapes).

WOOMERA

This township of several thousand inhabitants is of interest on two accounts—one is that it is an example of a settlement which was fully planned at its inception

TABLE 10.7

TREES AND SHRUBS RECOMMENDED BY THE DEPARTMENT OF NORTH-WEST, GOVERNMENT OF WESTERN AUSTRALIA, FOR PLANTING IN SPECIFIED AREAS OF THE NORTH-WEST OF WESTERN AUSTRALIA

NATIVE SPECIES	
Trees(a)	
Callitris glauca	white cypress pine
Casuarina decaisneana	desert sheoak
C. equisetifolia	beach sheoak
C. glauca	swamp sheoak
Erythrina vespertilio	grey corkwood or batswing coraltree
Eucalyptus bigalerita	Adelaide River white gum
E. camaldulensis	river red gum
E. dichromophloia	variable-barked bloodwood or gum-topped bloodwood
E. grandifolia	large-leaved cabbage gum
E. microtheca	coolibah
E. striaticalyx	—
Ficus decora	fig
F. racemosa syn. *F. glomerata*	cluster fig
F. platypoda	fig
F. rubiginosa	Port Jackson fig
Melaleuca leucadendron	broad-leaved paperbark
Nauclea orientalis syn. *Sarcocephelaus cordatus*	Leichhardt pine
Thespesia populnea	portea-tree
Shrubs(a)	
Acacia spp. incl.	wattle or acacia
A. elliptica	
A. impressa and many others	
Cassia spp. incl.	cassia
C. leursenii	
C. eremophila	
C. pleurocarpa	
C. pruinosa	
C. venusta	
Hibiscus spp.	hibiscus
Sesbania formosa syn. *S. grandiflora*	
Vitex trifolia	
EXOTIC SPECIES	
Trees(a)	
Albizia lebbek	albizzia or siris
Anacardium occidentale	cashew nut
Calophyllum spp.	Indian oil nut
Cassia fistula	golden shower or purging cassia
C. javanica	Java shower
Ceratonia siliqua	carob bean
Ficus spp.	edible figs
Jacaranda mimosifolia	jacaranda
Kigelia pinnata	sausage-tree
Melia azederach(b)	Cape lilac

(Continued on next page)

(a) Whether a perennial woody plant should be classified as a tree or a shrub depends primarily upon its growth form—if the plant attains a height of more than 20 feet and has a single main stem for half or more of that height, it is normally classified as a tree. A shrub, on the other hand, often has a number of stems from at or near ground level and may not attain more than 20 feet in mature height. The environment usually controls mature height and may influence growth form. In the case of the genus *Acacia* there are many species which are unquestionably trees but, on the other hand, there is also a large number which are never more than moderate size shrubs.

b) The Australian species of *Melia* is *M. azedarach* var. *australasica* (white cedar).

TABLE 10.7—*continued*

Poinciana gilliesii syn. *Caesalpinea gilliesii*	
Schinus molle	pepper tree
S. terebinthifolius	Brazilian pepper tree
Tamarindus indica	tamarind
Shrubs	
Euphorbia spp.	poinsettia
Caesalpinea pulcherrima	peacock flower or Barbados pride
Cassia alata	ringworm cassia
Citrus spp.	lime, lemon etc.
Hibiscus spp.	hibiscus
Psidium spp.	guava
Tecoma smithii	
T. stans	yellow elder

and adequate provision made for residential areas with minimum of through traffic, sites for schools, shopping centre, churches and sporting and social organisations; the other is that it is an excellent example of the wise, planned use of trees and shrubs to improve living amenities in an adverse environment. When the Commonwealth Government took measures to establish Woomera immediately after World War II, the Commonwealth Department of Works both designed the township and initiated an investigation to decide what species of shrubs and trees could be used for amenity planting. Mr R. H. Patterson, senior arboriculturist of the Department, enthusiastically led the early investigations and guided the initial plantings. Later he wrote an account of the project (1959). Subsequently the operation and maintenance of the township was transferred to the Department of Supply.

Woomera is close to the geographical centre of South Australia, but the meteorological area in which it is situated is known as the North-West Interior. As would be expected from its location about 200 miles from the Southern Ocean, altitude of about 600 feet and latitude of approximately 31°S., there is a moderately large temperature range from a mean minimum for the coldest month of 35-40°F to a mean maximum of the hottest of about 95°F. The days in winter are pleasant with a mean maximum of about 65°F, which is also approximately the mean annual value. Appendix I gives temperature data for Tarcoola and Marree, both in the arid zone of South Australia. At Woomera over a 20-year period the 50 percentile rainfall value is 7.47 inches, the 10 percentile 4.15 and the 90 percentile 11.10 inches. The lowest rainfall was 3.84 inches but for a longer period it is more likely to be comparable to those for Tarcoola (66 years) with 1.99 and Coober Pedy (54 years) with 1.20 inches.

The topography of the area may be described as very gently undulating. It can be divided, however, into very broad 'tablelands' and broad, shallow valleys and smaller drainage channels which may be as much as 100 feet lower than the tablelands. Since this is a part of Australia which has internal drainage most of the shallow valleys lead to basins which may hold water during years of above average rainfall but are only large salt pans at other times. The accumulation of salt, especially common salt and gypsum, in depressions is usually so high that either no vegetation can survive or only plants which are highly tolerant of such conditions, for example saltbush and bluebush. Where the vegetation has been damaged and where there is no layer of gibbers, small sand dunes arise and present problems in control. Around most of Woomera township there is only a shallow sandy surface layer in the soil so that there is not sufficient material for very large dunes.

Most of the soil in the township area is Woomera clay loam, which carries scattered gibbers over a very thin (1-3 inches) sandy loam, beneath which is a tight clay, red to brown in colour and with varying amounts of salt and gypsum. This clay is not readily penetrated by plant roots and when trees are established it is necessary to remove the clay from large holes and replace with the sandy loam from the surface. In the valleys the same type of clay soil is present but it is beneath a variable depth of sand and sandy loams. The soils generally are slightly to markedly alkaline, with pH values from 7.5 to over 9; by agricultural standards the soils are not poor and most minor elements are present, the most notable deficiency being boron.

In view of the adverse conditions it is not surprising that there is little vegetation more than 1-2 feet high on the tablelands. The most prominent plants are species of saltbush (*Atriplex* spp.), bluebush (*Kochia* spp.), many grasses and some members of the daisy family; shrubs and small trees are very rare. In the valleys and other depressions in addition to the plants already mentioned there may be scattered shrubs—the pea family is the most prominent with several species of *Acacia* and *Cassia*. Other shrubs include hopbush (*Dodonea attenuata*), needlewood (*Hakea leucoptera*), some paperbarks (*Melaleuca* spp.), one eucalypt (*Eucalyptus oleosa* var. *glauca*), white cypress pine (*Callitris glauca*) and pittosporum (*Pittosporum phylliraeoides*). Some of these, such as the cassias, may only grow to a few feet in height whilst the tallest are some of the paperbarks, acacias and white cypress pine, which may attain 20-25 feet under favourable conditions.

In this difficult environment, but with the liberal addition of water throughout the year, the authorities have been able to produce the results which are illustrated in Plate 42, whilst photographs of individual tree and shrub species are given in the plates which accompany Chapter 17. As would be expected from its location in the arid zone, Woomera has no local water resources, but obtains its supplies by pipe line from the Murray River via Port Augusta, an airline distance of over 250 miles. The principal catchments of the Murray and its major tributaries are up to 1,000 miles away. The water is first used for domestic purposes and is subsequently treated and used for watering trees and shrubs. After retreatment, which includes chlorination, the water is fit for human consumption except for the presence of significant quantities of detergent. Except perhaps during heat wave conditions in summer the quantity of water available from the township sewerage plant is more than sufficient for maintenance of the present scale of planting, but there is an obvious limitation upon the extent to which the plantations can be expanded.

A full list of tall and small growing trees and shrubs planted in the town of Woomera and nearby areas with comments, is given by R. H. Patterson in Table 6 and 7 of his report, whilst species of climbing plants that are responding well to extra water, in defined locations, such as arboretum, dwellings, etc., are listed in

TABLE 10.8

A SELECTED LIST OF TREES AND SHRUBS WHICH HAVE PROVED SUCCESSFUL AT WOOMERA WITH REGULAR WATERING(*a*)

	Very good	*Good*
NATIVE SPECIES		
Eucalypts	*E. brockwayi*	*E. camaldulensis*
	E. campaspe	*E. flocktoniae*
	E. dundasii	*E. griffithsii*
	E. forrestiana	*E. grossa*
	E. salubris	*E. lesouefii*
	E. sargentii	*E. stricklandii*
	E. torquata	*E. transcontinentalis*
		E. woodwardii(*b*)
Acacias (wattles)	*A. pendula*	*A. cyanophylla*
	A. salicina	*A. sowdenii*
		A. stenophylla
Other species	*Agonis flexuosa*	*Callistris glauca*
	Dodonea attenuata	*Casuarina cristata*
	Melaleuca pauperiflora	*Grevillea robusta*(*c*)
		Hakea laurina
EXOTIC SPECIES	*Schinus molle*	*Cupressus arizonica*
	Tamarix aphylla	*Rhus lancea*

(*a*) All the species listed here can be assumed to be at least moderately resistant to salt and to tolerate a high pH reaction. Analysis of local soils show total soluble salts mainly in the range of 2.5 to 3.5 per cent, with chloride as NaCl 1 to 3 0 per cent. Values for pH are 7.9, and up to 9.5 in the deeper horizons.

(*b*) With maturity this tree usually develops an unattractive, straggly form.

(*c*) This rainforest species needs better than average soil and watering if it is to grow into an attractive tree.

Table 8. As would be expected most of the small trees are eucalypts native to the arid and semi-arid zones of south-western Western Australia, whilst both these and most of the other native and exotic species are the same as those which thrive without watering, or with only limited watering, at Broken Hill. It should be appreciated, however, that R. H. Patterson visited Broken Hill when preparing his list of species for Woomera. His visit was about 12 years after the first plantings at Broken Hill during 1936-37, so that the results there were an excellent guide.

Trees and shrubs are classified by the author in Table 10.8.

Notes on the preparation of holes for tree planting, the use of asbestos watering pipes and the method and frequency of watering are referred to in Chapters 20 and 21 respectively.

The success of nearly all the species planted at Woomera illustrates the wisdom of selecting mainly species which have a proven record in comparable climatic areas, either naturally or following introduction over a period of many years, and to limit the use of previously untried species. Above all Woomera is an excellent example of what can be done, where water is available, to convert a comparatively barren arid area into one where small trees and shrubs provide shade and shelter and bring beauty to a residential and recreational centre.

FURTHER READING

Patterson, R. H. (1959). *The climate, soils, plant ecology, arboricultural activities and vegetative development; Woomera Research Project areas, North West Arid Interior, South Australia.* Department of Supply, Melbourne.

It is a matter of common observation that domestic animals seek shade during periods of high temperatures in the sun; whilst biologists who study desert fauna are aware that small native animals suspend activity during the hottest part of the day and seek protection underground, in caves or in the shade of vegetation. Not only should grazing animals have access to shade but when they are crowded together in stock yards and other places shade is even more important. The illustrations in this plate come from N.S.W., Q'ld. and W.A.

Top left: Although it was March when this photo was taken in Cunderdin, W.A., sheep had gathered in the shade of local eucalypts.

Upper centre left: Here cattle are in the shade of paperbarks (*Melaleuca* sp.) in the comparatively temperate area of south-eastern coastal Q'ld.

Lower centre left: Sheep yards at Griffiths, N.S.W., showing the Texas umbrella tree (*Melia azedarach* '*umbraculiformis*'). This tree is particularly suited to street and garden planting or for the use shown in this illustration. There are a number of examples of it in the Murrumbidgee Irrigation Area of N.S.W. but it is scarce elsewhere; it deserves wider use.

Bottom left: These river red gums (*Eucalyptus camaldulensis*) provide some shade to sheep in the nearest pens but more extensive planting is desirable to provide an adequate level of protection.

Lower centre right: The old sheep yards at Deniliquin N.S.W., where some shade has been available for a long time; notice the guards to protect the bark from animals.

Bottom right: An old eucalypt in the same stock yards as those above.

Plate 43 ANIMALS AND SHADE

Top left: This very fine example of oldman saltbush (*Atriplex nummularia*) growing at Broken Hill, N.S.W., shows that it has few equals for low shelter in the arid zone. Rainfall 9 inches, but the plants probably received additional watering when they were established.

Top right: Amongst exotics there are few species to equal the athel tree (*Tamarix articulata*). This fine belt is growing in semi-arid Vic. not far from the border of the arid zone; rainfall about 10 inches.

Upper centre left: Cooba (*Acacia salicina*) planted at Broken Hill, N.S.W.; rainfall 9 inches. This species is representative of the semi-arid zone, but the fact that it is well established at Broken Hill is evidenced by the natural regeneration taking place.

Lower centre: Lombardy poplar (*Populus nigra* var. *italica*) and weeping willow (*Salix babylonica*) comprise this very attractive belt on the edge of a water course in the Griffith irrigation area of semi-arid southern N.S.W. There is no question of the aesthetic appeal of these trees, but whether they are the most suitable selection of species, allowing for the value of the water, is a matter for a study of the economic aspects.

Bottom left: Useful shade in W.A. from native shrubs growing on the road verge or along the fence lines.

Bottom centre: An unusual species used for a low hedge; in this case silver cassia (*Cassia artemesioides*) at Woomera, S.A.; it receives regular watering.

Bottom right: Some of the small shrubs of the dry inland have very dense foliage and provide good protection for sheep during spells of very cold winds. This is umbrella bush (*Acacia ligulata*) growing naturally near Woomera, S.A.; rainfall about 7½ inches.

Plate 44 EXAMPLES OF SHADE AND SHELTER SPECIES

CHAPTER 11

Shade and shelter on the farm

ANIMALS suffer from exposure to extremes of climatic conditions but given suitable shade and shelter they will thrive better than on completely exposed sites. There is some difference of opinion amongst research workers as to whether the poorer condition of exposed animals is directly due to lack of shelter and shade or due to the poorer growth of pasture and browse species. The writer considers both factors are present and that the important practical point is that suitable levels of shelter are of benefit to farm production. There is ample evidence that shelter, especially in areas which experience cold winters, can result in increased production of 10-25 per cent and, exceptionally, at much higher rates for some wind-sensitive crops such as french beans or rice. Much of this information comes from well-designed research and confirms the popular view that under certain conditions crop and pasture production is higher on sheltered sites. There is also a strong case supporting the harmful effects arising from excessive exposure to extremes of climatic conditions, especially high mortality amongst newly-born animals or recently-shorn sheep as a result of cold, wet weather. It appears reasonable to assume that those which survive suffer loss of condition, but there is still limited statistical evidence. In recent years, however, there has been some useful information published, including some in Australia by officers of C.S.I.R.O. This includes the effect of shelter on sheep at critical times, for example after shearing (Hutchinson, 1968; Lynch, 1968), at lambing (Miller, 1968), and in very cold environments (Miller, 1968). Lynch and Marshall (1969) have published an account of the effect of sheet iron fences used to separate sheep in very small (0.3 acre) paddocks. These two authors state:

> It is concluded that the variation found in animal productivity was largely due to the difference in pasture availability rather than to an effect of shelter as such.
>
> This experiment demonstrates an increase in pasture availability and sheep productivity in response to shelter under certain conditions. The increase of pasture is reflected in animal productivity only if the stocking rate is raised to take advantage of it. However, when an overriding environmental factor such as drought limits pasture growth to less than potential animal intake, any increase in plant growth due to shelter also gives increased animal gain. Indeed, because of the curvilinear nature of the pasture availability/stocking rate relationship (Willoughby, 1959) slight increases in plant growth or availability at low levels may have considerable effects on animal live-weight.

SHADE

The provision of shade is necessary throughout the pastoral areas of Australia and it is particularly important in the dry country where summer temperatures are high.

There are few places in the arid and semi-arid zones where the mean maximum in the shade of the hottest month is less than 85°F and most are over 90°F, with significant areas exceeding 100°F. These are only *mean* temperatures and the maximum in the same month may be 10-15 degrees higher; consequently it follows that there are few localities which do not at least occasionally experience century figures. The data quoted have all been temperatures in the shade; those in the sun may be 30-50 degrees higher and research workers in Queensland have recorded values as high as 190°F at the tips of wool fibres on the backs of sheep in the western part of that State. It is no wonder that under such conditions stock eagerly seek any shade available. Suitable shade will result in reducing water loss from bodies and a lowering of respiratory energy usage. Protection promotes the welfare of the animals and, in the course of the year, leads to better financial returns to the farmer. All stock owners should ensure that there is some shade in every paddock, however small, and in the case of large areas there should be scattered shade trees so that animals do not have to travel too far for protection. It is suggested that no part of a grazing area should be more than about a quarter of a mile from good shade and preferably less for breeding stock; this means that on the larger grazing areas suitable shade should be spaced about every half a mile.

All animals require access to some shade but not only do the individual requirements vary with the temperature regime of a locality, which in turn is significantly influenced by latitude, but also with the breed and their development. In general, localities north of the Tropic of Capricorn have higher mean maximum temperatures for the hottest month than do those to the south; for instance in Western Australia Hall's Creek (18.2°S.) has 100.5°F but Kalgoorlie (30.7°S.) has only 93.2°F; whilst from south to north in Queensland and New South Wales Deniliquin (35.2°S.) has 88.0°F, Cobar (31.5°S.) has 94.2°F, Charleville (26.4°S.) has 97.6°F and Cloncurry (20.7°S.) has 100.4°F. It was fortunate that the first sheep extensively bred in Australia were Merinos, a breed which can tolerate hot and dry conditions better than some of the heavier breeds from Britain.

In the case of cattle it is only since World War II that serious consideration appears to have been given to breeds which are better adapted to the high temperatures, with a short monsoonal period, which characterise the semi-arid and subhumid areas of northern Australia; in this region the Santa Gertrudi breed appears promising.

Cows and ewes with young, and store cattle in the final stages of fattening also require more protection than animals at other stages of development.

The value of shelterbelts for providing shade will depend on the geographical location of a farm and the orientation of the belts. The furthest north of the climatological arid zone is about 19½°S., whilst the semi-arid zone extends to about 13½°S. latitude and from these boundaries southwards to the Tropic of Capricorn (23½°S. lat.) the sun at midday is vertically overhead twice during the year. Within this region but also southwards to about 30°S. latitude, east-west oriented shelterbelts will not provide worthwhile shade, though north-south belts will give adequate shade in mornings or afternoons. Although suitably-designed shelterbelts will provide some shade it is better to provide this amenity either by retaining suitable natural cover or, if that has all been cleared, by planting appropriate species either singly or in clumps.

Trees for shade fall into two main categories. The first is native trees already growing in the area; nearly all species will provide some shade—the eucalypts have a longer life than most other genera. The second category consists of species introduced into the locality—other Australian species, especially the eucalypts, are usually likely to prove satisfactory, whilst some of the exotics are valuable additions to the native flora. The athel tree (*Tamarix articulata*) and the pepper tree (*Schinus molle*) are amongst the hardiest and provide light but useful shade; of the true pines (*Pinus* spp.) the umbrella pine (*P. pinea*) has a naturally suitable crown for shade, whilst any of the others can be induced to develop larger crowns by cutting out the leader at a height of 10-20 feet when it is 1½-2½ inches diameter; this encourages the development of several laterals.

Where water is available, irrespective of the climatic zone, the willows (*Salix* spp.) and some of the poplars (*Populus* spp.) are particularly good for summer shade. These two genera also offer the advantage that, being deciduous, they can be established in winter by cuttings up to 10 feet long and these only take a few years to produce useful shade; it is possible, though hardly economic, to transplant trees up to 20 or 30 feet in height if there is suitable soil moisture. The use of very large planting stock requires the provision of supports for perhaps a couple of years, by which time the plants will have developed adequate lateral roots. It is sometimes practical to combine the provision of shade with timber production; one method is to fence the young trees until they are about 20 feet in height and gradually prune the lower branches to 6-8 feet before animals are admitted. If it is desired to provide shelter as well as shade then the branches on the outer trees of a woodlot should be retained. Should the animals use the woodlot as a night camp there may be excessive nitrogen added to the soil in their urine, with possible malformation of the main stem of future timber trees.

SHELTER

In contrast to shade, which mainly benefits animals, shelter affords protection to both animals and crops, including pasture. Although scientific data are limited on the effect of exposure on animals, there is considerable information available on the greater productivity of crops when given suitable shelter. Much of this information has been secured from field experiments in the cooler parts of Europe. The additional net yield from sheltered areas, as compared with unprotected sites has been as high as 18 per cent for grain crops, clover and turnips on effectively sheltered sites, but down to as low as only 2 per cent for sites with very little protection. When there was an associated drought the benefit of shelter was even more striking and yields were as much as 31 per cent higher. In an experimental field trial in Gotland (Denmark) the extra yield of potatoes varied from 11 to 22 per cent, according to the degree of shelter given. The point is then well established that crops in some climates benefit from suitable shelter. It is not to be expected that shelter will give similar benefits in all climates but wherever there are extremes of conditions, especially cold or very hot winds, suitable shelter will result in greater productivity. In Australia most fruit crops throughout the coastal zone, hops in Tasmania and many vegetables require some degree of protection.

In the case of stock, one of the readiest measures of the effect of exposure is the reduction in milk yield when cows are subject to full exposure, even for a few hours a day, as compared with production on sites with a high level of protection. In northern Europe cows may be kept in barns for several months of the year. One striking example of the value of shelterbelts comes from New Zealand. The shelterbelts for one farm, which protected it from the prevailing westerly winds, were cut down to allow for extension of the approaches to runways of the Whenuapai airport near Auckland. In the following 1958-59 season, in spite of a district group increase of 4¾ per cent in butterfat averages, the farm deprived of shelter showed a decline of 11 per cent. Losses of 10 per cent in milk yield are not uncommon and this is of even greater economic significance when net profits are being considered. On unprotected sites during spells of cold, squally weather, animals will stand with their heads downwind, looking miserable and dejected—as they are. During such weather the animals do not graze normally, and call on their bodily reserves to provide the heat and energy necessary for life, with a consequent loss in condition.

Shelterbelts, by reducing the velocity of wind, will reduce evaporation from a free water surface, for example a dam. This evaporation varies mainly from 30-65 inches a year in the coastal-tablelands zone and any substantial reduction in this makes more water available for stock and irrigation. In the dry country, on the other hand, potential evaporation can be very high; for instance at Alice Springs it is slightly over 100 inches, Broken Hill about 80, Kalgoorlie about 95 and Kambalda slightly less, Mt Isa nearly 110, the north-west of Western Australia within 300 miles of the coast mainly within the limits of 85-105, and Woomera about 105 inches; Oodnadatta stands above all these places with a potential of about 135 inches. Shelterbelts should not be planted so close to dams so that their roots draw water from the dam or reduce the efficiency of the dam wall, yet near enough to provide optimum reduction of wind velocity on the water surface. The planting distance should be in the range of 4-8 times the expected height of the mature trees comprising the belt and, in any case, not closer than 100 feet for even belts of low shrubs. Reduction of water loss from dams can be of great importance in time of drought, especially if strong, hot winds prevail.

The striking effect of tree plantings on reducing wind velocity may be seen in the data for several weather recording stations in the Australian Capital Territory. Canberra is built on the plains and low rolling topography of the Southern Tablelands of New South Wales. This area was either treeless or only had a very open cover of scattered eucalypts. A wise programme of tree planting was started in the early days of Canberra and records of total annual wind mileage are available from the days when the new tree plantings were still very small until present times, when most of the older suburbs have well established street, park and woodlot plantings. At the Australian Forestry School at Yarralumla, the average annual wind mileages for successive 5-year periods from 1930 to 1965 were 43,000; 39,000; 35,000; 31,000; 27,000; 25,000 and 27,000. At the Royal Military College, Duntroon, from 1911 to 1930, the comparable figures were 37,500; 29,500;

24,500 and 20,000. Compared with these figures is a value of about 68,000 miles a year recorded at the top of nearby Mount Stromlo during the period 1930 to 1939.

FACTORS IN A SHELTERBELT BALANCE SHEET

So far in this chapter it has been claimed that shelter increases plant growth and, in the case of animals, either because of this or combined with increased vigour there is greater productivity. Some of the credit and debit items in a balance sheet are elaborated below.

Advantages (credits)

1. Increase in plant yield, involving:
    - quantity;
    - quality in many cases, sometimes including palatability and reduction in mechanical damage.

    These advantages come, in part, from:
    - reduced loss of moisture from the plant and is marked during periods of strong winds and, especially, high temperatures. Under such conditions the stomatal openings (pores) of the leaves may partially close and this results in a decrease in the amount of carbon dioxide necessary for plant growth. Whilst the openings decrease in size it is not sufficient to prevent undue loss of large quantities of moisture, thus reducing the reserves in the soil. There is support for this in evidence showing significant increases in crop production in dry countries, for example 11 per cent for potatoes in Israel, when shelter has been provided;
    - increase in dew formation, which in arid lands may be of real importance in the water budget;
    - better insect pollination in the case of those plant species which are dependent upon such means.
2. Better control in the use of pesticidal and fungicidal sprays.
3. Increased animal productivity as a result of:
    - increase in the quantity and, sometimes, the quality of feed;
    - decrease in the effects of exposure to extremes of climatic conditions involving:
        - prevention of the deaths of newly-born lambs or recently-shorn sheep;
        - reduction of the energy loss in combating severe conditions, either of cold or heat.
4. An important tool in control of soil erosion.
5. With wide shelterbelts, that is 6 rows or wider, or woodlots acting as shelterbelts, the production of saleable timber and, indirectly, such other products as honey.
6. Last, but by no means least, the direct effect on man of aesthetic enhancement of his environment, as well as amelioration of climatic extremes within the sheltered area.

Disadvantages (debits)

1. Reduction in crop production arising from:

   decrease in soil moisture, nutriments and light intensity in areas immediately adjacent to the shelterbelts;

   sometimes reduction in production, as well as palatability, of crops from heavily shaded areas.

   (There is a recognisable reduction in production from a width up to height $\times$ 1 on one or both sides of a shelterbelt, but this is more than balanced by the increased production on the remaining area, so that there is greater *overall* production. Very tall east-west belts are undesirable in high latitudes.)

2. Cost of establishment and maintenance of the belt, and cost of additional fencing necessary. (Good farm management would plan gradual development of a complete shelterbelt system, using as long-lived species as possible; additional fencing, either of a temporary or permanent nature, would be necessary on *one* side of a belt.)

3. Decreased production due to:

   slower removal of excessive moisture from the soil after heavy rain;

   delayed drying of excessive moisture from crops during wet harvest periods;

   possibly increased local incidence of injurious insects and, especially fungal diseases.

   (In areas subject to these factors good planning would decrease the intensity of protection. Particularly in the case of fungal attack an open design belt could afford a satisfactory level of protection to crop and stock without significantly increasing the liability to fungal development.)

The following points have been raised by some writers on the subject but they are considered of limited significance.

4. Congregation of stock and transference of fertility. (This can occur in small paddocks on highly fertile soils but the breaking up of clods of dry manure and its distribution can be done by the use of harrows or other farm implements. This practice is generally desirable on very productive farms for areas used as night camps.)

5. Possible shortage of carbon dioxide in protected areas. (This is considered an academic objection rather than a practical one so far as it relates to Australian operations.)

THE PLANNING AND DESIGN OF SHELTER

Compared with the southern parts of the sub-humid and humid zones where shelter is needed for both animals and plant development, in the dry country protection is mainly of importance for animals, especially during periods of cold winds accompanied by rain or when there are strong, hot winds which may be associated with duststorms. Because of this accent on animal protection the tall continuous shelterbelts of the colder country are not necessary to the same extent in the warmer and drier parts of the arid and semi-arid zones. In areas of extensive grazing, adequate protection can usually be afforded by shrubs and small bushy trees which hold foliage-bearing branches to ground level. This type of growth is most suitably provided by the retention of natural vegetation, especially that with non-edible foliage.

Before considering further shelter planning in the dry country it will be useful to make a comparison between the size of farm holdings in different parts of Australia. Dairy farms in the more fertile coastal regions may be less than 100 acres in area and, under intensive management, may be subdivided into paddocks of less than 10 acres. In contrast with this are the vast cattle stations of the semi-arid north where holdings are usually referred to in square miles and there are a number in the range of 1-2 million acres; these large holdings usually have no boundary fences and only limited areas are fenced around the homestead. Intermediate between these extremes are many of the sheep stations of the semi-arid and better parts of the arid zones where individual holdings of 100,000-250,000 acres, are ring fenced and in many cases divided into several paddocks. In some districts, for instance, property owners refer to any paddock of less than 10,000 acres as being small. In general the larger the size of a holding the more likely is it to have a lower productivity per acre and it is obvious that on most of the larger stations very little capital expenditure on shelterbelt planting is economically justified. This does not imply that shelter is not needed. It is; but an adequate level of protection should be secured by the cheapest means; whenever a station has trees and shrubs on it, sufficient should be retained to afford the minimum level of protection desired. Unfortunately, in some areas there may be insufficient trees and shrubs for such purposes. This is often due to continued overstocking, especially during periods of drought, and heavy lopping to provide supplemental feed during such periods. In these cases lower stocking or the fencing of suitable areas to encourage natural regeneration may be possible but, in either case, there may be a long interval before adequate protection develops. There are a few districts where tree vegetation is naturally sparse, for example the black soil plains of south-eastern Queensland. In such areas it may be necessary to introduce species which will grow on such soils and also to develop more shelter on any more suitable soil types which may be present.

The position of shelter for animals in the lower rainfall and warmer parts of the dry country may be summarised:

1. Because of the milder winter temperatures, lower rainfall and usually greater number of days with calm conditions or only very light winds, the level of protection does not need to be as high as elsewhere.
2. In the areas of lower productivity, for example a carrying capacity of say one sheep to 15-20 acres, property owners cannot economically justify the capital investment per acre which may be essential for good management elsewhere.
3. When trees and shrubs are planted, protection from grazing animals by fencing and some watering during the first six months is usually necessary; these all add to costs and should be avoided if practicable.
4. Conservation of sufficient shrubs and trees which grow naturally on a property is usually the best way to provide minimum shelter at the lowest cost. This

involves overall farm management and may present difficulties on the largest holdings. In the case of medium size sheep stations divided into several paddocks, rotational grazing and reduction of stock during droughts are aids to conservation. Many shrubs and trees, especially those with edible foliage, will only regenerate when protected from grazing animals; it may be necessary to fence around small groups of seed trees in order to encourage growth of young plants.

Where shelterbelts are desirable, as in much of the temperate semi-arid zone, then design and planning can follow along conventional lines. This is dealt with in some detail in Chapter 6 of *Growing Trees on Australian Farms,* but is also discussed under the following headings:

ORIENTATION

It is customary to place shelterbelts at right angles to the prevailing wind, and so protect the largest possible area. Basically this is a sound principle but the position is not as simple as may seem in a preliminary consideration. Wind roses for both 9 a.m. and 3 p.m. are given in Fig. 3.5 for selected meteorological stations. Attention is drawn to several important factors exemplified by wind roses:

1. The wind directions at 3 p.m. may be quite different from those at 9 a.m. Coastal positions usually show the greatest contrast, for example Onslow, Western Australia, has a predominance of easterly, south-easterly and southerly winds at 9 a.m. throughout the year, but westerly, north-westerly and northerly at 3 p.m. On the other hand an inland location such as Alice Springs has much the same pattern both at 9 a.m. and 3 p.m.
2. The prevalence of any wind direction may vary considerably from season to season. Cloncurry, Queensland, has a predominance of south-easterly winds in winter, although it is still a common direction in summer. At Eucla, Western Australia, the common winds at 9 a.m. in winter are from the north-west and north, but at 3 p.m. they include directions from the east, south-east, south and south-west.
3. Wind roses do not give a true indication of the occurrence of cold winds accompanied by rain.

FIG. 11.1 (*opposite*)

The effects of windbreaks upon wind movement.

(a) The effect of a five-row shelterbelt on wind speed. A cross-section of the belt is shown on the left, and changes in wind speed about the belt on the right. Distances and heights on the graph axes are expressed in multiples of shelterbelt height. The area lying between lines of value Ut/U = 0.25 and 0.50 would experience wind velocities ranging between one quarter and one half of those which would occur in the absence of the shelterbelt. (U = wind velocity without shelterbelt; Ut = wind velocity with shelterbelt in position.)

(b) These two sketches show diagrammatically the effect of shelterbelt density upon wind velocity. The dense belt (B) results in considerable turbulence, which would often be less desirable than the more uniform air flow near the penetrable belt (A).

(c) These graphs show in more detail the effects of barrier density upon wind speed. The more extensive shelter produced by the penetrable screen would usually be more satisfactory than that behind the dense screen. The units on the *isotachs* show wind speed in metres per second.

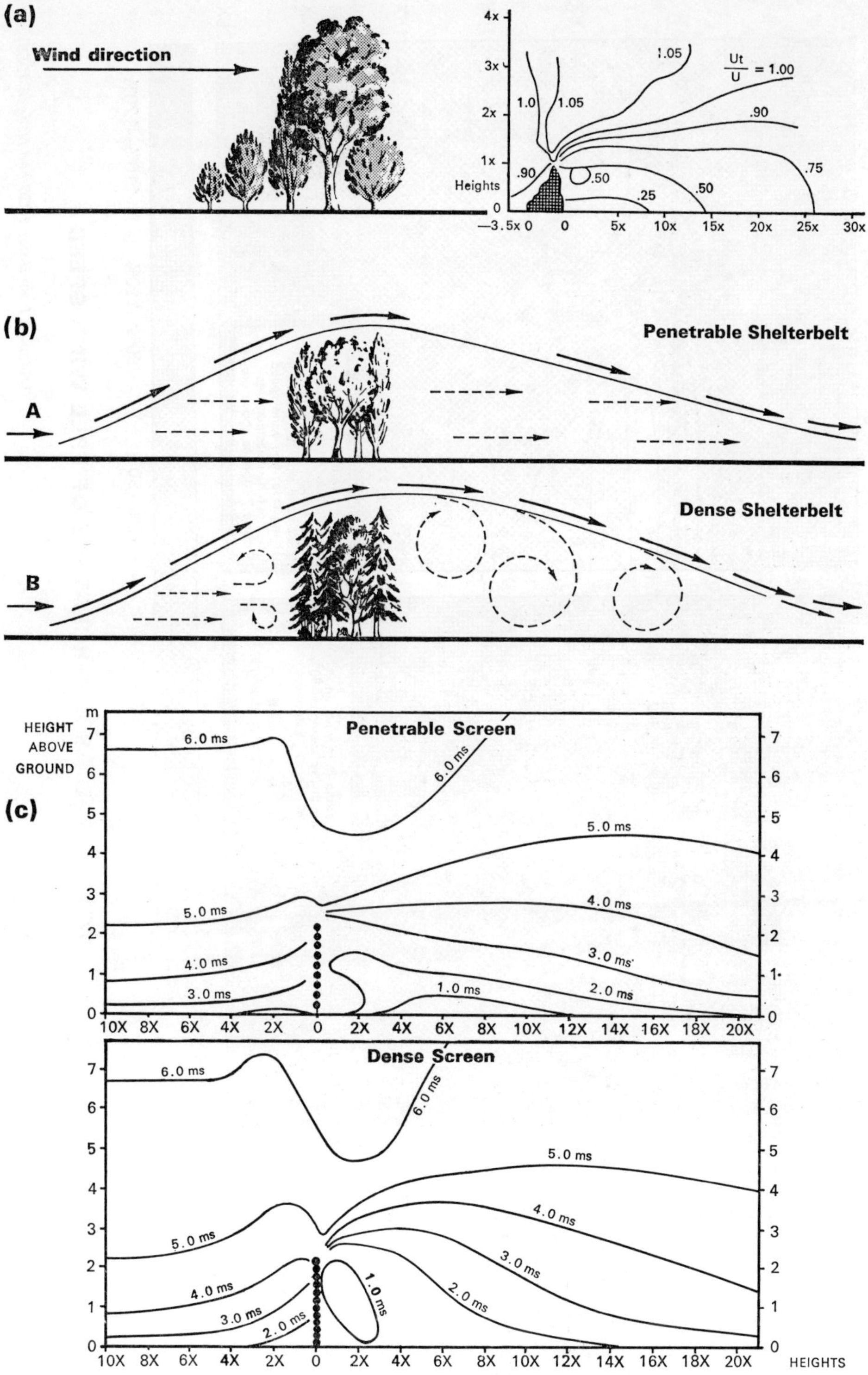
(a)
Wind direction
4x
3x
2x
1x
Heights
0
1.0
1.05
1.05
Ut/U = 1.00
.90
.90
.75
.50
.25
.50
−3.5x 0
0
5x
10x
15x
20x
25x
30x
(b)
Penetrable Shelterbelt
A
Dense Shelterbelt
B
(c)
HEIGHT ABOVE GROUND
m
Penetrable Screen
6.0 ms
5.0 ms
4.0 ms
3.0 ms
2.0 ms
1.0 ms
Dense Screen
10X 8X 6X 4X 2X 0 2X 4X 6X 8X 10X 12X 14X 16X 18X 20X
HEIGHTS

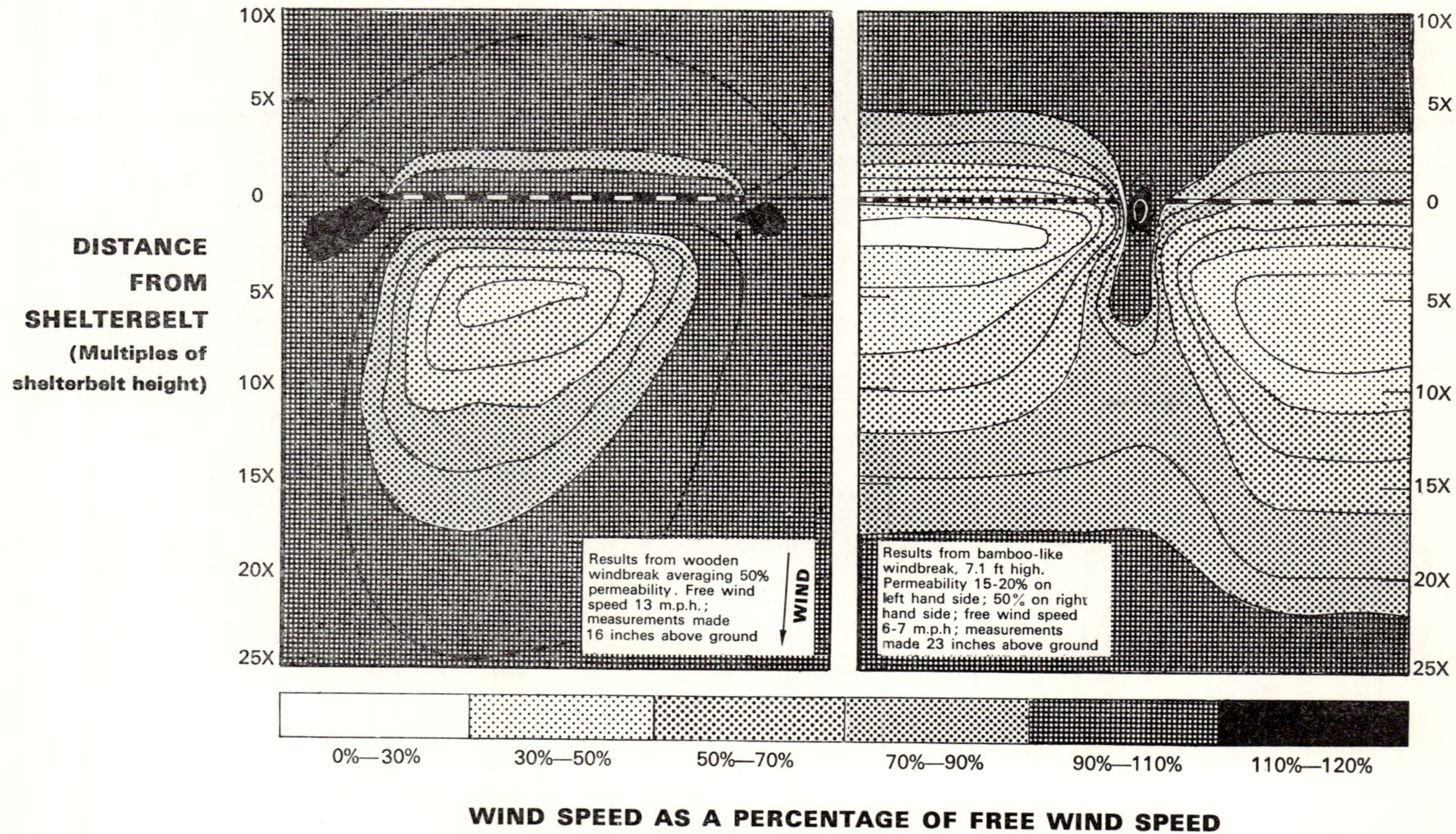

(*Adapted from Bates 1936 and Naegeli 1953*)

It should be remembered that wind roses are based on data collected at specific sites, often open areas of level ground such as may be found at aerodromes; whilst the resultant roses give the correct analysis of wind at these points, the wind in hilly country possibly only 10-20 miles distant may be quite different due to deflection of direction caused by the topography. Therefore, whilst wind roses are a useful general guide they should always be supplemented by local experience when consideration is being given to the design of shelterbelts for a specific farm. Wind roses are of greater value in the level and gentle topography of much of the dry country than in the mountain and tableland areas of coastal Australia.

Farmers will be aware of the wind pattern in their districts and in deciding upon the alignment of shelterbelts consideration should be given to:

1. Whether the primary objective is protection of animals or of crops, or both.
2. Whether shelter is wanted throughout the year or mainly during the coldest months.
3. The relative priority of protection to cows and ewes at calving and lambing time, compared with other times of the year, for example immediately after shearing.

HEIGHT

A very close connection exists between the height of a belt and the distance from the belt over which a reduction in wind speed can be measured on level ground. As a result it is usual to express the extent of shelter effects both upwind and downwind from a belt as a multiple of the height of the belt. On level ground a shelterbelt will measurably reduce the speed of the wind within the zone near the ground where farm animals and crops are, for about thirty times tree height downwind. Marked shelter, however, rarely extends beyond 15 or 20 heights downwind. In addition to this effect to leeward, useful shelter may be given on the windward side of a belt for two to five times belt height. Examples of the pattern of wind movement near a belt are shown in Figures 11.1 and 11.2.

DENSITY

In farm situations shelterbelts of moderate permeability give more suitable protection than solid barriers, such as a brick wall. The reason for this is that the former allows some air movement through it, with the result that the air speed on the downwind side is reduced to an acceptable level for an appreciable distance.

FIG. 11.2 (*opposite*)

Air flow near shelterbelts on flat ground.

The *left-hand diagram* shows that very short belts are inefficient in providing shelter.

When the wind is blowing at about right angles to the belt, a belt length of 12 heights is needed to attain maximum shelter; when the wind is at 45° to the belt, lengths of 24 heights are needed. The belt in the diagram is 19 heights long.

The *right-hand diagram* shows the effect of a fairly large gap (2.3 $\times$ H) in a windbreak.

In gaps the length of which is comparable with the height of the belt, or longer, high wind speeds may occur. These high speeds, however, are usually confined to the gap itself, there being little lateral extension of the 'draft zone'. Narrow, oblique or zig-zag openings, or short supplementary belts, minimise the size of the draft zone.

In addition there is no serious eddy effect. With a solid wall there will be a small area of comparative calm close to the barrier but beyond that there will be considerable turbulence and the maximum distance of effective protection will be less. Measurements near a number of European shelterbelts have shown that moderately penetrable shelterbelts yield an average reduction in wind speed of about 30 per cent throughout a leeward zone 30 × heights wide. If the belt density is increased the total shelter produced may not be necessarily less, but there will be a difference in its leeward distribution—there will be more shelter near the belt but a reduced effect further away.

Studies show that the optimum 'porosity' of model screens is about 40 per cent. It is obviously difficult to accurately measure the porosity of belts of trees, but belts of medium density with no large gaps and a uniform structure from the ground to the crown would approximate an ideal structure. Small, highly sheltered areas for newly-shorn sheep, young lambs and calves will be best provided by densely foliaged shrubs and trees with good performance to ground level. The lower level of protection desirable for pasture and most crops calls for a belt of only moderate density. Shelterbelts open at ground level are undesirable as wind speeds through the gaps may increase appreciably above that in the open. In the case of mature eucalypts the smaller branches usually die and fall from the lower one-quarter to one-third of the tree height, whilst in the case of open grown true pines (*Pinus* spp.) although branches are more persistent it is not uncommon to observe them dead to a height of 3-6 feet as a result of access by animals or, sometimes, by pruning. Trees which eventually lose their lower branches, and this is most of the larger broad-leaved species, should be planted in belts of two or more rows; the outer row or rows consisting of a low-growing bushy shrub. The selection of the most suitable shrub will vary with the region of the country but, in general, many of the non-edible wattles (*Acacia* spp.), paperbarks (*Melaleuca* spp.), eremophilas (*Eremophila* spp.) should be considered. In the very arid areas where it can be protected from grazing animals, a very good plant is oldman saltbush (*Atriplex nummularia*). Fig. 11.1 illustrates some of the effects of shelterbelts.

WIDTH

From the point of view of land utilisation and the efficient production of shelter, the belt should be no wider than is necessary to produce a screen of the required density. Wider belts may be desirable when it is planned to use them for both timber production and shelter. These two purposes are, however, not often fully compatible and more satisfactory results may be usually obtained by using a single-row narrow belt primarily for shelter and devoting a separate area to timber production.

CROSS-SECTIONAL PROFILE

Where shelterbelts are designed with protection as the sole objective a few-rowed belt can combine economy in the use of land and a suitable level of protection. Only under some conditions is it possible to obtain satisfactory protection throughout life with a single row belt and then the choice of species is restricted; amongst the most suitable trees are the native cypress pines (*Callitris* spp.) which are slow growing but long living, and the exotic true pines (*Pinus* spp.). With such species the lower branches should be protected from grazing animals and initial spacing between trees should not be so close that lateral development of branches is seriously inhibited. Where there is adequate soil moisture available in the semi-arid zone, such as on river flats or in irrigation areas, there are few tall-growing species to surpass Lombardy poplar (*Populus nigra* var. *italica*), which combines good maximum height with narrow form. Whilst maximum shelter is provided when the tree is in leaf, the deciduous habit allows winter sunshine to penetrate yet, at the same time, the branches reduce wind velocity appreciably so that the trees constitute a belt of moderate permeability even when leafless.

When two or more tall-growing species are used in a multirow belt not more than one species should be planted in any one row. If eucalypts are used in one row, they should be planted in the centre or on the leeward edge because of their normally bare lower stem; a pine or other species which retains foliage to the ground should be used on the windward side. In the case of low-growing bushy shrubs some mixture of species in an outer row is permissible for aesthetic purposes provided it is known that the rates of growth are reasonably comparable. It is desirable that these low-growing shrubs should be at least moderately tolerant of some shade and that they retain their lower branches and foliage under such conditions.

In order to obtain useful temporary shelter as quickly as possible, a close initial spacing of 2-4 feet may be chosen for trees which normally attain a mature height of 60-150 feet. In the dry country this may be done with many of the native eucalypts and the smaller exotic pepper tree (*Schinus molle*) and athel tree (*Tamarix articulata*) or, where there is sufficient soil moisture, Lombardy poplar. In suitable climatic areas radiata pine (*Pinus radiata*) may also be planted close together. This practice is very suitable for quick, short term shelter *which it is planned to completely remove* when an adjacent longer-term shelterbelt has attained sufficient size. The close initial spacing restricts branch growth and also affects root development, with resultant lessened resistance to windthrow. Even if such belts are thinned at an early age the subsequent branch pattern never equals that of trees which have been planted at wide spacings.

The drier the site the more important it is that each tree or shrub should have sufficient space for its root system to expand. In practice this means that in the arid zone single-row belts (possibly with the addition of one row of a low-growing shrub) are preferable to multi-row belts, unless very wide spacing is used both between trees in a row and between rows for the latter. In very general terms a tree with a mature height of 50 feet which could be spaced 8-10 feet apart in the sub-humid zone, should have the spacing increased to 12-16 feet in the semi-arid zone and 18-20 feet in the arid zone.

Multiple-row shelter belts which have the nature of a narrow plantation will provide satisfactory protection, especially if a row of low-growing shrubs is planted on one side, preferably the windward side. The spacing of the inner rows of trees should approximate to that for plantations, though wider spacings of 12-16 feet may be desirable in the two outer rows. The fewest management problems arise when only one tree species is used in a narrow plantation belt, though in long belts which have changing soil conditions it may be desirable to change the species, for example, on dry ridges from those on fertile flats. Compared with single-row shelterbelts where the death of even a few trees leave undesirable gaps, mortality of up to 10 per cent in a narrow plantation is not serious, either from the viewpoint of shelter effect or timber production, whilst unhealthy or badly formed trees can be removed in thinnings. In addition, eventual replacement of the shelter is made easier than is the case with single-row belts.

In all cases where low-growing shrubs are included in shelterbelt design, consideration should be given to their normal length of life compared with that of the tall-growing species. Some of the wattles (*Acacia* spp.) of the southern higher rainfall areas in particular, have a life which may be only 15-25 years and effective replacement of such shrubs alongside tall-growing trees presents considerable difficulties.

FIRE RISK

The risk of fire is one which should be taken into consideration when planning shelterbelts. There are many factors involved, some of them conflicting; for example the retention of branches close to ground level is desirable to maintain a high level of shelter but, on the other hand, such retention facilitates the movement of fire from grass or stubble into the crowns of the trees. The fire hazard of trees varies considerably between genera and species and the inflammability of the bark and foliage presents different problems in tree selection. The bark of eucalypts, for instance, varies between the highly-resistant type shown by ironbarks and the very inflammable stringybark group. The foliage of most eucalypts burns readily in the green condition once temperatures are raised high enough for combustion of the contained essential oils. Green conifer needles also usually burn readily but this is partly due to the associated drier needles still adhering to the branchlets. Most of the conifers, such as the true pines (*Pinus* spp.) cypresses (*Cupressus* spp.) and the native cypress pines (*Callitris* spp.) are killed by crown fires but will tolerate very light ground fires amongst trees which have been low pruned. In the southern parts of the semi-arid zone a notable exception is Canary Island pine (*P. canariensis*); young trees of this species will withstand ground fires which scorch branches up to half their height whilst older trees will develop adventitious buds when all the foliage and even the smaller branchlets have been destroyed. In the summer rainfall regions some of the southern pines such as slash pine (*P. elliottii*) and Caribbean pine (*P. caribaea*) will recover from ground fires of moderate severity. Only a few of the species suggested for the dry country have green foliage which will not readily burn. The most fire-resistant species are some of the introduced broad-leaved species such as elms (*Ulmus* spp.) and poplars (*Populus* spp.) but, unfortunately, most of these are unsuitable for planting in the dry country, except possibly in some of the better-watered parts of the southern semi-arid zone. They can, of course, be used more extensively in urban planting where some additional watering can be given during critical periods of the year.

The retention of branches to ground level and the use of low-growing bushy shrubs introduces a fire risk which cannot be fully overcome; the main measures to reduce the chance of loss of a large part of a shelterbelt system is to provide cultivated firebreaks or improved pasture with a low fire hazard, to design discontinuity in the length of long belts and, finally, to have species with different fire hazard in different sections.

LENGTH

Shelterbelts are most effective when they stretch without major gaps for distances exceeding twelve times the mature height, because the sheltered leeward area assumes the shape of a blunt triangle (Figure 11.2).

If the winds from which protection is required are known as to direction, the desirable minimum length is about 24 heights. If belt lengths are less than these, the ratio of sheltered area to belt length and cost becomes increasingly unfavourable. Gaps should be avoided as much as possible as high wind velocities occur in these. When a gap is unavoidable, the effect of it can be minimised by making the opening obliquely or crookedly through the trees. However, when the topography is very irregular, there may be little real advantage in long breaks.

LAYOUT

Layout of major shelterbelts must be considered in conjunction with the subdivision of the property, both for ease of management, such as cultivation and, because the cost of the fencing necessary for satisfactory establishment of the belt, can be reduced by suitable siting adjacent to existing paddock boundaries. Sometimes use can be made of existing patches of trees by joining them together with new plantings. When it is desired to give a moderate degree of shelter to extensive fields, this may best be achieved by planting a series of belts parallel with each other so that their zones of shelter overlap, planning each belt as an individual unit. It was at one time thought, notably in Russia, that a series of belts would have a cumulative effect which would ultimately result in the wind being reduced to negligible strength over large areas. This is not the case, however; each belt makes a separate and independent contribution to the shelter of the land within its zone of influence.

SHADE AND SHELTER ON THE FARM

The area of productive land near a shelterbelt lost through shade or root competition can be reduced by siting the belts beside roads, creeks or other unproductive areas. In moist, cool climates care should be taken that roads do not remain excessively wet or frosty; this is less likely to happen when the road is on the sunny or windy side of the trees. In hot climates a shaded road may be attractive.

Where it is not practicable to provide extensive shelterbelts in grazing areas, valuable protection can be given to stock for relatively little outlay by planting small shelters. If these are located in the corners of paddocks a minimum length of additional fence is required.

ESTABLISHMENT AND MAINTENANCE

Information on the raising of nursery stock, planting techniques and other silvicultural practices are discussed elsewhere in this book, as well as in Chapter 11-15 of *Growing Trees on Australian Farms*. Some of the more important factors which should be considered include:

(a) *Establishment*

Cultivation of grassy and weedy sites before planting is highly desirable in order to reduce root competition and, in very dry areas, the demand for soil moisture. When a very heavy grass sward is present it should be cultivated twice or more in the preceding six to nine months. If a pan is present or the site has many small rocks, it will be beneficial to early tree growth if the soil is deep-ripped.

In many districts phosphorus, lime and some of the minor elements are deficient in the soil and the application of the same manures as used for farm crops will assist early tree growth.

One to several waterings in the first few months may be desirable, especially in the arid zone, but continued heavy applications should be avoided. In very dry areas the use of a surface mulch improves soil conditions and economises in the use of water.

Protection from grazing animals and rabbits, where present, is necessary.

(b) *Early maintenance*

Control of grass and weeds should be continued until the young trees and shrubs have grown to a stage where such competition will not affect their growth. In the dry country if there has been adequate preliminary cultivation there should not be need for later control. Where it is necessary consideration should be given to the use of weedicides and attention is drawn to Appendix III (Properties of weedicides important in forestry and other tree growing activities) in *Growing Trees on Australian Farms*.

Early replacement is desirable of any plants which have failed to survive or which show very abnormal development; this is particularly important in single-rowed or few-rowed belts. There is much to be said for holding a small reserve of planting stock in the nursery or home garden so as to have plants available for use during the first planting season; such a reserve may be 2½-5 per cent of the total original planting. Insect attack and fungal disease are not common on young trees and shrubs but when these are present, possibly due to a particularly unfavourable season, use of the common pesticides will usually be adequate. Attention is drawn to Chapter 18 and 19 of *Growing Trees on Australian Farms*; other advice can be sought from the State Forestry Services or Departments of Agriculture.

In the case of narrow plantation belts or the inner rows of multi-rowed shelter belts it is desirable to improve tree form by pruning double leaders or unusually dominant laterals; this will improve log form and reduce the possibility of wind damage at a later stage.

(c) *Later maintenance*

Provided that there has been wise selection of species, once trees are well established they should rarely require much attention. If large branches or leaders have been broken by wind storms the broken stubs should be sawn or cut with an axe to provide a clean slope which will shed

This property is located on the border of the semi-arid and sub-humid zones in southern central New South Wales. The mean annual rainfall is 23 inches. Originally the lowlands carried an open woodland. The soils are fertile and suitable for cropping or for good pasture. The top illustration is an air photograph taken in 1969 and shows some of the shelter belts and woodlots planted by the then owner; in the middle background are eucalypts which have been retained on public roads or along watercourses. The smaller illustrations were taken in 1970.

Centre left: A wide belt of eucalypts with an outer row of pepper trees (*Schinus molle*).

Bottom left: A small woodlot of Canary Island pine (*Pinus canariensis*) which provides dense shelter at present and will furnish a useful construction timber in the future.

Bottom right: A narrow belt of eucalypts in the centre and a single row of white cypress pine (*Callitris glauca*) on the outside in order to provide good protection near ground level.

(*Top photo Division of Radiophysics, C.S.I.R.O.*)

Plate 45—TREES ON A WELL-MANAGED PROPERTY IN NEW SOUTH WALES

water; the exposed surface should be painted with tar or other suitable material to prevent decay and to encourage development of a callus—see Chapter 22 for details.

The trees should be examined periodically for signs of insect attack or fungal disease. If any tree appears to be dying and there are signs of fungal attack it may be desirable to fell it and burn all the foliage and branches. In the case of the inner rows of multi-rowed belts and narrow plantation strips the trees can be thinned and treated as in a normal plantation operation; the drier the locality or the season the more important it is that trees should not be allowed to become over-crowded as they develop. Chapter 15 of *Growing Trees on Australian Farms* describes routine thinning operations.

(d) *Renewal*

The need for renewal of shelterbelts should be kept in mind when a system is first planned. For financial and other reasons, including taxation benefits, the establishment of a complete system is likely to be spread over a number of years. For a large estate this might mean 10-20 years or even longer. Preferable to planting all the shelter for one paddock in any year it is better to do it in at least two stages, separated by several years. When the time comes for renewal it is possible to replace part of the shelter at a time so that the area is never without at least 50 per cent of its maximum designed shelter. The longevity of trees and shrubs may vary from possibly only 10 years for some of the smaller faster-growing wattles to 100 years for the larger eucalypts, the native cypress pines (*Callitris* spp.) and the exotic Canary Island pine (*Pinus canariensis*) when planted on suitable sites. This subject is also discussed in Chapter 18.

FURTHER READING

Caborn, J. M. (1965). *Shelterbelts and Windbreaks.* Faber and Faber, London.

Caborn, J. M. (1967). *Shelterbelts and Microclimate.* For. Comm. Gt. Britain. (Bull. 29).

F.A.O.(1962). *Forest Influences.* (Forestry and Forest Products Studies 15). F.A.O., Rome.

Hutchinson, J. C. D. (1968). 'Death of Sheep after Shearing'. *Aust. J. Exp. Agric. Anim. Husb.*, 8, 393.

Lynch, J. J. (1968). 'Sheltering Behaviour of Newly Shorn Sheep'. *Proc. Aust. Ecol. Soc.*, 3, 120.

Lynch, J. J. & Marshall, J. K. (1969). 'Shelter: a factor increasing pasture and sheep production'. *Aust. J. Sci.*, 32 (I).

Marshall, J. K. (1967). 'The effect of Shelter on the Productivity of Grasslands and Field Crops'. *Fld. Crop Abstr.*, 20, 1.

Miller, G. R. (1968). 'Some Responses of Hill Ewes and Lambs to Artificial Shelter'. *Anim. Prod.*, 10, 59.

Sturrock, J. W. (196?). 'The Control of Wind: the Value of Research and Farm Forestry'. *Farm Forestry*.

——— (1967). 'The effect of wind on crops: a preliminary report'. *Farm Forestry*, 9 (3).

——— (1969). 'The control of wind: the roles of research and farm forestry'. *Farm Forestry*, 11 (2).

Willoughby, W. M. (1959). 'Limitations to Animal Production Imposed by Seasonal Fluctuations in Pasture and by Management Procedures'. *Aust. J. Agric. Res.*, 10, 248.

World Meteorological Organisation (1964). *Windbreaks and Shelterbelts.* (Tech. Note 59). W.M.O., Geneva.

CHAPTER 12

Examples of shade and shelterbelt design

IF a farmer wants an efficient shelterbelt which will provide the level of protection desirable for his crops or stock, and which will remain healthy over a long rotation, he must select trees which will grow well in the locality; establish them in suitable positions, protect them against grazing, especially during the early years of life, and give them the maintenance and protection necessary. A number of the factors which should be taken into consideration when selecting suitable species is dealt with in Chapter 18, but it should be noted that a good early rate of growth and longevity are not always found together in the one species, and though many eucalypts, radiata pine and poplars (*Populus* spp.) do combine them, very few dry country species show fast early growth.

The location and design of shelterbelts should be committed to plans and the years for progressive establishment recorded. All young trees must be protected from grazing animals, including rabbits and hares—many hedges are ruined through allowing heavy grazing of the lower branches. Then it is not enough to simply plant trees and erect a fence and forget them. Weed growth should be controlled until the stage is reached when the trees dominate all competing vegetation, and a periodic watch kept for insect or fungal attack. Finally, a time comes, which may be in 20, 40 or 100 years, when it is necessary to renew the shelterbelts; this is usually best done by replacing short sections each year or at periodic intervals until the complete system has been renewed. Further information on the planning of shade and shelterbelt planting and management is given in Chapter 9 of *Growing Trees on Australian Farms.*

EXAMPLES OF THE DESIGN OF SHELTERBELTS AND SHADE TREES ON FARMS

Some aspects of an approach to the design of shelter and shade on four representative farms in the dry country are given in Figures 12.1 to 12.6. Farms have been selected from the different climatic areas of the 15-21 inch mean annual rainfall country of southern semi-arid Australia, the southern arid zone where the rainfall is only 9-15 inches, the semi-arid country of south-western Western Australia with a rainfall of 14-18 inches and the fourth example is from southern semi-arid Queensland where the rainfall is 19-25 inches. The extent to which a farmer is justified in spending money on tree planting varies, of course, with the returns he gets from his crops and stock. As farming extends into the arid zone productivity per acre decreases to a stage where conservation of existing trees and shrubs may be all that is justified, other than some amenity planting around the homestead.

When a farmer wishes to proceed from general considerations of design which are given in this chapter, to planning a protection system for his individual holding, he will need to allow for the numerous local factors which will influence his plan. Of particular importance are the intensity of farming operations, temperature at various seasons of the year and the direction of prevailing winds, especially those in winter. The examples given only show a limited number of tree and shrub species and before a selection is made consideration should be given to those which grow well in the locality or are recommended by the forest or conservation services of the State.

There are also legal aspects of planting tall shelterbelts along boundary fences, but these can usually be solved by mutual agreement between adjacent land owners. Because of the risk of change in ownership it is very desirable to have agreements in writing and copies kept in safe custody.

CASE A—3,840 ACRES OF SOUTHERN SHEEP-WHEAT COUNTRY

Locality

Southern semi-arid country, mainly south of 30-32°S. latitude but varying in the different States. The rainfall limitation of approximately 15-21 inches restricts it to a comparatively narrow belt of the climatically semi-arid zone in Western Australia, with Northam and Narrogin located on the boundary with the sub-humid zone; in South Australia the rainfall is mainly on the lower margin and Port Lincoln, Kadina and Yorketown fall within the type; whilst in New South Wales rainfall tends to the higher limits with Condobolin and Wagga Wagga representing the lower and higher margins. More representative places are Dubbo and Forbes.

Soils

The most extensive types are the red-brown earths and the non-calcic brown earths; both these types are inherently fertile soils and form the most reliable lands of Australia for cereal production. They are moderately heavy and are affected by cracking in the drying-out period in the seasonal cycle, though not to the extent of the black earths. On the higher rainfall side of the semi-arid zone podzolic types predominate. Where the surface soil to a depth of at least one foot consists of sandy loams and the clay of the subsoil is not too heavy most tree species will grow satisfactorily but where a heavy clay is near the surface and drainage is poor greater care needs to be taken with the selection of species. In such areas most of the true pines (*Pinus* spp.) and cypress pines will not grow as well as some of eucalypts and other hardwoods and if used at all should be restricted to slopes where the drainage is better than on the flats. In Western Australia lateritic podzolic soils occur throughout the semi-arid zone. These are pale-coloured, mildly acidic soils with a marked difference in texture between the surface and subsoil horizons and with associated lateritic features such as ferruginous grists and gravels with mottled and pale-coloured clay beneath. They require trees which are adapted to the soil conditions and in this regard it is best to give first preference to those species which grow naturally in the area; introduced species, whether from elsewhere in Australia or from overseas, should only be used on the advice of the forest service of the State.

Land use

This class of country represents the major wheat lands of Australia, as well as being the region with a high concentration of sheep. The better quality sites may alternate between wheat and sheep according to world demands for the products. Sheep management includes not only the fattening of store animals bought from the drier areas but the breeding and fattening of lambs for both internal and external markets; wool production is axiomatic with the other activities but some properties prefer to buy in breeding ewes rather than raising them. This type of sheep management allows for a wide range in stocking, depending

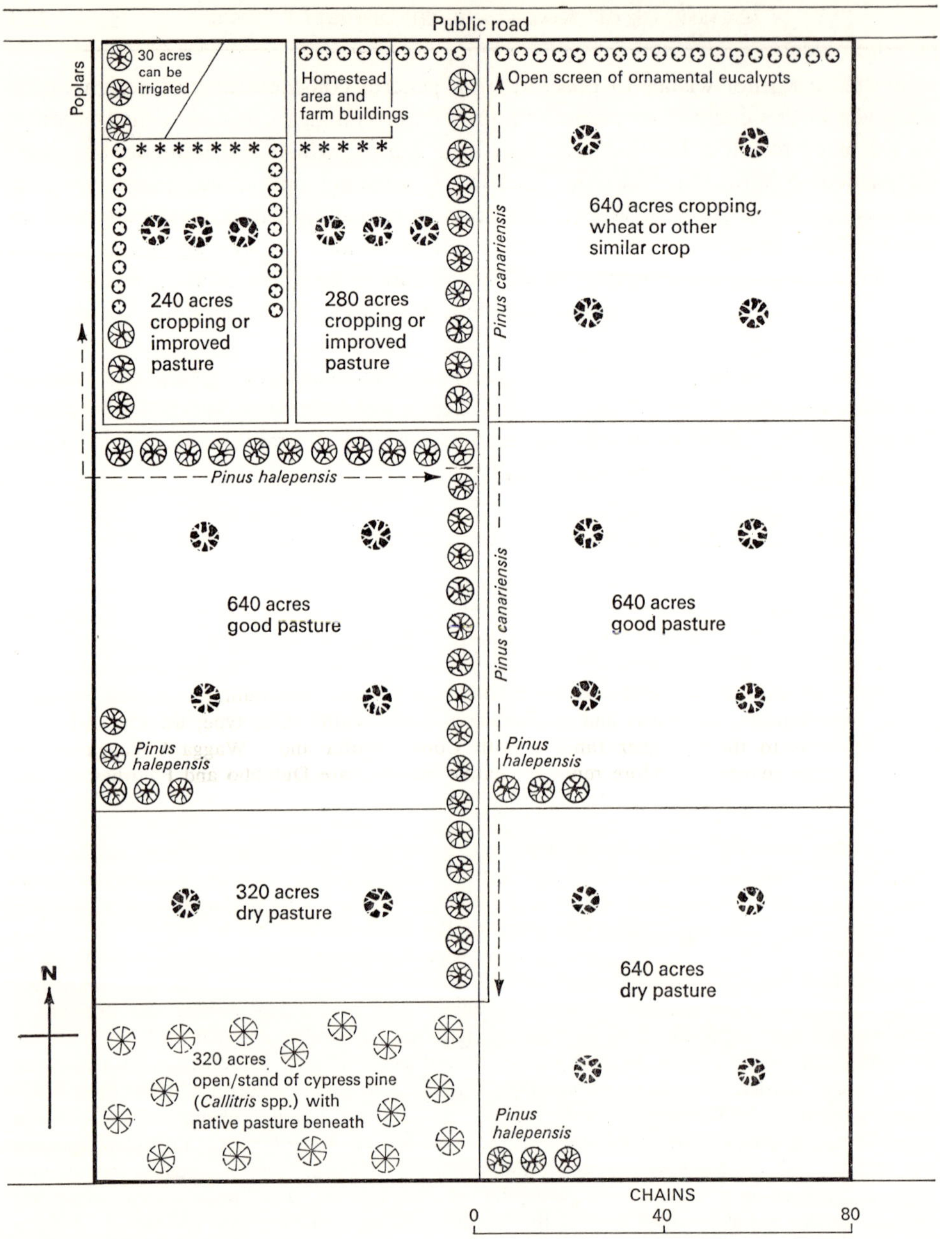

CASE A—3,840 ACRES

LOCATION—southern semi-arid sheep-wheat country. Rainfall 15–21 inches, mainly in winter.

Topography	Shelterbelts	Shade
Level to very gently undulating at southern end of farm.	(tall tree symbol) Tall and medium size trees, mature height, 60 feet and over O O Moderately low screen, height, 20–40 feet * * Low screen, height, 8–20 feet	(dark tree symbol) Evergreen trees (open tree symbol) Cypress pine forest (natural)

FIG. 12.1
Example of shelterbelt and shade layout on farms: Case A.

upon the season. In contrast with stocking in the arid zone which may be as low as one animal to 15-20 acres, the capacity here may be 2-5 sheep per acre on the better parts, depending upon season and type of stock. Cattle breeding and fattening is another land use.

Where small irrigation areas are available these may be used in the conventional way for intensive production of grapes for sultanas and raisins, citrus fruit or, alternatively, for special pastures and lucerne for fattening of stock. Within reach of specialised markets poplar growing may also be a useful aspect of diversified production. The more recent growing of cotton is usually restricted to large areas under single use management.

Climate

Rainfall. Mean annual 15-21 inches on 70-90 days a year, varying from a well-defined winter maximum in the semi-arid zone in Western Australia and South Australia south of about 30° and 32°S. latitude respectively, to more or less uniform in New South Wales near 33°S. Partly an effect of the more southerly latitude in Victoria, most of the 15-21 inch rainfall area in Victoria is classified as sub-humid and is outside the scope of this layout. Whilst droughts are not uncommon throughout this area the variability is not as great as is experienced in the arid zone. A composite set of data, based on representative meteorological stations in the three States concerned gives a 10 percentile rainfall value of about 13 inches, that is a 1 in 10 probability that the annual rainfall will not exceed that amount; a 50 percentile value of 18 inches and a 90 percentile value of 25 inches. The lowest annual fall is likely to be about 10 inches.

Temperatures. Mean annual 60-65°F; mean maximum of the hottest month 80-95°F; mean minimum of the coldest month mainly 37-43°F but as high as 45°F on the coast; frosts based on 32°F in the screen 1-15, but some coastal sites may be frost free; snow nil.

Wind. There are considerable differences in wind directions from west to east in this long belt, which extends through about 30 degrees of longitude or nearly 1,900 miles. Near the coast in the west, morning winds are mainly land winds with directions from south-west through west and north to north-east, but afternoon winds commonly vary from south-east to south-west. Inland, on the edge of the arid zone winds from the south and south-west are uncommon but all other directions occur. Along the South Australian coast of the Great Australian Bight south-west winds predominate in winter, and south and south-west in summer. In the southern part of the semi-arid zone in New South Wales south-west and west winds are important at all seasons, whilst further north, winds occur from all directions. In general, throughout this area protection of stock and crops is most important from cold and often strong winds from the south and west.

Timber utilisation

Because of the relatively low rainfall the semi-arid zone has not the potential for commercial timber production of the sub-humid and humid zones but, under suitable conditions, timber production can be a worth-while minor source of farm revenue. The possibilities include:

(a) Forest management and grazing of any existing cypress pine (*Callitris* spp.) areas.

(b) Combining shade and shelter with timber production from eucalypts; the open position of shade trees, for instance, encourages a good diameter growth although the resultant logs may be short compared with hardwood logs from coastal forests.

(c) Where there is accessibility to cheap water the growing of poplars (*Populus* spp.) under irrigation for special markets.

(d) Finally the designing of wide shelterbelts, that is 6-10 rows wide, of some of the exotic pines such as Aleppo pine (*P. halepensis*) and Canary Island pine (*P. canariensis*).

Both the cypress pines, the eucalypts and the exotic true pines will produce timber useful for fuel, fencing and barn construction, the exotic pines needing preservation. Where timber is grown with the intention of sale a careful examination of the possible markets should be undertaken before planting is done on other than a very small scale. This applies particularly to poplars and the exotic pines. Further reference is made to this aspect in Chapter 7 of *Growing Trees on Australian Farms.*

*Trees and shrubs suggested**

Tall shelter. Only one tall shelterbelt is suggested and this should be sited in a north-south position in the middle of the farm and could be 2-3 miles long. The tree recommended is Canary Island pine (*Pinus canariensis*) which has few equals in this environment for retention of foliage to near ground level, longevity (100 years or more) and fire resistance. It could be planted in a single row, with a spacing of 10-16 feet; it would take some years (say 8-10) for this spacing to close but well-balanced, wind-firm trees should result. It should not be planted across drainage depressions where soils may be waterlogged for part of the year, but replaced by a suitable eucalypt, for example one of the red gums or some of the boxes. If the property owner requires a eucalypt then one of the species which grow in the area could be used, but this would mean at least a 2-3 row belt, since provision would need to be made for a shrub species to provide low shelter as the eucalypt grew and shed its lower branches. Attention is drawn to plates 20 and 21 of *Growing Trees on Australian Farms.* For best use of farm labour and equipment, as well as obtaining the maximum benefit from taxation deductions, it is recommended that the planting of a major belt should be spread over a number of years or sections done every second or third year for a long period, say 10-20 years.

Medium shelter. As the farm is level or only gently undulating any shelter should provide protection for a maximum distance; it is suggested that a species such as Aleppo pine (*P. halepensis*) should be a major species for the approximately three miles of medium height shelterbelt included in the layout. It has a proven record of satisfactory growth in the environment and except on poorly drained and heavy soils should grow to 60 feet in height and remain healthy for 50 years or more. Early growth is moderately good. An alternative species is the much slower growing but longer lived white cypress pine (*Callitris glauca*); and it should be preferred on well-drained stony hills. Not only can a long length of shelterbelt be established over a period of many years but it is desirable that all the planting for any one paddock should not be done in the one year, so that eventual replacement can be done in progressive stages. There is also a strong case for the shelterbelts around any one paddock to be of different species, especially if one, for example Aleppo pine or white cypress pine is fire-sensitive. In this example consideration should be given to using some Canary Island pine in separate lengths of belts as shown in the layout. Where the property owner prefers eucalypts for his medium shelter a number of the dry country species listed in Table 17.1 can be used, though the addition of at least one row of a low-growing shrub is necessary to provide protection to ground level.

Low shelter. Here small trees and tall shrubs with a maximum mature height of 25-30 feet are considered; most of them will be native species, of which a wide choice is possible and only a few can be listed here. The genus *Acacia* (wattles) has many useful species, though some of the larger ones tend to lose their lower branches with maturity; species include the cooba or native willow (*A. salicina*) and myall or boree (*A. pendula*) in eastern Australia and raspberry jam (*A. acuminata*) and manna wattle (*A. microbotrya*) in Western Australia. Other genera are represented by the Western Australian peppermint (*Agonis flexuosa*), some of the taller growing species of *Cassia*; bottlebrushes (*Callistemon* spp.), and eremophilas (*Eremophila* spp.), certainly some of the paperbarks (*Melaleuca* spp.) such as *M. armillaris,* and *M. lanceolata,* boobialla (*Myoporum montanum*), pittosporum (*Pittosporum phylliraeoides*) and wilga (*Heterodendron oleiofolium*).

Ornamental screen-shelterbelt. The two-mile long northern road frontage offers possibilities for the planting of a widely-spaced single row of trees of attractive appearance which will not only be ornamental but also provide some protection from the occasional hot northerly

* The species listed here are only a few examples of the large number which could be used. Readers are referred to Chapter 17 and Table 17.1 of this book and, especially, to the catalogues and lists of trees and shrubs issued by the forest services of the States, as well as by some other organisations.

wind without unduly shading crops. It is suggested that long-lived native trees, preferably of not more than medium height would be most appropriate; spacing could be 25-50 feet. If the property owner prefers trees and shrubs only growing 15-30 feet in height then some of those suggested for low shelterbelts could be used. Eucalypt species are discussed in the next section, whilst many of the species listed for example Case B will also do very well on well-drained somewhat sandy soils in this area.

Shade. Wherever there are existing native trees, especially eucalypts, sufficient should be retained in each paddock to provide a minimum amount of shade; in addition it is assumed that in every paddock animals will have access to a north-south shelterbelt which will provide shade from the late afternoon sun. Local species of eucalypts which may occur on the property in Western Australia include brown mallet (*E. astringens*), wandoo (*E. wandoo*), York gum (*E. loxophleba*) and salmon gum (*E. salmonophloia*), with possibly Western Australian flooded gum (*E. rudis*) along drainage channels in the more western areas. South Australian species could include sugar gum (*E. cladocalyx*), peppermint box (*E. odorata*) and yellow gum or South Australian blue gum (*E. leucoxylon*). The New South Wales eucalypts are mainly those of the box group which attain their best development in the wheatlands of the State and adjacent areas of Victoria and Queensland; they are trees which commonly grow to about 80 feet in height and have a well-developed crown above a comparatively short and thick bole. The more important box species include grey boxes (*E. woollsiana* and *E. microcarpa*) and the closely allied narrow-leaved grey box (*E. pilligaensis*), bimble box (*E. populnea*) and fuzzy box (*E. conica*). On the drier side of the area black box (*E. largiflorens*) occurs and on the higher rainfall side white box (*E. albens*), yellow box (*E. melliodora*) and apple box (*E. bridgesiana*). A number of the ironbarks may also be present, especially in the more northerly areas: these include narrow-leaved red ironbark (*E. crebra*), red ironbark (*E. sideroxylon*) and silver-leaved ironbark (*E. melanophloia*) as well as brown bloodwood (*E. trachyphloia*). Most of the boxes are indicators of at least moderately-good soils suitable for wheat growing, whereas the ironbarks tend to less favourable sites.

Cypress pine (*natural forest*). Where the property owner has a small percentage of his farm with natural cypress pine forest he should consider the management of the area for both timber production and grazing; the forest service of the State is able to give expert advice on the subject. Some of the more important factors which affect management are:

(a) Cypress pine is so slow-growing that planting of seedlings is not considered an economic proposition.

(b) Natural regeneration is the usual way of obtaining establishment of a new crop. This needs a better than average rainfall, that is a 90 percentile value as shown in Appendix I, and protection from grazing stock. In practice this means that when a favourable year comes a portion of the forest should be fenced from grazing animals for several years until the young plants are large enough not to be damaged.

(c) Except for areas under regeneration grazing can be allowed in the forest, but periodic pasturing is preferable to continuous grazing.

(d) Young cypress pine stands and dense forests are susceptible to fire; danger can be reduced by the provision of firebreaks in all cases, but light grass fires can run through open mature stands with little damage. Apart from the provision of firebreaks the maintenance of mature stands in an open condition and heavy grazing at the beginning of the fire season will usually avoid serious damage.

Other aspects of timber growing are dealt with in more detail in *Growing Trees on Australian Farms*, especially Chapter 11 for the techniques of raising poplar cuttings. Some of the forest services of the States have publications which give advice on growing poplars for timber.

CASE B—100,000 to 250,000 ACRES, SOUTHERN SHEEP COUNTRY

Type locality

Southern dry country, mainly south of 30°S. latitude. It is typically an arid zone type but in places fringes onto the semi-arid zone. Most of the area is sufficiently inland from the southern coast of Australia not to be significantly affected by cold southerly winds; where these winds are experienced greater attention should be given to low shelter for homestead buildings and newly shorn sheep. Broken Hill, New South Wales, fits the environment envisaged.

Soils

There is a wide range of soil types in the southern arid zone. The most extensive is the solonised brown soils which are mainly restricted to 9-15 inches rainfall; they comprise deep sandy to shallow loamy soils overlying calcareous subsoils and are very susceptible to wind erosion. Where water is available for irrigation they can be used for the production of horticultural crops, principally grapes and citrus fruits, but problems of rising ground water and secondary salinity occur where subsoil drainage is inadequate. Next in importance are the desert loams and arid red earths; these are fertile, alkaline in the subsoil and require careful management of the vegetation cover if erosion is to be avoided. Skeletal soils are of more limited occurrence and only have a very low value for grazing. In the arid zone, in general, trees show better development on plains and very gentle slopes than on steeper slopes and hills where drainage is stronger and soil fertility lower.

Land use

This is typical arid sheep country where the main production comprises wool and breeding ewes for station use or for sale, or hoggets for fattening on land nearer to markets. On the better sites and where transport is an economic proposition some animals may be finished off ready for slaughtering. Because of the periodical droughts which are superimposed on a low mean annual rainfall it sometimes becomes necessary to significantly reduce stock numbers in order to avoid death of animals, and to ensure that the condition of fodder plants is not deteriorated to the stage when recovery will take a very long time.

Climate

Rainfall. Mean annual of 8-12 inches on 40-75 days a year. It has been mentioned in Chapter 3 that mean annual values have limited significance in the arid zone; a major factor in survival is the occurrence of years of very low rainfall, especially when two or more such years occur together. Taking the data for Broken Hill as an example, the 10 percentile rainfall value is only 5.15 inches, the 50 percentile value is 9.12 inches and the 90 percentile value 14.47 inches. Trees and shrubs which grow naturally in the area can be expected to survive the years of 10 percentile value reasonably well and most of them the lowest rainfall on record, which in the case of Broken Hill is 2.24 inches. In general, it is only in years of rainfall exceeding the 90 percentile values that good regeneration of native trees and shrubs can be expected; these are years when not only the trees regenerate but the whole vegetation cover improves from the effects of serious drought.

Temperature. Mean annual mainly about 65°F; mean maximum of the hottest month 90-95°F; mean minimum of the coldest month 40-45°F, the lowest being more common at the higher altitudes of 1,000 feet or more; frosts based on 32°F in the screen usually in the range of 1-10 a year; snow nil.

Wind. With the exception of a small belt where the arid zone reaches the coast near the border of Western Australia and South Australia, most of the area is sufficiently inland to experience a large number of days of calm or of only very light winds. In these inland areas the occasional cold winds from the southern quarter or the less common days of very hot, strong and drying winds from the interior are of greater importance in designing shelter protection for homesteads and paddocks where recently-shorn sheep may be held, than the prevailing winds. Because local knowledge is so important in all designing and the geographical

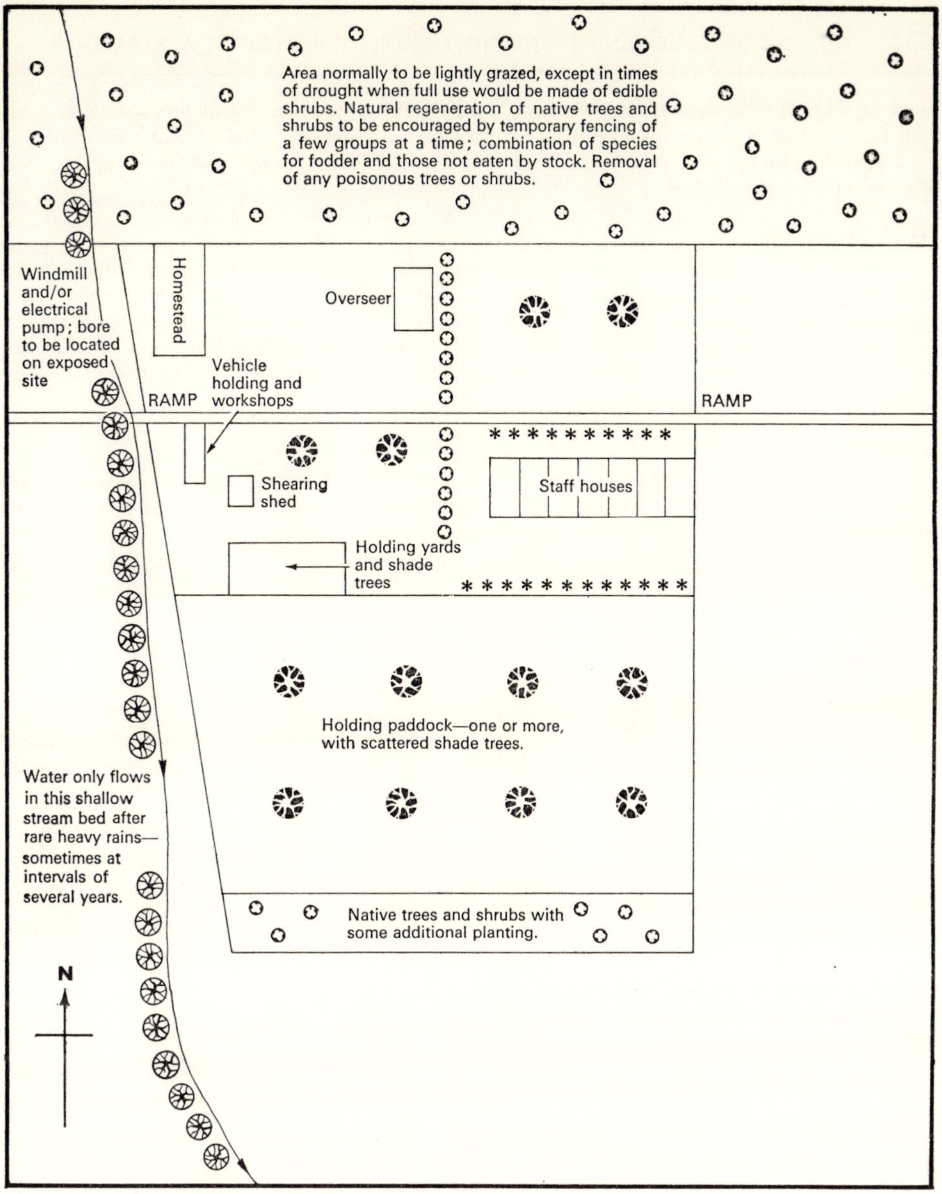

0 CHAINS 24

CASE B—HEADQUARTERS; AREA SHOWN 800 ACRES

(Total area 100,000–250,000 acres.)

LOCATION—southern arid zone, rainfall 8–12 inches, but with some years as low as 2–4 inches.

Topography	Shelterbelts	Shade
Almost level except for depression of drainage channel.	Tall and medium size trees along watercourse, mature height, 60 feet and over Moderately low screens, either planted or natural, height, 20–40 feet Low screens, height, 8–20 feet	Evergreen trees

FIG. 12.2

Example of shelterbelt and shade layout on farms: Case B.

area of Case B extends east and west for nearly 1,500 miles, the layout given only presents general suggestions which may need to be modified according to local critical wind direction and intensity. In general, protection from the west and south is more likely to be required than from the north and east.

Timber utilisation

This is not an area for growing timber for sale, but most of the larger native shrubs and trees have wood which is suitable for fuel and fencing material. Some of the best fuel wood will be from various acacias or eucalypts of the box and red gum groups. When of suitable form these trees will also produce fencing material; fissility is often poor with the groups of eucalypts named and consideration should be given to preservation of the sapwood of small diameter round material.

Trees and shrubs suggested

Because of the low productivity per acre of this class of property capital expenditure on shade and shelter needs to be kept at a comparably low level; fencing for protection of regeneration or of planted tree stock must be restricted to short lengths. It is suggested that:

(a) Protection of the homestead area should be based firstly on selecting the most suitable site and secondly on encouragement of the local species which are already growing there.

(b) There should only be a limited amount of planting, and this mainly restricted to protection and beautification of the residential buildings or providing shade in sheep and cattle yards.

(c) Operations away from the homestead should be mainly restricted to temporary fencing of small areas of potential natural regeneration of tree and shrubs species during and immediately following years of heavy rainfall.

(d) When introduced species are planted, either from elsewhere in Australia or from overseas, preference should be given to those which have the ability to survive without watering after the first few months. This means survival not only in average years but also in those of markedly below average rainfall.

(e) Where shelterbelts are desired, other than those only 6-12 feet high for homestead gardens, reliance should be placed on wide belts with trees widely spaced, for example belts say 5 chains wide with small shrubs perhaps 20-30 feet apart and trees which grow to a mature height of 25-50 feet at a spacing of 50-100 feet. This provides an open screen which reduces wind velocity but does allow some cooling effect when that is wanted; in addition it allows for the adequate expansion of the root systems which is so necessary for survival during dry years.

Because of the variable local conditions and the great number of species which can be used in this type of country, species are not nominated in the layout figure, but suggestions will be made of species which should be considered for natural regeneration or planting.

Nearly all species of acacias are important and undoubtedly mulga (*Acacia aneura*) heads the list; there are, however, many others such as cooba or native willow (*A. salicina*), myall or boree (*A. pendula*), ironwood (*A. excelsa*), raspberry jam (*A. acuminata*) and manna wattle (*A. microbotrya*) which have a high value.

Another important group are the sheoaks (*Casuarina* spp.). Probably the two most important are belah (*C. cristata*), which is widespread in the south-east of the dry country, and desert sheoak (*C. decaisneana*) which is native to very hot and dry areas.

Wherever the native cypress pines (*Callitris* spp.) exist they should be encouraged but their optimum development is in the higher rainfall semi-arid country.

Amongst the eucalypts there are many from which to select; river red gum (*E. camaldulensis*) has a very wide distribution along drainage channels and where it can gain access to deep underground water will grow in areas of very low rainfall and several species of the box group also grow on somewhat similar sites, especially those which are inundated at intervals. Outstanding, however, are many small tree species usually growing to 25 to 50 feet in height, native to the south-western dry country of Western Australia; some of the more important are Dundas mahogany (*E. brockwayi*), silver-topped gimlet (*E. campaspe*), Dundas blackbutt (*E. dundasii*), merrit (*E. flocktoniae*), Gardners mallet (*E. gardneri*), gimlet (*E. salubris*), salmon gum (*E. salmonophloia*), Salt River mallet (*E. sargentii*) and coral-flowered gum (*E. torquata*).

In addition to the four genera named there are many others which are of importance in this environment; only a few will be given here and readers are referred to Chapter 17 and Table 17.1 for further information, as well as that given in several of the State publications listed at the end of the same chapter. Some of these include the cassias (*Cassia* spp.), of which there are many species, hopbush (*Dodonea attenuata*), certainly some of the eremophilas (*Eremophila* spp.) which will grow in areas with an average rainfall as low as 6-7 inches and have emu bush, native fuchsia, budda and turpentine bush as common names, wilga (*Geijera parviflora*), some of the grevilleas (*Grevillea* spp.) such as beefwood (*G. striata*), inland rosewood or boonery (*Heterodendron oleifolium*), certainly some of the paperbarks, also known as teatrees, (*Melaleuca* spp.), especially *M. armillaris, M. halmatorum* and *M. lanceolata,* some of the species of *Myoporum* such as boobialla (*M. montanum*) and dogwood (*M. platycarpum*), pittosporum (*Pittosporum phylliraeoides*) and sandalwood (*Santalum lanceolatum*). Where a low, dense shelter is wanted for a homestead garden consideration should be given to oldman saltbush (*Atriplex nummularia*) which may grow to 8 feet or more.

Of exotics the two species which have shown reliable growth over a wide area of the southern dry country are the pepper tree (*Schinus molle*) and the athel tree (*Tamarix articulata*). On level sites free from stones these will grow without additional watering in areas with an average rainfall as low as 9 inches. Another species which will tolerate dry conditions is the European olive (*Olea europaea*). The most tolerant of the exotic true pines is probably Aleppo pine (*Pinus halepensis*) and, whilst moderately good specimens can be seen in the streets of Broken Hill, it is not recommended for general use in such dry country.

CASE C—3,600 ACRES OF SOUTH-WESTERN SHEEP-WHEAT COUNTRY

Locality

South-western semi-arid country of Western Australia, within the latitudinal range of 28-33°S. The specified rainfall of about 14-18 inches limits the area to a belt about 50 miles wide, which is located 50-100 miles inland at the latitude of Narrogin but approaches the coast near Geraldton. Locations in the southern part of the belt include Wagin (17 inches mean annual rainfall) and Lake Grace (14 inches); in the central parts are Northam (17 inches) and Cunderdin (16 inches), whilst Kellerberrin (13 inches) is somewhat dry. North of Perth are Coorow (16 inches) and Mingenew (16 inches), whilst Morawa and Mullewa (both 13 inches) are somewhat dry; Geraldton (18 inches) and Dongara (19 inches) represent the higher rainfall coastal locations.

Soils

The soils vary from lateritic podzols with a shallow leached horizon over a lateritic layer which may be 6-30 inches deep, and either a loose gravel or a compacted hardpan, to deep, sandy podzols with the clay layer 5-10 feet below the surface. Intermediate are red-brown earths and some non-calcic brown soils which are at least moderately fertile. There are also solonised brown soils of moderate texture, 12-24 inches thick, over deep accumulations of lime mixed with clay; they are moderately fertile and, where the rainfall is adequate, are suitable for wheat-growing.

From the viewpoint of growing trees the important factors are:

1. Where hardpans occur near the surface, that is within 18 inches, they should be deep-ripped prior to planting.
2. Near the coast where there are deep, sandy podzols the depth of the clay layer and the micro-topography are critical in determining what species will succeed. Reference should

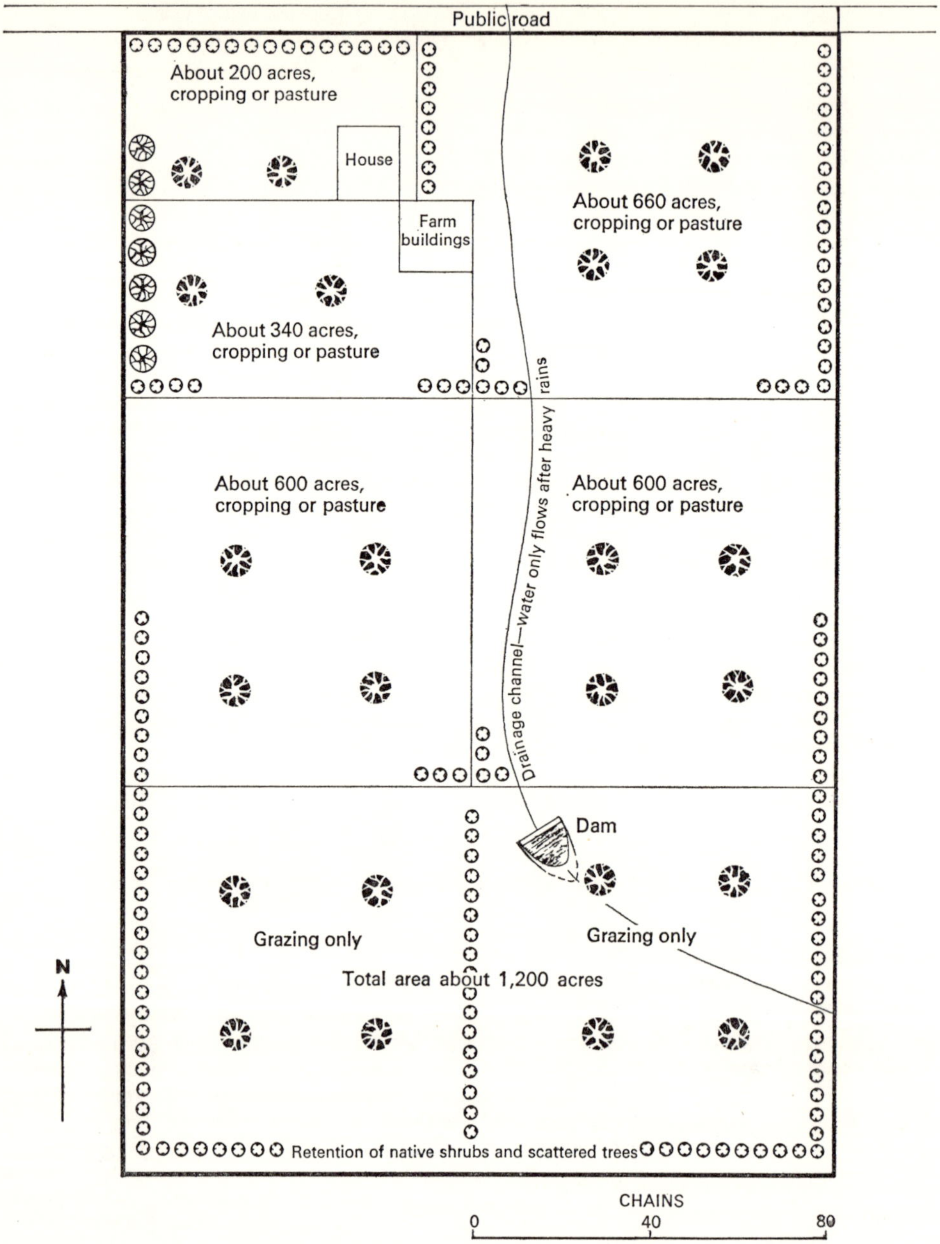

CASE C—3,600 ACRES

LOCATION—south-western sheep-wheat country in the semi-arid zone but bordering the sub-humid zone. Rainfall 14–18 inches, mainly in winter.

Topography	Shelterbelts	Shade
Level to gentle undulating, 3,150 acres Undulating to moderately hilly, 450 acres	Tall and medium size trees, mature height, 60 feet and over ○○ Moderately low screen, height, 20–40 feet * * Low screen, height, 8–20 feet	Evergreen trees ○○ Native trees and shrubs

FIG. 12.3
Example of shelterbelt and shade layout on farms: Case C.

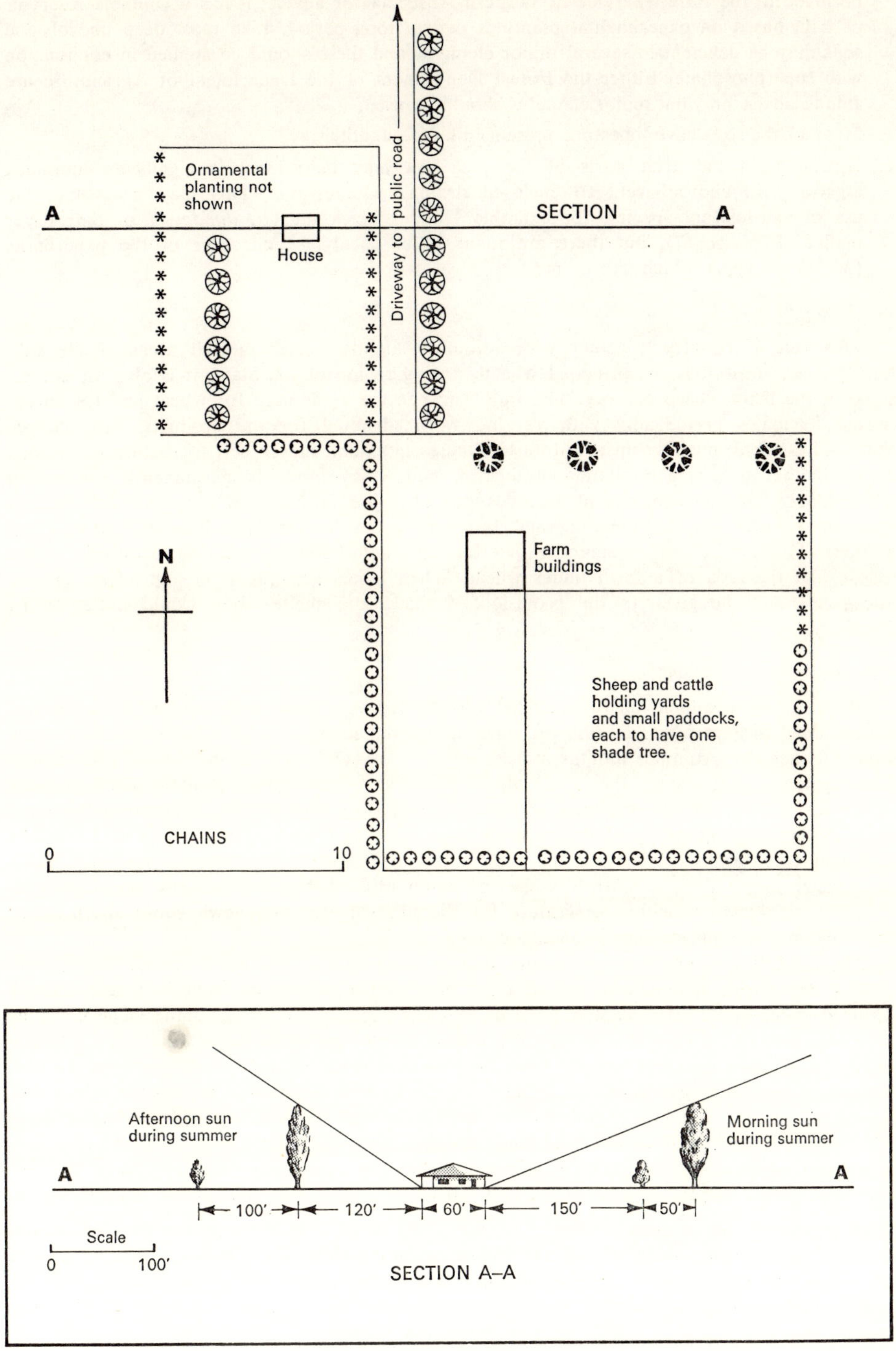

FIG. 12.4
Example of shelter and shade layout for homestead: Case C—*continued*.

be made to the Forest Service of Western Australia for advice; it has a valuable reservoir of data based on experimental plantings over a long period. Like most deep podzols the soils may be deficient in several minor elements and these should be applied in conjunction with superphosphate. Either the Forest Department or the Department of Agriculture are able to advise on what minor elements should be used.

3. Many of the soils have limestone present in large quantities.
4. Especially in the drier parts of the area salt may be present in significant amounts, especially in and adjacent to enclosed drainage depressions. This may necessitate the use of salt-tolerant species, for example Dundas blackbutt (*E. dundassi*) or Salt River mallet (*E. sargentii*), but there are many other eucalypts and some of the paperbarks (*Melaleuca* spp.) which can be used.

Land use

This belt of country, together with adjoining slightly higher rainfall areas of the sub-humid zone, constitutes the principal wheat-growing region of the State. It is also, of course, some of the better sheep country. The ratio between the areas used for wheat and for sheep-raising fluctuates periodically with varying world demand for the products. This chance that a wheat field, possibly up to a thousand acres in extent, may revert to pasture is a strong argument that at least a minimum number of shade trees should be permanently retained in all paddocks. Sheep management includes not only the fattening of store animals bought from drier areas but the breeding and fattening of lambs for both internal and external markets. This type of management allows for easy adjustment in stocking according to the season and the area of a farm under wheat. When lambs are raised on a property greater attention should be given to the provision of shade and shelter than would otherwise be necessary.

Climate

Rainfall. Mean annual of 14-18 inches on 60-70 days a year, with most of it in the five months May to September. During the remaining seven months the mean monthly fall varies from ¼-1 inch. The summer months are usually very dry and some of the precipitation comes from cyclones which reach their peak development in northern coastal areas from Broome to Carnarvon. At average intervals of several years they extend their influence southwards to reach the wheat country and inland as far as Kalgoorlie. This may bring falls of 2-4 inches in the northern areas of the belt, but such a fall in February, for instance, has little value for pasture unless there are worth-while follow-up rains. When the cyclones occur there is usually a significant drop in temperature for the time of year and newly-shorn sheep suffer and some may die, unless there is adequate shelter.

The rainfall variability is low at the southern end of the belt but increases appreciably towards the north. A composite set of data, based on a number of meteorological stations within the area and a 50 percentile value of 16 inches, indicates a 10 percentile value of about 10 inches and an all-time low of about 7 inches; the 90 percentile value is about 22 inches and the maximum on record varies from 25-30 inches, being highest in the northern part of the belt. These latter values should be kept in mind when considering the species which can be planted on different topographical sites; for instance in drainage channels and other low-lying areas eucalypts such as river red gum (*E. camaldulensis*) should be planted in preference to either the native cypress pines (*Callitris* spp.) or the exotic true pines (*Pinus* spp.). The latter two genera should be planted in areas where there is at least moderately good drainage.

Temperature. Mean annual 60-65°F in the southern parts of the belt and 65-70°F in the northern parts. Mean maximum of the hottest month mainly 85-95°F but higher in places, for instance Mullewa has 97.4°F; mean minimum of the coldest varies from about 40°F in southern inland areas to over 50°F for northern coastal sites. Frosts, based on 32°F in the screen, 1-6 a year but some coastal sites are frost free; snow nil.

Wind. Throughout the belt there is a marked change in direction between morning and afternoon winds. Early in the day easterly directions prevail but as the day advances the winds veer through southerly to south-westerly or westerly. Due to the proximity of the ocean, winds from these directions are often very cold during winter months and stock need protection. At times morning winds from the east can be both cool and very strong, so that some shelter is also needed from that direction. Northerly winds occur mainly in winter and although strong at times, they are not accompanied by very low temperatures.

Timber utilisation

The principal objective in planting trees in a low rainfall area should be for amenity purposes or soil conservation, whilst timber production should only be incidental. Since most of the trees planted will be open-grown, that is in single-row shelterbelts or as scattered shade trees, they will either retain branches to near ground level in the case of cypress and true pines or only have short clear boles for some of the taller-growing eucalypt species. Trees intended for shelter should not, of course, have any of their lower branches removed in order to encourage the production of clear timber since this would defeat the principal purpose for which they were planted. Some pruning can be done, however, of shade trees. Clear timber is not necessary for fencing material but is desirable for certain classes of sawn material.

With suitable selection of species and subsequent management the trees suggested for this farm can produce:

(a) Posts and other fencing material from naturally durable heartwood. Many species of eucalypts and acacias and some of the cypress pines produce suitable timber. In the case of round material 3-6 inches in diameter with a durable heartwood but a non-durable sapwood it may be practicable to preserve the sapwood. This is referred to in Chapter 23, whilst further information on the treatment can be obtained from the Forest Service. It is suggested that planning the preservation of the timber of *Pinus* is probably not justified in a region which has so many durable native species.

(b) Firewood for use on the farm or for sale where there are suitable markets within economic distance. Most of the native species are good for fuel but nearly all of them have dense timber which takes an appreciable time to dry unless cut to short lengths and split to thicknesses of 3-6 inches. The importance of drying firewood is discussed further in Chapter 13. Once a shelterbelt and shade planting programme has been completed over a period of years, the removal of damaged or unhealthy trees or the gradual replacement of old belts will normally be an adequate source of firewood. When firewood is wanted in the shortest possible time a small woodlot can be planted at a spacing of 8 x 8 feet or less and thinning operations commenced as soon as the trees reach a useable size, say 3-6 inches diameter.

(c) Clear sawn timber. This will only be available in small quantities from pruned shade trees, but where there is fast diameter growth useable logs should be available from 15 years of age upwards.

*Trees and shrubs suggested**

The following principles have been followed in considering species for this area:

(a) Only species are suggested which are likely to survive in the low rainfall of the area, that is not only during years of average rainfall but also during the dry years. It is poor ecology to plant any species for which the average rainfall is already at the lower limit of their ability to survive; under such conditions marginal soils and years of markedly below average rainfall will result in death or unhealthy growth. It is far wiser to select species which are known to be well adapted to the climatic pattern in which the farm is located.

(b) In general, the drier the area the wiser it is to select species which are known to grow satisfactorily in the district and on the soil types which occur on the farm concerned.

* The species listed here are only a few examples of the large number which could be used. Readers are referred to Chapter 17 and Table 17.1 of this book and, especially, to the catalogues and lists of trees and shrubs issued by the forest services of the States, as well as some other organisations.

(c) Species introduced from elsewhere in Australia or from overseas should only be planted extensively if they have a proven record of good performance in the district.

Because of the varying resistance of different species to disease, insect attack, fire and years of exceptional drought it is not wise to rely on only one or two species. but several should be planted. When a farmer or his wife is keenly interested in trees and wishes to introduce new species into the area it is desirable to establish initially only a few plants, preferably where there is some protection from the existing shelterbelt systems and spaced so that the distance from other trees is equal to the likely mature height of the taller growing species.

Retention of existing vegetation. On any farm which is being established in a developing area it is all too easy with present day heavy tractors and farm equipment to completely clear all native vegetation, irrespective of its value, the climate of the area or the need of stock for shade and shelter. It is strongly recommended that before any clearing is commenced the property owner should at least outline his plan for retention of native vegetation and the planting of additional shelter and shade. Even a heath with shrubs only 3-6 feet high can be of considerable value to sheep at lambing time or during cold and wet conditions immediately after shearing. It is suggested that any farmer involved in the development of a new property should initially reserve about 10 per cent of the area; this can be located so as to give protection from cold and strong winds and on the higher and poorer parts of the farm which are least suitable for cultivation. At a later stage it may be practical to reduce the area reserved to say 5 per cent or less. In the example shown the area is just under 5 per cent and it is assumed that where these areas of native vegetation are not available additional planting of moderately tall shelter is needed. In the southern 1,200 acres such belts could be located where the three north-south strips of native vegetation are shown or, with additional fencing, they could be replaced by two belts placed centrally north-south in the two 600 acre paddocks. Tree species could include those discussed below. If the land is held under Crown lease there may be restrictions limiting the removal of scheduled trees and shrubs.

Moderately tall and medium shelter. Most of the trees in these categories will be eucalypts and undoubtedly the outstanding species for the lower rainfall areas, especially where there are deep soils holding underground water, is river red gum (*E. camaldulensis*). The variety of this species which is used in Western Australia is particularly suitable for moderately open shelterbelts, as the lower branches are retained to maturity and the ultimate branchlets are drooping. In single rows it should be planted 20-25 feet apart if a good level of protection is needed; very wide spacing, for example 33 feet and greater, does not afford satisfactory shelter as it allows small wind funnels to develop between adjacent trees and the effect of these channels of increased wind velocity extends some distance down wind. When this and other species are used for ornamental purposes spacing can be increased to 33-66 feet. Most of the other tall eucalypts which can be planted in this climatic area carry a high crown at maturity and whilst useful for shade are less satisfactory for shelter unless there is an associated lower storey shrubby species. Although this is practical in higher rainfall areas it is not a good practice in dry country. Other eucalypts which can be considered include sugar gum (*E. cladocalyx*) of South Australia and the local brown mallet (*E. astringens*), wandoo (*E. wandoo*), flat-topped yate (*E. occidentalis*) and, amongst species which do not grow quite as tall, Dundas blackbutt (*E. dundassi*) and blue mallet (*E. gardneri*).

Because of their dense foliage most true pines should be planted in north-south lines so that the sun has access to both sides during the day; in east-west lines there is an appreciable strip which does not receive any sunlight during the winter half of the year. In the drier parts of the area *Pinus halepensis* var. *brutia* is considered the most promising exotic pine by the Forests Department of Western Australia. Aleppo pine (*P. halepensis*) is satisfactory where the rainfall is at least 16-18 inches, and it provides denser shelter, although the tree is heavily branched and of poorer form than the variety. In the higher rainfall areas maritime pine (*P. pinaster*) also merits consideration. None of these pines should be planted in or near drainage channels which may contain water for several weeks at a time.

Moderately low and low shelter. There are a number of native species suitable for use.

These include several of the cypress pines, such as black cypress pine (*Callitris endlicheri* syn. *C. calcarata*), white cypress pine (*C. glauca*) and Rottnest Island cypress pine (*C. preissii* subsp. *preissii*, formerly known as *C. robusta*). Amongst sheoaks there are northern sheoak (*Casuarina dielsiana*) and rock oak (*C. huegeliana*).

There is an almost bewildering number of shrubs 10-25 feet high suitable for low shelter, mainly acacias and eucalypts. Some of the principal acacias are raspberry jam (*Acacia acuminata*), manna wattle (*A. microbotrya*) and orange wattle (*A. cyanophylla*). Of the eucalypts there is a very large number. When a dense shelter is required then moort (*E. platypus* but particularly *E. platypus* var. *heterophylla*) is outstanding; others include Bald Island marlock (*E. lehmannii*) and a form of wandoo (*E. redunca* var. *melanophoia*). Highly favoured in the State, but not such a compact tree, is dwarf sugar gum (*E. cladocalyx* var. *nana*) and, when a hardy but somewhat open species is wanted, swamp mallet (*E. spathulata*) is very satisfactory.

Ornamental screen. To enhance the appearance of the property, especially when approaching it by road, it is suggested that consideration should be given to planting a line of trees or large shrubs along all or part of the road frontage, but particularly that part opposite the homestead. It is axiomatic that the trees should be of ornamental appearance but, in addition, it is desirable that they should be long-lived and resistant to ground fires. For this climate eucalypts fill the requirements best and most of those listed for shelter and shade are suitable. If ornamental appearance only is wanted then spacings of 33-66 feet are suitable, but if shelter effect is to be included, then the spacings should be closer and the list of possible species is more restricted. When the road frontage crosses topography, that is includes both drainage channels with deep or heavy soils and ridges with well-drained shallow soils, it is important to select different species for the different sites.

Should the property owner prefer to use shrubs and small trees of the height range of 15-30 feet he has a large number from which to select, especially among the hardy flowering eucalypts. One of the hardiest species in very dry areas is coral-flowered gum (*E. torquata*), which combines the ability to grow on very dry sites, produce very attractive flowers and has a moderately dense canopy which provides useful shade. As this book does not deal with species only of ornamental value readers are referred to Western Australian nursery lists for details of suitable species.

Shade. First of all consideration should be given to the retention of any existing native trees and shrubs. If suitable species occur and they are not over-mature, they may be included in the long-term plan, otherwise new plantings should be made. Since almost any long-lived tree with at least a moderately dense crown is useful for shade it follows that nearly all those suggested for shelter are also suitable for shade but, in addition to those named, there are a number of others which are particularly valuable for shade. For this area possibly the best is salmon gum (*E. salmonophloia*) which is reported to have a slow initial growth but certainly develops into the best shade tree in its climatic range. Other species include Dundas mahogany (*E. brockwayi*) and York gum (*E. loxophleba*). Preference has been indicated for native species, especially some of the hardier and longer-lived eucalypts, but an open-grown *Pinus* tree which has had its leader pruned at an early age in order to develop heavy branching, also provides good shade. One of the advantages of a tree like salmon gum which carries a high crown, is that wheat and other crops can be grown close to the stem without serious drop in production.

When planning shade requirements allowance should be made for the effect of north-south shelterbelts in the afternoons when maximum temperatures are attained. East-west belts also afford some shade at midday, although this may be limited from November to February. For shade around farm buildings and especially, in the holding yards and small paddocks pepper tree (*Schinus molle*) and the athel tree (*Tamarix articulata*) should not be overlooked.

Shade and shelter for the homestead

Fig. 12.4 provides a plan and an elevation to illustrate certain aspects of planning shade and shelter for homesteads in hot climates. The northern part of the area of Western Australia concerned has very long periods of sunshine; for instance, in January these are

325-350 hours and, on the inland side, may exceed 350. These figures are high by any standard. Towards the southern end of the area the hours are somewhat lower but still are mainly over 300. Even in mid-winter the values are 175 to over 200 hours. Earlier in these notes it has been recorded that the mean maximum temperature in the shade for the hottest month may be over 95°F. This means that there may be many days with shade temperatures of 100-110°F but objects in direct sunlight may attain 150°F or more. Unshaded walls absorb a considerable amount of heat especially during hot afternoons and this heat is only slowly released during the night, thus maintaining high internal temperatures. Shading of the walls from the sun, especially during the period from 3.00 to 6.00 p.m. when maximum temperatures are attained for the day, is very important if the occupants of the house wish to keep conditions as moderate as possible. This text is mainly concerned with the use of trees and shrubs, but wide eaves which protect the walls from the sun around midday and the placing of the long axis of a house east-west also reduces the extent to which eastern and western walls become heated.

The upper part of Fig. 12.4 gives a plan for a suggested layout. First of all the western side of the house; it will be noted that the moderately-tall row of trees is located from well south of the building to level with an east-west line through the northern side of the building. This belt will provide shade in the late afternoon for somewhat less than six months. If the owner desires late afternoon shade for a longer period of the year it would be necessary to plant one to three trees more at the northern end of the line, depending upon latitude. In most of the area concerned river red gum (*E. camaldulensis*) spaced at 20-25 feet is probably the most suitable species. Most of the other taller eucalypts which thrive in the area would be acceptable. *Pinus* species could be used but the writer does not support their general use in this climate area because of their greater inflammability and the extent to which dry needles accumulate in gutters. The fire danger can be reduced by the provision of a firebreak and the fact that open-grown pines retain green branches to near ground level is in their favour as a shelterbelt species. When river red gum or a pine species is used it would be possible to do without the low shelterbelt on the outside. This belt is designed not only to provide effective shelter near ground level but also to increase the shading effect just before sunset. Most of the species listed for low dense shelter with a mature height of 15-30 feet would be suitable, for example moort (*E. platypus* and especially the var. *heterophylla*) or a variety of wandoo (*E. redunca* var. *melanophloia*).

It will be noted that on the eastern side of the house the relative positions of the moderately-tall and the low shelterbelts have been reversed. The reason for this is that it is assumed the shading of the eastern wall is not as important as that of the western wall and, because the belt is planned to extend northwards to the road, there will not be undue shading in mid-winter. A continuation of the moderately-tall belt to enclose the southern end of the homestead area is desirable in order to prevent a wind tunnel effect between the two side rows. For the low shelterbelt on the eastern side it is suggested that some of the less dense species should be used; one of the best is swamp mallet (*E. spathulata*), but raspberry jam (*Acacia acuminata*) and manna wattle (*A. microbotrya*) would also be very suitable. Spacing between these plants could be 10-15 feet.

CASE D—3,840 ACRES OF EASTERN CATTLE-SHEEP-WHEAT COUNTRY

Locality

Southern semi-arid Queensland, mainly 200-400 miles from the coast and with particular reference to the latitudinal range of 23½-29°S. A typical location is Roma (mean annual rainfall of 20 inches), whilst further north are Augathella (20 inches), Injune (26 inches) and Blackall (19 inches); to the south are St George (20 inches) and Goondiwindi (23 inches). The type of country envisaged does not quite reach as far as Charleville (18 inches) or Bollon (17 inches), though the general principles suggested here could be applied in those districts. Around Cunnamulla (13 inches) there is a change in the natural vegetation and different practices would be desirable.

Soil

The soils fall into two broad groups. The first are heavy clays which occur on the more level sites and are usually of at least moderate fertility. They are suitable for cropping, especially wheat and other grains, but are not suitable for many species of trees. Those which do find the conditions satisfactory often show very good early growth. The other group of soils consists mainly of red and brown bleached soils exhibiting a light textured but often shallow upper layer, with an abrupt differentiation between it and the compact subsoil. These soils usually occur on gentle to moderate slopes and have freer drainage than those of the previous group. They are suitable for a wide range of tree and shrub species, especially where the sandy upper layer is of some depth. Deep-ripping, preferably along contour lines, is often a desirable practice for the encouragement of good early tree growth. On rocky rises the soils are skeletal and the number of trees which show good growth are fewer.

In the drier part of the region there may be an increased saline content of the soils in depressions and other low-lying locations; in which case salt-resistant species should be selected for planting.

Land use

On the more suitable soils in the higher rainfall parts of the area wheat is of importance, but in the drier locations it is of little significance. Sheep for wool and breeding stock are of major importance but progressively towards the hotter northern parts cattle replace sheep. Other aspects of land use are of comparatively minor importance.

Climate

Rainfall. Mean annual of about 21 inches, but with a range of 18-26 inches. This is a region with a summer maximum but there is a difference in distribution from south to north; Goondiwindi, for instance, has no month with an average of less than one inch, whilst Blackall in the north has four months below one inch—May being as low as 68 points.

An important aspect of the rainfall to land owners selecting trees for planting, as well as to farmers in general, is its very high variability. For a 50 percentile value of 21 inches the 10 percentile value (one year in ten probability) may be only 12-14 inches, with an all-time low of possibly only six inches. Roma, which has a somewhat less variable rainfall than other parts of the area, has a 50 percentile value of 21.88 inches, but a value of only 6.44 inches for the driest year and as high as 60.07 inches for the wettest year on record in a period of 92 years. In observing the growth of recently planted trees or the behaviour of natural vegetation in any district it is important to take into consideration the rainfall over the preceding 5-10 years. This is particularly desirable when noting introduced species, either from elsewhere in Australia or from overseas.

Temperature. Mean annual 67-73°F. The mean maximum of the hottest month varies between 93.3°F (Mitchell) and 99.0°F (Blackall), whilst the mean minimum of the coldest month ranges from 37.5°F to 43.2°F for the same two towns. Frosts, based on 32°F in the screen, shows a wide variation in average number per year. In the north this may be less than one, whilst in the central area at altitudes of about 1,000 feet Mitchell has 23.3 and Roma 18.4; on the southern edge at altitudes of only 600-700 feet the numbers vary from 1-10 a year. Frost is an important factor during the first few years of establishing new plantations.

Wind. Like many regions there is marked difference between the directions of morning and afternoon winds. In general, directions from north-west, via north-east to south-east predominate at the 9.00 a.m. readings, with many calms during winter; at the 3.00 p.m. readings directions are mainly from south-east via south-west to north-west, with few calms. The predominance of winds from the south-east, with some south-west, during all seasons of the year, but particularly in winter, should be taken into consideration when designing the retention of natural vegetation or the planting of shelter for stock. Crops sensitive to exposure may also benefit with protection from the west. In the case of homes, tall shelter appears desirable from the south and west, with hedges to give low protection from the north and east.

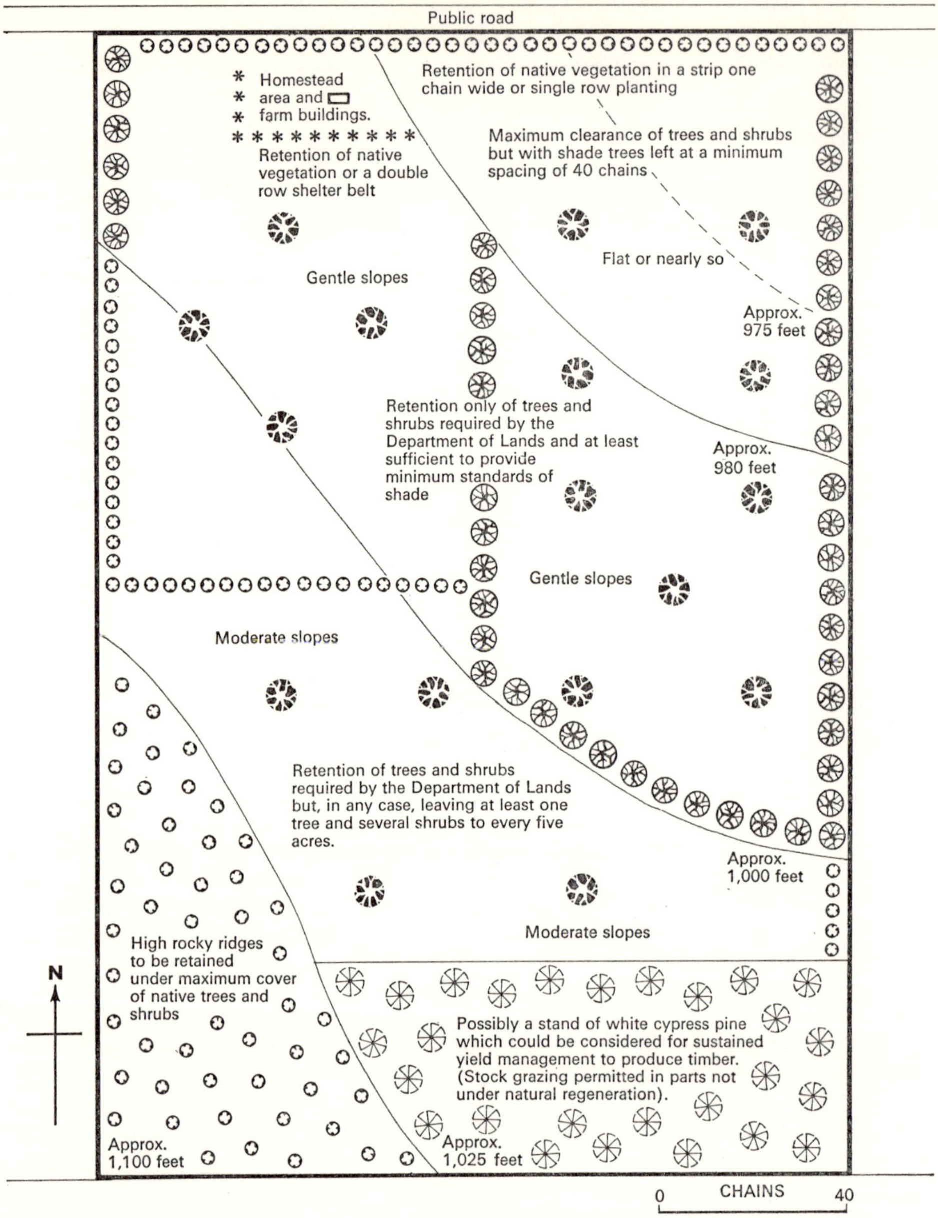

CASE D—3,840 ACRES

LOCATION—eastern semi-arid cattle-sheep-wheat country. Rainfall about 21 inches, with a moderate summer maximum.

Topography	Shelterbelts		Shade	
Flat, 640 acres Gentle slopes, 1,320 acres Moderate slopes, 1,400 acres Rocky ridges, 480 acres	(symbol)	Tall and medium size trees, mature height, 60 feet and over	(symbol)	Evergreen trees
	O O	Moderately low trees and large shrubs, either planted or natural, height, 20–40 feet	(symbol)	Cypress pine forest (natural)
	* *	Low screens, height, 8–20 feet		

FIG. 12.5

Part 1. When most of the native trees and shrubs are still present.

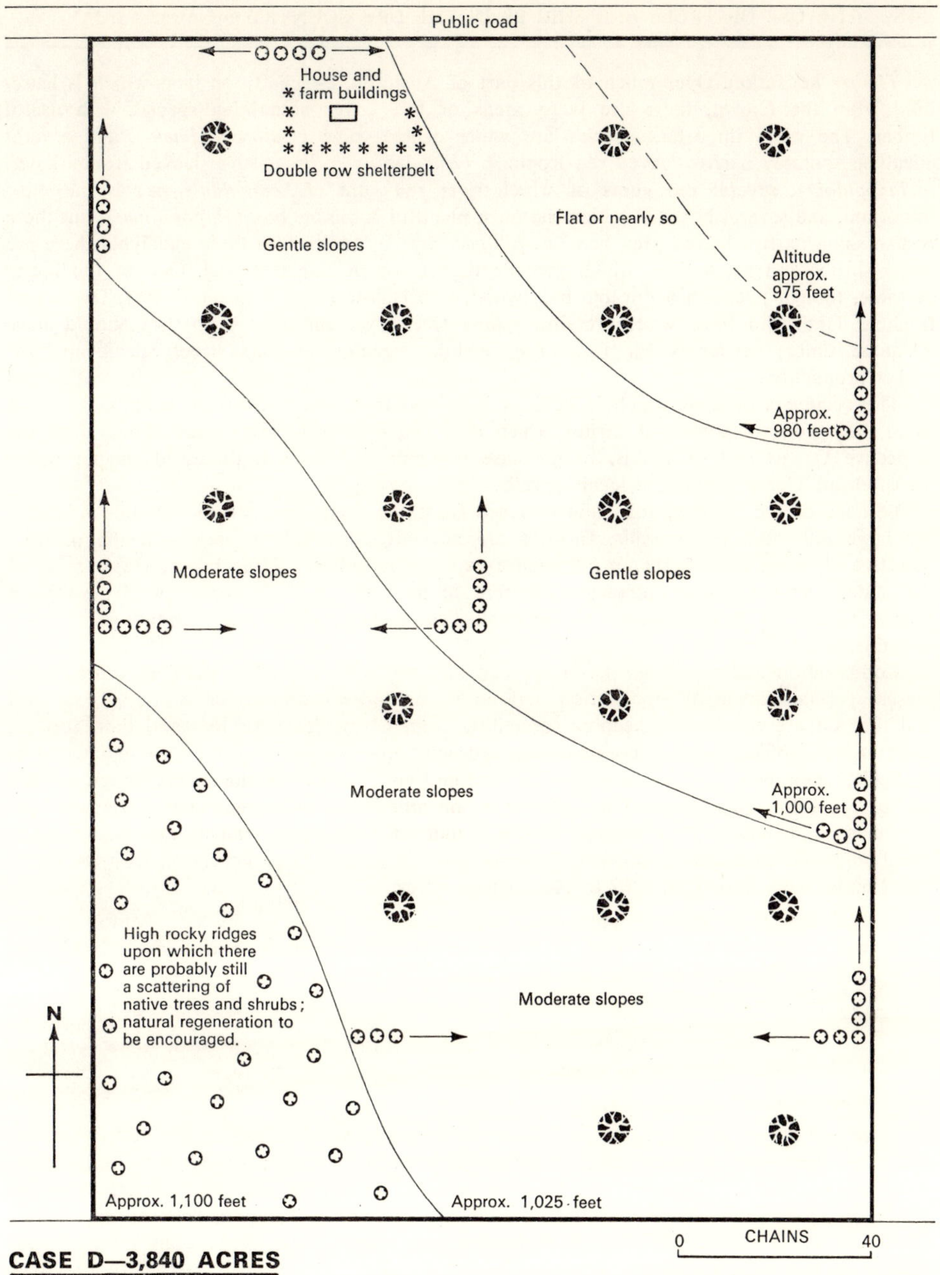

CASE D—3,840 ACRES

LOCATION—eastern semi-arid cattle-sheep-wheat country. Rainfall about 21 inches, with a moderate summer maximum.

Topography
Flat, 640 acres
Gentle slopes, 1,320 acres
Moderate slopes, 1,400 acres
Rocky ridges, 480 acres

Shelterbelts		Shade	
⊛	Tall and medium size trees, mature height, 60 feet and over	(symbol)	Evergreen trees
○○	Moderately low trees and large shrubs, either planted or natural, height, 20–40 feet		
* *	Low screens, height, 8–20 feet		

FIG. 12.6
Part 2. When most of the native trees and shrubs have been destroyed.

Timber utilisation. Over much of this part of Australia, especially on land which is leasehold from the Crown, there are large areas of tree cover containing species with useful timber. The more important species are white cypress pine (*Callitris glauca*) and several eucalypts, notably narrow-leaved red ironbark (*Eucalyptus crebra*), silver-leaved ironbark (*E. melanophloia*), several red gums of which river red gum (*E. camaldulensis*) is the most important, and several boxes of which the most plentiful is bimble box (*E. populnea*), but there is also some narrow-leaved grey box (*E. pilligaensis*). In addition to these eucalypts there are many shrubs and small trees which provide timber for fencing and fuel. One of the larger of these, though not with a durable heartwood, is brigalow (*Acacia harpophylla*). Owners of freehold land who have wisely retained some tree cover on their properties should have adequate timber resources for farm use, without needing to plant trees specifically for timber production.

The occupiers of land which is held as leasehold from the Crown are usually restricted as to the species of trees and shrubs which they can destroy without prior approval of the respective Department of Lands. In the case of Queensland this is discussed further under the sub-head 'Conservation and legal aspects'.

Because of the climate, soils and distance from the major markets or suitable locations for large pulp and paper mills, farmers are not advised to plant trees with the primary objective of profit from timber production except on the advice of the Forest Department of the State. The position is somewhat different in the case of land owners who have good natural forests on their property, for example cypress pine, which can supply logs to local sawmills.

Conservation and legal aspects. If we can divest the word 'conservation' of much of its present popular catch-call application and go back to the definition of 'conserve', we will find that we are considering keeping something from harm, decay or loss and that, further, there is no conflict between conservation and *wise use*. Throughout the dry country it is desirable, even essential, to adopt wise conservation practice in the management of the native vegetation. This conservation attitude is inherent in the official policies of Departments of Lands and Forestry. The attitude of the Department of Lands of Queensland is that of sound conservation, as exemplified by Condition 4 in Schedule II of Permit to Destroy Trees, issued under the Lands Acts, 1962 to 1965, which states:

> Sufficient trees shall be preserved, within the area embraced by this Permit, and in the manner, if any, specified in Schedule I of this Permit, *to ensure the provision of adequate shade, shelter and, where necessary, windbreaks.*

Condition 1 states:

> No trees shall be destroyed within forty-four yards of a main road, State forest or timber reserve or of the banks of any river, creek, lagoon or main watercourse or within the bed thereof.

Condition 2 lists a number of trees and shrub species which must not be destroyed anywhere on the property; these include *Acacia aneura, A. pendula, Apophyllum anomalum, Atalaya hemiglauca, Bauhinia* spp. *Callitris glauca, Casuarina cunninghamiana, Grevillea striata, Heterodendron oleifolium, Owenia acidula, Santalum lanceolatum, Sterculia* spp., deadfinish and any vinetree, and any other fodder tree, except during conditions of prolonged drought. In addition, Condition 3 gives the circumstances under which certain timber trees may be destroyed.

The drier the environment the more important it is that sufficient existing trees and shrubs should be retained to provide a desirable level of protection to the soil, crops and domestic animals. It is easier and less expensive to retain native trees and shrubs than it is to completely denude an area and then be faced with the necessity of planting in order to secure, after a delay of several to many years, a minimum level of shade and shelter. In addition, the complete denudation of some sites may do so much damage to the soils that it takes years or even decades to restore them to their original physical state. Owners of freehold land are recommended to prepare a plan for retention of trees and shrubs before commencing any large-scale clearing of their properties.

*Treatment and species of trees and shrubs suggested**

In the first place Case D will be considered on the basis of a farm which is only partly developed and still has most of the area under native vegetation. Later suggestions will be made where the area has been largely denuded of trees and recommendations made for minimum planting. In the case of selecting trees and shrubs for planting attention is drawn to the first two paragraphs under this heading for Case C.

Part 1. When most of the native trees and shrubs are present

Flats—about 640 acres of heavy soils

Originally this area would probably have had a moderate to high stocking of bimble box (*E. populnea*); if these have not all been cleared from the flats it will be desirable to leave one to two vigorous young mature trees to every 80 acres, in order to provide a minimum of shade for stock on the area even when it is being cropped. Depending mainly upon the strength and coldness of winter winds and the extent to which stock will be grazed on flats, it will probably be desirable to plant a narrow shelterbelt on or near the eastern boundary of the farm. The species selected must be able to tolerate the wetness of the soils in the rainy season, alternating with development of deep cracks when it dries. The bimble box is particularly suited to the soil type but there are several other eucalypts which are either native to the area or which have shown promising development in trial plantings of the Queensland Forest Department. These include river red gum and narrow-leaved grey box which are native to parts of the area, and red ironbark (*E. sideroxylon*) which has its principal occurrence further south. Other genera include river sheoak (*Casuarina cunninghamiana*) and belah (*C. cristata*); several wattles including brigalow (*Acacia harpophylla*), yarran (*E. homalophylla*), myall (*A. pendula*) and cooba (*E. salicina*). A narrow but effective belt could comprise a row of eucalypts combined with a row of one of the wattles or sheoaks.

Wherever there is any slope at all cultivation of the soil should be designed to ensure that there is a minimum risk of water erosion when there are periodic rains of high intensity. When a farmer has any doubt on the best practice he can secure advice from the Soil Conservation Service of the Department of Primary Industries for Queensland. Some of the soils are self-mulching and there may be little risk of wind erosion, but there may be a shallow, very fine-textured upper layer which is subject to wind erosion under conditions which combine a finely-cultivated surface, drought and strong winds. Whether wheat stubble should be burnt immediately after a crop is harvested or retained as a protective stubble mulch against wind and water erosion, is outside the scope of this book, but it is a subject which farmers should take into consideration.

Gentle slopes—about 1,320 acres

The average gradient here is very gentle, about 1:250, but there may be slopes where it is up to 1:100. Whilst erosion is unlikely to be a problem in pasture areas it could be serious if the land were cropped and there were a combination of poor soil conservation practice and unfavourable climatic conditions. These slopes should be regarded as the most productive grazing part of the farm and managed with this objective in view. The natural vegetation will include a large number of trees and shrubs. There will be several box eucalypts, especially bimble box and, in the southern part, narrow-leaved grey box; others may include river red gum. Amongst the wattles there are likely to be brigalow, yarran, myall and cooba. In addition there could be scattered kurrajong (*Brachychiton populneum*) and possibly bottle tree (*B. rupestre*), as well as many of the shrubs protected in Schedule II of Permit to Destroy Trees, previously mentioned.

It is suggested that as high a density as possible of native vegetation should be retained in belts about 5 chains wide. These belts can be conveniently located in north-south and east-

* The species listed here are only a few examples of the large number which could be used. Readers are referred to Chapter 17 and Table 17.1 of this book and, especially, to the catalogues and lists of trees and shrubs issued by the forest services of the States, as well as some other organisations.

west directions about a mile apart, as shown in Fig. 12.5, preferably with a fence line on one side. As it is assumed that stock will have access to the belts it should be kept in mind that a time will come when it is desirable to give encouragement to natural regeneration to replace the ageing and deteriorating original stand. Depending upon the species, this may be a very long time—perhaps 20 years for some of the smaller shrubs to 100 or more years for the larger trees. This can only be done by fencing the belts from stock and retaining such protection until the young plants have grown beyond the stage of serious risk from grazing animals.

The provision of other shade or fodder trees and shrubs will normally be covered by the retention of those species required by the Department of Lands for leasehold properties. In the unusual case where there are few of those species present any useful shade trees or shrubs should be retained in small groups or as individual specimens, at spacings not greater than half a mile.

Moderate slopes—about 1,400 acres

The term 'moderate' here is used in relation to the general topography of the area; elsewhere it would probably be called gentle. The average gradient is less than 1:100, but locally it may be much steeper due to the more broken nature of the topography and the presence of small rocky hills. With the exception of these poorer areas most of the slopes will provide moderate-quality grazing. The natural tree and shrub vegetation shows some contrast compared with that of the more gentle slopes. Amongst the eucalypts there may be several ironbarks, notably narrow-leaved red ironbark (*Eucalyptus crebra*), blue-leaved ironbark (*E. fibrosa* subsp. *nubila*) and silver-leaved ironbark (*E. melanophloia*); other eucalypts could include spotted gum (*E. maculata*) and Queensland peppermint (*E. exserta*). White cypress pine (*Callitris glauca*) may be locally of considerable importance. There will be a large number of smaller trees and shrubs, some of which will be useful as a drought fodder reserve or for small round fencing material.

The main potential of this type is for grazing, but if there are any good natural stands of white cypress pine it may be desirable to seek the advice of the Queensland Department of Forestry as to whether they could be profitably managed for timber production. Apart from the possibility of retaining some of the cypress pine and the necessity, in the case of leasehold land, of keeping specified species of trees and shrubs it is recommended that the area be ringbarked to the extent that the growth of pasture plants is encouraged but that there still remains an adequate scattering of trees and shrubs. When there is one medium-sized tree and several shrubs to every five acres or so there is a general shelter effect, combined with all the shade that is desirable. When the area shows variation between small dry rises and somewhat larger more fertile depressions it is obvious common sense to leave most of the trees on the poorer sites and give the maximum encouragement to grass in the depressions.

High rocky ridges—about 480 acres

Although with only a small altitudinal range this area will include many short steep slopes and rocky outcrops. Native trees and shrubs will include many of the species listed for the moderate slopes. Slight depressions will contain soils capable of growing grasses and other small fodder plants.

This type has a limited value for grazing, but the presence of shrubs of value in times of drought may add to its overall importance to the farmer. It is suggested that no felling or ringbarking should be done, with the exception that undesirable species may be destroyed in order to encourage the regeneration of edible shrubs.

The natural, undisturbed vegetation of this type has only a moderate density, but it affords a satisfactory level of shelter and all the shade that is needed. Where most of the original trees and shrubs have been destroyed it may be possible to fence small, selected groups where there are some seed trees, and so encourage natural regeneration. Conditions are not favourable for tree planting, nor is the area sufficiently productive to justify more than a minimum outlay in such work.

The homestead

It is suggested that if there is a good cover of natural vegetation near the homestead that a belt, two chains or more in width, should be retained on the south-western, southern and south-eastern sides. These should be a distance of 5-10 chains from the house, depending upon the mature height of the tallest trees. These belts should be incorporated in the fencing scheme, so that there is a fence on at least one side. Within this protected area ornamental and shade trees can be planted.

Part 2. When most of the native trees and shrubs have been destroyed

The larger a rural holding and the harsher the climate the greater is the problem, and also the expense, of re-establishing even a minimum level of shelter and shade. Any new planting will need protection from stock and, if present, rabbits and hares, and the cost of fencing is a major item of outlay. The outline proposals given in Fig. 12.5 merely provides a starting point at which farmers can commence a planting programme. Upon this they can build year by year until they have sufficient protection on their farm and have made the immediate surrounds of their home as attractive as possible. Due to variations in seasonal climatic conditions from year to year, and the benefits of distributing taxation deductions over a number of years, it is desirable to plan operations over a period of ten or more years. It is also advisable to commit to writing an initial plan indicating the priorities for the first few years, estimate the cost of materials and labour, ground cultivation, fencing and planting, and consider whether the species of plants selected should be grown on the property or purchased from the Department of Forestry or a private nursery.

Fig. 12.6 offers suggestions for a minimum programme. The location of the shelter belts follows the same general plan as in Fig. 12.5, but in this case they consist of only one or two rows each. It is recommended that planting should commence in a south-eastern or south-western corner of a large paddock but possibly only initially extend five chains in each direction. When the trees are well established, say in 3-5 years, additions can be made which will eventually extend the belts to the desired final points.

Species suggested. Here it is only possible to mention a few of the many trees and shrubs which can be planted in the area; additional information is given on species in Chapter 17 whilst everyone in the region should secure a copy of the descriptive list issued by the Department of Forests. The latter list is particularly useful in the information which it provides on the resistance or susceptibility of species to frost, which is an important factor in the region, and the types of soils upon which they grow satisfactorily.

Tall shelter, that is 50-100 feet in mature height

Callitris glauca, white cypress pine
Eucalyptus camaldulensis, river red gum
E. crebra, narrow-leaved red ironbark
E. moluccana, grey box
E. pilligaensis, narrow-leaved grey box
E. populnea, bimble or poplar box
E. sideroxylon, red ironbark
E. tessellaris, carbeen

Medium shelter, that is 25-50 feet in mature height

Acacia harpophylla, brigalow
A. homalophylla, yarran
A. pendula, myall
A. salicina, cooba
Bauhinia carronii, Queensland ebony
B. hookeri, mountain ebony
Casuarina cristata, belah

C. cunninghamiana, river sheoak*
Eucalyptus ochrophloia, yapunyah or napunyah
E. orgadophila, mountain coolibah

and the following exotic species—

Pinus halepensis, Aleppo pine*
P. halepensis var. *brutia**
Tamarix articulata, athel tree

Homestead. The desirable location of trees and shrubs to provide shade on buildings in the late afternoons, as well as early mornings, during summer has been discussed under this subhead for Case C and will not be repeated in detail. Allowance should be made for the difference in latitude and for the longer period of high temperatures. Many of the tree and shrub species listed for Case C can be used, but others should be considered. For tall and medium shelter, species can include some of those suggested for use on the farm in the preceding subsection. In addition there are many eastern small wattles which are ornamental and provide low shelter, for example Queensland silver wattle (*Acacia podalyraefolia*) and another silver wattle (*A. spectabilis*). No country homestead would be complete without one or two specimens of kurrajong (*Brachychiton populneum*) and bottle tree (*B. rupestre*), and the very ornamental wilga (*Geijera parviflora*). Amongst the smaller shrubs suitable for the drier and hotter parts are the cassias, for example silver cassia (*Cassia artemisioides*) and amongst the larger trees is silky oak (*Grevillea robusta*) and honey locust (*Gleditschia triacanthos*). In the case of the latter tree, which is an exotic, it should be appreciated that it is deciduous, prickly and should not be planted within 100 feet or more of sewers or any closed drains, as its root system extends for long distances and causes blockages. From Western Australia there are many dry country eucalypts which appear to thrive in this part of Australia. These include coral-flowered gum (*Eucalyptus torquata*), Dundas mahogany (*E. brockwayi*) and moort (*E. platypus* and its variety *heterophylla*).

* Species so denoted may grow to heights greater than 50 feet.

CHAPTER 13

Timber and minor forest products

TIMBER

WHILST the principal sources of timber for industry come from forests which are outside the dry country there are some forest types—cypress pine, dry eucalypt and riverain red gum—which occur solely or mainly in the semi-arid zone and supply products for use throughout Australia; (these three types are briefly described in Chapter 1.) The timber resources of the dry country are, however, of considerable local importance although, since the widespread introduction of fuel and heating oil the demand for wood fuel has lessened noticeably in the more populated areas. Fuel oil for steam production to generate electricity has replaced wood in towns and mining centres, whilst most commercial baking and cooking is done by electricity or fuel oil. Even in remote homesteads small diesel-powered units generate electricity for light and to run refrigeration plants, but wood fuel still serves for cooking and domestic heating. In the arid zone, additional to fuel, the most important use of timber is for fencing and in the semi-arid zone whilst the demand for fencing purposes is greater the production of railway sleepers is of considerable importance. Most sleeper timber is from eucalypts, especially those of the box and ironbark groups. With the exception of the cypress pines (*Callitris* spp.) most of the dry country timbers are hard to very hard, dense and correspondingly heavy and, whilst mainly durable, they are often not easy to saw and work by hand. Many of the eucalypts, notably those of the box and ironbark groups and some of the gums, have interlocked grain which makes them difficult to split and sometimes presents problems in obtaining a smooth machine finish on planed or moulded products.

The more important areas of the natural cypress pine and riverain red gum forests are managed on a sustained yield basis* by the various State forest services. With the exception of these two types and some of the dry eucalypt forests, most of the timber used in the dry country comes from scattered trees and woodlands on properties which are managed primarily for grazing or cropping. Some may come from shelterbelts and avenue plantings but these, at present, are a minor source of supply. The dry country, largely due to low rainfall is less suitable for the profitable growing of timber than the wetter zones of Australia and landowners are recommended not to engage in commercial production except with the advice of the forest service of the State. Poplar growing, mainly to produce logs suitable for match and other veneer purposes, on river flats or adjacent areas where irriga-

* Definition of this and other terms used in forest management are given in Chapter 7, especially page 76, of *Growing Trees on Australian Farms*.

tion is possible, can be considered. The production of poplar timber is, however, a specialised industry and should only be engaged in after careful study both of the technical aspects of silviculture and of profitable outlets for the sale of logs. Here also the forest services are able to give advice; Chapters 10 and 11 of *Growing Trees on Australian Farms* provides some information on poplars and the raising of cuttings.

FENCING MATERIAL (POSTS, STRAINERS, STAYS AND BATTENS)

Fencing material is an important part of the timber economy and although reported annual removals of industrial hardwood log timber for this purpose amount to around 100 million super feet or about 2 per cent of total removals of timber, the total usage is probably much greater, since there is a large amount of timber grown and used on the same farm which would not enter statistics and, in addition, there is an increasing amount of softwood thinnings. In all it appears that the annual usage is in the range of 20-30 million posts, to which must be added a proportional number of strainers and stays. This total for Australia compares with about 7.5 million in New Zealand during the early 1960s, an estimated 400-600 million for the United States but, because of different farming practices, only about 20 million for Canada. The capital investment represented by fencing varies from only one to two per cent on very large-size cattle stations to as high as 25-50 per cent on smaller-scale operations in the semi-arid and wetter zones. Even on some of the large-size sheep stations, that is 1-200,000 acres, fencing may be the major investment and represent several dollars per head of sheep carried.

The farmer who requires fencing material, particularly posts, has three principal types of timber which he can use. The first is split hardwood material, usually locally grown, which is often very durable but heavy to transport and frequently difficult to split except by using explosives. Due to shortage of farm labour and the consequent tendency to mechanise as many operations as possible, many farmers prefer to purchase posts rather than prepare them on their own properties. This may mean hardwood posts produced by local contractors or, where the properties are within reach of preservation plants, the use of treated softwood material. The second type is small diameter hardwood timber. If there is only a very narrow sapwood—say about ½ inch—the posts may be used as cut, but with wider sapwood it is desirable to preserve the sapwood against insect attack and decay. Some treatments can be conveniently done on the farm and are described in Chapter 23. The third type is the preserved round softwood post which is now available from timber preservation plants operating on thinnings from commercial plantations, especially those of *Pinus radiata.* There are a number of these plants in areas adjacent to the dry country and because of the light weight per post they may be transported up to several hundred miles from the preservation plants. The light weight is also an advantage in handling on the farm and, because of their roundness and small diameter (3-5 inches), the posts are convenient for mechanical driving in suitable soils—all of which reduces the labour content per mile of erected fence.

Before a farmer decides between alternative fencing material, both posts and strainers, it may be desirable to consider some of the characteristics of the timbers used. Hall (1969) has discussed this in Chapter 7 of *Growing Trees on Australian Farms* in which he states:

> The first criterion is the obvious one of durability, which is justifiably highly valued by farmers. The best box and ironbark posts or preserved pine can be expected to have a life of 40-50 years, which is comparable to that expected from concrete or galvanised steel of the quality used for fence posts. The real cost of fencing is not that of the initial cost of posts or wire but that of the annual cost per unit, say per chain, over its service life. In considering this it must be borne in mind that the cost of transporting a post and setting in the ground may be 1½-3 times the cost of the post in the forest or at the preservation plant and, consequently, the initial cost of a post is not necessarily the largest item in deciding annual cost.
>
> In considering the second criterion of strength it is important to separate two aspects—firstly that of resistance to the post bending or breaking and secondly of resistance of overturning. In both respects the large split hardwood post, possibly up to 6 x 8 inches in cross section, has a large margin of strength and, on the other hand, the small steel post is inadequate for use on its own except for very light fencing. The round post with a 3-4 inch diameter top has been proved by tests carried out by the Division of Forest Products, C.S.I.R.O., to have adequate strength in relation to bending (average maximum load of over half a ton per post) and resistance to overturning can be secured by correct depth of set, that is up to about ten times the diameter. Another criterion is lightness in weight which is of increasing importance as longer haulage becomes necessary. Light weight is also a help in reducing distribution costs within the farm, especially in hilly or mountainous country. A preserved round pine post of 4-inch top diameter may weigh only about 20 lb when ready for use, compared with up to 100 lb for a large split post of a heavy hardwood. Other desirable criteria are those of suitability for mechanical driving of the posts into the ground and the ability to hold staples.

With reference to staples it should be noted that the standard type used for hardwoods is not suitable for softwood timber and should be replaced by special staples which are now becoming available for the purpose.

Fencing battens can consist of durable or moderately-durable hardwoods or preserved softwoods. Many eucalypts and other hardwoods of the dry country are not sufficiently fissile for the easy production of split battens; also interlocked grain makes them unsuitable for reduction to small cross-sectional size. They can, of course, be sawn when there is access to a suitable saw bench. If sawn and preserved plantation grown softwood is considered, care should be taken to secure timber which does not contain a high proportion of large knots which will weaken the strength of the batten. The present trend in fence design is to reduce the number of battens used and to incorporate some form of netting or wire supports in combination with one or more barb wires.

As a matter of internal policy farmers should decide whether they will purchase all their wood fencing material or grow some or all of it on their properties; the latter action appears to offer a number of attractions, especially on the larger and more remote holdings. If a farmer decides to grow his own material he may be able to rely on existing native timber, with provision for its regeneration or, if the rainfall is adequate, to establish small woodlots or multiple row shelterbelts. The timber from such woodlots, unless the trees are allowed to grow to maturity, will usually need preservation; even in the case of hardwoods there will be a much higher proportion of non-durable sapwood than is the case with large size slowly-grown trees. Although softwood woodlots will not normally be planted where the rainfall is less than 18-20 inches, hardwoods can be grown at wide spacings in areas as dry as 12-15 inches a year in southern Australia and 15-20 inches further north.

POLES

Land owners in most of the dry country are only likely to be concerned with the provision of small poles for private telephone lines or, in some cases, for the framework of barns and other farm buildings. The necessary poles can often be obtained from the property or local woodlands; the trees are likely to be eucalypts or other hardwoods, often of rather poor form but with narrow sapwood and durable heartwood. In the higher rainfall parts of the semi-arid zone, farmers may be interested in the sale of poles from their natural forest or from plantations. When planning to grow timber with poles as a product the forest service of the State should be consulted as to probabilities of markets when the time for harvest arrives.

Trends in the usage of poles for telecommunications and power lines have been discussed by Hall (1964) whilst Chapter 7, pp. 80-82, of *Growing Trees on Australian Farms* refers to poles and piles in relation to farm forestry.

SAWN TIMBER

The use of logs for sawing is the dominant form of timber utilisation. In the year ending 30 June 1968, 70 per cent of the logs from the forest were classified as saw logs, with an additional 4 per cent as logs for sawn sleeper production; there was also approximately another 2 per cent of logs used for hewn sleepers. The very great part of this saw log production came from the commercial forests in the humid and sub-humid coastal and tableland forests of the eastern and southern parts of the continent.

The main commercial production of saw logs in the dry country is of cypress pine (*Callitris* spp.), river red gum (*Eucalyptus camaldulensis*) or various eucalypt hardwoods, especially those of the box and ironbark groups, for sleepers. From a statistical viewpoint any other production of saw logs is of minor importance.

Landowners who have cypress pine, river red gum or good eucalypt forest on their farms should consider managing for sustained yield, but large-scale plantings of trees with the objective of selling saw logs for profit should only be commenced after a careful consideration of all factors and the advice of the forest service of the State. Small supplies of saw logs for use on a property can be produced from woodlots or multiple row shelterbelts if there is not enough growing naturally.

Most farmers who have a saw-bench will be aware that the usual simple bench is only suitable for breaking down logs up to about 12 inches diameter at the large end. Even then the equipment must be in good condition and at least one of the sawyers must be experienced in the type of operation. When there is a small sawmill within reasonable distance it may be possible to arrange for large logs to be sawn on behalf of the farmer; one suggestion is that instead of sawing to final dimensions at the sawmill the log should be cut into flitches, say 4 inches thick, and that these are stored at the farm until timber is needed. The usual small farm saw-bench would then readily resaw the flitches to the sizes needed, but it should be noted that if the timber has become partly seasoned the wood will be harder to saw.

Information which could be of help to farmers intending to grow trees for saw log purposes will be found in Chapter 7, especially pp. 82-84 of *Growing Trees on Australian Farms.*

OTHER USES OF TIMBER—PARTICLE BOARD, FIBREBOARD, PULP AND PAPER

There are many other uses of timber but most of them are unlikely to be of interest to the dry country farmer as a possible outlet for his timber. Some of the products, like veneer, have been in use since at least the time of the Pharaohs of ancient Egypt whilst others, such as particle board, are a development since World War II.

The Australian manufacture of plywood, which is based on thin layers of wood called veneers, commenced in Melbourne in 1911 but the centre of the young industry soon shifted to Brisbane. Australian-grown sources of ply logs are mainly from the rainforest and hoop pine areas of Queensland or the exotic pine plantations of southern Australia; hardwoods, mainly eucalypts, are also used in Western Australia, Victoria and Tasmania. It has been previously mentioned that the growing of poplars for ply logs offers limited possibilities on river flats and irrigated areas in the dry country. To produce logs in the shortest possible time use should be made of (a) suitable fast-growing species; (b) favourable growing sites; (c) wide spacing for fast diameter growth; and (d) progressive pruning up the stem to the maximum practicable height.

Fibreboard, based on the disintegration of wood or other suitable material into its constituent elements, is one of the cheapest sheet materials and has a multitude of uses. The first local manufacture of soft fibreboard from bagasse (the residue of the cane after the manufacture of sugar) was in 1934 and of hard fibreboard in 1939. The industry is based on a few large plants using numerous species, mainly mixed hardwoods, and largely dependent upon cheap procurement of raw materials, including water. All plants are in coastal areas.

The pulp and paper industry in Australia commenced the production of pulp at the beginning of World War II, although paper had been made from imported pulp since the middle of the previous century. The annual requirement is now about 65 million cubic feet of locally grown timber (equal to about 16 per cent of total log production), as well as a large quantity of imported pulp. All this is insufficient to supply the country's needs and a large amount of paper, especially newsprint, is also imported. Even more than is the case with fibreboard manufacture, the pulp and paper industry is based on a small number of very large plants located on or near the coast and adjacent to major sources of timber and adequate water, accessible to cheap transport and preferably where effluent can be disposed of into the sea.

The fourth industry is that of particle board manufacture. This is a relatively thick sheet material with thinnings or slabs of radiata pine (*Pinus radiata*) as the principal raw material. At present its consumption of timber is about 2 per cent of total log production. All the plants are located in or near coastal regions.

FIREWOOD

In spite of a society which has a preference for fuels which, largely irrespective of cost, entail the least possible work, wood still remains an important form of energy for space heating in the small towns and more remote country areas of Australia. When considering the economics of any type of fuel one of the more important criteria is the cost delivered to the consumer. In this regard electricity and oil in the main ports, offset their initial price by lower cost of delivery per unit of heat. On the other hand, in rural areas where initial cost of wood fuel is low and the only transport is from the woodlot to the house on the same farm, the real cost of heat may be considerably lower than for competitive fuels. One argument in favour of farmers in remote areas growing their own fuel is that trees from shelterbelts or tops of other trees of negligible value for sale as industrial wood can be used and the second is that, with long-term inflation, there is certain to be a gradual increase in the cost of fuel oil and electricity. Another point is that the farmer who grows his own wood is self-contained, whilst anyone dependent upon imported fuel oil or even locally-produced oil which has to be transported long distances by sea, could find themselves severely rationed in times of national emergencies.

The heat (calorific) value of a fuel is measured in terms of *British thermal units* (Btu), usually expressed for a pound weight or, in the case of electricity, for a kilowatt hour. Table 13.1 gives the approximate calorific values for some of the common fuels.

When wet or green wood is burnt, some of the calorific value is used to drive off the moisture in the wood, leaving less available for heating purposes. This loss, added to the weight of the water, may equal half or more of the heating value of the dry wood. Before use, firewood should be air-dried in open stacks on well-exposed sites. The period necessary to dry from green to below 20 per cent moisture content, that is more or less air-dry, will vary with the species and density of the

TABLE 13.1

CALORIFIC VALUE OF COMMON FUELS USED IN AUSTRALIA

Brown coal briquettes	9,500 Btu/lb
Black coal	10–12,000 Btu/lb
Anthracite	13–15,000 Btu/lb
Fuel oil	19,000 Btu/lb
Electricity	3,412 Btu/kWh
Wood	
Completely dry	8,300 Btu/lb
Air dry—15% moisture content	7,082 Btu/lb
Air dry—20% moisture content	6,744 Btu/lb
Partly dry—50% moisture content	5,187 Btu/lb
Green—100% moisture content	3,631 Btu/lb

timber, the size of the pieces and the time of year. Pine in thicknesses up to 3-4 inches may dry sufficiently in 2-3 months during summer, even less in very hot dry areas, whilst hardwood of similar size will need much longer and logs, especially those several feet in diameter, may take years. The log, particularly in the case of hardwood, is not a suitable form for quick drying. Green timber intended for firewood should be sawn into short lengths, preferably stove lengths of 9-15 inches, and split to thickness not greater than about 6 inches in order to ensure reasonably rapid drying. The drying process can also be hastened by removing the bark where practical and piling in high, narrow stacks so that there is free air movement.

The comparative costs of heating units for different fuels and assumed costs of purchase is given in Table 7.2 on p. 20 of *Growing Trees on Australian Farms*. Cost is given in cents per therm of 100,000 Btus and the table shows that wood fuel is still by far the cheapest source of energy for space heating; this table was based on purchase of wood fuel and the relative cost position would be even more favourable to wood fuel when it is grown and cut on the home property.

MINOR FOREST PRODUCTS

This term has long been used to cover a wide range of products, other than timber, which are derived from shrubs and trees. Not all of these plants are, however, characteristic of forests and, indeed, many of them belong to the vegetation of woodlands and heaths. Statistics on recorded production are very incomplete but Table 13.2, based on Table 35 of the 1968 annual report of the Forestry and Timber Bureau, provides some indication of the production from State forests and Crown lands. It will be noted that some items show a moderately steady or even rising production during the last 30 years, for example tanning materials; whilst

TABLE 13.2

MINOR FOREST PRODUCTS, RECORDED PRODUCTION—AUSTRALIA

	Unit	1938–39	1946–47	1959–60	1964–65	1965–66	1966–67	1967–68*
Aromatic wood—								
Rosewood(*a*)	tons	194	221	11	—	—	—	—
Sandalwood(*b*)	,,	347	1,668	610	812	1,040	741	775
Charcoal—								
Forest charcoal(*a*)	tons	1,897	3,596	1,182	8,740	} 40,408	48,169	41,036
Industrial charcoal(*c*)	,,	—	—	38,783	41,848			
Essential oils, etc.—								
Eucalyptus oils	lb	807,903	1,384,475	913,187	545,536	471,977	420,447	370,717
Duboisia leaf(*a*)	,,	—	—	210,747	1,500,000	1,425,500	1,400,000	1,397,536
Foliage for rutin(*a*)	,,	—	—	26,726	101,694	1,577,310	1,042,540	...
Tea tree oil	,,	7,209	16,092	18,135	8,120	5,084	4,251	3,507
Boronia blossom	,,	1,876	328	2,198	1,275	2,763	4,322	4,223
Sandalwood oil	,,	...	...	...	...	7,658	3,995	3,561
Tanning materials—								
Tanbark	tons	3,279	2,735	5,076	3,049	2,835	2,208	...
Wandoo for tanning	,,	29,988	65,077	46,566	43,775	44,210	45,008	43,766
Miscellaneous—								
Canes and rattans(*a*)	,,	...	19,063	20	1	7	4	5
Mulga wood(*a*)	,,	20	97	30	96	50	86	...
Gums and resins(*a*)	,,	2,247	1,252	534	978	1,324	381	...
Honey	000's lb	...	...	...	...	36,200	31,800	39,700
Beeswax	,,	...	...	...	...	494	467	429

* Provisional. ... Not available. — Nil or negligible.

(*a*) Incomplete. Crown Land only figure in at least one State. (*b*) Includes roots and butts for oil. (*c*) Produced by Wundowie, W.A. for Charcoal Iron & Steel Industry (commenced 1948).

others show marked fluctuations, for example kino, gums and resins. Some items, such as rosewood for the aromatic content, have disappeared whilst new ones, for example eucalypt foliage for rutin, have appeared for the first time.

CHARCOAL

Charcoal may be defined as a form of carbon derived from incomplete combustion of animal or vegetable matter, that is wood or bones, and is one of the principal products from the destructive distillation of wood. It has a very wide range of uses at the present time but the most important ones in Australia are metallurgical processes in steel manufacture where coal resources are low, for example at Wundowie in Western Australia, and for barbecues. During World War II it was also extensively used in gas producers to provide the motive power for road vehicles; although dirty to use and needing frequent attention it took the place of petrol at a critical time. Producer gas is particularly effective in the heavier, slow r.p.m. type of stationary engines. Charcoal was also an important factor in the winning of World War I when the enemy introduced poison gases, wood charcoal in an 'activated' form being the vital element in gas masks.

The denser a timber and the drier it is at the time of carbonisation the higher is the yield of charcoal. Most of the heavier eucalypts give a good yield of a satisfactory quality material. One ton of charcoal can be expected from 4-8 tons of wood, depending upon the moisture content of the latter. In the case of dry wood the yield, by weight, may be 25 per cent charcoal, 25 per cent gases and 50 per cent liquids. The chemicals recovered from the liquid contents varies between hardwoods and softwoods; in the former there is a wide range, including acids, especially those of the acetic group, acetones, formaldehydes and phenols; in the latter pine oils, turpentine and tarry substances are predominant.

It is not difficult for farmers to produce charcoal on a small scale, that is using as small a kiln as a 44-gallon drum. The wood used can be any dense hardwood, preferably air-dry to below 20 per cent moisture content and, very important, sawn or split to pieces of approximately uniform thickness. Details of how to make charcoal are available in the pamphlets and leaflets of most of the forest services and of the Division of Forest Products, C.S.I.R.O.

GUMS, RESINS AND KINOS

Until the period between World Wars I and II the collection of these minor products was a common source of additional income in some country districts. The importance of them has now decreased. There are various reasons for this: one is that the imported, often synthetic, materials are either cheaper, more uniform in quality or more convenient to use, and another is that the natural habitats from which collections were made have been converted to other uses. This latter reason is particularly the case with the plant known as blackboy, grass tree or yacca (*Xanthorrhoea* spp.). It grows on rather poor land which was considered only of use for extensive grazing, but where the topography was suitable for cultivation it was found that with the addition of superphosphates and minor elements, useful pastures could be established. Prior to World War II exports were up to

1,000 tons a year, mainly from South Australia; its principal uses are in the manufacture of linoleum and the lower grades of varnishes. Even up to 1960, exports of this class of product, mainly yacca gum, was as high as 800 tons.

Gums are typically products of broad-leaved trees and resins of coniferous (softwood) species. Gums either dissolve or swell on contact with water and are used for a wide range of purposes, particularly in medicines as demulcents, in confectionery and as finishes in textile and paper manufacture. Resins, on the other hand, are not soluble in water and are used extensively in varnishes, whilst derived products include turpentine, and the preparation of insulating materials. Some gums are eaten in the natural condition and survival booklets issued to Allied air force crews during World War II carried information on edible gums which could be used in the event of them being forced down in unpopulated areas. Gums are also used as binding agents in rouges and face powders as well as the moustache pomades of 50-100 years ago. Whilst some resins are collected from natural secretions most of the commercial production of the world results from the tapping of *Pinus* species of trees, notably maritime pine in France and some of the southern pines in the warm south-east of the United States of America. These species are grown in Australia but it appears doubtful whether any large-scale tapping industry is likely to develop.

TANNINS

Tannins comprise either certain minerals, for example chromium salts, substances obtained from parts of plants, or synthetic products. They have the unique ability to combine with the proteins in raw hides to form leather. Natural tannins from plants are found in the wood, bark, leaves and fruits of many shrubs and trees. They are complex, water-soluble, dark-coloured compounds with an astringent taste and are able to form coloured solutions with ferric salts. The demand for natural tannins has shown a marked decrease in recent decades. This has been due to several causes: one is the increased production and favourable prices of synthetics, another is the changing pattern of footwear, especially the replacement of heavy leather soles by rubber, and still another is the very great reduction in the demand for harness leather as the horse has been replaced by the motor vehicle.

Australia has many sources of natural tannins, notably the bark of many species of *Acacia* and both the bark and wood of some eucalypts. Examples of tan-bark acacias are black wattle (*A. mollissima*) and golden wattle (*A. pycnantha*). A eucalypt (*E. wandoo*), yields tanning material from the wood. In the case of black wattle South Africa introduced this Australian species more than half a century ago and now exports black wattle tannin back to Australia. Tannin products are of some significance in the overseas trade of Australia; exports, mainly from wandoo, amount to over $250,000 a year, whilst imports are greater at $300,000, mainly wattle extract from South Africa.

Formerly the collection of tan bark was a useful source of extra income in some rural areas, but there is now only a very minor demand for the bark. The stripping of bark, of course, kills the trees and should only be done when there is no further use for them. It is considered unlikely that it will ever be profitable to grow trees, other than possibly wandoo, with tan bark or wood as the principal source of revenue, but where farmers already have suitable species established it could be worthwhile inquiring from local authorities as to whether there was a current demand for the bark.

ESSENTIAL OILS

The presence of essential oils in the leaves of eucalypts has been known since the early days of European settlement of Australia. A. R. Penfold (1958) has recorded Governor Phillips sent a sample to Sir Joseph Banks in 1788 and that two years later John White, the Surgeon-General, sent a quart for testing which had been distilled from the foliage of Sydney peppermint (*Eucalyptus piperita*). Very little commercial production was done until the middle of the following century when the first still started operations in Victoria. At the present time there is small-scale distillation in nearly all States, although the principal locations are in south-eastern New South Wales and northern Victoria.

The term essential oil is given to many volatile substances that occur in different parts of plant; they may be more or less homogeneous substances, but are usually a mixture of a wide range of alcohols, aldehydes, ketones, phenols, esters and terpenes. Until the commencement of the present century eucalypt essential oils were used in Australia mainly for medicinal purposes, especially the treatment of colds and as a mild disinfectant which could be used as a mouth wash. Subsequently use extended to solvents for substances such as tar, grease and paint; in perfumery, especially using geraniol; industrial uses also included mineral recovery and flotation purposes, but these have declined. There has been a considerably increased use as inexpensive, non-toxic disinfectants.

The essential oil in eucalypts is contained in the leaves, where it appears as translucent dots when the leaf is viewed against a suitable light (even a moderately strong daylight is sufficient). There is a great variation in the recovery during distillation of the leaves and although most commercial production is 20-35 lb per 1,000 lb of leaves, in the case of Camden woollybutt (*E. macarthurii*) the yield of geraniol may be only 2 lb, but this is a particularly valuable oil. Only some eucalypt species produce useful quantities of essential oil and, unfortunately, they are mainly species other than those felled for timber. The most important belong to the peppermint and mallee groups; some physiological strains of narrow-leaved peppermint (*E. radiata* and *E. australiana*) and broad-leaved peppermint (*E. dives*) are amongst the most important species in south-eastern Australia; the mallee group is represented by several species including green mallee (*E. viridis*).

Most of the leaves used in distillation come from Crown land and the harvesting is subject to licence. Commercial production is on a small scale and is sometimes done intermittently with other rural operations. Due to varying demand as a result of competition from cheaper oil from plantation-grown eucalypts in Spain and South Africa and the use of substitute materials, the industry offers little incentive to large-scale planting of suitable species or for the development of equipment suitable for harvesting coppice. Since World War II reported production has mainly varied between 0.4 and 1.4 million pounds.

DRUGS FROM TREES AND SHRUBS

Many Australian plants contain alkaloids and other drugs, some in sufficient quantity to justify commercial extraction for use in medicines. Rainforest trees and shrubs, in particular, have a high percentage of species which contain alkaloids but, on the other hand, very little is found in dry country vegetation.

It is only slowly that we are discovering what the Aborigines had known for a long time. They were aware that powerful narcotics are contained in the leaves of *Duboisia* spp., whilst other shrubs contained rotenone-bearing substances which could be used in fish poisoning. They also knew that the latex of several genera of trees and shrubs contained substances which assisted the healing of wounds and sores.

The quantities of *Duboisia* leaves, and *Eucalyptus* foliage for the extraction of rutin are recorded in the statistics of minor forest products. *Duboisia* leaves contain hyoscine and hyoscyamines which are used in medicine; collection was first recorded separately in 1959-60 and has varied from 0.1 to 1.4 million pounds of leaves a year. Rutin is of value as a source of vitamin P and is used in decreasing capillary fragility in man. Commercial extraction is from the leaves of eucalypts; the first species used was red stringybark (*E. macrorhyncha*) but subsequently silvertop stringybark (*E. laevopinea*) was found to be more suitable. Annual production since the early 1960s has varied widely from 0.1 to 1.6 million pounds of leaves.

FURTHER READING

Australia. Forestry and Timber Bureau. *Annual Report 1967*.

Hall, N. (1964). *Pole Supply from Australian Forests*. (Forestry and Timber Bureau Bulletin No. 39). Government Printer, Canberra.

Penfold, A. R. (1958). 'Essential Oils'. In *Australian Encyclopaedia*. Vol. III, pp. 409-411.

Penfold, A. R. & Morrison, F. R. (1951). 'Commercial Eucalyptus Oils', *Bulletin 2 of the Museum of Applied Arts and Sciences, Sydney*. (5th ed.).

CHAPTER 14

Honey and beeswax production in semi-arid and arid Australia

by J. H. Leigh

WITHIN the whole of Australia between 35 and 46 million lb of honey and 0.5 and 0.7 million lb of beeswax are produced annually. This honey has a gross value of 4-5 million dollars and the beeswax a value of ¼-½ million dollars. In terms of export earnings honey brings into Australia about ½ million dollars and beeswax about ¼ million dollars annually.

The ability of apiarists to successfully harvest honey and beeswax depends both on the management of their hives and/or the availability of adequate supplies of nectar and pollen at appropriate periods. Nectar provides the main source of carbohydrates whilst pollen is almost the sole source of proteins and essential vitamins and an essential prerequisite for breeding bee replacements.

HONEY FLORA

In Australia honey and beeswax are produced mainly from various species belonging to the genus *Eucalyptus*, this being particularly true of the semi-arid and arid areas which are the subject of this chapter. Most eucalypt species, for example the 'gums' yield both nectar and pollen which is attractive to bees and therefore gathered. Other species of eucalypt however, whilst amongst the highest yielders of excellent quality nectars, produce pollens which for as yet unknown reasons are unattractive to bees which either ignore these pollens or collect them only in very limited quantities, for example the 'box' and the 'ironbark' groups. Under conditions of inadequate pollen the important process of bee replacement through brood-rearing dwindles or ceases and hivestrength drops alarmingly. In many cases such pollen-deficient species grow in association with other plants which produce adequate pollen, but in the case of the summer and autumn-flowering species which grow in arid areas, associated pollen-bearing plants are usually absent. Amongst the most important pollen-producing plants are many trees belonging to the genus *Acacia* and numerous ground flora shrubs and herbs such as *Melaleuca* spp., *Hakea* spp., *Arctotheca calendula, Echium lycopsis, Hypochaeris* spp., *Asphodelus fistulosus, Brassica tournefortii* and *Leucopogon* spp. In Tables 14.1 to 14.5 an attempt has been made to list the more important honey and pollen-producing trees and shrubs for each State. These lists are by no means exhaustive. Included in these tables is the botanical name; official common name as adapted by the

Standards Association of Australia; other common names; distribution, communities and habitats in which each species is found; time and frequency of flowering; yield, colour and quantity of honey yielded; and quality and quantity of pollen produced. For this information the author is indebted to a number of leading apiculturists, whose help has been gratefully acknowledged at the conclusion of this chapter, as well as information drawn from a number of State Department of Agriculture publications which are listed under 'Further Reading'.

Apart from those species listed in these tables, and as has been partly mentioned earlier, there exists a wide variety of lesser species which because they are restricted to a geographically-small area, or because they yield infrequently or only in small quantities do not rate individual mention. Such plants are however often vital to the beekeeper since they provide 'breeding conditions' and 'maintenance flows'. Breeding conditions may be defined as a light flow of nectar and abundant pollen which encourage and enable bees to rapidly increase the hive population. Such conditions are most desirable at least six weeks before the first major honey flows. Often the abundance of pollen stored during good breeding conditions is sufficient to carry hives through a later-occurring large honey flow, even when such a honey flow comes from a species with unattractive and/or little pollen. Maintenance flows are somewhat similar but generally lack the abundance of pollen or occur at times when breeding is not important to the hive. As the name implies, the flow maintains the hive at its present existing level and the hive does not have to draw on its reserves of honey or pollen.

BUD INITIATION, FLOWERING AND NECTAR SECRETION

The frequency and reliability of flowering varies markedly between various eucalypt species. Depending on the amount and timing of rainfall, vegetative growth, bud development and carbohydrate storage may in some cases be followed by further vegetative growth or alternatively by flowering and nectar secretion. This sequence of events may be well exemplified by *E. camaldulensis* growing in South Australia. Good rains in winter and spring followed by a dry spring and summer result in good flowering and copious nectar secretion.

Bearing in mind the climate-growth cycle relationships which have been outlined above the most commonly observed flowering frequencies for each species can be given. Certain species may bud and flower each year, for example *E. largiflorens*; others may bud one year and flower twelve months later, for example *E. camaldulensis*; others in particular circumstances may bear two age groups of buds at one time, for example *E. wandoo* and *E. salmonophloia*; whilst others may only flower once in four to eight years. Moreover, it is not uncommon for flowering in a particular year to be prolific whereas the next flowering is rather sparse.

The various species of eucalypt vary markedly in the duration between bud initiation and the time the buds open and flower. In some species this period is as low as a few weeks whereas in others it may be up to two years.

The length of time over which individual trees flower or communities of one species flower also varies markedly from as little as a few weeks duration (*E. melliodora*) to a period of two to three months (*E. largiflorens*) or even longer.

Information of the sort outlined in the preceding paragraphs is of vital importance to apiarists who study these 'bud prospects' in planning their migratory itinerary for the coming year.

Many factors affect the quantity and quality of the nectar produced, amongst which are altitude, aspect, temperature, humidity, soil-moisture status, soil type and atmospheric pressure. Of these, soil-moisture status, temperature and humidity are probably the most important, with the other factors having an indirect effect. High humidities may not only affect the plants but they make it particularly difficult for bees to ripen honey and this is often more important than thin dilute nectar in determining how much the bees can harvest. A characteristic of many eucalypts is nectar secretion at night so that bees work fervently until mid-morning and then stop; alternatively nectar secretion is slow and evaporation during the day is high with the result that nectar can only accumulate at night.

As already mentioned, prevailing climatic and soil conditions just prior to and during flowering are important, for favourable growing conditions tend to promote a flush of vegetative growth at the expense of flowering and nectar secretion. In this regard the semi-arid and arid regions are more reliable than the higher rainfall areas for the extremely dry conditions which usually prevail at their flowering time in summer generally precludes the possibility of regrowth taking place. In comparatively rare cases trees may flower profusely without secreting a flow of nectar.

Experienced apiarists frequently take advantage of known environmental differences to extend their workings of a particular species and it may in certain years be possible to harvest three or four successive crops from a widely-distributed species. This is widely practised with *E. camaldulensis* which extends over long distances along the Murray-Darling River systems. Apiarists commence extracting operations in South Australia in early summer where because of higher temperatures in early spring, flowering commences first. As flowering diminishes they move their hives eastwards where flowering is later (associated with colder springs). As flowering tapers off at the second site the hives are again moved and the process is repeated until eventually the foothills are reached.

For reasons mainly of topography, altitude and aspect, and hence a greater variety of species and flowering times, apiarists in the higher rainfall regions have more options open to them as to which species and areas they can exploit, as compared to those apiarists working in the semi-arid and arid zones where the number of species is limited and where there is less variation in micro-climate and hence periodicity of flowering.

FACTORS AFFECTING QUALITY AND QUANTITY OF HONEY PRODUCED

The nectar secreted both within and between species may vary in its content of moisture, sugars, minerals, proteins and other nitrogenous substances, essential oils, pigments, higher alcohols and other undetermined constituents. Constituents in the nectar are not only affected by the particular species but are also affected by

There are many factors which should be taken into consideration when siting bee hives, and of particular importance in the dry country are availability of water and shade on the hives during the hotter parts of the afternoons. If there is no natural water supply or dams nearby, water must be carted to a site. Most of the shrub and small tree vegetation of the dry country, especially mallee, affords sites where hives can be placed to take advantage of afternoon shade.

Top: Hives working mallee in the Euabolong District of New South Wales.*

Bottom: These hives are situated near Hillston, New South Wales. The water is supplied to a tank in which floating bottles support sacking, thus affording a suitable surface upon which bees can alight and absorb water.†

(* *Photo A. Clemson.* † *Photo J. H. Leigh.*)

Plate 46 THE SITING OF BEE HIVES IN THE DRY COUNTRY

such factors as soil type, altitude and temperature. As a result, colour, flavour, aroma, viscosity and moisture content may vary markedly between floral sources and often very distinctive differences between certain honeys are evident. Included amongst the wide range of well-known and distinctive Australian honeys are a number produced in the semi-arid and arid zones, for example yellow box (*E. melliodora*), river red gum (*E. camaldulensis*) and red ironbark (*E. sideroxylon*).

Yields of honey per hive from individual species vary tremendously both with hive management and with the peculiarities of species and their environment. Australia-wide yearly averages from productive hives generally vary between 90 to 150 lb/hive. However, highly-productive hives may yield upwards of 100 lb of honey per hive from a species such as yapunyah (*E. ochrophloia*); 180 to 240 lb/hive from *E. camaldulensis*, 300 lb/hive from mixed *E. ochrophloia, E. populnea* and *E. camaldulensis* communities and as much as 480 lb/hive from certain New South Wales mixed mallee communities.

The highest commercial apiary hive yield in Australia has been established outside the zone discussed in this chapter but gives some idea of the immense potential which can be attained by well-managed bees given optimal conditions. In Western Australia 786 lb/hive from 400 hives was extracted over a twelve-month period from a range of eucalypt species, and included a big karri (*E. diversicolor*) flow.

THE HONEY INDUSTRY

The bulk of the honey extracted in Australia is produced by a relatively small number of full-time apiarists. Of the total annual production 75 per cent is produced by 443 apiarists who manage over 300 hives each (and sometimes as many as 1,000 to 1,500 hives). These 443 beekeepers would represent only 7.5 per cent of the total number of registered beekeepers.

Hives are usually moved over wide areas so as to ensure a continuous supply of nectar and pollen for the bees and in some cases the annual mileage covered by individuals would exceed 20,000 miles. Because of the migratory pattern of apiculture in all States it is difficult to accurately determine the relative importance of particular areas or particular species on a locality basis. For sociological reasons, such as the proximity to schools, hospitals, main roads and railways, etc., most apiarists live in towns and the smaller cities and exploit the flora accessible from those centres. For these reasons and the present availability of reliable flowering eucalypt forest in the higher rainfall areas, most beekeepers reside outside the area considered in this chapter. In New South Wales, for example, only 5 per cent of apiarists operating more than 300 hives live within the semi-arid and arid area. In more recent years however apiarists have made increasing use of semi-arid species. This has come about for a variety of reasons, chief amongst which has been the progressive clearing of State and privately owned forest land for more intensive farming, with subsequent seeding to improved pastures. To some extent beekeepers have been compensated by the nectar production from sown species such as clover and lucerne and to a greater extent by the increase in pasture weeds such as cape weed (*Arctotheca calendula*), Paterson's curse (*Echium lycopsis*), and various thistles. Other important factors enabling the use of the more remote semi-arid and arid regions has been the improvement of roads and trucks and the adoption of more efficient hive-loading and honey-extracting equipment.

There are, however, a number of problems specific to the semi-arid and more particularly the arid areas which will continue to limit the extent to which these vast areas will be used for honey and wax production. Low and unreliable rainfall results in erratic and infrequent budding and may even ultimately affect the flowering of budded trees. In addition many of the eucalypt and acacia pollens are of low biological quality and of low attractiveness to bees. The reasons for this are largely unknown, although currently under investigation. It has been suggested that it may be due to such factors as low mineral content, pollen too dry for bees to be able to collect, and pollen grains being too hard and thus being more difficult to digest. In relation to the last mentioned, pollen constipation has been observed both in South Australia and Western Australia. Because of the frequent lack of attractive high-quality pollen sources certain species of eucalypt, for example *E. ochrophloia* on the Channel Country, can be dangerous unless there has been sufficient heavy rains or floods to promote growth of pollen-rich ground flora species.

The operation of beekeepers in these semi-arid and arid zones when commuting long distances from their home bases requires a great deal of additional organisation. In two of the States, for example South Australia and Western Australia, beekeepers are required by law and by necessity to provide fresh water for their bees. In these hot environments populous hives will each consume as much as two pints of water per day and with extreme temperatures higher consumption probably occurs. This adds considerably to the beekeepers' work-load and is undoubtedly a major difficulty in areas where water is scarce. In Western Australia beekeepers have been allowed by special arrangements to take water from the Kalgoorlie pipeline and here again they still often cart it long distances. Another major problem in most areas is the destruction caused by ants which will attack and kill out many hives unless their nests are constantly sought out and destroyed. Termites are a lesser problem and destroy hive bodies, although risers on the bottoms of hives are usually made of termite-resistant timber.

Lastly, of paramount importance are the economics inherent in the utilisation of those areas which lie remote from the apiarists' main areas of operation. The main costs are those associated with transport and as transport costs continue to rise so the range over which the beekeeper can still profitably operate must diminish. It has been estimated that it costs between \$1.50 and \$2.00 per hive to carry a load of hives 300 miles. In addition to this it costs approximately three cents per pound to extract and tin honey from the hive. The honey derived therefrom has later to be dispatched or delivered to the markets and the hives moved to their next sites. Hence many apiarists consider that unless they are likely to be able to harvest a sustained flow (averaging about 300 lb/hive) of high grade honey (low Pfund reading) for which they can realise at least ten cents per pound, such long moves are likely to prove uneconomic.

Whilst it is impossible to accurately assess the extent to which the semi-arid and arid areas are being exploited because of the many factors already enumerated, the following appears to be the current situation. The semi-arid areas of each State are utilised moderately, the degree of use being greatest in Victoria and New South Wales. The arid areas are largely unused except when conditions within the traditional higher rainfall areas do not look very promising and there are known prospects of good flows in the arid areas. At present there is still a lack of regular beekeeping in these more isolated regions with the result that not enough basic information is available on nectar yields or more importantly on pollen production and pollen quality.

TABLE 14.1

HONEY AND POLLEN FLORA OF SEMI-ARID AND ARID QUEENSLAND

Botanical Name	*Official and Local Common Names*	*Distribution*	*Community(ies) and Habitat*	*Time of flowering*	*Importance as a source of honey; honey quality*	*Yield of pollen*	*Remarks*
E. albens Benth.	white box(*a*)	Darling Downs	Woodland on rich, well-drained soils	April-July	Major producer of white to extra light honey 19–46(*b*)	Medium	Heavy losses of bees have occurred when this species has been the sole source of supply
E. caleyi Maiden	Caley's ironbark(*a*), drooping ironbark	Southern Darling Downs	Woodland on sandy or stony soil	June-October	Major producer of extra white honey 21–55	Nil	
E. camaldulensis Dehnh.	river red gum(*a*), river gum	Condamine and Macintyre Rivers	Forest on river flats	Nov.-Jan.	Major importance; Honey white to light amber in colour and of mild flavour 32–63	Major	Flowers every two to three years
E. crebra F. Muell.	narrow-leaved red ironbark(*a*) narrow-leaved ironbark	South-east	Woodland on stony or sandy soils	Aug.-Dec.	Variable, extra white to extra light 14–35	Minor	An erratic producer of nectar yielding good harvests in only one of four or five years. Commonly associated with *Callitris* species.
E. dealbata A. Cunn. ex Schau.	tumbledown red gum, blue-leaf gum, leatherjack, mountain gum, silver gum	Darling Downs	Woodland on sandy soils or stony ridges	Nov.-Dec.	Medium producer of extra light amber to light amber honey	Major	
E. fibrosa F. Muell. subsp. *nubila* (Maiden et Blakely) L. Johnson	bluetop ironbark, blue-leaved ironbark(*a*), dusky-leaved ironbark, mountain ironbark broad-leaved red ironbark, red ironbark	Darling Downs	Woodland on stony ridges on poor silts and clay soils	June-Sept.	Medium, extra white to white 5–21.5	Minor	
E. melanophloia F. Muell.	silver-leaved ironbark(*a*), broad-leaved ironbark	East and south-east	Open forest on silts and also on light and heavy-textured soils	Dec.-Feb.	Medium white to light amber 28–52	Medium to major depending on location	The flowers of this species are easily affected by variations in the weather
E. melliodora A. Cunn. ex Schau.	yellow box(*a*), yellow jacket yellow ironbox honey box	Southern Darling Downs	Open forest and woodland on a variety of soils	Oct.-Dec.	Major, dense extra white to extra light amber honey of fine flavour	Nil	Usually flowers freely about every second year
E. microtheca F. Muell.	coolabah, coolibah(*a*), coolebah	South-western Darling Downs, South-eastern Maranoa Districts	Forest on river banks and flood plains on heavy soils	Dec.-Feb.	Medium producer of white honey 27–44	Medium	The coolibah requires good spring rains for a good honey crop

E. ochrophloia F. Muell.	yapunyah(*a*), napunyah, lapunyah	South-west	Forest on flood plains on heavy soils	May-Oct.	Medium, extra light amber 38–44	Minor	Dwindling often occurs when this species alone is being worked
E. panda S. T. Blake subsp. *panda*	Brogan's ironbark, tumbledown ironbark, corky ironbark	Southern Darling Downs	Woodland on sandstone country	April-Nov.	Minor to major light amber	Medium	
E. pillagaensis Maiden	narrow-leaved grey box(*a*), narrow-leaved box, gum-topped box, mallee box, ribbon box, molly box	Darling Downs and Maranoa District	Woodland on sandy or loamy soils	Feb.-April	Medium, white to extra light amber 21–42	Nil	Very serious dwindling often occurs with this species
E. populnea F. Muell.	bimbil box, bimble box(*a*), poplar box, grey box, shiny-leaf box	Widespread except south-east	Open woodland mainly on loam and sandy loam soils but also on heavy seasonally water-logged soils	Dec.-Jan.	Moderate producer of light amber	Minor	An erratic flowerer and is not held in high esteem as a honey producer
E. sideroxylon A. Cunn. ex Woolls subsp. *sideroxylon*	red ironbark(*a*), mugga	Darling Downs and southern Burnet District	Open forest hillsides	July-Nov.	Major, extra light amber 41	Nil to minor	
E. tessellaris F. Muell.	carbeen(*a*), Moreton Bay ash	Widespread throughout south-eastern Queensland	Open woodland on loam and sandy soils	Nov.-Feb.	Nil to minor medium amber	Nil to minor	Many beekeepers consider that this species is of little or no importance as a nectar or pollen producer
Angophora floribunda (Sm.) Sweet	rough-barked apple(*a*), apple-tree, apple	Darling Downs, Burnet, Maranoa and Warrego Districts	Woodland on creek banks and alluvial flats	Dec.-Jan.	Medium producer of dark amber honey 75	Major	This is the most important species of the Angophora group which blossoms best in drier years
Melaleuca lanceolata Otto	moonah(*a*), western tea-tree, black tea-tree	Darling Downs	Understorey shrub on black clays	Dec.-Jan.	Medium light amber	Major	

(*a*) Official names according to 'Nomenclature of Australian Timber' (Standards Association of Australia ASO2–1970).
b) Numbers refer to the Pfund Honey Colour Grader System for describing honey colour.

TABLE 14.2

HONEY AND POLLEN FLORA OF SEMI-ARID AND ARID NEW SOUTH WALES

Botanical Name	*Official and Local Common Names*	*Distribution*	*Community(ies) and Habitat*	*Time of flowering*	*Importance as a source of honey; honey quality*	*Yield of pollen*	*Remarks*
E. albens Benth.	white box(*a*)	Western Slopes	Open forest and woodland on rich well-drained soils	April-Aug.	Fair producer of good quality honey	Poor pollen producer	Unreliable-frequent losses of bees occur with this species because of poor pollen
E. blakelyi Maiden	Blakely's red gum(*a*), cabbage gum, red gum, forest red gum	Western Slopes, Lower Tablelands	Open woodland on good loam soils	Aug.-Dec. every third year	Good producer of good quality honey	Good pollen producer	Shy producer in some areas. Commonly associated with *E. melliodora* and *E. sideroxylon*
E. camaldulensis Dehnh.	river red gum(*a*), red gum, Murray red gum	Western Slopes and Plains	Forests and woodlands on river flats and flooded depressions	Nov.-Jan. Usually every second year	Good producer of good quality honey	A surplus of pollen	
E. conica Deane et Maiden	fuzzy box(*a*), apple box, blue box	North-western and Central Slopes	Woodland on river flats and open plains	Oct.-Nov.	Good producer, light amber	Of little consequence as a pollen producer	
E. dwyeri Maiden et Blakely	Dwyer's red gum(*a*), mallee gum	Western Plains and Slopes	Mallee on sandy or gravelly soils or elevated areas	January	Important yielding good quality honey	Most useful	
E. largiflorens F. Muell.	black box(*a*), flooded box, red box	Western Plains	Woodland on river flats and flooded depressions	Spring and summer	Infrequent honey producer; honey of good quality	Good pollen producer	Trees do not flower as a body but over some two to six months
E. melliodora A. Cunn. ex Schau.	yellow box(*a*), yellow jacket	Western Slopes and Lower Tablelands	Open woodland on all soils excluding poor sandy types	Variable but usually Sept.-Dec.	Good producer of excellent quality honey	Poor pollen producer	If no other flora is available bees have great difficulty in keeping up sufficient brood-rearing in the hives
E. microcarpa Maiden	inland grey box, grey box(*a*)	Western Slopes and Plains	Open woodland on clay loam and clay soils	Feb.-July	Unreliable producer of average quality honey	Good pollen producer	Closely resembles *E. woollsiana*
E. oleosa F. Muell. ex Miq.	red mallee(*a*), giant mallee	Western Plains	Mallee on sandy loams and sandy soils	Nov.-Jan.	Good producer of excellent quality honey	Useful	One of the most common mallee trees
E. pillagaensis Maiden	narrow-leaved grey box(*a*), Pilliga box	Western Plains	Woodland on loams and in flooded depressions	Jan.-April	Fair producer of good quality honey	Nil	
E. polybractea R. T. Bak.	blue mallee(*a*)	Western Plains	Mallee on a wide range of soils	Apr.-June	Good producer of good quality honey	Pollen yielded in adequate amounts	
E. populnea F. Muell.	bimbil box, poplar box, bimble box(*a*)	Western Slopes and Plains	Open mixed shrub-woodland on loam and heavy clay soils	Jan.-March	Fair producer of light amber honey	No pollen	Not a heavy honey producer on its own but useful when used in conjunction with Mallee gum and oil mallee

E. sideroxylon A. Cunn. ex Woolls subsp. *sideroxylon*	red ironbark(*a*), mugga ironbark, pink flowering ironbark	Widespread	Open woodland on variety of soils	March-Sept.	Good producer of light amber honey	No pollen	On its own may cause dwindling of bees because of lack of pollen; flowering may extend from autumn to early spring
E. viminalis Labill.	manna gum(*a*), ribbon gum, white gum	Widespread, mainly tablelands and slopes	Mixed woodlands on a variety of soils	March-May	Fair producer of good quality honey	Good pollen	Important for stimulating brood-rearing and getting bees in good heart for wintering
E. woollsiana R. T. Bak.	narrow-leaved grey box, narrow-leaved box, mallee box, grey box(*a*)	Western Slopes to mallee and desert area	Open woodland on loams and clay loams	Feb.-July	Unreliable producer of good quality honey	Useful	Closely resembles *E. microcarpa*
Acacia baileyana F. Muell.	Cootamundra wattle	South-western Slopes	Understorey tree or shrub to many woodland communities	July-Aug.	Nil	Produces an abundance of pollen	A valuable source of pollen for bees in winter
Acacia doratoxylon A. Cunn.	brown lancewood(*a*), currawong	Western Plains	Found mainly on stony hills	Sept.-Oct.	Nil	Appears to be the most prolific producer of pollen of all acacias	
Acacia pycnantha Benth.	golden wattle(*a*), broad-leaved wattle	South-west	Mostly on dry stony hills	September	Nil	Produces pollen abundantly	Moist air, dew, or favourable soil moisture needed to allow bees to gather pollen otherwise pollen is too dry
Brachychiton populneum (Schott) R.Br.	kurrajong(*a*)	Western Plains	Scattered trees rocky hills and on well-drained loams	March	Producer of dark strong-flavoured honey	Abundant pollen	Important for getting bees into excellent condition for wintering
Casuarina cunninghamiana Miq.	river sheoak(*a*), river oak	Western Plains	Woodland also on river banks	Spring	Poor	Useful	Important for providing a source of pollen for bees when drought has occurred in drier regions
Geijera parviflora Lindl.	wilga(*a*)	Western Plains	Mixed shrub woodland mainly on loam and clay-loam soils	Can flower for most of the year	Fair producer of dark, strong-flavoured honey	Limited	Useful for providing winter stores. Nectar collected during summer only
Hakea leucoptera R.Br.	needlewood(*a*)	Western Plains	Mixed shrub mostly on poor soils and sandy ridges	Usually every second year	Light producer of fair quality honey	Does not appear to be of any use for pollen	
Heterodendron oleifolium Desf.	inland rosewood(*a*)	Western Plains	Shrub woodland mainly on loam soils	Aug.-Feb.	Good producer of medium quality honey	Nil	Irregular flowering yielding for 3 to 4 weeks
Nitraria schoberi L.	dillon bush, nitre-bush	Western Plains	Degraded communities on loam and clay-loam soils	Sept.-Oct.	Moderate	Moderate	Regular flowering assures a reliable source of build up honey and pollen between mallee floras

(*a*) Official names according to 'Nomenclature of Australian Timber' (Standards Association of Australia ASO2–1970).

TABLE 14.3

HONEY AND POLLEN FLORA OF SEMI-ARID VICTORIA

Botanical Name	*Official and Local Common Names*	*Distribution*	*Community(ies) and Habitat*	*Time of flowering*	*Importance as a source of honey; honey quality*	*Yield of pollen*	*Remarks*
E. behriana F. Muell.	Bull mallee(*a*), broad-leaved mallee box	North-west, north and south	Mallee, dry sandy soil	Dec.-Jan. Unreliable	Producer of poor quality, dark amber honey	In abundance	Not common usually as a minor species in association with other mallee species
E. calycogona Turcz.	red mallee	North-west	Mallee, dry sandy soil	Sept.-Nov.	Unreliable yielder of medium amber, medium density poor-flavoured honey	Pollen yielded in medium quantities	
E. camaldulensis Dehnh.	river red gum(*a*), red gum, Murray red gum	Widespread	Forest common along river banks and alluvial valleys	Dec.-Jan. Reliable only every third year	Good producer light amber honey	Major	Subject to severe insect devastation especially by leaf and gall insects
E. dumosa A. Cunn. ex Schau.	white mallee(*a*), dumosa mallee, congoo mallee	North-west	Mallee on sands and sandy loam soils	June-Feb.	Reported to be a good producer of light to medium amber dense honey of fair flavour	Quantities of pollen are reported to be good	Shy yielder in dry years but gives large yields after good rains prior to flowering
E. foecunda Schau.	hooked mallee black mallee	North-west	Mallee on sandy loam soil	Dec.-March	Produces large amounts of good quality honey	Major	As a nectar and pollen producer this is one of the best of the Mallee eucalypts known to beekeepers
E. gracilis F. Muell.	yorrell(*a*), white mallee	North-west	Mallee, grows best on sandy soils 1 to 2 ft deep overlying limestone	Mainly Sept.-Oct. but may extend Apr.-Nov.	Producer of probably best mallee honey. Honey is light amber, fairly dense and of mild flavour	Yielded in very small quantities	Beekeepers are very wary of working this species because of the heavy bee losses sustained because of lack of pollen
E. incrassata Labill. var. *costata* N. T. Burbidge	angular-yellow mallee	North-west	Mallee on sands and sandy loam soils	Feb.-May	Good producer of good honey	Adequate	
E. largiflorens F. Muell.	black box(*a*), swamp box, river box, dwarf box, Murray box	North-west Plains	Woodlands in swamps and along water-courses	Jan.-Feb. Generally a heavy flowering every two years	Good producer of good quality honey	Major	One of the best bee forage trees in the districts in which it grows. Frequently adjoining *E. camaldulensis* forest
E. leucoxylon F. Muell.	yellow gum(*a*), white gum, white ironbark	North-west, north south-west	In open woodland and mallee areas on a variety of soils	Aug.-Nov.	Heavy producer of excellent quality light amber honey	Nil	A peculiarity of this species is that under certain climatic conditions it secretes nectar which the bees will not collect
E. macrorhyncha F. Muell.	red stringybark(*a*)	North-east, north, south-east	Open mixed woodland mainly on dry stony ridges	February	Somewhat unreliable but produces dark honey of good flavour	Good	Keenly sought after honey for blending purposes because of its visible lustre

E. melliodora A. Cunn. ex Schau.	yellow box(*a*)	Widespread	Open woodland usually on fertile well-drained soils	Nov.-Feb.	Heavy producer of excellent quality honey	Nil	Flowers mainly every second year but can flower every year or every third year
E. microcarpa Maiden	grey box(*a*)	Widespread	Open woodland found mostly sandy loams and clay loams	Feb.-May	Good producer of excellent quality medium amber honey which candies quickly	Yields abundant pollen	Crops vary season to season
E. oleosa not F. Muell. ex Miq.	red mallee(*a*), water mallee, oil mallee	North-west	Mallee on clay and sandy soils	March-Apr.	Yields medium amber honey of mild flavour	Medium	
E. polyanthemos Schau.	red box(*a*)	Eastern	Open woodlands on poor slaty or shaley ridges in moderately dry areas	Sept.-Jan.	Produces pale, very dense 'oily' flavoured honey with a dull appearance	Pollen yielded in small quantities	
E. polybractea R. T. Bak.	blue-leaved mallee, blue mallee(*a*)	North-west	Mallee on loam and clay-loam soils	April-June	Honey of good quality	Adequate pollen	
E. viridis R. T. Bak.	green mallee(*a*),	North-west, north	Mallee on sandy soil	Nov.-Jan. Flowers profusely at irregular intervals	Occasionally yields a honey of good colour and flavour	Medium	

(*a*) Official names according to 'Nomenclature of Australian Timber' (Standards Association of Australia ASO2–1970).

TABLE 14.4

HONEY AND POLLEN FLORA OF SEMI-ARID AND ARID SOUTH AUSTRALIA

Botanical Name	*Official and Local Common Names*	*Distribution*	*Community(ies) and Habitat*	*Time of flowering*	*Importance as a source of honey; honey quality*	*Yield of pollen*	*Remarks*
Eucalyptus anceps (R.Br. ex Maiden) Blakely	sand mallee, Kangaroo Island mallee, white mallee	Lower Eyre Peninsula, Yorke Peninsula, Murray Mallee	Mallee on sands and sandy loam soils	Jan.-Feb.	Occasionally produces moderate quantities of medium amber honey	Poor producer of average quality pollen	
E. angulosa Schau.	big red mallee, ridge-fruited mallee, giant angular mallee	Eyre Peninsula, Upper South East	Mallee on sands and sandy loam soils	Aug.-Nov.	Medium amber	Light	Where regrowth has occurred nectar production is very high
E. behriana F. Muell.	bull mallee(*a*), broad-leaved mallee box	Scattered trees occur over Eyre Peninsula, Southern Flinders Range, Upper South East	Mallee on sands and sandy loam soils	Nov.-Jan.	Variable producer of honey	Poor producer of medium quality pollen	Yields very heavy honey crops in occasional years
E. calcicultrix F. Muell. in Miq.	whipstick peppermint, black mallee box	Yorke Peninsula, Flinders Range	Mallee on sands and sandy loam soils	May-Nov.	Unreliable producer of medium amber honey	Useful	
E. calycogona Turcz.	slender mallee, black mallee, square fruited mallee	Eyre and Yorke Peninsulas, Murray Mallee, Upper South East	Mallee on sands and sandy loam soils	Aug.-Nov.	Amber	Useful	
E. camaldulensis Dehnh.	river red gum(*a*), red gum, Murray red gum	Widespread	Forest and woodlands along rivers and drainage lines	Dec.-Jan.	High producer of light amber, mild flavoured honey	Heavy	Flowers profusely every two to three years
E. cladocalyx F. Muell.	sugar gum(*a*)	Scattered occurrence over Eyre Peninsula, Flinders Range	Mixed woodland, woodlands on a variety of soils	Jan.-Feb.	Occasionally produces good yields of light amber honey	Produces good yields of poor quality pollen	
E. cooperana F. Muell.	merrit(*a*)	Eyre Peninsula	Woodland on mallee most commonly on sandy or medium loams	July-Nov.	Variable producer of light amber honey	Produces variable quantities of average quality pollen	
E. diversifolia Bonpl.	soap mallee(*a*), desert mallee, stony mallee, white mallee, S.A. coastal mallee	Eyre and Yorke Peninsulas, Upper and Lower South East	Mallee mainly on poor shallow sandy soils overlying limestone	July-Sept.	Variable producer of amber honey	Heavy producer of good quality pollen	
E. dumosa A. Cunn. ex Schau.	dumosa mallee(*a*), white mallee(*a*), Congoo mallee, scrub mallee, Waikerie mallee	Eyre Peninsula, Murray Mallee, Gawler Northwards to Flinders Ranges, Upper South East	Mallee on sands and sandy loam soils	June-Feb.	Light amber	Medium	

E. gracilis F. Muell.	yorrell(*a*), Kong mallee, green mallee, red mallee	Eyre and Yorke Peninsulas, Murray Mallee, Upper South East, Flinders Range	Woodland or mallee on sands and sandy loam soils	April-Aug.	Variable producer of light amber honey	Useful	
E. incrassata Labill.	yellow mallee(*a*), big white mallee, lerp mallee, mallee box, ridge-fruited mallee	Murray Mallee, Upper South East	Mallee on sands and sandy loam soils	Feb.-May	Medium producer of medium amber honey	Produces medium quantities of medium quality pollen	Frequently occurs on the edge of *E. camaldulensis*
E. lansdowneana F. Muell. et J. E. Brown	crimson mallee, crimson mallee box, red-flowered mallee, red mallee	Southern Eyre Peninsula	Mallee on sands and sandy loam soils	Dec.-Feb.	Medium amber	Variable	This mallee is restricted to small areas only
E. largiflorens F. Muell.	black box(*a*), Murray box, river box	River Murray area	Woodlands along flood-plains and in flooded depressions on grey clay loam soils	Oct.-Dec.	Variable producer of medium amber honey	Variable producer of average quality pollen	Unreliable as a nectar producer. Frequently occurs adjoining *E. camaldulensis* forest
E. leptophylla F. Muell.	march mallee, little red mallee, narrow-leaf red mallee	Eyre and Yorke Peninsulas, Murray Mallee, Upper and Lower South East	Mallee on sands and loam soils	Mar.-July	Light amber	Little	Unreliable
E. leucoxylon F. Muell.	yellow gum(*a*), S.A. blue gum	Widespread	Mixed open woodland on clay and sandy loam soils	May-Jan.	Good producer of medium amber honey	Nil	Flowers freely approximately every second year
E. leucoxylon F. Muell. var. *macrocarpa* J. E. Brown	water gum, red-flowered yellow gum	Lower Eyre Peninsula	Woodland	Aug.-Sept.	Medium amber	Nil	Cold wet weather may reduce bee activity and lower honey flow
E. oleosa not F. Muell. ex Miq.	red mallee(*a*), oil mallee, giant mallee	Widespread	Mixed mallee community on grey and brown loam soils, less commonly on sands	May-June sometimes Nov.-Dec.	Produces medium quantities of light amber honey	Produces negligible quantities of average quality pollen	
E. pileata Blakely	capped mallee, Ravensthorpe mallee, blue mallee	Eyre and Yorke Peninsulas, Murray Mallee	Mallee on sands and sandy loam soils	Nov.-Jan.	Medium amber	Heavy	
E. rugosa (R.Br.) Blakely	Kingscote mallee	Eyre Peninsula, Upper South East, Murray Mallee	Mallee on sands and sandy loam soils	Nov.-Feb.	Average producer of medium amber honey	Produces negligible quantities of average quality pollen	
E. viminalis Labill.	manna gum(*a*), ribbon gum	Mount Lofty Ranges, Lower South East	Woodland on a wide range of good and medium quality soils	Jan.-May	Variable producer of medium amber honey	Medium to heavy producer of good quality pollen	Nectar production unreliable

(*a*) Official names according to 'Nomenclature of Australian Timber' (Standards Association of Australia ASO2–1970).

TABLE 14.4—*continued*

Botanical Name	Official and Local Common Names	Distribution	Community(ies) and Habitat	Time of flowering	Importance as a source of honey; honey quality	Yield of pollen	Remarks
Acacia spp.	acacia, wattle	Widespread	Trees and understorey shrubs	July-Oct.	Nil	Useful	Nearly all species of acacia are a useful pollen source, but quality of pollen is doubtful
Bursaria spinosa Cav.	sweet bursaria(*a*), prickly box, christmas bush, native blackthorn	Widespread	Tree or understorey shrub most common along streams and gullies	Dec.-Feb.	Average producer of dark amber honey	Produces medium to heavy quantities of good quality pollen	Very useful plant in an off-season
Callistemon spp.	bottlebrush	Widespread	Understorey shrub	Spring	Medium amber	Useful	May produce a small surplus of honey
Casuarina stricta Ait.	drooping sheoak(*a*)	Widespread	Woodland mainly on poor soils and rocky slopes	April-May		Useful	In some parts of the State a good plant for winter breeding
Hakea leucoptera R.Br.	needlebush, needlewood(*a*), pinbush, water tree	Scattered throughout the Mount Lofty Range, Murray Mallee, Upper and Lower South East	Understorey shrub mostly on dry sandy ridges and poorer soils	Sept.-Jan.	Occasionally yields small amounts of light amber honey	Produces medium to heavy quantities of good quality pollen	May yield large amounts of pollen in some years. Produces small amounts of nectar and is good for spring build-up
Loranthus spp.	mistletoe	Widespread	Epiphytic parasite on numerous tree and shrub species	May-Jan.		Useful	Little nectar at times
Melaleuca spp.	paperbark, tea-tree, broombush tea-tree, dryland tea-tree	Scattered throughout Eyre and York Peninsulas, Murray Mallee, Upper and Lower South East	Understorey shrubs mostly on sands and sandy loam soils	Sept.-Mar.	Variable producers of light amber to dark amber honeys	Good producers of good quality pollens	Similar in characteristic to Leptospermum although the honey does not jell. Honey from some species goes quite dark in second-hand containers
Myoporum platycarpum R.Br.	sugarwood(*a*), sandalwood, false sandalwood, dogwood	Murray Mallee	Scattered trees mainly on sandy soils in mallee	Sept.-Nov.	Moderate producers of light amber honey	Nil	Not a pollen-bearing tree

(*a*) Official names according to 'Nomenclature of Australian Timber' (Standards Association of Australia ASO2–1970).

TABLE 14.5

HONEY AND POLLEN FLORA OF SEMI-ARID AND ARID WESTERN AUSTRALIA

Botanical Name	*Official and Local Common Names*	*Distribution*	*Community(ies) and Habitat*	*Time of flowering*	*Importance as a source of honey; honey quality*	*Yield of pollen*	*Remarks*
E. anceps (R.Br. ex Maiden) Blakely	peaked crown top mallee, Kangaroo Island mallee	Stirling Range and southern mallee	Mallee on sands and loam soils	Jan.-April; growth of buds takes 2-3 years	Important yielding good quality honey	Good pollen producer	Owing to long flowering period best suited to small apiaries
E. annulata Benth.	open-fruited mallee	Local occurrence in mallee zone, Bremer Range and southwards	Tree or thickets on sands and loam soils	Oct.-Mar.	Important yielding white to extra light amber honey	Useful pollen producer	
E. calycogona Turcz.	gooseberry mallee	Widespread and locally common in mallee area	Mallee on sands and loam soils	Variable; usually Aug.-Nov.	Important yielding extra light amber high quality honey	Possibly deficient	*E. celastroides*, *E. calycogona* and *E. gracilis* have crossed extensively and thus flowering times vary
E. camaldulensis Dehnh.	river red gum(*a*), river gum	North-west and north of State	Forest along watercourses and flood plains	Nov.-Dec.	Important producer of good quality honey	Good	
E. campaspe S. Moore	silver-topped gimlet(*a*)	Eastern Goldfields from north of Bremer Range to Coolgardie	Woodland on flats	Nov.-Jan.	Useful but not fully known	Considered potentially useful	This species has not yet been used very extensively by apiarists
E. celastroides Turcz.	mirret, mealy blackbutt, blue snap and rattle	Widespread and locally common in mallee areas	Scattered trees in mallee on sands and loam soils	Variable but usually Sept.-Nov.	Important yielding good quality extra light amber honey	Possibly deficient in pollen	
E. clelandii Maiden	Cleland's blackbutt	In Goldfields district	Open woodland on stony hills and on flats	Aug.-Feb.	Unknown	Unknown	Value to bees unknown
E. cooperana F. Muell.	merrit(*a*)	Locally common in mallee zone and particularly in eastern Goldfields	Mallee and open woodland mainly on sandy to medium loams	Flowers Feb. to Nov. but most useful Sept. to Nov.	Important, but flavour poor; light amber honey	Produces an abundance of pollen	
E. concinna Maiden et Blakely	—	East of Hyden	Mallee on sands and loam soils	Full flowering Feb.	Unknown	Unknown	Worked by bees
E. corrugata Luehm.	rough-fruited mallee	Widespread in mallee areas	Mallee on sands and loam soils	Nov.-Apr.	Honey potential not known	Useful	Has not been worked by apiarists but is worked by bees and birds
E. cylindrocarpa Blakely	woodline mallee	Common in Widgiemooltha area and extendng eastwards	Woodland or mallee on sands and loam soils	December	Unknown	Unknown	Worked by bees

(*a*) Official names according to 'Nomenclature of Australian Timber' Standards Association of Australia ASO2–1970).

TABLE 14.5—*continued*

Botanical Name	*Official and Local Common Names*	*Distribution*	*Community(ies) and Habitat*	*Time of flowering*	*Importance as a source of honey; honey quality*	*Yield of pollen*	*Remarks*
E. dundasii Maiden	Dundas blackbutt(*a*)	Bremer Range to Norseman area	Woodland mainly on loam soils	Feb.-Apr.			
E. eremophila (Diels) Maiden	Eastern Goldfields mallee, horned mallee	Widespread in mallee zone	Open thickets in mallee zone on loamy soils	Oct.-Jan. but may commence in June	Yields abundantly	Good	Frequently on the margins of *E. salmonophloia*
E. gardneri Maiden	blue mallet(*a*), Gardner's mallet	Wetter mallee areas in the south-west	Open woodland on gravelly laterite soils	May-June	Good producer of excellent honey	Excellent	
E. gracilis F. Muell.	yorrel(*a*), snap and rattle, small-budded mallee	Widespread and locally common in mallee zone	Tree or shrub often in thickets on sands and sand loam soils which are frequently calcareous	Flowers from April to Nov. with honey flow from Aug.-Nov.	Good producer of good quality extra light amber honey	Deficient in pollen	Bees must have another source of pollen to be maintained on this honey flow
E. griffithsii Maiden	grey gum	Coolgardie Goldfields, Bullabulling to Widgiemooltha	Woodland or more commonly a mallee on a variety of soils	Sept.-Jan. May hold buds for a very long period	Potentially useful producer	Unknown	Value unknown
E. leptophylla F. Muell. ex Miq.	slender-leaved white mallee	Widespread in wheat belt and mallee zone	Mallee on sands and loam soils	Dec.-Mar.	Low yielding, honey of good quality	Useful	
E. lesouefii Maiden	Goldfield's blackbutt	Eastern Goldfields area	Open woodland on alluvial flats	Flowers Oct. to Dec. but may shed buds without flowering	Unknown	Unknown	Value to bees unknown
E. loxophleba Benth.	York gum(*a*)	Widespread	Wooded grassland on granite and diorite sandy loams	Every 2 years, buds taking about 10 months to mature. Flowers May-Dec. with main nectar flow from Sept. to Dec.	Very important as producer of excellent extra light amber to light amber honey	Poor	Usually flows on alternate year to *E. rudis*. Usually grows associated with *Acacia acuminata*
oleosa F. Muell. ex Miq. var. *glauca* Maiden	redwood, glauca, boongul	Widespread in mallee zone, very common in eastern Goldfields	Mallee and open woodland on sands and medium loams	Sept.-Dec.	Important producer of good quality extra light amber honey	Abundant pollen produced	Individual trees tend to start and finish flowering within about 4 weeks, but within an area trees will flower over several months

E. oleosa F. Muell. ex Miq. var. *longicornis* (F. Muell.) C. A. Gardn.	red morrel, morrel, longicornis	Widespread from wheat belt to eastern Goldfields	Open woodland usually on loam soil overlying limestone	Nov.-Mar.	Important producing good quality extra light amber honey	Good	Often associated with other tree species particularly *E. salmonophloia*
E. oleosa F. Muell. ex Miq. var. *obtusa* C. A. Gardn.	oleosa	Eastern Goldfields and southwards	Open woodland or mallee on sands and loam soils	Feb. (Bremer Range) and May (Goldfields)	Important producer of extra light amber honey		
E. pileata Blakely	capped mallee, crown-topped mallee	East of Hyden down to Ravensthorpe	Mallee on sands and loam soils	Variable, usually Dec.-April	Useful producer of good quality honey	Useful	Has not been worked extensively
E. platypus Hook.	round-leaved moort(*a*)	Southern mallee areas, particularly Ravensthorpe district	Open woodland in Mallee areas on moist sandy loam flats and depressions	June-Jan.	Excellent producer of high quality honey	Excellent	One of the better flows
E. rudis Endl.	Western Australian flooded gum(*a*), flood gum, river gum	South-west	Forests occurring on river banks and areas subject to flooding	Aug.-Nov. earlier in more southern areas than northern	Good producer of light amber honey	Useful	Can be valuable and helps build up hives
E. salmonophloia F. Muell.	salmon gum(*a*)	Widespread wheat belt and mallee areas	Mixed open woodland on heavy red clays and lighter soils	Unpredictable most usually Dec.-Mar.	Unreliable but when it flows yields a mild, extra-light amber honey	Poor	May carry two age-groups of flower buds
E. salubris F. Muell.	gimlet(*a*)	In wheat belt, Goldfields districts and mallee areas	Open woodland on red clay flats	Oct.-Jan.	Produces extra-light amber honey	Abundant pollen producer	Often associated with *E. salmonophloia*
E. sheathiana Maiden	ribbon-barked mallee	Local occurrence in mallee zone	Open woodland or mallee on alluvial soil	Apr.-June	Useful	Useful	
E. stricklandii Maiden	Strickland's gum	Eastern Goldfields	Woodland on stony hilltops	Nov.-Jan.	Useful		Yields honey when enough trees are available
E. torquata Luehm.	coral-flowered gum(*a*)	Occurs in eastern Goldfields from Bullabulling through Coolgardie to Norseman	Woodland on stony hills	Sept.-Jan.	Useful, occasionally produces a crop of extra light amber honey	Useful	Very showy flowering
E. wandoo Blakely	wandoo(*a*), white gum mallee	Local pockets in mallee areas	Mallee on sands and sandy loam soils	March	Important producer of mild flavoured extra-light amber to light amber honey	Rather poor	Mainly cleared in wheat belt except for isolated clumps of var. *elata* Benth.
Acacia pulchella R.Br.	prickly moses	Widespread	Understorey shrub	Nov.-Dec.	None produced	Very good	Acacias generally are a valuable source of pollen, but West Australian species do not appear to yield nectar

(*a*) Official names according to 'Nomenclature of Australian Timber' (Standards Association of Australia ASO2–1970).

TABLE 14.5—*continued*

Botanical Name	*Official and Local Common Names*	*Distribution*	*Community(ies) and Habitat*	*Time of flowering*	*Importance as a source of honey; honey quality*	*Yield of pollen*	*Remarks*
Leucopogon spp.	white bell	Widespread	Heaths and deep sands	Spring; variable according to rainfall and temperature	Important but honey rather strong flavoured	Very important source of pollen	
Melaleuca spp.		South-west semi-arid	Low-lying sandy soil		Good flows of extra light amber to light amber honey. Flavour is variable from poor to excellent	Useful	

(*a*) Official names according to 'Nomenclature of Australian Timber' (Standards Association of Australia ASO2–1970).

ACKNOWLEDGMENTS

The author wishes to sincerely thank Dr K. M. Doull (Senior Lecturer in Entomology, Waite Agricultural Institute), Mr R. B. Gulliford (Senior Apicultural Officer, New South Wales Department of Agriculture), Mr D. F. Langridge (Livestock Research Officer, Victorian Department of Agriculture) and Dr F. G. Smith (Senior Apiculturist, W. A. Department of Agriculture) and Mr A. L. Eastley (C.S.I.R.O., Division of Plant Industry) for unpublished information and constructive comments. One plate was kindly supplied by Mr A. A. Clemson (Chief Apicultural Officer, New South Wales Department of Agriculture). Lastly, I would like to thank Mr N. Hall and Mr G. M. Chippendale (Forest Research Institute, Canberra) for assistance relating to the taxonomy of the eucalypt species described.

FURTHER READING

Anonymous (1949). *The Honey Flora of Victoria.* Govt. Printer, Melbourne.

Australian Honey Board (1969). *Sixth Annual Report 1968-1969.*

Blake, S. T. & Roff, C. (1958). *The Honey Flora of South Eastern Queensland.* Govt. Printer, Brisbane.

Coleman, R. S. (1962). 'Commercial bee-keeping: 2. Honey flora of Western Australia'. *Western Australian Department of Agriculture Bulletin*, No. 3038, pp. 15.

Cooke-Yarborough, R. E. (1963). 'The honey industry in New South Wales'. *Review of Marketing and Agricultural Economics*, 31 (No. 1), 5-39.

Goodacre, W. A. (1951). *The Honey and Pollen Flora of New South Wales.* Govt. Printer, Sydney.

Penfold, A. R. & Willis, J. L. (1961). *The Eucalypts.* Leonard Hill Limited, London.

Purdie, J. D. (1968). 'Honey and pollen flora of South Australia'. *J. Dep. Agric. S. Aust.*, *71* : 207-16.

Roff, C. (1969). 'Honey production in Paroo District'. *Qd. Agric. J.*, 95 : 163-6.

Roff, C. & Rhodes, J. (1968). 'Colour grades of some Queensland honeys'. *Qd. Agric. J.*, 94 : 137-9.

Smith, F. G. (1969). 'Honey Plants in Western Australia'. *W.A. Dept. Agric. Bull.*, No. 3618, p. 78

CHAPTER 15

Saltbush and other chenopod browse shrubs

by J. H. Leigh

APPROXIMATELY six per cent (174,000 square miles) of the total land area of mainland Australia is occupied by a shrub-steppe dominated by various annual and perennial species of the family *Chenopodiaceae*. These communities lie mostly south of the Tropic of Capricorn in an environment receiving predominantly winter rainfall ranging from 6 to 20 in. per annum (Figure 15.1). The most extensive areas occur in South Australia (99,000 sq miles), western New South Wales (31,000 sq miles) and south-eastern Western Australia (30,000 sq miles). Smaller areas (approximately 6,000 sq miles) are found in Queensland and Northern Territory. Victoria contains less than 1,000 sq miles of this vegetation type.

The shrub-steppe is composed characteristically of xeromorphic halophytes (drought and salt-tolerant plants). Of these the most widespread and economically important genera, with the number of endemic species in parentheses, are: *Atriplex* (40), *Bassia* (80), *Chenopodium* (20), *Enchylaena* (1), *Kochia* (40), *Rhagodia* (15) and *Salsola* (1). Of less importance and more restricted occurrence are species of the genera *Arthrocnemum* (8), *Salicornia* (2) and *Pachycornia* (2). These shrubs may occur as extensive mono-specific stands presenting a simple uniform appearance, or as mixed stands of two or more species. Generally the plants are less than 1½ metres in height, although *Atriplex nummularia* may attain a height of 3 metres. Bushes may be congregated in groups or occur as individuals separated by areas of bare soil. Most species may also be present as an understorey to low woodland species of *Eucalyptus, Casuarina, Acacia* or *Myoporum*.

DESCRIPTION AND DISTRIBUTION OF SPECIES

The original distribution of individual species was determined by factors such as the physical and chemical properties of the soil, and the soil moisture status, particularly as influenced by topography. A secondary but nevertheless significant factor affecting the present-day occurrence of the shrubs has been the introduction by man of domestic and feral animals.

From the resource stability and the pastoral viewpoint the most valued shrubs are *Atriplex vesicaria, Kochia sedifolia, K. pryamidata, K. astrotricha, K. aphylla* and *Atriplex nummularia.*

Atriplex vesicaria (bladder saltbush)

This perennial shrub with woody brittle stems, reaches a height and diameter of 60-70 cm. Its leaves are roughly oval in outline, up to 25 mm in length and

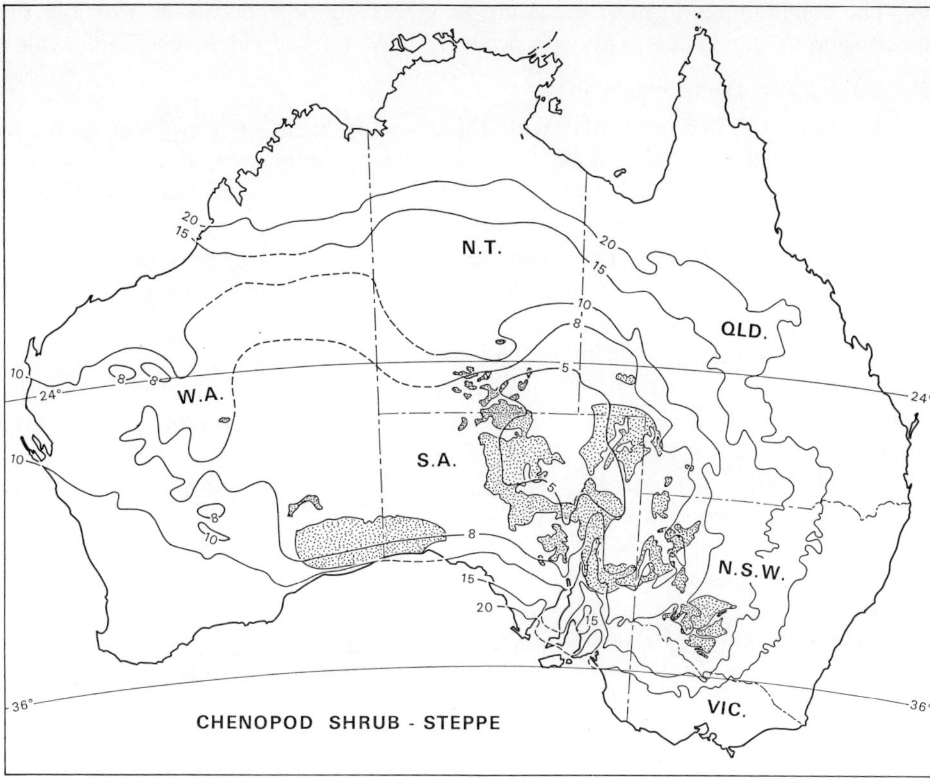

FIG. 15.1

Map of Australia showing the distribution of chenopod shrub-steppe and rainfall isohyets (inches/annum). After R. M. Moore and R. A. Perry in 1969.

have a silvery-grey appearance due to the vesicular hairs on the surface. The seed is enclosed within two bracteoles, on each of which there generally occurs a large membranous bladder-like appendage, the whole being up to 12 mm in diameter.

The most extensive tracts of *A. vesicaria* occur over parts of western New South Wales and the northern and eastern regions of South Australia. Other areas are found on the Nullarbor Plain, extending into Western Australia, in north-western Victoria, south-western Queensland and southern-central Australia.

This shrub grows on a wide range of soil types and in a variety of habitats. In the riverine plain of central south-western New South Wales it is found on flat terrain of brown and grey soils of heavy texture. Here it may occur in pure stands or in association with *Kochia aphylla* or *A. nummularia.* In South Australia and south-western New South Wales, *A. vesicaria* is supported by red-brown soils, some with a clay sub-soil, others with a sub-soil containing limestone nodules. In the latter case the associated shrubs are usually *Kochia sedifolia* or *K. astrotricha.* In north-western New South Wales and northern South Australia *A. vesicaria* is found on the loams of the gibber plains and on the stony desert soils of the undulating and hilly country surrounding the Barrier and Grey Ranges.

The communities in most areas are in a severely degraded state. Possibly the most vigorous stands of *A. vesicaria* occur in the vicinity of Hay, New South Wales.

Kochia sedifolia (pearl bluebush)

A much-branched perennial with thick woody stems, *K. sedifolia* grows to a height and diameter of about 1 m. The young branches and foliage have a dense covering of white hairs which imparts a pale-blue, almost white appearance to the bush. The leaves are soft, thick, almost cylindrical in outline and 3-10 mm in length. The short tube of the fruiting body is encircled by a reddish or light-brown horizontal wing 8-10 mm in diameter, finely veined, unbroken or with one radial slit.

K. sedifolia extends over much of western New South Wales and South Australia and is one of the major shrubs on the Nullarbor Plain. Either in pure stands or in association with *A. vesicaria, K. pyramidata* or *K. astrotricha*, or as an understorey to *Myoporum platycarpum* (sugarwood) or *Casuarina cristata* (belah [N.S.W.], black oak [S.A.]), this shrub occupies possibly the largest area of any of the Chenopod browse shrubs in Australia.

K. sedifolia prefers calcareous soils in which the limestone lies close to the surface. These soils are particularly susceptible to erosion if the surface cover is removed, either by wind in the case of the level alluvial desert loams or by both wind and water in the case of the undulating stony desert soils.

Kochia pyramidata (black bluebush)

Growing to a height of almost 2 m, *K. pyramidata* is the largest of the Australian *Kochia* species. Its greenish-grey, fleshy leaves are more or less obovoid and only 2-4 mm long. The characteristic fruiting body has a pyramid-shaped spongy extension above the narrow horizontal wing encircling the short basal tube. Before maturity the fruits are bright green; when mature and dry they turn black.

The distribution of *K. pyramidata* is similar to that of *K. sedifolia*, being confined to the calcareous red and red-brown soils. *K. pyramidata* is perhaps the more prevalent species of the two in the eastern areas and is a more common constituent of the shrub component of *Eucalyptus* (mallee), *C. cristata—Heterodendron oleifolium* (inland rosewood), and *Acacia aneura* (mulga) communities.

Kochia astrotricha (southern bluebush)

Similar in habit and appearance to *K. sedifolia*, although greyish rather than blue, *K. astrotricha* can be distinguished by its flatter leaves, which taper to a conspicuous petiole. The lower leaves may be up to 15 mm long and 3 to 4 mm broad. The winged fruits are 6 to 15 mm across, with the prominently-lobed wing surmounting a large hard obconical tube 3 to 5 mm long.

The major areas of *K. astrotricha* occur on sandy soils in the far west of New South Wales and in South Australia (where the similar *K. planifolia* also occurs). Minor areas of *K. astrotricha* are also found in Central Australia, Queensland and Western Australia. This shrub may be associated with *A. vesicaria* and various other *Kochia* species.

The relative abundance of each of the *Kochia* shrubs so far discussed is largely determined by the depth at which the limestone layer occurs beneath the soil surface. With the layer close to the surface *K. sedifolia* is the predominant species, with *K. pyramidata* and then *K. astrotricha* taking precedence as the depth to the layer increases.

Kochia aphylla (cotton bush)

K. aphylla is a much-branched perennial shrub growing to a height and diameter of about 1 m, with slender striate branches which become rigid and spiny with age. The leaves, short (1 to 2 mm long), thick, and crowded along the branchlets, fall off at an early stage. A membranous wing, pink when young, later maturing to brown, surmounts the short tube of the fruiting body, the whole being up to 12 mm across. Clusters of white cotton-like galls, caused by small grubs, often occur along the branches, hence the name 'cotton bush'.

Of widespread distribution, *K. aphylla* occurs throughout much of the arid and semi-arid interior. With a preference for clay and clay-loam soils, this species reaches its maximum density in the southern riverine plain of New South Wales. Here it may be the dominant species or it may be associated with other Chenopod shrubs, particularly *A. vesicaria.* Elsewhere, associated plants may be *K. pyramidata* or *Eremophila maculata.*

Atriplex nummularia (oldman saltbush).

A. nummularia is a blue-grey shrub growing to a height and diameter of about 3 m, with large oval or almost circular leaves 1-2 cm long. The leaves are covered with a scaly layer; their margins may be uninterrupted or slightly toothed. The plants are usually one-sexed, the female flowers occurring in dense clusters in the leaf axils at the ends of the branchlets. The fruiting body consists of two flat paper-like bracteoles, hemispherical in outline, toothed or smooth-edged, pressed closely together but united only at the base, where the seed is enclosed, the whole being 5-8 mm across.

The largest endemic species of *Atriplex, A. nummularia* is more common in the south-east, becoming restricted to local drainage areas in the more inland regions. This shrub occurs in association with *Acacia pendula* (boree or myall) (an association considered to be once far more widespread than it is today), as an understorey to *Eucalyptus largiflorens* (black box) or in association with *A. vesicaria.* Although it shows a preference for clay and clay-loam soils, *A. nummularia* can be found on most soil types.

An additional number of Chenopod browse plants associated with the above species are widespread but of little economic value; others may be of pastoral value but of such minor occurrence that their contribution to the overall vegetation is relatively small. Some of these species will be discussed briefly.

Atriplex semibaccata (creeping saltbush)

With weak prostrate stems spreading to 1 metre or more and rarely exceeding 30 cm in height *A. semibaccata* barely qualifies as a 'shrub'. The leaves are up to 25 mm long, roughly oblong in outline and may be toothed along their outer margins. They are green above, mealy white below. The fruiting bodies may be flat, green, diamond-shaped and toothed but generally they are succulent red berries about 5 mm long. This species can be confused with *A. leptocarpa* which has a similar habit and appearance and the fruiting bodies of which are tubular, up to 6 mm long, with the upper portion, consisting of two triangular lobes, somewhat compressed.

A. semibaccata is a widespread and fairly common species encountered in a range of situations from treeless plains to dense woodlands.

A. leptocarpa is less common and usually prefers soils with a light-textured surface.

Chenopodium auricomum (northern bluebush or golden goosefoot)

C. auricomum is an erect perennial shrub, often 1 m or more in height, with bluish-green leaves of variable shape. The flowers form dense globular golden clusters along short branchlets at the ends of the stems. This is a highly-regarded species found mainly in the northern regions of New South Wales and South Australia, in south-central Australia and south-western Queensland. It is usually confined to drainage areas on red and brown clays.

Chenopodium nitrariaceum (nitre goosefoot)

An erect perennial shrub growing to a height and diameter of about 2 m, *C. nitrariaceum* has pale green or grey oblong leaves up to 2.5 cm long clustered along the branchlets, which may often end in a spine. Small yellow or brown flowers occur densely packed at the ends of the branches. The seeds, which are minute round and black, are enclosed in a mealy covering. A widely distributed species, *C. nitrariaceum* prefers clay and clay-loam soils. Its most frequent habitats are drainage areas and river flats.

Kochia brevifolia (a bluebush)

This is a semi-erect much-branched shrub which may reach a height of 1½ m. The branches are somewhat hoary; the leaves are green, succulent, planoconvex or almost cylindrical, usually hairless and less than 5 mm long. The flattened fruiting body is 6 to 8 mm across, including the 5 broad membranous, equal-sized wings, narrowed towards their bases.

K. brevifolia is tolerant of saline conditions and occurs in a wide variety of situations from salt ground near the sea and inland water to undulating loamy soil in the dry interior areas.

Kochia georgei, K. tomentosa, K. villosa, K. triptera (bluebushes)

K. georgei is a low shrub with stout stems and white-woolly branchlets. The rather cylindrical leaves are 6-12 mm long and covered with long silky hairs. The fruiting body, probably larger than that of any other Australian species of *Kochia*, consists of a large obconical tube 4-5 mm long surmounted by a broad lustrous membranous wing 15-20 mm in diameter and often wavy on the margin. Before drying the wings are particularly colourful, changing from bright pink to red and finally to brown.

K. tomentosa, a low white woolly shrub with narrow triangular or oblong leaves usually closely appressed to the stems, has a fruiting body of a short narrow tube surmounted by a pale brown wing up to 10 mm across, with one radial slit.

K. villosa has narrow flat cylindrical fleshy leaves which may be silky-haired or hairless, the stems may be striated and often red tinged. The fruiting bodies of this short branching shrub have a small rounded base and a flat membranous wing up to 8 mm across, cleft with one radial slit.

A low hairless shrub, *K. triptera* has its narrow, cylindrical fleshy leaves crowded along the stems. The leaves may be up to 20 mm in length and gradually pointed. The horizontal membranous wing of the fruiting body is entire or it may be slightly lobed. Before drying it may be bright pink or red. The broad tubular base bears 3, occasionally 4 or 5, narrow vertical wings.

Of the above four species *K. georgei* and *K. villosa* are perhaps the more widespread, *K. villosa* occurring in the areas of higher rainfall to a greater extent than any of the others. All species have an apparent preference for red and red-brown soils, although *K. villosa* may also be found on grey clays.

Rhagodia spinescens (thorny saltbush)

This shrub may grow to a height of 2 metres, but is more often a low branching bush of 1 metre or less. Its oblong, sometimes almost round leaves taper to a conspicuous petiole and vary from green to mealy-white. The flowers usually occur in clusters at the ends of the branchlets, many of which terminate in a spine. The fruits are succulent red or orange berries 5 mm or less in diameter. *Rhagodia nutans* (climbing saltbush) is a closely related species with weak procumbent or climbing stems and much smaller fruits (usually less than 3 mm).

Enchylaena tomentosa (ruby saltbush) also bears succulent fruits of 5 mm diameter or less. These may be red, orange, yellow or purple before drying. The narrow, nearly cylindrical silky-haired leaves are up to 15 mm long. They are borne on procumbent or ascending woolly stems, which may reach a height of about 1 metre. *R. spinescens, R. nutans* and *E. tomentosa* are widespread, particularly on soils of light texture. All form part of the understorey beneath low woodlands of *Eucalyptus, Acacia* and *Casuarina.*

Salsola kali (soft roly-poly, buckbush)

An annual or biennial herb or sub-shrub, *S. kali* is a cosmopolitan plant which is widely distributed throughout most vegetation types. Rigid stems and branches reach a height and diameter of 1 metre, the bush assuming a uniform hemispherical shape. The leaves are variable, those at the base of the plant being narrow, cylindrical up to 3 cm long and prickly-pointed, whilst the upper leaves are shorter, flattened, curved and tapered. The fruiting body consists of a small central tube encircled by a horizontal, semi-transparent wing about 5 mm in diameter. The wing consists of five overlapping segments, three of which are broader than the others.

Arthrocnemum, Pachycornia and *Salicornia* spp. (samphires, glassworts)

These low-growing salt-tolerant plants are structurally similar in that they are comprised of many jointed superposed sections, the younger sections being green and succulent, the older uniting and hardening to form the stems. Although the density of these species is greatest around the margins of salt lakes, they also extend into dry, non-saline situations, usually in association with *Atriplex* or *Bassia* spp.

These then are the most common and/or the most economically important Chenopod shrubs within the dry interior of Australia. Associated with them there is a multitude of other shrubs, sub-shrubs and herbs of the same family. These range from valuable browse plants, through plants of little forage value but useful as stabilisers of erodible soil, to spiny aggressive species which may reduce or even eliminate the more useful types. Within the genus *Bassia* there are species which fall into each of the above categories.

DROUGHT RESISTANCE

A characteristic feature of the Chenopod shrubs, and one which explains their presence in the low-rainfall areas, is their resistance to drought. The ability of these plants to persist and produce growth in conditions under which few other species could exist is due to several factors, some of which are known, others of which are not yet fully understood and, in certain cases, the subject of scientific controversy.

The semi-succulent, flattened leaves of *Atriplex* species have the main photosynthetic tissue concentrated around the vascular bundles. It is thought that this structure, characteristic of *Atriplex* species, is of some importance in relation to drought resistance. In addition the leaves have bladder-like vesicular hairs on the surface. The bladders collapse, to form a mat of fused vesicles which give to the leaf a scaly or mealy appearance. It has been suggested that this vesicular tissue acts as a screen to reduce the intensity of excessive illumination and as insulation against excessive transpiration. It is also thought that the tissue may act as a medium to allow the rapid absorption of atmospheric moisture into the mesophyll of the leaf.

Many species of *Atriplex* have a high concentration of dissolved salts in their cell sap, which may be a consequence of a high rate of salt concentration under conditions of high soil water deficit. Many *Atriplex* species require sodium as an essential element and the addition of sodium chloride to a culture solution results in increased yield and succulence. In addition it is known that *Atriplex* species can actively synthesise oxalate compounds and it has been suggested that this mechanism enables the plant to increase the osmotic concentration of its cell sap in response to externally increasing water stress.

The leaves of other species (*Kochia, Bassia, Enchylaena*) are cylindrical and have no vesicles. The surfaces of these leaves are covered by fine hairs, which presumably reduce transpiration.

Some *Atriplex* species (notably *A. vesicaria*) have a drought-evading mechanism by which they shed their wilted leaves during excessively dry periods. The bushes may reach a state of almost complete defoliation without dying.

The ability of the arid Chenopod shrubs to exploit the available soil moisture is another factor contributing to their survival. With the exception of *A. vesicaria*, which has a shallow, wide-spreading root system, most species possess a deep tap root. Roots of *A. nummularia* may reach a depth of 4-5 m. In addition to a deep rooting system *Kochia* species, notably *K. sedifolia* and *K. pyramidata*, develop shallow deciduous feeder roots which are shed during dry conditions.

Top: Branches with leaves, flowers and fruiting bodies. (First six) Fruiting bodies; longitudinal section of fruiting body.

Below: Close-up of branchlets showing leaves and fruiting bodies.

(*Photos and drawings courtesy Division of Plant Industry, C.S.I.R.O.*)

Plate 47 CHENOPOD BROWSE SHRUBS (*Atriplex leptocarpa*)

Left hand page

Top: Branches bearing male (*second from left*) and female flowers and fruits. Fruiting bodies (*left*); longitudinal section of fruiting body (*upper centre*); seed (*right below*); longitudinal section of seed (*right above*).

View of contrasting *A. vesicaria* communities.

Bottom left: *A. vesicaria—Kochia aphylla* community growing on heavy clay soils on the Riverine Plain, N.S.W.

Bottom right: *A. vesicaria* growing on calcareous earth soils 150 miles north of Alice Springs.

Right hand page

Top left: *Acacia aneura—A. vesicaria* community growing on loam soils near Mt. Magnet, W.A.

Top right: Close-up view of branchlet showing leaves and fruiting bodies.

Centre: View of an individual *A. vesicaria* bush.

(*Photos and drawings courtesy Division of Plant Industry, C.S.I.R.O.*)

Plate 49 (*above and facing*) CHENOPOD BROWSE SHRUBS (*Atriplex vesicaria*)

Top: Branches with leaves, flowers and fruiting bodies. Fruiting bodies (*first* 7); longitudinal section of fruiting body; see (*above*).

Bottom left: Individual bush of *A. semibaccata* growing on heavy clay soils near Deniliquin, N.S.W.

Bottom right: Close-up of branch showing leaves and fruiting bodies.

(*Photos and drawings courtesy Division of Plant Industry, C.S.I.R.O.*)

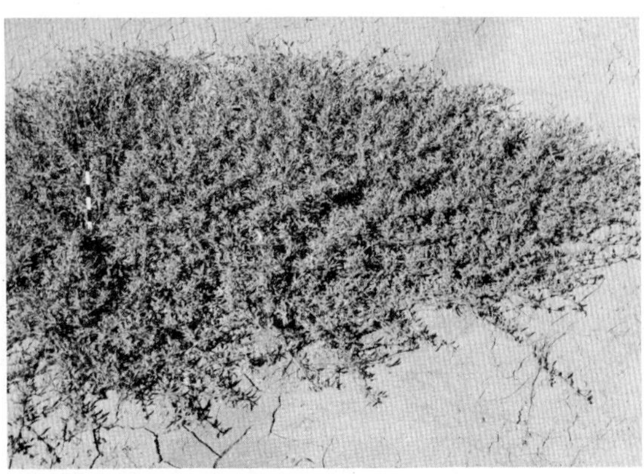

Plate 50 CHENOPOD BROWSE SHRUBS (*Atriplex semibaccata*)

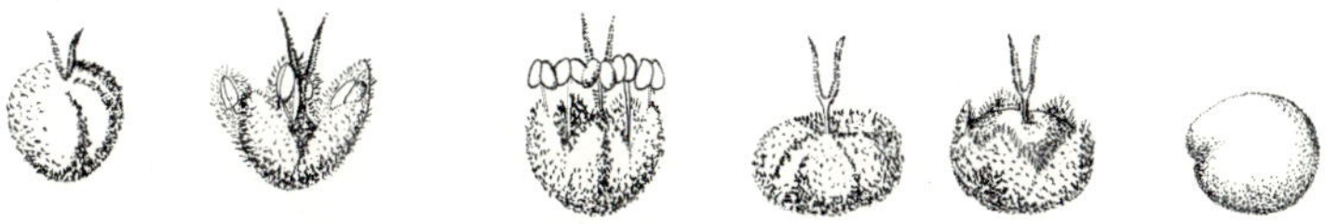

Top: Individual leaves; branches showing leaves and fruiting bodies. Flowers; fruiting bodies and seed.

Bottom left: General view of a vigorous *C. auricomum* community growing on heavy, cracking clay soil 70 miles north-west of Alice Springs, N.T.

Bottom right: Heavily-grazed degraded community of *C. auricomum* growing 80 miles north-west of Alice Springs, N.T.

(*Photos and drawings courtesy Division of Plant Industry, C.S.I.R.O.*)

Plate 51 CHENOPOD BROWSE SHRUBS (*Chenopodium auricomum*)

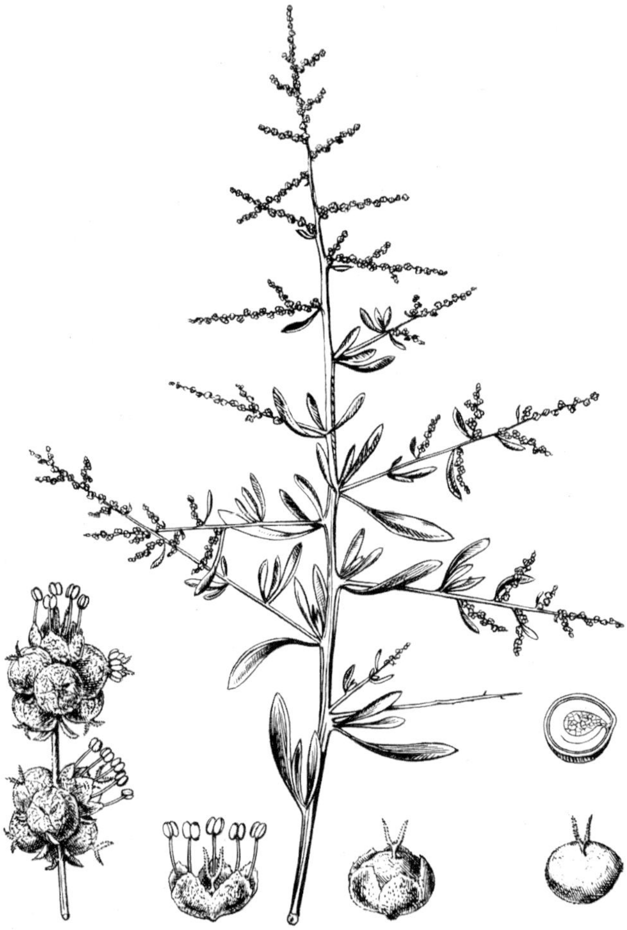

Top: *l. to r.*: Portion of branchlet with flowers; flower; branch with leaves, flowers and fruiting bodies; fruiting body; exposed fruiting body (*below*); cross-section of seed (*above*).

Bottom left: General view of a *Eucalyptus largiflorens—C. nitrariaceum* community near Deniliquin, N.S.W.

Bottom right: Close-up view of branch showing leaves, flowers and fruits.

(***Photos and drawings courtesy Division of Plant Industry, C.S.I.R.O.***)

Plate 52 CHENOPOD BROWSE SHRUBS (*Chenopodium nitrariaceum*)

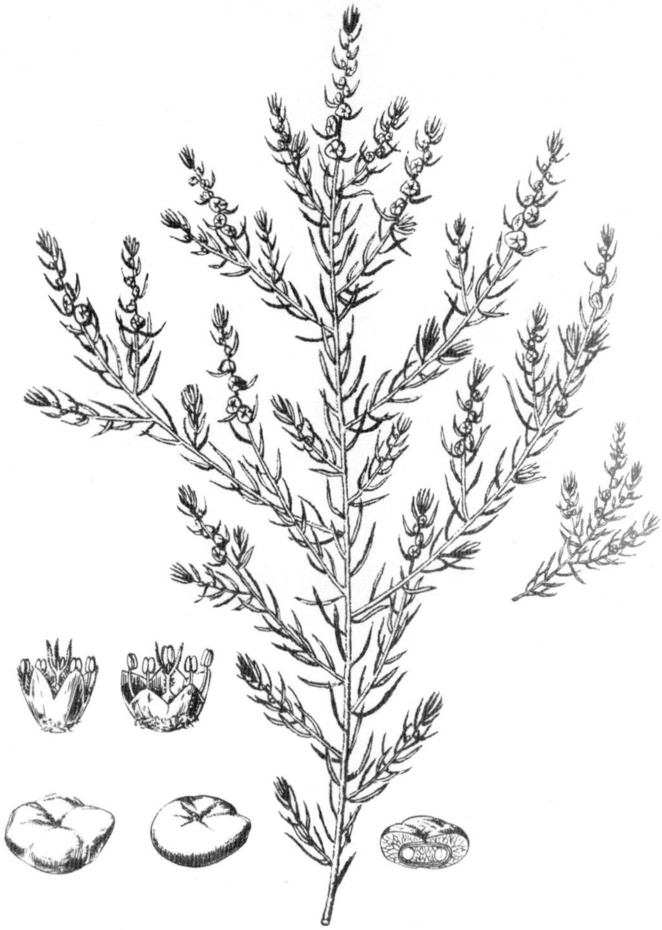

Top: Branches with leaves, flowers and fruiting bodies. Flowers (*left upper*); fruits (*left lower*); longitudinal section of fruit (*right*).

Bottom left: View showing habit of an individual bush of *E. tomentosa* growing in a *Heterodendrum-Casuarina* shrub-woodland community near Ivanhoe, N.S.W.

Bottom right: Close-up view of branchlets showing leaves and fruiting bodies.

(*Photos and drawings courtesy Division of Plant Industry, C.S.I.R.O.*)

Plate 53 CHENOPOD BROWSE SHRUBS (*Enchylaena tomentosa*)

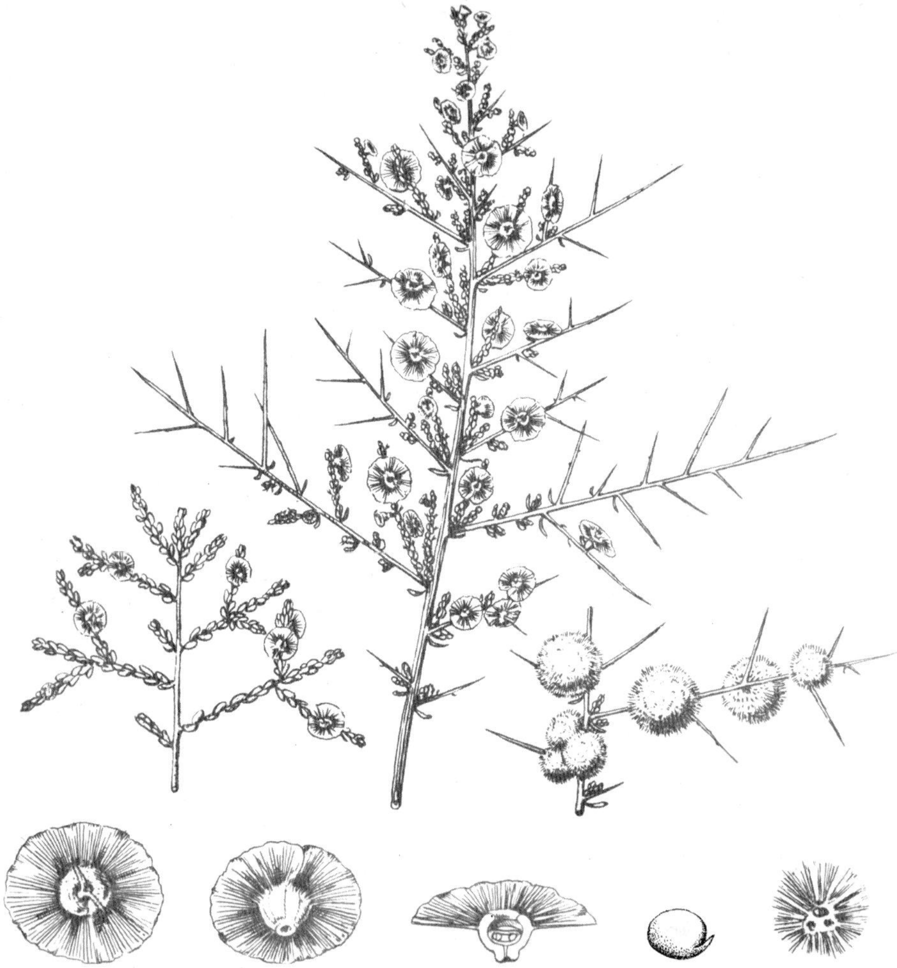

Left hand page

Top: Branches with leaves, flowers and fruiting bodies; portion of branch with galls (*lower right*). Fruiting bodies, longitudinal section of fruit, seed, longitudinal section of gall.

Bottom left: View showing habit of an individual bush.

Bottom right: View of *K. aphylla* community growing on heavy clay soils near Wanganella, N.S.W.

Right hand page

Top: Close-up view of branch showing leaves and fruiting bodies.

Bottom left: Close-up view of branch showing leaves and cotton-like galls.

(*Photos and drawings courtesy Division of Plant Industry, C.S.I.R.O.*)

Plate 54 (*above and facing*) CHENOPOD BROWSE SHRUBS (*Kochia aphylla*)

Branches with leaves, flowers and fruiting bodies. Fruiting body.
(Photos and drawings courtesy Division of Plant Industry, C.S.I.R.O.)

Plate 55 CHENOPOD BROWSE SHRUBS (*Kochia planifolia*)

Top: Branches showing leaves, flowers and fruiting bodies.

Bottom left: View of branch showing leaves and fruiting bodies.

Bottom right: Fence line view showing how grazing control has allowed the establishment of *K. brevifolia* near Kulin, W.A.

(***Photos and drawings courtesy Division of Plant Industry, C.S.I.R.O.***)

Plate 56 CHENOPOD BROWSE SHRUBS (*Kochia brevifolia*)

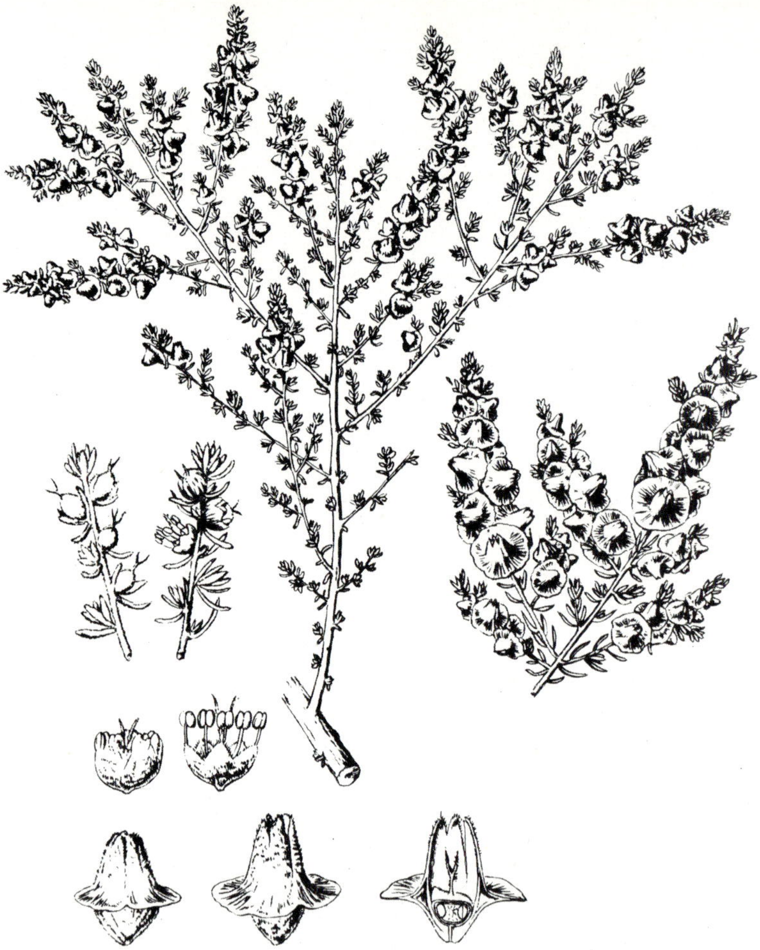

Top: Portions of branchlets with leaves and flowers (*left*); branches with leaves, flowers and fruiting bodies (*centre right*). Flowers (*upper left*); fruiting bodies (*lower left*); longitudinal section of fruiting body (*right*).

Bottom left: View of *K. pyramidata* community growing on sandy loam soil near Mossgiel, N.S.W.

Bottom right: Close-up view of branches showing leaves and fruiting bodies.

(Photos and drawings courtesy Division of Plant Industry, C.S.I.R.O.)

Plate 57 CHENOPOD BROWSE SHRUBS (*Kochia pyramidata*)

Top: Branch with leaves, flowers and fruiting bodies. Flower (*left*); fruiting body (*front view, back view and longitudinal section*).

Below: General view of a *K. sedifolia*—*K. pyramidata* community grown on loam soil near Balranald, N.S.W.

(*Photos and drawings courtesy Division of Plant Industry, C.S.I.R.O.*)

Plate 58 CHENOPOD BROWSE SHRUBS (*Kochia sedifolia*)

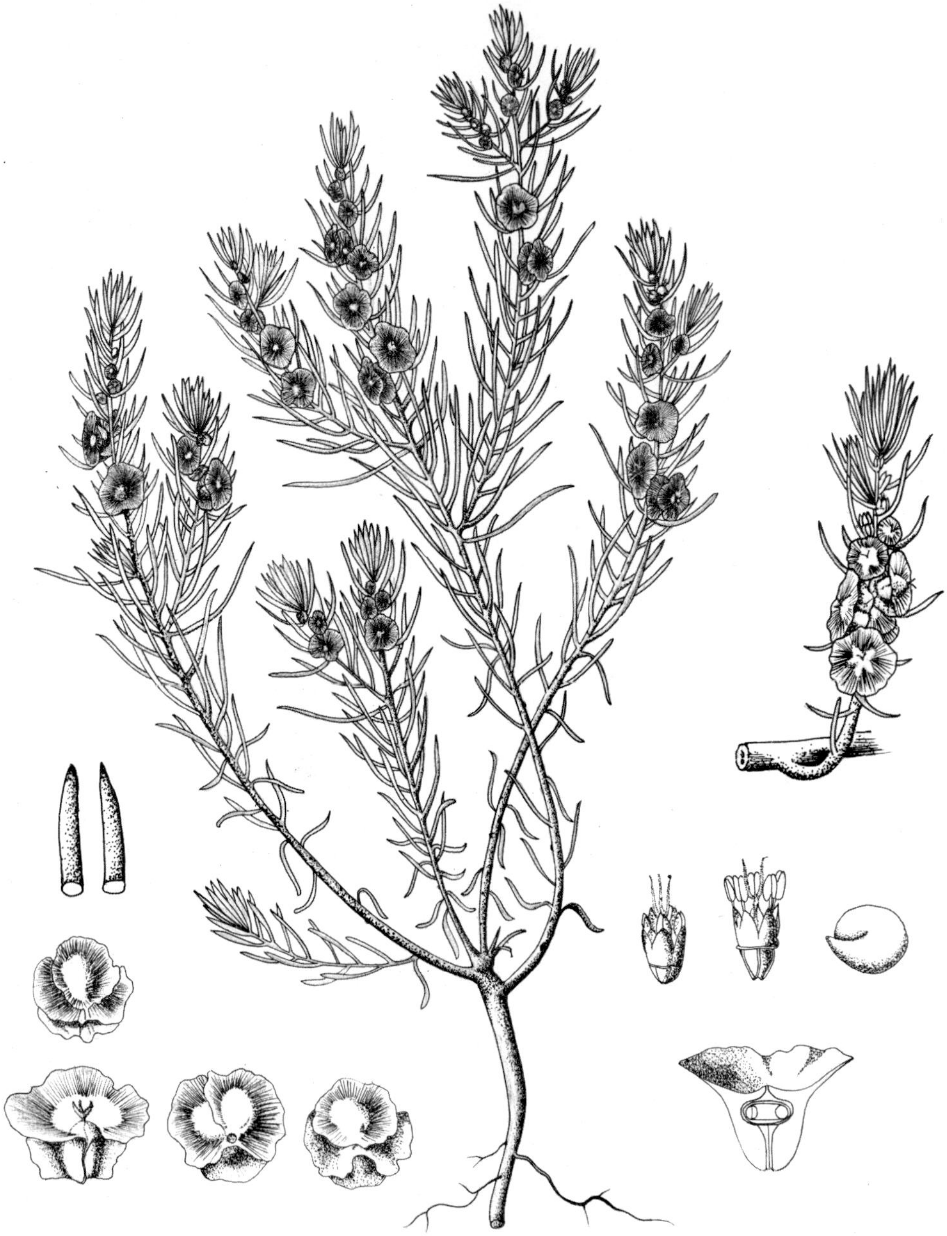

Whole plant showing stems, leaves, flowers and fruiting bodies; close-up view of fruiting bodies (*right*). Leaves (*lower left*); fruiting bodies (*front, side and back views—lower left*); flowers and seed (*upper right*); longitudinal section of fruiting body (*lower right*).

(*Photos and drawings courtesy Division of Plant Industry, C.S.I.R.O.*)

Plate 59 CHENOPOD BROWSE SHRUBS (*Kochia triptera*)

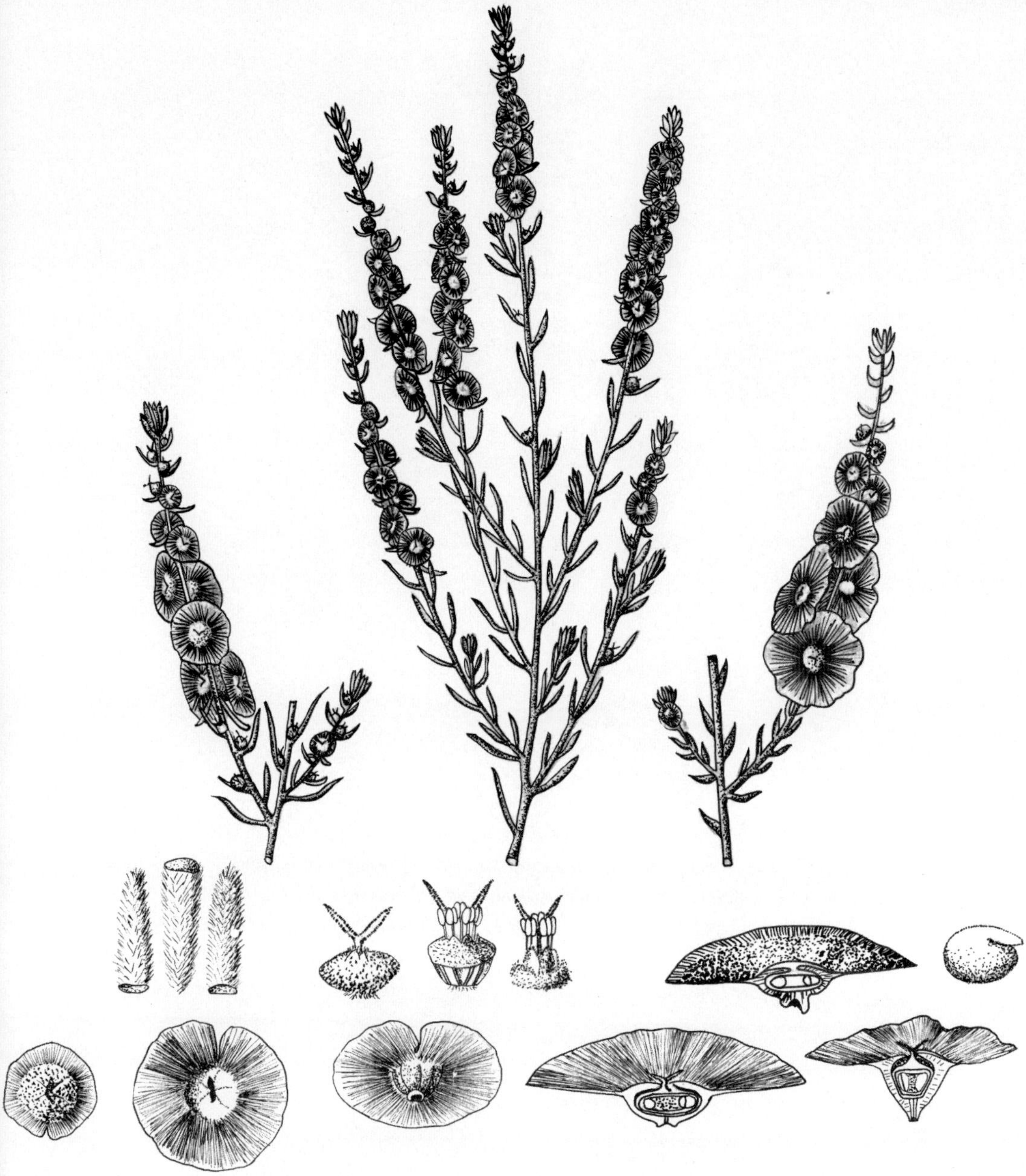

Branchlets showing leaves, flowers and fruiting bodies.

Centre row, leaves; flowers; longitudinal section of fruiting body; seed.

Lower row, fruiting bodies (*front and rear views and in longitudinal section*).

(*Photos and drawings courtesy Division of Plant Industry, C.S.I.R.O.*)

Plate 60 CHENOPOD BROWSE SHRUBS (*Kochia villosa*)

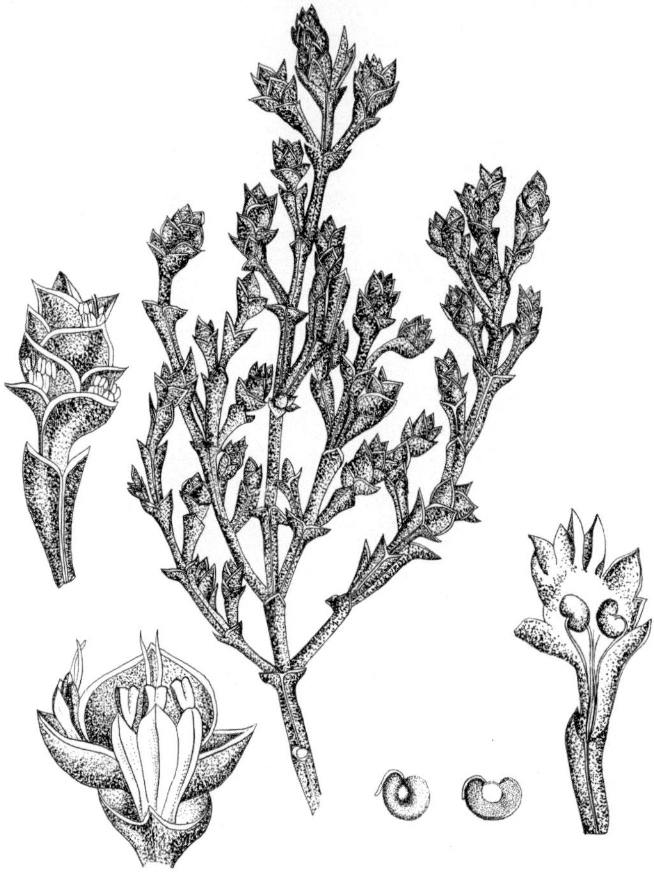

Top: Branches showing leaves and flowers. Flowers and seeds.
Bottom left: View of individual bush of *Pachycornia triandra*.
Bottom right: Close-up of branches of *P. triandra* showing leaves and flowers.

(Photos and drawings courtesy Division of Plant Industry, C.S.I.R.O.)

Plate 61 CHENOPOD BROWSE SHRUBS (*Pachycornia triandra*)

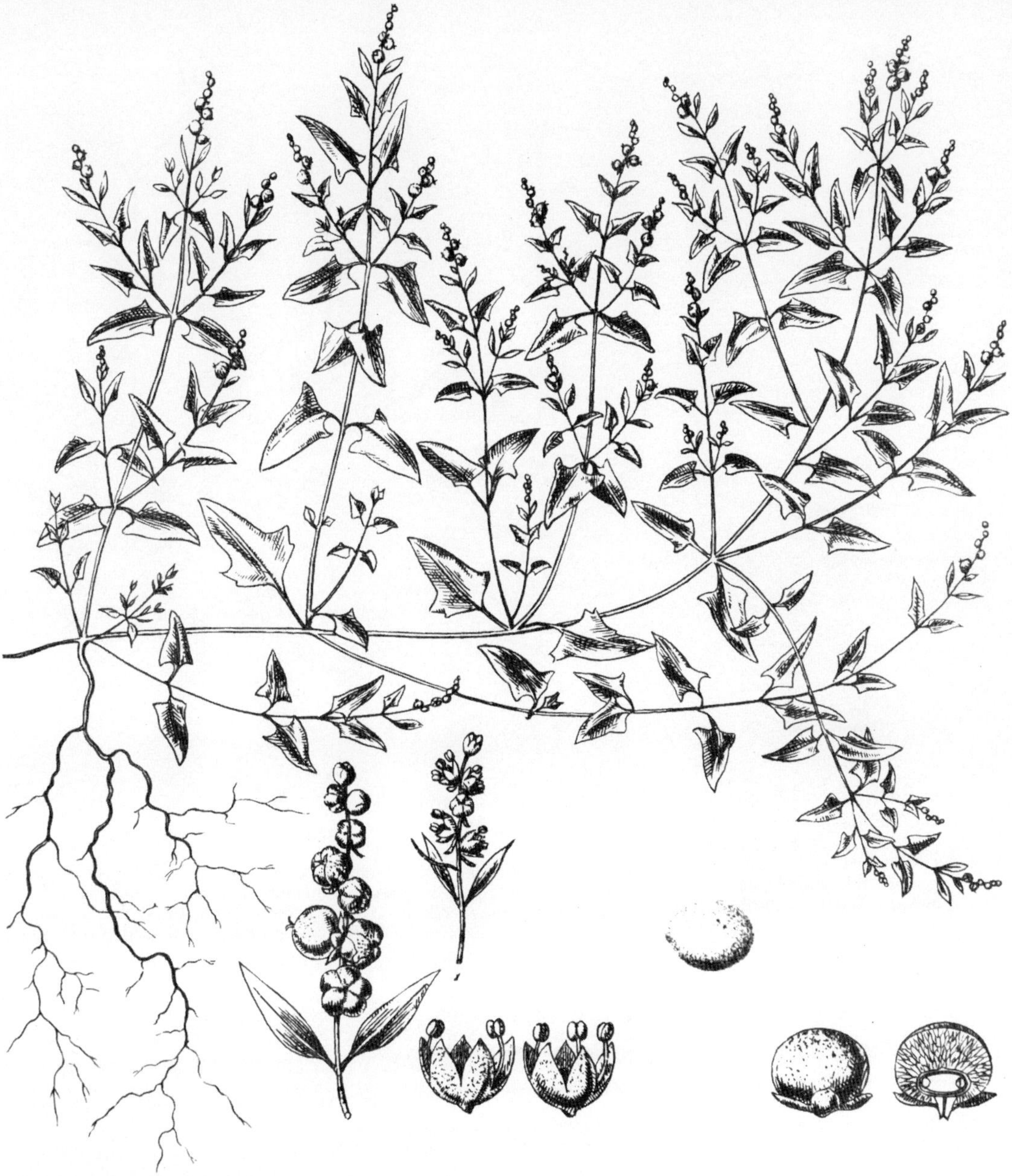

Top: Branches showing leaves, flowers and fruiting bodies. Branches with fruiting bodies and flowers; flowers, seed; fruiting body; longitudinal section of fruit.

Bottom left: Close-up of branch showing leaves and fruiting bodies.

(*Photos and drawings courtesy Division of Plant Industry, C.S.I.R.O.*)

Plate 62 CHENOPOD BROWSE SHRUBS (*Rhagodia nutans*)

Top: Branches showing leaves, spiny tips and fruiting bodies (*upper centre*); branchlets showing contrasting leaf shapes and fruiting bodies. Flowers (*left*); fruiting bodies (*centre left*); longitudinal section of fruiting body (*centre right*); seed (*right*).

Centre left: Close-up view of branchlets showing leaves, spines and fruiting bodies.

Bottom left: General view of *Acacia victoriae—R. spinescens* shrub-woodland community growing along drainage line near **Balranald, N.S.W.**

Bottom right: View of an individual bush.

(*Photos and drawings courtesy Division of Plant Industry, C.S.I.R.O.*)

Plate 63 CHENOPOD BROWSE SHRUBS (*Rhagodia spinescens*)

PALATABILITY

No single index of acceptability to livestock in general can be given for a particular species, for acceptability is dependent on a number of interacting factors. The more important of these are class of livestock, and level of availability and growth stage both of the species in question and that of other associated species.

In general it has been observed that most Chenopod shrubs are of moderate to low palatability and that stock prefer associated grasses and forbs when these are available.

In the Riverine Plain of south-western New South Wales it was found that *K. aphylla* and *A. vesicaria* were ungrazed or only lightly grazed (10 per cent of the diet or less) when grasses and forbs were present even though these grasses and forbs were often only minor constituents of the pastures. In times of feed scarcity and an absence of associated species it was found that *Atriplex vesicaria* made up 70 to 90 per cent of the diet eaten. *Kochia aphylla* seldom exceeded 10 per cent of the diet eaten.

Observations made of species growing around watering points in a range of communities in north-western South Australia (and in which species were marked 1 to 5 in order of decreasing palatability) confirm the general unattractiveness of many of the widespread Chenopod shrub species. Rankings given to some of the species which have been discussed previously were *Atriplex vesicaria* 2, *Salsola kali* 2, *Kochia planifolia* 2, *Kochia sedifolia* 3, *Kochia aphylla* 4, *Kochia georgei* 4, *Chenopodium nitrariaceum* 4, *Enchylaena tomentosa* 4, *Rhagodia spinescens* 4, *Kochia pyramidata* 5, *Kochia tripteria* 5, *Atriplex rhagodioides* 5, *Arthrocnemum* spp. 5. Such ratings can only serve as a general guide for even an unpalatable species like *Kochia pyramidata* will be heavily grazed under drought conditions when nothing else is available. In fact, under extreme conditions sheep will even eat dry sticks and twigs. Stage of growth may also have a marked effect on the degree and extent to which individual species may be grazed. For example, the new shoots of *Kochia aphylla* are readily eaten whereas old shoots are generally shunned except during drought conditions.

Whilst palatability comparisons as presented above may be generally valid, it is also true that in certain circumstances considerable differences in palatability within individual species may exist, even when the plants are growing on the same soil type. To date no information is available as to what factors or constituents are responsible for these differences.

REACTION TO GRAZING

The effects of grazing on the various plants differ between species but it is generally agreed that some degree of grazing is beneficial. The point at which defoliation by grazing has a detrimental effect on the plant is largely determined by the individual species concerned. The degree to which *A. vesicaria* is grazed has a direct bearing on its chances of survival. Moderate grazing leads to a more compact, leafier bush; complete defoliation will almost certainly result in its death. On the other hand *A. nummularia* and *Kochia* species appear able to withstand periodic total defoliation. None of the plants will survive extended terms of continual defoliation.

The stems of the two major *Atriplex* species, *A. vesicaria* and *A. nummularia* are particularly brittle, hence they may be severely damaged by trampling by livestock. Large stands of *A. nummularia* have been destroyed by the combined effects of trampling, especially by cattle, and grazing.

PRESENT CONDITION AND TREND

The extent of the deterioration of Australia's Chenopod shrublands cannot be accurately assessed because of the absence of permanent and precise records which date back to first settlement. Much of the deterioration occurred in the early phases of settlement, but even at the present time this deterioration is still continuing. Various estimates of the extent of deterioration have been made, the most recent being given at an Academy of Science Arid Lands Symposium. Present pastoral value was related to expected pristine condition. It was estimated that 25 per cent of Australian Chenopod shrublands were severely degenerated (< 40 per cent of original pristine conditions), 40 per cent were moderately deteriorated (40-60 per cent of original condition), 25 per cent showed minor degeneration (60-80 per cent of original condition) and only 10 per cent showed little or no deterioration (80-100 per cent of original condition).

The texture-contrast soils often associated with the shrublands have frequently been a significant feature in the deterioration of these lands. Once the lighter permeable surface has been lost through erosion the hard clay sub-soil is exposed. This sub-soil has an extremely low infiltration capacity and hence an increasingly arid root environment develops leading ultimately to a depletion of the shrubs.

In the semi-arid regions elimination of the original dominant Chenopod shrubs has led to the development of mixed grass-forb pastures which have been shown to be productive and stable under a wide range of stocking conditions. In lower rainfall regions removal of the shrubs has resulted in the invasion of the denuded areas by spiny, non-edible species of *Bassia* in some cases, and in other instances by unpalatable annuals, including some *Atriplex* species. In severe cases the layer of top-soil has been removed by the action of wind and large scalds have developed.

REGENERATION AND CULTIVATION

Many of the severely deteriorated Chenopod communities are unlikely to ever return to their original condition. Natural regeneration of most species is often exceedingly slow both with and without the exclusion of stock, and usually occurs only after extended wet periods.

In most instances reclamation requires some hindrance to run-off and various mechanical treatments designed to reduce run-off and impound water have been tested. Several of the species discussed earlier have been sown in the reclamation of eroded soils and in schemes aimed at increasing the productivity of existing pastures. *Atriplex vesicaria* has been used successfully in the revegetation of soft scalds, in conjunction with ploughing and checkerboard furrows. Attempts, however, to regenerate this species on stony desert soils on which it was once common have met with little success. This is due in part to the poor infiltration capacity of the soil, from most of which the top-soil has been removed. With the removal of the top-soil much of the soil nutrients were also lost, resulting in an environment inhospitable to the establishment of any vegetation.

Atriplex nummularia has been widely utilised as a means of increasing the productivity of existing pastures. In the southern riverine plain of New South Wales its introduction into a perennial grass pasture (*Danthonia caespitosa—Stipa variabilis*) was shown to be of limited benefit in terms of animal production. Nevertheless, as shade and shelter, especially from the wind, and as a producer of high-quality forage in areas lacking in these aspects, *A. nummularia* has much to commend it.

This species can be propagated readily by a number of methods. It grows readily from seed, which is relatively easy to collect from mature bushes, and it may be planted as seedlings grown in veneer or metal tubes. Cuttings also strike fairly readily under moist conditions. Whichever method is used, the soil in which the plants are to be placed should be cultivated to reduce competition from other species during the establishment phase.

Despite the potential increase in germination suggested by laboratory tests incorporating the leaching out of the salt contained in the bracteoles surrounding the seed, and the mechanical removal of the bracteoles, trials have shown that this potential is not always realised in the field.

In Western Australia, *Atriplex* species, *Kochia brevifolia* and *Salicornia* species have been successfully introduced into saline country. *K. brevifolia* has been found to be especially suitable for this purpose.

MINERAL CONTENT AND FORAGE VALUE

In comparison with non-Chenopods, these shrubs are high in nitrogen, sodium, potassium and chloride salts. Typical analyses showing the range in these constituents, determined over a number of localities, are presented in Table 15.1. Species of *Atriplex* accumulate ions to approximately the same concentration despite wide variations in the mineral composition of the soils on which they grow. High sodium levels are usually accompanied by high nitrogen levels. From an animal husbandry viewpoint these high mineral contents may have a number of deleterious effects. High sodium levels reduce the water use of Chenopods per unit of dry weight and increase the water needs of grazing animals. Salt intakes approaching

TABLE 15.1

PUBLISHED VALUES FOR SODIUM, POTASSIUM, CHLORIDE AND CRUDE PROTEIN CONTENT AND DRY MATTER DIGESTIBILITY OF SEVERAL CHENOPOD SHRUBS (EXPRESSED AS A PERCENTAGE OF DRY WEIGHT)

	Sodium	*Potassium*	*Chloride*	*Crude Protein*	*Digestibility*
Atriplex angulata	3.3	1.8	–	17.5	58
A. nummularia	3.8–8.2	1.14–3.9	7.8–13.6	17.0–21.9	68–74
A. semibaccata	3.5	4.2	–	10.0–19.9	–
A. vesicaria	3.2–6.5	1.2–7.2	3.9–14.3	11.1–18.4	52–54
Kochia brevifolia	–	–	–	24.8	–
K. pyramidata	6.6	1.5	3.4	15.1–22.0	58
K. sedifolia	–	–	–	17.2	–

200 g NaCl/sheep/day have been measured for sheep grazing *Atriplex vesicaria* in summer. Penned sheep fed on a diet of *A. vesicaria* leaf were shown to consume 11.3 litres of water/day, sheep grazing on a natural *A. vesicaria* community drank 7.5 litres/day, whereas similar sheep grazing on grassland consumed less than 3.2 litres/day. The increased water intake is necessary for the urinary excretion of the high intakes of salt. Substantial reductions in feed intake have also resulted when sheep consuming Chenopod leaf have had only saline water to drink. This situation, where sheep grazing saltbush pastures in summer have no feed other than saltbush leaf and have access only to highly saline bore water is by no means uncommon. It has been suggested that in such situations, because of the need to make frequent visits to water, that the grazing range of sheep is likely to be severely decreased resulting in over-exploitation of the vegetation growing nearest to the water source.

The range in crude protein content and dry matter digestibility of leaf samples of a number of Chenopod species analysed over a number of years are shown in Table 15.1. The crude protein values are high relative to other pasture grasses and legumes and are maintained throughout the year. The ability of these species to remain green under conditions of moisture stress together with their high crude protein contents has led numerous workers to suggest that these Chenopod shrubs could serve as valuable protein supplements to sheep and cattle when the associated grasses were dry and possibly deficient in protein. Weight losses of pen-fed sheep have been reduced when *A. nummularia* and *A. vesicaria* have been given as a supplement to a protein deficient diet, but similar supplements have not been shown to improve weight change or wool growth of animals grazing natural grassland communities. Obviously protein levels of the species available within the pasture have been adequate.

Field experiments designed to assess the value of *A. nummularia, A. vesicaria* and *Kochia aphylla* have been conducted near Deniliquin in south-western New South Wales. Comparisons were made of wool-growth rates and body-weight changes of flocks of sheep grazing established native perennial grass communities in which Chenopod shrubs were either dominant constituents or absent.

The presence of the three Chenopod shrub species was of little or no value in reducing seasonal fluctuations in animal production and it was concluded that in terms of wool production the introduction of these shrubs into a perennial grassland pasture was of doubtful economic value. The main benefit from the retention of the shrubs in areas in which they are already growing is to ensure an amount of feed available for use during dry periods and also for the possible reduction in wind erosion.

Recent pen-feeding work may provide an important clue to explaining why diets containing appreciable quantities of Chenopod forage fail to provide barely more than sufficient nutrients for the maintenance of energy and nitrogen equilibrium. This failure may be related to markedly different digestion processes observed between Chenopod rich diets and diets of the more conventional grasses and legumes. Studies have shown that with *A. nummularia*, relative to grasses and legumes, the stomach plays a less important role in the digestion of organic matter and fibre, ruminal absorption of fatty acids is impaired, and protein is extensively degraded to ammonia in the rumen. Accordingly the protein value of an *A. nummularia* diet may be much lower than is indicated by either its crude protein content or its digestible crude protein content.

CONCLUSIONS

From the review of existing published information and current work it is clear that there is an urgent need for further information on a number of lines. It is essential that permanent reference sites be located on existing stands of each of the major Chenopod shrubs so that trends in vegetation change and production can be accurately assessed over a range of habitats and management practices. Information is also needed on the influences of density of Chenopod shrubs on the species composition and production of associated ground flora species as well as on wind and water erosion. Lastly further critical work should be conducted on the digestion and utilisation by ruminants of the herbage of a range of Chenopod shrubs so that their true forage value can be assessed.

ACKNOWLEDGEMENTS

The author wishes to thank Mr W. E. Mulham (C.S.I.R.O., Riverina Laboratory, Deniliquin) for assistance with the preparation of this chapter.

Permission was kindly granted by the Director of the National Herbarium, Melbourne to reproduce in part the line drawings from Baron F. von Mueller's *Iconography of Australian Salsolaceous Plants* and one drawing (*Pachycornia triandra*) from A. J. Ewart's *Plants Indigenous to Victoria*. The Director of the National Herbarium, Sydney kindly allowed the reproduction of *Kochia planifolia* from F. Turner's *The Forage Plants of Australia*.

Photographs were kindly loaned by Messrs C. V. Malcolm (W.A. Department of Agriculture, Perth), W. E. Mulham, R. A. Perry (C.S.I.R.O., Rangelands Research Unit, Canberra) and C. J. Totterdill (C.S.I.R.O., Division of Plant Industry, Canberra). Lastly I would like to thank Miss S. C. Haldane and Mrs A. J. M. Simpson (both C.S.I.R.O., Division of Plant Industry, Canberra) for redrawing and reproducing the illustrative material presented.

FURTHER READING

Ashby, W. C. & Beadle, N. C. W. (1957). 'Studies in the halophytes. III: Salinity factors in the growth of Australian saltbushes'. *Ecol.*, 38 : 344-52.

Barker, S. & Lang, R. T. (1969). 'Effects of moderate sheep stocking on plant populations of a Black Oak—Bluebush association'. *Aust. J. Bot.*, 17 : 527-37.

Beadle, N. C. W. (1948). *The vegetation and pastures of western New South Wales with special reference to soil erosion.* Government Printer, Sydney.

Beadle, N. C. W. (1952). 'Studies in the halophytes. I: The germination of seed, and establishment of the seedlings of five species of *Atriplex* in Australia'. *Ecol.*, 33 : 49-62.

Beadle, N. C. W., Whalley, R. D. B. & Gibson, J. B. (1957). 'Studies in halophytes. II: Analytic data on the mineral constituents of three species of *Atriplex* and their accompanying soils in Australia'. *Ecol.*, 38 : 340-44.

Black, J. M. (1948-57). *Flora of South Australia*, Parts 1-4. Government Printer, Adelaide.

Black, R. F. (1954). 'The leaf anatomy of Australian members of the genus *Atriplex*. I: *Atriplex vesicaria* Heward, and *A. nummularia* Lindl'. *Aust. J. Bot.*, 2 : 269-86.

Black, R. F. (1958). 'Effect of sodium chloride on leaf succulence and area of *Atriplex hastata* L'. *Aust. J. Bot.*, 6 : 306-21.

Blake, S. T. (1937). 'The plant communities of western Queensland, and their relationships, with special reference to the grazing industry'. *Proc. Roy. Soc. Qld.*, 49 : 156-204.

Brownell, P. F. (1968). 'Sodium as an essential micronutrient element for some higher plants'. *Pl. Soil,* 28 : 161-4.

Burbidge, N. T. (1945). 'Germination studies of Australian Chenopodiaceae with special reference to the conditions necessary for regeneration. I: *Atriplex vesicarium* Heward'. *Trans. Roy. Soc. S. Aust.*, 69 : 73-84.

Burbidge, N. T. (1946). 'Germination studies of Australian Chenopodiaceae with special reference to the conditions necessary for regeneration. II: (a) *Kochia sedifolia* F.v.M.; (b) *K. pyramidata* Benth.; (c) *K. georgei* Drebs'. *Trans. Roy. Soc. S. Aust.*, 70 : 110-119.

Charley, J. L. & Cowling, S. W. (1968). 'Changes in soil nutrient status resulting from over-grazing and their consequences in plant communities of semi-arid areas'. *Proc. ecol. Soc. Aust.*, 3 : 28-38.

Condon, R. W. & Knowles, G. H. (1952). 'Saltbushes'. *J. Soil Conserv. Serv. N.S.W.*, 8 : 149-57.

Eichler, H. (1965). *Suppleme it to J. M. Black's Flora of South Australia.* Government Printer, Adelaide.

Ewart, A. J. (1910). *Plants Indigenous to Victoria.* Victorian Government Printer, Melbourne.

Gates, C. T. & Muirhead, W. (1967). 'Studies of the tolerance of *Atriplex* species. I: Environmental characteristics and plant response of *A. vesicaria, A. nummularia* and *A. semibaccata*'. *Aust. J. exp. Agric. Anim. Husb.*, 7 : 39-49.

Greenway, H., Gunn, A., & Thomas, D. A. (1966). 'Plant response to saline substrates. VIII: Regulation of ion concentrations in salt-sensitive and halophytic species'. *Aust. J. biol. Sci.*, 19 : 741-56.

Guthrie, F. B. (1897). 'The fodder-value of saltbush'. *Agr. Gaz. N.S.W.*, 8 : 574-8.

James, J. W. (1956). 'Scald reclamation experiments in the Bourke district'. *J. Soil Conserv. Serv. N.S.W.*, 12 : 44-54.

Jennings, D. H. (1968). 'Halophytes, succulence and sodium in plants—a unified theory'. *New Phytol.*, 67 : 899-911.

Jessup, R. W. (1949). 'The saltbushes'. *J. Agric. S. Aust.*, 53 : 157-61.

Jessup, R. W. (1951). 'The soils, geology and vegetation of north-western South Australia'. *Proc. Roy. Soc. S. Aust.*, 74 : 189-273.

Jones, R. M. (1966). 'Scald reclamation studies in the Hay district. Part I: Natural reclamation of scalds'. *J. Soil Conserv. Serv. N.S.W.*, 22 : 147-60.

Jones, R. M. (1966). 'Scald reclamation studies in the Hay district, N.S.W. Part II: Reclamation by ploughing'. *J. Soil Conserv. Serv. N.S.W.*, 22 : 213-30.

Jones, R. M. (1967). 'Scald reclamation studies in the Hay district, N.S.W. Part III: Reclamation by ponding banks'. *J. Soil Conserv. Serv. N.S.W.*, 23 : 3-17.

Jones, R. M. (1967). 'Investigations of seedling establishment on a reclaimed scald'. *J. Soil Conserv. Serv. N.S.W.*, 23 : 187-91.

Jones, R. M. (1968). 'Studies on the stimulation of germination and emergence of old man saltbush (*Atriplex nummularia* Lindl.)'. *J. Soil Conserv. Serv. N.S.W.*, 24 : 271-8.

Jones, R. M. (1969). 'Scald reclamation studies in the Hay district. Part IV: Scald soils—their properties and changes with reclamation'. *J. Soil Conserv. Serv. N.S.W.*, 25 : 104-20.

Jones, R. M., Leigh, J. H. & Mulham, W. E. (1968). 'Preliminary evaluation of bore water in relation to sheep tolerance in the central Riverine Plain'. *Fld Stn Rec. Div. Pl. Ind. C.S.I.R.O. (Aust.)*, 7(2) : 95-103.

Knowles, G. H. A. (1949). 'Soil erosion and soil conservation in the Hay district'. *J. Soil Conserv. Serv. N.S.W.*, 5 : 171-77.

Knowles, G. H. A. (1954). 'Scald reclamation in the Hay district'. *J. Soil Conserv. Serv. N.S.W.*, 10 : 149-56.

Knowles, G. H. & Condon, R. W. (1951). 'The perennial saltbush'. *J. Soil Conserv. Serv. N.S.W.*, 7 : 123-31.

Knowles, G. H. & Condon, R. W. (1951). 'Some western pasture species—bluebush, cottonbush, and other perennial chenopods'. *J. Soil Conserv. Serv. N.S.W.*, 7 : 189-98.

Leigh, J. H. & Mulham, W. E. (1964). 'The performance of introduced dryland species on three soil types in the southern Riverine plain'. *Fld Stn Rec. Div. Pl. Ind. CSIRO (Aust.)*, 3(2) : 9-20.

Leigh, J. H. & Mulham, W. E. (1965). *Pastoral plants of the Riverine plain.* Jacaranda Press, Melbourne.

Leigh, J. H. & Mulham, W. E. (1966). 'Selection of diet by sheep grazing semi-arid pastures on the Riverine plain. I: A bladder saltbush (*Atriplex vesicaria*)—cottonbush (*Kochia aphylla*) community'. *Aust. J. exp. Agric. Anim. Husb.*, 6 : 480-7.

Leigh, J. H. & Mulham, W. E. (1966). 'Selection of diet by sheep grazing semi-arid pastures on the Riverine plain. II: A cottonbush (*Kochia aphylla*)—grassland (*Stipa variabilis-Danthonia caespitosa*) community'. *Aust. J. exp. Agric. Anim. Husb.*, 6 : 468-74.

Leigh, J. H. & Mulham, W. E. (1967). 'Selection of diet by sheep grazing semi-arid pastures on the Riverine Plain. III: A bladder saltbush (*Atriplex vesicaria*)—pigface (*Disphyma australe*) community'. *Aust. J. exp. Agric. Anim. Husb.*, 7 : 421-25.

Leigh, J. H., Wilson, A. D. & Mulham, W. E. (1968). 'A study of merino sheep grazing a cottonbush (*Kochia aphylla*)—grassland (*Stipa variabilis-Danthonia caespitosa*) community on the Riverine Plain'. *Aust. J. agric. Res.*, 19 : 947-61.

Leigh, J. H. & Noble, J. C. (1969). 'Vegetation Resources'. In Slatyer, R. O. & Perry, R. A. (eds.). *Arid Lands of Australia*, pp. 73-92. A.N.U. Press, Canberra.

Leigh, J. H., Wilson, A. D. & Williams, O. B. (1970). 'An assessment of the value of three perennial chenopodiaceous shrubs for wool production of sheep grazing semi-arid pastures'. *Proc. XI International Grasslands Congress*, 55 : 9.

Macfarlane, W. V., Howard, B. & Siebert, B. D. (1967). 'Water metabolism of Merino and Border Leicester sheep grazing saltbush'. *Aust. J. agric. Res.*, 18 : 947-58.

Malcolm, C. V. (1961). 'Bluebush seed: Cleaning, drying and storing'. *J. Agric. West. Aust.*, 2 (Fourth Series) : 37-9.

Malcolm, C. V. (1963). 'Don't put sheep in yet'. *J. Agric. West. Aust.*, 4 (Fourth Series) : 549.

Malcolm, C. V. (1964). 'This is bluebush'. *J. Agric. West. Aust.*, 5 (Fourth Series) : 109.

Malcolm, C. V. (1964). 'Samphire: Well worth encouraging'. *J. Agric. West. Aust.*, 5 (Fourth Series) : 345.

Malcolm, C. V. (1969). 'Use of halophytes for forage production on saline wasteland'. *J. Aust. Inst. agric. Sci.*, 35 : 38-49.

Meadley, G. R. W. (1947). 'Western Australia'. In *The use and misuse of shrubs and trees as fodder.* (I.A.B. Joint Publ. No. 10, 29-39).

Moore, C. W. E. (1953). 'The vegetation of the south-eastern Riverina, New South Wales. I: The climax communities'. *Aust. J. Bot.*, 1 : 485-547.

Mueller, F. von (1889-91). *Iconography of Australian Salsolaceous Plants.* Government Printer, Melbourne.

Muirhead, W. A. & Jones, R. M. (1966). 'Species trials in the semi-arid south west of New South Wales. Part II: Native shrub trials in the Hay district'. *J. Soil Conserv. Serv. N.S.W.*, 22 : 138-46.

Newman, J. C. & Condon, R. W. (1969). 'Land Use and Present Condition'. In Slatyer, R. O. & Perry, R. A. (ed.). *Arid Lands of Australia*, pp. 105-129. A.N.U. Press, Canberra.

Osborn, T. G. B. & Wood, J. G. (1923). 'On some halophytic and non-halophytic plant communities in arid South Australia'. *Trans. Roy. Soc. S.A.*, 47 : 388-99.

Osborn, T. G. B. (1925). 'Ecology of vegetation of arid Australia. I: Introduction and general description of the Koonamore Reserve for the study of saltbush flora'. *Trans. R. Soc. S. Aust.*, 49 : 290-7.

Osborn, T. G. B., Wood, J. G. & Paltridge, T. B. (1932). 'On the growth and reaction to grazing of the perennial saltbush, *Atriplex vesicarium*. An ecological study of the biotic factor'. *Proc. Linn. Soc. N.S.W.*, 57 : 377-402.

Osmond, B. (1963). 'Oxalates and ionic equilibria in Australian saltbushes (*Atriplex*)'. *Nature* (Lond.), 198 : 503-4.

Osmond, C. B. (1966). 'Divalent absorption and interaction in *Atriplex*'. *Aust. J. biol. Sci.*, 19 : 37-48.

Osmond, C. B. (1967). 'Acid metabolism in *Atriplex*. I: Regulation of oxalate synthesis by the apparent excess cation absorption in leaf tissue'. *Aust. J. biol. Sci.*, 20 : 575-87.

Palmer, J. H., Trickett, E. S. & Linacre, E. T. (1964). 'Transpiration response of *Atriplex nummularia* Lindl. and upland cotton vegetation to soil-moisture stress'. *Agr. Meteorol.*, 1 : 282-93.

Perry, R. A. (1960). *Pasture lands of the Northern Territory, Australia.* (C.S.I.R.O. Land Research Series No. 5, pp. 1-55).

Perry, R. A. (1962). 'Natural pastures of the Alice Springs area'. In *Lands of the Alice Springs area, Northern Territory, 1956-57.* (C.S.I.R.O. Land Research Series No. 6, 240-57).

Perry, R. A. (1970). 'Arid Shrublands and Grasslands'. In Moore, R. M. (ed.). *Australian Grasslands.* A.N.U. Press, Canberra.

Richardson, A. E. V., Trumble, H. E. & Shapter, R. E. (1931). *Factors affecting the mineral content of pastures.* (Coun. Sci. Industr. Res. Aust. Bull. No. 49).

Rowan, J. N. & Downes, R. A. (1963). *A study of the land in north-western Victoria.* (Soil Cons. Auth. Vic. T.C.2).

Russell, F. C. (1947). 'The chemical composition and digestibility of fodder shrubs and trees'. In *The use and misuse of shrubs and trees as fodder.* (I.A.B. Joint Publ. No. 10, 186-225).

Trumble, H. C. & Woodroffe, K. (1954). 'The influence of climatic factors on the reaction of desert shrubs to grazing by sheep'. In Cloudsley-Thomas, J. L. (ed.). *Biology of deserts.* Institute of Biology, London.

Turner, F. (1891). *The Forage Plants of Australia.* N.S.W. Government Printer, Sydney.

Valder, G. (1896). 'Trials of saltbush at Wagga Wagga'. *Agr. Gaz. N.S.W.*, 7 : 609-11.

Weston, R. H., Hogan, J. P. & Hemsley, J. A. (1970). 'Some aspects of the digestion of *Atriplex nummularia*'. *Proc. Aust. Soc. Anim. Prod.*, 8 : (in press).

Williams, O. B. (1962). 'The Riverina and its Pastoral Industry'. In Barnard, A. (ed.). *The Simple Fleece*, pp. 411-434. Melbourne University Press in association with the Australian National University.

Williams, O. B. (1964). 'A relationship between forage availability and wool growth-rate for high and low producing sheep'. *Proc. Aust. Soc. Anim. Prod.*, 5 : 280-3.

Williams, O. B. (1963). 'The effect of leaching and drying on the viability of old man saltbush seeds'. *CSIRO Div. Pl. Ind. Fld Stn Rec.*, 2 (No. 1) : 55-8.

Williams, O. B. (1968). 'Pasture management in the pastoral zone—a review'. *Wool Technol. Sheep Breed.*, 15 : 45-8.

Wilson, A. D. (1965). 'Adult performance of sheep reared on native and irrigated pastures at Deniliquin'. *Fld Stn Rec. Div. Pl. Ind. CSIRO (Aust.).*, 4 : 91-4.

Wilson, A. D. (1966). 'The value of *Atriplex* (saltbush) and *Kochia* (bluebush) species as food for sheep'. *Aust. J. agric. Sci.*, 17 : 147-53.

Wilson, A. D. (1966). 'The intake and excretion of sodium by sheep fed on species of *Atriplex* (saltbush) and *Kochia* (bluebush)'. *Aust. J. agric. Res.*, 17 : 155-63.

Wilson, A. D. (1966). 'The tolerance of sheep to sodium chloride in food and drinking water'. *Aust. J. agric. Res.*, 17 : 503-14.

Wilson, A. D., Leigh, J. H. & Mulham, W. E. (1969). 'A study of sheep grazing a bladder saltbush (*Atriplex vesicaria*)—cottonbush (*Kochia aphylla*) community on the Riverine Plain'. *Aust. J. agric. Res.*, 20 : 1123-36.

Wilson, A. D. & Leigh, J. H. (1969). 'Use of water intakes in the estimation of saltbush intake by grazing sheep'. *Fld Stn Rec. Div. Pl. Ind. CSIRO (Aust.)*, 8 : 1-3.

Wood, J. G. (1925). 'The selective absorption of chloride ions and the absorption of water by the leaves in the genus *Atriplex*'. *Aust. J. exp. Biol. med. Sci.*, 2 : 48-56.

Wood, J. G. (1936). 'Regeneration of the vegetation on the Koonamore Vegetation Reserve, 1926-1936'. *Trans. Roy. Soc. S. Aust.*, 60 : 96-111.

Wood, J. G., Woodroffe, K. & Trumble, H. C. (1947). 'South Australia'. In *The use and misuse of shrubs and trees as fodder.* (I.A.B. Joint Publ. No. 10, 25-9).

Woodroffe, K. (1941). 'Shrub pastures under low rainfall. An experiment on grazing management'. *J. Aust. inst. Agric. Sci.*, 7 : 117-21.

Zimmer, W. J. (1946). 'The flora of the far north-west of Victoria'. *Forest Comm. Vic. Bull. No. 2 (2nd Edn).*

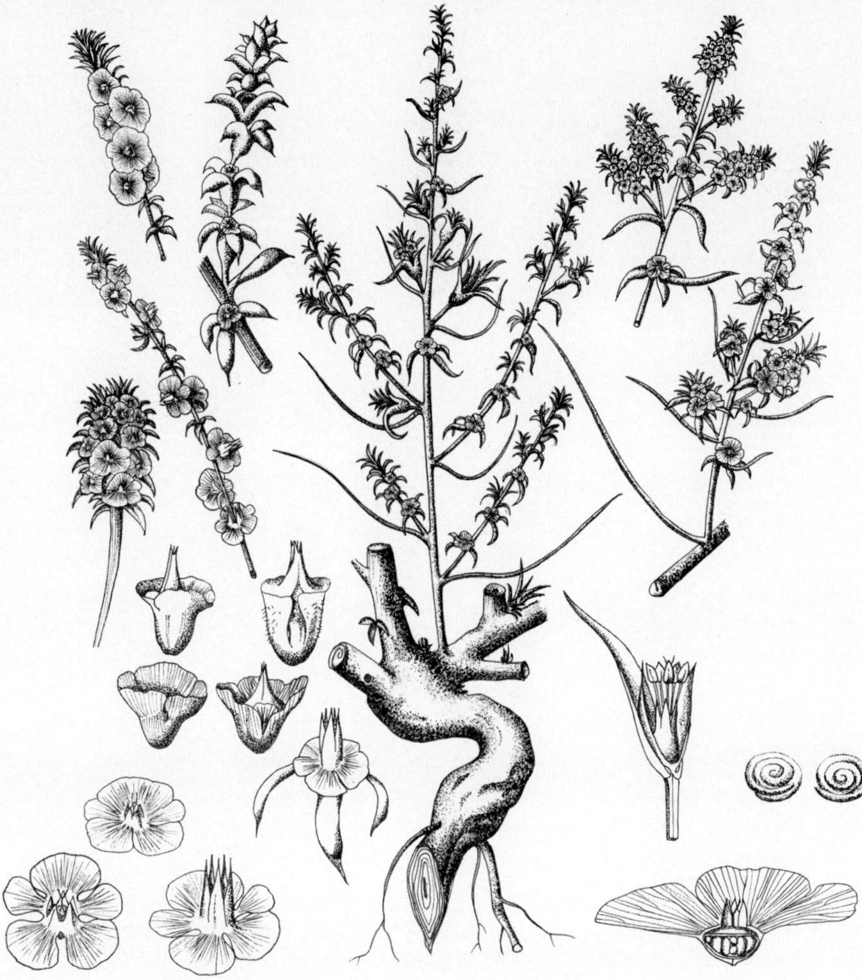

Top: Branches showing leaves, flowers and fruiting bodies. Fruiting bodies (*left*), longitudinal section of flower and fruiting body (*right*) and seed.

Bottom left and right: Close-up of branchlet showing leaves and fruiting bodies.

(*Photos and drawings courtesy Division of Plant Industry, C.S.I.R.O.*)

Plate 64 CHENOPOD BROWSE SHRUBS (*Salsola kali*)

CHAPTER 16

Plants harmful to animals

TREES and shrubs are of considerable importance as fodder plants in the dry country. Throughout the world it is probable that more animals feed on these plants or on associations in which they play an important part than on true grass or grass-legume pastures. Areas in which trees and shrubs are the principal if not the only source of fodder, include large areas of semi-desert shrublands and bushlands throughout Africa, the maquis of Mediterranean and adjacent eastern countries, the chaparral of California and the forest ranges of the southern States of the United States of America, caatinga and monte in South America as well as large areas in India and much of the arid and semi-arid country of Australia. In many parts of the arid zone of Australia the pastoral industry is largely dependent upon this class of fodder, whilst it is still of importance in the drier parts of the semi-arid zone.

For animal feed it is convenient to divide Australian trees and shrubs into two categories. The first is that of browse plants which are characteristic of the arid zone and in many parts constitute the principal or sole source of fodder. They are mainly low shrubs upon which even sheep can graze readily, and are typically represented by saltbush (*Atriplex* spp.) and bluebush (*Kochia* spp.); both genera belong to the *Chenopodiaceae*. The second category is that of drought reserve fodder, which are trees and shrubs only fed to stock at times when other food is very scarce. Due to the height of these plants it is necessary to lop or fell them to give the animals access to the leaves, or other edible parts such as the pods of certain species of *Acacia*.

Everist (1969) summarises some aspects of fodder trees and shrubs:

> Trees have two characteristics which make them particularly useful as drought fodder. Firstly, they are able to draw on moisture and minerals deep in the soil out of reach of grasses; secondly, the leaves of most trees retain their nutritive value even when they are mature.
>
> For those who have them, edible trees on a property can make the difference between survival and disaster for stock during drought. It is important to be able to recognise the trees which are edible, know how to use them and learn how to preserve them so that they will be available during future droughts.

The management of fodder trees and shrubs presents a number of problems, some of which are easy to overcome whilst others present greater difficulties. Amongst the easier problems are recognition of the edible and, equally important, of the poisonous species and how to lop the forage species so that the plants survive for future use. More difficult aspects of management, especially when costs per acre must be kept low, are those of regeneration and change in species composition. In this book Chapter 15 deals with browse plants, and Chapter 17 provides notes on a large number of species. There are many State publications on the subject and some of these are listed at the end of this and other chapters.

A useful definition of a toxic plant is given by Gardner and Bennetts (1956):

> For general purposes a toxic plant may be defined as one which detrimentally affects the health of man or animals when eaten in such amounts as would be taken normally or under special circumstances, as for example, restriction of diet or extreme hunger.

The term harmful plants not only includes toxic or poisonous vegetation but also that which causes deleterious effects on the alimentary system by mechanical action, for example thorns, very harsh fibres, etc., or indirectly by introducing enzymes which promote the release of toxic substances from glucosides, for example prussic acid, or other harmful agents. This chapter is only an introduction to the subject and details of poisonous plants, botanical descriptions and veterinary treatment where possible, should be obtained from the books listed under further reading or in the many leaflets and other publications issued by the Departments of Agriculture in the States.

HISTORICAL

From the time of the early explorers there have been reports of the toxic effects of native plants on European man. There is no doubt, of course, that the Aboriginals of the country, who relied heavily on wild plants for their food, had a sound and comprehensive knowledge of the effects of many plants not only on human life but also on animals and fish. They had discovered for themselves the efficacy of rotenone-bearing *Derris* and *Tephrosia* as fish poisons, besides plants of the genera *Barringtonia* and *Faradaya* which contain substances toxic to fish. In addition, they used a wide range of plant products to relieve headache, toothache, fever, dysentery and other afflictions. Comparatively recent research into the constituents of many of these plants has shown them to contain appreciable quantities of useful alkaloids. The Aborigines were fully aware of the powerful narcotic properties of *Duboisia*.

Webb (1948) notes that in August 1770, Captain Cook recorded the first investigation by Europeans of the poisonous action of an Australian plant:

> We found here the palm of three different sorts . . . The third sort, which, like the second, was found only in the northern parts, was seldom more than ten feet high, with small pinnated leaves, resembling those of some kind of fern; it bore no cabbage, but a plentiful crop of nuts, about the size of a large chestnut, but rounder; we found the hulls of these scattered round the places where the Indians had made their fires, we took for granted that they were fit to eat; those however who made the experiment paid dear for their knowledge of the contrary, for they operated both as emetic and cathartic with great violence. Still, however, we made no doubt but that they were eaten by the Indians, and judging that the constitution of the hogs might be as strong as theirs, though our own had proved to be much inferior, we carried them to the stye; the hogs eat them, indeed, and for some time we thought without suffering any inconvenience, but in about a week they were so much disordered that two of them died, and the rest were recovered with great difficulty.

Webb observes that the plant to which Captain Cook referred was probably *Cycas media* R.Br., the kernels of which were eaten by the Aborigines only after maceration, washing and baking.

From the early days of settlement the pastoral industry has had the experience of stock dying suddenly, often without any obvious cause other than the suspicion, sometimes justified by later investigations, that the mortality was caused by eating a specific plant. This type of occurrence was not as common in southern coastal areas as when stock were moved inland and to more northern regions. Unfortunately, in those days a part of the native flora was still undescribed and there were no floras readily available. This, combined with the often somewhat vague descriptions of suspected plants, made it difficult to connect stock mortality to a specific plant. By the latter half of the nineteenth century botanical knowledge had increased and there were more Australian authorities, so that the latter years of the century and the first four to five decades of the twentieth century saw a large volume of literature on the subject. At the present time many of the new accounts appear in the journals of the Departments of Agriculture of the States. Since World War II there has been considerable work on alkaloids and other drugs, especially from rainforest species, because of their possible use in medicine.

HARMFUL AGENTS

Harmful agents fall into two groups; firstly those of a chemical nature and include alkaloids, some of the glucosides, especially those which are responsible for prussic acid poisoning and mineral poisons such as oxalic acid and selenium and, secondly, those which have an effect by mechanical means. Included in the first group are those enzymes which act on the cyanogenetic glucosides and cause the release of prussic acid, and pigments which cause photosensitivity. After eating certain plants stock may become abnormally sensitive to light, so that sunlight causes more or less severe inflammation of any exposed skin. This condition appears to be associated with liver damage, whereby the liver is not able to deal with the pigments which cause photosensitisation. Trefoil dermatitis has occurred in sheep and cattle in Western Australia when they have been grazed on pastures containing burr clover (*Medicago denticulatus*), whilst facial eczema is a serious and well-known disease of sheep in New Zealand. Examples of mechanical agents are harsh fibres which cause injury to the alimentary canal and may result in enteritis or other complaint, thorns and other hard, sharp-pointed parts which the digestive juices of animals cannot deal with before injury is done. The effect which a particular poison or other harmful agent has on stock varies with a number of factors and some of these are:

1. *Plant factors*

   (a) *Species and variation of toxicity within a species.*

   Some species nearly always appear to contain a dangerous level of harmful substances and have a serious effect on stock, even if eaten in only small quantities. On the other hand, there are plants which fairly consistently only have small amounts of poison and are only dangerous if eaten in large quantities at one time. The toxicity of some plants appear to vary with the environment; this applies particularly to those which have a wide geographical distribution. In the case of *Duboisia leichhardtii* from the humid zone it has been found that the alkaloid hyoscine appeared dominant in the north part of the range but is replaced by hyoscyamine in the southern part.

(b) *Parts of the plant.*

The toxicity of different parts of a plant may vary considerably; for instance the leaves may be harmless whilst the fruit or seed are very poisonous.

(c) *Stage of development.*

There is sometimes a changing level of toxicity between unripe and mature fruit or between fast-grown young leaves and mature foliage. In some of the herbs, *Oxylobium* and *Gastrolobium* for example (Gardner and Bennetts, 1956) the young leaves are invariably the most poisonous part of the plants.

(d) *Extent and frequency of occurrence.*

There are less chances of fatalities amongst stock from species which only occur sparsely compared with those which may be common.

(e) *Palatability.*

It is obvious that if a plant is attractive to animals there is a greater chance of large intakes in a short time, than if it is non-palatable.

2. *Animals and animal management factors*

(a) *Species.*

With many plants there is a considerable variation in the sensitivity shown by different types of animals. Horses, for instance, appear to be more susceptible to some substances which have little effect on sheep or cattle; whilst pigs appear to be very susceptible to some plants, but this may be partly associated with their partiality for fruits rather than leaves.

The larger animals, of course, have access to greater volumes of plant growth than sheep or other smaller creatures.

(b) *State of general health.*

Like human beings, stock in poor health, especially if liver disease is present, are more susceptible to ill effects from harmful substances than healthy animals.

(c) *Drought.*

Not only does prolonged drought reduce the general health of animals but it often induces them to eat plants which they would not normally consume.

(d) *Travelling of stock.*

When animals have been travelling for long periods without access to feed they are more likely to eat heavily of any species which may be accessible. This can mean a large intake of a plant which, when eaten in small amounts at a time, would not normally have ill effects. Except in the case of cumulative poisons the body is able to excrete them if there is only a small intake at a time. Another aspect is whether animals are allowed to rest after eating. Gardner and Bennetts (1956) state that sheep, for example, if allowed to rest after eating some poisonous plants may show few ill effects but if disturbed or frightened by dogs may show heavy mortality.

(e) *Timing of access to water.*

It has been recorded that the effects of plant poisoning may become more evident when animals have been watered soon after feeding.

NOTES ON SOME TREES AND SHRUBS SPECIES

The species listed below are described and, in many cases, illustrated in Chapter 17.

Acacia

Many of the species of this important genus are rightly considered important as drought fodder reserves and, whilst some have fallen under suspicion of being harmful to stock, the foliage of only a limited number have been proved to contain poisons, for example cyanogens. The bark and fruit of several acacias contain harmful substances and wounds made by the green wood can prove very dangerous. Hurst (1942) refers to the ill effects felt by Leichhardt's party when near the Roper River in the Northern Territory, following drinking a decoction made from *Acacia* seeds. Another party in Central Australia suffered serious effects from water in which 'silver wattle' sticks had been placed.

The most important acacia for fodder is mulga (*A. aneura*) and animals can exist on the foliage for long periods, especially if the diet is supplemented with small quantities of bone flour or meat and bone meal, plus salt and molasses (Everist, 1969). In the case of some species, for example *A. estrophiolata* and *A. excelsa*, impaction may result if stock feed only on these species for periods as short as a few weeks. *A. georginae* has not as good a reputation as some acacias and was considered to contain poisons, but it now appears (Hurst, 1942) that the fruit contains an enzyme which is capable of releasing prussic acid from certain substances, for example a glucoside found in *Eremophila maculata*. *A. salicina* has been suspected of poisoning hungry cattle and this may be associated with large intakes of the foliage which contains tannins and the pods which have saponins.

Atalaya hemiglauca

The effect of this species on stock, especially horses, has attracted considerable attention. White (1927) writes:

> It is rather a strange fact that only in central and northern Queensland is whitewood accused of causing staggers in stock. The tree is abundant throughout western Queensland and New South Wales, but one never hears of the southern trees as the cause of any trouble though it is freely eaten everywhere by horse, sheep and cattle.

The leaves contain a high percentage of saponin, more in material from north-west Australia than from the south-east, and continued intake causes serious damage to the liver, but other plants besides *Atalaya hemiglauca* have probably contributed to the 'walkabout' symptoms shown by horses in the Kimberley region of northern Western Australia.

Brachychiton populneum

Although the kurrajong is highly regarded as a reserve fodder tree, there have been reports of ill effects and, in rare cases, of death when sheep have been fed on the seeds.

Casuarina

The foliage of the sheoaks is sometimes eaten by cattle, although it is not an attractive form of fodder as it is very harsh and fibrous. Impaction is reported to occur if animals eat a large amount.

Codonocarpus cotinifolius

In central Australia it has been suspected of poisoning cattle. The small tree is, however, nowhere very plentiful and it is probable that if it is only eaten as a small proportion of the daily intake, there are unlikely to be any harmful effects.

Eremophila

Many of the species are readily eaten by both sheep and cattle and are well regarded as a drought fodder in spite of the fact that there have been numerous reports of deaths attributed to them, notably *E. maculata*. Chemical analysis of the leaves and fruit may not reveal any prussic acid but there is evidence that an enzyme may act on glucosides to release the acid. The enzyme may originate from some part of the eremophila plant or from another plant which has been eaten at the same time. Greater details on this genus are available in Everist (in the press) and Hurst (1942). Most of the stock fatalities have occurred in animals which have been travelling for a long time without feeding or during drought conditions; in general, the species seem rarely to have harmful effects on stock under normal grazing conditions.

Eucalyptus

As most eucalypts are trees or tall shrubs from which foliage is not accessible unless the plants are lopped or felled, animals do not commonly have access to the leaves except in some mallee areas. In addition, the leaves are usually not attractive to animals because of their content of essential oil. Some species are lightly grazed, however, during times of drought or when stock have been driven for long periods without opportunity to feed. Everist (1969), Hurst (1942) and others state that the following species may be eaten: *E. cladocalyx, E. papuana, E. pilligaensis, E. populnea, E. ochrophloia* and *E. thozetiana.* Hurst (1942) reports that *E. cladocalyx* has been responsible for deaths of cattle, horses and sheep following eating lopped or fallen foliage. On testing the leaves gave a positive result for prussic acid.

Melia azedarach var. *australasica*

The fruits are poisonous to pigs and there is some evidence that they affect poultry, so that the trees should not be planted in or close to pig pens or poultry yards. Everist (1969) states that the leaves are sometimes fed to dairy cattle and slightly taint the milk.

Nerium oleander (exotic)

The leaves of both this and the yellow oleander (*Thevetia neriifolia*) exude a poisonous milky juice.

Robinia pseudoacacia (exotic)

The leaves are stated to be toxic if eaten in large quantities and the Queensland Department of Primary Industries does not recommend it for further planting (Everist, 1969).

FURTHER READING

Chippendale, G. M. (1963). *Topfeed.* (Extension article 5). Animal Industry Branch, Northern Territory Administration.

Everist, S. L. (1969). *Use of Fodder Trees and Shrubs.* (Advisory Leaflet 1024). Division of Plant Industry, Department of Primary Industries.

Everist, S. L. (1969). Some thoughts on drought. A paper presented at the Seminar on Drought, University of New England, May 1969.

Everist, S. L. (1972). *The Poison Plants of Australia.*

Gardner, C. A. & Bennetts, H. W. (1956). *The Toxic Plants of Western Australia.* West Australian Newspapers Ltd, Perth.

Herbert, D. A. (1926). *The Poison Plants of Western Australia.* (Bull. 96). Department of Agriculture.

Hurst, E. (1942). *The Poison Plants of New South Wales.* Poison Plants Committee of New South Wales.

Imperial Agricultural Bureaux (1947). *The Use and Misuse of Shrubs and Trees as Fodder.* (Joint Publication No. 10).

Webb, L. J. (1948). *Guide to the Medicinal and Poisonous Plants of Queensland.* (Bull. 232). Council for Scientific and Industrial Research, Melbourne.

Webb, L. J. (1969). 'Australian Plants and Chemical Research'. In *The Last of Lands.* Jacaranda Press.

White, C. T. See the very long list given in Hurst above, where there are about 125 references to his writings, mainly in the *Queensland Agricultural Journal* or the *Annual Reports of the Government Botanist*, Queensland.

All State Departments of Agric., in their journals and leaflets.

In addition to the above attention is drawn to the very extensive references and bibliographies in some of the above books, especially Everist (in the press), Gardner and Bennetts, and Hurst.

CHAPTER 17

Notes on some of the more important trees and shrubs

THIS chapter provides notes on 178 of the more important tree and shrub species of value for conservation or planting in the dry country. The choice of species has necessarily been arbitrary, but an attempt was made to include both the better-known trees and shrubs and some which are less well known but appear to have an important role in the arid and semi-arid zones. Many important genera which are strongly represented in the Australian flora presented serious problems in the limitation of species to be described. The genus *Acacia* (wattles) has hundreds of species in the dry country, whilst *Casuarina, Grevillea* and *Hakea*, to mention but a few, are also strongly represented. No book of this type could adequately cover the acacias and it was only by the severest limitation of numbers that it was possible to keep the list down to 35 species. With other important genera, for example *Cassia* and *Eremophila*, it was only possible to describe some of their characteristics and mention a few individual species. Detailed information is not readily available on many of the plants and that given in the notes has been taken from the references listed or from personal observation.

Australians who live in or are accessible to the dry country are urged not to be content with the species described here but to investigate for themselves other trees and shrubs which grow well in their localities. This is not only important for pastoralists but a rich source of species for ornamental planting.

The few exotics dealt with in the text have been restricted to those which have a proven performance in the dry country, for example the pepper tree (*Schinus molle*) and the athel tree (*Tamarix articulata*). In the higher rainfall parts of the southern semi-arid zone there are also a few conifers which deserve mention.

NOMENCLATURE

In writing about plants it is desirable that the latest botanical nomenclature should be used, but it is not essential. What is important is that there should not be any doubt as to what species is being described. There are changes in botanical names from time to time as knowledge increases and experts revise genera and, sometimes, families. If a botanical name and author are correctly cited any subsequent name change does not vitiate the information given, as it will be known to what species the new name applies. Where there has been a name change in the past few decades it has been the practice to also give the name which applied previously, whether it was a true synonym or a name which had been, botanically speaking, incorrectly but widely used for the species concerned. The principal references used for botanical names have been:

(a) Eucalypts—

Johnston, R. D., & Marryatt, R. (1965), *Taxonomy and nomenclature of eucalypts*, (Forestry and Timber Bureau Leaflet 92), Department of National Development, Canberra, as amended by later advice from State herbaria.

(b) Other Australian genera—

the floras of the states, especially those which have been recently published, for example—

Beard, J. S. (no date but during mid 1960s), *Descriptive Catalogue of West Australian Plants,* Society for Growing Australian Plants (issued by the King's Park Board, Perth).

Eichler, H. (1965), *Supplement to J. M. Black's Flora of South Australia,* (Second edition, 1943-1957), Government Printer, Adelaide.

Standards Association of Australia (1970). *Nomenclature of Australian Timbers,* (A.S. 02-1970), S.A.A., Sydney.

(c) Exotic trees not included in (b) above—

Streets, R. J. (1962), *Exotic Forest Trees in the British Commonwealth,* Clarendon Press, Oxford.

Which common name to use has presented problems with many species, especially for those plants which occur throughout the dry country. The policy followed is that where a species is restricted to one State, the local name has been used, but where it is found in several or all States the most generally-used name has been adopted. Considerable use has been made of Australian Standard 02-1970 *Nomenclature of Australian Timbers*, which provides 753 standard trade common names and lists over 1,000 species of trees and large shrubs. In most, but not all, cases the common name of the timber also applies to the living plant. In a few cases where there did not appear to be an acceptable established common name the botanical name alone has been used. For exotic trees reference has been made to *Exotic Trees in the British Commonwealth* (1962), Australian Standard 0118-1968 *Nomenclature of Commercial Timbers imported into Australia* or the various handbooks of horticulture, and overseas manuals of cultivated plants. The general practice in the text is that for the first mention of a species in any chapter both the botanical name and the standard common name are given, and an example would be *Acacia aneura* (mulga). Subsequent naming is by either name, according to the context.

The danger of using only a common name, without any further qualifying description, is well exemplified in the case of the name 'blue gum' when applied to a eucalypt. To Australians in general and most people overseas, 'blue gum' would bring to mind *Eucalyptus globulus* Labill., for which the standard trade common name is 'southern blue gum'. The name 'blue gum', however, if used in South Australia may refer to *E. leucoxylon* F. Muell. (yellow gum); in Victoria to *E. bicostata* Maiden *et al.* (eurabbie) or to *E. globulus* Labill.; in northern coastal New

South Wales to *E. saligna* Sm. (Sydney blue gum) and in Queensland to *E. tereticornis* Sm. (forest red gum). In view of the confusion which can arise by the use of common names alone, it is not surprising that forest services, in issuing catalogues and lists of trees and shrubs available from their nurseries or making recommendations for planting, adhere to the use of botanical names as the primary means of reference.

A complete botanical name consists of three parts, for example *Eucalyptus globulus* Labill. The first is the generic name, in this case *Eucalyptus*; the genus being a group of plants with a number of common characteristics which separate them from other genera in the same family (*Myrtaceae*). The second part, *globulus*, is the specific epithet which is given to a smaller group which, whilst having the general characteristics of the genus, have certain features which separate them from other species. The third part of the name, often omitted in general references, is that of the author who named the species. This may be given in full but is often shortened, for example Labill. is an abbreviation of J. J. H. de Labillardiere who first named the botanical material of *E. globulus*. It has happened that two different names have been published for the same plant, but under International Rules of Botanical Nomenclature the correct name is the one first validly published. A wrong name is sometimes in use for a long time before this is realised and a name change becomes necessary. An example of the confusion which can arise was exemplified when the name *E. camaldulensis* Dehn. (red river gum but also known as Murray red gum) took priority over *E. rostrata* Schlecht. This was a case where strict adherence to International Rules changed a name which had been sanctioned by custom for over a century.

The use of a botanical name, without the addition of the name of the author, is not always sufficiently precise to identify a species, since there are many examples where two authors have given the same specific name to different trees. One example is that of *E. viminalis* Hook. (1846); in this case the first name is valid and is the correct nomenclature of manna gum and the second was invalid and has been replaced by *E. tessellaris* F. Muell. (1859), which is the carbeen or Moreton Bay ash of north-eastern Australia. A relatively minor complication is the general adoption amongst botanists of Recommendation 73C of the International Code of Botanical Nomenclature concerning the manner in which specific epithets should be formed from personal names, but this has usually only involved minor changes in spelling. For example, what was *E. andrewsi* is now spelt *E. andrewsii* and *E. muelleriana* is now spelt *E. muellerana.*

Below the level of species there can be divisions into sub-species and varieties, which can be distinguished on the basis of features such as growth habit, colour of the flowers or, in the case of species, the colour of the autumn foliage. Varieties and forms can be of considerable importance in the selection of trees for shelter belts, as well as in commercial timber production.

The rank of *variety* is used for naturally occurring forms and the term *cultivar* for forms arising in cultivation and is of particular importance in connection with ornamental trees and shrubs.

A taxonomic term which may be given after a botanical name, especially a synonym, is *auctt.* and indicates that the name so qualified, whilst not the correct botanical nomenclature, is in common use by botanical writers. An example of this is in *Ulmus procera* Salisb. (*U. campestris* auctt.), English elm, where the first name is now recognised as the correct nomenclature but that the second name (*U. campestris*) is still in common use. When the term is qualified, as in *Melia azedarach* L. var. *australasica* (A. Juss.) C.DC., syn. *M. dubia* auctt. austral., white cedar, it means that the name *M. dubia* is still in common use by Australian botanical writers.

IDENTIFICATION

The correct identification of trees and shrubs is of great importance and often presents problems to the farmer and forester and, at times, even to the specialist botanist. Taking eucalyptus as an example, information which may be necessary in order to make a correct determination includes plant form, area of collection, bark type, juvenile and mature leaves and, particularly important, buds and fruits (capsules). In forwarding material for identification to a forest service or state botanist, mature leaves, buds and fruits should be included together with a brief note as to place of collection, nature of the bark and size of the tree.

Literature covering the description of the trees and shrubs in Australia, keys for their identification and standard lists of names forms a small library in itself. Some of the more important publications in this field are listed below.

AUSTRALIA AND GENERAL

Blakely, W. F. (1965). *A Key to the Eucalypts.* Forestry and Timber Bureau, Canberra.

Food and Agriculture Organisation of the United Nations (1960). *Choice of tree species.* (Forestry Development Paper 13). F.A.O., Rome.

Forestry and Timber Bureau (1970). *Forest Trees of Australia.* 3rd edition. Government Printer, Canberra.

Forestry and Timber Bureau (1960). *Notes on the commercial forest regions of Australia.* Aust. For. Timb. Bur. Leafl. 61 (2nd edition) and map. For. Timb. Bur., Canberra.

Hall, N., Johnston, R. D., & Marryatt, R. (1963). *The natural occurrence of the eucalypts.* Aust. For. Timb. Bur. Leafl. 65 (2nd edition). For Timb. Bur., Canberra.

Johnston, R. D. & Marryatt, R. (1964). *Taxonomy and nomenclature of eucalypts.* Aust. For. Timb. Bur. Leafl. 92. For. Timb. Bur., Canberra.

Standards Association of Australia (1970). *Nomenclature of Australian timbers.* (A.S. 02-1970), Standards Association of Australia, Sydney.

Streets, R. J. (1962). *Exotic Forest Trees in the British Commonwealth.* Clarendon Press, Oxford.

The list of trees and shrubs available from State forest nurseries, issued annually (most of these contain valuable notes of the more important species, together with recommendations on their suitability for various purposes).

AUSTRALIAN CAPITAL TERRITORY

Burbidge, N. T., & Gray, M. (1965). *The plants of the Australian Capital Territory.* (Division of Plant Industry Tech. Paper 21). C.S.I.R.O., Melbourne.

Pryor, L. D. (1962). *Trees in Canberra.* Department of the Interior, Canberra.

NEW SOUTH WALES

Anderson, R. H. (1968). *The Trees of New South Wales.* 4th ed. N.S.W. Government Printer, Sydney.

Beadle, N. C. W., Evans, O. D., & Carolin, R. D. (1962). *Handbook of the Vascular Plants of the Sydney District and Blue Mountains.* Authors, Armidale, N.S.W.

Stannard, M. E. & Condon, R. W. (1968). 'Fodder trees of western New South Wales'. *Journal of the Soil Conservation Service of N.S.W.*, 24(2).

QUEENSLAND AND NEW SOUTH WALES

Everist, S. L. (1969). *Use of fodder trees and shrubs.* (Advisory Leaflet No. 1024). Division of Plant Industry, Queensland.

Francis, W. D. (1951). *Australian Rain-forest Trees.* For. Timb. Bur., Canberra.

VICTORIA

Ewart, A. J. (1930). *Flora of Victoria.* Government Printer, Melbourne.

Lord, E. E. (1964). *Shrubs and Trees for Australian Gardens.* Griffin Press, Adelaide.

Willis, J. H. (1962). *A Handbook of Plants in Victoria.* Vol. 1. University Press, Melbourne.

WESTERN AUSTRALIA

Beard, J. S. (no date but during mid 1960s). *Descriptive Catalogue West Australian Plants.* Society for Growing Australian Plants (issued by the King's Park Board, Perth).

Brockway, G. E., *et al.* Reprints of the *Journal of Agriculture of Western Australia,* issued as bulletins of the Department of Agriculture, W.A. (The authors are officers of the Forests Department and the series covers many aspects of tree planting for farmers in that State.)

Gardner, C. A. (1952-63). *Trees of Western Australia.* A series of Department of Agriculture leaflets and bulletins reprinted from the *Journal of Agriculture of Western Australia,* Government Printer, Perth.

SOUTH AUSTRALIA

Black, J. M. (1943-57). *Flora of South Australia* (in four parts) (2nd ed.) Government Printer, Adelaide.

Boomsma, C. D. (1966). *Tree planting guide for rural South Australia.* Woods and For. Dep. Bull. 13 (2nd ed.). Adelaide.

Eichler, H. (1965). *Supplement to J. M. Black's Flora of South Australia.* Government Printer, Adelaide.

TASMANIA

Curtis, W. M. (1956-67). *The Students Flora of Tasmania.* Government Printer, Hobart.

NORTHERN TERRITORY

Chippendale, G. M. & Jephcott, B. R. (1963). *Topfeed.* (Extension Article No. 5). Animal Industry Branch, Northern Territory Administration.

ILLUSTRATIONS

Illustrations of most of the species described in this chapter are given in Plates 65-83. Most of the photographs have been taken by the author, but thanks are tendered to the many organisations and individuals who co-operated in the project. Brief notes on the plates indicate whether, in the case of native species, the illustration is of a tree growing naturally or whether it is a planted specimen; also it is stated whether the plant is immature or mature. The latter has also been shown for exotic trees and shrubs.

Every effort has been made to ensure that the illustrations are true to name. In the case of some photos, notably those from the Division of Land Research, C.S.I.R.O., they are of plants from which botanical material was collected and identified. Other photographs are of labelled specimens in forest service experimental plantings, whilst others were named by State officers who accompanied the author on some of his travels. In still other cases the author collected botanical material from trees and shrubs which he photographed and arranged for identification at State herbaria.

NOTES ON SPECIES

The notes include reference to:

1. The natural occurrence of the species and, in the case of exotics, the parts of Australia where they have been established satisfactorily or can be expected to grow as specimen trees; the size to which they grow under favourable conditions.

2. The climatic conditions under which they grow naturally or as a result of successful introduction. When growing conditions are not optimum, either because of temperature, rainfall or soil, poorer growth, especially in height, can be expected. Planting in the open will produce bushier trees of lower height growth than trees grown in a plantation or woodlot. Different species show a wide variation in their tolerance to adverse environmental conditions. For example, trees in narrow shelterbelts and, especially, ornamental specimens on watered lawns will withstand much lower rainfall than trees closely spaced in plantations. A combination of two or more adverse factors, such as heavy soil and low rainfall, is much more serious than when only one is unsuitable. In general, it is easier to acclimatise a species in a warmer climate than its natural habitat than the reverse, but there may be a greater liability to disease, especially when the warmer climate has a wet summer and the natural habitat has a dry summer. When trying to establish trees from a warmer climate in a cooler one, frosts during early life may be a critical factor. Frost hardiness is all-important when planting large areas, but in the case of specimen trees around homesteads protection from already established hardy species, or some form of covering in winter time, may allow trees which are frost-tender in early life to pass safely through the stage at which they are liable to be seriously damaged.

3. The principal wood properties in the case of trees which produce useful timber.

4. The more important uses of the trees on farms and for planting in rural towns.

Space precludes these notes from being comprehensive accounts of individual species, and anyone proposing to manage natural stands or to plant the species referred to can, in most cases, secure more detailed information in publications from the forest and conservation services of their State. More detailed information on the natural occurrence of many of the eucalypts is given under headings of size, forest type, geographical range, climate, topography and general items in *The Natural Occurrence of the Eucalypts*, Leaflet No. 65, 2nd edition, 1963. Other information and photos of tree, bark, leaves, buds and fruits of 118 eucalypts and 15 other native trees are given in *Forest Trees of Australia*, 3rd edition, 1970.

EXPLANATION OF TABLE 17.1

This table presents a list of the species described in Chapter 17 and a summary of their more important characteristics and uses. In such a table it is obviously impossible to give an infinite range of categories and qualify each classification and, hence, amplification should be sought from the notes for the individual species. The classification used for columns 10-25 is:

1 = indicates that the species is very suitable for the purpose shown.

2 = indicates the species can be used for the purpose shown, but may be subject to limitations which are given in the notes or, for nectar and pollen, in Tables 14.1-14.5.

– = reported as unsuitable or of only limited value for the purpose shown; usually mentioned in the notes.

(P) = the species is illustrated with a photograph.

The absence of a symbol in any column does not necessarily mean that the tree or its product cannot be used for the purpose listed.

Entries for columns 10-25 have been based mainly on recommendations of State authorities such as Departments of Agriculture, Conservation and Forestry or, in the case of timber uses, the appropriate Commonwealth or State authority.

Botanical and common names (col. 1)

Because a botanical name is not an adequate description of a species without the appropriate author citation, this has been given here as well as in the notes on the species. Most trees and shrubs have several common names and these vary not only from State to State but also within a State; the most generally acceptable name has been used in col. 1. In a large number of cases it has been possible to use the standard trade common name given in Australian Standard AS 02-1970, *Nomenclature of Australian Timbers*.

CLIMATIC (cols 2-7)

Summary of natural distribution for the main occurrence (col. 2)

Taken in conjunction with the maps of Australia and States (endpapers and maps in Appendix I) this column provides the briefest possible summary of the natural occurrence of Australian species. In the case of exotic trees and shrubs it indicates the principal climatic area where, allowing for cols 5-7, the species can be expected to grow reasonably satisfactorily. Abbreviations used are:

A = arid climatic zone
S-A = semi-arid climatic zone
S-H = sub-humid climatic zone
H = humid climatic zone
N, E, S and W = north, east, south and west respectively
WA, SA, V, NSW, Q and NT = Western Australia, South Australia, Victoria, New South Wales, Queensland and the Northern Territory respectively

Meteorological reference stations (cols 3-4)

Selected stations located within the principal area of occurrence are listed as 'representative', whilst those which are less typical are shown as 'others'. As far as possible meteorological stations used are those which are shown on maps of States on pp. 526-531, and for which meteorological data is given in Appendix I. These two columns have not been entered for exotic species.

Rainfall (cols 5-6)

For Australian species the values entered are those representative of the principal natural occurrence of the species and sometimes different limits have been recommended in the text when it is proposed to plant the tree concerned. Further information on rainfall is given in Chapter 3 and Appendix I. The abbreviations for season of greatest rainfall are: W = winter; U = uniform; S = summer.

The values given are general indications of rainfall requirements and allowance must be made for temperature, soil type and depth, and topography. There are some species, for example river red gum (*Eucalyptus camaldulensis*), which will grow in very arid areas if planted on the low-lying ground of drainage channels where additional moisture is available from deep sub-soil storage.

Frosts (col. 7)

The heading of the column is self-explanatory. Temperatures of 32°F in the screen represents 'white' frosts, and grass temperatures are usually several degrees lower, some times as much as ten degrees.

THE LIVING TREE (cols 8-16)

Range of mature height (col. 8)

This is the range of height which a species is expected to attain for the climate and other conditions described in the notes, and for uses recommended.

Tolerance to salt in the soil (col. 9)

Where a species is known to be salt tolerant it has been classified. Most other trees and shrubs can be assumed to be non-tolerant, even though the '—' sign has not been used.

The more important uses of the living tree (cols 10-14). *Shelterbelts* (col. 10) and *low shelter* (col. 11)

The basis of classification adopted here is the suitability of mature trees for single or few-rowed belts. For a species to secure a '1' rating it must retain branches to near ground level and is, hence, suitable for use in single-row belts without an associated lower-storey species. The '2' rating signifies that, in general, the species would require the assistance of a lower-storey shrub to provide satisfactory protection to near ground level. It will be appreciated that, largely irrespective of form, most trees and shrubs in wide belts, that is 5 chains or greater, provide at least a moderate level of protection. The height distinction between a shelterbelt and low shelter is approximately 25-30 feet. To qualify for this latter use a species must have a bushy habit from ground level and provide satisfactory protection without recourse to pruning or trimming.

Shade (col. 12)

Any tree can provide some shade but for a species to be classified '1' it must develop at least a moderately broad crown with suitable foliage. This type of tree provides satisfactory shade when the sun is high in the sky, whereas a narrow, pyramidal-shaped species gives the maximum shade when the sun is low in the sky. Both types of trees should be used in any complete farm plan.

Ornament (col. 13)

This category includes trees and shrubs which are suitable for public and private amenity planting. It is possible that a species which merits a '1' classification for the arid zone should not receive the same rating for use in the moister parts of the semi-arid zone.

Street planting (col. 14)

A species is only given a rating in this column if it is of small or medium size, that is mainly in the height range of 15-20 feet, does not have a very inflammable bark and is tolerant of conditions in paved streets and footpaths. Attention is drawn to various limiting factors discussed in Chapter 18.

Drought fodder (col. 15)

This is a general heading to indicate that the foliage or other parts of the plant may be eaten by stock. Some plants are grazed regularly and constitute an appreciable part of the food intake even in normal times, for example mulga (*Acacia aneura*), whilst others are only eaten under stress of severe droughts. Chapters 15, 16 and the notes on species in 17 discuss this subject.

Acacia acuminata (raspberry jam)

A. aneura (mulga)

A. aulococarpa (brown salwood)*

A. beckleri (Barrier Range wattle)

A. acuminata—growing naturally near Toodyay, about 50 miles north-east of Perth, W.A.; rainfall about 21 inches.

A. aneura—a large specimen, growing naturally in the arid zone of N.S.W. near the border of S.A.; rainfall about 8 inches.

A. aulococarpa—growing naturally in northern Australia near the boundary of the N.T. and W.A.; rainfall about 20 inches.

A. beckleri—planted at Woomera, S.A. but receiving some regular watering.

A. argyrodendron—growing naturally near Natal Downs Station about 70 miles south of Charters Towers, Q'ld.; rainfall about 24 inches.

A. bidwillii—growing naturally in Q'ld.

A. cambagei—growing naturally near Richmond, about 70 miles west of Hughenden, Q'ld.; rainfall 18 inches.

A. coriacea—growing naturally near Lucy Creek Station, about 180 miles north-east of Alice Springs, N.T.; rainfall about 9 inches.

(* *These photos by M. Lazarides, Division of Land Research, C.S.I.R.O.*)

A. argyrodendron (black gidyea)*

A. bidwillii (corkwood wattle)*

A. cambagei (gidgee)*

A. coriacea (wirewood)*

Plate 65 AUSTRALIAN BROADLEAVED SPECIES—I *Acacia acuminata* to *A. coriacea*

Acacia cunninghamii
(Cunningham's wattle)*

A. estropholiolata
(ironwood)*

A. georginae
(Georgina gidgee)*

A. cyanophylla
(orange wattle)

A. cunninghamii—growing naturally about 30 miles south of Emerald, Q'ld.; rainfall about 24 inches.

A. estropholiolata—growing naturally near Alcoota Station about 70 miles north-east by north of Alice Springs, N.T.; rainfall about 11 inches.

A. georginae—growing naturally near Mt. Riddoch Station about 70 miles north-east of Alice Springs, N.T.; rainfall about 11 inches.

A. cyanophylla—growing naturally near Gin Gin, about 50 miles north of Perth, W.A.; rainfall about 29 inches.

A. homalophylla—growing naturally west of Dubbo, N.S.W.; rainfall about 20 inches.

A. iteaphylla—planted at Woomera; receives some regular watering.

(A colour photo of *A. harpophylla* (brigalow) is given in Plate 1.)

(* *These photos by M. Lazarides, Division of Land Research, C.S.I.R.O.*)

A. homalophylla
(yarran)

A. iteaphylla
(Flinders Range wattle)

Plate 66 AUSTRALIAN BROADLEAVED SPECIES—II *Acacia cunninghamii* to *A. iteaphylla*

A. kempeana
(witchetty bush)*

A. pendula
(myall)

A. ligulata
(umbrella bush)

A. microbotrya
(manna wattle)

A. kempeana—growing naturally near Murray Downs Station, about 180 miles north of Alice Springs, N.T.; rainfall about 11 inches.

A. pendula—growing naturally near Gilgandra, 40 miles north of Dubbo, N.S.W.; rainfall about 22 inches.

A. ligulata—growing naturally near Woomera, S.A.; rainfall about 7½ inches.

A. microbotrya—growing naturally near Toodyay, about 50 miles north-east of Perth, W.A.; rainfall about 21 inches.

A. linophylla—planted in a Forests Department plot, Wongan Hills, about 100 miles north-east of Perth, W.A.; rainfall about 15 inches.

A. notabilis—growing naturally near Mount Olga, about 220 miles south-west by west of Alice Springs, N.T.; rainfall about 9 inches.

A. oswaldii—growing naturally near Mantuan Downs Station, about 90 miles south-west of Emerald, Q'ld.; rainfall about 25 inches.

(* *These photos by M. Lazarides, Division of Land Research, C.S.I.R.O.*)

A. linophylla
(horse mulga)

A. notabilis
(notabilis wattle)*

A. oswaldii
(umbrella bush)*

Plate 67 AUSTRALIAN BROADLEAVED SPECIES—III *Acacia kempeana* to *A. pendula*

Acacia salicina (cooba)*

A. sowdenii (western myall)

A. tetragonophylla (dead finish)

A. victoriae (prickly wattle)

A. peuce
(waddywood)†

A. spectabilis
(Pilliga wattle)

A. shirleyi
(lancewood)*

A. salacina—growing naturally on Laglan Station about 75 miles north-west of Clermont, Q'ld.; rainfall about 24 inches.

A. sowdenii—planted in a Forests Department seed orchard, Kalgoorlie, W.A.

A. tetragonophylla—growing naturally in western N.S.W. near the border with S.A.; rainfall about 8 inches.

A. victoriae—growing naturally in western N.S.W. near the border with S.A.; rainfall about 8 inches.

A. peuce—growing naturally in south-eastern N.T.; annual rainfall about 7 inches.

A. spectabilis—growing naturally near Chinchilla, southern Q'ld.; annual rainfall about 29 inches.

A. shirleyi—68 miles south-east of Montejennii Station, which is about 110 miles south-west by west of Daly Waters, N.T.; rainfall about 20 inches.

(* *These photos by M. Lazarides, Division of Land Research, C.S.I.R.O.*
† *This photo by Professor L. D. Pryor, Australian National University.*)

Plate 68 AUSTRALIAN BROADLEAVED SPECIES—IV *Acacia peuce* to *A. victoriae*

Agonis flexuosa
(Western Australian peppermint)

Angophora melanoxylon
(coolibah apple)

Atriplex nummularia
(oldman saltbush)

Bauhinia carronii
(Queensland ebony)

Brachychiton gregorii
(desert kurrajong)

Brachychiton rupestre
(Queensland bottle tree)

Agonis flexuosa—planted in arboretum at Woomera, S.A.; watered.
Angophora melanoxylon—growing naturally near St. George, southern Q'ld.; rainfall about 18 inches.
Atriplex nummularia—planted in rocky area, Broken Hill, N.S.W.; not watered, annual rainfall 9 inches.
Bauhinia carronii—planted at Dalby, southern Q'ld.; annual rainfall 25 inches.
Brachychiton gregorii—growing naturally in Kambalda town, where it was retained during road construction.
Brachychiton rupestre—growing naturally about 3 miles south of Mitchell, southern Q'ld.; rainfall 21 inches.
Apophyllum anomalum—growing naturally in N.S.W.
Atalaya hemiglauca—growing naturally in N.S.W.
Brachychiton populneum—growing naturally south of Forbes, southern N.S.W.; rainfall about 19 inches.
(* *These photos by the Soil Conservation Service of N.S.W.*)

Apophyllum anomalum
(broom bush)*

Atalaya hemiglauca
(whitewood)*

Brachychiton populneum
(kurrajong)

Plate 69 AUSTRALIAN BROADLEAVED SPECIES—V *Agonis flexuosa* to *Brachychiton rupestre*

Capparis mitchellii
(wild orange)

Carissa ovata
(carissa)*

Capparis mitchellii—growing naturally near Warren, N.S.W.; rainfall about 20 inches.

Carissa ovata—growing naturally near Natal Downs Station, about 70 miles south of Charters Towers, Q'ld.; rainfall about 24 inches.

Cassia artemisioides—growing naturally on Undoolya Station near Alice Springs, N.T.; rainfall 10 inches.

Cassia nemophila var *coriacea*—growing naturally about 50 miles south of Broken Hill, N.S.W.; rainfall about 9 inches.

Casuarina campestris—growing naturally near Wongan Hills, about 100 miles north-east of Perth, W.A.; rainfall about 15 inches.

Casuarina cristata—growing naturally in arid western N.S.W., where the rainfall is near the minimum for the species, that is about 10 inches.

(These photos by M. Lazarides, Division of Land Research, C.S.I.R.O.)*

Cassia artemisioides
(silver cassia)*

Cassia nemophila var *coriacea*
(cassia)

Casuarina campestris
(sheoak)

Casuarina cristata
(belah)

Plate 70 AUSTRALIAN BROADLEAVED SPECIES—VI *Capparis mitchellii* to *Casuarina cristata*

Casuarina dielsiana
(northern sheoak)

Dodonea attenuata
(hopbush)

Eremocitrus glauca
(limebush)*

Eremophila longifolia
(berrigan)

Eremophila maculata
(a native fuschia)

Casuarina decaisneana
(desert oak)*

Casuarina huegeliana
(rock sheoak)

Codonocarpus sp.
probably *C. cotinifolius*
(native poplar)

Eremophila latrobei
(a native fuschia)*

Casuarina dielsiana—young planted tree, Kalgoorlie.
Dodonea attenuata—growing naturally, Broken Hill.
Eremocitrus glauca—growing naturally, 40 miles east of Emerald, Q'ld.; rainfall 24 inches.
Eremophila longifolia—growing naturally, Broken Hill.
Eremophila maculata—growing naturally, Broken Hill.
Casuarina decaisneana—growing naturally near Alice Springs, N.T.; rainfall 10 inches.
Casuarina huegeliana—growing naturally about 100 miles north-east of Perth, W.A.; rainfall 15 inches.
Codonocarpus sp.—a few miles east of Charleville, southern inland Q'ld.; rainfall 18 inches.
Eremophila latrobei—north of Alice Springs.
(* *These photos by M. Lazarides, Division of Land Research, C.S.I.R.O.*)

Plate 71 AUSTRALIAN BROADLEAVED SPECIES—VII *Casuarina decaisneana* to *Eremophila latrobei*

Eucalyptus alba
(white gum)

E. argophloia
(Queensland western white gum)

E. astringens
(brown mallet)

E. behriana
(bull mallee)

E. brockwayi
(Dundas mahogany)

E. camaldulensis
(river red gum)

E. alba—growing naturally in the Humpty Doo area of the N.T.; rainfall about 55 inches.
E. argophloia—growing naturally near Burra Burri State School, north of Chinchilla, southern sub-humid Q'ld.; rainfall about 28 inches.
E. astringens—young planted trees south of West Wyalong, N.S.W.; rainfall about 19 inches.
E. behriana—growing naturally near West Wyalong, N.S.W.; rainfall about 19 inches.
E. brockwayi—young planted tree in Forest Department seed orchard, Kalgoorlie, W.A.
E. camaldulensis—young planted trees with well developed drooping branchlets, in a plot south of Merriden, W.A.; rainfall about 12 inches.
E. campaspe—planted tree at Woomera, S.A.; watered.
E. cladocalyx var. *nana*—young planted tree in Forest Department plot near Wongan Hills, about 100 miles north-east of Perth, W.A.; rainfall 15 inches.

E. campaspe
(silver-topped gimlet)

E. cladocalyx var *nana*
(dwarf sugar gum)

Plate 72 AUSTRALIAN BROADLEAVED SPECIES—VIII *Eucalyptus alba* to *E. cladocalyx* var *nana*

Poisonous to animals (col. 16)

Some plants are always harmful to animals and, if this is known, they are classified as '1'; other plants, however, which are usually edible may be toxic if eaten at certain times of the year or consumed in excessive quantities at any one time and these are classified as '2'.

Timber (cols 17-21)

These categories are intended only as general indications of suitability for the various uses shown. To be classified under sawlogs (col. 17) the trunks should normally attain at least one foot in diameter and lengths over 10 feet, whilst the wood should be suitable for use as boards or framing timber. In the case of poles (col. 18) form and natural durability are requisites for a species to be listed. There are a number of trees which produce boles of suitable form but which require preservative treatment; these are not listed. Natural durability of the heartwood is also necessary for being classified as suitable for fencing material (col. 19). Where information states that a species is suitable for turnery or other special uses it has been shown in column 20, but this does not preclude the use of many species which are not so listed. Nearly all wattles (*Acacia* spp.) and most eucalypts (*Eucalyptus* spp.) are first class firewoods, but where a species has been reported as unsuitable it has been shown '—'.

OTHER USES (cols 22-25)

Honey (cols 22-23)

Flowers are of interest to apiarists for two principal purposes: (a) nectar for honey production; and (b) pollen as a food for the workers. The subject is dealt with in Chapter 14, whilst columns 22 and 23 are a summary of information in Tables 14.1-14.5, and additional notes in the descriptions of the species.

Edible parts (col. 24)

In addition to fruits in the horticultural usage of the word, some trees have edible seed. The best Australian example is bunya pine (*Araucaria bidwillii*), whilst the exotic stone pine (*Pinus pinea*) is a source of edible seed in its native area.

Other uses (col. 25)

Where a tree or shrub has a use not listed in the previous columns it is classified in column 25. A good example is green-leaved mallee (*Eucalyptus viridis*), the leaves of which are used for the distillation of essential oils.

TABLE 17.1

A LIST OF 178 OF SOME OF THE MORE IMPORTANT TREES AND SHRUBS SUITABLE FOR USE IN THE ARID AND SEMI-ARID ZONES OF AUSTRALIA, TOGETHER WITH A SUMMARY OF CERTAIN CHARACTERISTICS

1 = indicates that the species is very suitable for the purpose shown.
2 = indicates that the species can be used for the purpose shown, but may be subject to limitations which are given in the notes or, for nectar and pollen, in Tables 14.1 to 14.5.
– = reported as unsuitable or of only limited value for the purposes shown; usually mentioned in the notes.
(P) = is illustrated with a photograph.

It is important to read the explanation of the table before using information contained therein.

Botanical name, common name	*Summary of natural distribution for the main occurrence*	*Climatic summary for the principal range of the species in Australia*			
		Meteorological reference stations		*Rainfall*	
		Representative	*Others*	*Range of mean annual values, inches*	*Season*
1	2	3	4	5	6
AUSTRALIAN BROADLEAVED					
Acacia acuminata Benth. (P) raspberry jam	S-A SW of WA, extending to S-H	Merredin, Northam	Geraldton, Kalgoorlie	10–30	W
A. aneura F. Muell. ex Benth. (P) mulga	Widespread, mainly temperate A	Mt Magnet, Alice Springs, Cobar	Barcaldine	8–15	U
A. argyrodendron Domin. (P) black gidyea	Mainly SA, Q		Charters Towers, Hughenden	17–23	S
A. aulococarpa A. Cunn. ex Benth. brown salwood (P)		Fitzroy Crossing, Rockhampton	Grafton	20–60	U–S
A. beckleri Tindale (P)	SA and S-A	Broken Hill	Dubbo	9–20	W–U
A. bidwillii Benth. (P) corkwood wattle		Fitzroy Crossing, Bowen	Charters Towers, Katherine	20–60	S
A. brachystachya Benth. umbrella mulga	Widespread in temperate A and S-A	Kalgoorlie, Broken Hill, Nhill		8–15	U
A. cambagei R. T. Bak. gidgee (P)	Widespread in E, 17–32°S lat.	Camooweal, Hughenden, Barcaldine, Bourke		10–25	U–S
A. coriacea DC. (P)	Widespread, mainly A, 17½–26°S, some to 30°S	Marble Bar, Alice Springs, Hughenden, Bourke		10–20	U–S
A. cunninghamii Hook. (P) Cunningham's wattle	H, S-H, some A E Aust., north of 35°S lat.	Brisbane, Inverell	Dalby	20–60	S
A. cyanophylla Lindl. (P) orange wattle	H, S-H, some A SW WA	Geraldton, Northam, Perth	Albany, Merredin	15–40	W
A. estrophiolata F. Muell. (P) ironwood	Mainly S of NT and some in N of SA	Alice Springs		10–15	U

	Size	Salt	*The more important uses which have been reported*															
			The living tree							*Timber*					*Other uses*			
															Honey			
Frosts based on 32°F in screen No./yr	Range of mature height	Tolerance to salt in soil	Shelter and windbreaks	Low shelter	Shade	Ornament	Street planting	Drought fodder	Poisonous to animals	Sawlogs	Poles	Fencing material	Turnery, special uses	Firewood	Nectar	Pollen	Edible parts	Other uses
7	8	9	10	11	12	13	14	15	16	17	18	19	20	21	22	23	24	25
0–8	15–25		1	1	2	1	2					1	1					
0–12	15–25		2	2	2	2		1				2	1	1				
0–1	15–20				2			1										
0	20–80				2			2		1		–				2		
1–10	3–10			1	2	1							2					
0	20–40				2	1		2					2					
0–8	8–20		2		2	2		2				2	2	1				
0–8	25		2		2	2						1	2	1				
0–12	10–15		2		2			2				2		1				
0–10	15–30				2	2		2							–	2		
0–6	10–25		2		2	1												
8–12	40–50		2		2	1		1				1						

TABLE 17.1—*continued*

Botanical name, common name	*Summary of natural distribution for the main occurrence*	*Climatic summary for the principal range of the species in Australia*			
		Meteorological reference stations		*Rainfall*	
		Representative	*Others*	*Range of mean annual values, inches*	*Season*
1	2	3	4	5	6
A. excelsa Benth. ironwood	NT, S-A and S-H of Q, N of NSW	Bourke, Hughenden	Rockhampton, Dalby	15–25	U–S
A. georginae F. M. Bail. (P) Georgina gidgee	Mainly central NT and adjacent Q	Alice Springs, Barrow Creek		9–11	U–S
A. harpophylla F. Muell. ex Benth. brigalow (P)	S-A and S-H Q and NSW, 22–33°S	Clermont, Roma, Narrabri	Rockhampton, Dalby	12–40	U–S
A. homalophylla A. Cunn. ex Benth. yarran (P)	Mainly S-A NSW, less in Q	Bourke, Condobolin	Dalby	15–25	U–S
A. iteaphylla F. Muell. ex Benth. Flinders Ranges wattle (P)	S-A, S-H SA	No typical station for which full data is available		12–20	W
A. kempeana F. Muell. (P) witchetty bush	Widespread 21–29°S	Mt Magnet, Alice Springs, Charleville		8–18	U–S
A. ligulata A. Cunn. ex Benth. (P) umbrella bush	Widespread in S of NT, W of NSW, most of S-A N of V	Alice Springs, Broken Hill, Mildura, Cunnamulla		7–16	W–U
A. microbotrya Benth. (P) manna wattle	S-A and S-H of SW, WA	Dandaragan, Northam	Geraldton, Merredin	11–20	W
A. notabilis F. Muell. (P) notabilis wattle	Scattered S of NT, SA and NSW	Port Augusta, Broken Hill		10–12	W–U
A. oswaldii F. Muell. (P) umbrella bush	Widespread in A and S-A Q and NSW; also SA and WA	Kalgoorlie, Broken Hill, Narrabri, Barcaldine		9–25	W–U–S
A. pendula A. Cunn. ex G. Don (P) myall or boree	A, S-A, S-H of NSW and Q	Condobolin, Moree, Roma	Charleville, Dalby	12–25	U–S
A. peuce F. Muell. (P) waddywood	A NT, some Q, SA, mainly 23–27°S lat.	Finke	Alice Springs	6–10	U
A. podalyriifolia A. Cunn. Queensland silver wattle	S-H, H of E Aust., mainly S of 23½°S lat.	Brisbane, Toowoomba	Inverell	20–60	U–S
A. pycnantha Benth. (P) broad-leaved wattle golden wattle	SA, V, NSW mainly S of 32°S; also S-H and H	Adelaide, Bendigo, Wagga Wagga		15–30	W–U
A. rigens A. Cunn. ex G. Don nealie or needle wattle	A and S-A of SA, V, NSW; S of Q	Port Lincoln, Broken Hill, Griffith	Cobar, Gilgandra	9–19	W–U

	Size	*Salt*	*The more important uses which have been reported*															
			The living tree							*Timber*					*Other uses*			
															Honey			
Frosts based on 32°F in screen No./yr	Range of mature height	Tolerance to salt in soil	Shelter and windbreaks	Low shelter	Shade	Ornament	Street planting	Drought fodder	Poisonous to animals	Sawlogs	Poles	Fencing material	Turnery, special uses	Firewood	Nectar	Pollen	Edible parts	Other uses
7	8	9	10	11	12	13	14	15	16	17	18	19	20	21	22	23	24	25
0–12	25–50		2		1	1		2		2		2	2	2				
2–12	12–15		2		2			2	?									
0–18	25–60		1		1	1			?	2		2	1	1				
0–12	20–40		2		2	2				2		2	1	1				
0–10	6–10			1		2												
0–12	10			2	2	2		2										
2–12	4–15			2		2												
0–5	12–20			1	2	2												
0–12	4–20			2	2													
1–12	6–15				2	2		2						1				
2–18	20–40		2		1	1	2	1				2	1	1				
1–10	25–50		2		1	1	2	2			2	1		1				
0–20	15			2	2	1												
2–10	15–30		2	2	2	1	2						2	1	–	1		1
2–12	6–20		2		2	2							2	2				

TABLE 17.1—*continued*

Botanical name, common name	*Summary of natural distribution for the main occurrence*	*Climatic summary for the principal range of the species in Australia*			
		Meteorological reference stations		*Rainfall*	
		Representative	*Others*	*Range of mean annual values, inches*	*Season*
1	2	3	4	5	6
A. salicina Lindl. (P) cooba or native willow	Widespread, esp. Q and NSW	Deniliquin, Roma, Hughenden	Bowen, Brisbane	15–40	W–U–S
A. shirleyi Maiden (P)	S-A, NT, Q		Daly Waters, Tennant Creek	15–20	S
A. sowdeni Maiden (P)	S of WA and SA in A and S-A	Eucla, Port Augusta	Cook	7–18	W–U
A. spectabilis A. Cunn. (P) Pilliga wattle	Mainly S-H, some H SE Q and N of NSW	Coonabarabran, Roma	Chinchilla, Moree	20–30	U–S
A. stenophylla A. Cunn. ex Benth. river cooba	A and S-A E Aust.	Bourke, Condobolin	Barcaldine, Mildura, Newcastle Waters	8–25	W–U–S
A. tetragonophylla F. Muell. (P) dead finish	Widespread, mainly in A	Alice Springs, Broken Hill, Mt Magnet, Windorah		7–12	U–S
A. victoriae Benth. (P) prickly wattle	Widespread in A and S-A zones	Alice Springs, Broken Hill, Camooweal	Adelaide	8–20	W–U–S
Agonis flexuosa (Spreng.) Schau. peppermint (P)	SW, WA, mainly S-H and H	Albany, Perth	Kalgoorlie	15–40	W
Angophora melanoxylon R. T. Bak. coolibah apple (P)	S-A, S-H of NSW; S Q	Charleville, Narrabri	Bourke	12–25	U–S
Apophyllum anomalum F. Muell. currant bush or (P) warrior bush	A and S-A of NSW and Q; also S-H	Cobar, Narrabri, Roma, Barcaldine	Rockhampton	12–25	U–S
Atalaya hemiglauca (F. Muell.) F. Muell. ex Benth. (P) whitewood	Widespread N and NE Aust., incl. S-H	Bourke, Roma, Katherine, Fitzroy Crossing	Rockhampton, Alice Springs	12–40	U–S
Atriplex nummularia Lindl. (P) old man saltbush	Widespread in A S of Aust.	Kalgoorlie, Alice Springs, Broken Hill, Roma	Clermont, Mildura, Bourke	8–20	U–S
Bauhinia carronii F. Muell. (P)	A, S-A, S-H of E Aust., mainly Q		Bourke, Chinchilla, Roma	8–25	S

	Size	*Salt*	*The more important uses which have been reported*															
			The living tree							*Timber*					*Other uses*			
															Honey			
Frosts based on 32°F in screen No./yr	Range of mature height	Tolerance to salt in soil	Shelter and windbreaks	Low shelter	Shade	Ornament	Street planting	Drought fodder	Poisonous to animals	Sawlogs	Poles	Fencing material	Turnery, special uses	Firewood	Nectar	Pollen	Edible parts	Other uses
7	8	9	10	11	12	13	14	15	16	17	18	19	20	21	22	23	24	25
0–18	20–60		2		1	1	2			2		2	1	1				
—	30–45									2	2	–						
0–5	10–15			1	2	2												
5–20	10–15			–		1	2											
0–12	35–60		2		2	2						1	1	1				
0–12	10–15				2			2										
0–12	10–15			2	2	2		2										
0–1	20–30		2		1	2	1											
2–8	35–50		2		1	1								2	2	1		
0–18	8–20		2		2	2		1	?									
0–18	20–40				1	1	2	1	?			–						
1–18	6–10			1				1										
0–15	25			1	2	1	1	2										

TABLE 17.1—*continued*

Botanical name, common name	*Summary of natural distribution for the main occurrence*	*Climatic summary for the principal range of the species in Australia*			
		Meteorological reference stations		*Rainfall*	
		Representative	*Others*	*Range of mean annual values, inches*	*Season*
1	2	3	4	5	6
B. hookeri F. Muell.	NE of Q incl. S-H and H	Barcaldine, Clermont, Charters Towers	Townsville, Rockhampton	20–50	S
Brachychiton gregorii F. Muell. (P) desert kurrajong	Mainly A S of WA; also S of NT, N of SA	Alice Springs, Kalgoorlie, Mt Magnet		8–12	W–U–S
B. populneum (Schott.) R.Br. (P) kurrajong	Mainly S-A and S-H of E of Aust.	Charleville, Cobar, Wagga Wagga	Brisbane, Sydney, Orange	20–35	U–S
B. rupestre Benth. (P) Queensland bottle tree	Mainly S-A and S-H of Q; some NSW	Bourke, Clermont, Roma	Rockhampton, Brisbane	18–35	S
Callistemon viminalis Cheel	S-H, H E Aust.	Brisbane, Lismore	Warialda	20–60	S
Canthium latifolium F. Muell. ex Benth. native currant	Widespread A 20–30°S	Alice Springs, Bourke, Meekatharra		8–11	U–S
Capparis mitchellii (Lindl. ex F. Muell.) Benth. (P) wild orange	Scattered NSW, Q	Alice Springs, Bourke, Roma, Barcaldine	Rockhampton, Dalby	8–35	U–S
C. umbonata Lindl. wild orange	N of WA and NT, A, S-A and S-H	Fitzroy Crossing, Tennant Creek	Katherine, Barrow Creek	12–40	S
Carissa lanceolata R.Br. conkerberry	N of Aust., mainly 14–24°S	Fitzroy Crossing, Tennant Creek, Mt Isa, Hughenden		9–22	U–S
Cassia artemisioides Gaud. ex DC. silver cassia (P)	Widespread A, S-A, 20–34°S	Kalgoorlie, Alice Springs, Cobar, Barcaldine		8–18	U–S
C. eremophila A. Cunn. ex R.Br. syn. *C. nemophila* A. Cunn. ex Vogel	Very widespread A, S-A and S-H, S of 20°S lat.	Kalgoorlie, Alice Springs, Mildura, Roma	Geraldton, Port Augusta, Bourke, Clermont	8–20	U–S
C. nemophila Vogel var. *coriacea* (Benth.) Symon syn. *C. sturtii* R.Br. (P)	Widespread, mainly A	Mt Magnet, Alice Springs, Barcaldine	Giles, Adelaide, Hughenden	8–20	W–U–S
Casuarina campestris Diels (P) a sheoak	S–A, SW, WA		Geraldton Mullewa	12–20	W

	Size	Salt	*The more important uses which have been reported*															
			The living tree							*Timber*					*Other uses*			
															Honey			
Frosts based on 32°F in screen No./yr	Range of mature height	Tolerance to salt in soil	Shelter and windbreaks	Low shelter	Shade	Ornament	Street planting	Drought fodder	Poisonous to animals	Sawlogs	Poles	Fencing material	Turnery, special uses	Firewood	Nectar	Pollen	Edible parts	Other uses
7	8	9	10	11	12	13	14	15	16	17	18	19	20	21	22	23	24	25
0–2	30–40		2		1	1	2	2										
1–12	12–25		2		1	1	1	?										
0–20	25–50		2		1	1	1	1		–					2	1		
0–8	50–60		2		1	1	1	1										
0–30	15–60			1	2	1	1											
0–12	6–12			2	2	2												
0–18	15–30		2		1	1		2										
0–10	15				2	2		2										
0–2	4–6			1		2		1										
1–12	4–8			1		1		2										
1–18	3–6			1		1		2										
0–10	4–7			1		1												
0–5	5–10			1		2												

TABLE 17.1—*continued*

Botanical name, common name	*Summary of natural distribution for the main occurrence*	*Climatic summary for the principal range of the species in Australia*			
		Meteorological reference stations		*Rainfall*	
		Representative	*Others*	*Range of mean annual values, inches*	*Season*
1	2	3	4	5	6
C. cristata Miq. (P) belah	Widespread, mainly S of 26°S in S-A	Port Augusta, Broken Hill, Condobolin, Emerald	Mildura, Narrabri	9–25	U–S
C. decaisneana F. Muell. (P) desert oak	Very A S of NT and adjacent WA	Alice Springs, Giles		7½–10½	U–S
C. dielsiana C. A. Gardn. (P) northern sheoak	S-A, S-H SW WA		Mullewa	12–20	W
C. huegeliana Miq. (P) rock sheoak	S-A, some S-H, SW, WA	Dandaragan, Narrogin, Northam	Geraldton, Mullewa	12–30	W
C. luehmannii R. T. Bak. (P) bull oak	S-A and S-H E of Aust., mainly temperate	Echuca, Condobolin, Narrabri	Stanthorpe, Roma, Clermont	15–25	W–U-
C. stricta Ait. drooping sheoak	Mainly S-A and S-H of V and S of NSW	Ballarat, Canberra, Melbourne	Adelaide, Bombala, Condobolin, Dubbo	15–30	W–U
Codonocarpus cotinifolius (Desf.) F. Muell. (P) native poplar	Scattered, mainly S half	Kalgoorlie, Alice Springs, Bourke	Port Hedland, Merredin, Mildura	10–20	W–U-
Dodonea attenuata A. Cunn. hopbush	Widespread S of 20°S incl. S-H	Kalgoorlie, Alice Springs, Bourke	Narrogin, Adelaide, Canberra	10–25	U–S
Eremocitrus glauca (Lindl.) Swingle limebush	S-A of Q and NSW, some A and S-H	Broken Hill, Dalby, Moree	Barcaldine, Charleville, Rockhampton	10–20	S–U
Eremophila latrobei F. Muell. native fuchsia	Widespread, esp. A 20–30°S	Mt Magnet, Giles, Alice Springs, Windorah	Mt Isa, Tennant Creek, Cunnamulla	8–25	U–S
E. longifolia (R.Br.) F. Muell. berrigan or emubush	Very widespread	Mt Magnet, Alice Springs, Cobar, Barcaldine	Giles, Mt Isa, Nhill, Dubbo	8–20	U–S
E. maculata (Ker.-Gawl.) F. Muell. native fuchsia	Scattered 18–34°S esp. A	Kalgoorlie, Alice Springs, Bourke	Meekatharra, Charleville	8–20	U–S
Eucalyptus alba Reinw. ex Blume white gum (P)	S-A to H N and NE	Darwin, Townsville	Derby, Rockhampton	30–60	S

	Size	Salt	*The more important uses which have been reported*															
			The living tree							*Timber*					*Other uses*			
															Honey			
Frosts based on 32°F in screen No./yr	Range of mature height	Tolerance to salt in soil	Shelter and windbreaks	Low shelter	Shade	Ornament	Street planting	Drought fodder	Poisonous to animals	Sawlogs	Poles	Fencing material	Turnery, special uses	Firewood	Nectar	Pollen	Edible parts	Other uses
7	8	9	10	11	12	13	14	15	16	17	18	19	20	21	22	23	24	25
1–12	20–70		2		2	2	2	?			2	2		1				
12	35–50		2		2	2	2	2		2	1	1		1				
0–2	15–30		2		2	2												
0–10	20–30		2		2	1												
5–50	25–60		2		2	2	2			2		2	1	1				
0–40	20–30		2		1	1	2	2						1		2		
1–12	8–30		–		2	1	–	?										
1–20	5–15			1	2	2		2										
0–2	10–18			1		1												
0–12	3–10			1		1		1										
0–12	5–25			1		1		1	?									
0–12	3–7			1		1		2										
0–1																		

TABLE 17.1—*continued*

Botanical name, common name	*Summary of natural distribution for the main occurrence*	*Climatic summary for the principal range of the species in Australia*			
		Meteorological reference stations		*Rainfall*	
		Representative	*Others*	*Range of mean annual values, inches*	*Season*
1	2	3	4	5	6
E. angulosa Schau.	Mainly S-A, SW, WA	Norseman, Salmon Gums	Kalgoorlie, Esperance	9–27	W–U
E. argophloia Blakely (P) Queensland western white gum	S-H SE Q	Chinchilla		24–28	S
E. astringens (Maiden) Maiden (P) brown mallet	Mainly S-H SW WA	Narrogin		14–22	W
E. behriana F. Muell. (P) broad-leaved mallee	SA, V, NSW; mainly 33–37°S S-A	Lameroo, Nhill, Temora		16–20	W–U
E. brockwayi C. A. Gardn. (P) Dundas mahogany	Restricted in S-A of SW WA	Norseman		9–20	W
E. camaldulensis Dehnh. (P) river red gum	Widespread except coastal E and S	Mildura, Hamilton	Alice Springs	8–50	W–U–S
E. campaspe S. Moore (P) silver-topped gimlet	S-A southern WA	Kalgoorlie		8–12	U
E. cladocalyx F. Muell. (P) sugar gum	S-A, S-H SA	Clare, Kingscote, Port Lincoln		15–25	W
E. clelandii (Maiden) Maiden (P) Cleland's blackbutt	Kalgoorlie area WA	Kalgoorlie		8–12	U
E. conica Deane et Maiden (P) fuzzy box	S-A and S-H NSW, some Q	Narrabri, Forbes	Dalby, Inverell	20–25	U
E. crebra F. Muell. (P) narrow-leaved red ironbark	S-A, S-H, H Q, mainly S of 18°S lat.; some NSW	Dalby, Narrabri	Clermont, Charters Towers	20–30	U–S
E. dichromophloia F. Muell. (P) variable-barked bloodwood	Most N Aust. from 28°S	Tennant Creek, Alice Springs, Charters Towers, Darwin		10–55	U–S
E. dumosa A. Cunn. white mallee	Mainly A and some S-A of southern Aust.	Griffith, Kalgoorlie, Ouyen	Alice Springs, Cobar, Condobolin	10–18	W–U
E. dundasii Maiden (P) Dundas blackbutt	Restricted area S-A, S of Kalgoorlie WA	Norseman		10–25	W
E. eremophila (Diels) Maiden (P) sand mallee	Restricted in S-A of SW WA	Kalgoorlie, Merredin, Norseman		9–18	W–U
E. exserta F. Muell. (P) Queensland peppermint	S-A to coast of Q	Gladstone Maryborough	Roma, Clermont	17–45	S

	Size	Salt	*The more important uses which have been reported*															
			The living tree							*Timber*					*Other uses*			
															Honey			
Frosts based on 32°F in screen No./yr	Range of mature height	Tolerance to salt in soil	Shelter and windbreaks	Low shelter	Shade	Ornament	Street planting	Drought fodder	Poisonous to animals	Sawlogs	Poles	Fencing material	Turnery, special uses	Firewood	Nectar	Pollen	Edible parts	Other uses
7	8	9	10	11	12	13	14	15	16	17	18	19	20	21	22	23	24	25
0–2	10–20			2	2	2									1	2		
15–20	120		2		1	1				1	2	1		1				
2–4	40–80		2		1	2	1			1	1	2		1				
0–20	15–25		2		2									1	2	1		
1–12	50–80		2		1	1	1			1	1	2		1				
1–15	30–150		2		1	1				1	2	1		1	1	1		
0–2	25–40	2	2		2	1	1					–		1	?	?		
0–15	50–100		2		2	1	2			1	1	2		1	?	?		
0–2	20–40				2	2	1					2						
10–50	50–70		2		2	1				1	2	1		1	1	2		
0–15	60–90		2		2	2				1	1	1		1	2	2		
0–12	25–80				2	2						2		1				
1–10	10–40			2	2	2								1	1–2	1		
1–12	30–60	1	2		2	1	1					2		1				
1–12	15			2		1	1								1	1		
0–18	35–80		2		1	2				1		1		1				

TABLE 17.1—*continued*

Botanical name, common name	*Summary of natural distribution for the main occurrence*	*Climatic summary for the principal range of the species in Australia*			
		Meteorological reference stations		*Rainfall*	
		Representative	*Others*	*Range of mean annual values, inches*	*Season*
1	2	3	4	5	6
E. flocktoniae (Maiden) Maiden (P) merrit	Mainly S-A of S of WA and SA	Norseman, Port Lincoln	Moonta	10–20	W–U
E. forrestiana Diels (P) Forrest's mallee	Restricted in S-A of SW WA	Salmon Gums		12–20	W–U
E. gamophylla F. Muell. (P) blue mallee	Very A WA and NT, N SA, mainly 20–27°S	Alice Springs, Finke	Meadow (WA)	6–10	U
E. gardneri Maiden (P) blue mallet	S-A and S-H of SW WA	Narrogin, Corrigin	Merredin	16–22	W
E. gillii Maiden (P) curly mallee	A NE of SA and SW of NSW	Broken Hill	Marree	6–12	W–U
E. gongylocarpa Blakely (P) marble gum	Very A WA, S of NT and N of SA; 24–30°S	Giles, Finke	SW of Alice Springs	6–10	U
E. gracilis F. Muell. (P) yorrell	Mainly S-A, S of Aust.	Southern Cross	Kalgoorlie, Port Lincoln	10–18	W–U
E. intertexta R. T. Bak. gum-barked coolabah	Mainly A 23–32°S	Alice Springs, Cobar	Condobolin	10–18	U–S
E. kondininensis Maiden et Blakely Kondinen blackbutt	S-A SW, WA	Lake Grace, Norseman	Corrigin	10–15	W–U
E. largiflorens F. Muell. (P) black box	Mainly A and some S-A NSW; less SA, V, Q	Ivanhoe, Bourke	Broken Hill	10–15	W–U–S
E. lehmannii (Preiss. ex Schau.) Benth. Bald Island marlock	H, S-H and S-A of S WA	Albany, Esperance		20–40	W
E. lesouefii Maiden (P) Goldfield's blackbutt	S-A southern WA	Kalgoorlie, Norseman		10–13	W–U
E. leucoxylon F. Muell. (P) yellow gum	S-A and S-H V; some SA	Bendigo, Horsham	Clare	20–30	W
E. longicornis (F. Muell.) F. Muell. ex Maiden (P) red morrell	A and S-H SW WA	Narrogin, Kalgoorlie		10–15	W–U
E. loxophleba Benth. (P) York gum	Mainly S-H and S-A SW WA	Merredin, Narrogin, Geraldton	Norseman	10–19	W
E. melanophloia F. Muell. (P) silver-leaved ironbark	S-A to coast Q; some S-A NSW	Roma, Narrabri	Clermont, Charters Towers	15–30	S–U

	Size	Salt	*The more important uses which have been reported*															
			The living tree							*Timber*					*Other uses*			
															Honey			
Frosts based on 32°F in screen No./yr	Range of mature height	Tolerance to salt in soil	Shelter and windbreaks	Low shelter	Shade	Ornament	Street planting	Drought fodder	Poisonous to animals	Sawlogs	Poles	Fencing material	Turnery, special uses	Firewood	Nectar	Pollen	Edible parts	Other uses
7	8	9	10	11	12	13	14	15	16	17	18	19	20	21	22	23	24	25
0–8	34–40				2	1												
12	10–18			2		2	2											
8–12	8–15			2		2		2										
2–6	30–35		2		1	1	1			2		1		1	1	1		2
1–5	15–20		2	2		2												
8–12	30–45				2													
0–4	25–40		2		2	2	2					2			1–2	2		
5–12	30–60	2	2		2	2				2		2		1				
0–3	35–60	1			2	2	2											
1–5	30–60				2	2				2		1		1	1–2	1		
0–5	4–30			1		2	2											
0–2	20–45				2	1	1											
1–15	30–60		2		1	1				1	2	1		1	1	–		
1–4	30–70	2	2		1	1	1			2	1	1			1	2		
0–6	15–40		2		2	2	2			2					1	–		
1–18	20–90		1		2	1	2			2		1		1	1	1–2		

TABLE 17.1—*continued*

Botanical name, common name	*Summary of natural distribution for the main occurrence*	*Climatic summary for the principal range of the species in Australia* — *Meteorological reference stations*: *Representative*	*Others*	*Rainfall*: *Range of mean annual values, inches*	*Season*
1	2	3	4	5	6
E. mellidora A. Cunn. ex Schau. (P) yellow box	S-A and S-H V, NSW, some Q	Tamworth, Wagga Wagga, Bendigo	Inverell	15–35	W–U–S
E. microtheca F. Muell. (P) coolibah	Widespread N of 32°S except coastal NT, Q	Tennant Creek, St George, Alice Springs	Katherine, Onslow	8–40	U–S
E. occidentalis Endl. flat-topped yate	S-A and S-H SW WA	Katanning, Esperance		13–35	W
E. ochrophloia F. Muell. (P) yapunyah	N of NSW, S of Q	Thargomindah, Charleville, Bourke		10–20	U–S
E. odorata Behr et Schldl. (P) peppermint box	Restricted S-A and S-H SA	Adelaide, Port Lincoln	Yongala	15–30	W
E. oleosa F. Muell. ex Miq. var. *angustifolia* Maiden (P)	S-A, S-H S Aust.	Lameroo	Swan Hill	12–15	W
E. orgadophila Maiden et Blakely mountain coolibah (P)	S-H SE Q	Oakey, Warwick	Dalby, Toowoomba	25–35	S
E. papuana F. Muell. (P) ghost gum	Most of N Aust.	Alice Springs, Katherine, Charters Towers	Atherton, Darwin, Emerald, Rockhampton	10–60	S
E. pilligaensis Maiden (P) narrow-leaved grey gum	S-A of NSW, adj. Q	Narrabri		18–25	U–S
E. platypus Hook. (P) moort	SW, WA mainly S-A		Albany, Lake Grace, Esperance	15–35	W
E. polycarpa F. Muell. (P) long-fruited bloodwood	WA, NT N of 18°S; most of Q and edge NSW	Katherine, Charleville, Roma	Broome, Darwin, St George	10–55	S
E. populnea F. Muell. (P) bimble box	Mainly S-A NSW and Q S of 22½°S	Condobolin, Roma	Emerald, Bourke	14–22	U–S
E. '*pterocarpa*' (P)	S-A of southern WA	Norseman		10–13	U
E. redunca Schau. var. *melanophloia* Benth. (P) black-barked marlock	Mainly S-A SW of WA		Albany	25–35	W
E. salmonophloia F. Muell. (P) salmon gum	SW WA, mainly S-A	Kalgoorlie, Merredin		8–20	W–U

	Size	Salt	*The more important uses which have been reported*																
			The living tree							*Timber*					*Other uses*			
															Honey			
Frosts based on 32°F in screen No./yr	Range of mature height	Tolerance to salt in soil	Shelter and windbreaks	Low shelter	Shade	Ornament	Street planting	Drought fodder	Poisonous to animals	Sawlogs	Poles	Fencing material	Turnery, special uses	Firewood	Nectar	Pollen	Edible parts	Other uses
7	8	9	10	11	12	13	14	15	16	17	18	19	20	21	22	23	24	25
5–20	60–100		2		1	1				2	2	1		1	1	–		
0–12	25–70		2		2	2						2		1	2	2		
1–3	60–80		2		1	2	1			2	1	1		1				
1–10	40–70		2		2	2				2		1			1–2	2		
0–30	20–50		2		1	2					2	1		1	2	2		
1–5	30–40		2		2	2	2							1				
10–20	35–50		2		2	1												
0–12	20–60				2	1	1					2		2				
5–20	40–80		2		1	2				2	2	1		1	1–2	–		
0–3	25		1	1	2	2									1	1		
0–6	40–70		2		2	2				2	2	1		1				
2–18	35–70		2		1	1				2		1		1	1–2	2		
0–2	30–40		2		1	1	2											
0–2	30		1	1	2	2												
1–6	40–80		2		1	1	1			1	1	1		1	1–2	–		

TABLE 17.1—*continued*

Botanical name, common name	*Summary of natural distribution for the main occurrence*	*Climatic summary for the principal range of the species in Australia*			
		Meteorological reference stations		*Rainfall*	
		Representative	*Others*	*Range of mean annual values, inches*	*Season*
1	2	3	4	5	6
E. salubris F. Muell. (P) gimlet	Kalgoorlie area WA	Kalgoorlie, Merredin	Narrogin	8–18	W
E. sargentii Maiden (P) Salt River mallet	S-A SW WA	Merredin, S of Kalgoorlie, E of Northam		13–18	W
E. sideroxylon A. Cunn. ex Woolls subsp. *sideroxylon* red ironbark (P)	S-A, S-H NSW, V	Bendigo, Condobolin, Gunnedah		15–25	W–U–S
E. spathulata Hook. (P) swamp mallet	S-A SW WA	Cunderdin, Katanning		12–15	W
E. stricklandii Maiden (P) Strickland's gum	Kalgoorlie area WA	Kalgoorlie, Norseman		9–11	W–U
E. tectifica F. Muell. Darwin box	N of WA, NT, NW Q S-A to H	Katherine, Darwin	Daly Waters, Fitzroy Crossing	15–60	S
E. tessellaris F. Muell. (P) carbeen	S-A to coast Q, small extension NSW	Roma, Charters Towers	Moree, Brisbane	15–60	S
E. thozetiana F. Muell. ex R. T. Bak. mountain yapunyah (P)	S of Q	Roma, Clermont	Charleville, St George	12–25	S
E. torquata Luehm. (P) coral-flowered gum	To S and W of Kalgoorlie, WA	Kalgoorlie	Norseman	9–11	W–U
E. trachyphloia F. Muell. (P) brown bloodwood	S-A N NSW; S-H and H Q	Narrabri, Bundaberg	Brisbane	20–50	U–S
E. viridis R. T. Bak. (P) green mallee	S-A NSW, V; also some SA and Q	Cobar, Condobolin, Bendigo		16–20	W–U–S
E. wandoo Blakely (inland form) wandoo	S-H, S-A SW WA	Narrogin	Merredin	16–20	W
E. woollsiana R. T. Bak. grey box	Mainly S-A NSW, some Q	Wagga Wagga, Narrabri	Moree, Gunnedah	15–25	U
Ficus platypoda (Miq.) A. Cunn. ex Miq. (P) fig	Mainly N of 28°S; A to H	Fitzroy Crossing, Alice Springs	Atherton, Mackay, Brisbane	10–60	U–S
Flindersia maculosa (Lindl.) F. Muell. (P) leopardwood	A and S-A N NSW S Q	Bourke, Cobar	Broken Hill, Charleville	12–22	U–S

	Size	*Salt*	*The more important uses which have been reported*															
			The living tree							*Timber*					*Other uses*			
															Honey			
Frosts based on 32°F in screen No./yr	Range of mature height	Tolerance to salt in soil	Shelter and windbreaks	Low shelter	Shade	Ornament	Street planting	Drought fodder	Poisonous to animals	Sawlogs	Poles	Fencing material	Turnery, special uses	Firewood	Nectar	Pollen	Edible parts	Other uses
7	8	9	10	11	12	13	14	15	16	17	18	19	20	21	22	23	24	25
1–6	40–60	2	2		1	1	1			2	1	1			1	1		
1–6	25–35	1	2		1	1	1											
1–25	70–100	2	2		1	1				2	2	1		1	2	–		
1–5	15–20			1	2	1	1											
1–6	20–25				2	1	1								1–2			
0–1	30–45		2		2	2						2		1				
0–18	40–100		2		1	1	1			1	2	1		2	–	–		
1–18	30–80		2		1	2				1	2	1						
6	20–25	1	2		1	1	2							2	1–2	2		
0–15	60–80		2		1	2				2		1		1				
2–16	15–35		2	2											2	2		1
3–6	50–70		2		2					2	1	1		1	1	2		1
5–25	25–80		2		1	1				2	2	1		1				
0–12	30				1	1												
1–8	15–30					2		1				–						

TABLE 17.1—*continued*

Botanical name, common name	*Summary of natural distribution for the main occurrence*	*Climatic summary for the principal range of the species in Australia*			
		Meteorological reference stations		*Rainfall*	
		Representative	*Others*	*Range of mean annual values, inches*	*Season*
1	2	3	4	5	6
Geijera parviflora Lindl. (P) wilga	NSW, Q, mainly S-A, S-H 21–35½°S	Clermont, Condobolin, Roma	Gunnedah, Dalby, Rockhampton	9–30	U–S
Grevillea pinaster Meissn. (P)	S-A, S-H SW of WA		Geraldton, Mullewa	10–18	W
G. striata R.Br. (P) beefwood	Scattered C and E Aust. N of 32°S	Meekatharra, Alice Springs, Cobar	Narrabri	9–25	W–U
Hakea divaricata L. Johnson syn. *H. intermedia* Ewart et Davies	A, S-A, mainly S Aust.	Alice Springs, Giles		8–15	W–U–S
H. leucoptera R.Br. (P) needlewood	Widespread S NT, S Q, SA, V, NSW	Alice Springs, Mildura, Charleville	Broken Hill, Nhill, Bourke	9–20	W–U–S
H. petiolaris Meissn.	S-A, S-H SW of WA		Perth	15–30	W
H. vittata R.Br.	Mainly S Q, NSW, V	Mildura, St George	Horsham	10–30	U–W
Heterodendrum oleifolium Desf. (P) inland rosewood	Widespread; to S-H Q	Kalgoorlie, Alice Springs, Narrabri, Hughenden	Carnarvon, Tarcoola, Broken Hill, Roma	9–20	W–U–S
Lagunaria patersonia Ait. (P) Norfolk Island hibiscus tree	Restricted coastal Q (also Norfolk Island)	Townsville, Bowen		35–60	S
Leptospermum laevigatum (Sol. ex Gaertn.) F. Muell. (P) coast teatree	S-H, H S of Aust.	Melbourne		20–30	W–U
Melaleuca armillaris Sm. (P) bracelet honey myrtle	S-H, H S of Aust.	Bairnsdale, Bateman's Bay, Sale		23–35	W–U
M. halmaturorum F. Muell. ex Miq. South Australian swamp paper-bark	Mainly S-A SA, V, also some A and S-H	Kingscote, Port Augusta, Port Lincoln	Adelaide	15–25	W
M. lanceolata Otto = *M. pubescens* Schau. (P) dryland teatree	SE Aust., mainly S-A, S-H	Port Lincoln, Ouyen, Melbourne, Dalby	Griffith	12–30	W–U–S
M. pauperiflora F. Muell. (P)	A, S-A of SW WA	Kalgoorlie, Salmon Gums	Tarcoola	7–12	U–S

	Size	*Salt*	*The more important uses which have been reported*															
			The living tree							*Timber*					*Other uses*			
															Honey			
Frosts based on 32°F in screen No./yr	Range of mature height	Tolerance to salt in soil	Shelter and windbreaks	Low shelter	Shade	Ornament	Street planting	Drought fodder	Poisonous to animals	Sawlogs	Poles	Fencing material	Turnery, special uses	Firewood	Nectar	Pollen	Edible parts	Other uses
7	8	9	10	11	12	13	14	15	16	17	18	19	20	21	22	23	24	25
1–18	20–30		2		1	1	2	1							1	2		
0–2	6–10			1		2												
1–12	20–40				2	1		2				1		–				
10–12	15–20				2	2												
1–12	8–18				2	2		2					2		2	2		
0–2	10–15			1		2												
1–15	10–15					2		2										
0–30	10–25		2		2	2		1	?						1	–		
0–10	25–60		2		1	1	2					–						
1–10	15–30			1	2	1							1	1				
0–10	8–15	2		1		2								1				
0–5	10–25	2	1	1	2	2												
0–15	15–20	1	1	1	2	2									1–2	1		
0–5	4–10			1		2												

TABLE 17.1—*continued*

Botanical name, common name	*Summary of natural distribution for the main occurrence*	*Climatic summary for the principal range of the species in Australia*			
		Meteorological reference stations		*Rainfall*	
		Representative	*Others*	*Range of mean annual values, inches*	*Season*
1	2	3	4	5	6
Melia azedarach L. var. *australasica* (A. Juss.) C.DC.	Coastal to S-A Q	Rockhampton, Brisbane	Taroon, Gayndah	25–60	S
Muehlenbeckia cunninghamii (Meissn.) F. Muell. lignum	Widespread, esp. S-A E Aust.	Mt Magnet, Tennant Creek, Mildura, Charleville	Adelaide, Melbourne, Narrabri, Clermont	8–25	W–U–S
Myoporum montanum R.Br. (P) boobialla or water bush	Scattered in A, S-A	Alice Springs, Ouyen, Cobar		10–18	U–S
M. platycarpum R.Br. dogwood	Mainly S-A 30–36°S	Kalgoorlie, Broken Hill, Griffith, Narrabri	Horsham, Murray Bridge	9–25	W–U
Owenia acidula F. Muell. (P) emu apple bush	Mainly NSW, Q 20–31°S; also NT, ext. to WA	Bourke, Roma, Barcaldine	Tibooburra, Narrabri, Biloela	8–25	U–S
Petalostigma quadriloculare (P) F. Muell. petalostigma or quinine-tree	N and NE Aust., S-A to coast	Katherine, Tennant Creek	Darwin, Brisbane, Townsville	15–60	S
Pittosporum phylliraeoides DC. (P) pittosporum or inland pittosporum	Widespread A, S-A, some S-H; S of 20°S	Alice Springs, Condobolin, Mullewa, Roma	Adelaide, Horsham, Dalby, Clermont	8–35	U–S
Santalum lanceolatum R.Br. (P) sandalwood or plum bush	Widespread, esp. N and E	Alice Springs, Meekatharra, Fitzroy Crossing, Narrabri, Roma	Broken Hill, Tennant Creek	9–25	U–S
Ventilago viminalis Hook. (P) supplejack	Widespread in N, NE; mainly A, S-A, some S-H	Fitzroy Crossing, Tennant Creek, Bourke, Clermont	Broken Hill, Narrabri	10–25	U–S
AUSTRALIAN CONIFERS					
Actinostrobus pyramidalis Miq. (P) native cypress	S-A and S-H of SW, WA	Perth		20–35	W
Callitris endlicheri (Parl.) J. Garden black cypress pine (P)	Mainly S-A and S-H of E of Aust.	Dalby, Narrabri, Wagga Wagga		20–30	U
C. glauca R.Br. ex R. T. Bak. et H. G. Sm. (P) white cypress pine	Common in S-A of NSW and S Q; scattered elsewhere	Condobolin, Narrabri, Roma	Alice Springs, Barcaldine, Bourke	12–30	U–S

	Size	*Salt*	*The more important uses which have been reported*																
			The living tree							*Timber*					*Other uses*			
															Honey			
Frosts based on 32°F in screen No./yr	Range of mature height	Tolerance to salt in soil	Shelter and windbreaks	Low shelter	Shade	Ornament	Street planting	Drought fodder	Poisonous to animals	Sawlogs	Poles	Fencing material	Turnery, special uses	Firewood	Nectar	Pollen	Edible parts	Other uses
7	8	9	10	11	12	13	14	15	16	17	18	19	20	21	22	23	24	25
0–5	50–150		2		1	1	1		?						2			
0–20	3–10							2										
0–20	8–10			1	2	1												
1–20	10–25			2	1	2									2	–		
0–15	12–25		2		1	1		1										
0–10	12–30				2	2			?									
0–20	10–25		2	2		1	1	1										
0–15	15–20				2	2		1										
0–12	15–30			2	2	2		1										
1–3	15–25		1	1	2	2												
10–30	40–70		1		2	2	2			2	2	1		1				
0–18	20–100		1		2	2	2			1	2	1		1				

TABLE 17.1—*continued*

Botanical name, common name	*Summary of natural distribution for the main occurrence*	*Climatic summary for the principal range of the species in Australia*			
		Meteorological reference stations		*Rainfall*	
		Representative	*Others*	*Range of mean annual values, inches*	*Season*
1	2	3	4	5	6
C. preissii Miq. subsp. *preissii* Rottnest Island cypress pine (P)	Scattered S-A, S-H S WA, SA, V, NSW	Perth	Adelaide, Nhill	(Main WA occurrence only) 20–35	W
EXOTIC BROADLEAVED					
Ceratonia siliqua L. carob bean (P)	S-A, S-H S Aust.	Meteorological stations are only shown for native species. Climatic information for exotics is given in terms of rainfall, frost and the zones in which they grow at least reasonably satisfactorily.		15–25	W–U
Cercis siliquastrum L. Judas tree	S Aust. S-H			18–30	W–U
Cytisus proliferus L.f. false tree-lucerne	S Aust. S-A, S-H, H			18–50	U
Fraxinus oxyphylla Bieb. desert ash	S Aust. S-A, S-H			17–35	W–U
Gleditschia triacanthos L. honey locust	S-H			20–50	W–U–S
Nerium oleander L. oleander (P)	S Aust. A, S-A, S-H, H			15–50	W–U–S
Olea europaea L. olive (P)	S Aust. A, S-A, S-H			10–25	W–U
Parkinsoniana aculeata L. Jerusalem thorn	A, S-A			10–18	W–U
Rhus lancia L.f. karreeboom (P)	A, S-A, S-H, S Aust.			15–30	U–S
Robinia pseudoacacia L. black locust	S-H, some S-A			20–30	W–U–S
Schinus molle L. pepper tree (P)	A, S-A S Aust.			12–25	W–U–S
Tamarix articulata Vahl. athel tree (P)	S Aust. A, S-A			10–20	W–U–S
EXOTIC CONIFERS					
Cupressus arizonica Greene Arizona cypress	S Aust. S-A, S-H			20–30	W–U

	Size	Salt	The more important uses which have been reported															
			The living tree							Timber					Other uses			
															Honey			
Frosts based on 32°F in screen No./yr	Range of mature height	Tolerance to salt in soil	Shelter and windbreaks	Low shelter	Shade	Ornament	Street planting	Drought fodder	Poisonous to animals	Sawlogs	Poles	Fencing material	Turnery, special uses	Firewood	Nectar	Pollen	Edible parts	Other uses
7	8	9	10	11	12	13	14	15	16	17	18	19	20	21	22	23	24	25
0 0–5	20–30		1		2	2												
1–10	20–30				1	1		1										1
0–5	20				2	1												
0–10	10–15			1	2	1		1										
1–30	25–50		2		1	1	1											
1–30	60		2		1	1				1		1		1				
0–30	6–12			1		1			1									
1–15	15–25			1	2	2												1
1–10	20–30		2	1														
0–10	15–30			1	1	1	2	2									1	
1–30	40–50				1	1				1		1	1	1				
1–10	25–40		1	2	1	1												
1–15	25–40		1		2	1	2						2	2				1
1–20	50–70		1		2	2								2				

TABLE 17.1—*continued*

Botanical name, common name	*Summary of natural distribution for the main occurrence*	*Climatic summary for the principal range of the species in Australia*			
		Meteorological reference stations		*Rainfall*	
		Representative	*Others*	*Range of mean annual values, inches*	*Season*
1	2	3	4	5	6
C. sempervirens L. Roman cypress	S Aust. S-A, S-H			18–30	W–U
Pinus canariensis C. Sm. (P) Canary Island pine	S Aust. S-A, S-H			14–50	W
P. coulteri D. Don big-cone pine				20–30	W
P. halepensis Mill. (P) Aleppo pine	S Aust. S-A, S-H			18–30	W
P. halepensis Mill. var. *brutia* (Ten.) Elwes et Henry	S-A S Aust.			12–20	W
P. pinaster Ait. maritime pine	S Aust. S-A, S-H			20–35	W
P. pinea L. (P) stone pine	S Aust. S-A, S-H			18–30	W
P. sabiniana Douglas digger pine	S Aust. S-A, S-H			20–30	W
P. torreyana Parry ex Carr. Torrey pine	S Aust. SA, S-H, H			22–30	W–U

	Size	Salt	*The more important uses which have been reported*															
			The living tree							*Timber*					*Other uses*			
															Honey			
Frosts based on 32°F in screen No./yr	Range of mature height	Tolerance to salt in soil	Shelter and windbreaks	Low shelter	Shade	Ornament	Street planting	Drought fodder	Poisonous to animals	Sawlogs	Poles	Fencing material	Turnery, special uses	Firewood	Nectar	Pollen	Edible parts	Other uses
7	8	9	10	11	12	13	14	15	16	17	18	19	20	21	22	23	24	25
0–25	25–60		1		2	1				2				2				
1–30	80–120		1		2	2				1	1	2		2				
2–30	60–70		2		2	2				2				2				
0–15	40–60		2		2	2				2				2				
0–15	40–80		1		2	2				1	2			2				
2–20	50–100		2		1	2				2				2				
1–20	50–60				1	2								2				
1–20	40–70		1		2	2				–				2				
1–30	30–100		1		2	2				1				2				

AUSTRALIAN BROADLEAVED

Acacia P. Miller **ACACIAS or WATTLES**

This very large genus of trees and shrubs belong to the pea family (*Leguminosae*, subfamily *Mimosaceae*) and has a wide distribution throughout the warmer and drier parts of the world. It has been estimated that there are about 600 species which grow naturally in Australia, most of them restricted to this country but a few being international, and some hundreds which are native to Africa. They are typically species of the arid and semi-arid zones but, in Australia, they are also plentiful in the sub-humid and humid zones, whilst a few occur in rain-forest. The generic name is derived from the Greek, *ake*, in reference to the spiny development of the stipules; a feature which is particularly prominent amongst African species. The common name of wattle goes back to Anglo-Saxon times and refers to flexible twigs and branchlets which can be closely interwoven to form the walls of primitive dwellings, to be followed by daubing with wet clay to give better protection against the weather. The term acacia is sometimes used for species of other genera, especially *Robinia*. The black locust (*R. pseudo-acacia*) of north America is also known as false acacia.

In size most of the Australian species are 10-50 feet in height, but the range is from 5 to over 100 feet. The greatest height is attained by blackwood (*A. melanoxylon*) in the moist valleys of cool temperate rainforest in Victoria and Tasmania. The foliage of the genus is commonly either feathery bipinnate leaves or flattened phyllodes, usually of firm texture, 2-10 inches long and ¼-1½ inches wide. The bipinnate leaves prevail amongst species of the humid zone but phyllodes in the arid and semi-arid zones. The flowers, although individually quite small, are one of the best known features of the genus and many species are planted in gardens for the very impressive display of the species. When flowering many species appear to be almost a mass of light yellow, yellow or deep orange-yellow flowers. The flowers appear in small globular heads or spikes. Plate 1 shows a small plant of golden wattle (*A. pycnantha*) flowering as a planted specimen at Broken Hill.

Most wattles are fast growing in early life but vary considerably in longevity from as low as about 15 years to well over 100 years. The attractive Cootamundra wattle (*A. baileyana*) when planted in a dry and otherwise unfavourable site commences to deteriorate after 12-15 years whilst, on the other hand, a large rainforest species such as blackwood has a life span which certainly exceeds 100 years. Accurate data is not available on many of the dry country species, but there is little doubt that most of the small trees 20-50 feet in mature height have a life span of 50 years and possibly longer.

Not surprising for such a large genus, members of it have a wide variety of uses. Due to the wide geographical range of the species and the variety in size, it is not difficult to select suitable wattles for a wide range of ornamental use. Many species have a considerable value as drought fodder in the dry country, some even for use as general browse plants; the most notable species being mulga (*A. aneura*). Acacia foliage and pods, in general, rarely have toxic effects on stock although the green pods of a few species may be harmful. Nearly all the larger growing species have attractive, dark-coloured heartwood; one of the best known species which is commercially available both as plywood and as sawn timber, is blackwood. The wood of some of the smaller growing species is valued for small ornamental and turnery work, notably mulga. True gum arabic is produced from an African species (*A. senegal*), but there are several others which produce a product which is nearly as good. The bark of many species has a high tannin content, notably golden wattle, black wattle (*A. mollissima*) and green wattle (*E. decurrens*). The second species has been extensively planted in South Africa for tannin from the bark and the wood for use in mines.

The value of nectar and pollen varies considerably with the species, for instance currawong (*A. doratoxylon*) is considered a valuable tree but with many others there is little nectar produced and the pollen is of limited value as a bee food.

It has only been possible to describe 34 species in the notes below. For information on the large number of other wattles which have a value for ornament or other farm amenity planting, readers are referred to the reading list given at the end of this chapter and to various nursery catalogues, especially those of forest services of the States.

A. acuminata Benth. **RASPBERRY JAM**

The unusual common name of this acacia is based on the strong odour of the freshly cut wood, which resembles that of crushed raspberries. The main occurrence of the species is in the semi-arid zone of south-western Western Australia, especially on land regarded by farmers as being suitable for wheat production or as good-quality sheep country. It has been cleared from much of this area. The species also extends to the fringing arid zone in the north and east and the sub-humid zone in the west.

Raspberry jam grows as a large bushy shrub or small tree, usually 15-25 feet in height but sometimes smaller on adverse sites. Typically it has a short main stem which divides into numerous ascending and spreading branches which develop a broad crown. Only rarely is the trunk long enough for sawing but both it and the larger branches are often suitable for fencing material in the round. The wood is very heavy, very hard and durable in the ground. The usually attractive grain, often with fiddleback, combined with the dark reddish colour make the wood suitable for small ornamental boxes and similar uses. Formerly it was also used for machinery bearings, sheave blocks and heavy turnery.

A. aneura F. Muell. **MULGA**

Mulga is one of the most widely distributed acacias, extending from the coast of W.A. to within a few hundred miles of eastern coastal Queensland. There are three major areas of occurrence—central south-western arid Western Australia, central South Australia extending into north-western New South Wales and southern Queensland as far north as about 24°S. latitude, and a third area in the southern Northern Territory where it is associated with northern spinifex (*Triodia* spp.). These three areas are estimated to total about 566,000 square miles and, in addition, there is probably another 500,000 square miles in which it occurs as a minor species.

This acacia, as at present accepted by botanists, comprises at least 12 forms which can be distinguished on either growth habit or leaf size and shape. Associated with these differences are variations in palatability to sheep and cattle. There are several closely related acacias which resemble *A. aneura* and are known by the same common name, for example *A. adsurgens* and *A. luerssenii*, whilst some have the name mulga incorporated in their common names, for example umbrella mulga (*A. brachystachya*) and horse mulga (*A. linophylla*). Typically mulga is a bushy shrub or small tree 15-25 feet high, though it may attain 35 feet, and a short trunk which often divides at 4-8 feet into a number of diverging branches.

Most of the area in which mulga grows experiences a rainfall of 8-12 inches and with some 12-15 inches, high variability and associated prevalence of drought. For instance a district with a 10-inch mean annual rainfall may have a 10 percentile value of 4-5 inches and the lowest recorded of 2-3 inches. Sheep in the south and cattle in the north graze most of this country and pastoralists have a high regard for mulga as an emergency drought feed. There are also a number of useful publications which refer to mulga and attention is drawn to writings by Everist, Chippendale, and Stannard and Condon, which are listed at the end of this chapter. Everist gives advice on the management of mulga and the feeding of it to stock, with information on supplementary rations such as boneflour, salt and molasses.

In the arid zone it is usually sound husbandry to manage mulga so that it is available as a reserve supplementary fodder in times of drought. The principles of natural regeneration are discussed elsewhere in this book, whilst State agricultural and conservation departments are able to provide specialised advice on securing the best results from mulga in their climatic areas.

The timber is hard, heavy and durable, with a dark-coloured heartwood and a narrow pale-coloured sapwood. It can be used in the round for fencing material and is a favoured Australian timber for the manufacture of small ornamental items, especially those of a souvenir type.

A. argyrodendron Domin **BLACK GIDYEA**

This acacia is a small tree, 15-20 feet high, which has a restricted natural occurrence in semi-arid Queensland, mainly between 20-22°S. latitude. It often forms dense thickets, for example on the flood plains of the Belyando River and the Torrens Creek which are located south of Charters Towers and between that town and Hughenden respectively. It has a good local reputation as sheep and cattle fodder; both the fallen leaves and the heavy production of pods are readily eaten.

A. aulococarpa A. Cunn. ex Benth. **BROWN SALWOOD**

Brown salwood, also known as hickory wattle, is a small to moderate size tree which occurs across northern Australia from the Kimberley region of northern Western Australia, through the Northern Territory, to the east coast of Queensland and as far south as north-eastern New South Wales. Under optimum conditions approaching those of rainforest it attains to 80 feet in height and 2½ feet in diameter; on the other hand in the drier sub-humid zone it may be reduced to a small tree only 20 feet in height. Climatically this zone of occurrence is tropical to sub-tropical, with maximum rainfall in summer and high temperatures throughout all or most of the year. Rainfall varies over a wide range from about 20 to 60 inches a year.

The foliage is only eaten by stock under drought conditions. The timber is light brownish in colour, about 50 lb/cu ft, of moderate strength and has a durability which is adequate for use where exposed to the weather but not in contact with the ground. The heartwood is suitable for a wide range of uses—framing and boards both exposed and protected from the weather, flooring, joinery and furniture. The narrow, light-coloured sapwood is susceptible to *Lyctus*. The light yellow pollen is used by bees at the early stages of flowering but the species is soon abandoned as the flowers commence to wilt, and during dry periods. Apiarists consider the pollen a poor bee food.

A. beckleri Tindale **BARRIER RANGE WATTLE**

The common name suggested for this species has been given because the holotype was collected from a glen to the Nothungbulla Gorge, Hodgson's Basin, near the Barrier Ranges in the arid south-west of New South Wales. This wattle is a small, straggling or erect bushy shrub 3-10 feet in height. It branches freely from at or near ground level and the spread of the bush may be appreciably greater than its height. It occurs naturally in parts of the arid and semi-arid country of southern Australia—at Gawler, not far from Adelaide, and in the Flinders Ranges north of Spencer Gulf, South Australia, and in the Broken Hill district and the Dubbo-Medooran area of New South Wales. Summers are hot with mean maxima of the hottest month 85-95°F, and winter days mild but with a few frosts. Rainfall varies from 9-20 inches, with moderate to high variability; in the drier parts Broken Hill has a 10 percentile value of 5.15 inches and a lowest on record rainfall of 2.24 inches. The species has been found on sandy and granitic soils, and on permatite hills.

Barrier Range wattle is an attractive, compact, earth-hugging species which should be considered for ornament, low hedges which do not need trimming, and in shelterbelts combined with taller-growing species which thin out near the ground. When flowering in late winter or early spring the flowers have a fragrant odour. It is recommended principally for use in areas of 8-15 inch rainfall, but can be expected to grow satisfactorily on sandy soils and well-drained sites where the rainfall is as high as 25 inches.

A. bidwillii Benth. **CORKWOOD WATTLE**

More than one of the acacias has the common name of corkwood wattle, another being *A. sutherlandii*. It was named after J. C. Bidwill (1818-1853), a horticultural botanist who acted as Government Botanist in New South Wales (1847) and was later Commissioner of Crown Lands in Queensland. This wattle extends across northern Australia from the Kimberley region of northern Western Australia through the Northern Territory to northern coastal Queensland near Bowen. High temperatures prevail throughout most or all of the year and the rainfall of 20-60 inches occurs mainly in summer; winter and spring are usually very dry. Climatic zones include the semi-arid, sub-humid and humid. The size of the tree varies from only 20 feet in height in adverse sites to 40 feet or more in optimum locations. It is often found on clay loams and clays, especially those with a gravelly content. Corkwood wattle has limited value as a drought fodder but is of use for shade and ornament.

A. brachystachya Benth. **UMBRELLA MULGA**

This acacia, also known as turpentine mulga in Queensland, has a geographical range which is generally comparable with that for mulga (*A. aneura*) except that in that State it occurs mainly west of the Paroo River in the south-western corner. Typically it is a rather bushy shrub, 8-20 feet in height, which branches from just above ground level to develop a wide, spreading crown seldom wider than the shrub is high. The foliage is sometimes eaten by stock but is not as palatable as that of true mulga. The wood, however, is somewhat comparable but, because of the generally smaller size, is not used as extensively for posts or ornamental work.

A. cambagei R. T. Bak. **GIDGEE**

This species is widespread in eastern Australia north of about Broken Hill in New South Wales and has its greatest occurrence in arid and semi-arid Queensland where it mainly grows on clays and clay loams where the rainfall varies from 10-25 inches. This is mainly a region of summer rainfall with periodic very severe droughts. The common name is of Aboriginal derivation and has several alternative spellings, for example gidyea and gidya. It is also sometimes called stinking wattle since the foliage of this species and of Georgina gidgee (*A. georginae*) have a very unpleasant odour, particularly during humid weather. Gidgee is a small tree, often up to 25 feet in height and when growing in thickets have straight trunks up to 15 feet in length and 9 or more inches in diameter. The specific name honours R. H. Cambage, at one time Chief Mining Surveyor of N.S.W., and an enthusiastic botanist.

The foliage has little value for stock feed, but the timber is heavy, hard, and has a close grain which is often interlocked. The heartwood is dark-red to almost purplish in colour and the sapwood is very narrow. Like most of the acacias the timber is very durable and is used for fence posts, mainly in the round; and is a first-class firewood. Sometimes it is used for ornamental work.

A. coriacea DC. **WIREWOOD**

This acacia is an excellent example of the value, in many cases, of using botanical names rather than local names. The present writer is adopting 'wirewood', since it was applied to the species during the last century; there are, however, two other names in current usage—desert oak in Queensland and dogwood in the Northern Territory. The first of these two names is already well established for *Casuarina decaisneana* and 'oak' is often used in compound names for species of that genus; the second name, whilst not used for many species in Australia, is well established for a well-known North American ornamental tree.

Wirewood has an extensive occurrence from north-western tropical Western Australia through the southern half of the Northern Territory into Queensland to within a few hundred miles of the east coast. There are some extensions into northern South Australia and New South Wales. It is usually found on sandy soils, often of some depth, in regions of 10-20 inches annual rainfall associated with high variability. It is a shrub, usually 10-15 feet in height, but in heavily-stocked areas may be kept browsed to much lower heights. In some of the western spinifex areas it may occur as the most common large shrub. Apart from limited value as a browse plant its main use is protecting soil and providing light shelter and shade in arid locations. When the timber is large enough it may sometimes be used in the round for fencing material.

A. cunninghamii Hook. **CUNNINGHAM'S WATTLE**

This wattle was named in honour of Alan Cunningham (1791-1839), a pioneer botanist and explorer of eastern Australia. His body is buried in the Royal Botanic Gardens, Sydney. The tree is also known as black wattle. The standard trade common name of the timber is curracabah. In New South Wales it occurs both on the central and northern divisions of the Western Slopes and in the northern coastal areas. In Queensland it extends inland for more than 200 miles from the coast. The climatic zones range from humid to sub-humid and even into the semi-arid. Rainfall is mainly in summer and has a wide range of 20-60 inches. This wattle grows both on the edges of rainforest and in open savannah woodlands on the inland side of the Great Dividing Range; at times it may form pure stands, especially on sandy or stony soil.

Cunningham's wattle has some value for shade and ornament and is a useful conservation species. Stock eat the phyllodes in times of scarcity but the species is not of major importance as a drought fodder. It has no value for nectar and, like several other members of the genus, the pollen has only a moderate food value. Bees do not gather pollen during dry periods.

A. cyanophylla Lindl. **ORANGE WATTLE**

Orange wattle is a large, bushy shrub or small tree which is a native to the semi-arid and humid zones of south-western Western Australia, with some extension into the adjacent arid zone; mainly it is a species of 15-40 inches annual rainfall. Height varies, of course, with the conditions under which it is growing but it is usually in the range of 10-25 feet, and in open formation the spread of the branches may give the plant a greater width than height.

The attractive orange-yellow blossoms, which appear in early spring, result in the tree being esteemed for ornament; it is also useful for shade for small animals and fowls around homesteads. Like most moderately hardy, wide-spreading plants it is of value for soil and sand stabilisation, and has the added advantage that it is moderately salt tolerant.

A. estrophiolata F. Muell. **IRONWOOD**

This acacia is only one of several trees which are called 'ironwood'; one is an acacia (*A. excelsa*), whilst another tree is a species toxic to animals (*Erythrophloeum chlorostachys*) which is a tree of northern Australia. Under favourable conditions the acacia described here is a tree which may attain 40-50 feet in height and 12-18 inches in diameter. The graceful habit of the mature tree is due to the slender terminal branchlets and the somewhat dense, pendulous willow-like foliage; the habit of the young trees, on the other hand, is quite different—it is erect with shorter and broader leaves. The partiality of cattle to the foliage and even the smaller branchlets results in shaping of young plants and the trimming back of all accessible foliage on the larger trees.

The main occurrence is in the southern half of the Northern Territory, especially around Alice Springs and the ranges towards the border of Western Australia and South Australia; it also extends to the Everard and Musgrave Ranges in the far north-west of the latter State. Ironwood prefers well-drained sandy soils; typically it occurs as scattered trees in open woodlands, sometimes with species such as mulga (*A. aneura*) and Georgina gidgee (*A. georginae*).

In its native area it is highly regarded as a supplementary fodder, whilst as one of the larger trees in a hot and arid area it is also of value for shade and shelter. Like most of the acacias the timber is heavy, hard and has a dark-coloured heartwood with a narrow, light-coloured sapwood. It is strong and sufficiently durable for fence material, though the area in which it occurs is mainly a location of very large, unfenced cattle stations.

A. excelsa Benth. **IRONWOOD**

This acacia, which has been known as ironwood since the nineteenth century, is a species of semi-arid and sub-humid eastern Australia, especially Queensland where it extends right across the State and southwards from about 19°S. lat. to cross the southern border into the central northern parts of New South Wales. There is some extension into the eastern region of southern Northern Territory. The principal occurrence has a rainfall of 15-25 inches but with very high variability—for a location with a mean annual value of 20 inches the 10 percentile value may be 8-10 and the all-time low only 5 inches. There is a marked summer maximum precipitation throughout most of the area.

Ironwood is one of the largest species of the genus *Acacia* and, under favourable conditions, may attain 80 feet in height and 3 feet in diameter; more commonly it is 25-50 feet high and 12-18 inches in diameter. Well-grown mature specimens are of attractive appearance as a result of drooping terminal branchlets and moderately dense foliage. The bark is hard, dark in colour and deeply furrowed.

Wherever this species grows naturally on their holdings land owners would be well advised to retain sufficient to provide adequate shade for stock and, where desirable, shelter. The foliage is of some value as a drought fodder but should only be regarded as a supplement to other feed. It is a species which should be considered for farm planting in its climatic area. The timber is very heavy, exceptionally hard, strong and tough but there are variable reports on its durability. In general, it is considered sufficiently durable for fencing; it has been used for mining timbers where strength and toughness are important. Many people appreciate it as an attractive timber for ornamental purposes since it has a dark brown heartwood and a light yellowish sapwood but, because of its hardness and difficulty in working, it is not now used extensively for this purpose.

A. georginae F. M. Bail. **GEORGINA GIDGEE**

Georgina gidgee has a more restricted area of natural occurrence than the closely related gidgee (*E. cambagei*); it grows naturally in the south-eastern quarter of the Northern Territory from the border of South Australia northwards to Barrow Creek and eastwards into the adjacent areas of south-western Queensland, mainly around Urandangie and the basin of the Georgina River. In parts of the South Barkly area of the Northern Territory it may be the dominant shrub over hundreds of square miles. The climate is very dry and very hot; the mean annual rainfall is 9-11 inches with a marked variability, for instance Barrow Creek in the Northern Territory with a 50 percentile value of 10.44 inches has extremes of 2.76 and 38.86 inches. Temperatures are high at all seasons, with mean maxima of the two hottest months over 100°F.

This acacia is a bushy shrub, often only 12-15 feet high, with the main stem branching a few feet from ground level, and usually a dense crown. It is reported to be slow-growing and long-lived; one writer has suggested that it may attain 100 years, but this is probably somewhat optimistic. Georgina gidgee shows a preference for soils containing lime and, whilst growing on chalcedonic ridges, is not common on deep sands. In its native area it is of value for drought fodder, though there are variable reports on its toxicity; further information on this aspect is given in papers by Chippendale and Everist listed at the end of the chapter.

A. harpophylla F. Muell. ex Benth. **BRIGALOW**

Brigalow is characteristically a species of the semi-arid and sub-humid areas of Queensland, chiefly southwards from the latitude of Rockhampton (c. 23½°S.) to the border of New South Wales. In the latter State it is mainly in the northern part of the Western Plains division, but there are minor occurrences further southwards. Most of the area in Queensland experiences an annual rainfall of 15-30 inches, but the range is 12-40 inches; precipitation is in summer and variability is high. It is warm to hot throughout the year and the mean maximum of the hottest month is 95-100°F.

This acacia is a small to moderate-size tree, mainly in the range of 25-50 feet but, under favourable conditions, attaining 60 feet in height. At its best the main stem may yield a useful log up to 15 feet long and 9-15 inches in diameter, but usually the trunk is short and the tree exhibits an irregular branching habit. In its principal area in subtropical Queensland, mainly 200-400 miles from the coast, brigalow grows in almost pure dense thickets but, because in most years the rainfall there is adequate for dryland cropping or partially-improved pastures, large parts have been completely cleared. Much of this development has been done since World War II. Like many other species of the same genus, brigalow grows from root suckers and this, combined with the longevity of the seed, presents difficulties when complete eradication is desired.

When grown on favourable sites with sufficient growing space brigalow is a useful ornamental species and also merits consideration for shelter and shade. Farmers who have it growing on their property should hesitate before planning any programme of complete eradication. The timber is heavy, hard and remarkably strong; it is hard on tools but planes to a fine finish and takes a high polish. Durability is adequate for use as fencing material and it is a good fuel and source of charcoal. Formerly it was used for fishing rods and is still used, to a limited extent, for cabinet and other ornamental work.

A. homalophylla A. Cunn. ex Benth. **YARRAN**

This acacia has its principal occurrence in the semi-arid zone of New South Wales, but extends into southern Queensland and is found, less commonly, in the adjoining arid and sub-humid zones. The most southern occurrence is along the Murray River in Victoria. The mean annual rainfall in Queensland is chiefly 20-25 inches and in New South Wales 15-20, though the species does occur where it drops to about 10 inches. The specific name is derived from two Greek words which refer to the smoothness of the leaf (in this case the phyllodes, which are not true leaves but fulfil their function).

Yarran is one of the moderately large acacias in its natural habitat and the height range is 20-40 feet. It may have a fairly straight main stem up to one-third of the total height but is more often heavily branched from within a few feet of ground level. It is found on a variety of topographical sites which include river flats, plains and dry ridges; soils are typically somewhat heavy, such as clays, clay loams and red soils. Like many other tree species in the dry country, especially acacias, it may grow as solitary specimens on dry ridges but form small thickets on good soils, especially where the site has not been subject to heavy browsing during the regeneration stage.

Under favourable conditions the young trees are very ornamental, with a moderately dense crown and the typical yellow, rounded flower heads of one group of the genus. Where a landowner has it growing on his property he should plan the retention for shade and shelter of suitably located trees and thickets. The timber is very heavy, very hard and strong and the heartwood is dark brown in colour, with a faint violet odour. It is well suited for cabinet work and fancy articles where an attractive appearance is desired and heavy weight is not critical; formerly it was used for native spears and boomerangs.

A. iteaphylla F. Muell. ex Benth. **FLINDERS RANGE WATTLE**

The natural occurrence of this wattle includes, as the common name implies, the Flinders Ranges but it is also found in the Gawler Ranges; both areas are in central South Australia near Spencer Gulf. The specific botanical name means 'willow-leaved', in allusion to the narrow-shaped phyllodes which have a resemblance to those of the true willows (*Salix* spp.); this shape is, however, characteristic of many species of the genus. The mean annual rainfall is probably mainly within the range of 12-20 inches, but it should be noted that the Forests Department of South Australia only recommends it for planting in areas receiving more than 17 inches. The mean maximum of the hottest month is 90-95°F, whilst there may be a few light frosts in winter.

This bushy shrub has a mature height of 6-10 feet and is suitable for unpruned or lightly-pruned hedges, or for use in two-tier shelterbelts where the taller species loses its lower branches at an early stage of development. It also merits consideration for ornament and where some additional water is available it will grow in areas where the rainfall is only 7-10 inches.

A. kempeana F. Muell. **WITCHETTY BUSH**

Witchetty bush is a small shrub, attaining about 10 feet in height, with many stems from near ground level and a moderately dense crown. It is found in the dry country from the coast of Western Australia to central Queensland and has its principal occurrence in the southern half of the Northern Territory; there is a small extension into the far north-west of South Australia. This is an area of 8-10 inches rainfall but near Geraldton in W.A. and Charleville in Queensland it may be up to 18 inches; associated with low rainfall is marked variability and very high summer temperatures.

It is most commonly found on limestone soils and occurs with mulga (*A. aneura*) and spinifex (*Triodia* spp.). Although a small shrub, it is of considerable importance in its area for soil protection and of minor value as an emergency fodder in times of drought.

A. ligulata A. Cunn. ex Benth. **UMBRELLA BUSH**

Umbrella bush is found throughout the southern half of the Northern Territory, most of South Australia except the moister south-eastern parts, north-western Victoria and the drier western districts of New South Wales. This acacia is not listed for Queensland but possibly occurs near the border with South Australia and the Northern Territory.

It is a bushy shrub which varies in height from 4-15 feet or more, with branches usually retained to near ground level. Umbrella bush, which is also known as small cooba in New South Wales, is closely related to cooba (*A. salicina*) but it has more rigid branches than that species; formerly it was considered a variety. The pods are distinctive in that they are markedly constricted between the seeds and become brittle on drying. The name dune wattle is also used, no doubt due to its preference for sandy soils and ability to stabilise drifting sands. In favourable years it regenerates readily and shows fast initial growth—two factors of value for soil conservation.

Umbrella bush grows naturally in some of the hottest and driest parts of Australia and will exist where the average annual rainfall is only 7-8 inches, but when planted for low windbreaks on sandy soils a rainfall of 10-16 inches is desirable.

A. microbotrya Benth. **MANNA WATTLE**

Manna wattle is a moderately dense bush, usually in the height range of 12-20 feet, which is native to the semi-arid and part of the sub-humid zone of south-western Western Australia. This area includes the main wheatbelt between Perth and Kalgoorlie and sheep-wheat country northwards from Perth to Geraldton. Rainfall shows a wide range of 11-20 inches, but the Forest Service of the State does not recommend it for planting in areas drier than about 15 inches. On suitable lowland sites which receive drainage from adjacent slopes it could possibly be planted in slightly drier areas, but the advice of the Forest Service should first be sought.

This useful wattle is recommended for planting in southern Australia for low shelterbelts or where a lower storey is required for two-tier belts, for ornamental use and, because of its good early growth, for shade for sheep whilst larger and longer-living trees are developing. Although it grows naturally on some poor sites, for example granitic slopes, it shows a preference for loamy soils.

A. notabilis F. Muell. **NOTABILIS WATTLE**

This acacia has a scattered occurrence in South Australia south of the Flinders Ranges and in the Broken Hill district of New South Wales. The climate has hot to very hot summers and a rainfall which is about 10 inches at Broken Hill but approaches 20 inches in the wetter districts in the south of South Australia.

It varies from a small shrub perhaps only 4-5 feet high to a large plant 15-20 feet in height; it has a bushy habit and normally retains branches and foliage to near ground level. It offers possibilities for low shelter as well as soil protection in dry areas.

A. oswaldii F. Muell. **UMBRELLA BUSH**

This acacia is a densely branched shrub or even a small tree branched from near ground level, usually 6-15 feet in height but sometimes attaining 20 feet or more. The common name adopted here was originally applied in recognition of the shape of the tree, especially when stock had eaten the foliage from the lower branches, and the shade which it provided. Other common names included nelia (Queensland), umbrella acacia (Victoria), miljee (New South Wales Aboriginal name), midget and ram's horn tree (Queensland). The specific botanical name was given in honour of Ferdinand Oswald, once living in Adelaide but who later returned to Germany. It usually grows as scattered specimens in open eucalypt forest or among other acacias.

Umbrella bush is found throughout the semi-arid and arid areas of eastern Australia south of about 20°S. latitude, and extends into the sub-humid zone on the Queensland coast near Rockhampton. It is also found in the mid-latitudes of South Australia from Adelaide to the Transcontinental railway and to parts of the dry country of Western Australia, especially around Kalgoorlie.

Reports on its palatability to sheep and cattle vary and it is probable that this is due to whether drought or over-grazing are responsible for shortage of feed. Sheep eat the foliage in spite of the sharp points on the phyllodes and it seems likely that these caused irritation in the throats of animals and were responsible for one of the common names, coughing bush. The pods, which are often produced in abundance, are readily eaten by sheep, and formerly the Aborigines ground the seeds into a paste which they consumed. The wood is a very good fuel.

A. pendula A. Cunn. ex Benth. **MYALL**

Myall is one of the most attractive species of the large genus *Acacia*. The specific botanical name of *pendula* was given by the author of the botanical name in reference to the drooping branchlets which reminded him of the weeping willow. The ornamental appearance is further enhanced by the silver-grey phyllodes ('leaves') which are also pendulous. Myall is a small tree, usually 20-30 feet in height, but sometimes attaining 40 feet and diameters of 1½-2½ feet on favourable sites. The trunk may be one-quarter of the total height but usually divides into several stems within a few feet of the ground. The tree is also known as weeping myall and boree, but the latter name is deprecated since it is used for another species (*A. cana*) which grows in Queensland. Myall prefers good soils—rich alluvial, clays or black soils, with access to adequate ground water; its occurrence is usually a sign of good land for farming.

This acacia has its principal occurrence in the semi-arid zone of New South Wales and Queensland, mainly south of 23½°S. latitude; it also extends into the sub-humid zone and, to a lesser extent, the arid zone. The summers are hot to very hot and rainfall is 12-25 inches, with a more or less uniform incidence in the south but varying to a marked summer maximum in the north.

Myall should be considered both for retention in its native habitat and for planting whereever the climate and soils are suitable. It has many desirable attributes—it provides a useful drought fodder, is a very ornamental species, is suitable for shade and shelter planting on farms and has an attractive timber. Due to the liking of stock for myall foliage natural regeneration may need some protection during early years, whilst trees planted for low shelter should also be fenced. With age the crown of the tree is carried higher and shelterbelts may need an associated species to provide protection to ground level where such is needed. The handsome dark-coloured wood is very heavy, very hard, close-grained and has a distinctive odour of violets which is better developed than in most woods. It is useful for making ornamental objects and is, of course, an excellent fuel.

A. peuce F. Muell. **WADDYWOOD**

This small tree, 25-50 feet in height, is remarkable in growing to such dimensions in one of the most adverse climates in Australia. The natural occurrence consists of a few small areas in the south-east of the Northern Territory, including part of the Simpson Desert, the extreme south-east of Queensland and a small adjacent area of South Australia. In this region the mean maximum temperatures of the hottest month are 95-100°F, and individual daily maxima frequently exceed 110°F. The rainfall is some of the lowest in Australia; Birdsville, for instance, has a 50 percentile value of 4.51 inches and an all-time low of 1.30 inches, whilst Finke has a mean annual rainfall of 6.17 inches. Potential evaporation is exceptionally high at 115 to over 130 inches a year.

In general appearance the tree has a resemblance to species of *Casuarina* and in the Northern Territory is sometimes known by that name. The specific botanical name comes from the Greek *peuke*, a pine-tree, referring to some similarity of general shape with trees of the genus *Pinus*. Waddy wood shows variation in form from a narrow-crowned, erect single-stemmed tree, to one which divides into two or more stems near ground level, and has pendulous branchlets with drooping phyllodes ('leaves') which may vary from 2-12 inches in length.

When planted around homes in Alice Springs this acacia has shown promising early growth, but nothing is known of its further development. The species certainly deserves consideration for ornamental and shade planting in very arid areas. L. D. Pryor (1967) suggests that in view of its extraordinary performance in its native habitat small trials are justified as a plantation species. There are about 3,000 seeds per pound and plants can be readily raised by soaking the seed in hot water, as is successful with most other species of the genus.

Everist (1969) reports the foliage is eaten by cattle and that the wood is used for fencing material and stock yard construction. The timber is very heavy, hard and has proved durable to termites even after 40 years use; the heartwood is very dark in colour.

A. podalyrifolia A. Cunn. **QUEENSLAND SILVER WATTLE**

This ornamental shrub, growing to about 15 feet in height, is a native of south-eastern and eastern Queensland, mainly near the coast, but there are a few localities where it grows naturally in the northern parts of the Western Slopes of New South Wales. It flowers profusely for a longer period than most other wattles and is one of the most popular cultivated species. It can be used both for ornament and temporary shelter but, because it is short-lived, other shelter should be planned to replace it after 10-15 years.

In the dry country Queensland silver wattle should be restricted to the higher rainfall parts, that is above 20 inches, of the semi-arid zone or used in the wetter areas. It will grow on nearly any soil but is somewhat sensitive to frost, especially in early life.

A. pycnantha Benth. **GOLDEN WATTLE**

When in flower this wattle well merits the epithet of 'golden', since the complete plant may be a mass of attractive golden-yellow blossoms. The specific botanical name also refers to the flowering, since it comes from two Greek words, *pyknos* (dense), and *anthos* (a flower), hence dense-flowered. Other common names include South Australian golden wattle (South Australia) and broad-leaved wattle (New South Wales). The species varies from a well-branched bushy shrub, 15-20 feet high, to a small tree 30 feet or more in height with a somewhat open crown. Stem diameters rarely exceed 12 inches.

Since the middle of the nineteenth century golden wattle has been extensively planted in south-eastern Australia outside of its natural area, so that a distinction should be made between its natural range and where it has been introduced. Originally it grew in South Australia, south of the Flinders Ranges, much of the moderate rainfall area of Victoria and only those parts of New South Wales adjoining those two States. Now it has been acclimatised throughout Victoria with the exception of the driest north-west and the Alps, and southern and central coastal New South Wales, including the Sydney and Blue Mountains areas. In its native area the summers are mild to hot and rainfall mainly 15-30 inches, but extending to over 40 for areas where it has been acclimatised. It has a reputation for being drought hardy and this is supported by fine specimens at Broken Hill which do not get additional watering. Young plants are somewhat frost tender and care needs to be taken when planting them in areas where frosts occur of greater severity than 27°F in the screen.

Nowadays golden wattle is mainly prized as an ornamental species in areas with hot, dry summers and a rainfall of 15-30 inches. It has also been used with sugar gum (*E. cladocalyx*) as the lower storey of two-tier windbreaks. Formerly it was highly prized for the bark, which is one of the richest sources of tanning materials in the world. The amount of tannin from the bark of the trunks is usually 30-40 per cent, but higher returns have been obtained. The older and thicker bark from the trunks of mature trees usually contain the highest percentage of tannin and can be stripped as soon as the plants carry a worthwhile return. This may be at 7-10 years from planting on good sites. The trees, of course, die after the bark has been stripped and the wood is available for fuel. Nowadays the establishment of golden wattle in plantations, usually by the direct sowing of seed on prepared soil, offers little financial attraction due to the competition from other tanning materials. There is still a limited amount of bark peeling done as a side line to other rural activity.

A. rigens A. Cunn. ex G. Don **NEALIE**

Nealie occurs in South Australia, mainly from the Flinders Ranges southwards to Eyre Peninsula and the Murray River area, north-western Victoria and arid and semi-arid New South Wales where it is fairly common in the Roto-Condobolin-Wyalong-Griffith area. It extends to the border with Queensland. This is a region of hot to very hot summers. The rainfall is 9-19 inches, either mainly in winter or more or less uniformly distributed.

This acacia may be a small shrub 6-10 feet high or may develop tree form and attain heights of 15-20 feet. The specific botanical name comes from the Latin, *rigens*, referring to the stiffness of the narrow, needle-like phyllodes; some forms of the species have broader, silvery 'foliage' and these trees are useful for ornamental purposes. The tree, in general, is just one of many which has a value for conservation in the dry country. Another common name is needle-bush acacia (South Australia) in reference to the phyllodes of the typical form. The timber is dark brown and resembles that of myall (*A. pendula*), but the stems are usually too small to be of value for any purpose other than firewood.

A. salicina Lindl. **COOBA**

Cooba has several other common names and probably the most important is that of native willow, which refers to the drooping branchlets and foliage which gives a general resemblance to some of the true willows (*Salix* spp.). The specific botanical name is based on the Latin, *salix, salicis* (a willow). The general form of the tree has a marked similarity to that of myall or weeping myall (*A. pendula*), *q.v.* In New South Wales the occurrence of these two species overlap and care needs to be taken when identifying them at a distance on the basis of general form; a closer examination, especially when pods are available, will readily separate the two species. Like myall, it is found on alluvial and clay soils on river flats, but it also extends to sandy soils on higher land. The common name of doolan is used in Queensland, whilst Broughton willow is the official name in South Australia.

The principal occurrence of cooba is in the semi-arid zone of New South Wales and Queensland, south of about 20°S. latitude; it extends, however, both into the arid zone on the inland side and to the sub-humid and even humid zones of coastal Queensland. It is also found in north-western Victoria, scattered throughout much of South Australia, extends as far north as Alice Springs in the Northern Territory and is reported to have a small occurrence in Western Australia. The climatic range is considerable but probably the best development is in the semi-arid and adjacent sub-humid zones of New South Wales and Queensland, where summers are hot to very hot and the annual rainfall 15-22 inches in New South Wales but up to 40 inches in Queensland. The total range of rainfall is, however, about 8 to 60 inches. Because of this variation in climate care needs to be taken in the source of seed when it is planned to introduce the species to an area outside of its natural habitat.

Cooba is typically a small, somewhat bushy tree with a wide crown, 20-40 feet in height. The trunk is usually fairly straight and up to one-quarter of the total height, above which it divides into several stems. In the north of its range, which is within the tropics and with a rainfall of 25 inches or more, it attains heights of 50-60 feet and diameters of 2-2½ feet.

The drooping habit and pendulous silver-grey 'foliage' presents an attractive appearance and in suitable areas cooba should be considered for both park and garden planting. Together with myall it is one of the most ornamental acacias which occur in the dry country of eastern Australia. It provides good shade and is also useful for shelter on farms. The tree often suckers freely from the roots and this adds to the value of the species for many forms of soil conservation. As a fodder it is eaten by stock, but not as freely as many other species. The wood is dark brown and attractively marked, close-grained, tough and only moderately heavy (about 48 lb/cu ft); it has been used for furniture and ornaments, as well as making boomerangs.

A. shirleyi Maiden **LANCEWOOD**

In northern Australia lancewood is the common name of several other species of trees and large shrubs with very hard, strong wood, although in the Northern Territory this wattle is the only species of the genus known by that name. In Queensland, on the other hand, the name is also applied to *A. doratoxylon* to which it is very similar except for the fine structure of the flowers. The occurrence in the Northern Territory is near and west of Newcastle Waters, which is located about 80 miles south of Daly Waters. The latitude of the principal area is about 17-18°S., whilst altitude is mainly 700-1,000 feet. The range of rainfall is 15-25 inches, with a summer incidence and high variability. In the Northern Territory it does not extend eastwards on to the black soils of the Barkly Tablelands.

The occurrence in Queensland is disconnected with that in the Northern Territory, but it is still within the same limits of rainfall. The type specimen comes from the Eidsvold area which is about 25°S. latitude and has a rainfall of 30 inches. In that State the altitudinal range is mainly 500-1,000 feet. Both in the Northern Territory and Queensland the species occurs chiefly in the semi-arid zone.

This lancewood is a small tree, usually 30-45 feet high, and when growing in dense thickets has a straight, erect stem and a crown carried high above the ground. It is often found on rather stony soils.

Everist (1969) reports that the phyllodes ('leaves') are eaten fairly readily by stock but that there are reports of local difference in palatability. In the Northern Territory, however, it is not regarded as a topfeed species. The timber is not sufficiently durable for use in the ground but when growing in dense scrubs the form of the stems is suitable for use as rails.

A. sowdenii Maiden **WESTERN MYALL**

This small wattle grows in the height range of 10-15 feet in the arid and semi-arid south-west of South Australia and adjacent areas of Western Australia, mainly within 150 miles of the southern coast. The 50 percentile rainfall value varies from 6.55 inches (Rawlinna) to 9.79 (Eucla), with the corresponding 10 percentile values 3.36 and 6.69 inches respectively. The lowest rainfall on record for Port Augusta, over a period of 106 years, is only 2.21 inches. Temperatures vary appreciably, for instance the mean maximum of the hottest month for Eucla on the coast is only 78.3°F, but 100 miles inland the value is about 90°F. Frosts vary from 0-5 a year.

Western myall has a very bushy habit, usually with many branches from near ground level. In southern Australia, where the rainfall is 8-18 inches and the soils are sandy or sandy-loams, it is a promising species for use as low, unpruned hedges or in conjunction with taller growing eucalypts where wind protection is desired to ground level. The Forests Department of South Australia recommends it for use in areas as dry as 8 inches, whilst the Forests Department of Western Australia have used it in numerous trial plots.

A. spectabilis A. Cunn. **PILLIGA WATTLE**

This is a wattle of very ornamental appearance which usually grows in the height range of 10-15 feet. It has several other common names, such as golden, silver and glory wattle. The shrub is of very open habit, with small diameter branches which are often contorted and twisted. Pilliga wattle has its principal occurrence in the central subdivision of the Western Slopes of New South Wales, but it extends northwards into the adjacent semi-arid and sub-humid zones of southern Queensland. The mean annual rainfall is in the range of 20-30 inches, with a uniform or moderate summer maximum. Summers are hot and the mean maximum temperature of the hottest month is around 90°F, whilst winters experience 5-20 frosts a year.

In the late winter and early spring Pilliga wattle produces an abundance of golden flowers in long, drooping racemes. It is a species for ornamental use and has negligible value for shelter. A preference is shown for sandy soils but, in the Darling Downs district of south-eastern Queensland, it has been found to grow satisfactorily on much heavier soils if they are well drained.

A. stenophylla A. Cunn. ex Benth. **RIVER COOBA**

This acacia grows to larger dimensions than most of the dry country members of the genus —well-grown specimens may attain 60 feet in height, with trunks of moderately good form and up to two-fifths of the total height; diameters may be 2-2½ feet. It is more commonly seen, however, as a somewhat bushy tree 35-45 feet in height, with the trunk divided into several large stems at 5-10 feet from ground level. It prefers clay and other alluvial soils and is not usually seen on sandy areas; in the drier part of its habitat it grows on river banks and along intermittent watercourses.

The specific botanical name is derived from the Greek words meaning narrow-leaved; the phyllodes are distinctive in being very long and relatively narrow, that is 6-16 x 0.1-0.25 inches. These long, pendulous, greyish coloured phyllodes present a distinctive appearance and together are responsible for the alternative common names of willow and native willow; names which, however, are also applied to several other acacias with drooping foliage. In Queensland the preferred name is belalie, whilst eumong is listed for Victoria; this latter name, however, is not recommended since it is used for a number of acacias, especially those with a drooping habit.

River cooba has a wide geographical range from tropical northern Western Australia to temperate semi-arid north-western Victoria—a distance of over 1,700 miles. It is scattered in the Northern Territory, mainly between 17-21°S. latitude, and in the arid and semi-arid zones of western Queensland. In South Australia it occurs mainly in the dry far north, but is also found in the vicinity of the Murray River in the south-east of the State. The principal occurrence, however, is probably in the semi-arid zone of New South Wales. Summer temperatures are high throughout the area of distribution but winter frosts occur in the southern half; rainfall is mainly 12-18 inches in the south but as high as 25 inches in parts of Queensland, there is a sparser occurrence where it drops as low as 8 inches. Variability is high.

The wood is very dark reddish-brown to almost black and attractively marked; only moderately heavy for an acacia, hard, close-grained, works satisfactorily and takes a polish readily. Where it occurs naturally it is used for fencing and other farm purposes, and is an excellent firewood; but it is suitable for furniture and ornamental work where weight is not critical.

A. tetragonophylla F. Muell. **DEAD FINISH**

Dead finish is a shrub with irregular branching and usually 10-15 feet in height, which grows on sandy soils on flats or along intermittent watercourses or is found scattered in mulga (*A. aneura*) or other acacia vegetation. It has a wide distribution in the arid zone of Australia, mainly south of 21°S. latitude, from the coast of Western Australia to the border of the semi-arid zone in central New South Wales and Queensland. It is absent from Victoria and the south-east of South Australia. Over much of the area the rainfall is only 7-10 inches and has very high variability; mean maximum temperatures of the hottest month are around 100°F. The name dead finish is used in most States but is also applied to other species, for example *Albizia basaltica* in Queensland.

Reports on the value of the foliage as a drought fodder vary from State to State; it is however, agreed to be fairly palatable but of low nutritive value. In Queensland it is stated that the phyllodes are shed during severe drought, lessening the value of the shrub at a time when the demand for drought fodder reaches a peak. Like most other bushy plants in the arid zone it has value for shade and in soil conservation.

A. victoriae Benth. **PRICKLY WATTLE**

It is a wide-spreading, bushy shrub often with several stems from near ground level, and a height of 10-15 feet. Grazing of the foliage on the lower part of the plant may leave it open near ground level, but in protected areas it retains foliage low down. The first part of the common name refers to the spines which are usually found at the base of the phyllodes, whilst the specific botanical name was based on the Victoria River or Upper Barcoo in Queensland, where the type was collected. Other common names include elegant wattle (South Australia) and gundabluey (Queensland). Prickly wattle occurs in all mainland States other than Victoria, and is moderately common in the Northern Territory as far north as about 17°S. latitude; in New South Wales it is mainly a species of the arid zone but in Queensland it is also scattered in the semi-arid zone. The rainfall varies from 8-10 inches in southern Northern Territory, to 10-13 inches in New South Wales, and as high as 20 inches in the north-eastern parts of its occurrence in Queensland.

Prickly wattle is usually found on sandy soils and occurs both on ridges and depressions, and along intermittent watercourses. It will tolerate clipping and is recommended by some Forest Services for low shelterbelts in areas as dry as 10 inches of rainfall a year. In favourable years it has excellent natural regeneration and shows good early growth, so that it may even become a nuisance around waterholes; these attributes, however, are desirable for a plant used in soil conservation in sandy areas of the dry country.

Agonis flexuosa (Spreng.) Schau. **WESTERN AUSTRALIAN PEPPERMINT**

This is a small tree, usually in the height range of 20-30 feet, which has its principal occurrence in the humid and sub-humid districts of the south-west of Western Australia where the rainfall is mainly 25 inches or more. On favourable sites and with adequate growing space it will, however, do reasonably well where the rainfall is as low as 15 inches. It is reported to be somewhat frost-sensitive when young.

Western Australian peppermint is an ornamental tree with a well-developed crown, drooping small branches and pendant foliage. The leaves give an odour of peppermint on being bruised. In its State it has been successfully used as a street tree in Perth and is recommended for shade plantings in parks.

Since the printing of *Growing Trees on Australian Farms* a new author citation has become available, whilst the species has been listed in AS 02-1970 of the Standards Association of Australia, with the name of Western Australian peppermint. In its own State it is usually simply called peppermint, whilst in the eastern States the name of willow myrtle has become well established.

Angophora melanoxylon R. T. Bak. **COOLIBAH APPLE**

Coolibah apple, also spelt coolabah, is an irregular, heavily-branched tree with a wide crown, attaining 35-50 feet in height. The trunk is short and stout. It is a species of the northern parts of inland semi-arid New South Wales and the adjoining areas of southern Queensland; in the latter State it also extends to the sub-humid coastal areas. The rainfall is mainly 12-25 inches, with a moderate summer maximum.

Beekeepers are interested in this tree as a good source of pollen, though it is of minor value for honey production. The honey is dark in colour, with a strong flavour. The timber is sometimes used for fuel but otherwise has little value.

Apophyllum anomalum F. Muell. **BROOM BUSH**

This is a small bush, usually 8-20 feet high, in which numerous, thin, green, cylindrical branchlets take on the function of leaves to supplement the sparse foliage. These branchlets have a sharp mustard-like flavour, giving rise to one of the other common names—mustard bush. Other names which are used include warrior bush and currant bush. It has a wide occurrence in Queensland from the edge of the arid zone to the drier parts of the coast, for example near Rockhampton on the Tropic of Capricorn, and in the semi-arid area of New South Wales, particularly the northern part. In New South Wales the rainfall is 12-20 inches with a slight to moderate maximum in summer; the Queensland rainfall is mainly 15-25 inches with a summer maximum which is most marked in the north. Variability is high to very high.

Broom bush grows on a wide range of soils from sands to clay loams and is regarded as a hardy species, especially in relation to drought. Under even moderately reasonable conditions it develops into a bush quite suitable for shelter and ornament in the dry country and also has some value for shade. The thin green branchlets are eagerly consumed by sheep and cattle, but are not ranked high in nutritive value.

Atalaya hemiglauca (F. Muell.) F. Muell. ex Benth. **WHITEWOOD**

Whitewood is a tall bush to a small tree, with mature heights of 20-40 feet; the trunk, which usually divides at a low height into two or more large ascending stems, may attain 18 inches diameter. The crown is often moderately dense and will tolerate pruning which can enhance the value of the tree for shelter and ornament.

This tree has an extensive natural distribution from the coastal areas of north-western Western Australia to those of Queensland, especially in the drier region of Rockhampton; it occurs on both sides of the Tropic of Capricorn, that is from about 14-32°S. latitude. In New South Wales the main occurrence is in the northern parts of the inland area. Whitewood is not found in Victoria and only to a limited extent in the northern part of South Australia. Most of the area has high to very high summer temperatures and mild to warm winters. The rainfall is 12-25 inches in the south and as high as 30-40 inches on the monsoonal northern fringe; variability is high, with the probability of extended periods of severe drought.

Whitewood is one of the more important native species in its area and should be retained wherever it grows naturally. It is worthy of consideration for amenity planting around homesteads or in public parks on both sandy and clay soils in its climatic area. Young plants are readily grown from seed and root suckers are produced. It also ranks high as drought fodder, but there are variable reports on the toxic effects of the seed and sucker shoots. An old report on it causing staggers in horses in northern Western Australia has been proved to be ill-founded, as the causative agent was another plant which grew in association with whitewood. The timber is soft and non-durable, whilst the whitish colour gives rise to the common name.

Atriplex nummularia Lindl. **OLDMAN SALTBUSH**

Saltbush is the common name applied to members of this genus and is in reference to the ability of the species to grow in alkaline soils. *A. nummularia* is the largest representative and may attain 10 feet in height, hence the common name of oldman saltbush. In Queensland, however, this name is also applied to *Rhagodia spinescens* and *R. parabolica*. It is a small, densely-branched, often sprawling bush and long branches which touch the ground may root and an original plant, if not eaten by stock, may have a radius of 20 feet or more. With a minimum of attention it can be trained into a satisfactory shelterbelt 6-10 feet high, which gives protection to ground level; it is doubtful if there is a more suitable plant for such purposes on alkaline soils in the 8-12 inch rainfall area of southern Australia.

This saltbush occurs mainly in the arid zone of Australia south of the Tropic of Capricorn, with some extension into the adjacent parts of the semi-arid zone. Summer temperatures are normally high to very high, but some oldman saltbush grows sufficiently near the southern coast to experience cold changes with marked falls of temperature; winters are mild, with 1-10 frosts. It is most plentiful in areas of 8-14 inches rainfall but, in the northern part of its range, may be found where the rainfall is as high as 18-20 inches. Oldman saltbush occurs with other saltbushes and sometimes in open areas with sparsely scattered shrubs. It grows on a wide range of soils from sands to clays on flats and, as previously indicated, these may have so high a salt content that most other plants are unable to exist.

Oldman saltbush is one of the most palatable browse plants of the arid country and is one upon which sheep can exist almost indefinitely with a minimum of other fodder. Under continuous heavy grazing it will rapidly disappear, so where it is an important component of the fodder resources of an area, grazing should only be maintained at a level at which deterioration does not occur. Other species of the genus have been discussed in Chapter 15, browse plants.

Bauhinia carronii F. Muell. **QUEENSLAND EBONY**

This bauhinia is a shrub or small tree which may attain 25 feet or more in height on favourable sites. The principal natural occurrence is in Queensland, where it is found in the arid, semi-arid and sub-humid zones, but it extends to the northern edge of the Western Plains in New South Wales and to the north-eastern corner of South Australia. Mean annual rainfall is mainly in the range of 8-25 inches; it has a summer incidence and high variability. Queensland ebony is of ornamental appearance and can be used for public planting, as well as for shade and shelter. The leaves and pods are readily eaten by stock. Bauhinia trees are leafless during late winter and early spring and, usually during severe droughts.

B. hookeri F. Muell. **HOOKER'S BAUHINIA**

The chief occurrence of this bauhinia is in northern Queensland from the coast inland to the higher rainfall parts of the semi-arid zone, where it is common along drainage lines and the banks of streams in the 25-30-inch belt. There is some extension into the northern parts of the Northern Territory. Rainfall is mainly in summer, with a very dry period in winter and spring.

Hooker's bauhinia, also known as pegunny, is mainly regarded in Queensland as a species for planting in coastal areas, but in the higher rainfall parts of the inland it grows as an attractive tree up to 40 feet in height. It has a fairly dense, spreading crown which provides good shade. The tree is frost tender when young but is resistant to droughts.

Brachychiton gregorii F. Muell. **DESERT KURRAJONG**

This is a small tree, commonly 12-25 feet in height, with a short bole which may be only 5-8 feet long and 9-15 inches in diameter. Above this short bole the tree develops a number of large branches which form the framework of a crown which may be wider than the height of the tree. The density of the foliage is good for the environment in which the tree grows.

Desert kurrajong has its natural occurrence in the arid and semi-arid south-western part of Western Australia and in the Northern Territory south of Barrow Creek (21° 30′S. lat.). A few specimens have also been reported from the far north-west of South Australia. Most of this area has a rainfall of 8-12 inches, with high variability; summer temperatures are high to very high and winters experience a few light frosts.

In the Northern Territory it occurs mainly in areas which are not heavily stocked and there are no reports on its palatability to stock but, like most species of the genus, it is probably acceptable in times of drought. Under favourable conditions around homesteads, where a little additional watering may be possible, it is a species which could be planted for shade and ornament under very severe climatic conditions.

B. populneum R.Br. **KURRAJONG**

The kurrajong is a small to moderate size tree of eastern Australia and attains its best development in the moister parts of the semi-arid zone and the adjacent sub-humid zone of New South Wales and southern Queensland, with a rainfall of 20-25 inches in the former State and 25-35 inches in the latter. While it extends to the coast it avoids the colder parts of the tablelands and higher mountains and it is only found scattered in a few localities in the arid zone. It will grow on a wide range of soils including poor podsols, but shows best development on good light loams which are reasonably well drained.

The height range is 25-50 feet and a large tree may have a crown in width equal to or exceeding half the height. The trunk is straight and bears a very dense, deep crown with glossy green leaves which reminded the author, Robert Brown, of the specific botanical name of poplar leaves, hence *populneum*. The common name is of Aboriginal origin. Kurrajong stands pre-eminent for fodder and shade, is a striking ornamental species and can be used for avenue and park planting and in shelterbelts. For the latter use, however, a lower storey is required where protection to ground level is needed. Of attraction to the farmer who wishes to be his own nurseryman, it can be readily raised from seed and, with some pruning, can be transplanted at an advanced size. For feeding the foliage to stock it is necessary to lop the branches, but if this is done judiciously the tree will readily put out new growth and return to its previously attractive appearance. Everist (1969) has an illustration of the regrowth of kurrajong five years after lopping.

The bark of the tree contains a strong fibre which the Aborigines used for making fishing nets. Like nearly all species of the genus, the timber is soft, weak and of no economic value. Nectar and pollen production from the flowers is sufficient to be useful to beekeepers for maintenance of their hives.

B. rupestre Benth. **QUEENSLAND BOTTLE TREE**

This *Brachychiton* species has the typical swollen bole to which the common name of bottle tree is applied. In open country heights may reach 50-60 feet and even higher on the edge of rainforest; the boles of mature specimens may reach 6-8 feet in diameter, or even more. Above the top of the bole the tree divides into a number of heavy branches which form a large crown which may have a width up to three-fifths of the tree height. The leaves show different forms—those on the flowering branchlets are simple in shape but those on seedling plants and vigorous sterile shoots are compound. The specific name, *rupestre*, is derived from the Latin, *rupes*, a rock, referring to the tree sometimes being found growing amongst rocks.

The Queensland bottle tree has its principal occurrence in eastern Queensland, mainly south of 20°S. latitude, and includes both semi-arid and humid coastal areas; it is fairly common in some of the drier coastal scrubs south of Rockhampton. There is a limited extension into northern inland New South Wales. The rainfall is mainly 18-35 inches, with most of the precipitation in summer. It grows on clays and clay loams and is often associated with brigalow (*Acacia harpophylla*) and gidgee (*A. cambagei*).

The leaves are considered of comparable value for stock fodder as those of kurrajong (*B. populneum*), whilst the soft tissues in the swollen trunks contain a large percentage of water and are sometimes eaten by cattle in times of severe drought. Because of the large size mature trees attain, felling is usually necessary to make foliage available for stock, but with young trees lopping should be practised. The Aborigines made use of a gummy exudation and ate both young roots and the seeds. This species is recommended for planting in parks and other places where there is sufficient growing room—say at least 60 feet between specimens. Like the kurrajong it can be grown readily from seed and, with pruning, transplanted at an advanced size, that is 6-8 feet in height.

Callistemon viminalis Cheel **DROOPING BOTTLEBRUSH**

Drooping or red bottlebrush is a small tree or large shrub of eastern New South Wales and Queensland. It is most common in coastal areas but extends to the tablelands and the higher rainfall areas of the western slopes. Under favourable conditions along the banks of coastal streams it attains 60 feet in height but on drier inland sites it may only grow to 15-20 feet. In youth the branchlets tend to be erect but with maturity they develop a drooping habit, hence the first part of the common name. The term bottlebrush is used for the genus and is in allusion to the dense showy spikes of flowers, which have long, conspicuous filaments.

Within the dry country drooping bottlebrush will grow in areas with an annual rainfall as low as 20 inches, which is either uniform in distribution or has a moderate summer maximum, and have up to 20 frosts a year.

Drooping bottlebrush has a use for hedges and low shelterbelts, street planting and as individual specimens for ornament. It is fast growing in early life and in southern Queensland has proved very suitable for use on the heavy black soils of the region, especially in comparatively damp situations along drainage channels.

Canthium latifolium F. Muell. ex Benth. **NATIVE CURRANT**

This canthium is taken as an example of several species of the genus which are small shrubs or trees and occur in the arid zone; most of them eaten by stock. Native currant is found from the coastal areas of Western Australia around 20-26°S. latitude, through the southern half of the Northern Territory to the south-western districts of Queensland and the north-western areas of New South Wales. There is also a sparse occurrence in the northern half of South Australia. The rainfall is 8-11 inches, with high variability; summer temperatures are particularly high and a number of places have mean maxima of 95-105°F for the hottest months, whilst potential evaporation is 90-130 inches a year. *Canthium oleifolium* (myrtle tree) is a small tree with a short trunk which divides into several upright stems carrying stiff, spreading branches. It grows in arid Queensland and New South Wales and is readily eaten by stock.

Native currant is a small shrub, usually within the range of 6-12 feet. The native name refers to the way in which the fruits are borne like those of garden currants. The rather large, almost orbicular leaves are distinctive amongst the shrubs and bushes of the arid interior. Nowhere it is plentiful but in some areas, such as the southern part of the Northern Territory it is widespread as scattered specimens growing on deep red sand in spinifex communities or in mulga (*Acacia aneura*) associations.

Capparis mitchellii (Lindl. ex F. Muell.) Benth. **WILD ORANGE**

The name of wild orange was taken from the resemblance of the large fruit (about 2 inches diameter) to a small orange or mandarin. This name is used in New South Wales and the Northern Territory, in South Australia it is called native orange and in Queensland it is known as bimbil or bumble. The fully mature fruit is reported to be edible and formerly eagerly sought by Aborigines. The generic name is from the Latin for caper, *capparis*, whilst the specific name is in honour of Sir Thomas L. Mitchell, a former Surveyor-General of New South Wales at a time when it represented the continent. Wild orange is a small tree 15-30 feet high, which is heavily branched from near ground level and supports a crown which may be appreciably wider than the tree is high. The foliage is moderately dense and provides a useful level of shade.

Wild orange extends from central Australia to near the eastern coast in the drier parts of Queensland, for example near Rockhampton. The principal occurrence is in the semi-arid zone of New South Wales and Queensland (mainly south of the Tropic of Capricorn), but it extends to the arid zone in both States, as well as the southern half of the Northern Territory and, sparsely, to the northern half of South Australia. Between 23½ and 30°S. latitude it also extends into the sub-humid zone of eastern Australia. The rainfall has a wide range at 8-35 inches, but most of the occurrence is in a region of 12-25 inches. Summer temperatures are high, with mean maxima of the hottest month 90-95°F.

Young plants of this tree appear as very small-leaved, tangled, thorny shrubs which only later develop main stems; the young thorny shoots may even scramble up adjacent trees before the tree habit develops. The prickles are usually absent on the mature flowering branches. Wild orange grows on a wide range of soils from skeletal types to loamy clays, and sometimes on limestone. It is a useful tree for shade and where plentiful can provide shelter. It is not generally recognised as an ornamental species but it merits wider consideration in this respect. Where it grows naturally it should be retained both for use as a drought fodder and for soil protection and amenity purposes.

There are several other species in the genus which are either shrubs or small trees and grow in the same type of habitat; some are eaten by stock. *C. loranthifolia* (narrow-leaved bumble of Queensland) is found in that State, New South Wales and probably the north-eastern corner of South Australia. *C. spinosa* var. *nummularia* (caper bush) is a small shrub of the southern half of the Northern Territory, but it is not common. *C. canescens* (wild orange or wild pomegranate) is a small tree of Queensland, somewhat like *C. mitchellii*, which grows mainly in the sub-humid areas nearer the eastern coast.

C. umbonata Lindl. **WILD ORANGE**

This wild orange or native pomegranate occurs sparsely in the semi-arid and arid zone of northern Australia, mainly 14-22°S. latitude. This includes the Kimberley area of Western Australia, much of the Northern Territory north of Barrow Creek and the far west and north-west of Queensland. Summer temperatures are high throughout the area but on the higher inland sites there may be 1-10 frosts a year. Rainfall is mainly 12-22 inches, but extends to 40 inches, and has a monsoonal pattern.

It is a small, slender tree up to 15 feet in height, with long drooping leaves and fruit which are similar to those of *C. mitchellii*. Both the fruit and foliage are eaten by stock, but in the Northern Territory the species is rated as of only moderate value for fodder.

Carissa lanceolata R.Br. **CONKERBERRY**

Conkerberry is a small, finely branched, dense shrub 4-6 feet high, which is found from the north-western coast of Western Australia to near the coast of Queensland, mainly in the latitudinal range of 15-22°S. but extending as far south as 23° 38′ at Alice Springs in the southern half of the Northern Territory. Most of this occurrence is in the semi-arid zone, but it is also found in many places of the arid zone. Over most of the area the annual rainfall is 9-22 inches, but on the northern monsoonal fringes it may be as high as 30 inches, with a prolonged dry winter. In Queensland it is also known as boorum bush.

Locally it tends to form dense low thickets and in spite of the pair of sharp spines at the base of each leaf is readily eaten by stock, especially cattle. After browsing it freely produces new shoots. Conkerberry is mainly found on sandy soils or sandy loams, is considered a hardy species and will tolerate periods of prolonged drought.

Another representative of the genus is *C. ovata*, which has the same general distribution in the semi-arid zone and in Queensland is most common in areas of 20-25 inch rainfall. In that State it may be associated in scrub areas with brigalow (*Acacia harpophylla*) and gidgee (*A. cambagei*). In the drier parts it is considered a useful drought fodder plant but in the wetter areas it may become a weed. In Queensland it is known as currant bush, which is also applied to other species which bear fruit like garden currants, but this is a name which is applied to several other shrub species in Queensland, for instance *Apophyllum anomalium*.

Cassia L. **CASSIAS**

This is a large genus of more than 400 species scattered throughout the tropical and warm temperate regions of the world, but excluding southern Europe. Many are cultivated and useful products are obtained from some; for example the laxative senna is obtained from the leaves of certain Northern Hemisphere species, extracts of the bark and roots of others have been used for the treatment of skin diseases and stomach disorders, whilst the nuts of still other species are ground as a substitute for coffee or for the control of asthma.

Cassia includes trees, shrubs and herbs but most of the Australian species are small trees or bushy shrubs in the height range of 4 to 40 feet. They are widely distributed throughout the continent and whilst most common in the dry country they also extend to the higher rainfall areas of coastal New South Wales and Queensland.

For the purpose of this book the principal interest in the genus is that it contains some of the most attractive small bushy shrubs of the dry country and to recommend them for more extensive use for ornament and low shelter. Many of the arid zone species grow from 4-8 feet in height, have very beautiful yellow flowers and retain a dense bushy habit to ground level so that they can be used either alone or in combination with trees, such as eucalypts, which become open near the ground. Unfortunately most of the smaller species are comparatively short-lived. Brief notes are given on only a few species, but there are many others which should be considered locally.

C. artemisioides Gaud. ex DC. **SILVER CASSIA**

This is a small bushy shrub commonly about 4 feet in height but sometimes attaining 6-8 feet. The leaves are silver-grey and the flowers golden-yellow, the latter usually borne profusely on the ends of branchlets during late winter. Chippendale (1963) states that this species is fairly quick-growing but short-lived. When used as an ornamental or as a temporary low hedge there would need to be plans for periodical replacement. Fortunately, like most leguminous plants, it can be readily raised from seed; the latter benefit from soaking prior to sowing.

Silver cassia is widespread in the arid and semi-arid areas of Australia. Summers are very hot and winters mild to cool. Representative meteorological stations include Kalgoorlie (Western Australia), Alice Springs (Northern Territory), Cobar (N.S.W.) and Barcaldine (Queensland); the annual rainfall is 8-18 inches and varies from a weak winter maximum to a well defined summer maximum. Variability is high, for example Barcaldine, with a 50 percentile value of 18.58 inches, has a lowest recorded rainfall of only 5.74 inches for 79 years, and Broken Hill, with a 50 percentile value of 9.12 inches, has a lowest value of 2.24 inches for 77 years. Although a native to the dry country it will also grow satisfactorily on well-drained soils in the warmer parts of the sub-humid and humid zones.

C. eremophila A. Cunn. ex R.Br. **DESERT CASSIA**
syn. *C. nemophila* A. Cunn. ex Vogel

Like many other species of the genus this cassia has a very wide range in the arid, semi-arid and sub-humid zones south of about 20°S. latitude. The climate for these areas has been previously described. Desert cassia is a small shrub, mainly in the height range of 3-6 feet but sometimes taller under particularly favourable conditions. It is reasonably quick growing at first but is not long-lived. The foliage and pods of this species and its several varieties contain useful amounts of protein and phosphorus, and is low in fibre content, but opinions differ as to their value as a stock fodder. Animals like the shrub and certainly in some areas it is grazed to the extent that regeneration is affected and the number of mature plants reduced.

C. nemophila A. Cunn. ex Vogel var. *coriacea* (Benth.) Symon
syn. *C. sturtii* R.Br.

This is a very attractive, yellow-flowered, moderately dense, small bush which has an extensive range in the dry country, with some extension into the southern sub-humid zone. The height is only a few feet, but it provides adequate shelter for sheep and also merits retention near homesteads for both low shelter and the beauty of its flowers.

Casuarina L. **SHEOAKS**

This is a genus of about 35-40 species which vary in size from small to large trees. They are found in the Southern Hemisphere from India and Malaya to the islands of the South Pacific Ocean, and find their greatest development in Australia where they occur both in the arid areas of the centre and the higher rainfall coastal regions. Most of them have a pyramidal habit and whilst some of them, *C. stricta* for instance, are of ornamental appearance under favourable conditions, their principal value is for shelter, conservation or timber supply. The timber generally is dark in colour, fissile due to the large medullary rays, moderately strong and heavy and very tough; it is a first-class fuel and was formerly extensively used for bakers ovens.

Notes on seven species are given below, but there are a number of other sheoaks which merit consideration, especially in the 14-20 inch rainfall belt south of about 26°S. latitude.

Descriptions and illustrations of river oak (*C. cunninghamiana*) and forest oak (*C. torulosa*) which grow in the eastern coastal areas are given in *Forest Trees of Australia.*

C. campestris Diels **A SHEOAK**

This sheoak is a small, very bushy shrub which only grows to 5-10 feet in height, but merits consideration for use where low shelter is needed for sheep or home gardens, or as the lower part of two-storey shelterbelts. It is a casuarina of the semi-arid zone of south-western Western Australia and climatic conditions are, in general, comparable to those described for *C. huegeliana.*

C. cristata Miq. **BELAH**
syn. *C. lepidophloia* F. Muell.

Belah is one of the more important casuarinas in semi-arid south-eastern Australia, but its distribution also includes arid south-western Western Australia, the central latitudes of South Australia with some extension into the north, the northern edge of Victoria, the arid, semi-arid and sub-humid zones of New South Wales and adjoining areas of Queensland as far north as about 22½°S. lat. The size varies from a small tree 20-30 feet high in areas of 9-12 inches rainfall, to a fine, moderately large tree 50-70 feet high and with trunks 2-2½ feet in diameter, in regions of 20-25 inch rainfall. Belah prefers relatively low-lying sites which collect water during heavy rains but which do not remain sodden for long periods; it is also found on other flat land and plains, but rarely on dry ridges. It prefers moderately heavy soils, such as clay loams, but may grow on red soils of the arid zone, and will tolerate a moderate amount of alkalinity. The trunk is usually short, often less than one-quarter of the total height, above which it divides into several large, ascending stems which carry ascending and spreading branchlets. The ultimate branchlets are green and in function replace the leaves which are reduced to minute scales.

The tree is of value for ornamental planting, such as avenues or as specimen trees, for rather light shade and some types of shelter on farms, for timber and, to a limited extent, as an emergency drought fodder. Where it occurs naturally farmers should retain sufficient for their requirements but planting is only recommended on good sites, that is, where the rainfall and soil moisture is adequate. This means, in general, relative lowlands with at least moderately good alluvial or loamy soils, and a rainfall minimum of 15 inches. It will grow under conditions of lower rainfall but in these circumstances sites should be restricted to depressions and the sides of seasonal drainage channels where underground water reserves will be greater than on slopes and ridges. Spacing should be 20-30 feet and if a shelterbelt is wanted this should be of two or more rows with trees staggered in alternate rows.

Belah can be readily grown from seed, but should be planted into its permanent position whilst still quite small, that is 6-8 inches high. Early height growth, as with most of the genus, is comparatively fast, that is 2-3 feet a year for the first 5 years or so. It sometimes suckers from exposed roots.

The timber is typical of the genus in general. The narrow sapwood is pale in colour, whilst the heartwood is dark brown to reddish brown. The broad medullary rays, which reminded early European settlers of those shown by the true oaks (genus *Quercus*), enables logs to be readily split and, in spite of only moderate durability in the ground, this was one factor in the timber being used for split posts, rails and other farm purposes. Other properties include moderately heavy weight, toughness and moderate strength. In the early days it was used for tool handles, shingles and cabinet work, but at present it is chiefly regarded as one of the best fuel woods in Australia.

C. decaisneana F. Muell. **DESERT OAK**

This is one of the remarkable species of the centre of Australia as it grows to 50 feet in one of the hottest and driest parts of the continent. It occurs mainly south and west of Alice Springs in the southern half of the Northern Territory, and in adjacent areas of Western Australia and South Australia, that is, around Giles in the former State and in the Musgrave Ranges of the latter. The recording station of Giles has a 50 percentile rainfall value of 8.60 inches based on 13 years of observation, the 10 percentile value is 2.52 inches and the lowest recorded is 1.49 inches. The mean maximum temperature of the hottest month is about 95°F, whilst readings of 100-110°F are common. All of this is associated with a potential evaporation which may exceed 130 inches a year.

Desert oak has a graceful pendulous habit when mature but the young plants have a more upright habit. The trunk is straight and may be more than half of the total height. Cattle will sometimes eat the foliage when hard pressed for fodder. In the Northern Territory it grows on deep sandy soils south of the MacDonnell Ranges and on other undulating or lowland sites where the maximum seepage into the sub-soil can be expected during the rare heavy falls of rain.

C. dielsiana C. A. Gardn. **NORTHERN SHEOAK**

This sheoak is a native of Western Australia, mainly within 200 miles of the west coast and between latitudes 26-30°S., with the altitudinal range from near sea level to about 1,000 feet. Rainfall is 12-20 inches, with a well defined winter maximum and only moderate variability. Summers are long and very hot, with the mean maximum of the hottest month approaching 100°F.

Northern sheoak is a large shrub or small tree with heights of 15-30 feet and provides good shelter when young, but with age the lower branches die. The Forests Department recommends it for shade and for use in multi-row shelterbelts. In its natural occurrence it grows on well-drained, poor, rocky soils, but it is considered to have possibilities for heavy soils in areas where the rainfall may be as low as 12 inches.

C. huegeliana Miq. **ROCK SHEOAK**

This sheoak is a native of the semi-arid south-west of Western Australia, with some extension eastwards into the jarrah forests of the sub-humid zone. Much of its occurrence is in the wheatbelt. The summers are hot with mean maximum temperatures of the hottest month 85-95°F, whilst mean minima of the coldest month are as low as 40°F. Areas near the coast may be frost-free, but inland there are 1-10 a year. Rainfall is in the range of 12-30 inches and shows a well-developed winter maximum.

Rock sheoak derives its common name from its occurrence on difficult sites around granite outcrops where even eucalypts are very sparse. The Forest Service of the State considers that it merits greater attention than it has hitherto received for shade planting on poor, shallow sandy soils. It is less suitable for shelterbelts, unless associated with a lower-storey shrub, because the moderately dense crown is carried on the upper two-thirds or half of the stem. It also deserves consideration for ornamental planting on difficult sites. Mature height is 20-30 feet.

C. luehmannii R. T. Bak. **BULL OAK**

This casuarina has a natural occurrence in the eastern States which overlaps much of that for belah (*C. cristata*), but is less common in the arid zone. The Queensland distribution is mainly in the southern inland semi-arid and sub-humid zones, but it reaches the coast in places from Maryborough to Rockhampton. In New South Wales it is found both on the Western Plains and the Western Slopes and it approaches the coast in some of the drier areas, such as the valley of the Hunter River. It occurs in the semi-arid north-west of Victoria and adjoining districts of South Australia—Lameroo, Bordertown and Francis. For most of the area the rainfall is 15-25 inches, but in northern New South Wales and Queensland it may be as high as 30 inches. Summer temperatures are warm in the south and hot in the north, that is the mean maxima of the hottest month vary from 85-95°F. Rainfall variability is comparatively low for Australia and potential evaporation only 50-70 inches.

Commonly bull oak grows on plains, gentle undulating slopes and, especially in the drier areas, around the edges of swamps and depressions where surface drainage is poor, but it also extends, as a poorer tree, to slopes and ridges. especially in the moister parts of its distribution. The soils show a wide range from poor, well-drained sands to sandy loams and moderately firm poorly drained clays. With the variation in topographical site and soils it is not surprising that there is a considerable range in mature heights of 25-60 feet or more, with diameters of 1-2½ feet. On all except the poorest sites the main stem is straight and often up to half the total height, above which it divides into several large ascending stems. These in turn carry the rather open crown of smaller ascending branches, with the green branchlets which function as leaves. This species only provides light shade and has limited value for shelter except where it occurs naturally in dense belts.

The heartwood is dark red in colour, strong, heavy, close-grained and with the very wide and deep medullary rays typical of the genus; durability is not high for use in contact with the ground. It is sometimes used for fencing and formerly, to a limited extent, for fancy turnery. As a fuel wood it ranks high, burning quietly and with great heat.

The specific botanical name is in honour of J. G. Luehmann (1843-1904), who was Government Botanist in Victoria (1896-1904).

C. stricta Ait. **DROOPING SHEOAK**

This sheoak occurs in the higher rainfall parts of South Australia from the Flinders Range to the south-east, most of Victoria (except the north-west and the higher mountains), and the Western Slopes, Southern Tablelands and South Coast of New South Wales. Rainfall is mainly within the limits of 15-30 inches, with a winter or uniform incidence. Summer temperatures are mild to hot, but winters experience 1-40 frosts. The tree grows on moderately poor soils and rocky slopes and avoids sandstones and quartzites.

Drooping sheoak is a small tree, usually 20-30 feet in height at maturity, with a short trunk and, in open situations, a well-branched dense crown. The smallest branchlets are drooping, hence the first part of the common name. The tree is rated as a useful fodder species and the foliage is eagerly sought by stock. One result of this is that little natural regeneration is found in areas which are grazed. The trees stand lopping well and this allows most of the smaller branches of the crowns to be cut off in times of severe drought without the trees dying. On even moderately-good soils open-grown specimens are useful for shade and ornament but, like most sheoaks, they are not the most suitable type of tree for use in single-row shelterbelts. The heartwood is reddish, fissile and has been used for tool handles and bullock yokes; it is, like nearly all members of the genus, an excellent fuel.

Codonocarpus cotinifolius (Desf.) F. Muell. **NATIVE POPLAR**

Several common names have been applied to this small tree, but because of the resemblance of the leaves, both in colour and shape, to those of some species of true poplars (genus *Populus*) that of native poplar is being used here. It is also used in South Australia and the Northern Territory but in New South Wales, however, horseradish or mustard tree is given to it because of the very pungent taste of the bark and leaves. The generic name alludes to the shape of the fruit, being based on the Greek, *codon* (a little bell), and *carpos* (a fruit); hence the common name in Victoria of bell-fruit. The species epithet was given because of the likeness of the foliage to that of species of the genus *Cotinus*.

This plant is nowhere plentiful but it is found scattered over a wide area of Australia from the coasts of Western Australia to the southern half of the Northern Territory, parts of South Australia, north-western Victoria, western New South Wales and south-western Queensland. Most of this arid country with the exception of the occurrence in Victoria, part of South Australia and the most south-western distribution in Western Australia.

Native poplar is a small tree which is 8-15 feet high in the arid interior but may be 30 feet in the more temperate south-eastern parts of its distribution. It has a slender, pyramidal form and with light greyish green, broadly ovate leaves 1½-3 inches long, on long petioles which renders it conspicuous amongst dry country vegetation. It is readily raised from seed and is not difficult to transplant, and in favourable years there is free natural regeneration. Because early growth is good and in spite of a short life, possibly only 10 years on poor sites, it is recommended for temporary shelterbelts and ornamental planting around homesteads in areas of 10-14 inch rainfall where it is not subject to insect attack and the soils are moderately good sandy loams. Stock will rarely eat it, whilst the timber is soft and valueless.

Dodonea attenuata A. Cunn. **HOPBUSH**

This dodonea has a wide distribution in temperate Australia, mainly south of 26°S. latitude, and whilst it occurs in such dry localities as Kalgoorlie, Alice Springs, Ooldea and Broken Hill, it also extends throughout most of the semi-arid and sub-humid zones of the southern States. The distribution in Queensland is mainly in the southern and south-western parts of that State. Because of its overall distribution the name hopbush is preferred to that of desert hopbush, which latter name is justifiably used in the Northern Territory. It was called hop-bush by the early settlers because of the similarity of the papery wings on the fruit to those of the hop used in brewing beer (*Humulus lupulus*). The generic name honours Rembert Dodoens (1517-85), a Flemish botanist and professor of medicine at the University of Leyden; the specific name is in allusion to the comparatively narrow leaves.

Hopbush is a small, compact shrub with many stems from ground level which carry a good covering of attractive green leaves; height is 5-15 feet. It should be considered for dense, low-level shelter of homesteads in areas of 15-20 inch rainfall; on low-lying deep sandy sites it will thrive where the rainfall may be as low as 10 inches. It is eaten by stock to a limited extent and planted specimens should receive protection from grazing. There are other members of the genus which merit consideration in the dry country, for example *D. lobulata* in areas as dry as 12 inches, and *D. stenophylla* for localities like those for hopbush.

Eremocitrus glauca (Lindl.) Swingle **LIMEBUSH**

This shrub belongs to the same family as the lemon and lime (*Rutaceae*), and the small fruits, ½-1 inch in diameter, have a thin skin and an acid pulp which has been used for making jams and preserves. Limebush or wild lime is a shrub, usually 10-18 feet high, which has a narrow but dense crown. It has a distinctive growth habit; the juvenile and sucker stages are intricately branched growth with thorns up to 2 inches long. Normal stem formation and branching habit develops later, but the thorns may be retained for a considerable time on the lower part of the stems of mature shrubs.

Its main occurrence is in the semi-arid and sub-humid zones of Queensland, mainly south of about 22°S. latitude, and the semi-arid zone of central and northern New South Wales, with some extension into the arid zone. Because of the spiny habit of young plants it is considered a pest in some districts, but this characteristic may be useful for some forms of low shelter, and later the plant develops into quite an attractive ornamental species.

Eremophila spp.

The generic name of this exclusively Australian group of plants is derived from two Greek words, *eremos* (a desert), and *philos* (love of); hence desert-loving. The 90 or more species are characteristic of the arid and semi-arid zones of the continent and Western Australia, with over 80 species, is the State in which the genus reaches its greatest development. Most of the eremophilas are densely-branched shrubs, mainly 4-12 feet in height, but they range from plants only 1-2 feet high to small trees which exceed 20 feet and have a trunk 6-8 feet long. Many of them have graceful, drooping greenish leaves and pendulous, bell-shaped flowers which are borne on the plants during several months of the year. The fruit of some species is a drupe, that is it has a central hard 'stone' surrounded by a thick flesh. The fruit of some species is eaten by emus; hence one of the common names—emu bush; also the similarity of some fruit to that of a plum gives rise to another common name—native plum.

Nearly all eremophilas are attractive shrubs useful for ornamental planting both around homesteads and in parks of the dry country, as well as for low shelter. The foliage of some species is occasionally eaten by stock, whilst the flowers may produce moderate quantities of pollen and nectar. The size of the shrubs is usually too small for the timber to be of any practical use, but a few species may produce butt logs which may be over 6 feet in length and 1 foot in diameter.

Eucalyptus clelandii
(Cleland's blackbutt)

E. conica
(fuzzy box)

E. dichromophloia
(variable-barked bloodwood)*

E. dundasii
(Dundas blackbutt)

E. eremophila var *grandiflora*
(sand mallee)

E. exserta
(Queensland peppermint)

E. flocktoniae
(merrit)

E. forrestiana
(Forrest's mallee)

E. crebra
(narrow-leaved red ironbark)†

ucalyptus clelandii—a young tree at Kalgoorlie, W.A.
. *conica*—growing naturally near Gilgandra, N.S.W.
. *dichromophloia*—growing naturally near Rolleston in semi-arid Q'ld.
. *dundasii*—planted at Kalgoorlie, W.A.; rainfall 9½ inches.
. *eremophila* var *grandiflora*—planted at Broken Hill.
. *exserta*—growing naturally near Maryborough, Q'ld.
. *flocktoniae*—planted at Broken Hill, N.S.W.; rainfall 9 inches.
. *forrestiana*—planted at Broken Hill, N.S.W.; rainfall 9 inches.
. *crebra*—growing naturally in Q'ld.

** This photo by M. Lazarides, Division of Land Research, C.S.I.R.O.*
This photo by the Department of Forestry, Queensland.)

Plate 73 AUSTRALIAN BROADLEAVED SPECIES—IX *Eucalyptus clelandii* to *E. forrestiana*

Eucalyptus gardneri
(blue mallet)

E. gillii
(curly mallee)

E. gongylocarpa
(marble gum)*

E. gracilis
(yorrell)

E. largiflorens
(black box)

E. lesouefii
(Goldfield's blackbutt)

E. gamophylla
(blue mallee)†

Eucalyptus gardneri—young planted tree near Wongan Hills, about 100 miles north east of Perth, W.A.; rainfall about 15 inches.

E. gillii—planted at Broken Hill, N.S.W.; rainfall 9 inches.

E. gongylocarpa—growing naturally south of Camel Well, W.A.

E. gracilis—planted at Kalgoorlie, W.A.; rainfall 9½ inches.

E. largiflorens—growing naturally near Echuca, Vic.; rainfall about 17 inches.

E. lesouefii—a young tree planted at Kalgoorlie, in the seed orchard of the Forest Department of W.A.

E. gamophylla—growing naturally on the McDonald Downs Station, N.T.; rainfall about 10 inches.

E. leucoxylon—growing naturally near Bendigo, Vic.; rainfall about 20 inches.

E. longicornis—planted at Kalgoorlie, W.A.

E. loxophleba—growing naturally in W.A.

(Photo by Dr. N. H. Speck, formerly of the Division of Land Research, C.S.I.R.O. † Photo by R. A. Perry, Division of Land Research, C.S.I.R.O.)*

E. leucoxylon
(yellow gum)

E. longicornis
(red morrel)

E. loxophleba
(York gum)

Plate 74 AUSTRALIAN BROADLEAVED SPECIES—X *Eucalyptus gamophylla* to *E. loxophleba*

Eucalyptus odorata
(peppermint box)

E. orgadophila
(mountain coolibah)

E. oleosa var. *angustifolia*
(narrow-leaved giant mallee)

Eucalyptus odorata—planted at Broken Hill, N.S.W.; receives some additional watering.

E. orgadophilla—growing naturally in Q'ld.; rainfall about 30 inches.

E. eleosa var. *angustifolia*—planted at Broken Hill, N.S.W.; some extra watering.

E. melanophloia—growing naturally near Kilcoy, Q'ld.; rainfall about 37 inches.

E. melliodora—growing naturally; rainfall about 24 inches.

E. microtheca—growing in Q'ld.

E. ochrophloia—growing naturally north-west of Charleville; rainfall about 18 inches.

E. papuana—growing naturally near Humpty Doo homestead, N.T.; rainfall about 54 inches.

(* *This photo by the Department of Forests, Q'ld.*)

E. melanophloia
(silver-leaved ironbark)

E. melliodora
(yellow box)

E. microtheca
(coolibah) *

E. ochrophloia
(yapunyah)

E. papuana
(ghost gum)

Plate 75 AUSTRALIAN BROADLEAVED SPECIES—XI *Eucalyptus odorata* to *E. papuana*

Eucalyptus platypus
(moort)

E. 'pterocarpa'

E. populnea
(bimble box)

E. redunca var. *melanophloia*
(black-barked marlock)

E. salubris
(gimlet)

E. sargentii
(Salt River mallet)

Eucalyptus platypus—a young tree planted in a plot of the Forests Department, W.A.; rainfall about 18 inches.

E. 'pterocarpa'—a young planted tree at Broken Hill. This plant does not receive regular watering.

E. populnea—naturally grown near Gilgandra, N.S.W.; rainfall about 22 inches.

E. redunca var. *melanophloia*—a young planted tree in the Yuna sample plot of the Forests Department of W.A. The rainfall is probably about 16 inches.

E. salubris—planted at Broken Hill; not watered.

E. sargentii—planted at Broken Hill; not watered. This is in a skimp dump area and there is a large amount of salt showing on the surface.

E. pilligaensis—naturally grown in Baradine State Forest area, N.S.W.; rainfall 24 inches.

E. polycarpa—naturally grown in Q'ld.

E. salmonophloia—a young planted tree at Woomera, S.A.; receives additional water.

(* *This photo acknowledged to the Department of Forestry, Q'ld.*)

E. pilligaensis
(narrow-leaved grey box)

E. polycarpa
(long-fruited bloodwood)*

E. salmonophloia
(salmon gum)

Plate 76 AUSTRALIAN BROADLEAVED SPECIES—XII *Eucalyptus platypus* to *E. salmonophloia*

Eucalyptus stricklandii
(yellow-flowering blackbutt)

E. thozetiana
(mountain yapunyah)

E. torquata
(coral-flowered gum)

E. trachyphloia
(brown bloodwood)

E. viridis
(green mallee)

E. viridis
(green mallee)

Eucalyptus stricklandii—planted at Woomera; receives regular watering.

E. thozetiana—naturally grown about 30 miles west of Tara which is south of Chinchilla in southern Q'ld.; rainfall about 24 inches.

E. torquata—naturally grown between Kalgoorlie and Kambalda, W.A.; rainfall about 10 inches.

E. trachyphloia—naturally grown in Baradine State Forest, N.S.W.; rainfall about 24 inches.

E. viridis—four years from planting on a farm south of West Wyalong, N.S.W.; rainfall about 18 inches.

E. viridis—large naturally grown mature specimens between Dubbo and Gilgandra, N.S.W.; rainfall about 21 inches.

E. sideroxylon—naturally grown at Glenmaggie, Vic.; rainfall about 23 inches.

E. spathulata—a young planted tree at Broken Hill, N.S.W.; not watered, rainfall 9 inches.

E. tessellaris—naturally grown young tree midway between Mitchell and St. George, southern Q'ld.; rainfall about 19 inches.

E. sideroxylon
(red ironbark)

E. spathulata
(swamp mallet)

E. tessellaris
(carbeen)

Plate 77 AUSTRALIAN BROADLEAVED SPECIES—XIII *Eucalyptus sideroxylon* to *E. viridis*

Ficus platypoda
(a fig)*

Grevillea pinaster
(a grevillea)

Hakea leucoptera
(needlewood)

Flindersia maculosa
(leopardwood)

Flindersia maculosa
(leopardwood)

Lagunaria patersonia
(Norfolk Island hibiscus tree)

Ficus platypoda—growing naturally at Palm Valley about 70 miles from Alice Springs, N.T.; rainfall about 9 inches.

Grevillea pinaster—a young plant in a sample plot of the Forests Department of W.A.; rainfall probably about 20 inches.

Hakea leucoptera—naturally grown about 40 miles north-west of Broken Hill.

Flindersia maculosa—rainfall about 18 inches. The upper photos show a mature tree whilst the lower shows the unusual growth habit of young plants.

Lagunaria patersonia—planted tree; rainfall about 10 inches but the roots of this tree undoubtedly extends to the watered areas of the adjacent garden.

Geijera parviflora—naturally grown tree; rainfall about 22 inches.

Grevillea striata—naturally grown tree; rainfall probably about 16 inches.

Heterodendrum oleifolium—naturally grown specimens, southern Q'ld.; rainfall about 21 inches.

(** Photo by M. Lazarides, Division of Land Research, C.S.I.R.O. † Photo by G. M. Cunningham of the Soil Conservation Service of N.S.W.*)

Geijera parviflora
(wilga)

Grevillea striata
(beefwood)†

Heterodendrum oleifolium
(inland rosewood)

Plate 78 AUSTRALIAN BROADLEAVED SPECIES—XIV *Ficus platypoda* to *Lagunaria patersonia*

Leptospermum laevigatum
(coast teatree)

Melaleuca armillaris
(bracelet honey myrtle)

Melaleuca lanceolata
(dryland teatree)

Melaleuca pauperiflora
(a melaleuca)

Leptospermum laevigatum—planted in a plot of the Forests Department, W.A.; rainfall about 18 inches.

Melaleuca armillaris—planted at a Department of Agriculture Experimental Farm near Condobolin, N.S.W.; rainfall 16 inches.

Melaleuca lanceolata—(formerly known as *M. pubescens*) —planted at Broken Hill. These bushes are on a very salty site at the foot of a skimp dump. Some years have been very dry, for example 1967 had only 2·67 inches.

Melaleuca pauperiflora—growing naturally, W.A.; rainfall about 10 inches.

Melaleuca halmatorum—in the cemetery at Broken Hill.

Melaleuca glomerata—growing naturally near Palm Valley, about 70 miles south-west by south of Alice Springs, N.T.; rainfall about 9 inches.

Myoporum montanum—close up of the dense foliage of a small shrub about 7 feet high, growing naturally near Woomera, S.A.; located in a drainage area and probably benefits from run-on.

(* *Photo by M. Lazarides, Division of Land Research, C.S.I.R.O.*)

Melaleuca halmatorum
(South Australian swamp paperbark)

Melaleuca glomerata
(white teatree)*

Myoporum montanum
(boobialla)

Plate 79 AUSTRALIAN BROADLEAVED SPECIES—XV *Leptospermum laevigatum* to *Myoporum montanum*

Owenia acidula
(emu apple bush)*

Petalostigma quadriloculare
(quinine tree)

Santalum spicatum
(sandalwood)

Pittosporum phylliraeoides
(inland pittosporum)

Owenia acidula—growing naturally west of Rannes, which is some 55 miles south-west by south of Rockhampton, Q'ld.; rainfall about 20 inches.

Petalostigma quadriloculare—growing naturally a few miles west of Miles, southern Q'ld.; rainfall about 24 inches.

Santalum spicatum—growing naturally between Kalgoorlie and Kambalda, W.A.; rainfall about 10 inches.

Pittosporum phylliraeoides—planted in a private garden area near Cunderdin, W.A.; rainfall 14 inches.

Santalum lanceolatum—about 50 miles north-west of St. George, southern Q'ld.; rainfall about 20 inches.

Ventilago viminalis—near Castlevale Station, some 100 miles south-west of Emerald, Q'ld.; rainfall about 24 inches.

(* *These photos by M. Lazarides, Division of Land Research, C.S.I.R.O.*)

Santalum lanceolatum
(sandalwood)

Ventilago viminalis
(supplejack)*

Plate 80 AUSTRALIAN BROADLEAVED SPECIES—XVI *Owenia acidula* to *Ventilago viminalis*

Separate notes are provided for three of the more important species—one of the native fuchsias (*E. latrobei*), berrigan (*E. longifolia*) and budda (*E. mitchellii*), and brief reference is made here to native fuchsia (*E. maculata* (Ker.-Gawl.) F. Muell.) and turpentine bush (*E. sturtii* R.Br.). The native fuchsia is a small shrub which grows in the semi-arid and arid zones throughout most of Australia from 17-36°S. latitude. There are variable accounts of its effect on stock and Everist (1969) states it yields large quantities of prussic acid and is dangerous to hungry and travelling stock. Sheep which are normally grazed in areas where it grows, however, can eat large amounts without ill effect and in Queensland it is commonly regarded as an excellent fodder plant. Turpentine bush, also known as budda in Queensland, is a shrub or small tree which occurs in southern Northern Territory, western Queensland and New South Wales and, to a very limited extent, in the north of South Australia.

E. latrobei F. Muell. **NATIVE FUCHSIA**

Native fuchsia is a small, much-branched shrub in the height range of 3-10 feet, which occurs throughout most of the arid zone but with some extension into the semi-arid zone, especially in Queensland. It has the same common name as *E. maculata* and should not be confused with that species. The range of rainfall is mainly 8-15 inches, with high variability, but it may be as much as 25 inches in some of the northern areas. The rainfall occurs mainly in summer but varies from a marked monsoonal incidence in the north to more or less uniform in the south. The mean maximum temperatures of the hottest month are mainly over 90°F and even reach 107.5°F at Marble Bar, Western Australia. Frosts may be absent or as high as 12-15 a year for high inland locations.

Chippendale (1963) reports that in the Northern Territory native fuchsia occurs on sandy soils with a clay sub-soil or on lateritic sand country. It is commonly one of the principal species in mulga (*Acacia aneura*) or Georgina gidgee (*A. georginae*) associations. He also states that it is often grazed quite heavily, both by sheep and cattle, and in heavily stocked areas may be eaten back to the extent that the plants are seriously damaged. In the Northern Territory it is considered a useful topfeed species; previous reports on stock poisoning by this species is now known to be due to Georgina gidgee. Native fuchsia should be retained in native vegetation in the dry country and can also be considered for use as low shelter and ornament for homesteads and in parks and reserves of the arid and semi-arid zones.

E. longifolia F. Muell. **BERRIGAN**

This eremophila has several other common names; that of berrigan is used in New South Wales, Queensland and the Northern Territory and has been adopted here in preference to emu bush, which is applied to a number of shrubs and small trees. Still another name is native plum tree in reference to the nature of the fruit.

Berrigan has an extensive geographical range throughout most of the arid and semi-arid zones of Australia, covering about 37 degrees of longitude (114-151°E.) and some 21 degrees of latitude (16-37°S.). Rainfall mainly varies from 8-20 inches, with high variability, and temperatures are high to very high for most of the year.

It is a shrub to a small, well-branched tree, usually wth dense foliage, 5-15 feet in height and with a stem 4-8 inches in diameter. At times it may be 20-25 feet high, with a moderately heavily-branched crown and a main stem 6-8 feet long and 12-15 inches diameter. Berrigan grows on a wide range of soils, including those of a calcareous type. It is recommended for ornament and low shelter both around homesteads and in public parks in the dry country. The foliage is freely eaten by sheep and cattle and, whilst there are reports of it being poisonous if fed as an exclusive diet, it does not appear to be harmful when part of a mixed diet.

E. mitchellii Benth. **BUDDA**

Budda grows mainly in the semi-arid and sub-humid zones of Queensland and New South Wales, with its greatest occurrence from 19-33°S. latitude. In the former State it extends to the coast in localities such as Rockhampton and Townsville, whilst in New South Wales it occurs, less commonly, in the arid zone. Rainfall in New South Wales is 10-15 inches and in Queensland mainly 15-25 inches.

The size of budda varies with the locality, but is mainly within the range of a shrub 8-12 feet high or a small tree up to 25 feet with a trunk 12-15 inches in diameter. When of tree size the trunk may be long enough for fence posts, before the stem divides into a number of heavy branches which support a moderately dense crown. The plant grows on a variety of soils which range from sandy ridges to shallow loams over clay. These may sometimes be moderately alkaline.

On favourable sites budda is an ornamental species which deserves consideration for homestead and park use in its climatic area. It is not regarded as a browse shrub but is sometimes eaten by stock. The sapwood is pale yellowish in colour, whilst the heartwood is brownish, tough and reputed to be very resistant to white ants; due to this latter attribute it has been used for fence posts in the round. It is an excellent fuel and has a sweet and pleasant odour when burnt; because of this latter characteristic it has been used as a substitute for true sandalwood and is sometimes known by that name. The specific botanical epithet is in honour of Sir T. L. Mitchell, Surveyor-General of N.S.W., 1827-1855.

Eucalyptus alba Reinw. ex Blume **WHITE GUM**

White gum is mainly a species of the humid and sub-humid zones of northern and north-eastern Australia, but it has a limited extension into the semi-arid zone, especially in the north-west of Western Australia. The climate is mainly tropical monsoonal, with an annual fall of 30-60 inches, most of which occurs between December and April, but the species also grows in drier areas if it has access to drainage water. Temperatures are high throughout the year; for example the mean maxima of the hottest month, usually November, are in the range of 90-98°F and the mean minima of the coldest month 55-60°F. Only rarely at the higher inland altitudes do temperatures fall below 40°F.

Best growth is on flats which may be waterlogged for several months of the year; in such areas the trees grow to 60 feet in height, but on poor rocky locations it may barely attain 20 feet. This is a useful tree for use on seasonally inundated land in tropical areas or where there is an adequate rainfall throughout the year. Well-grown specimens are ornamental and provide a useful shade. The wood is hard, heavy and durable and is used locally; the bark contains 30-32 per cent tannin.

E. angulosa Schau. **A MALLEE**

This mallee attains about 20 feet in height and in its native habitat often has a rather straggly form. When planted on suitable sites in south-eastern Australia the young trees are of better shape and carry dense foliage. The natural occurrence is from Esperance on the south coast of Western Australia to Comet Vale, about 60 miles north of Kalgoorlie. Most of the area falls in the semi-arid zone, but north of Kalgoorlie it is arid and south of Salmon Gums it changes to sub-humid. Both temperature and rainfall show a large range and care should be taken in the source of seed for a particular project. The Forests Department of Western Australia recommends seed from the northern part of the range for sand dune planting in arid areas, and from near Esperance in the south for coastal dune planting. The mean maximum temperature at Kalgoorlie for the hottest month is 93.2°F but for Esperance is only 76.6°F; the mean minimum temperatures for the coldest month are 42.9°F and 45.4°F respectively. Rainfall also shows a large range from 9 inches at Kalgoorlie to 27 inches at Esperance.

This eucalypt is particularly suitable for use in sand and soil conservation but can also be planted as the lower element in two-tier shelterbelts or on its own for 15-20 foot high shelter.

E. argophloia Blakely **QUEENSLAND WESTERN WHITE GUM**

This very attractive tree is comparatively little known outside of its very restricted native habitat, which is mainly 5-15 miles north and north-east of Chinchilla in southern Queensland. This is about latitude 26-26½°S. and some 150 miles from the coast; altitude is around 1,000-1,100 feet. The summers are hot with the mean maximum temperature of the hottest month about 90°F, whilst nights in winter are cool with the mean minimum of the coldest month about 40°F. Snow does not occur but there are 10-15 frosts a year. The mean annual rainfall is 24-28 inches, with high variability and a moderate summer maximum.

Queensland western white gum grows on or around the edges of flats in country of low topographic relief. The soils are clays and clay loams of at least moderate fertility and most of the habitat has been developed for cropping or pasture, so that there is very little of the original forest now left. The tree has an erect habit and attains 120 feet or more in height; when grown in forest formation there is a clear, straight trunk half or more of the total height of the tree, but when the tree develops in open situations it retains good lateral branch development until early maturity.

The timber is reported to be very strong, hard and durable, and to be used locally for fencing and general construction. The heartwood is a deep red in colour.

The Forests Department of Queensland recommends this species for shade, windbreaks and woodlots. It is an ornamental species which deserves consideration for parks and reserves in its climatic area. There is little doubt that it is a tree which merits being better known than it is at present.

E. astringens (Maiden) Maiden **BROWN MALLET**

As a result of clearing for agriculture and of heavy stripping of bark prior to 1914, the original mallet areas are now much reduced and the present stands remain by virtue of the fact that the sites are not attractive to agriculture. As a consequence most of the trees are less than 50 feet in height, but formerly the larger specimens in the original forests on the better soils attained 70-80 feet in height, with trunks up to 2½ feet diameter and 30 feet in length. The specific botanical epithet refers to the astringent taste of the bark which has the remarkably high tannin content of 40-50 per cent (dry weight basis).

Brown mallet is native to a relatively small area of the semi-arid zone in south-western Western Australia, east of Narrogin and south-east of Northam, where the rainfall is 14-22 inches with a moist winter and a dry, hot summer. The Forests Department of W.A. recommends it for street and ornamental planting in sandy to loamy soils, preferably over clay, in areas where the rainfall is 16 inches or greater. The shade provided by the tree is good in early life but tends to become rather thin with age and brown mallet is not recommended for shelterbelts, except in association with other species. It is not difficult to raise from seed and tubed stock is usually the best proposition for farmers wishing to plant it as specimen trees or in mixed shelterbelts. It has shown good promise in semi-arid south-eastern Australia but it is somewhat fire tender and therefore liable to damage from grass fires when young.

The timber is heavy (about 60 lb/cu ft), very hard, very tough, strong but only of moderate durability for a eucalypt.

E. behriana F. Muell. **BROAD-LEAVED MALLEE BOX**

Usually in mallee form and 15-25 feet high it sometimes grows as a single stem tree up to 35 feet in height; bole diameters as a mallee may be only 3-4 inches but the trees attain 7-9 inches. Broad-leaved mallee box is nowhere very plentiful but it has a wide distribution from Kangaroo Island and Eyre Peninsula in South Australia, through north-western Victoria to the Wyalong district of southern New South Wales. Rainfall is mainly 16-20 inches, with a marked to slight winter maximum.

It can usually be separated from associated mallees by the relatively wide leaves and the small paniculate inflorescences which resemble those of grey box (*E. woollsiana*), with which it overlaps in New South Wales. The specific epithet honours Dr Hermann Behr who was a botanical collector in South Australia during the middle of the nineteenth century. Where this species is growing on a property the mallee form can be retained for shelter and conservation, whilst the tree form provides light shade. It prefers sandy soils, but also grows on sandy loams which may be slightly alkaline. The timber is usually too small except for use as fuel or for farm purposes; the heartwood is reddish but there is little information on its durability in the ground but from its botanical grouping probably is at least good enough for temporary fencing.

E. brockwayi C. A. Gardn. **DUNDAS MAHOGANY**

This fine dry country eucalypt has a very restricted natural occurrence around Norseman, Western Australia, which is about midway between Kalgoorlie and the coast to the south. The rainfall for Norseman is 11 inches and the Forests Department of W.A. considers this about the minimum for satisfactory growth; in eastern States it has done well at Broken Hill with only 9 inches and a 10 percentile value of 5 inches. For these latter conditions, however, it must be on sandy loams of some depth and spaced sufficiently far apart to have an adequate volume of soil from which to draw moisture. Experience elsewhere in the east indicates that it will thrive on slopes and other well-drained sites where the rainfall may be as high as 20 inches. The specific botanical name honours Mr G. E. Brockway, who has taken a keen interest in the dry country eucalypts. He retired from the Forests Department of W.A. in the 1960s.

Dundas mahogany is a medium size tree, 50-80 feet in height and with diameters up to 2½ feet. The trunk is moderately straight and usually does not exceed half of the total height, whilst the crown is of medium size and rather open in maturity. It is recommended, therefore, mainly as an ornamental species and merits consideration for use in wide streets or in parks; it is a useful dry country shade tree but for shelterbelts it should only be considered for mixture in multi-row belts. On even moderately favourable sites it is relatively fast growing and can be expected to attain 20 feet in 6-8 years. The timber is tough, reddish in colour, straight grained but not resistant to termites.

E. camaldulensis Dehnh. **RIVER RED GUM**

River red gum is the most widely spread of all eucalypts, being absent only from the tablelands and coastal areas of eastern, south-eastern and south-western Australia. Even in the desert areas it may be found along drainage channels. With this wide distribution is a variation in rainfall of 8-50 inches, but it should be noted that in the lower rainfall areas, especially below about 15 inches, the tree relies on stored water from seasonal flooding or a high water table. Some of the finest specimens are to be found along the banks and on the flood plains of the Murray River, especially where it forms the boundary between Victoria and New South Wales. Overseas it has a good reputation for being adaptable to a wide range of soil and temperature conditions, but care should be taken in the source of seed.

In size river red gum varies from small, straggly specimens 30-40 feet high, to large trees up to 150 feet high. Trunks may be relatively short but with diameters up to 8 feet or more, whilst the crown is heavy and wide-spreading. Wherever it is planted for ornament it should be allowed plenty of space, that is at least 100 feet if the site is suitable and it is proposed to grow the trees to maturity. It is an excellent shade tree and can be used in shelterbelts of mixed species. Under favourable conditions of sufficient ground moisture and suitable soils early growth is comparatively fast and may be several feet a year for the first 10-20 years. This eucalypt seeds freely and in favourable years natural regeneration is good, but for farmers planting only a few trees it is best to use tubed stock.

River red gum does not flower freely every year, but when it does pollen and nectar production is profuse. Honey produced from the nectar is clear or pale in colour, mild in taste and has a very good flavour. The timber is red, hard, heavy and very durable but strength properties are somewhat variable due to irregularity of grain. It is a favoured railway sleeper timber and makes an excellent flooring, especially in the form of parquetry. As a firewood it has few equals.

E. campaspe S. Moore **SILVER-TOPPED GIMLET**

This eucalypt is a small tree with a height range of 25-40 feet, which grows on sandy loams within a radius of about 100 miles of Kalgoorlie, Western Australia. The name gimlet refers to the shallow flutings on the bark, a characteristic also shared with *E. salubris*. The climate is arid to semi-arid, with a rainfall of about 9 inches and a 10 percentile value of 5.74 inches; summers have very high mean maximum temperatures (93.2°F for Kalgoorlie), but in winter temperatures may sometimes drop to near freezing point.

Although silver-leaved gimlet is native to sandy loams it will grow on clay soils and is tolerant of a moderate level of alkalinity. The Forests Department of W.A. recommends it for planting in areas where the rainfall is at least 12 inches. Experience in the east shows that it will grow satisfactorily at Broken Hill where the rainfall is 9 inches, with a high variability. When planted at adequate spacing it develops a moderately dense crown and a well-branched habit. Silver-leaved gimlet is recommended for street and park planting in dry areas where a small tree is desirable. When used in mixed shelterbelts care needs to be taken in ensuring that the mature heights of the different species are comparable, so that one tree does not get suppressed by an adjacent, taller-growing species. The timber is pale brown, straight grained, strong and reasonably durable against fungi but is not resistant to termite attack; it is used for farm purposes and makes a satisfactory fuel.

E. cladocalyx F. Muell. **SUGAR GUM**

There are few eucalypts which have been more extensively planted for shade, shelter and ornament than sugar gum. A native of three, small disconnected occurrences in South Australia it has been planted in diverse climatic zones in every other mainland State, including the southern inland region of Queensland. Under favourable growing conditions it will attain 100 feet in height and 3-5 feet in diameter. In South Australia the climate shows a striking range for the mean maximum temperature of the hottest month from 72.5°F for Kingscote on Kangaroo Island to about 90°F in the Flinders Ranges. The mean annual rainfall is 15-25 inches, with a marked winter maximum. Planted elsewhere in Australia it shows an ability to tolerate a wide range of conditions from sites in the arid zone where the rainfall may be less than 10 inches, but where it needs watering for satisfactory growth, to southern inland Queensland where the rainfall is 20-30 inches, with a summer maximum and high variability. The popularity of this species is undoubtedly due to its good early growth, attractive appearance and ability to tolerate high summer temperatures, periodic droughts and withstand hot, strong winds. It is a good species for shade and ornament in Australia south of the tropics but, due to its habit of carrying its crown high up the stem, has been, in the opinion of this writer, markedly over-rated for shelterbelt planting. This defect can be met by using a shrub or small tree of bushy habit in multi-row design. It has good coppice ability and will readily tolerate pollarding, but cannot stand salt winds, and is sensitive to frost when young.

The timber is hard, heavy and moderately strong, but needs preservation to give long life in the ground. It is useful for construction but is too heavy for interior fittings. Under suitable forest management the tree is a good species for the production of both small and large diameter poles.

E. cladocalyx var. *nana* Hort. is a dwarf horticultural variety which may grow to 25 feet in height. It has a much bushier habit than the species; hence is more suitable for low shelterbelts.

E. clelandii (Maiden) Maiden **CLELAND'S BLACKBUTT**

Cleland's blackbutt is a small tree, 20-40 feet in height, which is found on red, sandy loam in the Kalgoorlie district of Western Australia. The rainfall is about 10 inches a year and summer temperatures are high; for example the mean maximum of the hottest month in Kalgoorlie is 93.2°F. This tree was named after A. F. Cleland, civil engineer in the Kalgoorlie area, and Sir John Cleland, a noted collector of plants in the dry country. Blackbutt refers to the short stocking of dark-coloured, rough bark at the base of the trunk.

Although drought hardy and tolerant of somewhat poor soils it responds well to better conditions and shows good growth on light sands in areas of higher rainfall, for example that of Perth with 36 inches. In general, however, it should not be planted in areas with more than 20-25 inches of rain unless the soils are light and the sites freely drained. Under favourable conditions it may attain a height of 20 feet in 10 years. Because of its hardiness and small size it has been recommended for street use where there is limited space. Since the filaments of the anthers are white the flowers are not as spectacular as many of the other eucalypts from the Goldfields area of Western Australia.

E. conica Deane et Maiden **FUZZY BOX**

The main occurrence of this box is on the moister side of the semi-arid zone of New South Wales, with a limited extension into southern Queensland. The climate is characterised by a rainfall of 20-25 inches, in places up to 30 inches, with hot to very hot temperatures; the mean maxima of the hottest month are 85-95°F, whilst winters have cool nights with up to 15 frosts at low altitudes and as many as 60 at the highest.

In appearance fuzzy box is not unlike some of the grey boxes (*E. woollsiana* and *E. pilligaensis*) with which it grows, but is usually not quite as large at 50-70 feet in height and 1½-2½ feet in diameter; on poor sites it may be much smaller and of poor form. It is commonly found on somewhat heavy alluvial soils, clay loams and better quality sandy loams. Like many of the boxes this species is of attractive appearance when growing on reasonably favourable sites and where it grows naturally on farms some should be retained for shade. It may be planted for ornamental purposes and is of value to beekeepers, since it produces good quality nectar during the flowering season. The timber has the general properties of most of the larger tree-size boxes—it is heavy, hard, strong and very durable but does not split or work readily.

E. crebra F. Muell. **NARROW-LEAVED RED IRONBARK**

This ironbark is a small to medium-sized tree which usually grows 60-90 feet in height. It has an extensive occurrence in eastern Australia, with a latitudinal range of nearly 20 degrees from near Cairns in tropical Queensland to a little south of Sydney, and from the humid and sub-humid coastal areas to the higher rainfall parts of the semi-arid zone. The annual rainfall of the more important inland occurrence is 20-25 inches, but it is higher in coastal areas. Inland areas experience very high summer temperatures and up to 15 frosts a year.

The wood has the general properties of most of the better quality ironbarks—it is hard, strong and durable but does not split or work readily. It is a very good firewood. Several forest services recommend it for planting on farms, especially in woodlots for timber production. It also has a use in multi-row shelterbelts, especially on moderately heavy soils where the rainfall is 25-30 inches.

E. dichromophloia F. Muell. **VARIABLE-BARKED BLOODWOOD**

This bloodwood is so named because the bark shows considerable variation from being completely shed to persisting on the trunk and the bases of the larger branches. It occurs throughout most of the northern half of Australia, including both the arid interior and humid tropical coastlands. Size and form shows a considerable range from a small tree 25-40 feet high and of a shrubby habit with a short, poorly-shaped main stem, to a moderately large tree 80 feet in height and with a reasonably straight bole 15-20 feet in length. Both temperature and rainfall show a large range; the mean maximum temperatures of the hottest month may be up to 100°F whilst the mean minima of the coldest month may be below 40°F for the higher inland sites. Rainfall has a range of 10-15 inches in the drier areas to over 55 inches in the north where the pattern is monsoonal, with an almost rainless season of several months.

Variable-barked bloodwood is one of the eucalypts which can be considered for use in the northern parts of the dry country. The crown is somewhat open but it does provide useful light shade. The timber is a good firewood and has sufficient durability for use in fencing.

E. dumosa A. Cunn. **WHITE MALLEE**

This species is variously called blue, red, white, Congoo and Waikerie mallee. *E. dumosa* is typically a mallee, 10-25 feet high with several stems 3-6 inches in diameter. Occasionally it develops into a small, bushy tree 30-40 feet in height and with a trunk 1-1½ feet in diameter. It occurs throughout the middle latitudes of South Australia, in north-western Victoria and the southern and central semi-arid zone of New South Wales. Rainfall is mainly 12-18 inches.

Like most of the dry country eucalypts it is drought-resistant and can be useful for shelter and soil conservation in low rainfall areas both inland and near the coast. The timber is small in size but is useful for garden and horticultural stakes, whilst the 'mallee' roots are a good firewood. The nectar is suitable for commercial honey production.

E. dundasii Maiden **DUNDAS BLACKBUTT**

This is a small tree attaining 50-60 feet under suitable conditions but sometimes finishing height growth at 30-40 feet. Even in very dry areas it has a dense crown in early life but with maturity the foliage is sparser but still sufficient to provide light shade. Although planted in many places it has a restricted natural occurrence, being limited to a small area of the semi-arid zone around Norseman in Western Australia.

The rainfall in its native area is 10-13 inches, with a moderate winter maximum and a 10 percentile value for Norseman of 7.03 inches. Planted at Broken Hill and without watering other than in the first few months it has, however, shown an ability to grow under even lower rainfall and has survived a year with less than 3 inches (1967). It can be expected to grow satisfactorily on permeable soils where the rainfall is as high as 20-25 inches in the hot summer-mild winter areas of southern Australia. Once established the tree will tolerate light frosts.

Like most eucalypts Dundas blackbutt can be readily raised from seed in nursery beds and should be planted as tubed stock. This species ranks in a high place for street, ornamental and shade planting in very dry areas with a winter or uniform rainfall, that is the southern half of the dry country. It is tolerant of salt in the soil and is one of the hardiest and most attractive-looking eucalypts at Broken Hill. When planted for shelter it should be appreciated that whilst it provides good single row protection in youth, as it grows towards maturity it thins out appreciably in the lower half and needs an associated bushy shrub when maximum protection to ground level is needed.

E. eremophila Maiden **SAND MALLEE**

This eucalypt is commonly in mallee form. It is usually about 15 feet in height, with a number of stems which are only 2-4 inches in diameter. The natural distribution is mainly in the semi-arid zone of south-western Western Australia from the eastern Goldfields area around Kalgoorlie through agricultural lands to within about 100 miles of the east coast, and southwards to Esperance. The range of rainfall is about 9-18 inches, but most of the area is only 9-14 inches. Sand mallee grows on a range of soils from those of a sandy nature to clay loams.

It is an ornamental species suitable for use in small parks and offers possibilities for hedges and low shelterbelts for homesteads in dry localities.

E. exserta F. Muell. **QUEENSLAND PEPPERMINT**

The name peppermint refers to some general similarity in appearance of the bark with that of species of the peppermint group of eucalypts, although botanically Queensland peppermint belongs to the red gum group. It is a species of the semi-arid, sub-humid and, to a limited extent, the humid zones of Queensland south of about 17°S. latitude. The rainfall varies from about 17-45 inches and the size of the tree shows a corresponding variation from a small, rather straggly tree, 35-50 feet in height, to a straight boled tree 80 feet or more in height and with a trunk diameter up to 5 feet.

It grows on a wide range of topography, which varies from coastal lowlands and hill slopes to gentle undulating areas of the inland and the small but often steep hills and ridges which occur on them. The soil types range from poor sands and skeletal formations to soils derived from volcanic material; whilst best development is on somewhat heavy loams of alluvial origin. On suitable sites in the summer rainfall area it grows into an attractive tree, but there are sometimes other species which are more suitable for shelter because they retain branches and foliage nearer to ground level than Queensland peppermint. Well grown open specimens are very satisfactory for shade. The timber is pale pinkish brown, heavy, hard and durable but tending to be brittle; it is suitable for general construction, especially house framing, and in the round or split for fencing material.

E. flocktoniae (Maiden) Maiden **MERRIT**

Merrit is usually a small tree 30-40 feet high, but it sometimes grows in mallee form. It mainly occurs near the Great Australian Bight from the vicinity of Esperance in Western Australia to Eyre Peninsula in South Australia; this is chiefly within the semi-arid zone. The rainfall shows a fairly wide range from 10 to over 20 inches and possibly the drought resistance and hardiness of the species varies with the source of seed.

This is an ornamental eucalypt which deserves consideration for planting in streets, parks or private gardens in southern areas where the rainfall is over 10 inches and the soil sandy or loamy. The common name is of Aboriginal origin, whilst the specific botanical epithet is in honour of Miss M. Flockton (1861-1953) who was official artist to the Sydney Botanic Gardens and drew most of the illustrations in the monumental work *A Critical Revision of the Genus Eucalyptus* by J. H. Maiden, published in the years 1903-33.

E. forrestiana Diels **FORREST'S MALLEE**

This eucalypt is typically a small bushy shrub, usually 10-18 feet high, which rarely grows in mallee form. The specific botanical epithet and the first part of the common name were given in honour of Lord Forrest who, when Premier of Western Australia in 1900, afforded Dr Diels and Dr Pritzel facilities for botanical exploration in the south-west of the State. Forrest's mallee has a restricted natural occurrence in the Salmon Gums-Esperance area south of Kalgoorlie. The rainfall at Salmon Gums is 13 inches but increases to over 20 at Esperance; summer temperatures are warm to hot, but much milder than at Kalgoorlie.

Forrest's mallee grows on alluvial clays which may be somewhat saline and in its natural environment carries dense green foliage and striking scarlet buds which enhances its attractive appearance. Where it has been planted outside of its native area, often on sandy soils and in a different climate it all too often grows as a small, spindly tree lacking the ornamental appearance of the plant in its natural habitat. Wherever it is introduced it should be planted under conditions as close as possible to those which exist at Salmon Gums and southwards. Its small size is suitable for use in gardens and for small hedges.

E. gamophylla F. Muell. **BLUE MALLEE**

Blue mallee is native to one of the driest and hottest parts of Australia. It grows in the southern part of the Northern Territory, mainly south and east of Alice Springs, in the adjacent parts of Western Australia and the north-western corner of South Australia. The rainfall is 6-10 inches, with a high variability. For instance Alice Springs with a 50 percentile value of 10.09 inches has an all time low of only 2.37 inches; potential evaporation is the highest in Australia at 100 to over 130 inches a year. Mean maximum temperatures of the hottest month are 95-100°F, whilst 10 or more frosts occur each winter.

Growing naturally it is always in mallee form and the spreading habit of the plant is excellent for soil protection and ground shelter. Height is in the range of 8-15 feet. Its drought and frost resistance, associated with its ability to grow on deep, red sands, including sand dunes, makes blue mallee a desirable species for desert areas. In addition, it has ornamental glaucous foliage and is relatively quick growing during early life. Chippendale (1963) reports that 'this species is not palatable, and under normal conditions is not known to be grazed. However, cattle confined on spinifex country eat the foliage avidly, and break the trees down'.

E. gardneri Maiden **BLUE MALLET**

Blue mallet occurs in the semi-arid and parts of the sub-humid zone of south-western Western Australia where the rainfall is 16-22 inches. Part of this distribution overlaps that of brown mallet (*E. astringens*). It is a moderately small tree, usually in the range of 30-35 feet in height and 1-2 feet in diameter, and grows on sandy soils and light loams, including those of a lateritic type. The name *gardneri* honours Mr C. A. Gardner, Government Botanist of Western Australia, 1928-60.

Introduced into south-eastern States this eucalypt develops a large, dense crown and grows well on a variety of soils in areas where the rainfall is 15-25 inches. If planted on loamy sands with adequate soil moisture it will thrive in drier areas. It is a suitable species for shelterbelts and park planting but if used in gardens should be placed where it has sufficient growing space, that is not less than 25 feet between adjacent trees. Early growth is good and trees may attain 20 feet in 6-8 years.

The timber is strong, hard, heavy, but only moderately durable. It is used for mining purposes, round construction and firewood. The bark has a tannin content of 22-30 per cent.

E. gillii Maiden **CURLY MALLEE**

This is a mallee or small tree which grows to about 20 feet in height and usually exhibits crooked stems and branches. The principal natural occurrence is in South Australia, where it extends from the Flinders Ranges (about 32°S. latitude) north-east to Lake Frome and eastward to the vicinity of Broken Hill in south-western New South Wales. The rainfall is 6-12 inches, with very high variability; the mean maximum of the hottest month is 90°F or higher.

When planted the form is better than that of the natural occurrence and at Broken Hill it has shown reasonable growth without watering. In very dry and hot areas it should be considered for ornament, low shelter and soil conservation.

E. gongylocarpa Blakely **MARBLE GUM**

This hardy eucalypt is a small tree, 30-45 feet in height, which grows in the corner of the Northern Territory south and west of Alice Springs, the adjacent area of Western Australia, including the Victoria Desert, and the north-western corner of South Australia. This is some of the driest and hottest country of the continent and has been described briefly for *E. gamophylla*, which occurs in the same general area.

Marble gum was only described as a separate species in 1936, though it had been observed by the Horn Expedition in 1894, when it was thought to be a form of *E. eudesmoides*. It is considered to be of importance for the reason that it attains tree dimensions under some of the most adverse climatic conditions in Australia. The Horn Expedition reported it as growing on inland sandhills, so that the species merits trial on well drained sandy soils in the arid zone.

E. gracilis F. Muell. **YORRELL**

Yorrell is a small tree, usually 25-40 feet in height, which develops a good crown in open positions but when growing close together the crown is high and comparatively narrow. In other words, the species requires plenty of space and light for satisfactory growth. One of the varieties is a mallee which is found in New South Wales. This eucalypt has a natural occurrence in the semi-arid zone of southern Australia which includes the Southern Cross-Narrogin-Kalgoorlie area of Western Australia, the middle latitudes of South Australia, including some of the arid zone, the north-western corner of Victoria and the south-west of New South Wales. For most of the area the rainfall is 10-18 inches, with a slight to marked winter maximum. Summer temperatures are moderate to hot; winter frosts are few in number and mainly light. Yorrell prefers sandy soils and is moderately tolerant to salt.

It is recommended for shelterbelts, shade and street planting in areas where the minimum rainfall is as low as 11-12 inches. The heartwood is dark brown, heavy, very hard and very strong but, unfortunately, is susceptible to termite attack and is therefore unsuitable for use in the ground.

E. intertexta R. T. Bak. **INLAND RED BOX**

This tree shows a wide range in mature size, but is usually in the range of 30-60 feet in height and 1½-3 feet in diameter. Even in very dry areas it may have a deep, well-developed dense crown, the trunk is typically of poor form and about one-third of the tree height. The native habitat includes the southern part of the Northern Territory and adjacent areas in Western Australia and South Australia, and the Western Plains of New South Wales. Most of this is in the arid zone, but the species extends to semi-arid country in New South Wales and part of South Australia. In the drier parts of its distribution it grows on relatively low-lying sites which may retain water after heavy rains. It is moderately salt tolerant and at times is found near the edges of salt pans.

It is recommended for dry country planting by several Forest Services, mainly where the rainfall is greater than 14 inches. On favourable sites it develops into an attractive tree; it is suitable for use in parks and for shade and shelter. When used for shelter it should be either in multi-row belts or with an understorey to provide protection near ground level.

The timber of inland red box is heavy, hard, strong but only of moderate durability in the ground; it is often difficult to split because of the interwoven grain (hence the specific name of *intertexta*). It is a good fuel and can be used for most farm requirements.

E. kondininensis Maiden et Blakely **KONDININ BLACKBUTT**

The discovery of this species was made at Kondinin in 1923 by C. A. Gardner, at one time Government Botanist of Western Australia. Kondinin is a small settlement in the semi-arid zone of Western Australia, about 150 miles south-east by east of Perth and some 170 miles north of Albany. Since the original discovery the species has been found southwards at Lake Grace and Pingrup, as well as about 250 miles eastwards in the Fraser Range. The rainfall of the western part of the distribution is 14-15 inches, with a well-defined winter maximum and a comparatively low variability for the dry country. Corrigin, which is not far from Kondinin, has a 50 percentile value of 14.96 inches and a 10 percentile value of 10.52 inches; for Norseman, which is not far from the Fraser Range, the values are 10.45 and 7.03 inches respectively. Summers are moderately hot with mean maxima of the hottest month about 90°F.

Kondinin blackbutt is a small to moderate size tree with a wide, well-developed crown when growing in open situations; mature height is 35-60 feet. It grows on medium to heavy loams and near Kondinin occurs on heavy clay flats surrounding salt lakes. It is rated as one of the most salt tolerant species of Western Australia and is recommended for shade and ornamental planting in areas of 12-20 inch rainfall, especially where salt is present in the soil and many other eucalypt species cannot exist.

E. largiflorens F. Muell. **BLACK BOX**

This box has its main occurrence in western New South Wales, where it is found on low country near drainage channels throughout the arid zone and the drier part of the semi-arid area. There are small extensions into southern Queensland, along drainage channels in north-western Victoria and near the Murray River in eastern South Australia. In these areas river red gum (*E. camaldulensis*) grows on the lowest flats which may be inundated for weeks or even months during periods of high river flow, and black box occurs on adjacent areas which may be only a few feet higher but where water does not remain for long periods. The alternative common name of flooded box refers to sites on which the tree is often found. Most of the region is one of 10-15 inches rainfall, but the species makes use of ground water from additional infiltration. The soils are typically heavy, grey clays which have at least moderate fertility.

Black box is usually in the height range 30-60 feet and has diameters of 1-2 feet, but on adverse sites it may be smaller. The trunk is of poor form, whilst the crown is wide-spreading and usually rather open.

This eucalypt has use for shade and soil conservation in areas of southern Australia where the rainfall is 12-15 inches and the soil may be water-sodden for short periods; in areas of greater rainfall there are more attractive species. Individually the trees flower freely, although not every year. The flowers produce acceptable quantities of both pollen and nectar. Beekeepers report that all or most of the trees in an area do not flower at the same time, and this discourages the shifting of large numbers of hives. The timber is very heavy, hard, durable but not readily sawn or split; it is useful for fencing and makes an excellent fuel.

E. lehmannii (Preiss ex Schau.) Benth. **BALD ISLAND MARLOCK**

One form of this eucalypt grows as a dense bush or small tree in the height range of 20-30 feet, with branches nearly to ground level during early years. In Western Australia it is considered only slightly inferior to the yardstick, *E. platypus* var. *heterophylla*. It is recommended by the Forests Department of several States; that of Western Australia suggests it particularly for use as a low shelter and also lists it as an ornamental species for street planting where its low height can be an advantage, and for shade because of its dense crown. It grows best on sandy soils or sandy loams where the rainfall is 18 inches or better.

The other form of Bald Island marlock is that of a mallee, 4-18 feet in height, which is found on granitic, lateritic or quartzite hills near the south coast of south-western Western Australia from King George Sound (Albany) to Cape Arid (near Esperance) three hundred miles or more to the east. Both latitudinal and altitudinal ranges are small, the former about 33½-35°S. and the latter from near sea level to 1,000 feet. The summers are mild, the mean maximum of the hottest month near the coast being about 75°F, and winters have 0-5 frosts. The mean annual rainfall of most of the natural occurrence is in the range of 20-40 inches.

E. lesouefii Maiden **GOLDFIELD'S BLACKBUTT**

This eucalypt has a small distribution in the Kalgoorlie to Norseman area of Western Australia. The area is marginal between arid and semi-arid, with a rainfall of 10-13 inches and high temperatures in summer. It is found mainly on alluvial flats which are often moderately alkaline. Goldfield's blackbutt is a small tree, usually 20-45 feet in height, with a rough, dark-coloured bark for a few feet at the base. The specific botanical epithet is in honour of Mr Ernest le Souef, who was a Director of the Zoological Gardens, Perth, in the early years of the twentieth century.

When planted on favourable sites it is fast growing in early years and may attain heights of up to 20 feet in six years, after which growth becomes slower. It is recommended for ornamental planting both in parks and the larger private gardens in southern Australia where the rainfall is 10-20 inches.

E. leucoxylon F. Muell. **YELLOW GUM**

This eucalypt has several common names; in South Australia it is known as South Australian blue gum but as the species has its principal occurrence in Victoria that name is not acceptable as a standard common name; it is also known as white ironbark in Victoria in reference to the fact that although it has a gum-type bark, botanically it belongs to the ironbark group. The natural occurrence is mainly in sub-humid country, but it is also found to a limited extent in the semi-arid zone. The rainfall is in the range of 20-30 inches, with a winter maximum and a warm to moderately hot summer.

The size and form of the tree varies with the site—on good soils in the higher rainfall area it may be up to 90 feet in height, with a straight bole and a moderately dense crown, but on marginal sites it may be reduced to a shrubby tree possibly only 30 feet in height. There are two horticultural varieties with pink flowers; these are often selected for ornamental planting in parks and large gardens, in preference to the white-flowering form. In southern Australia, where the rainfall is above 18 inches, yellow gum is also of use for shade and shelter planting; in the latter case it provides good protection to ground level in early life but later needs the addition of a lower-storey shrub. On favourable sites early growth of the tree is good, sometimes averaging 4 feet a year for the first ten years. It has limited pollen production but nectar is produced about every second year when the tree flowers freely; useful flowering often commences with planted stock at the third to fourth year. The timber is of high quality —heavy to very heavy, hard, strong and very durable and like that of many of the closely allied ironbarks. It is a favoured sleeper timber but is not used as much as formerly for bridge timbers, as a result of the change in constructional materials to steel and concrete. It is used in the round for poles and strainers and, whilst suitable for both light and heavy construction, it is not readily sawn and nailed.

E. longicornis (F. Muell.) F. Muell. ex Maiden **RED MORREL**

Red morrel has a moderately wide distribution in the semi-arid and sub-humid zones of south-western Western Australia, occurring in an area which starts about 50 miles from the coast and extends over 300 miles to near Kalgoorlie in the east; whilst north and south it is found from the vicinity of Geraldton to near the south coast. It is a tree with a height range of 30-70 feet and boles 1-2 feet in diameter, with good form, and is remarkable in that it attains good heights in rainfall as low as 10 inches. It is one of the largest trees and best timber species in the 10-15 inch rainfall belt of south-western Western Australia. It also grows with rainfall as high as 25 inches. Commonly it is found on sandy or red and brown loams which may have calcareous subsoils. The specific botanical epithet means long-horned, a reference to the usually long opercula of the buds.

This is one of the eucalypts which should receive consideration when selections of species are made for planting in the dry country of southern Australia where rainfall is 10-20 inches. It is drought and frost resistant and will tolerate a moderate level of salinity in the soil. It is suitable for use in parks and as shade trees on farms, but for shelter is preferably used in multi-row belts or associated with a lower-storey shrub species. The timber is of good quality and has been mainly used locally for mining, in round construction and as fuel.

E. loxophleba Benth. **YORK GUM**

York gum is found in a large part of south-western Western Australia from the sea coast, especially between Perth and about 100 miles north of Geraldton, to 50-100 miles east of Kalgoorlie and Esperance. The main occurrence is in the sub-humid and moister parts of the semi-arid zones, but there is some extension into the arid zone near Kalgoorlie. It is a small tree, usually in the height range of 15-40 feet and 1-1½ feet in diameter, with an umbrella-shaped crown which provides useful shade. Form is typically poor, though sometimes the trunk may be moderately straight and up to half of the total height. Most of the occurrence is in a region of 10-19 inches rainfall, with a winter maximum. Temperatures in summer are high, with mean maxima of the hottest month 90-94°F. York gum has a preference for sandy soils including sandy loams, but will grow on both lateritic soils and those of somewhat heavy texture. It is tolerant of moderate levels of alkalinity. York in the common name refers to the town of York where it was plentiful.

The Forests Department of W.A. recommends it for shade and shelter purposes in areas where the rainfall is 16 inches or greater, but on deep soils it will probably grow satisfactorily in slightly drier localities. The timber is heavy, hard and tough, with an interlocked grain and was formerly used for wheelwright and similar work.

E. melanophloia F. Muell. **SILVER-LEAVED IRONBARK**

Silver-leaved ironbark is a species of the higher rainfall parts of the semi-arid zone and most of the sub-humid zone in northern New South Wales and in Queensland nearly as far north as the Gulf of Carpentaria. In the latter State it reaches the coast in many localities. The climate shows a wide range of conditions; rainfall is mainly within the limits of 15-30 inches, with a summer maximum and a dry winter which is most marked in the northern latitudes. The size of the tree shows considerable variation from not much more than a large shrub about 20 feet high, to a moderately well-shaped tree up to 90 feet in height. Trunk form is rarely good.

The qualification of silver-leaved in the common name refers to the glaucous foliage; the fruits are also glaucous and the tree is one of the easiest eucalypts for the layman to identify. It grows best on good silts, but will tolerate both light and heavy-textured soils. Because of the glaucous foliage and the tendency to retain branches and foliage low down on the trunk it is recommended for ornamental planting and shelterbelts in summer rainfall areas.

E. melliodora A. Cunn. **YELLOW BOX**

This tree is widely distributed in New South Wales and Victoria, especially on the inland side of the Dividing Range; it also occurs on the tablelands of New South Wales, with a minor extension into south-eastern Queensland. Climatically this region includes the moister parts of the semi-arid zone and much of the sub-humid zone. The rainfall is 15-35 inches and summer temperatures are warm to hot. Yellow box is a tree of medium size, 60-100 feet in height and usually a short trunk 2-4 feet in diameter. With full maturity the tree develops a large, heavily-branched, wide-spreading crown which affords excellent shade in pasture country.

The specific name, *melliodora*, means honey-scented; and is associated with the copious nectar production of the species. The trees flower freely, usually about every second year, and the species is regarded as the best honey producer amongst the eucalypts. The honey is pale straw in colour, very dense and aromatic, with a pronounced flavour and is one of the highest quality native honeys. So far as is known bees do not collect pollen from yellow box and when other pollen-yielding sources are absent during the flow of yellow box nectar, the worker force of the hives is reduced following upon restricted reproduction by the queen bees.

Yellow box prefers light to moderately heavy alluvial soils, but is found on sandy loams. It does not grow on poor soils, especially sands, and the occurrence of well-grown specimens is usually an indicator of better-quality agricultural soils. In the drier parts of its range, that is 15-20 inches rainfall, it shows a preference for sites on flats near watercourses where there is additional sub-soil moisture available.

In addition to its value for honey production, yellow box is a first-class shade tree and can be used for ornamental purposes in parks where there is sufficient growing space, that is 60-80 feet between trees. When used for shelter it is preferable to incorporate it with other species in multi-row belts. The light pinkish to yellowish brown timber is heavy, hard and very durable. The grain is frequently interlocked and the timber is difficult to work, especially split, so that it is more often used in the round for posts and poles and, formerly, hewn for heavy construction. It is pre-eminently a high class firewood. The wood should be allowed to become air-dry before being used for fuel; this may take 2-3 years or longer for material one foot in diameter.

E. microtheca F. Muell. **COOLIBAH**

Coolibah is a species of the arid and semi-arid zones of Australia, especially the Northern Territory and Queensland, but there is a considerable extension into northern Western Australia. It also occurs in north-western New South Wales and, to a limited extent, in northern

South Australia. The range of rainfall is 8-40 inches, but the widest distribution is in areas of 10-25 inches. Except in the lowest rainfall areas there is a marked summer maximum with a very dry winter. Temperatures are high throughout most of the year and in summer the mean maximum of the hottest month may attain 100°F; there are a few winter frosts in localities away from the coast. Associated with this wide geographical and climatic range is a variation in size from a tree 50-70 feet in height, with a moderately well-formed bole up to 3½ feet in diameter, down to a poorly-formed, small tree with a straggly crown and a trunk only a few feet in length to the first large branches. Soils show a wide variation, although the tree is often found on heavy soils such as alluvials and alkaline clays. Especially in the drier part of its range coolibah is characteristically a species of very open savannah formation in seasonally inundated country around the edges of swamps or lagoons, or in open belts along water courses when soil conditions are favourable.

It is considered a useful species for shade and shelter on suitable sites throughout its natural range, and has proved satisfactory in dry country to the south. The timber is dark in colour, very heavy, hard and has an interlocked grain; it is durable and notably termite-resistant. Nowadays it is only used to a limited extent for fencing and fuel.

E. occidentalis Endl. **FLAT-TOPPED YATE**

This eucalypt attains mature heights of 60-80 feet and diameters up to 2½ feet. Like some other Western Australian eucalypts the larger branches are at a very small angle to the main stem and they grow to about the same height, giving the tree a nearly level or umbrella-like crown. This type of tree provides good light shade and allows crops to be grown close to the base, but is not suitable for most forms of shelter. The name *occidentalis,* meaning western, was given to it by Endlicher who did not, however, ever see the living tree, and the name has no significance. The first part of the common name alludes to the form of the crown, whilst the second part refers to a slight resemblance of the tree to that of yate (*E. cornuta*).

The natural occurrence of flat-topped yate is in the moister parts of the semi-arid zone of south-western Western Australia, and parts of the adjacent sub-humid zone. The principal distribution is east of the main jarrah belt and north of Albany; there is a sparse occurrence as far east as Esperance and Salmon Gums. The range of rainfall is 13-35 inches, but optimum development is in areas of 16-20 inches; variability is comparatively low and data for Narrogin, which is on the western edge of the distribution, shows 20.11 inches for the 50 percentile value, 14.95 inches for the 10 percentile and 10.50 inches for the lowest on record. The tree grows in low-lying sites which are typically poorly drained and the soils are clay or heavy loams, sometimes with a sandy surface.

Flat-topped yate is recommended by the Forests Department of W.A. for shade planting; it is also suitable for park use where a high, light shade is required and where grass can grow to the base of the tree. The timber is of high quality and can be used where a very heavy and strong wood is required. In Western Australia it is rated as inferior to yate, but then the latter is one of the hardest and strongest timbers in the world.

E. ochrophloia F. Muell. **YAPUNYAH**

There is so much similarity between the common names applied to this species and *E. thozetiana* that confusion can arise. In Australian Standard 02-1970 the standard trade common name of yapunyah is applied to *E. ochrophloia* and mountain yapunyah to *E. thozetiana,* whilst the local names of napunyah and lapunyah are used for both species. Yapunyah has a limited occurrence in the dry country on either side of the border of New South Wales with Queensland. This is both in the arid zone and the drier parts of the semi-arid zone. The rainfall is 10-20 inches, with a weak summer maximum and a high variability; summer temperatures are high.

Yapunyah is a moderately small-size tree, attaining 40-70 feet in height, with a good crown in view of the dry habitat. It grows mainly on silty or heavy alluvial soils along drainage channels and on flats subject to accumulation of water after irregular heavy rains. This eucalypt is one of the better species for shade and shelter on heavy soils in dry summer rainfall areas. The timber is very heavy, very hard, tough and compares favourably with ironbarks for strength.

E. odorata Behr et Schldl. **PEPPERMINT BOX**

This is a species of South Australia, where it occurs as scattered trees from the southern Flinders Ranges in the north southwards to Adelaide and Kangaroo Island, and westwards to southern Eyre Peninsula. There is a small isolated occurrence in the south-east of the State which extends into Victoria. The climatic zones include semi-arid and sub-humid. Rainfall is 15-30 inches, mainly in winter. Growth habit varies from a small tree, usually 20-50 feet in height to a shrub without a well-defined main stem; on some sites it may show mallee form. It grows on a variety of soils and on the Adelaide plains may be found on clay loams over comparatively impervious clay sub-soils. In the lower rainfall areas it is found on red-brown earths and sandy loams.

Where it grows naturally some specimens should be retained for shade and shelter. When planted at wide spacings it develops a bushy habit and retains branches to near ground level. It offers possibilities for single row shelterbelts in southern winter rainfall areas where precipitation is 18-25 inches.

E. oleosa F. Muell. ex Miq. var. *angustifolia* Maiden **NARROW-LEAVED GIANT MALLEE**

The botanical name given above is that in *A Key to the Eucalypts,* W. F. Blakely (2nd ed., 1955), but the variety is not currently recognised by most botanists, as they consider it falls within *E. oleosa* subsp. *oleosa.* It is recorded in this book because when planted at Broken Hill it has proved to be one of the best eucalypts for growing without watering. The strain used there grows into a small tree 30-40 feet in height, with a short main stem which divides into several large branches. The high crown is well developed and provides good shade, but the tree is not suitable for use in single-row shelterbelts. It is quite an ornamental eucalypt and should be considered for planting in southern inland Australia where the rainfall is as low as 9 inches. In order to ensure that the strain used is similar to that growing at Broken Hill it would be necessary to take great care in seed procurement.

Eucalyptus orgadophila Maiden et Blakely **MOUNTAIN COOLIBAH**

This eucalypt is found on both sides of the mainly low and undulating country of the Great Dividing Range in eastern Queensland. The main area is 80-280 miles from the coast and extends for about 500 miles from Warwick in the south to Clermont and Emerald in the north, with a small extension between Hughenden and Pentland further north. The range of latitude is 20-28½°S. and that for altitude 500-1,900 feet. The climate is warm temperate to inland tropical, with a considerable range in the mean maximum temperatures of the hottest month of 83-99°F. Winters are cool to mild, with the mean minima 39-50°F and whilst much of the area has only rare frosts some of the coldest places may have 10-20 a year. The mean annual rainfall is mainly 18-30 inches, but attains up to 35 inches near Toowoomba, with maximum precipitation in summer.

Mountain coolibah is a small, ornamental tree, usually in the height range of 35-50 feet, with a short trunk which supports a large heavily-branched crown. Open grown planted trees, however, grow well to 30 feet or more and retain their lower lateral branches for many years.

The Forestry Department of Queensland recommends this species for shelterbelts, shade and ornament on both red and black soils in its region. Like most eucalypts, however, in shelterbelts it would need to be associated with a lower-storey dense shrub to give good protection when the trees attain maturity.

E. papuana F. Muell. **GHOST GUM**

Ghost gum is probably best known to people who live outside its habitat as the subject of paintings by many Aboriginal artists, notably Albert Namatjira. It grows over most of Australia north of the Tropic of Capricorn, especially in the Northern Territory and Queensland, but with extension into the Kimberley region of northern Western Australia. The rainfall varies from 10-60 inches and is of a monsoonal type in the north, but in the arid interior has a variable incidence from year to year. The climatic zones include arid, semi-arid and sub-humid. Size varies with growing conditions but even in the Alice Springs area may attain 40-50 feet in height; elsewhere height can be 20-60 feet and form poor to good.

This eucalypt has a clean white trunk and an open crown which collectively gives a particularly attractive appearance to the tree. It deserves to be retained in most areas of central Australia where it occurs naturally, and to be planted in northern areas of 10-25 inches rainfall, especially on sandy flats which receive run-off from adjacent slopes. It is also worthy of use in higher rainfall areas.

E. pillagaensis Maiden **NARROW-LEAVED GREY BOX**

Narrow-leaved grey box is found on the eastern side of the semi-arid and the western side of the sub-humid zones in northern New South Wales and adjacent areas of southern

Queensland. It is also known as Pilliga box, after a district of that name, where it is fairly common, in the former State. Most of the distribution is within the limits of 18-25 inches rainfall, with a weak summer maximum but, on an average, no month with less than 1-1½ inches.

It is a moderate size tree, usually 40-80 feet in height, and on all fairly good sites it has satisfactory form. Specimens which have grown in the open from early life may retain branches on the lower part of the stem, but otherwise the crown tends to be carried on the upper half of the tree. Like the closely associated grey box (*E. woollsiana*) in the drier parts of its range it grows on plains and shallow depressions with rather heavy alluvial soils and clay loams, but is found on slopes with sandy loams in the higher rainfall districts.

In its climatic area it is a good shade species and can be used in multi-row shelterbelts. The timber is possibly not quite equal to some of the other boxes, but is still of high quality—heavy, hard, strong and usually very durable. It is used in the round and split for fencing and is, of course, an excellent fuel.

E. platypus Hook. **MOORT**

Moort is a large shrub or small tree up to about 25 feet in height. Typically it has a very bushy habit with many branches and dense foliage, so that it is capable of providing a high level of shelter near the ground. In nature the tree often grows in dense thickets and exhibits poor form, but when planted in south-eastern semi-arid Australia the shape is satisfactory. Moort occurs in southern coastal Western Australia from Albany to Esperance, with the principal occurrence in the Ravensthorpe district; it extends inland for only a limited distance, for example the Stirling Ranges about 50 miles north of Albany. The climatic zone is sub-humid and the rainfall shows a wide range from about 15 inches to over 35 inches; temperatures are moderate throughout the year, with mean maxima for the hottest month of 73-77°F. Frosts are absent or very light.

This species should receive high priority for low, dense shelter in southern semi-arid and sub-humid regions where the rainfall is 15-25 inches. Introduction into eastern States show that it will tolerate much higher maximum summer temperatures than it does in its area of natural occurrence. It is recommended for low shelter, either on its own or in association with a taller-growing species; if the lower branches are removed it affords excellent shade and it is suitable for use in parks and other ornamental planting.

E. platypus Hook. var. *heterophylla* Blakely **Same common name as species**

This variety of moort was originally described from material collected near Ongerup, some 100 miles south-west of Narrogin and about latitude 34°S. The bushy habit and the retention of branches close to the ground is even more marked than in the species and the Forests Department of Western Australia strongly recommends it for farmers and others who require a dense low shelter. This writer considers that there is possibly no other eucalypt which is better for this purpose in its climatic area. Whilst it occurs naturally in a moderately temperate climate it is showing good growth in sample plots as far north as the Geraldton district, where the mean maximum of the hottest month may exceed 90°F. When planted in favourable sites it is possible that its ultimate height may reach about 35 feet.

E. polycarpa F. Muell. **LONG-FRUITED BLOODWOOD**

This bloodwood is a tree of northern and north-eastern Australia, where it is found throughout most of Queensland, the northern part of the Northern Territory and the Kimberley area of Western Australia. There is a minor extension into northern inland New South Wales. Much of the occurrence is in the semi-arid zone, but the species occurs throughout the sub-humid and humid zones of the north and there is some extension into the arid area. There is, consequently, a considerable variation in rainfall from about 10 to over 55 inches; this falls in a monsoonal pattern throughout most of the area and has still a well-developed summer rainfall pattern in the southern parts of the occurrence. The mean maximum temperature of the hottest month is in the range of 90-100°F, whilst the mean minima of the coldest month vary from 68°F at Darwin to about 40°F for southern inland districts. Whilst most of the area is frost-free there are 1-6 a year in the latter areas.

Long-fruited bloodwood at its best is a shapely tree up to 70 feet in height, with a moderately straight bole one-third or more of the tree height and 2-2½ feet in diameter. More generally it is 30-50 feet in height, with an open and rather straggly crown and a bole which is short and of poor form.

It is not an important timber tree but throughout the extensive area over which it occurs, it provides light shade and some shelter. The timber can be used for fencing material and, when the logs are of sufficiently good form, it is sometimes sawn. Like most bloodwoods it is a good fuel.

E. populnea F. Muell. **BIMBLE BOX**

This box is a small to moderate-sized tree, usually in the range of 35-70 feet in height and 1-2½ feet in diameter. It commonly has a straight trunk about half of the tree height, with a medium-sized, moderately dense crown. The leaves are of firm texture, dark glossy green and of a shape similar to many true poplars (*Populus* spp.); hence the botanical name of *populnea* and an alternative common name of poplar box.

The principal occurrence of bimble box is in the semi-arid zone of central and northern New South Wales, and northwards into Queensland as far as about 20°S. latitude. In the latter State it extends to the sub-humid zone and reaches the coast near Rockhampton on the Tropic of Capricorn. For most of the area rainfall is 14-25 inches, with a more or less uniform distribution in the south but changing to a marked summer maximum in the farthest north. Especially in the driest parts of its distribution it tends to grow on flats which receive run-off from adjacent slopes or even become waterlogged for short periods; these are clays or heavy black soils which become cracked during the dry season. In higher rainfall areas it grows on gentle slopes where the soil may be a reddish loam of light texture over a clay sub-soil.

Bimble box is one of the more ornamental members of the box group and merits public planting for parks where there is sufficient growing space. Where it occurs naturally it certainly justifies retention at picnic spots near rivers and by farmers for shade purposes. When protected from grazing it regenerates freely from seed during years of sufficient rainfall and, like the true poplars, it also suckers. The timber is pale brown, heavy, hard, strong and durable but does not saw or work readily. Utilisation is limited to local pastoral needs.

E. 'pterocarpa'

This is a eucalypt which has not been botanically described, although it was listed and its uses given by G. E. Brockway in the *Journal of Agriculture of Western Australia*, Vol. 8 (third series), No. 2 (1959). The article was subsequently issued as Bulletin No. 2616, *Tree Establishment in the Wheatbelt. E. 'pterocarpa'* has some similarities to *E. corrugata* Luehm. Its natural occurrence is on sandy and gravelly soils in the Norseman district of Western Australia.

It has been planted on a small scale in the dry country of Western Australia and at Broken Hill, New South Wales, and gives promise of being an excellent eucalypt for shade and for use in multi-row shelterbelts in areas where the rainfall may be as low as eight inches and the summers very hot. The mature height is at least 30-40 feet and the tree has a wide-spreading, moderately-dense crown. The Forests Department of W.A. recommends it for greater use on farms in the dry country than has hitherto been the case. It would probably be undesirable to plant it on heavy soils without experience of its behaviour on such types, or where the rainfall exceeds about 20 inches a year. Seed should only be procured from reliable sources.

E. redunca Schau. var. *melanophloia* Benth. **BLACK-BARKED MARLOCK**

The species *E. redunca,* as at present described, shows considerable variation and it is possible that systematic botanists will eventually revise the group of eucalypts now included in that species. Of interest to farmers is the variety *melanophloia* for its bushy habit and retention of branches to ground level, making it one of the better species of the State for low shelter. It is not, however, quite equal to *E. platypus* var. *heterophylla.* It is a native of the southern part of the south-western semi-arid and sub-humid zone of Western Australia where the climate is moderately temperate. The evidence of Forest Department plots is that it will grow satisfactorily where the rainfall is as low as 12-14 inches and the dry, hot summers have mean maximum monthly temperatures as high as 95°F.

E. salmonophloia F. Muell. **SALMON GUM**

Salmon gum is one of the finest dry country eucalypts of Western Australia. Even in areas of rainfall as low as 9 inches, for example Kalgoorlie, it attains 40-80 feet in height and diameters up to 2 feet. Rainfall varies, however, from about 8 to 20 inches and shows high variability. The tree has a proven performance in the dry country and is recommended by the Forests Department of its State for areas as dry as 10 inches, whilst the Forestry Commission of New South Wales advises a minimum of 12 inches. Salmon gum occurs throughout the wheat belt and eastern Goldfields area of Western Australia, with the easternmost occurrence about 70 miles from Perth and the westernmost beyond Kalgoorlie. Northwards it extends to near Geraldton and southwards to within 50 miles of the coast. Both the specific botanical epithet and the common name refers to the salmon colour of the bark fracture; the external colour of the bark varies from brownish-pink at the end of summer to almost white in winter.

This tree grows on a wide range of better-quality soils, which vary from somewhat light-textured loams to heavy clays. In the early days of land classification in W.A., salmon gum and gimlet (*E. salubris*) were often taken as indicators of first-class agricultural soil. This was probably of most significance in the higher rainfall areas.

The Forests Department of W.A. recommends salmon gum for shade and shelter, and whilst young trees are suitable for single-row shelterbelts, this is not the position when they attain maturity and the lower part becomes open. It can, of course, be used in multi-row windbreaks or in single-row belts with under-planting. It is a handsome species and should be considered for parks and avenue planting where there is sufficient space. The timber is of good quality and reasonably durable, so that it is suitable for farm woodlots and can be sawn for constructional purposes. The planting of salmon gum should be restricted to southern Australia, whilst specimen trees can be grown in parks where the rainfall is as low as 10 inches, for shade and shelter a minimum rainfall of 12-14 inches is desirable.

E. salubris F. Muell. **GIMLET**

The common name of gimlet arises from the fluted or spirally-twisted trunks which is most marked in young trees. This tree should not be confused with silver-topped gimlet (*E. campaspe*). *E. salubris* is one of the best-known eucalypts of the eastern agricultural and eastern Goldfield districts of south-western Western Australia, but salmon gum (*E. salmonophloia*) takes first position. The area of natural occurrence is mainly in the semi-arid zone from about 100 miles inland from Perth to slightly east of Kalgoorlie. The rainfall has a range of 8-18 inches and a winter maximum; summer temperatures are high, with mean maxima of the hottest month 85-95°F.

Gimlet is a tree 40-60 feet in height and carries a moderately dense crown near the top. In its natural habitat the presence of gimlet trees is considered an indication of good loamy soils, often with a large proportion of clay; in such areas it is frequently growing in association with salmon gum. When they grow separately it is usually found that gimlet occupies the heavier soils. It is also found in New South Wales that best development is on heavy soils. Both in Western Australia and when planted in areas such as Broken Hill it has proved to be tolerant of salt in the soil.

It is suitable for ornamental planting in areas where the rainfall may be as low as 10 inches, but there is difference of opinion as to its value for farm shelter. Whilst quite young trees may retain branches to near ground level they soon lose them, so that the species is quite unsuitable for single-row shelter, but it can be used for multi-row belts. The timber is pale brown, very hard, tough and strong but only moderately durable as it is subject to termite attack.

E. sargentii Maiden **SALT RIVER MALLET**

This is still another useful dry country eucalypt from the south-western semi-arid zone of Western Australia. Its natural distribution is mainly 125-200 miles east and south-east of Perth in an area where the rainfall is 13-18 inches, with a marked winter maximum. The common name refers to occurrence around the Salt River near Lake Mears where it enters the Avon River, and is particularly suitable because the tree is one of the most salt-tolerant eucalypts. In Western Australia where the salt level of some soils is increasing and the more sensitive species are dying out, Salt River mallet survives the longest. This salt tolerance is also a well-established characteristic in eastern States, for example at Broken Hill in an area with an average annual rainfall of 9 inches but where the lowest on record is only 2.64 inches.

Salt River mallet is a moderately bushy small tree, 25-35 feet in height and of attractive appearance. It is suitable for street and park planting where a hardy species of this size is desirable and is more adapted for shelterbelt planting than many of the other eucalypts from dry country Western Australia. For extensive shelterbelt use it should probably be restricted to a minimum rainfall of about 15 inches, but for ornamental use where ample growing space is available and a little additional watering in drought years is possible, it could be considered for deep soils where the rainfall may be as low as 9 inches. Its salt tolerance is an important attribute for some dry country planting.

E. sideroxylon A. Cunn. ex Woolls subsp. *sideroxylon* **RED IRONBARK**

This ironbark has a wide distribution, but the main area is in the higher rainfall parts of the semi-arid zone in New South Wales, with extension into sub-humid central northern Victoria. There are minor occurrences in southern Queensland and on the coast of New South Wales and Victoria. The rainfall of the inland areas is 15-25 inches and it varies from a moderate winter maximum in the south to a moderate summer maximum in the north. The specific botanical epithet means that the wood is like iron. Certainly the wood is very heavy and hard, but not more so than many other species of the genus, notably ironbarks and boxes.

Red ironbark is a tree of only medium size; commonly it is 70-100 feet in height and 2½-4 feet in diameter. The form of the trunk in old trees is often poor, whilst the length does not exceed half the total height. Young planted trees are usually of acceptable form and retain branches to near ground level for many years. Naturally-grown trees are found on clay soils, often with a very shallow surface loam or even none at all.

The tree is recommended for shade, shelter and woodlot purposes on heavy soils in southern Australia where the rainfall is 20 inches or more. There are varieties which have pink flowers and these are often favoured for ornamental use. In the western areas of New South Wales the flowering period is mainly during winter, a period which is generally unfavourable to bees because of reduced pollen supplies and the damper weather prevailing. Red ironbark does not produce any worthwhile quantities of pollen. The timber is very strong and very durable, but difficult to work. It is hewn and sometimes sawn for railway sleepers and, formerly, for heavy construction such as bridges.

E. spathulata Hook. **SWAMP MALLET**

This is a useful species for low shelterbelts of moderate density. It occurs naturally in the central and southern wheat belt area of south-western semi-arid Western Australia where the rainfall is mainly in the range of 12-15 inches. The summers are hot to very hot with mean maxima of the hottest month about 85°F in the southern part of the range but 90-95°F in the northern part. The mean minimum of the coldest is about 40°F and there are 1-5 frosts a year. The rainfall is not as variable as for some parts of dry Australia and the 1 in 10 years lowest value is about 10 inches; the all-time records may be several inches lower. It grows on medium to heavy loams and generally on low-lying areas.

Swamp mallet is botanically described as a shrubby mallee or slender marlock (the latter is defined as a shrubby eucalypt with a reduced rootstock). When planted on farms, however, it usually has only one main stem at ground level, though this soon branches profusely. The many fine branches and the small leaves provide a desirable type of moderately open wind screen. Growing naturally it is often only 15-20 feet in height, but when planted on good soils and, especially, with a higher rainfall of 15-20 inches it may attain a height of 35 feet. It is recommended for low shelterbelts in areas of 12-20 inches rainfall and for use in street or other public planting where a bushy eucalypt of low mature height is desired.

E. stricklandii Maiden **STRICKLAND'S GUM**

The specific botanical epithet, *stricklandii*, and the first part of the common name is in honour of Sir Gerald, later Lord, Strickland, a former Governor of Western Australia and later of New South Wales, but best known as Governor of Malta during World War II. Less commonly the tree is called yellow-flowering blackbutt; the latter part of the name is due to the dark grey to almost black, persistent bark at the base of the tree (which is otherwise a gum) and the former part to the striking yellow of the anther filaments of the large flowers.

Strickland's gum is a small tree, usually 20-25 feet high, which has an open, widespreading crown with rather sparse foliage. In its natural area it grows on stony, schistose hills but it has been found to grow moderately well on both sands and clay soils, which may contain appreciable salt. The rainfall is 9-11 inches. Like many dry country species this is a eucalypt which if planted in a high rainfall area or is heavily watered in a dry one, does not develop its true form and colouring. It is recommended for ornamental and possible shelterbelt use in southern areas with a rainfall of 10-15 inches.

E. tectifica F. Muell. **DARWIN BOX**

This eucalypt occurs mainly north of latitude 17½°S. in a belt which includes the northern part of Western Australia, most of the Northern Territory north of Daly Waters and the north-western corner of Queensland. The climatic zones include both the higher rainfall parts of the semi-arid and the humid area as far as Darwin. Katherine, in the wetter part of the semi-arid zone of the Northern Territory, has a 50 percentile rainfall value of 40 inches, and a lowest on record rainfall of 17 inches. This is in the monsoonal area where the peak rainfall is December to March, whilst June to September may be almost rainless. Temperatures are high at all seasons and the mean maximum just before the break of the monsoon in November may exceed 100°F in the semi-arid zone and even 93°F at Darwin, which is on the coast. In Western Australia it is known as grey box whilst the name of Macarthur River box refers to the location from which the first botanical specimens were secured.

It is a tree 30-45 feet in height with a form which is comparable to those of many of the grey boxes of south-eastern Australia—a trunk about one-third of tree height, of moderate form and a crown of moderate density. North of the Leopold Range in Western Australia, its occurrence is so closely connected with basalt and andesite soil types that the latter can be mapped on the basis of the species distribution. Elsewhere, both in Western Australia and the Northern Territory it grows on a variety of soils, sands and sandy loams on river flats, as well as extending to hill slopes.

Where it grows naturally a sufficient number should be retained for shade purposes both around homesteads and at stopping points on roads and tracks. The timber is very heavy, hard and is reputed to be sufficiently durable for use in the ground.

E. tessellaris F. Muell. **CARBEEN**

Carbeen is found throughout most of the semi-arid zone of Queensland, all the sub-humid and is scattered throughout the coastal humid areas south of about 17°S. latitude. It also occurs in the adjacent northern parts of New South Wales. Rainfall shows a wide variation from 15-60 inches, but most of the distribution occurs within the limits of 15-25 inches. Maximum precipitation is in summer and in the northern part of the range there are several months in the year when the average monthly falls are only 0.25-0.50 inches. Temperatures are high throughout most of the year but a few frosts may occur at the higher inland locations. The specific botanical name refers to the tessellated appearance of the bark on the lower part of the tree. Another common name is Moreton Bay ash, which refers to the Brisbane area of Queensland; but this is also used for several other trees.

According to the environment carbeen varies from a stunted tree, say 40 feet in height, to a moderately large tree up to 100 feet in height and 3 feet in diameter. The crown is rather open, with slender branches and drooping branchlets. The tree occurs mainly on sandy soils or alluvials, but it shows a wide tolerance to differences in texture and fertility. The occurrence of large *E. tessellaris* of good form is generally regarded as a sign of good pasture country or, in some areas, of land suitable for cropping.

The tree is of attractive appearance and because it is resistant to heat, moderate drought and some frosts it is very suitable for ornamental and shelter planting on suitable sites in the summer rainfall zone.

E. thozetiana F. Muell. ex R. T. Bak. **MOUNTAIN YAPUNYAH**

This species has long been known as Thozet's box but in the Australian Standard 02-1970 the name of mountain yapunah was adopted for the timber and is being used in this text. Under the notes for *E. ochrophloia* there is reference to the use of the names napunyah and lapunyah for the two species. On favourable sites in Queensland this box grows as an erect, graceful tree with an open crown; it attains 70-80 feet in height and up to 3 feet in diameter. On poorer and drier sites further inland it may be only a small tree 30 feet high, with a short trunk of poor form. The most important occurrence is in the semi-arid and part of the sub-humid zone of Queensland south of about 21½°S. latitude, but in the south-west it extends to the arid zone. There is a very limited distribution in northern New South Wales and about 50 miles north-east of Alice Springs in the Northern Territory. The botanical name, *thozetiana*, is after Monsieur A. Thozet, who collected it during his last botanical journey to the Expedition Range in Queensland.

The rainfall for most of the distribution is 12½-25 inches, with a moderate to well-developed summer maximum, and high variability. Summer temperatures are high with mean maxima of the hottest month 95-100°F. Mountain yapunyah grows on a variety of topography and soils, and though the latter are mainly light sandy loams they may sometimes be of a skeletal nature.

Some mountain yapunyah trees should be retained on farms for shade purposes, whilst it merits consideration for use in low summer rainfall areas. The timber is dark in colour, very heavy, hard, strong and very durable; it saws well for such a heavy timber (68 lb/cu ft air dry). It seasons without undue distortion and can be used for general construction and even interior fittings where a heavy timber is acceptable.

E. torquata Luehm. **CORAL-FLOWERED GUM**

The coral-flowered gum is a small tree, often only 20-25 feet in height but sometimes attaining over 30 feet. The trunk is short, the crown broad and well branched with a good cover of foliage even under adverse conditions, so that it is an excellent species for shade. This is one of the Western Australian species which is most extensively planted in the dry areas of the southern parts of the eastern States for street, park and ornamental purposes; good examples of it can be seen in Mildura, Broken Hill and Woomera. It is native to a small area west and south of Kalgoorlie; Norseman represents the more southern limits. The specific name *torquata* which was given to it by J. G. Luehmann of Melbourne, is derived from *torquis*—a collar or ring, in reference to the very characteristic ribbed swelling at the base of the buds and fruits.

The rainfall of its native area is 8-11 inches, with a moderate winter maximum and a 10 percentile value for Kalgoorlie of 5.74 inches. Planted at Broken Hill it has shown its ability to grow satisfactorily without watering and has survived a year with less than 3 inches (1967). It will also tolerate alkaline soils better than most species of the genus.

It can be readily raised from seed but should be planted as tubed stock. This species ranks in a high place for any form of ornamental planting in most areas with as low as 9 inches of rainfall a year. It will tolerate up to about 20 inches if the soil is well drained and the summers hot but with the winters having not more than a few light frosts. Because of the wide-spreading crown it should be widely spaced either in avenues or as ornamental specimens, that is 25-35 feet apart. When planted in parks for shade purposes in very dry areas, that is 9-14 inches a year, and it is not intended to water the specimens after the first few months, it is recommended that trees should be spaced 40-50 feet apart so that the root systems can spread over a large area.

One of the attractions of the tree is its masses of pink-coloured flowers, hence the common name, and it should be appreciated that whilst the species will flower in a wetter and cooler climate the flowers do not attain the magnificence of colour to be seen in the hot and dry parts of the continent.

E. trachyphloia F. Muell. **BROWN BLOODWOOD**

Brown bloodwood is an eastern Australian species which ranges from near Cairns on the north coast of Queensland to central inland New South Wales. The climate varies from tropical sub-humid and humid to inland semi-arid, and rainfall has the wide range of 20-50 inches. Throughout the area summer temperatures are warm to hot with mean maxima of the hottest month about 85°F; mean minima of the coldest month vary from 60-70°F in the north to about 40°F in the south. In the warmer areas frosts are absent but in the south there may be up to 15 a year.

This is a tree of only moderate size, commonly 60-80 feet high, and with a reasonably straight trunk, but varies in size and form with the site. In the better rainfall coastal areas it has a moderately dense crown. The timber is typical of bloodwoods in general, including the presence of gum veins. It is used for mine props, fencing and as a fuelwood; logs of suitable form can be sawn for structural timber. The species merits consideration for planting for shade and timber production in northern areas of 20-30 inch rainfall; it also has a value for ornamental planting in large parks.

E. viridis R. T. Bak. **GREEN MALLEE**

This eucalypt is one of a limited number of mallees which grow in the semi-arid zone of south-eastern Australia. Its principal occurrence is on the lower Western Slopes and parts of the adjacent Western Plains of New South Wales. In addition it is found in the Bendigo district of central north-western Victoria, the Inglewood area of southern Queensland and there is a limited occurrence on the southern end of the Flinders Ranges in South Australia. Rainfall is typically 16-20 inches, but it is as low as 12 inches in the driest parts. Green mallee varies in height from 15-35 feet, with stem diameters from $2\frac{1}{2}$-$7\frac{1}{2}$ inches. It grows mainly in mallee form and only occasionally develops as a small tree. The canopy tends to be rather light and, especially in the case of the taller growth, is carried on the upper third of the height. Young plants and those on sites where the total height may be only 15 feet carry foliage nearer to the ground. Both the specific botanical name and the common name refer to the colour of the leaves.

Green mallee grows on soils which may be sandy or sandy loams, with some clay in the sub-soil. On some sites there may be limited surface drainage. This is a serviceable species for soil conservation and dense thickets are useful shelter for stock. When planted on better soils and protected from fire and damage by stock it develops as a small bushy tree excellent for low shelterbelts, or for planting in the outer rows of multi-row belts with taller growing eucalypts which lose their branches from the lower parts.

It ranks as one of the more important eucalypts for the distillation of essential oils. The return from leaves and terminal branchlets is about 1.5 per cent of the weight at time of collection. The valuable cineol content of the crude oil may be as high as 80 per cent. Since the young leaves yield a higher return than the older, adult foliage it is customary to practice a form of coppice rotation for commercial production. Not only does this give a higher proportion of juvenile leaves but the tonnage harvested per acre is higher.

E. wandoo Blakely **WANDOO**
syn. *E. redunca* Schau. var. *elata* Benth.

Wandoo is a species of the semi-arid and sub-humid zone of south-western Western Australia, where it occurs in a belt about 300 miles long on the inland side of the jarrah forest (*E. marginata*). It reaches the Indian Ocean about 50 miles north of Perth and extends southwards to within the same distance of the coast. The rainfall is mainly 16-20 inches, and shows a marked winter maximum; summers are dry and hot.

It is usually a small tree 50-70 feet in height and 2-3 feet in diameter. The trunk is straight and may be half of the total height before it divides to form a crown of moderate density and size. The tree shows a preference for rather heavy soils over clay or a pan, but will grow on loamy sands and on lateritic soils. The Forests Department of W.A. recommends it as a shade tree for farms in areas where the rainfall is not less than 15 inches. It is, however, a tree of attractive appearance with a clean, light-coloured bark which deserves attention for use in parks and elsewhere in southern winter-rainfall areas where it will have sufficient space to grow without restrictions. The heartwood is light reddish brown, close-textured, with wavy or interlocked grain. It is very heavy, very hard, very strong, stiff and durable. Formerly it was mainly used in the round or for heavy construction but nowadays it is used for high quality flooring and heavy joinery. Wandoo is also a source of tannin extract obtained by reducing the wood and bark to chips and leaching out the extract with water.

E. woollsiana R. T. Bak. **GREY BOX**

In size and appearance this is a typical inland grey box; it is commonly 50-80 feet in height and 1½-3 feet in diameter and on all fairly good sites is of excellent form, with a straight bole about half of the tree height. On poor sites it may be reduced to a small, somewhat straggly tree only 25-40 feet high. The reference to grey in the common name is to the colour of the bark, whilst the botanical name is in honour of the Rev. W. Woolls (1814-93), who was a keen amateur botanist.

Grey box has its principal occurrence in the wheatbelt area of New South Wales, but it also extends into southern Queensland. In the southern part of its range it mixes with another grey box (*E. microcarpa*) which occurs mainly in northern sub-humid Victoria, and in the northern part of New South Wales and Queensland with narrow-leaved grey box (*E. pilligaensis*). Most of the *E. woollsiana* grows in the moister parts of the semi-arid zone. The rainfall is 15-25 inches and varies from a weak winter maximum in the south to a moderate summer maximum in the north. It is a species of gentle slopes and plains and, occasionally, in the driest part of the range, of the beds and banks of drainage channels where water only flows periodically. The soils are typically rather heavy alluvials, clay loams and the better-quality sandy loams.

In New South Wales this grey box is considered a general purpose tree on heavy soils where the rainfall is 18 inches or greater. The crown is well-branched and is carried fairly high on the tree, so that the trees are well suited for shade, and yet wheat can be grown close to the trunks. Most farmers in the wheatbelt wisely leave scattered specimens throughout large wheat-fields as a provision for shade of stock when they are returned to the paddocks. The timber is very heavy, very hard, tough and durable but is not easily sawn. Limited quantities are mainly used in the round or hewn for sleepers or split for fencing material. It is an excellent fuel.

Ficus spp. **NATIVE FIGS**

This is a world-wide genus which is well represented in the rainforests of eastern Queensland and north-eastern New South Wales. There is no better shade tree in the warm coastal parts of eastern Australia for farms and large parks than Moreton Bay fig (*F. macrophylla*

Desf. ex Pers.). On suitable sites the crown of an old, open-grown specimen may have a total spread of up to 150 feet. The canopy is dense and light rain will not penetrate to farm equipment stored below; at the same time there is little growth of grass. This species and some others, such as Port Jackson fig (*F. rubiginosa* Desf. ex Vent.) and Hill's fig (*F. microcarpa* L.f. var. *hillii*), will grow moderately well with rainfall as low as 20 inches if the soil is deep and at least of some fertility and, preferably, on relatively low-lying country which receives some additional run-off. Under these conditions they are useful species, though they never attain the magnificent development seen on alluvial flats of coastal eastern Australia.

Ficus platypoda (Miq.) A. Cunn. ex Miq. is one of the few species of *Ficus* which occurs naturally in the dry country. It is scattered throughout Australia north of about 28°S. latitude from the arid country around Alice Springs and the coast of Western Australia, for example in the North West Cape region, to the humid east coast of Queensland. It grows to about 30 feet in height and frequents rocky sites. Land owners who wish to select figs for planting in the dry country should consult the forest service or other government organisations in their State.

Flindersia maculosa F. Muell. **LEOPARDWOOD**

Leopardwood is a small tree 15-30 feet in height and usually has a somewhat open crown at maturity. The development of the tree is unusual, as in the early stages it forms a tangled mass of long thin branches. Later a leading shoot develops and is protected from stock by the other juvenile branches; when it grows above the central mass it develops in the normal way, with many fine lateral branches carrying moderately dense foliage. By the time the tree reaches maturity the early tangled mass has died and dropped away. The bark of the tree has a spotted appearance, hence the specific botanical name and the common name. The genus is mainly an Australian one and is named in honour of Captain Matthew Flinders, a pioneer of early coastal exploration in southern Australia.

The principal distribution of this species is in the semi-arid and arid zones of central and northern New South Wales, with extension into Queensland as far north as about the Tropic of Capricorn. Rainfall is mainly 12-22 inches, but it occurs as far west as the Broken Hill district where the rainfall may be as low as 8 inches. It grows on plains with soils which are often a sandy loam over clay. Leopardwood is a fodder species and while it is rated as an excellent fodder in Queensland it is regarded as only second class in New South Wales. Because of the growth habit of the young plants and the somewhat slow growth at that stage this is not a species to be recommended for general planting, but where it exists naturally landowners in very dry areas would be well advised to conserve some specimens. The wood is usually too small in size to be of much use, apart from the fact that it is not durable out of doors.

Geijera parviflora Lindl. **WILGA**

This is one of the most ornamental small trees on the inland side of the Dividing Range in eastern Australia and few native species are more rewarding to plant or conserve for ornament, shade and drought fodder. It reaches its best development in the moister parts of the semi-arid zone and in the adjacent sub-humid zone in New South Wales north of about 34½°S. latitude and of Queensland south of about 21°S. There is a small occurrence in north-eastern South Australia, adjacent to the border with New South Wales, and in the far north-west of Victoria. Wilga is a small tree commonly 20-30 feet in height, usually with a short single main stem 9-18 inches in diameter. It has a broad, dense crown which may be appreciably wider than the tree is high. The thick foliage consists of green leaves 3-6 inches long and ¼-⅓ inches wide, and the numerous small white or cream-coloured flowers appear in terminal or sub-terminal panicles 2-3 inches long. The final branchlets are drooping and together with the pendulous foliage give some resemblance to a true willow (*Salix* spp.). It is believed that the common name of wilga is of Aboriginal origin, whilst the generic name is after J. D. Geiger, a botanical author, and the specific name is based on two Latin words meaning small-flowered.

Wilga grows naturally in areas where the rainfall has the wide range of 9-30 inches. The maximum may even be higher in parts of coastal Queensland, for example Rockhampton near the Tropic of Capricorn. The tree shows very fine development and has its greatest value for farmers in areas of 15-25 inches. The rainfall incidence is more or less uniform in the south but has a well-developed summer maximum in the north of its occurrence. Wilga grows on a wide range of soils from deep sandy loams to moderately heavy alluvials and shows excellent growth on low-lying sites which may become wet during the rainy season.

This tree is generally rated as one of the most palatable fodder species in eastern Australia, but authorities in both Queensland and New South Wales comment upon the variability between growth forms. In the former State, tree wilga, which grows mainly in the drier areas, is readily eaten and dry sheep can exist on 1½-2 lb of leaf a day, but a small, bushy form known as lavendar bush, which mainly grows in the higher rainfall areas, is left untouched. In New South Wales, with trees of apparently the same type and not botanically distinguishable, some will be eagerly sought by sheep and others left untouched. The edible forms are very obvious by the manner in which animals browse the lower foliage uniformly to a level which they can reach.

To repeat; wilga is one of the most attractive and useful native trees in its habitat. It can be used for the larger home garden where its ability to withstand trimming can be an advantage and for parks, but its wide-spreading crown is not satisfactory for street planting. On the farm it is one of the best shade species of the inland and, as already indicated, is a useful drought fodder. As a honey species in New South Wales it is reported by Goodacre (1939), that 'some trees flower through almost the whole of the year, but honey appears to be collected from the species during the summer only. The honey is dark in colour and has a very strong flavour. It is useful in providing winter stores; the colonies winter well on this class of food'.

Grevillea pinaster Meissn. **A GREVILLEA**

This small, very bushy grevillea is a native of the sub-humid and semi-arid country near the coast of Western Australia from Perth northwards to Geraldton and the southern edges of Shark Bay. The summers are long and hot, with mean maximum temperatures of the hottest month about 90°F near the coast but appreciably higher even a short distance inland. Rainfall is mainly within the range of 10-18 inches, with most of it falling in winter.

G. pinaster only grows to a maximum height of about 10 feet on the sandy soils where it occurs naturally, but it has a very dense habit with branches to ground level. It merits consideration for low shelter, both for homesteads and for animals, as well as planting in multi-row shelterbelts in areas where climate and soils are suitable.

G. striata R.Br. **BEEFWOOD**

The name beefwood was given to this tree by early settlers because of the resemblance in colour of the freshly-cut heartwood to that of raw beef. In New South Wales the common name of western beefwood is used because of its occurrence in the western part of the State, but the qualification of western is not considered suitable for a species which extends to several States. The specific botanical epithet is from the Latin, *striata,* in reference to the longitudinal markings on the leaves. Beefwood is found in an extensive part of the dry country of Australia—it occurs in Queensland and western New South Wales, especially the central and northern parts of the latter State, the southern half of the Northern Territory, and those parts of South Australia adjacent to the Northern Territory and New South Wales. Over much of this area the rainfall is 9-15 inches, but in the easternmost parts of its occurrence the rainfall may be as high as 25 inches. Seasonal distribution varies from more or less uniform to a well developed summer maximum in Queensland. Summer temperatures are high throughout the area, with the mean maxima of the hottest month mainly 95-100°F.

Beefwood is a moderately small tree, 20-40 feet in height and 1-2 feet diameter, with a main stem which may be half of the total height. The crown is comparatively narrow and the canopy light. The tree is characterised by the very rough, dark coloured bark and narrow leaves which are 6-18 inches long. Like many species of the genus it has smallish but very attractive flowers. Little information has been recorded of the honey value of the tree but other species of *Grevillea* are frequented by bees and honey-eating birds during the flowering season. The timber is of moderate weight, fissile and very durable. It is first class material for posts and other items, such as house blocks, which are in contact with the ground. In the past it has been used for shingles and slabs of pioneer homes, where fissibility and durability were important critera. It is not suitable for fuel.

The tree is of ornamental appearance and has possibilities for use in parks and large homestead gardens, especially in areas of 15-25 inches rainfall. The foliage is only eaten by stock during serious drought conditions.

Hakea spp.

Hakea is an Australian genus which has its greatest development in the dry country, especially Western Australia, with about 85 species, most of them endemic. They are shrubs or small trees in the height range of 3-40 feet. The species are readily distinguished from members of other genera by the very hard, woody capsule, often up to an inch in length in many species, and which divides into halves to free the two winged seeds. The leaves are usually small, sometimes cylindrical, with hard sharp points. Needlewood is the common name used for several of the large species, especially some of those eaten by stock. Nearly all hakeas have attractive flowers and many are cultivated for their ornamental appearance.

Probably the most important hakea to pastoralists in the dry country of eastern Australia south of the Tropic of Capricorn is *H. leucoptera*, and this species is described separately. Short notes and illustrations are given for several other species which occur over large parts of the continent. *H. chordophylla* is a large shrub or small tree up to 20 feet in height which mainly belongs to the tropics, as it is found from the Kimberley district of northern Western Australia, through the Northern Territory to north-western parts of Queensland where it is known as bootlace oak. To a limited extent it is eaten by stock. *H. divaricata* syn. *H. intermedia*, occurs in most States, as well as the southern part of the Northern Territory. *H. petiolaris* grows to about 15 feet in height on granitic soils in the Darling area (near Perth) of Western Australia. It has been used in trial plantings by the Forests Department of that State, whilst it is recommended by the Forests Department of South Australia and New South Wales; the common names given by the two latter States are broad-leaf hakea and broad-leaf seaurchin respectively. *H. vittata* is a small, well-branched shrub often 10-15 feet in height, the leaves of which are eaten by stock.

Hakea leucoptera R.Br. **NEEDLEWOOD**

Needlewood varies in size from a small bush, only 8-10 feet high with several main stems from near ground level, to a small tree 15-18 feet high with a well developed trunk several feet in length. The crown has numerous fine branches and, in spite of the small, cylindrical leaves, casts a useful shade. The name needlewood is well applied as the small leaves have very sharp rigid points.

This hakea is scattered throughout the dry country of central and eastern Australia, from a few degrees north of the Tropic of Capricorn to nearly 37°S. latitude in sub-humid north-western Victoria. The climate is mainly arid and semi-arid. Most of this is an area of 9-15 inches rainfall, with high variability, but it extends to over 20 inches. Needlewood grows on a wide range of soils from those of a poor type, for example deep, well-drained sands, to sandy loams and firm clay loams. The timber is like that of many other species in the same family (*Proteaceae*), in that it has an obvious figure due to medullary rays and is not difficult to split; it is reddish-brown, hard, close grained and tough. In the past it has been used for making small ornamental items such as smoking pipes, and in Sydney in 1895 a company was formed to use the timber for that purpose but it did not operate for long.

Where needlewood is growing on a property some should be retained for shade and, on the better sites in 12-20 inch rainfall, it should be considered as an ornamental species. Root suckers sometimes develop. The harsh foliage is a poor drought fodder but is occasionally eaten by animals.

Heterodendrum oleifolium Desf. **INLAND ROSEWOOD**

Inland rosewood varies from a bush which may be only 10 feet high, to a small tree with a height of 15-25 feet. The larger branches are often of poor form and rather irregular, but the crown provides a useful level of shade. This plant has a number of common names, but because it is mainly an inland species, the name adopted here appears logical. Other names include boonareet (Queensland) or boonery (New South Wales) and minga bush (Western Australia). The generic name is based on two Greek words, *heteros* (variable), and *dendron* (tree), probably in allusion to the foliage; whilst the specific name is from two Latin words, *olea* (an olive tree), and *folium*, because of some resemblance of the leaves to that of the olive tree.

The natural occurrence includes most of the arid and semi-arid country of Australia south of about 20°S. latitude, as well as some sub-humid areas in the three eastern States. West to east it extends for nearly 2,400 miles, for example from Carnarvon on the west coast of Western Australia to near Rockhampton on the east coast of Queensland. Rainfall has the appreciable range of 8-30 inches, but most of the occurrence is 10-25 inches. Temperatures are mainly hot to very hot and suitable recording stations are listed for the species.

Sometimes the tree is of rather straggly form but it responds well to pruning and lopping. It is quite suitable for shade and shelter near homesteads and for retention as shade for stock, or planting in parks of the dry country. The foliage is readily eaten by sheep and cattle but at some times of the year contains a significant amount of prussic acid; it should therefore only be fed to stock in limited quantities, and then only when mixed with other fodder.

Lagunaria patersonia D. Don **NORFOLK ISLAND HIBISCUS TREE**

This is a moderately large tree which grows to 60 feet or more on favourable coastal localities, but only 25-30 feet on dry inland sites away from its natural habitat. In early life the erect main stem carries lateral branches which gives the tree a broadly pyramidal form, but with maturity in higher rainfall areas the crown is wide-spreading and supported on a framework of heavy branches. The tree grows naturally on Norfolk Island and, with somewhat different botanical characteristics, on the north-eastern coast of Queensland. It is from the former area that the species was introduced into southern Australia. The generic name was given in honour of Andreas Laguna, a Spanish physician and botanist of the sixteenth century, whilst *patersonia* commemorates the name of Colonel Paterson, a Lieutenant-Governor of N.S.W. during the nineteenth century. On Norfolk Island the tree is known as white oak but as the resemblance to true oaks (*Quercus* spp.) is only slight and as the tree belongs to the same family as the hibiscus, the name used in this note appears the more desirable.

On its native island this tree experiences a fairly uniformly distributed rainfall of about 45 inches and warm temperatures but elsewhere it has proved to be tolerant of greater heat and is moderately drought-resistant. It is planted in southern inland Australia where the rainfall is as low as 14 inches. In such areas it has proved moderately successful but on good loamy soils in areas with a minimum of about 20 inches, especially in parks and large gardens with watered lawns, it is an excellent shade species. This writer, however, suggests that for areas of 14-20 inch rainfall there are many species native to the dry country which would be more suitable.

Leptospermum spp. **TEATREE**

This is a genus of shrubs or, less commonly, small trees, which is characteristically Australian. A few species occur overseas. The common name of teatree, with a qualifying epithet, is applied to many of them. They have a wide occurrence throughout Australia, but are most common in coastal and tableland areas of southern and eastern parts of the continent. Soils upon which they grow naturally include both coastal sands and poorly-drained heavy soils. Many are cultivated for their ornamental appearance, especially their white, pink or red flowers and whilst some will grow in the dry country they are of less importance than species of the closely allied genus *Melaleuca*. One example of a species which is native to temperate southern coastal areas, in this case mainly Victoria, but will grow satisfactorily in hot semi-arid and sub-humid conditions where the rainfall may be as low as 18 inches, is coast teatree (*E. laevigatum*).

Melaleuca spp. **PAPERBARK**

This large genus has its greatest development in Australia, especially Western Australia which has about 100 species, most of them endemic. In that State most of the melaleucas are small shrubs, sometimes only a few feet in height, but many of the eastern species are large shrubs or small trees 15-30 feet in height, and in the Northern Territory broad-leaved teatree (*M. leucadendron* (L.) L.) attains heights over 100 feet, with millable boles several feet in diameter. Many of the species show a preference for river banks and the edges of tidal waters; others grow on sandy heaths. The preferred common name for the larger representatives of the genus is paperbark, with some qualifying epithet, but unfortunately teatree is also used both for this genus and for species of *Leptospermum*. In southern Australia the name honey-myrtle is well established for many of the shrub-size species.

In addition to the four species of *Melaleuca* which have been described separately there are many others of importance for planting in the dry country; included amongst these are river teatree (*M. bracteata*), also known as black teatree, and white teatree (*M. glomerata*). Both have the typical bushy form of many species of the genus, especially in early life, and are suitable for dense, low shelter or for use in multi-row shelterbelts. The former melaleuca occurs in eastern Australia, where it extends to warm coastal areas, as well as to the dry country of the Northern Territory; whilst the latter species is found in both the arid and semi-arid zones of southern Australia from Western Australia to New South Wales. Most species of the genus will grow in damp situations and on heavy soils, as well as on sandy loams. Nearly all of them will tolerate pruning and shaping if commenced at an early age, but for the purposes recommended in this book they are better left to develop their normal form.

M. armillaris Sm. **BRACELET HONEY MYRTLE**

Like many species of this genus, bracelet honey myrtle is a very bushy shrub, 12-18 feet high, with many stems from at or near ground level, so that it is an excellent plant for low, dense shelter. In appearance it is attractive, especially when in flower. It is a native to the coastal areas of New South Wales and south-eastern Victoria, mainly on lowland sites near streams. On introduction to the dry country it has proved hardy in hot areas and with a rainfall as low as 18 inches and, where some additional watering is possible, with an even lower rainfall.

M. halmaturorum F. Muell. ex Miq. **SOUTH AUSTRALIAN SWAMP PAPERBARK**

The principal occurrence of this shrub or small tree is near brackish water in the coastal areas of South Australia from the Eyre Peninsula in the west to Beachport in the south-east, but it extends inland to drier areas in the vicinity of salt lakes. It is also found in the western districts of Victoria. The specific botanical name, which refers to kangaroos, was given because the type specimens came from Kangaroo Island. The climate is mainly semi-arid, but with extension into both the arid and the sub-humid zones. The annual rainfall of the principal occurrence is 15-25 inches, with a well-defined winter maximum and a comparatively small variability. Summers are warm to hot, with mean maxima of the hottest month rising to over 90°F for inland sites; frosts are absent immediately adjacent to the coast but there are 1-5 a year for inland areas.

Naturally it usually occurs in dense thickets but when planted in single rows on suitable sites it develops the dense bushy habit typical of most species in the genus, and provides protection to ground level. Mature height ranges from 10-25 feet. The forest service of S.A. recommends it for planting on saline sites with a shallow water table, but it will also grow satisfactorily on more favourable sites.

M. lanceolata Otto **DRYLAND TEATREE**
syn. *M. pubescens Schau.*

This melaleuca is variously called black teatree (South Australia), moonah (New South Wales) and Rottnest Island teatree (Western Australia), but the name dryland teatree is used by the forest service of S.A. and is adopted here. It seems particularly suitable since *M. lanceolata* is one of the hardiest species in the genus and planted at Broken Hill it thrives on unfavourable sites with a mean rainfall of only 9 inches. It is a large, bushy shrub, usually 15-20 feet in height at maturity and, when given sufficient growing space, with an almost equal width.

This description is for the species as it occurs in southern Australia, mainly 31-37°S. latitude. There it grows in the arid and semi-arid areas of South Australia and New South Wales, as well as the sub-humid zone of South Australia and Victoria. The rainfall is 10-25 inches, with warm to hot summers. The species is also reported from the coasts of south-western Western Australia and the Darling Downs region of Queensland.

Not only will dryland teatree grow in areas as arid as Broken Hill, but it will tolerate appreciable salt in the soil. For low shelterbelts which only require minimum attention, it has few equals in the drier parts of southern Australia. It is particularly recommended for areas of 9-16 inches rainfall, but is also useful where the rainfall may be as high as 25 inches. In the latter areas, however, it should be planted on either deep sandy loams or on free draining sites.

M. pauperiflora F. Muell. **A MELALEUCA**

This is a small, dense shrub rarely exceeding 10 feet in height. Its natural occurrence is in a long, narrow belt of arid and semi-arid country in southern Western Australia and South Australia. There is limited extension into the sub-humid zone. For most of the area the mean annual rainfall is 7-15 inches, falling mainly in winter.

In inland locations the summers are hot, that is the mean maximum temperature of the hottest month is 90°F or higher, but on the coast conditions are mild. This melaleuca is a hardy species which merits consideration for low shelter in the very dry areas of southern Australia; it can also be expected to grow well on sandy soils in areas with a rainfall up to 20 inches or more.

Melia azedarach L. var. *australasica* (A. Juss.) C. DC. **WHITE CEDAR**
syn. *M. dubia* Cav.

White cedar is a species of sub-tropical high rainfall areas but it has shown the ability to grow in the moderately dry inland. One Forest Service, for example, recommends it for parks and gardens in warm frost-free areas with a rainfall as low as 14 inches. The author however, believes that it should be restricted to moderately good soils and rainfall greater than 20 inches. Under such conditions, especially if surrounded by lawns which are watered, white cedar is an attractive tree for shade and ornament. In the dry country height may be limited to 50 feet, but in high rainfall sub-tropical areas it may grow to 150 feet, with diameters of 4 feet or more above the buttresses. It bears a profusion of lilac-coloured flowers, in large terminal panicles, hence the name of Persian lilac tree for the species *M. azedarach.* It grows readily from seed and shows good development during early life, even under somewhat adverse conditions.

The principal role of this species is as a deciduous shade and ornamental species in northern coastal New South Wales and south-eastern Queensland, as mentioned on p. 134 of *Growing Trees on Australian Farms.* The fruits are poisonous to pigs and there is some evidence that they affect poultry, so that the trees should not be planted within pig pens or near fowl runs.

The generic name is from the Greek name for the manna ash (*Fraxinus ornus*) because of some resemblance of the compound leaves to those of the ash. The species name is from the Persian *azad-darakt*, applied to a closely-allied species.

Muehlenbeckia cunninghamii (Meissn.) F. Muell. **LIGNUM**

Lignum is a small bush which has a scattered occurrence throughout most of the dry country of the continent south of about 17°S. latitude. It is usually found on areas subject to seasonal inundation, where soils are of fine texture and may include clay pans. The mean annual rainfall is mainly within the limits of 8-20 inches, but lignum sometimes extends to areas where the rainfall is higher. The shrub typically has tangled, sprawling stems and branches and forms a compact mass 3-10 feet in height and often of greater diameter. There are few leaves on the green twigs, but the latter are sometimes eaten by sheep and cattle.

Myoporum montanum R.Br. **BOOBIALLA**

Although the native name of boobialla is applied to other species of the genus, it is preferred to that of native myrtle since it is not in any way related to the true myrtles (*Myrtus* spp.). *M. montanum* grows in the arid and semi-arid areas of southern and south-eastern Australia, and occurs in the southern part of the Northern Territory, throughout most of South Australia north of Adelaide, north-western Victoria and western New South Wales. The range of rainfall is 8-20 inches; summer temperatures are hot to very hot, with mean maxima of the hottest month 85-100°F.

This boobialla is a very bushy shrub, often only 8-10 feet high, with numerous fine branches from ground level which are well clothed with foliage. In all open situations the width of the bush is often greater than the height. When allowed sufficient growing space it is a useful ornamental species for homestead gardens and small parks, as well as being a first class species for low shelter which does not need attention. In addition, it will tolerate light trimming.

M. platycarpum R.Br. **SUGARWOOD**

The name sugarwood is an allusion to the sweetish exudation which sometimes takes place from the sapwood, whilst the specific botanical epithet means flat-fruited. This is a shrub or small tree in the height range of 10-25 feet, which grows in the dry country of South Australia, north-western Victoria and western New South Wales. The rainfall may be as low as 9 inches, but best development is in areas of 20-25 inches. It grows mainly on sandy soils.

Branches are not retained to ground level as is the case with *M. montanum,* hence the species has not the same use for low shelter. It is of some ornamental value, but where it occurs naturally its main use is for shade and soil conservation. The wood is pale yellow in colour, soft and not of importance. The foliage is palatable to stock.

Owenia acidula F. Muell. **EMU APPLE BUSH**

Emu apple bush usually grows in the height range of 12-25 feet; sometimes it is a large bush with several main stems from near ground level or a small tree with a single trunk which may be up to 8 feet long and 1 foot in diameter. It has a well-branched crown which is often appreciably wider than the tree is high, and drooping branchlets carry thick foliage which provides excellent shade. Because the leaves are attractive to stock the lower part of the crown is usually trimmed to an even height from the ground. The genus was named after Sir Richard Owen (1804-1892), a comparative anatomist and superintendent of the natural history collection of the British Museum, whilst the species name refers to the sharp sub-acid taste of the globular, plum-like fruit. These fruit are $\frac{3}{4}$-1 inch in diameter and are edible; they are freely eaten by Aborigines and it is stated that they are very refreshing when drinking water is scarce. Amongst other common names are emu bush because emus are attracted to the ripe fruit, emu apple (Queensland) and gruie and colane (New South Wales)

The principal occurrence of this tree is in the semi-arid zone of Queensland, mainly south of the Tropic of Capricorn, and in New South Wales. It extends, however, to both the arid and sub-humid zones, the range in rainfall is 9-25 inches, and it shows a slight to a marked summer maximum. Temperatures are high throughout most of the year. This plant grows on a wide range of soils from light sandy loams to firm clays, but nowhere is it at all plentiful.

One writer has claimed this tree as the most attractive small tree of its climatic area; this might be some exaggeration, but it is undoubtedly one of the more ornamental trees of the eastern semi-arid zone. In areas of 12-25 inches rainfall it is worthy of consideration for ornamental and shade planting in parks and any large gardens; it is not a good street species because of the space needed for crown development. Propagation of the tree is not easy since seed germination is difficult and whilst root suckers develop freely, they are not easy to transplant. It is probable that the seed would respond to some form of pre-germination treatment.

Petalostigma quadriloculare F. Muell. **QUININE TREE**

This species occurs in the Kimberley district of northern Western Australia, the higher rainfall parts of the Northern Territory, and through Queensland to the northern slopes and coastal parts of New South Wales. It is a shrub or many-branched small tree up to about 30 feet in height in the higher rainfall coastal areas, but commonly only 12-18 feet in semi-arid inland regions. The foliage has been suspected of causing loss of stock when the tree has been lopped or felled for drought fodder.

Pittosporum phylliraeoides DC. **INLAND PITTOSPORUM**

This pittosporum has a very wide range; it is found from the west coast of Western Australia to near Rockhampton on the east coast of Queensland—an air distance of about 2,400 miles, and from about 20°S. latitude to near the drier parts of the south coast of the continent. In terms of habitat this covers about two million square miles. The rainfall has a wide range of 8-25 inches, with some extension to 35 inches in sub-humid Queensland near the Tropic of Capricorn. Inland pittosporum is a small tree, usually 10-25 feet in height, with a single stem 8-12 inches in diameter. Occasionally it attains 30-35 feet. It is a tree of graceful appearance, enhanced by the drooping ultimate branchlets and the narrow pendant leaves. This species is an excellent example of the better definition often obtainable by using botanical names in preference to common names; in Queensland it is known as meemeei, cumby cumby and cattle bush, in New South Wales as western pittosporum, berrigan and butterbush, whilst South Australia calls it native apricot. The name of inland pittosporum has been adopted here for the reasons that the tree is a pittosporum and that its characteristic habitat is the inland. The generic name comes from two Greek words, *pitte* (to pitch), and *sporos* (a seed), the seeds of many species being covered with a dark-coloured, sticky substance. The species name is in reference to the resemblance of the leaves to those of *Phyllirea angustifolia* L., a shrub of the Mediterranean region which is a member of the olive family.

Inland pittosporum grows on a range of soils which vary from sandy loams along dry drainage channels in central Australia to better quality clay loams in the semi-arid country. It is one of the more ornamental species of the inland and is well suited for planting around homesteads or in small parks in the dry country, especially where the rainfall is greater than about 14 inches. It has been recommended for planting as a fodder reserve but the financial advantage of such operations are open to doubt. There is, however, no question of the desirability of retaining naturally grown trees on a property and possibly managing such areas to encourage regeneration in favourable years. The orange-red fruits are contained in a yellow capsule and covered by a viscid substance. They are easy to germinate.

Santalum lanceolatum R.Br. **SANDALWOOD**

The standard trade common name of this species is northern sandalwood, but it is also widely known as plumbush due to the dark purple to almost black fruit which have a resemblance to small plums. It is a shrub up to about 20 feet in height and occurs on sandy or loamy soils in all mainland States, chiefly in the semi-arid zone. It is readily eaten by stock and is regarded as one of the most palatable and nutritious of native shrubs. Where it occurs naturally it should be retained and natural regeneration encouraged. It is worthy of use as an ornamental but its parasitisation of the roots of other plants presents problems, especially in establishment.

Ventilago viminalis Hook. **SUPPLEJACK**

This is a small tree with a height range of 15-30 feet. At times it will have two or more rather crooked main stems due to the habit of the leading shoots initially scrambling up a conveniently-located tree; at other times the young plant shows a normal habit with a single leading shoot and, consequently, develops only one main stem. With maturity the crown becomes moderately broad and dense; the tree will tolerate careful lopping and if this is done judiciously the crown will show increased density. Supplejack suckers freely from roots which have been slightly injured by soil cultivation or other means and, sometimes, a number of suckers will develop to form a dense shrubby growth. The generic name is from two Latin words, *ventilo* (I blow or winnow), *ago* (I drive gentle), in allusion to the winged seed, whilst the species name is from *viminalia* for all trees and shrubs yielding twigs suitable for wickerwork, for example most of the true willows (*Salix* spp.).

Supplejack occurs in the north-eastern half of the dry country and some of the adjacent sub-humid zone from the Kimberley region of northern Western Australia, through the southern two-thirds of the Northern Territory, to western New South Wales and much of arid and semi-arid Queensland, south of about 20°S. latitude. The rainfall is mainly 10-25 inches, but in the northern and north-eastern fringes it may exceed 30 inches. A summer maximum occurs throughout most of the area. For the northern part of the distribution temperatures are high at all seasons of the year.

It prefers sandy soils but will grow satisfactorily on any moderately well-drained light loam. It is readily eaten by cattle and sheep and grows well after light browsing, but if the stocking is unduly heavy seedlings and suckers may be killed. Supplejack is a moderately ornamental tree, especially vigorous specimens which have had some shaping. It should be retained in areas where it grows naturally and merits use in parks and the large gardens of homesteads, especially in summer rainfall areas where the mean annual fall is 15-25 inches.

AUSTRALIAN CONIFERS

Actinostrobus pyramidalis Miq. **NATIVE CYPRESS**

This small conifer is a native to a comparatively small area of south-western Western Australia, mainly around Perth and in the adjacent Darling Ranges. It is suggested that when ordering plants reference should only be made to the botanical name, since the use of 'cypress' can lead to confusion with the native cypress pines (*Callitris* spp.) The climate of the occurrence is warm sub-humid, with most of the area subject to coastal influence. The mean maximum temperature of the hottest month is 80-90°F, and the mean minima of the coldest 40-50°F; frosts are absent on the coast but there may be 1-3 on inland sites. Annual rainfall is 20-35 inches, with most of it occurring during the winter season.

Actinostrobus pyramidalis is a small tree, usually in the height range of 15-25 feet, with a compact, columnular habit which makes it very suitable for low shelterbelts in its climatic area. It is also recommended by the forest service of the State for ornamental planting. Like most conifers best growth is on sandy loams and it should not be planted on tight clays; it is tolerant of a moderate level of salt.

Callitris endlicheri (Parl.) Joy Garden **BLACK CYPRESS PINE**
formerly known as *C. calcarata* (A. Cunn. ex Mirb.) F. Muell. nom. illeg.

Black cypress pine has its principal occurrence in New South Wales, where it is found in the eastern parts of the semi-arid zone, throughout the sub-humid area (except on the coldest tablelands and in the high mountains) and extending to the drier parts of the coastal region. There are small extensions into adjoining areas of Queensland and eastern Victoria. The rainfall is mainly 20-30 inches and where the tree grows in wetter areas it segregates to the sunnier and better-drained sites. It prefers deep sands but grows on a wide range of soils, including those of a skeletal nature.

This cypress pine has the typical, elongated pyramidal shape characteristic of the genus and in most situations it retains its branches and foliage to near ground level. It is, consequently, an excellent species for long-lived shelterbelts where the landowner is prepared to wait a number of years for a belt to be effective, but desires trees which should last upwards of a 100 years. Where additional watering and manuring is not practicable the growth rate may be less than 18 inches a year for the first 10-20 years. On favourable sites adjacent to homesteads where superphosphate and some additional watering can be given at critical times the height rate may be increased to 2-2½ feet a year for the first 20 years. Compared with white cypress pine (*C. glauca*) it does not extend as far into the arid zone but will tolerate greater cold than that species, for example frosts which are both moderately heavy and frequent (as many as 30-40 a year) and even light falls of snow in occasional years. In addition to its use on farms it should be considered for ornamental planting, especially as avenues in parks. Mature height is 40-70 feet.

The timber is generally comparable to that of white cypress pine, and can be used for the same purposes, though it is not quite as durable. When a farmer has patches of this species growing naturally on his property he should manage them for shelter and timber production but, because of the relatively slow rate of growth it is not recommended for planting with the sole objective of timber production.

C. glauca R.Br. ex R. T. Bak. et H. G. Sm. **WHITE CYPRESS PINE**

This is the most common native cypress pine and it has an extensive natural distribution in eastern Australia, mainly south of the Tropic of Capricorn, but it is found scattered in all other States and the southern part of the Northern Territory. Associated with the wide range in climate, for instance rainfall from 8-28 inches, is a corresponding variation in size from about 20 feet in height in the arid interior to nearly 100 feet under optimum conditions. In the main cypress areas of semi-arid Queensland and New South Wales it is typically 60-70 feet in height and 1-2 feet in diameter at maturity. Even under adverse conditions the trunk is usually straight, whilst canopy tends to be rather light except with vigorous growing young specimens. The botanical nomenclature of this important species has had a checkered history and a summary has been given on p. 296 of the third edition of *Forest Trees of Australia.*

Consideration of it as a timber-producing tree should be restricted to areas in which it is already established as a natural forest; these will be mainly in areas of 16-25 inches rainfall and with either a uniform or moderate summer maximum. As an ornamental species, especially for avenues and shelterbelts, it should be considered for most areas with a rainfall greater than about 12 inches, where summers are hot and winters with not more than about 10-15 frosts a year. Of the meterological stations for which data is given in Appendix I, the more typical of it as a timber tree are Condobolin and Coonabarabran in New South Wales and Roma in Queensland.

White cypress pine can be readily raised from seed and tubed stock should be used for ornamental planting; the restocking of native forests is normally done by means of natural regeneration and the general conditions desirable for such operations are referred to in Chapter 19. When a landowner proposes to regenerate a native cypress pine forest he is recommended to seek the advice of the forest service of his State. Because of the wide natural range of the species care needs to be taken in the source of seed.

The timber is renowned for its high resistance to decay and termite attack and its very small shrinkage on seasoning. It is largely used for building scantlings, flooring, linings and weatherboards, with small diameter round material for fencing material and small poles. Where farmers have a small area of natural forest they should consider retaining at least some of it under forest management. The persistent lateral branches are an advantage for shelterbelt use but do result in knotty timber, even when grown in forest formation. In spite of its slow early growth—1-1½ feet a year—it is recommended in its climatic area for single row shelterbelts and avenues where its hardiness and long life of 100 years or more are advantageous. A spacing between trees of about 6 feet is recommended for homestead shelter but 8-10 feet for other farm belts. This species is fire sensitive in early life but older trees are not damaged by light grass fires.

C. preissii Miq. subsp. *preissii* **ROTTNEST ISLAND CYPRESS PINE**
also known as *C. robusta* (R.Br.) Mirb.

This tree has its principal natural occurrence in the coastal areas of south-western Western Australia, including Rottnest Island, but is variously reported as being found in all other southern States. These notes will refer only to the Western Australian form. This cypress pine has the typical growth habit of the genus and, when grown in open positions, retains branches to ground level and provides a very dense shelter. Mature height is 20-30 feet. It grows on sandy soils, often with limestone present, and will tolerate salt-laden winds. Climatically it occurs in the sub-humid and humid zones but experimental plantings indicate that it will grow satisfactorily in areas of hot to very hot summers and a rainfall at least down to 20 inches.

Rottnest Island cypress pine is a tree, par excellence, for low single-row shelterbelts; it also makes attractive ornamental specimens and has a long life. The principal disadvantages are that it shows somewhat slow early growth and, because of the retention of dry foliage and branchlets, it is sensitive to fire.

EXOTIC BROADLEAVED

Ceratonia siliqua L. **CAROB BEAN**

The carob bean is a small, heavily-branched tree which may attain to about 30 feet in height. The main stem is usually of poor form, often only a few feet in height, but the tree carries a dense canopy of thick leaves which provides excellent shade. It is a native of the Mediterranean region and is typically a species of that type of climate with hot, dry summers and cool to mild damp winters. Australian experience is that its best development is likely to be in a rainfall of 15-25 inches.

It is slow growing but may attain about 20 feet in 12-15 years on good sites; there is no detailed record of its longevity in Australia but it is probably at least 40 years. It can be either grown by nursery stock or sown direct. Goor (1968) reports that there are about 2000 seeds to the pound and that they retain their viability for five years. Before sowing they need to be pre-treated with concentrated sulphuric acid for one to two hours, followed by soaking in water for 48 hours.

In this country the main interest in it is as an ornamental for parks and private gardens where good shade is desired in the semi-arid zone of southern Australia. It should also be considered as a shade species for pigsties, fowl yards and dog kennels. Where it occurs naturally it is esteemed for the high nutritive value of the pods and seeds for animal feed. Goor states that large trees of high-producing varieties may yield up to a ton of pods in a good seed year. The male and female flowers are borne on separate trees and if anyone is considering growing carob bean for cattle food it is necessary to ensure that there is an adequate number of male trees, say about 5 per cent, scattered through a plantation.

Cercis siliquastrum L. **JUDAS TREE**

This is a small, deciduous tree which flowers before the leaves appear; the small, pink, pea-like flowers are usually produced in abundance all over the tree, including trunk and branches. One variety has white flowers. In the moderately dry country of Australia it grows to about 20 feet but in its native habitat of southern Europe and western Asia fully mature trees may attain 40 feet. The generic name comes from the ancient Greek word, *cer-cis*, whilst the species epithet refers to the fruit type—a siliqua. The common name has long been applied to this tree due to the belief that it was upon one of these that Judas hanged himself.

It is suitable for planting on a wide range of soils in the sub-humid and the moister parts of the semi-arid zones where the rainfall is greater than about 18 inches. It is an ornamental species, suitable for small parks and gardens but somewhat slow-growing in early life.

Cytisus proliferus L.f. **FALSE TREE-LUCERNE**

This is a small bush, 10-15 feet high, which is well branched from ground level. It is a native of the Canary Islands and grows satisfactorily in the higher rainfall parts of the semi-arid zone, as well as in sub-humid and humid coastal areas. It has four claims for consideration—it is fast growing in early life and will attain 10 feet in 3-4 years even if given only slight attention, it is not very inflammable and hence can be planted closer to buildings than many shrubs, it is very palatable to stock and, finally, it is of ornamental appearance. In addition, it is readily grown from seed; these should receive pre-treatment before sowing, for example placing in very hot water and allowing the seeds to remain in the water while it cools. Like many shrubs which have a fast early rate of growth, it has a comparatively short life as a vigorous, healthy plant. On somewhat unfavourable sites this may be as low as 10-15 years, but is sufficient to permit the establishment of a slower growing but more permanent shelterbelt. On good soils and with optimum rainfall the life may be 20 years or more.

Fraxinus oxyphylla Bieb. **DESERT ASH**
syn. *F. oxycarpa* Willd.

This ash is generally considered the hardiest species of the genus for growing in the sub-humid zone and the moister parts of the semi-arid zone of southern Australia. On most fertile soils it will grow satisfactorily with a rainfall of not less than 17 inches. It can be expected to attain to 50 feet in height and in open positions develops a crown which may be nearly as wide as the tree is high. In areas where the weather becomes cool in late autumn, possibly with a few frosts, the leaves turn to a clear, golden yellow. Desert ash is an introduced species; its natural occurrence is in southern Europe and eastwards as far as Iran and Turkestan.

It will withstand both prolonged, dry seasons and moderate frosts, and shows good growth up to middle age if conditions are suitable. Since it is a deciduous species it has the advantage that large nursery stock, that is say 6 feet in height, can be raised and planting done during the dormant season. Formerly it was extensively planted in streets, but any tree of this size and vigour should not be planted within 50-100 feet of sewerage and drainage locations. In some cities, for example Adelaide, it is banned from such sites. Nevertheless, this is a fine species for shade and ornament in parks and other public areas. It should not be planted in small gardens, but has a place in any extensive areas of ornamental plantings.

Gleditsia triacanthos L. **HONEY LOCUST**

This is an introduced deciduous tree which has proved hardy to moderately dry conditions and frost in the semi-arid and sub-humid zones of southern Australia. The minimum rainfall desirable is about 20 inches. Under the most favourable conditions in its native area of the central and south-eastern parts of the United States of America it attains heights of 140 feet, with large boles; locally-grown trees may only be 60 feet in height but have large crowns. It should not be planted on marginal sites, for example poor soils with minimum rainfall, since in such cases 'stagheadedness' may develop. Honey locust has the advantage of being deciduous, which enables large plants to be established, and yellow-coloured autumnal foliage. The species carries vicious spines 3-4 inches long. The variety *inermis* Willd. has the same growth characteristics of the species but is unarmed.

The variety is recommended for suitable sites as a tree for occasional use in moderately large parks and other public places where a large, deciduous species is wanted. When the roots are disturbed, for example by cultivation, it suckers freely and this may be a disadvantage, although in some types of shelterbelts it can be a desirable characteristic. It seeds freely most years and is easily raised from seed; root suckers can also be used. Like most broad-leaved species it will tolerate light selected pruning to improve crown shape or to remove dead or dying branches, but should not be heavily pruned to limit height growth.

Nerium oleander L. **OLEANDER**

This is a well-known small shrub, commonly 6-12 feet high, which has been introduced into Australia from the Mediterranean region. It has been extensively planted in areas which vary in climate from Woomera and Broken Hill in the arid zone to the coastal cities of the humid zone. In the dry country it has few equals amongst exotic shrubs for low shelter and ornament. If additional watering is not available it should be limited to a minimum rainfall of about 15 inches. It is fast growing, has a moderately long life for a small shrub (probably 20 years or more), and horticultural strains are available with a wide range of flower colours. Both the flowers and the leaves are poisonous if eaten, though not to the touch, and many people with young children prefer not to plant oleander around their homes.

Olea europaea L. **OLIVE**

This is the common or European olive which is the source of the olive oil of commerce. It is a native of the Mediterranean region but has been introduced into many other parts of the world which have a comparable climate, that is somewhat dry, with a winter rainfall and a long, hot summer. Whilst it is cultivated to a limited extent in Australia for the fruit, its role discussed in this text is as a hedge and shelter plant in the southern arid and semi-arid regions. One forest service recommends it for planting in a minimum rainfall of 15 inches, but it will grow under even drier conditions. At Broken Hill, which has only 9 inches, it exists without watering. It will tolerate lime in the soil and even slight salinity.

The common olive is a shrub which will grow with many branches from close to ground level, and is an excellent low shelter species. Mature height is 15-25 feet. It may also develop a single main stem which carries a dense crown which is also useful for shade and will tolerate a moderate amount of pruning and trimming. The olive prefers sandy loams of moderate depth, but will grow on soils which may be rocky and well drained. In any suitable environment it is a long-lived tree and 40-50 years is probably a conservative estimate for its life as shelter. The golden olive (*O. africana*) which grows to about 15 feet in height, is also planted in Australia and will tolerate drought and high summer temperatures.

Parkinsoniana aculeata L. **JERUSALEM THORN**

This is a small, spiny tree which occurs naturally in a region extending from southern United States of America and Mexico to northern South America. It has been introduced into many countries, especially those with large areas of arid and semi-arid lands such as Israel, India and Cyprus for soil conservation work and as low shelter. Mature height may reach 30 feet but it is often only 20-25 feet. It offers possibilities for hedges and low shelterbelts in dry southern Australia where the rainfall is 10-18 inches. Because of its prickly nature it is not recommended for large scale conservation work where a native species would be more suitable. Lord (1964) reports that it is grown in humid sub-tropical Brisbane and is usually a semi-creeping shrub.

Rhus lancea L.f. **KARREEBOOM**

Karreeboom is a native of southern Africa, especially the Orange Free State and the Transvaal, but extends northwards as far as the Zambesi River (*circa* 18°S. latitude).

There it grows as a shrub or small tree up to 30 feet in height along river banks or where there is underground water available; only to a limited extent does it grow on freely-drained sites. It is of interest in Australia as it has been planted at Broken Hill, both as a hedge plant and as an ornamental; it grows very satisfactorily with only a limited amount of additional watering. The dense foliage and the shining dark green of the leaves give an attractive appearance amongst the prevailing lighter colours of other plants. It is not known what is the minimum rainfall in which it will grow in the dry country of Australia, but it may be about 18 inches where there is no additional watering and about 12 inches on sites which receive run-off. Palmer and Pitman (1961) state that karreeboom is both drought and frost resistant, and grows on chalky or sandy soils, as well as on rich red earths. Also it is resistant to local species of termites. The tree is readily raised from seed and, while it will grow on well-drained sites, is recommended in South Africa for poorly-drained areas. Propagation may also be done by layering. Early growth is fast.

In some parts of its native area the foliage is considered a good fodder but, as the species shows a wide range in leaf form depending upon locality of natural occurrence, it would be desirable to be cautious in feeding the strain grown in Australia to hungry stock. The small fruit are edible and, after pounding and fermenting, produce a beer which is popular with the indigenes; they are also reported to be a good poultry food. The trunk is rarely of suitable form for utilisation but the wood is sufficiently durable for use as posts.

Karreeboom is used as a beautiful and suitable species for street planting in some of the towns and cities of South Africa, for example Kimberley at 28¾°S. lat. and about 4,000 feet altitude, with a mainly summer rainfall of 16 inches a year. It is recommended for more extensive trials in Australia.

Robinia pseudoacacia L. **BLACK LOCUST**

This exotic deciduous tree is indigenous to the mountains and highlands of eastern United States of America where the 40-55 inches rainfall occurs mainly in summer. It has, however, proved adaptable to winter rainfall climates and will tolerate both rather dry areas, that is 20-25 inches a year, and slightly calcareous and alkaline soils. It has been planted extensively outside its natural range in eastern U.S.A., in Canada and many countries overseas, mainly for ornamental purposes and soil conservation. On good soils and in areas of optimum rainfall it attains 80 feet or more in height, a trunk 2-4 feet in diameter and shows a well-developed crown. In the 20-30 inch rainfall belt of Australia open-grown specimens usually grow to 40-50 feet. There is variation in the form of the tree, which is attributed to the original seed source, though the early specimens grown in Canberra, for instance, show satisfactory form for use as an ornamental.

Black locust, also known as false acacia due to the resemblance of the leaves to some bipinnate-leaved species of *Acacia*, is recommended as an ornamental and shade species in the better parts of the semi-arid zone, as well as in humid areas. The light green foliage and

Actinostrobus pyramidalis
(native cypress)

Callitris preissii subsp. *preissii*
(Rottnest Island cypress pine)

Although Australia has an appreciable number of softwood genera many of them are of very limited distribution and of even less economic importance. With the exception of two genera—*Actinostrobus* and *Callitris*—the others are restricted to the comparatively high rainfall areas of eastern Australia and Tasmania. Some are reduced to small, stunted plants which grow at high altitudes.

Actinostrobus is comprised of two species which are native to W.A.; whilst the number of species in *Callitris* (cypress pines) is subject to the classification adopted by botanists, but one authority cites 16 as the number. The cypress pines spread over a large part of the continent and Tasmania and are particularly common in the higher rainfall parts of the semi-arid zone and in the sub-humid zone.

Actinostrobus pyrimadalis—young planted trees in a plot of the Forests Department near Eneabba, W.A.; rainfall about 18 inches. Eneabba is about 150 miles north of Perth.

Callitris preissii subsp. *preissii*—young planted trees from the same area as *Actinostrobus pyramidalis* above. This species is usually referred to under its synonym of *Callitris robusta*, but it is better known by its common name.

Callitris endlicheri—planted in a plot of the Forests Department near Wongan Hills, about 100 miles north-east of Perth, W.A.; rainfall about 15 inches.

Callitris glauca—growing naturally east of Deniliquin, N.S.W.; rainfall about 16 inches. The crop underneath the tree is potatoes.

Callitris endlicheri
(black cypress pine)

Callitris glauca
(white cypress pine)

Plate 81 AUSTRALIAN CONIFERS

Rhus lancea
(karreeboom)

Nerium oleander
(oleander)

Olea europaea
(olive)

Ceratonia siliqua
(carob bean)

Rhus lancea, planted at Broken Hill, N.S.W., receiving some additional watering.

Nerium oleander, planted at Woomera, S.A., receiving additional watering.

Olea europaea, planted at Broken Hill, N.S.W., not watered but the roots may extend to an adjacent area which receives watering.

Ceratonia siliqua, planted in city park, Dubbo, N.S.W.; watered.

Schinus molle, planted, old Wentworth Gaol (historical) western N.S.W.; not watered.

Tamarix articulata, planted, Woomera, S.A., receiving additional watering.

Schinus molle
(pepper tree)

Tamarix articulata
(athel tree)

Plate 82 EXOTIC BROADLEAVED SPECIES

Pinus pinea
(stone pine)

Pnus pinea
(stone pine)

Pinus pinea
(stone pine)

Pinus halepensis
(Aleppo pine)

Pinus halepensis
(Aleppo pine)

Pinus canariensis
(Canary Island pine)

Australia has no true pines (*Pinus* spp.), but there are a number of introduced species which have done well in one or more of the climatic zones.

Top three: These show stages in the development of stone pine. The top left is a tree only a few years old in a sample plot of the Agricultural Junior High School at Cunderdin, W.A.; rainfall 14 inches. Below it is a tree starting to show mature form in the A.C.T.; rainfall 24 inches, whilst the top right shows early maturity at Grenfell, N.S.W.; rainfall 24 inches.

Lower centre left: Young trees established as part of a shelter belt at an experimental farm of the N.S.W. Department of Agriculture, Condobolin; rainfall 16 inches. Apart from the characteristics of the species the good branch development shown is due to a combination of suitable initial spacing and protection from grazing.

Bottom left: Old trees growing on a house block at Broken Hill; they have benefited from watering but are hardly suitable for a small block of land.

Bottom centre: This specimen provides useful shade at a kindergarten centre in Broken Hill. The species tolerates high summer temperatures but would not thrive in the low rainfall (9 inches) without the benefit of lawn watering.

the masses of white flowers in the spring add to the attactive appearance of the tree, whilst the non-inflammable bark and foliage add to its suitability for planting in the vicinity of homesteads and in parks. Overseas it has been used to control coastal sands and erosion. There are two characteristics which may not be disadvantageous under all conditions—one is that it has strong spines and the other is its vigorous suckering habit. It does best on deep, light soils but, if given sufficient growing space, will thrive on most light, well-drained soils; it should not be planted on heavy, wet soils. Especially in low rainfall areas it should be allowed sufficient growing space, say 50 feet between trees, and not planted within at least that distance from sewers and drains. The timber is heavy, hard, strong, tough and durable and is highly regarded for use in the ground, as railway sleepers and many other purposes, including turnery and tool handles. It is an excellent fuel.

Schinus molle L. **PEPPER TREE**

The pepper tree, so named because the berries have been dried, ground and used as pepper, is a native of Peru, but has been introduced into many countries of the world. It is one of the outstanding exotics for use in the dry country and is at its best in areas of 12-18 inches rainfall in southern Australia and 15-25 inches in the north. It will grow without additional watering in areas as dry as 10 inches and should not be planted in wet soils or where heavy frosts occur. *Schinus molle* is a small to moderate-size tree, 25-40 feet in height, with drooping branchlets and long drooping foliage. The generic name is from *schi-nus*, Greek for the mastic-tree, and the species name is from *mulli*, the Peruvian native name. As an ornamental species it is probably not fully appreciated in Australia because of its extensive use in the very dry country, but it is one of the more attractive exotic trees of the arid zone. It retains its branches to near ground level and is suitable for low shelter.

This tree is recommended for planting around homesteads and in parks, as well as for shade and shelter of animals. It will grow on most light soils and will tolerate heavy trimming.

Tamarix articulata Vahl. **ATHEL TREE**
syn. *T. aphylla* (L.) Karst.

The athel tree, also known as tamarisk, is usually seen in Australia as a heavily-branched, small tree 25-40 feet in height, sometimes with not more than a very short (4-8 feet) trunk. In view of the fact that it is normally propagated from cuttings it is possible that most of the trees planted in this country have originated from one source. Although the form as a forest tree is poor it is very suitable for shelter and light shade; the crown is often wider than the height of the tree. In single-row shelterbelts branches are retained to ground level.

It is a native of the deserts and steppes of Asia, especially Asia Minor and north-west India; some varieties have straight boles up to 20-25 feet in height. In these regions it will grow with a rainfall as low as 4 inches. In this country it shows an ability to grow moderately well in areas as dry as Broken Hill and Kalgoorlie (both with 9 inches) where the 10 percentile values may be only 5 inches. One forest service recommends it for use in areas as dry as 7 inches but probably its optimum development is where the rainfall is 14-20 inches, with hot to very hot summers and winters with 1-15 frosts. It will grow on almost any soil type and is one of the exotic species most resistant to saline conditions.

This tree can be grown from its minute seed but is more readily raised from cuttings which are taken in the spring from the wood of the previous season; with suitable attention these will be well rooted and suitable for planting in the following winter. It does not produce suckers but coppices readily and shoots are developed freely from any large branches which have been lopped. This ability to shoot after a heavy lopping, which is rare with arid zone species, has advantages where it is desired to control height or size near buildings or power and telephone lines. Plate 83 shows that characteristic.

It is the outstanding exotic species for moderately low shelterbelts in the arid zone. Overseas it is recommended for erosion control and sand dune fixation, using long, well-rooted cuttings which are planted deeply in the sand. In its native area galls which form on the tree are collected for tannin production, whilst the white, hard and durable wood is used for ploughs, wheels and firewood; when there are logs of suitable form it may also be sawn for furniture manufacture.

Several other tamarisks are grown in the dry country, especially *T. gallica* L. (French tamarisk), whilst other species are flowering shrubs suitable for use in areas which are not as dry as those which the athel tree will tolerate.

EXOTIC CONIFERS

Cupressus arizonica Green **ARIZONA CYPRESS**
also known as *C. glabra*

The home of this tree is in the mountains and along the drainage channels which extend into the deserts of eastern central Arizona. In the mountains the rainfall is mainly 15-20 inches; there are many frosts and high maximum day temperatures occur. In Australia it has shown itself able to grow satisfactorily in the drier parts of the southern coastal areas and in the moister regions of the semi-arid zone. It should preferably not be planted in areas with a rainfall less than about 20 inches nor where there is a summer maximum precipitation. Height is likely to be 50-70 feet.

Like most species of *Cupressus* and *Callitris* it has numerous, persistent branches that are normally retained to ground level in open situations, which makes it suitable for dense shelter-belts. The inner foliage dies off as the branches grow, producing a highly inflammable mass so that the tree should not be planted close to the windward side of buildings. If used for farm shelter it should be restricted to short belts for the shelter of calves and lambs or milking cows near homesteads. Because of the risk of fire completely killing a belt it should be only planted on improved pasture or adjacent to cultivated areas where a reasonable level of fire protection can be afforded. Commonly-planted forms have wide, spreading crowns, but some narrow-crowned individuals do occur which are suitable for use where space is limited or a columnular effect is sought. As with other horticultural forms of this genus, vegetative reproduction should be used if plants of a particular habit are required or colour of foliage, for example the 'golden' types. In general, the species is hardy to high summer temperatures, dry conditions and can tolerate lime in the soil. Arizona cypress will grow on rather poor soils if they are moderately well-drained but will not tolerate badly-drained tight clays. Although the timber is used in its native habitat it should not be planted in Australia with the objective of timber production.

C. sempervirens L. var. *stricta* Ait. **ROMAN CYPRESS**

The species, *C. sempervirens*, is suitable for the better rainfall parts of the southern semi-arid zone and adjacent higher rainfall areas; it has been described on p. 165 of *Growing Trees on Australian Farms*. The variety *stricta* has a narrow, erect form and is mainly used where formal effects are desired in parks and avenues. The common name of Roman cypress is due to its use in the Rome of two thousand years ago. It is a tree which is often planted around cemeteries.

This cypress can be planted in southern Australia where the rainfall is 18 inches or better, on most soils other than very tight clays.

Pinus canariensis C.Sm. **CANARY ISLAND PINE**

This exotic conifer has a number of good qualities which are of interest to landowners in the higher rainfall parts of the semi-arid zone of southern Australia. One attribute is its ability to retain good form, especially an erect trunk, on poor soils and with a rainfall which may be less than 18 inches; another, and a very important one for farm use, is its fire resistance once it has attained the pole stage and associated with this is its ability to produce shoots from both the trunk and stump; still another is the longevity of the species—at least 100 years on even moderately favourable sites and, finally, the high quality of the timber. On the debit side is its slow early height growth for the first 5-10 years, after which it may average 2 feet a year until heights of 80-120 feet are attained, depending upon the local environment.

The tree is a native of the Canary Islands but has been planted in Australia since the end of the nineteenth century. It is suitable for the higher rainfall parts of the semi-arid zone in southern Australia, as well as the sub-humid zone. Meteorological stations which represent climates in which it will grow include Condobolin and Wagga Wagga (New South Wales), and Bendigo (Victoria) and Mt. Gambier (South Australia) in the sub-humid zones of the two latter States.

Canary Island pine seedlings develop a strong tap-root at an early stage and, unlike most of the exotic pines, are unsuitable for planting bare-rooted. On a small scale, establishment should be with tubed stock but where conditions are suitable and ants and other vermin can be controlled, direct sowing of the seed can be practised; several seeds should be sown in a spot and when several inches high surplus seedlings can be thinned by hand.

The principal role of this pine in the dry country is for single-row shelterbelts in areas where the rainfall is 16 inches or better; the trees should be spaced at least 10-12 feet apart if good branch development is desired to maturity. The timber is hard, heavy and strong. The heartwood is durable but Australian experience in higher rainfall areas has been that the plantation trees are slow to develop heartwood. As an ornamental tree it can possibly be planted on good soils in areas where the rainfall is as low as 12-14 inches; where some water is available it can be grown as specimen trees in areas as dry as 9-10 inches.

P. coulteri D. Don **BIG-CONE PINE**

Big-cone pine is a very suitable epithet since the cones are the largest in the genus, the length being 10-14 inches and diameter up to 6 inches, with the largest weighing 4-5 lb in the green condition. The seed is also large, about half an inch in length, and edible. The natural occurrence of the pine is on dry mountain slopes of California at altitudes of 3,000-7,000 feet. It has been introduced into Australia as an ornamental and does reasonably well in the sub-humid zone and in the higher rainfall parts of the semi-arid zone. It is suitable for use in large parks for shade and ornamental purposes, but requires plenty of growing space, that is 50-60 feet between specimens. Open-grown trees may attain 60-70 feet, with heavy branching and a wide spread. It is recommended only for use in southern Australia, that is the winter rainfall area, where annual precipitation is in the range of 20-30 inches. Growth is rather slow. The tree seeds freely every second year or so and, like most *Pinus* species, is easy to raise from seed.

P. halepensis Mill. **ALEPPO PINE**

Aleppo pine is a native of countries bordering the Mediterranean from Portugal and Morrocco in the west to Asia Minor and the Levant in the east. A distinctive feature of this pine throughout its natural occurrence is its association with limestone—other pines grow well only on acid or neutral soils. In Australia, however, it grows on a wide range of soils from moderately heavy clays to rather poor sandy podzols but does best, as would be expected, on a good sandy loam.

In this country Aleppo pine has been planted in the southern part of the continent and mainly in areas of 18-30 inches rainfall, with a winter maximum. Some of the earliest plantings were of very poor form but other strains used in later plantings show a fairly straight main stem and very numerous persistent branches. It shows little promise as a timber tree but in areas where it has been planted is one of the best of all species for shelterbelts of 40-60 feet height. Because of the bushy habit of open grown specimens single row shelterbelts require plenty of growing space, that is about 30 feet on each side of the row. Single trees grown in the open and pruned to 10-15 feet in height provide useful shade. Further information is given on p. 140 of *Growing Trees on Australian Farms.*

P. halepensis Mill. var. *brutia* (Ten.) Elwes et Henry

Since the nomenclature followed here is taken from *A Handbook of Coniferae and Ginkgoaceae* this pine will be shown as a variety of *P. halepensis*, but the name of *P. brutia* is also widely used. The most important natural occurrence is on the island of Cyprus, where it constitutes the principal species from sea level to nearly 5,000 feet. It is also found on the hills of southern Turkey and in northern Syria and northern Iraq. *P. brutia* grows on a variety of soils which range from those with a high lime content to acid podzols. The rainfall is mainly 20-40 inches, but it is found on coastal lowlands where it is as low as 10-15 inches.

The height varies from 40 feet on poor rocky ridges and slopes to 80-100 feet in mountain valleys and, occasionally, with elite trees to 120 feet. The form of the tree is much better than that of Aleppo pine (*P. halepensis*); the branches are smaller and more regular, whilst the trunk is straight even on sites which are only moderately favourable. The foliage is not as dense as in the species but is sufficient to provide a good level of shelter from wind.

Small plantings, mainly on an experimental scale, have been made in Western Australia and elsewhere in Australia and good form and a reasonable rate of growth has been retained where the rainfall is down to nearly 12 inches. In these dry areas, often associated with rather poor soils, this pine gives the best promise of any exotic *Pinus* species. The scale of planting desirable for commercial timber production on sites unsuitable for *P. pinaster* or *P. radiata* in Australia has not been determined, but there is little doubt that it has a future for farm shelterbelts in areas of southern Australia which experience very high mean maximum temperatures of the hottest month, that is up to 95°F, and a rainfall of 12-20 inches. For planting in areas of 12-15 inches rainfall it is recommended that the advice of the forest service of the State should be sought before commencing operations.

P. pinaster Ait. **MARITIME PINE**
syn. *P. maritima* Poir.

This is an important farm and timber species in the winter rainfall area of southern Australia. For commercial timber production a minimum rainfall of about 25 inches is desirable and for single row shelterbelts this can be reduced to about 20 inches. For marginal sites and rainfall Aleppo pine (*P. halepensis*) is more suitable.

Maritime pine has several strains, some of which are noted for pronounced butt sweep, and care needs to be taken in the source of seed. This and other aspects of the species are discussed on pp. 170-172 of *Growing Trees on Australian Farms*. Because of numerous and persistent branches this tree has a role for single row shelterbelts with a life span of over 80 years. Grown in the open it may take 30 years to reach 50 feet and another 20 to reach 80 feet, after which height growth will be slower.

P. pinea L. **STONE PINE**

This is the most useful pine for shade in the 18-30 inch winter rainfall area of southern Australia. It can tolerate both high maximum temperatures and light winter frosts and will grow on a wide range of soils, including those of rather heavy texture and sandy loams. It should not, however, be planted on very tight, impermeable clays or soils with a very high lime content.

Under good conditions the tree attains 50-60 feet in height and develops long, sturdy lateral branches well up to the top so that a mature tree appears, in outline, like a heavy umbrella, with the width equal to or exceeding the total height. Pryor (1962) records that in Canberra it has reached 56 feet in height and 92 feet in crown spread. The role of this tree is for use as single specimens for shade on farms or for high shade and ornament in parks. For landscape work Pryor draws attention to the need to use such a species sparingly. It has only a moderate rate of height growth for a pine in early life, but is very long lived—probably 60-100 years—and is readily raised from seed. Further information on this species is given on p. 172 of *Growing Trees on Australian Farms*.

P. sabiniana Dougl. **DIGGER PINE**

This species is botanically closely related to big-cone pine (*P. coulteri*) with its long needles and large cones. It is hardy to drought, will tolerate light frosts and grows on dry and shallow soils. Digger pine is a native of the arid foothills of the Sierra Nevada and coastal ranges of California at altitudes of 400-4,000 feet.

In Australia it has a possible role for shade and shelter on farms and for screening and general background planting in southern winter rainfall areas, where the average annual precipitation is 20-30 inches. Early growth is moderate but form, as a timber tree, is often poor. Experimental plantings have been made in the Australian Capital Territory and at Wirrabarra in South Australia. It seeds regularly and is readily raised from seed. It should not be grown for timber production.

P. torreyana Parry ex Carr. **TORREY PINE**

This pine has a very limited natural occurrence, being restricted to a small area on the San Diego coast and to Santa Rosa Island of southern California. In that dry environment it varies from a small, heavily-branched tree only 30-50 feet high, to a semi-prostrate shrub on exposed coastal sites. When planted overseas in more favourable locations it has shown much better form and a moderately fast growth rate, for example in the Australian Capital Territory it has shown an average of 3 feet in height and, for widely-spaced specimens, up to ½ an inch in diameter a year. In rainfalls of 50 inches or more in New Zealand it has attained heights of over 100 feet in 35 years and timber volumes per acre which were comparable to those of *P. radiata*.

Torrey pine has greyish-coloured needles, in contrast to the greenish hue of most *Pinus* species. It is worthy of small-scale planting for shelterbelts and background effect in winter rainfall areas of southern Australia, where the mean annual precipitation is 22-30 inches and only light to moderate frosts occur. Like most pines it shows best growth on moderately fertile sandy loams but will grow on a wide range of soils other than heavy, impermeable clays or those with a marked saline content. Like Canary Island pine (*P. canariensis*), it does not tolerate bare-rooted planting and should be established with tubed stock. Seed years, on Canberra experience, are somewhat irregular and good crops may only occur at intervals of 4-5 years.

FURTHER READING

The following references are additional to those on pages 309-10. Some of them also have an application to Chapters 18-21.

Audas, J. W. (1952). *Native trees of Australia.* New and enlarged ed. Whitcombe and Tombs, Melbourne.

Brockway, G. E. (1941). 'Forests of the arid gold fields region of Western Australia'. *Emp. For. J.*, 20 : 16-24.

Food and Agriculture Organisation of the United Nations (1958). *Fuelwood plantations in India.* (F.A.O. For. Occ. Pap. 5). Rome.

Goor, A. Y. & Barney, C. W. (1968). *Forest Tree Planting in Arid Zones.* Ronald Press.

Jones, D. K. (1953). *Carob (Ceratonia siliqua) culture in Cyprus.* (F.A.O. Working Party on Medit. Pasture and Fodder Developm.). Algiers.

Letourneux, C. (1957). *Tree planting practices in tropical Asia.* (F.A.O. For. Developm. Pap. 11). Rome.

Magini, E., & Tulstrup, N. P. (1955). *Tree seed notes.* (F.A.O. For. Developm. Pap. 5). Rome.

Metro, A. (1954). *Eucalypts for planting.* (F.A.O. For. and For. Prod. Stud. 11). Rome.

Penfold, A. R., & Willis, J. L. (1961). *The eucalypts: botany, cultivation, chemistry, and utilization.* Leonard Hill (Books) Ltd., London. Interscience, New York.

Pryor, L. D. (1967). 'Acacia peuce: a tree for arid regions'. *Unasylva*, 21 (1).

United States Forest Service (1948). *Woody-plant seed manual.* (U.S. Dept. Agr. Misc. Publ. 654). Washington, D.C.

World Eucalypt Conference (Second), Sao Paulo, Brazil (1961). *Reports and Documents.* 2 vols.

CHAPTER 18

Factors in the selection and use of trees and shrubs

ALL trees are suitable for at least one or two purposes and a number can be used for a wide range of amenity and conservation planting. Perhaps the requirements are fewest in the case of trees for shade on large farms, rather more is required from species for use in hedges and shelterbelts, still more from trees to be planted around homesteads or in town and city parks and the most stringent specifications are for those to be used in street planting. With reference to the latter the Superintendent of Parks and Gardens, City of Springvale, Victoria has written (1968):

> There is no perfect street tree. If there were such a tree it would measure up as follows. Its roots would thrive in any kind of soil and be sufficiently strong to support trunk and crown of tree without encroaching on, or damaging in any way, kerb, channel, footpath, foundations of front fences, underground drainage pipes, sewers, and other essential services such as water, gas, and power mains. Its top would grow rapidly to the required height and span without encroaching on overhead wires, passing vehicles or pedestrians and remain at the correct size without further attention.
>
> Its trunk would be of sufficient height also to ensure maximum visibility for both drivers of vehicles and pedestrians alike. It would have leaves that did not shade nearby gardens in winter, or cause shadow on the footpath at night. Its leaves, bark, flowers and seed pods would remain firmly fixed, thereby causing the minimum of litter on nature strips and in gardens. It would be immune to frost, and would stand up to the severest wind, storms, drought, and floods, but possibly best of all it would be able to withstand all forms of vandalism, and remain a thing of beauty at all times.

In spite of this pessimistic outlook and engineer-like specification, from the infinite variety of shrub and tree species it is possible to make suitable selections for nearly all purposes. The surprising thing is that in the very difficult environment of city streets with restrictions imposed by public utilities, species are available which will meet the principal requirements. The more important criteria which should be considered in selecting trees for amenity planting are considered under a number of headings. Somewhat different priorities apply in the case of determining species for commercial timber production and the reader is referred to *Growing Trees on Australian Farms* for necessary information.

1. *Legal requirements*

There are two main aspects: one is that of complying with restrictions imposed by local authorities on planting in any street, road or drainage area; the other is the limitations of the height of obstacles at the approaches to aerodromes, impeding the view on roads or interfering with power and telecommunication lines. In addition most Departments of Lands have restrictions on what species of trees and shrubs can be removed from Crown leasehold land in the dry country. The decade 1960-70 has seen most metropolitan local authorities, especially sewerage organisations, gazette regulations which limit the choice of shrub and tree species. The general pattern of these regulations is to ban certain species and to limit the distances from sewer mains and connections which other species may be planted. In 1960 the Sewerage Regulations in one Victorian city banned desert ash (*Fraxinus oxycarpa*), Moreton Bay fig (*Ficus macrophylla*), all poplars (*Populus* spp.) and all elms (*Ulmus* spp.); whilst an amendment in 1967 increased the total to 14 species. It is drawn to attention that the banning of a tree like Moreton Bay fig, whilst quite justified, is mainly an indication of the judgment of those who planted such a large species in streets and, therefore, adjacent to sewers and other utilities. Moreton Bay fig is a magnificent tree when allowed to grow to full size under favourable conditions, and as a shade tree for large parks in warm coastal areas it has few equals. It is suggested that such a species should not be planted within at least 100 feet of anything with which the roots may interfere; Plate 26 in *Growing Trees on Australian Farms* shows a fine specimen on the banks of the Manning River near Taree, New South Wales, which has a total crown spread of nearly 150 feet. It is unusual for the roots not to spread at least to the edge of the crown; in arid areas the root spread may exceed ten times the diameter of the crown. Although the banning of scheduled species is the pattern in the large cities of the sub-humid and humid zones of Australia, the general principle should be followed for all public planting in the dry country, as well as by any private establishment which has its own sewerage system. There only needs to be a microscopic hole in an earthenware or concrete pipe, or at a joint, for a root hair to enter and subsequently develop a root system within the pipe which will ultimately block the cavity.

Obstruction to air navigation is under the control of the Department of Civil Aviation and anyone wishing to plant major shelterbelts near aerodromes would be well advised to consult the office of that Department in their State. The avoidance of trees interfering with vision on road or with lines is mainly a matter of common sense. Local authorities have the legal power to remove branches, top trees and otherwise cut back any vegetation which affects public use. There is nothing attractive about a large tree which has had one side cut off or has been roughly topped, so avoid the position arising by not planting trees on the inside of road curves or where their size may interfere with public utilities. If it is particularly desired to plant trees near such positions either use smaller growing species or place them at a sufficient distance from the road boundary.

2. *Size and form*

The size to which a tree will grow depends not only on the species concerned and the climate and soil where it is planted, but also upon whether it is established in woodlot formation or planted as solitary specimens in exposed situations. Many eucalypts native to the humid zones, for instance, readily attain heights up to 200 feet on favourable sites in mountain valleys, but when planted as solitary specimens in a drier climate they may only grow to half the height. A common exotic, radiata pine (*Pinus radiata*), shows the same behaviour—some specimens on good sites in New Zealand have grown to over 200 feet, but if planted in the open on a dry basaltic hill may not attain more than 60-80 feet. This striking contrast in size and form is well exemplified by Plates 8 and 19 of *Growing Trees on Australian Farms*.

Anyone planting a tree or even a garden shrub should first of all consider the probable mature height it will attain in their locality, before deciding on the position where it can be established; height is important in planning a shelterbelt and it is critical in locating trees around homesteads. From the viewpoint of safety of buildings and people no large tree should be planted closer to a building than the equivalent of the estimated maximum height to which it is likely to grow. In this regard it should be noted that maximum shelter effect occurs at distances greater than this. Another aspect of mature size and relative position to a home or road is the amount of shade cast at various times of the year; unless heavy shading is wanted on buildings during winter in the southern hemisphere large trees should not be planted in the arc north-west, north, north-east from the house. Trees planted directly east and west may be advantageous in reducing summer sun on buildings when the sun is low in the sky; this can be very beneficial in shading against the western sun in summer, since maximum shade temperatures are usually attained in the mid-afternoon as the sun starts to drop in the sky. The size of trees in large parks does not present the problems which occur in planting near homes or in streets. Those responsible for public planting are aware that particular care needs to be taken to avoid establishing large evergreen species in suburban streets which are oriented east and west, since in such cases the homes on the southern side may receive excessive shade throughout the winter months. This position is more critical in southern parts of Australia than in areas nearer the tropics where the sun angle at noon is high even in mid-winter.

Form is also important as various tree uses have different requirements. For instance, a tree for a shelterbelt should desirably retain branches clothed with living foliage to ground level; this is a condition which is likely to occur in open-grown native cypress pine (*Callitris* spp.) and the exotic cypresses (*Cupressus* and *Chamaecyparis* spp.) and to a less, but still worthwhile extent, in the true pines (*Pinus* spp.). Broad-leaved trees (hardwoods) in general do not retain lower branches in maturity as well as the conifers (softwoods). Only rarely do fully-grown eucalypts provide good shelter to ground level, but this position can be remedied by a row of dense shrubs; this combination can be a very practical one for many parts of Australia. In the case of most trees for street planting and some for use in parks and around homesteads the retention of branches to near ground level is not desirable. A clean bole for the lower 10-20 feet may be preferable in order to allow visibility for traffic or to provide shaded recreational areas, as well as to enable grass cutting and cleaning operations to be done with ease. When shade from single trees is required, especially when the sun is high overhead, hardwood trees are usually more suitable than conifers, because of their tendency to grow relatively broad crowns in relation to height. In the case of conifers when used in avenues free from sewers and other utilities, pruning of the lower branches will secure the desired visibility without detracting from the general effect; hoop pine (*Araucaria cunninghamii*) and Norfolk Island pine (*A. heterophylla*) are particularly suitable for this type of treatment.

3. *Root systems*

The root system of a plant is necessary not only to absorb moisture and nutriments from the soil but also to provide the necessary stability against wind action. Typically it will extend to at least the edge of the crown and often much further, and any artificial restriction on this limit is prejudicial to both health and stability. Apart from inherent differences between species, root systems vary with soil type and depth. The depth to which roots penetrate increases with permeability of the soil and with availability of moisture at depth; lucerne (*Medicago sativa*) when growing on porous sandy loams along river banks has been known to have roots going downwards for 20-25 feet, though the above ground part of the plant may have only been 1-2 feet high. It is well known that trees growing on a shallow soil over a tight, almost impenetrable clay are much more subject to windthrow than those growing on a deep sandy loam. The rooting system of trees is relatively shallow when their heights are considered; a tree growing to over 100 feet may have a rooting system which only penetrates to a depth of 3-4 feet; even less on some soil types.

Especially when rainfall is low and there is no regular watering, plant roots may grow laterally to distances many times the height of the tree above ground, and many suburban areas with their combination of sealed surfaces and buildings which prevent the infiltration of rain can be dry. It is no wonder that tree roots seek moisture from sources such as sewers and drainage pipes and grow vigorously under watered lawns and gardens. It is this habit which has led to the banning of many trees from further planting in streets, but the fact is that any plant should not be established where there is not sufficient room for growth both above and below ground.

Root pruning is, in general, an undesirable practice because of the risk of introducing fungal infections and, in some cases, reducing the stability of trees. Where shelterbelts are planted around orchards the finer roots are likely to be attracted to the cultivated and manured surface soil between the fruit trees and some pruning can be done to control excessive robbing of the orchard area. One method is to dig a shallow drain (12-18 inches deep) 10 to 20 feet from the trees and sever all surface roots; this usually needs to be repeated every year or second year.

4. *Suckering habit*

Growth from underground shoots or arising from damaged roots is an undesirable characteristic for trees to be planted in streets or around homesteads, although it can be put to good use under suitable conditions for shelterbelts. Some trees, notably certain species of the exotic poplars and aspens and some native acacias, readily produce suckers, especially if soil cultivation cuts the smaller surface root. Where grazing or mechanical means are available to control the undue spread of the sucker growth, the presence of suckers, often associated with seedling plants, greatly improves the effectiveness of a belt. Attention is drawn to Plate 24 in *Growing Trees on Australian Farms* of a single-row belt of English elm (*Ulmus procera*). Alongside the road, protected from grazing, there is a dense growth of seedlings and suckers which affords a high level of protection from wind, even when the trees are without leaves. This contrasts with the other side of the belt where there are no branches and foliage for several feet from ground level, and a complete lack of seedlings and suckers as a result of grazing.

5. *Suitability to the local climate*

There is a strong case that this requirement should take precedence over all other factors. This is undoubtedly true in the selection of species for plantations in which timber production is the principal object and where vigorous growth with the minimum of attention is critical for profitable operations. It is almost equally true for farm shelterbelts, but less so for city parks and street planting where individual trees can receive attention. In the case of single trees planted for ornament on watered lawns in sheltered sites both rainfall and maximum temperatures are not critical for most species; probably the limiting factor is minimum temperature, that is frost, which may even kill young plants.

Generally most species will tolerate a warmer and moister climate than their natural habitat, but many do not adjust well to variation in the rainfall pattern, especially from a dry summer to hot and humid conditions. In such cases many species will show an increased susceptibility to fungal diseases and form will be poor. A deterioration both in form and colouring of flowers can be expected in trees and shrubs taken from a very hot arid climate and planted in a comparatively cool, moist area.

In most cases satisfactory results are obtained in plant introduction by establishing in comparable climates, but there are many anomalies which cannot be readily explained. With Australian species a tree which grows quite satisfactorily in the semi-arid country, but should not in view of its native habitat, is white cedar (*Melia azedarach* var. *australasica*). It is found in the rainforests of northern coastal New South Wales and both tropical and sub-tropical Queensland, as well as in New Guinea—areas of high rainfall and humidity—and usually grows in sheltered fertile sites. Yet, unexpectedly, it grows quite satisfactorily as a street and park tree in inland semi-arid New South Wales and Queensland, proving itself both moderately drought hardy and tolerant of high temperatures. The all-important value of artificial watering is well shown at Woomera, South Australia, where in an area with an average rainfall of about 6-7 inches and maximum summer temperatures which are mainly over 90°F, there is a specimen of Moreton Bay fig (*Ficus macrophylla*). This is an exception and should not be regarded as an example to be followed in planting.

Most of the trees used or recommended for planting in the dry country, especially the arid zone, are native species either from the same locality or from other areas with comparable climates. Of the exotics nearly all come from dry regions, especially countries around the Mediterranean or in Asia Minor. It is in the capital cities, especially Melbourne and Sydney, that there can be seen many examples of acclimatisation of a wide range of species from other parts of Australia or from overseas. On the one hand Melbourne has many fine examples of broad-leaved trees from cool temperate Europe, such as English elm (*Ulmus procera*), and on the other hand, desert mallees and other eucalypts from the eastern Goldfields of Western Australia (Kalgoorlie region) where the average annual rainfall is only 8-10 inches. Because of the nearer approach of Melbourne to the English climate than to that of Kalgoorlie the elms show to greater advantage there than do the Goldfields eucalypts.

6. *Tolerance to air pollution*

Fortunately air pollution is not the serious problem in the sparsely-populated dry country that it is in the large cities and industrial centres of the world. In the few locations where there are ore treatment works and other large industrial operations of a metallurgical nature, a study should be made of the directions of prevailing winds; this will allow plans to be made which will limit planting in locations which are likely to be seriously affected. Overseas, especially in cool temperate latitudes, it is found that deciduous trees are more suitable than evergreen species; the latter being at a disadvantage in that normally their leaves are retained for 3-4 years compared with the 6-8 months for deciduous species. Amongst the latter group the London plane (*Platanus acerifolia*) is outstanding in its ability to tolerate heavy air pollution, repeated pollarding and only a small area of soil not covered by pavement. Unfortunately in Australia there are few deciduous species which will thrive in the dry country; the principal example is white cedar (*Melia azedarach* var. *australasica*), which will grow satisfactorily in the semi-arid zone.

7. *Soil conditions*

The nutritional requirements of most trees and shrubs, especially the pines and eucalypts, is relatively low compared to the needs of agricultural crops. The soils of the arid zone and, to a certain extent, the semi-arid zone usually have sufficient fertility for tree development. In contrast to this are the leached soils of the higher rainfall zones where many not only have a marked deficiency in phosphorus and sometimes lime, but also show very low amounts of many minor elements.

If soil texture is suitable and the moisture levels satisfactory most trees will grow reasonably well even if fertility is low, unless there is serious deficiency in one of the more important minor elements. With a deep, moderately fertile sandy loam or loam nearly all species of trees can be expected to thrive; attention is needed on shallow soils over a tight clay or ironstone pan or where there are unduly low levels of either major or minor elements. In addition, drainage is important and either surface drains or the breaking of hard pans by deep ripping, boring or explosives may be necessary. On some sites it may be preferable to select species which will grow under the prevailing conditions and there are a number of trees which will tolerate seasonally wet soils. Amongst native species several of the paperbarks (*Melaleuca* spp.) will thrive on soils which are almost continually wet, whilst a number of eucalypts will grow on sites which are seasonally inundated; example of the latter include river red gum (*Eucalyptus camaldulensis*), black box (*E. largiflorens*) and a number of the northern Australian species which grow on flats which become waterlogged during the monsoon but dry out afterwards. Amongst exotics most poplars demand a moist sub-soil such as is found in river alluvials, whilst some willows will thrive with most of their roots in water; other exotics, especially the true pines (*Pinus* spp.), require well-drained soils.

Most Australian soils are deficient in phosphorus and the addition of rock phosphate or superphosphate to crops and improved pasture is general practice. In recent decades this practice has extended to pine plantations, preferably about time of establishment or when only a few years old; heavy dosages of 5-10 cwt per acre may be economically justified on soils with a serious deficiency of this element. Many soils in the sub-humid and humid zones are also known to be low in calcium and the application of lime may be necessary in order to ensure good growth. It is well known that nitrogen is one of the most important elements needed for plant growth and nearly all soils will benefit from the application of a nitrogen-based manure, but not only is it comparatively expensive but large doses at one time may have prejudicial effects on tree growth.

Many Australian soils show a deficiency in one or more minor elements, especially zinc and boron, but there are several others, including magnesium. The correct diagnosis of deficiencies in nutrition is not easy for small-scale operators and most farmers should rely on advice from officers of the Department of Agriculture or the Forest Service in their State. More information, including half-tone and colour plates, on soil infertility and the diagnosis of soil deficiency is given in Chapter 16 of *Growing Trees on Australian Farms*.

The accumulation of salt in the soils of the arid zone, especially in depressions with no external drainage, may be so high as to stop all plant growth, or only allow the development of those with a high tolerance to salt. Where vegetation is absent these areas are known as 'salt pans'. When the salt level is not quite as high it is possible to grow some halophytic plants. Outstanding in their tolerance to salt in the soil are members of the family *Chenopodiaceae*; the numerous Australian representatives are mainly herbs or small shrubs and amongst the latter are the saltbushes (*Atriplex* spp.) especially oldman saltbush (*A. nummularia*), the bluebushes (*Kochia* spp.) the bassias (*Bassia* spp.) and the berry saltbushes (*Rhagodia* spp.). Of larger size are some of the dry country eucalypts, especially those native to Western Australia, such as Dundas mahogany (*E. brockwayi*) Dundas blackbutt (*E. dundasii*), Salt River gum (*E. sargentii*), gimlet (*E. salubris*) and yellow-flowered blackbutt (*E. stricklandii*). Some of the species of paperbark (*Melaleuca*) also show tolerance to salt in the soil.

Sloping terrain is less likely to have as high a salt content as low-lying flats, since there is greater opportunity for salt to be washed out when there are the rare heavy rains. Before planting trees in the driest parts of the continent it is desirable to check the salt content of the soil and a satisfactory field method is by use of simple equipment developed by the Commonwealth Scientific and Industrial Research Organisation, known as the C.S.I.R.O. saltmeter. It should also be appreciated that much of the water available in the dry country has a significant salt content and that heavy, regular watering of trees in very arid regions may increase the salt in the soil, especially on flat sites and where there is a tight sub-soil.

8. *Resistance to wind*

Trees should have stability in strong winds and also be resistant to breakage. The first aspect involves root habit and soil type and has been previously discussed; in addition, adequate drainage is important since overthrow is more likely to occur on soils which have been waterlogged during heavy storms than on better-drained sites. Particularly in areas liable to gale force winds trees with soft, weak timber may have tops blown out and large branches broken; most of the species suggested for the dry country, especially acacias and eucalypts, have strong timber and are not subject to serious damage except during cyclonic conditions. Trees with double leaders, especially those originating below half height, are liable to have one leader broken off and when this happens it is not usually long before the other leader follows. The occurrence of low double leaders can be controlled by pruning in the year that a double leader develops.

A tree which has grown in an open position throughout life will be more wind-firm than one which has grown in a woodlot and then has been suddenly exposed by heavy thinning. Particularly in areas subject to gale force winds it is recommended that woodlots should be given a series of light thinnings rather than a few heavy thinnings; by doing this there is an opportunity for the remaining trees to develop their root systems and trunk progressively during the series of thinning operations. In the case of shelterbelts it is undesirable to plant trees close together with the idea of thinning them at a later stage; plant at the desired final espacement, especially in the case of single or few row belts or in the outer rows of wide belts.

Where the dry country extends to the coast in areas subject to periodic cyclones, particular care should be taken in the selection and siting of trees and shrubs. The principal area affected is the sub-tropical and tropical latitudes of Western Australia, especially from Carnarvon northwards through Onslow and Port Hedland. The general recommendations given in this and other sections apply and these and other advice are summarised as follows:

(a) Only use species with strong wood which is not likely to break under the impact of violent winds.

(b) Select trees with moderately open crowns in preference to those with very dense crowns; the latter are permissible in relatively short but very wide-spreading species, for example some of the figs (*Ficus* spp.).

(c) The problem of the greater leverage effect by wind on tall trees can be avoided or ameliorated in either of two ways: the first is not to plant tall-growing species, but this is an attitude of defeatism and should only be adopted for locations very close to buildings or other valuable property, or for very shallow soils; the second is to plant shorter-growing shrubs and small trees on the windward side, so that some protection is afforded to the tallest trees.

(d) Adopt an open spacing to encourage a widespread and strong root system.

(e) Only plant on soils which have sufficient depth to permit development of an adequate root system.

(f) Be particularly careful to check that the young nursery stock is not root-bound, especially that the principal side roots on planting are placed radially out from the centre and not allowed to be wound around the root system.

(g) If necessary, do some crown pruning of trees which attain large size.

(h) Before initiating any planting scheme check on any local information which may be available, especially the effect of previous cyclones on different species.

9. *Longevity and rate of growth*

Longevity is a desirable attribute in nearly all amenity planting, but fast growth to near maximum height and longevity are not usually associated in the one species. There are, however, some exceptions, and among native species most eucalypts and some of the sheoaks (*Casuarina* spp.) may combine these two factors to a significant degree. Amongst the longest-lived trees in the dry country are the cypress pines (*Callitris* spp.) which can be expected to survive for at least 100 years, but early growth is slow and may not average more than one foot a year for the first ten years. Amongst exotics which combine fast early growth and at least a moderately long life are some of the poplars, especially Lombardy poplar (*Populus nigra* var. *italica*) and radiata pine (*Pinus radiata*), but these are mainly species of the more humid zones. One of the longest-living exotics for the semi-arid zone is Canary Island pine (*P. canariensis*), which shows a very slow early growth (1-1½ feet a year for the first 5-10 years) but can be expected to live for 100 years or longer.

Detailed information is not available on many of the species suggested for the dry country but the following will serve as a guide—the larger woody plants have a longer life than the small shrubby species, especially if the latter are soft-wooded; trees on favourable sites will survive longer than those growing under adverse conditions. From this it is possible to generalise by stating that small, soft-wooded shrubs may only have a life of about 10 years, most of the dry country acacias at least 20-50 years, whilst cypress pines may have a minimum life of about 100 years and it is probable that most of the sheoaks survive for periods between that indicated for the acacias and for the cypress pines. The tree-size eucalypts have very long lives and it is probable that most of them have a minimum of about 100 years; it is certain that some of the larger commercial species in the humid zones last for several hundred years unless damaged by fire or storm.

Where shelter is needed as quickly as possible but is also required to last indefinitely, generally the best course is to plant a row of some quick-growing but usually short-lived species about 50-100 feet from the buildings or other objects to be protected and, at the same time, establish a more permanent belt far enough behind the temporary belt so that there is no mutual interference to development. Removal of the short-lived belt should be carried out as soon as the more permanent belt has attained sufficient height to give the level of protection desired.

Species for the short-lived belt will vary with the part of the dry country concerned, but amongst native species some of the faster-growing acacias should be considered and from the exotics the athel tree or tamarisk (*Tamarix articulata*) and the pepper tree (*Schinus molle*); some of the smaller dry country eucalypts planted as advanced stock 4-6 feet high also offer possibilities. These trees will mainly be in the height range of 20-40 feet and the athel tree offers the advantage that it will tolerate heavy trimming (see Plate 84). For the more permanent shelter the native cypress pines are particularly suitable, also some of the sheoaks; where conditions are milder in the southern semi-arid zone then Canary Island pine has few equals for a tall, long-lived belt.

10. *Freedom from insect attack and disease*

A vigorous tree will normally remain a healthy tree and is less likely to suffer from diseases than moribund specimens, and if attacked by insects will show a better recovery after the peak of the infestation has passed. It is, therefore, axiomatic that care should be taken in the selection of species to be planted in a given environment; in particular climate, soils and drainage conditions should be favourable to healthy growth. If a species which grows in the locality is being considered for planting and is known to be subject to disease or insect attack the existing specimens can act as foci from which further infection can spread. It is usually undesirable to use the species concerned. In the case of the introduction of species new to the locality, either from elsewhere in Australia or from overseas, it is desirable to consult the Forest Service or Department of Agriculture of the State for advice on the probable health of the species. Sometimes introduced species do surprisingly well; this can sometimes be explained by the absence of the insects or pathogens which attack the trees in their homeland. Leaf-eating or sucking insects are very common on most eucalypts in Australia and while not killing them have a prejudicial effect on rate of growth; when planted overseas many species show growth far exceeding that prevailing in Australia due to the absence of the particular type of insects which attack them here. The reverse position also applies when some northern hemisphere broad-leaved trees which have been introduced into Australia show much healthier growth than is common in their area of natural occurrence—elms and chestnuts are two examples.

The following rules will assist property owners to maintain healthy trees:

(a) Select tree species which are known to be healthy in the locality; if introduced species are considered then seek expert advice from the appropriate governmental organisations.

(b) Use healthy seedlings or transplants; if using stock grown on the property then grade out all plants which are undersized or of unsatisfactory colour.

(c) Encourage young plants by cultivation of the soil, drainage and manuring where indicated; after establishment do such tending as may be desirable.

(d) Several times a year examine trees for signs of insect attack or disease. If detected, decide what remedial measures are justified, seeking expert advice where necessary. Fungal diseases are usually more serious than insect attack and more difficult to detect in the early stages; non-parasitic foliage diseases are not as serious as fungal infections of the roots or stem.

(e) Wounds left by pruning large branches or due to vehicles knocking against the stems should be covered with tar or other antiseptic. Prompt attention is particularly desirable in warm, humid weather. Be very careful in guarding against root damage. The pruned stubs of small branches in most conifers are usually naturally soon covered with resin and do not need special attention.

(f) Fell any dead or dying trees, burning all infected material and removing from the site other material such as logs.

(g) The spraying of small to medium-sized trees can be readily done with orchard-type equipment but care should be taken in the selection and use of insecticides and fungicides. Most modern chemicals are very poisonous and many have a strong residual effect; the use of some of the earlier formulations has been banned by overseas countries and is discouraged in Australia. The application of sprays and care in their use has been discussed by M. Tanton (1969). Also Departments of Agriculture and Health have publications on the safety aspects of handling insecticides and fungicides.

11. *Adaptability to pruning**

The degree to which trees and shrubs will tolerate pruning and pollarding is an important consideration for street trees; it also has value for those planted in parks, as low hedges designed for regular trimming and specimen trees around homesteads, but is of less importance for tall shelterbelts.

Many broad-leaved species will tolerate heavy pollarding for very long periods if initiated at an early age and they will develop leaf-bearing branches from most parts of the framework; conifers, on the other hand, do not readily develop adventitious buds and new growth will usually only come from young wood. Because pollarding is typically a technique which operates best with certain deciduous species in a cool temperate climate it follows that it will have restricted application in the dry country of Australia. One of the species with which it could be used, though not recommended, is the exotic athel tree (*Tamarix articulata*). The eucalypts will tolerate some light pruning and shaping if commenced at an early age and some species respond favourably to coppicing, in which the main stem is severed 6-12 inches from ground level, resulting in a few to several main stems developing from the sides of the cut. Coppicing is illustrated and described in Chapter 17 of *Growing Trees on Australian Farms.* A native evergreen species which will tolerate a form of pollarding and very heavy pruning to control size and

* *Pruning*: in forestry the removal of live or dead branches or multiple leaders from trees in order to improve form or to ensure an increment of knot-free wood outside the pruned stubs. In horticulture and gardening the term is used to denote either shaping a tree for a particular purpose, for example fruit production, or to maintain a desired cross-section in a hedge.

Pollarding: the initial cutting of the main stem of a tree, usually restricted to deciduous broad-leaved species, above the height at which animals can browse or which is convenient for street trees, followed by regular pruning back of annual growth to a framework of a few large branches which have been allowed to develop immediately below the original severance of the main stem. It is a practice which is used in cool temperate climates, especially with some willows (*Salix* spp.) for the commercial production of osiers or in a modified form, with street trees such as London plane (*Platanus acerifolia*), where several large branches are left below the point of initial pollarding to carry the annual shoots.

form, is brush box (*Tristania conferta*), a coastal species which is extensively used for street planting in Sydney. Amongst conifers Canary Island pine (*Pinus canariensis*) is exceptional in that it will sometimes shoot from stumps in the manner of a broad-leaved species.

The following principles should be observed:

(a) Do not plant tall-growing trees or shrubs under power or telecommunication lines or wherever it may be necessary to control development. The ugliness of tree mutilation is shown in Plate 16 of *Growing Trees on Australian Farms* and Plate 85 of this book.

(b) When some form of shape training is necessary for street trees, as with brush box or London plane, such shaping should commence at an early age.

(c) Where visibility or recreation space is desired under trees, the lower laterals should be pruned at as early a stage of development as the size of the crown will permit. The gradual removal of branches from the ground upwards as they attain about one inch in diameter is usually a good practice. Sufficient crown should always be retained to maintain vigorous growth and discourage epicormics on the pruned stem; this means that crown depth should never be less than about half to two-thirds of total height.

(d) Winter is usually the best season of the year in which to prune trees, although conifers can be pruned in spring when growth is commencing if there is no danger of attack by the *Sirex* wasp.

(e) Never prune the bottom of trees in shelterbelts, although pruning to about two feet from ground level may be permissible with tall shelterbelts in areas where ground fires are difficult to control.

(f) After pruning broad-leaved species and the larger branches of conifers, that is over 3-4 inches diameter, a suitable wound dressing should be applied. Tar or creosote-based paints are effective antiseptics but discourage the development of callus growth over the exposed stub; zinc oxide-based paints are more suitable. Grafting waxes are expensive and usually do not last long enough, but see Chapter 22 for details.

12. *Evergreen or deciduous habit*

Nearly all the species listed for the dry country are evergreen and amongst native trees and shrubs there is only the rare deciduous species, for example white cedar. In the semi-arid zone, however, there are a few exotic deciduous trees which will grow satisfactorily if sufficient water is available. Chief among these are poplars and willows, which can be used on river flats or in irrigated areas.

Deciduous trees have certain advantages and disadvantages when compared with evergreens. Possibly the most important factor on the credit side is their ability to provide shade in summer but allow sunlight to penetrate in winter; this is a very desirable attribute in street trees and those for small parks, whilst this characteristic can be used with advantage in designing shade and shelter around homesteads, especially on their northern and western aspects. Another credit feature for areas of air pollution, though of limited importance in the dry country of Australia, is that each year deciduous trees commence with new and physiologically efficient leaves. On the aesthetic side most deciduous trees have attractive light green foliage when they first come into leaf and this makes a pleasing contrast with the darker and sometimes somewhat drab leaves of the evergreens. The principal disadvantage applies mainly to street planting and small parks where the leaves need to be gathered during and after leaf fall each autumn.

Where conditions are suitable for planting deciduous trees the property owner or local authority needs to make a decision on the balance between advantages and disadvantages.

13. *Fire risk*

The risk of trees being killed by fire or of causing fire in buildings is an aspect of species selection which is all too often overlooked by property owners. Fire risk not only involves the characteristics of a species but concerns distance from buildings and position in relation to the prevailing winds during fire weather. The risk is also affected by whether trees are in woodlots or are isolated specimens on well-watered lawns; it is also modified by low pruning.

The natural characteristics which have a bearing on inflammability are mainly the type of leaves and bark, and whether the crown is carried high on the stem or is continuous from ground level. Leaves which contain large quantities of essential oils, such as is common amongst many of the eucalypts, light up readily in hot weather and burn with a fierce heat. Species which retain their leaves for several years or accumulate dead leaves on their branches, as is common with most conifers such as the true pines (*Pinus* spp.), are more dangerous than deciduous trees or those which drop their leaves after only two to three years. Once a fire starts, in most conifers the whole of the crown is usually burnt and the tree killed, an exception being Canary Island pine which shows an unusual ability to throw out new shoots from adventitious buds which have been safely insulated from the fire by the thick bark.

Inflammable bark, well exemplified in the stringybark group of eucalypts, presents a high hazard; it will light readily from a low grass fire, burns vigorously up the vertical trunk and usually sets the crown alight. Probably worst of all is the tendency for pieces of burning bark to be carried very long distances during strong winds—in major bush fires spotting may develop half a mile or more ahead of the main front during unfavourable conditions.

The following notes summarise the principal precautions which should be taken in relation to fire risk from trees:

(a) Where a recognisable fire risk is present, especially in street planting, hedges fronting public thoroughfares and on the windward side of country buildings, it is desirable to select species which have a low inherent fire hazard.

(b) In the case of long shelterbelts, for example half a mile or more in length, avoid continuity of inflammable species; either design alternation of species or have suitable breaks in the continuity.

(c) Do not plant trees too close to buildings; the minimum distance is height $\times$ 1, but a greater space is desirable. *Under no condition should the crowns of trees overhang the roofs.*

(d) Where low pruning is permissible, that is in other than shelterbelts and hedges, low pruning to 6 feet or more may be desirable.

(e) Even when trees are located up to height $\times$ 3-4 from buildings it is necessary to clean out gutters at least twice a year.

(f) See that all openings into buildings, especially into the roof space, are adequately screened to prevent the entry of sparks during fires.

(g) Do not store firewood or other inflammable material against exterior walls or under floors of houses.

(h) Plough firebreaks or have a completely non-inflammable ground vegetation alongside hedges and shelterbelts. In the case of well-grazed improved pastures there is little fire risk but with unimproved or neglected pastures there is a much greater chance of grass fires setting shelterbelts alight.

14. *Reduction of the risk of damage in public places*

Those who are responsible for the establishment and maintenance of public planting are well aware of the chance of damage and nothing is more disheartening than to have several semi-mature trees in a street or avenue destroyed beyond hope of recovery. Damage may be broadly classified into that originating from carelessness, well exemplified by vehicles backing into trees in parking areas, and what can be comprehensively called vandalism. In general most planting in public places will be done with as large a stock as is practical, thus shortening the time during which the trees are most vulnerable.

Prevention of damage from vehicles, for instance, can be largely controlled by planting trees an adequate distance from thoroughfares or by constructing kerbs or small stone walls between parking areas and where the trees are planted. In public picnic spots where vehicle entry is normally allowed, damage by vehicles is usually not serious, whilst the provision of barbecue sites encourages people to light fires only in approved locations, thus reducing the risk of grass fires or damage to trees by hot air from fires lit beneath the crowns or against the trunk.

Guarding against the worst forms of vandalism is more difficult and apart from having rangers on duty, involves public relations and education of the public—aspects which are outside the scope of this book. One suggestion is made: in the case of historical trees consideration should be given to the provision of guard rails at a sufficient distance to prevent compaction of the soil immediately around the tree and discourage access of those people who have an irrational urge to carve their names in the bark and often through the cambial layer into the wood.

15. *Tree breeding*

Outside the scope of the private planter is work in tree breeding, especially the selection of forms which are particularly suitable for public planting. The vegetative propagation of selected forms offers considerable attraction for street planting, since by this means it is possible to establish a uniformity of growth which is not possible with most seedling stock.

16. *The risk of introduced plants becoming weeds*

Australia, as a large land mass which has been physically separated from other parts of the world for a very long time and with a wide range of environments, provides numerous examples of plants introduced either deliberately or accidentally, which have adapted themselves too well to their new habitats and become weeds, often with a serious effect on the economy. Farmers and rural councils have been and still are involved in considerable costs in controlling some of the more aggressive introductions. The cost of blackberry (*Rubus fruticosus*) control alongside roads in Victoria was reported in the press (1970) as being $160,000 a year. Probably the most important example of a group of introduced plants becoming major pests are the prickly pears. There are several species, mainly belonging to the genus *Opuntia.* There is uncertainty about their early introduction but it is believed that they were brought into Australia as curiosities. Everist (1959) says:

> The important point for us is that they were introduced deliberately and in ignorance of what they might do in their new home.

It is believed that the plants were taken to Yandilla in Queensland from New South Wales about 1843. By 1884 Queensland opinion was apprehensive about the spread of prickly pears. Everist (1959) continues the account of subsequent development:

> By 1900 an area of 10,000,000 acres was affected and, following the breaking of the 1902 drought the plant went mad. Within 20 years the 10,000,000 acres had expanded to 60,000,000, of which half was heavily infested and the other half scattered and four-fifths of this area was in Queensland. New country was being invaded at the rate of 1,000,000 acres every year when *Cactoblastis* finally stopped the pest.
>
> The destruction of the pest was even more spectacular than its increase. Within ten years, *Cactoblastis* had brought under control all the serious infestations. . . .

Fortunately for Australia the prickly pears were very susceptible to insect control, but there are few weeds which would respond in the same way.

In the case of most weed species in agricultural crops, cultivation and the use of selective weedicides give satisfactory control. On the other hand, the spread of non-palatable or even species poisonous to stock on rough grazing country may present serious economic problems to property owners. Most of the introduced weed species in Australia are found in the higher rainfall coastal areas and only to a limited extent in the dry country, but there is always the risk that unwise introduction may result in plants, especially some of the hardier shrubs, becoming truly naturalised. If such shrubs are non-edible or poisonous this presents a serious problem, especially in the arid zone where low carrying capacity of stock does not justify high per acre expense in weed control. The mesquite of North America (*Prosopis juliflora*) is reported to have spread naturally in one locality of Western Australia.

When selecting trees and shrubs, particular care should be taken in the case of exotic species that there is little risk of them spreading as a weed. Not only is it illegal to smuggle plant seeds into Australia but there is always the possibility of the introduction of another major weed. If anyone wishes to experiment with growing new plants and to import seed he should not only make enquiries from the Commonwealth Department of Customs but ask agricultural and forestry authorities in his State for advice regarding the species concerned. Also, in many cases, enquiries would possibly reveal that such plants had already been introduced and, if so, whether they had been successful or not.

17. *Tree planting in salt land areas and some recommended species and establishment practices* (by A. J. Hart, Silviculturist, Forests Department of Western Australia).

It is now generally accepted that problem areas occur in the drier zones of Australia due to the occurrence of salt, either on the surface of the soil or in the subsoil. The latter condition is usually associated with irrigation regions, such as the Murray River citrus fruit and grape areas and, to a much less extent, some parts of south-western Western Australia.

The influence of salts, mainly those of Mg, Na and K, is causing concern in water catchment areas of the drier parts of the continent and, in such areas, is regarded as a pollutant. It causes severe loss of production and, in the Carnamah district of Western Australia, land which once yielded 8 bags of wheat an acre has now become non-productive. The original vegetation included York gum (*Eucalyptus loxophleba*) and teatree (*Melaleuca* spp.).

For many years the Department of Agriculture of Western Australia has been active in surveys of salt-affected land. On the basis of 16,248 observations over an area of 40,465,455 acres in 1955,* 181,556 acres or 0.45 per cent was seriously affected by salt. In 1962, on an area increased to 44,548,077 acres, 13,246 observations indicated that the area affected was 305,270 acres or 0.69 per cent. The increase in the seven-year period was 123,714 acres or 68 per cent. The fact that the total area under observation in 1962 was nearly 10 per cent greater than in 1955 has been allowed for in the determination of the increase of 68 per cent. The area mainly affected was in the Central Agricultural Division of southern Western Australia.

CAUSES OF SALTING

Observation in Western Australia is in accord with world-wide experience in the causes of salting:

(a) When natural vegetation is removed there is a reduction or cessation of the evaporation of moisture by plants, with the result that there is a rise in the water table, and as this water is evaporated any included salts are deposited on the surface of the soil.

(b) Where there is an impervious clay pan in an enclosed drainage system, run-on from adjacent ground transports both water and salts, thus increasing the total salt in the area.

(c) Any condition which results in a fluctuating water table brings salt solution to the surface and when the water evaporates, salt accumulates at the surface. This condition can also occur with incorrectly designed irrigation drainage systems.

* Taken from data published in *W.A. Journal of Agriculture*, 1956, (5) 1 and 1962, (5) 7.

TREE PLANTING INVESTIGATION, YUNA AND WONGAN HILLS, 1950

The earliest plantings of trees near saltland areas in Western Australia were made by the Forests Department in co-operation with the Department of Agriculture on private property at Yuna in the northern wheatbelt area in 1950, and at the Wongan Hills Agricultural Research Station at the same time. Spacing was 15 x 15 feet.

The experiment at Yuna was well documented and supported by soil analysis giving the pH and the total soluble salts found in typical areas. At 8 years of age there was 100 per cent survival and tree species used and their mean height at that age were:

1. *Eucalyptus brachycorys*	10 feet 2 inches
2. *E. sargentii*	23 ,, 3 ,,
3. *E. salubris*	11 ,, 9 ,,
4. *E. brockwayi*	13 ,, 7 ,,
5. *E. gracilis*	16 ,, 4 ,,
6. *E. camaldulensis*	14 ,, 10 ,,
7. *E. salmonophloia*	15 ,, 7 ,,
8. *E. occidentalis*	27 ,, 7 ,,
9. *Tamarix articulata*	11 ,, 1 ,,

By November 1963, some 13 years after planting, some losses had occurred but the more important species surviving were *Eucalyptus sargentii* (27 feet), *E. camaldulensis* (over 25 feet) and *Tamarix articulata* (15 feet).

The Wongan Hills planting was on heavier soils but with similar topography and a comparable rainfall (15 inches). At approximately 20 years of age the heights of the following species were as shown below:

1. *Tamarix gallica*	10-20 feet
2. *Eucalyptus cladocalyx*	50 ,,
3. *E. occidentalis*	35-40 ,,
4. *E. sargentii*	25-30 ,,
5. *E. spathulata*	25-30 ,,
6. *E. camaldulensis*	40-45 ,,

In the case of *E. cladocalyx* and *E. camaldulensis* at Wongan Hills, it was considered that the soil in which they were planted was not significantly affected by salt.

Of the species used in the two locations it was considered that the following are the most useful:

1. *Eucalyptus sargentii*
2. *E. occidentalis*
3. *E. spathulata*
4. *Tamarix gallica*
5. *T. articulata*

TREE PLANTING INVESTIGATIONS, AVONDALE AND WONGAN HILLS, 1967

Further experiments were initiated in 1967 with the objectives of determining the tolerance to salt (mainly Mg, Na and K) of additional species and to test establishment procedures. Soil conditions were recorded, especially zones of salt deposition and the amounts of salt. The species used were *Eucalyptus camaldulensis, sargentii, spathulata, cladocalyx, planchoniana* and *platypus* var. *heterophylla* and *Tamarix articulata.* Three types of nursery stock were used: normal potted stock, tubed stock in clear and black polythene tubes up to 15 inches long, and bagged stock in one and three-pound-type black polythene bags.

Two methods of establishment were adopted: mounded planting spots to raise the seedlings above the level of winter inundation, when salts are flushed away, thus minimising contamination of introduced soil 'gache', and ploughed mounded planting spots using long tubed stock and bags in order to place the roots in the profile where salt concentration was least.

The average annual rate of growth after 3 years was as follows:

Eucalyptus camaldulensis	2	feet	5	inches	to	3	feet	6	inches
E. sargentii	3	,,	2	,,	,,	3	,,	7	,,
E. spathulata	3	,,	0	,,	,,	4	,,	2	,,
E. platypus var. *heterophylla*	2	,,	10	,,					
Tamarix articulata	1	,,	4	,,					

An example of salt conditions in the trial plot at Wongan Hills Research Station is given in Table 18.1.

TABLE 18.1

SALTS IN SOIL OF TRIAL PLOT AREA IN WESTERN AUSTRALIA

(Hole No. 2 at the Wongan Hills Research Station)

Position in profile	*pH*	*Total soluble salts (percentage)*	*Sodium (Na) as NaCl (percentage)*
Surface	7.82	1.654	1.475
36 inches	8.54	0.184	0.160
54 inches	8.60	0.173	0.093

The salts were determined by electrochemical titration methods and percentages calculated on oven-dry weight of soil samples.

RECOMMENDED SPECIES AND METHOD OF ESTABLISHMENT

At the time of writing, the species currently recommended by the Forests Department for planting on salt affected land are:

Eucalypts

Eucalyptus sargentii
E. platypus var. *heterophylla*
E. spathulata
E. camaldulensis

E. occidentalis
E. kondininensis
E. gracilis

Other species

Casuarina glauca
Tamarix articulata
Callitris glauca

The best method of establishment appears to be the formation of a mound about 12 inches high by ploughing a year in advance and planting late in the season to avoid unnecessary waterlogging. Areas which have previously had superphosphate applied do not require additional manure, as the phosphate levels are usually adequate.

POSSIBLE RESULTS OF TREE PLANTING ON SALTED AREAS

Tree planting is not the only means of rehabilitating salt-affected areas, as many cases can be cited where suitable land management can reverse the process and restore land to near normal fertility; the latter may improve, depending upon the location and cause of salting. The surface accumulation of salt can be caused by a number of factors and the following are considered to apply in Western Australia:

(a) Overstocking of pastures with consequent denudation of the surface soil.

(b) Excessive clearing of tree and other vegetative cover along watercourses and drainage systems, which results in ground-water tables rising above those prevalent before clearing. This accelerates the evaporation-accumulation cycles of salts at the surface of the soil.

Degradation of the site is accentuated by the reluctance of many landowners to carry out the unspectacular and seemingly unprofitable rehabilitation measures necessary on affected areas. A striking example of the effect of tree planting on ground water is that at the Wongan Hills Agricultural Research Station, where a plot of about 7 acres was established in 1950 across the catchment of a well. In eight years the trees had so reduced ground water that the well dried up.

To restore an affected area to its pre-saltation condition it appears necessary to fence the area from normal land usage and either repasture, using fertilisers and making drains where necessary, or plant trees. The establishment of suitable relatively fast-growing species in windbreaks or distributed for shade will assist by:

(a) Reduction in wind velocity over the affected areas and, hence, reducing evaporation.

(b) Directly shading soil and thus lowering temperature at ground level, and so again reducing evaporation.

(c) Allowing incorporation into the soil of leaf litter, thus improving soil conditions.

The views of many people closely associated with the land in Western Australia was given by G. E. Brockway, a former officer of the Forests Department, in an Australian Broadcasting Commission talk in 1967 when he stated:

The degree to which trees and shrubs will tolerate pollarding and heavy pruning is an important consideration in the selection of species for street planting. Most broad-leaved species will tolerate heavy pollarding much more readily than conifers. The best solution, of course, is to use smaller sized trees and shrubs in street and other public places.

Top left: A very successful example of initial pollarding, followed by periodic pruning of the athel tree (*Tamarix articulata*).

Top right: A very fine example of sugar gum (*Eucalyptus cladocalyx*) in a watered park. This tree was probably pollarded at an early age.

Centre left: A small-growing eucalypt which has been lightly pruned from a suitable stage in the streets of Broken Hill, N.S.W.

Centre right: An example of heavy pollarding on an athel tree. This photo was taken in an arid area with only 7-8 inches of rainfall, but the tree is watered regularly.

Bottom left: These are English elm (*Ulmus procera*) in a southern humid zone town. Work like this is very expensive and only under favourable conditions do the trees respond satisfactorily.

Bottom centre: Another example of a good response to heavy pollarding by the athel tree.

Bottom right: This is an example of too large a species being planted beneath overhead lines; the best solution would be to remove the large eucalypt and replace with a small shrub.

Plate 84 POLLARDING AND HEAVY PRUNING OF TREES

It would perhaps have been possible to call this plate 'the misuse of trees in public places', but the result is ugliness, whilst in the case of the bottom right illustration it is cruelty to the young tree, which will later result in deformation and ugliness.
When power lines are placed over very large trees, as in the top left illustration, it is undoubtedly best to remove the existing plants and replace with smaller species. Centre left and upper right are examples of the abuse which river red gum (*E. camaldulensis*) will take and still survive.

Bottom left: Undoubtedly in this case the young tree was cut back at an early age, as evidence by the numerous vigorous branches growing starting about 5-6 feet from ground level. With regular light pruning this tree should remain of reasonably attractive appearance for many years.

Bottom right: This vigorous blue gum (*Eucalyptus* sp.) will be seriously deformed if the guard is not removed. Should there still be danger from straying cattle then the present guard should be replaced with one of much greater diameter.

Plate 85 UGLINESS AND CRUELTY

> I made mention in an earlier talk of the fantastic ability of our perennial native vegetation to extract virtually the last drop of water from the soil during the summer months.
>
> Millions of acres of this native cover has now been removed and replaced by annual cereal crops and pastures whose demand on soil moisture are less intense and certainly less prolonged. Much of this now unused water finds its way to the lower-lying areas and a rise in the water table there has resulted.

Perhaps the biggest factor involved in this problem is that, in the rural community, this affects to a large extent the very economy of the primary industries as the soil is the basis for the industry concerned, apart from the side effects to other users of the water catchment channels.

FURTHER READING

Champion, Sir H., & Brasnett, N. V. (1958). *Choice of tree species.* (F.A.O. For. Developm. Pap. 13). Rome.

Departments of Agriculture of all States have leaflets on insecticides and fungicides.

Hamilton, C. D. (1969). 'Tree Planting around the Homestead'. Chap. 4 in *Growing Trees on Australian Farms.*

Heather, W. A. (1969). 'Tree diseases'. Chapter 19 in *Growing Trees on Australian Farms.*

Meadly, G. R. W. (1956). 'Mesquite (*Prosopis juliflora* DC.)'. *Journal of Agriculture of W.A.*, 5(1), 65-76.

Tanton, M. (1969). 'Protection against insects and other animals'. Chapter 18 in *Growing Trees on Australian Farms.*

Whittet, J. N. (1958). *Weeds.* (Farmers Handbook Series). New South Wales Department of Agriculture.

Wilkie, A. E. (1968). 'Selecting Trees for Planting in Streets'. In *Proceedings of Symposium on Street Trees* conducted by the Australian Institute of Parks and Recreation, Victorian Division, La Trobe University, June 1968.

Further reading for section 17 by A. J. Hart.

Brockway, G. E. (1967). 'Tree Establishment in the Wheatbelt'. Talk No. 8, *Trees for Saltland Areas.*

Burvill, G. H. (1956). 'Land Clearing Control in the Eastern Wheatbelt'. *W.A. Journ. of Ag.*, (5) 1.

Burvill, G. H. (1956). 'Saltland Survey, 1955'. *W.A. Journ. of Ag.*, (5) 1.

Hart, A. J. (1968). Interim Report on Saltland Plots and Others. Internal report of the Forests Department of Western Australia.

Lightfoot, L. C. *et al.* (1964). 'Saltland Survey, 1962'. *W.A. Journ. of Ag.*, (5) 7.

Malcolm, C. V. (1963). 'Profitable Returns from Saltlands'. *W.A. Journ. of Ag.*, (4) 11.

Pennefather, R. R. (1950). *The Salinity Problem in Western Australia.*

Stoneham, T. C. (1962). 'Winter Management of Saltlands'. *W.A. Journ. of Ag.*, (3) 8.

(Note: The *Journal* of the Department of Agriculture of Western Australia is issued in series, so that there are several Volumes 5; for instance, that in 1956 is in the third series and that in 1964 is in the fourth series.)

CHAPTER 19

Plant propagation

by R. W. Boden

PLANT propagation may range from an interesting and absorbing hobby to a full-time highly skilled profession; it can involve a few plants of several species or the mass production of millions of plants of a single species, as in some forest nurseries.

The methods and techniques described in this chapter relate mainly to the raising of small numbers of plants for use on the property in homestead beautification, windbreaks, shelterbelts and woodlots. The techniques are also suitable for school or rural youth groups who may be involved in local tree planting projects in schools, parks and reserves.

Whether it is better to purchase plants from a reliable private or government nursery or to raise planting stock locally, depends on a number of factors which should be considered for each situation. Some of the advantages of purchasing nursery stock lie in the ability to select from a wide range of species, relative cheapness, and possibility of obtaining plants at relatively short notice. The last mentioned may not be important if a planting programme involving site preparation in advance is undertaken. The importance of early site preparation is discussed in Chapter 20.

There are many benefits which can arise from raising plants locally: nursery stock will be more readily available when needed; it will be acclimatised and should therefore suffer less transplanting shock; local species difficult to obtain from nurseries may be raised; the seed source or provenance is known; the costs may be lower; and the operation is mentally rewarding and satisfying.

Before embarking on a plant propagation venture it is essential to have an adequate and reliable water supply, suitable shelter, time to care for the plants, and an interest in nature. Although the amount of time needed to care for plants in a small nursery may not be great it must be available consistently—one day without water or shade at a critical stage of growth may mean complete loss and considerable disappointment.

PROPAGATION METHODS

SEED AND VEGETATIVE METHODS

There are two main propagation methods—seed and vegetative. The latter includes the use of cuttings, grafts, layers and budding and is common horticultural practice where large numbers of plants which are identical among themselves and to the parent are needed; common examples are roses, fruit trees, and attractive ornamentals. Some forest and fodder trees, including poplars and willows, can also be raised vegetatively to ensure uniform wood or leaf properties. In the case of many Australian native trees, such as eucalypts and acacias, consistently reliable techniques for vegetative reproduction have not yet been found and the grower must rely on seed propagation.

SEED SELECTION AND SEED SOURCE

When purchasing shrub and tree seed it is recommended that this be done either through a State forest service or other governmental agency, or from a reputable seed merchant. It is usually false economy to secure seed from the cheapest source unless there is adequate information on the locality (provenance) and type of plants from which the seed has been collected. When a species has a natural occurrence over a wide geographical range, which is the case with most native shrubs and trees of the dry country, it is highly probable that there will be considerable variation in growth if seed from all provenances is planted in one locality and treated similarly. This behaviour is well known in forest practice, where great differences occur between provenances of maritime pine (*Pinus pinaster*) and also Douglas fir (*Pseudotsuga menziesii*). Some provenances show slow growth and poor form whilst others show fast growth and good form. The botanical differences between provenances are usually not sufficient to justify the creation of new species and rarely of subspecies.

Less is known about most Australian native plants than some exotic species; but there is little doubt that as knowledge increases the importance of provenance will be revealed. In the case of mulga (*Acacia aneura*) which has a wide geographical range, mainly within the arid zone, it is known that there is a large number of growth forms and amongst other factors, trees differ in the presence or absence of substances poisonous to stock. In general, it is highly probable that any species which has a large range in latitude, altitude and the amount and incidence of rainfall will show greater variation than a species which has a restricted natural occurrence. For example, among the eucalypts of Western Australia, Dundas blackbutt (*Eucalyptus dundasii*) and Dundas mahogany (*E. brockwayi*) have a limited range within 100 miles of Norseman, combined with a small altitudinal range and a high degree of uniformity in climate. Progeny raised from seed collected from the natural habitats of these two species would not be expected to show the variation which occurs with seed of mixed origin of species such as river red gum (*E. camaldulensis*), which is found over most of Australia. Another species which shows considerable variation, complicated by hybridisation, is manna gum (*E. viminalis*) which, while it has a much smaller geographical coverage than river red gum, has a considerable range in altitude and climate. White cypress pine (*Callitris glauca*), for instance, has a natural occurrence which extends from Western Australia to within 100-150 miles of the eastern coast, and from the edge of Victoria to central northern Queensland. Mean annual rainfall for it varies between 9 and 25 inches, with the seasonal incidence from more or less uniform to a marked summer maximum. Some botanists also group the northern cypress pine in the same species, though it occurs in a tropical climate with marked summer monsoonal rain; whether this is taxonomically correct is not the concern of this book, but what is important is that progeny from seed collected, say, near Narrabri in New South Wales and near Darwin in the Northern Territory may show different growth rates and produce timber which is quite distinctive.

In addition to differences due to genetic variation within a species it can be expected that a species which grows naturally only in the arid zone will show differences in vigour and form when planted in a milder climate. This aspect is discussed in more detail in the chapter on factors in the selection of shrubs and tree species for the dry country.

SEED COLLECTION AND STORAGE

When a property owner wishes to raise seedlings of shrubs or trees which grow locally, he should consider the possibility of collecting seed. The question of identification of the parent plants is important, and if there is any doubt reference should be made to one of the numerous books on Australian plants, or material consisting of at least fruits and leaves (buds also in the case of eucalypts) sent to a State herbarium or other botanical agency. Local agricultural and conservation authorities can be helpful, especially with plants which grow within their districts.

Fruit should be collected from the larger and better-formed plants or, where age is known, from those showing faster growth rate as well as suitable form and resistance to insect and fungal attack. *Never collect from plants of unhealthy appearance or those apparently lacking in vigour*; sometimes such trees fruit more prolifically than healthier and more vigorous plants, but those with the heaviest crop are not necessarily the most desirable seed source. Since most trees are cross-pollinated it means that although selection is made of a better specimen for seed collection this only represents one of the two parents; the other parent could be of poor form and resulting seed could carry these characteristics. Where possible collect from a group of good trees rather than from a single well-grown specimen amongst a group of poorer plants. Some species produce a good seed crop almost every year whilst others are shy seeders and carry only a moderate crop at intervals of several years. This is well recognised in forest practice and where large plantations are planned, large quantities are collected in good seed years and stored under appropriate conditions for use in years of poor production. Although it is nearly always the fruit which is collected it is common practice to refer to seed collecting; the latter term will be used in this text.

The most suitable time of the year to collect seed will vary with the species and weather prevailing during the season, and also with the geographical location within Australia. The time of maturation and seed dispersal, as well as other information on species for the temperate humid zones is given in Appendix IV to *Growing Trees on Australian Farms.*

Seed contained in fleshy fruit can usually be collected anytime after the fruit ripens, but there may be risk of loss from birds or bats. On the other hand, seed contained in capsules (eucalypts), pods (acacias) or woody cones (cypress pine or, a better example, some of the exotic pines) should usually be collected as soon as the fruit is ripe but before they open and the seed is shed. This opening may occur within a very short time when a spell of warm weather is followed by a heatwave with accompanying very low humidity. It is often possible to enclose branches in a coarse fabric bag to catch the seed if premature shedding is likely. Polythene bags should not be used as they are liable to cause overheating. Following seed-shed, the capsules and cones may close as a result of cool, humid conditions. It is therefore desirable to examine such fruits before making a collection in order to check whether they do contain seed. In the cooler parts of southern Australia and Tasmania, pine cones may not open for several years after ripening but the seed viability will have fallen.

Small shrubs present no problems in access for seed collection and in the case of taller shrubs and small trees large individual fruit or twigs bearing small fruit can often be reached using secateurs or pruning saws on long poles. Collecting from tall trees can be difficult but several methods can be employed: one is to climb the trees, using a safety belt (the type used by P.M.G. linesmen is very suitable, or one can be improvised by sound *new* rope or heavy leather belts with suitable fastenings), and cut off fruit-bearing branches; another is to fell the trees required. The latter may result in far greater quantities of seed than an individual property owner may need and involves loss of the parent tree, but is the only practical method for large commercial collections, other than extensive lopping by experienced tree climbers. Where small quantities of seed are needed a skilled shooter can bring down small branchlets. Some success has also been achieved using a commando saw. In this method a light weighted line is thrown over the branch carrying seed and a commando saw with a heavier rope is then attached to it and drawn up onto the branch. With two people operating on the ends of the rope it is possible to bring down fairly large branches very quickly. Commando saws are available fairly cheaply from army surplus stores. This method is limited only by the height of the tree and the skill of the thrower. (Fig. 19.1.).

In the case of trees with small capsules, such as eucalypts, a watch should be maintained on the progress of ripening and as soon as there are any signs of initial opening the collection should be made; completion of ripening will occur in the capsules after harvesting. In the case of *Pinus* species the most recently matured cones are those of full size nearest the outer ends of the branches; those nearer to the trunk are older. It is usually easy to extract seed from a closed woody fruit. Placing eucalypt capsules or acacia pods, for instance, on a canvas or plastic sheet in a dry, airy room for a few days to a week or more will usually result in preliminary drying as well as completion of maturation of the seed. Then if the fruits are placed in a calm, sunny position, 90 per cent or more of seed will be shed within half a day. Do not place fruit directly on to metal which is exposed to the sun, since this may result in high temperatures which will damage the seed. Also the exposure of large moist, woody cones directly to sunlight under glass may cause 'case-hardening' which inhibits subsequent opening, whilst if seed is subjected to moist, warm conditions viability is usually rapidly destroyed. A small seed extractor can be made from boxes by placing the fruit on netting a few inches from the bottom and placing a sheet of glass or plastic over the top. The fruit should be shaken each day and the seed removed so that it is not exposed to high temperatures longer than is necessary. Fruit with only a thin fleshy cover encasing

FIG. 19.1

The use of a Commando saw in the collection of eucalypt seed.

It is often desirable to collect seed from individual *Eucalyptus* trees, and still preserve the parent for future collections. Various methods of doing this have been tried, but the use of a "commando saw" has proved the most successful to date, enabling the collection of quite large samples of capsules to be made relatively easily.

The "commando saw" is similar to a chain saw in both appearance and principle. Pairs of teeth are individually jointed, enabling the saw to be folded into a circle of 4″ diameter. Fully extended, it is 4 feet long with eyes at each end for attaching ropes or handles.

A weighted nylon cord is first thrown over the desired branch and the saw, with heavier lines tied at each end, is then drawn up. Using the handles provided as grips on these ropes, two operators can quickly bring down large branches. Although more difficult, one person can operate the saw by using an adjacent tree as a pulley.

The only limitation in the use of the saw is in the operator's initial casting ability with the light line. Branches sixty feet up are well within the amateur's range.

The saw can also be used in falling, or as a cross cut, but operates best when the pull is downwards and outwards, as when removing limbs.

The tool comes complete with handles, file and tooth setter, packed in a stitched leather pouch 7″ x 7″ with belt loops.

the usually single seed are best treated by drying in an airy but shady position; small quantities of fruit with a very thick flesh such as apples or oranges, can have the seed removed by hand and subsequently dried before storage.

Seeds borne in capsules, pods or cones are usually of a type which can be readily stored in tight, screw-top containers or sealed plastic bags under cool conditions. Most eucalypt, acacia and pine seed will retain viability for many years if the containers are kept in a room which is neither heated in winter nor unduly exposed to the sun in summer. On a commercial scale such seed may be stored under refrigeration at temperatures of 35-40°F and will last for many years. The air in the containers must be as dry as possible when sealed and also free from insects. The first can be obtained by packing the seed on a dry warm day in containers as small as possible for the given quantity of seed. In humid areas it may be useful to enclose small packets of silica gel similar to those used in film and photographic packaging. Some proprietary insecticides can be used to keep seed insect-free during storage. Dry powders rather than liquid preparations should be used. Seed fumigation is practised in large organisations but is difficult and dangerous for the inexperienced to use on a small scale.

NURSERY SITE AND EQUIPMENT

SITE AND SHELTER

There are several points to consider when selecting a site for a nursery. These points apply regardless of the size of the operations.

Firstly, it is desirable to have a level or slightly sloping, well-drained site. If sloping, it should have an easterly or north-easterly aspect to receive morning sun and should never face north-west, west or south-west. Shade or shelter is important in protecting young plants from wind, severe heat and frost. Shelter can be provided by locating the nursery near a windbreak, paling fence or building; shade by erecting a lattice frame or using one of the proprietary shadecloths suspended over a timber or pipe frame. Shadecloth is made with a range of mesh sizes which give precise shade intensities—one giving 40-50 per cent shade is usually suitable for most plants. Shadecloth must be fixed firmly between battens or by tying to prevent wind flapping which will cause wear. It can be attached to a roller to allow removal during mild weather, and when hardening the plants in preparation for field planting.

WATER SUPPLY AND APPLICATION

Both the reliability of supply and the quality of the water are important in nursery operations. Where there is any doubt about the latter factor the local officer of the Department of Agriculture is likely to have information on the subject. Saline water may harm young seedlings and should not be used.

Water can be applied in several ways—by hand using watering cans, under pressure from fixed sprinkler of varying types and under gravity or gravity plus pumping as flood irrigation.

Hand watering is slow and inefficient and should only be practised where few plants are to be raised or during the seed germination phase when light watering is needed to prevent soil wash and damage to young seedlings.

Fixed sprinkler irrigation can be of a wide range of types, depending on the scale of operations. For large open seed beds, fixed or movable aluminium piping with variable sprinkler heads similar to the equipment used in irrigated pasture and field crops can be adopted. In most cases the scale of operations will mean that a small fixed sprinkler attached to the shade frame might be suitable, or normal garden-watering equipment on a flexible hose. Water pressures of 80-100 psi are needed for successful operation of most sprinklers. Sprinkler heads are generally of the rotating type with relatively large droplets, or misting heads to give a fine spray. Misting jets are valuable in propagation but must be used in wind-protected situations for effective operation. Irrigation equipment suppliers should be contacted for advice on fixed and flexible sprinkler installations.

Flood or sub-irrigation watering is an age-old practice which is still employed. In nursery work it has considerable application as it is cheap, efficient, effective and encourages deep rooting. It also has the advantage that it can be used where water pressures are low. Basically the principle is that plants in containers are placed in shallow waterproof tanks either of concrete or lined with polythene film. Tanks should be designed with either a side or central drainage channel deeper than the tank base to give rapid drainage and prevent water lying around the bases of the containers. The depth of flooding is determined by the size of plants, soil mixture and height of containers. With established plants it is desirable to flood to a depth so that water is lifted by capillary attraction to within 1-2 inches of the soil level. By keeping the upper soil surface dry, problems of collar rot and excessive weed growth can be reduced.

PROTECTION

In country districts it is important to locate the nursery in an area protected from rabbits, hares, stock and domestic poultry which can effectively and quickly destroy the contents of a seed box. Rats and mice will also dig for buried seed, and possums are very partial to young seedlings.

Lastly, the nursery should be sited in an accessible area where it is passed several times a day during other farm operations. The need for additional watering, shading or protection will not then be missed. In the young stages, in particular, plants are very sensitive to desiccation and should be checked at least twice daily.

SOILS AND POTTING MIXES

When seedlings are raised in open seed beds prior to field planting it is important to have a well-cultivated freely-drained loamy soil, which neither dries out too rapidly or becomes hard and compacted with watering. The incorporation of organic matter such as sawdust, straw or shavings will improve the structure and water retentive properties of the soil. Where such materials are used it is necessary to provide nitrogenous fertilisers such as urea, blood and bone, or farmyard manure to compensate for nitrogen used by soil micro-organisms in breaking down this added organic matter. If time permits a green manure crop of peas or lupins can be sown some months before the seed bed is needed, and cultivated as part of the seed bed preparation.

If conifers (for example, native cypresses and true pines) are to be raised for the first time it is advisable to inoculate the seed bed with the appropriate mycorrhizal fungus. These fungi grow in conjunction with the seedling roots and assist in plant nutrition. Leaf litter gathered from beneath mature trees of the species to be grown should be spread over the bed and dug in. Two or three seasons may be necessary for inoculation to become complete.

Where weeds, soil nematodes or fungal pathogens are persistent, nursery beds may be fumigated to achieve partial sterilisation. Methyl bromide is one chemical which can be used alone or in mixture with up to 30 per cent chloropicrin applied as a gas beneath a plastic cover, or by injection, when soil temperatures are at least 50°F and preferably above 60°F, and when the soil is in a moist but not wet condition. The covers are left in position for two days or more and sowing should not be carried out for a fortnight. Application rates vary from less than 1 lb to 4 lb per 100 square feet, depending upon soil texture, temperature and the purpose of the treatment. *Methyl bromide is very dangerous and must be handled with care* and the manufacturer's instructions followed at all times.

Soil for use in containers, pots, tubes, etc., should be prepared to John Innes or University of California recommendations. The former recommends proportions of sand, loam and peat moss with added fertiliser; the latter relies entirely on sand and peat moss with fertiliser—no soil is included. Details of mixtures and preparation for both types are given in *Seed and Potting Composts*, by W. J. C. Lawrence and J. Newell, and *The U.C. System for Producing Healthy Container Grown Plants*, ed. K. F. Baker. University of California.

John Innes Seed Compost

- 2 parts loam (by volume)
- 1 part peat moss ,, ,,
- 1 part clean sand ,, ,,

To each cu yd is added

- 2 lb superphosphate
- 1 lb ground limestone

John Innes Potting Compost

- 7 parts loam (by volume)
- 3 parts peat moss ,, ,,
- 2 parts clean sand ,, ,,

To each cu yd is added

- 5 lb John Innes Base
- 1 lb ground limestone

The John Innes Base consists of 2 parts by weight hoof and horn meal (13 per cent nitrogen), 2 parts by weight superphosphate, 1 part by weight potassium sulphate.

The U.C. mixes cover a wide range of sand/peat combinations, but the following should suffice for most tree and shrub propagation:

- 2 parts peat moss (by volume)
- 2 parts coarse sand
- 1 part fine sand

A fertiliser mixture containing:

- $2\frac{1}{2}$ lb superphosphate 22 per cent
- $2\frac{1}{2}$ lb blood and bone
- 4 oz potassium nitrate
- 4 oz potassium sulphate
- $2\frac{1}{2}$ lb calcium carbonate (lime)
- 8 oz magnesite

is added to each cubic yard of soil mix. Trace elements are also needed for healthy plant growth and can be provided by incorporating a proprietary trace element mix at the recommended rate, or by using a complete soluble fertiliser in the liquid feeding programme during the growing stages of the plants.

It is essential that the sand and peat are well mixed before adding the finely-ground fertiliser, which should be made up freshly for each soil batch. Peat moss should be sieved through a $\frac{1}{4}$-inch sieve and moistened to reduce dust problems.

Both soil mixes should be treated either chemically or with heat to control weeds and soil-borne pests prior to use. Heat treatment in large nurseries is usually carried out with a steam and air mixture at 140°F for 30 minutes and this is commonly called pasteurisation. The equipment to carry out this treatment may not be readily available and where only small quantities of soil are required the home oven can be used. Soil is spread in thin layers 1-1½ inches thick in flat trays and heated for ½ hour at 180°F. The soil should be moist. If larger quantities are needed it is necessary to resort to methyl bromide treatment as described above. Fungicidal soil treatment using Difoliton has proved useful where some plant diseases are prevalent.

PLANT CONTAINERS

A wide variety of containers can be used including tins; cans; paper cartons; clay pots; metal or wood veneer tubes; and polyethylene bags.

Of these the clay pot and beer can are the most familiar but also the most unsuitable. The clay pot loses moisture readily, is heavy when wet, expensive and breaks easily and the tapered shape encourages root spiralling. The beer can is cheap and has quite good proportions; however, the difficulty of removing the plant often results in root damage.

The two types of container recommended and used widely by nurseries and forestry authorities are the metal or wood veneer tube and the polyethylene bag. The metal tube is a piece of flat tin 6 in. x 8 in. with two flanges 1-2 in. long at

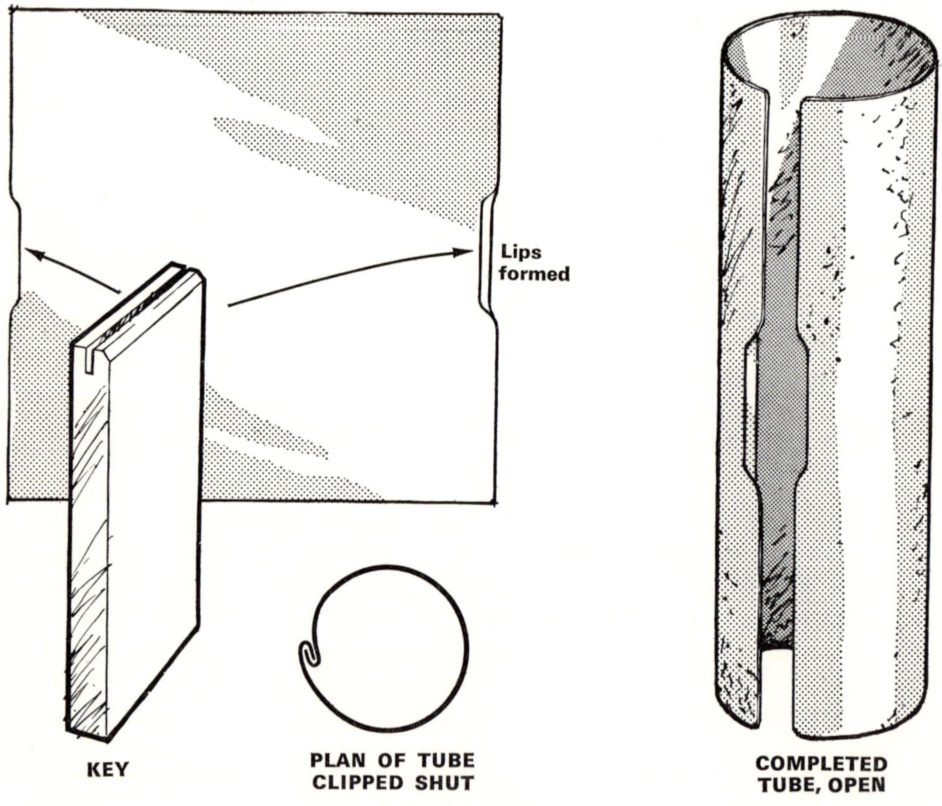

FIG. 19.2

Steps in making tubular metal plant containers from galvanised iron sheet or tinplate. These tubes are being displaced, for raising eucalypts, by peat pots in which seed is sown directly and which are planted in the ground with the seeding.

the centres of the 8 in. sides. The metal is rolled around a galvanised pipe to form a tube and the flanges bent to clip together. The tube has good proportions of width to length, encourages deep rooting, is light in weight, can be easily removed in the field and re-used. A similar tube can be made from wood veneer rolled around a pipe and held by staple or rubber band. The wood veneer tube is cheap, light and easily removed, but it can seldom be re-used. Some workers consider that the wood veneer tube can be left on at planting as it will eventually decay in the soil. This is not recommended unless the tube is in a decaying state at planting, as intact tubes tend to form thin air pockets and barriers to lateral root movement during the critical establishment phase.

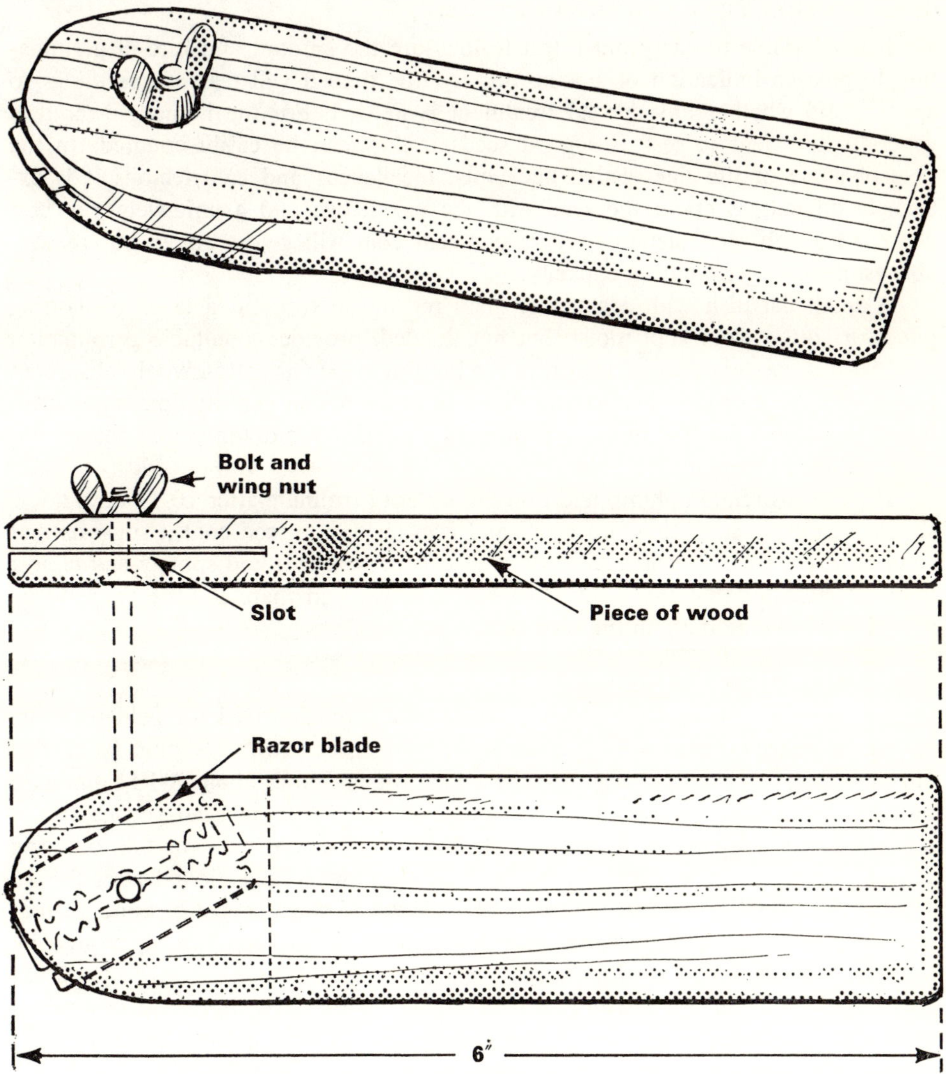

FIG. 19.3

Cutting tool for removing plants from polyethylene bag containers (design V. Russell).

The polyethylene bag has many advantages including low cost in initial price, shipment and storage, and is easily removed at planting. It is slow to fill with soil mix, but once filled can be handled without losing the soil which occurs with the bottomless tube. Polythene bags 2 in. x 2 in. x 8 in. are suitable for most one-year seedlings. They must be kept upright in baskets or boxes or where small numbers of plants are raised they can be stood in reinforcing mesh. The bags are cut and removed at planting—cutting should be carried out with a tool which does not penetrate deeply and cut the roots. An example of such a tool is shown in Fig. 19.3.

RAISING PLANTS TECHNIQUE

GERMINATION TEST AND SOWING TIME

It is advisable to carry out a test to measure the amount and rate of germination to give an indication of sowing density, the time it will take for seedlings to appear, and whether any special treatment is needed. For the home grower it is generally sufficient to test 100 good seeds if they can be easily counted. In the case of some of the fine-seeded eucalypts, melaleucas and callistemons it is extremely difficult to count the seed with the naked eye and a unit weight is best chosen. Enough seed just to cover a one-cent coin will give a rough but reasonable estimate of germination capacity.

A small flat dish with a sheet of glass resting on top and a layer of blotting paper or cotton wool kept moist, but not flooded, provides a suitable germination tray. Seed is spread or sprinkled on to the blotting paper or cotton wool (absorbent type), and the dish kept in a warm, dimly-lit place. When germination commences it will be noticed that the number germinating per day rapidly increases for several days and then tapers off.

If the seeds remain plump and firm but do not germinate after 10-14 days a few should be cut open. If they are white and firm inside the seed coat it suggests that some seed pre-treatment is necessary, whilst if they are soft and spongy and almost liquid it indicates that the seeds have begun to rot and there is little value to be gained from sowing these in the seed box.

TIME OF SOWING SEED

The sowing time is determined to some extent by the need for field planting. As the latter will generally be in autumn or winter, seed sowing should be carried out in spring, for example August-September, which allows about 8-10 months to produce a plant big enough to go to the field site. Autumn seed sowing is not recommended as generally it means that plants will be too small for the following winter planting, and will have to be held in the nursery and looked after for 15 months or more. This will mean restricting growth to suit the size of the container or transplanting to a larger one. The aim should be to keep the plant growing steadily, without major growth checks, throughout its life; to this end it should not be held too long in the nursery. *When buying plants never purchase any which have obviously been in the nursery for a long time or have a large amount of top growth in proportion to root volume.* Ideally the height of top should not be greater than twice the height of the container.

SEED PRE-SOWING TREATMENT

Seed pre-sowing treatment may be necessary for two main reasons. Firstly, to hasten germination where seed dormancy is present, and secondly, to protect the seed from insect and fungal attack during the germination period.

Seed dormancy occurs in many species and is a mechanism which prevents germination from occurring as soon as the seed is ripe. The advantage of this in the wild is that seed which might germinate in autumn or early winter when seedlings would face conditions of severe cold, are prevented from doing so until spring when conditions are more favourable. In arid areas some species will not germinate until there has been a substantial fall of rain which leaches chemical inhibitors from the seed coat; this requirement prevents germination occurring at times when soil moisture conditions for growth would not be suitable. In some species dormancy occurs only in some of the seeds of each crop, thus ensuring that germination of the species as a whole will be spread over a long period of time.

Dormancy is generally due to conditions of the seed coat or the embryo itself, or a combination of both. Where the seed coat is involved it may be hard and impermeable to water and oxygen, or contain chemical inhibitors to germination. Where the embryo is involved it may require an after-ripening period even though the fruit itself is apparently ripe. Induced or secondary dormancy is a condition sometimes brought about by unfavourable conditions being experienced during the germination process. A knowledge of seed dormancy and the methods to overcome it is important in horticulture, as it enables the propagator to obtain an even and rapid germination in the seed box or nursery bed.

The best known form of seed dormancy in Australian native plants from arid areas is the hard seed coat common to most acacias and other leguminous plants. To obtain rapid and even germination it is necessary to crack the hard coat either by filing, abrasion, acid or hot water. The last method is the simplest and most efficient and involves placing the seeds in near-boiling water and allowing them to cool for about 12 hours, or overnight. Seed should then be sown immediately after removal from the water.

Species which grow naturally in arid regions with cold winters, generally benefit from cold moist stratification for periods of from 2-6 weeks prior to sowing. Seed is mixed with moist but not wet, sand or sand/peat and stored at about 38°F in a polythene bag for 2-6 weeks, prior to sowing.

Chemical inhibitors which retard germination are important in the ecology of some desert plants, but unfortunately little is known at this stage, of the pretreatments required for many Australian species native to arid areas. It is believed that genera such as *Pittosporum*, *Eremophila* and *Flindersia* have particular requirements.

SEED PROTECTION

Protection from fungal, insect and other pests which attack the germinating seed may be required. Fungicidal coatings of proprietary materials containing thiram are useful and relatively easy to apply. The main fungal problem likely to be encountered is damping off, and the most successful treatment is prevention by using sterilised soil and containers, and maintaining good growing conditions.

Insect problems are largely associated with ant removal of seed during the germination phase. Proprietary insecticides can be applied to the area around seed boxes but should not be dusted or sprayed on to young seedlings.

Birds, mice and possums may disturb seed beds and boxes in a search for seed, and here the most effective control is to cover the box with fine wire netting.

SEED SOWING METHODS

Prior to sowing, the seed box or pot should be filled with a finely-sieved weed-free soil mixture. The method of sowing depends on the size of the seed. Large seed can be placed singly on the soil surface, whilst small seeds such as eucalypts are sprinkled finely over the soil surface using a V-shaped piece of cardboard and tapping gently with the finger, or adapting a salt shaker for the purpose. It is important to remember that there are many seeds per ounce in some of the fine-seeded species, for example *Eucalyptus microcarpa* (grey box) may range from 5,000 to 30,000 per ounce. In order to avoid fungal problems developing seed should not be sown densely. Fine seed should be pressed into the surface of the soil with a flat board, taking care not to let the seed become stuck to the board when it is lifted, and barely covered with a thin layer of sieved sphagnum or peat moss. Larger seeds may be covered one to two times their minimum diameter.

Most seed sowing is best carried out in spring when days are warm but not hot. The most suitable temperatures for many species is in the range of 68-70°F. Where nights are cold, seed boxes should be covered with hessian or even brought into a shed. Seed boxes should be kept moist but not wet, and protected from the direct rays of the sun, and from rain which might wash the seeds out of the box. If hessian, paper or cardboard is used it must be removed when seeds start to germinate.

Germination usually occurs within a week to ten days for many species; however, others may take several weeks, particularly if seed pre-treatment requirements have not been met adequately. As seedlings develop they should be kept moist and gradually exposed to more direct sunlight to prevent them becoming drawn up and spindly. Frequent light waterings are essential at this stage as the seedlings' root systems will be only poorly developed. Overhead watering should be avoided late in the day as fungal problems can be more prevalent if leaves of seedlings are wet during the night.

Damping-off, which is a fungal disease, sometimes becomes a problem, particularly if seed boxes and the sowing mix have been contaminated. Damping-off, which is recognised by browning of the stem near the soil surface and collapse of the seedling, will often commence in a small patch and spread rapidly. Control is difficult and results variable: proprietary fungicides containing thiram, or captan can be used, but the secret to damping-off control is prevention by using sterilised soil and seed boxes and pre-treated seed.

When seedlings have reached the first or second true leaf stage they should be pricked out of the seed box into containers such as tubes, bags or pots. During the pricking-out stage it is important to prevent the seedling roots from drying out at any stage and to handle the seedling by the leaf, not the stem. Great care should be taken when placing the young seedling root in the soil mix to avoid bending or twisting it. Once twisted the root may stay in this position throughout the life of the plant and severely retard its growth. If the root is too long for the container it is far better to cut it short than to twist it to fit the container. Soil should be pressed firmly around the seedling, and the container thoroughly watered to remove air pockets. For the first few days whilst the seedlings are becoming established the container should be shaded and watered lightly at regular intervals.

Direct seeding into nursery beds or containers is a technique used to overcome the transplanting shock and set-back which always occurs, no matter how great the care taken during pricking-off. Direct seeding also allows the root system to develop without the possibility of twisting during potting. Many drought-hardy species have an early high root-shoot ratio, that is the root grows much longer than the top in the young stages. This is partly a survival mechanism to assist in the establishment of the young seedling and, where possible, it is best to allow it to develop unhindered.

The main disadvantages with direct seeding are that there are more containers in need of constant care at the one time, and a knowledge of germination percentage is important to ensure that each container will eventually have at least one good strong seedling. If these disadvantages can be overcome, direct seeding is strongly recommended to aid in rapid establishment without check to growth.

Seed-bed and line-out propagation can be practised for some species but, generally speaking, it is better to have a plant in a container to avoid having to bare the roots when moving the plant to the field site. One major exception to this rule is in the propagation of kurrajong (*Brachychiton populneum*) by the Henley method, so named after its originator. It has been found with kurrajong that seeds are best sown in a seed bed and when plants are about one foot high they are transplanted to rows in a nursery. The long carrot-like root should be trimmed at transplanting stage. Plants are grown for several years in the nursery, transplanting when necessary if crowding occurs. Kurrajong can be grown to several inches in diameter before transplanting to its permanent site in the field, lopping back the top and root on each occasion in order to encourage the development of a strong trunk.

During the growing stage in the container, seedlings should be watered regularly, either by flood irrigation or overhead watering, depending on the irrigation method adopted. Liquid feeding with a proprietary soluble plant food applied in the irrigation water is desirable at regular intervals to maintain growth. Generally once a fortnight is sufficient but the frequency must be adjusted to suit the growth of the plants. It is important that the top growth does not develop at the expense of the root system and become soft and weak in the stem.

Occasionally fungal or insect attack may occur in the seedling stage. Mildew is common in humid weather with some species and can generally be controlled by using phaltan and watering to avoid water coming in contact with the leaf surfaces. If caterpillars or aphids become a problem it may be necessary to apply an insecticidal spray, taking care to follow instructions explicitly. Use only insecticides recommended for plants and not those used for insect control on animals.

Weed control in the nursery bed or line-out area is important in producing healthy plants. Weeds compete for water and nutrients and may act as hosts for insect pests or diseases. In container culture of plants, weed control is achieved by using sterilised soil, but in the open bed other means may be necessary. Where relatively few plants are being grown, hand chipping of weeds is advised, whilst in large nurseries pre- and post-emergent chemical weedicides can be used. The use of chemical weedicides is a somewhat complicated procedure where weed control without damage to desirable species is required. Tree species vary in their tolerance to different herbicides and it is advised that the guidance of local forestry or agricultural officers be sought before embarking on a chemical weedicide programme.

Vegetative propagation has less application in raising plants for the dry country than seed propagation; however, some species can be raised from cuttings with a degree of success. Trees which can be raised from hardwood cuttings are those which are dormant in winter, for example poplar, willow and tamarix. Evergreen species can seldom be grown successfully from cuttings without glasshouse and propagation conditions which provide fairly precise control of water and temperature. Hardwood cuttings are taken in winter and are usually from wood $\frac{1}{4}$-$\frac{1}{2}$ in. in thickness, about 8 in. long and trimmed to just above and just below a bud. Hardwood cuttings taken in winter are stored in moist straw or sawdust to allow them to form wound tissue on the cut ends prior to lining out in early spring. Soil to receive cuttings should be thoroughly and deeply cultivated. Cuttings are placed in a narrow slit trench and pushed into the soil to leave only one or two buds around ground level. The soil is firmed back around the cuttings and the whole bed watered. Apart from trimming to a single stem if more than one bud develops, and keeping the bed well watered and free of weeds there is relatively little maintenance required.

FURTHER READING

Hartmann, H. T. & Kester, D. E. (1968). *Plant Propagation; principles and practices.* 2nd ed. Prentice-Hall.

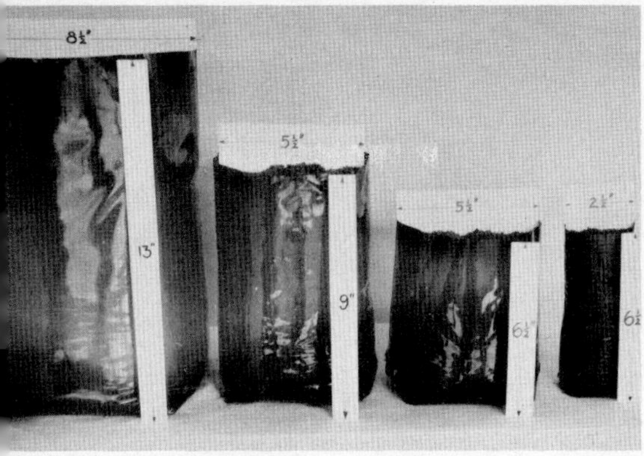

Top left: Polyethylene bags suitable for use as seedling containers.*
Top right: A range of peat pots with eucalypt seedlings. The pots decay after planting.

Centre left: Polyethylene bag cut away in preparation for field planting. Note the vertical rooting and the lack of spiralling.*

Bottom left: An example of lack of balance between the root system and the above ground part of the plant, due to the pot size being too small.

Centre: Spiral root system developed on *Eucalyptus melliodora* grown in a tapered terra cotta pot. The seedling died one year after it was planted in the field.*

Bottom centre: Spiral root system developed on *Eucalyptus stjohnii* grown in a 4-gallon metal drum. The plant survived 3 years in the field before dying of root strangulation.*

Bottom right: Sterilisation of the soil used for raising seedlings often results in more and better plants. *Eucalyptus regnans* 3 months old. Soil unsterilised on the left; sterilised on the right.†

(* *Photos R. Boden.*
† *Photo A. G. Brown.*)

Plate 86 THE GROWING OF NURSERY STOCK—I

Top left: Shadecloth in position over young eucalypt seedlings. The cloth is held tightly at each end in order to prevent sagging.†

Top right: The same frame with the shadecloth removed to show framework and seedlings.†

Upper centre left: Seedling bays facing north or north east, with hessian sides and provision for overhead hessian cover (slatted frames could be substituted).

Lower centre left: Timber or aluminium slatted cover to give partial shade. Solid side and back walls to protect seedlings from wind.

Bottom left: Details of overhead wooden slat construction; width of slats about 2 inches, spacing can be varied according to the degree of shading required (in this case about 1¼ inches).

Bottom right: Polyethylene bags filled with soil mixture and placed in sub-irrigation trays in preparation for direct seed sowing.*

(* *Photo R. Boden.* † *Photos V. Russell.*)

Plate 87 THE GROWING OF NURSERY STOCK—II

The availability and quality of water, as well as finance, decide whether the operation can be done. With the exception of the bottom left illustration, the photography was done at Woomera, in the arid zone of S.A. and a mean annual rainfall of about 7½ inches.

The two top illustrations and the top centre left show the basin which can be made around a tree to hold water, whilst the top left shows a perforated tube set 18-24 inches into the ground to assist in seepage.

Lower centre left: In this case there is a four-row shelter belt and it was found practical to reticulate the area with a 2 inch pipe. Were it not for the risk of vandalism it would have been more economic to have used plastic piping with a small hole drilled opposite each tree, and the flow controlled by a tap.

Bottom left: This was taken in a town where the mean maximum temperature of the hottest month is 93·3°F. The two drainage pipes are good but undesirable is the placing of a heavy layer of bitumen almost right up to the stem of the tree.

Bottom right: The re-use of water is of great importance in the arid zone. Here the froth is due to the domestic use of detergents which do not readily break down and are, in any case, very undesirable in water which is to be used again for any purpose whatever.

Plate 88 REGULAR WATERING OF ORNAMENTAL TREES IN VERY DRY COUNTRY

CHAPTER 20

Tree establishment

THE success of tree planting up to the end of the first year can be judged by a high percentage survival and initial vigorous and healthy growth. Factors which are controllable by man, additional to the wise selection of species for the environment, are suitable preparation of the planting site, the use of appropriate nursery stock or seed and the necessary protection of plants from harmful agencies.

CLEARANCE OF VEGETATION

Preparation of the planting site includes the clearance of competing vegetation, cultivation of the soil and, if necessary, the addition of fertilisers, including minor elements. The object in the control of existing vegetation is to ensure that the young trees do not have competition for soil moisture, nutriment and shade from other plants. The amount of vegetation to be removed will vary with the environment and the type of tree planting proposed; in the sparse natural vegetation of the arid zone this may not involve more than the removal of scattered shrubs and small trees located close to the planting site. In this very dry area, however, it will usually be better practice to retain all existing shrubs and to allow for this in the design of new planting. In higher rainfall areas where the existing vegetation consists of moderately dense scrub or open forest it is usually desirable to fell or root out all growth for a minimum distance from the new planting of once to twice the height of the natural vegetation. The trees can be felled by hand or with chain saws, lopped and allowed to dry for a few months before burning shortly prior to the commencement of the planting season, or they can be removed completely; in the case of trees up to 6-9 inches diameter this can be done with the usual mechanised farm equipment available; larger growth may call for heavy bulldozers, with a root rake and tree bar, which are capable of pushing over trees two or more feet in diameter. This type of operation is illustrated in Plate 34 of *Growing Trees on Australian Farms* and described in Chapter 12. Inexperienced plant operators should take great care against the risk of branches, especially dead ones, falling on them when pushing down trees. For heavy timber operations it is necessary to have a strong canopy over the tractor in order to protect the operator from falling branches.

In the arid zone grass is usually so sparse that competition is not serious but wherever there is a dense grass sward this should be destroyed for one to several feet from an individual planting spot some two to six months prior to planting. Where it is very dense, as with improved pastures in the higher rainfall parts of the semi-arid zone, it may be desirable to cultivate the site two or three times before planting, so that all grass roots are destroyed.

Trees may be killed by girdling or poisoning. This practice can be initially economical in the case of the occasional very large specimen, for example a 4-5 feet diameter red gum, but subsequently as the branches dry and eventually drop there can be resultant damage to young trees and also the possibility of serious injury to man and beast. This latter risk is greater than is generally acknowledged and within recent years more than one person has been killed, one fatality being due to a branch from a large river red gum. There is also an additional risk in the fire hazard and it is recommended that girdling should not be done adjacent to farm buildings and wherever there is a concentration of men or animals. Girdling means cutting through the bark to sever the cambium immediately beneath, preferably including one or two inches of the sapwood; the severance must be completely around the tree since even a small strip of active cambium may be sufficient to keep it alive. Trees may also be poisoned by girdling them with downward cuts to form a channel which will hold a liquid, and applying either an arsenical preparation or one of the modern weedicides. The use of chemicals have a role in large forestry operations but their application is not recommended on farms, especially where there is a risk of access by animals or children. In addition, it should be noted that the use of certain chemicals are prohibited in some States but in others only within water catchment areas. Further information on weedicides is given in *Growing Trees on Australian Farms* (Chapters 12 and 18 and Appendix III).

The chemical control of ground vegetation does not present the same safety problems as the poisoning of large trees, especially if grazing animals can be precluded from the area for a few weeks immediately after application. The time of spraying will depend upon local conditions; applications are most effective when plants are showing maximum growth activity and this period is largely influenced by local moisture and temperature conditions, as well as species of plants to be controlled. In most of the southern semi-arid country the season is likely to be spring or autumn, the latter being the more suitable where winter planting is the practice. In the northern monsoonal rainfall areas the appropriate period would be at the start of the monsoon, but is not likely to be as necessary, due to the natural drying of the ground vegetation during the very marked dry season. The use of sprays in the arid zone is only likely to have limited practical value. Application after planting presents the problems of preventing spray falling onto the young trees or of using a selective weedicide which may kill the weeds but have a retarding effect on the growth of the desired young trees or shrubs. Plate 36 of *Growing Trees on Australian Farms* shows the use of an attachment to avoid spray on young trees.

SOIL CULTIVATION

On all soils, other than deep sands and sandy loams, some form of pre-planting cultivation is desirable for shelterbelts and ornamental tree establishment. There are numerous methods of treatment, some of which are suitable for small-scale operations on farms and others which require heavy mechanical equipment. Surface cultivation usually combines control of grass roots and ground vegetation with preparation of the surface layers of soil. The most often required form of cultivation on farms, especially where there are heavy grass swards, is the ploughing or rotary hoeing of strips 4-6 feet wide for single row shelterbelts or the complete area in the case of multiple-row belts or plantations. When only widely-spaced single trees are involved the use of a rotary hoe or hand digging may be satisfactory. In the latter case the combination of a weedicide over an area 4-6 feet in diameter, and a pit 12-24 inches in diameter and 9-15 inches deep may give the desired results.

Pits for planting ornamental trees in public places usually call for greater size, since the plants used may be advanced stock possibly 6-8 feet high for evergreens and up to 15 feet or more in height for deciduous species. For these it may be desirable for the pit to be 2-3 feet deep and of at least equal diameter. For the largest pits a back-hoe is effective, preferably working in two directions at right angles to give a large '+'; this allows for expansion of the root system which, in turn, promotes the wind stability of the tree and allows for the fine roots to draw moisture from a greater volume of soil. Where back-hoes are not available the use of post and strainer-hole borer attachments on farm tractors can be very effective, especially if three or four holes are made adjacent to one another. When the soil is replaced in the holes it may be desirable to add a quantity of a sandy loam if the *in situ* material is of a heavy texture and, in any case, care should be taken that the original top-soil is not placed at the bottom of the hole and beyond reach of the roots of the young tree when first planted. When heavy soils and clay are replaced care should be taken that large voids are not left in the hole; a mixture of sand or a good sandy loam with the clay will help obviate this.

In all cases where the soil is stony or where an impervious pan is present, for example a lateritic layer, deep ripping is desirable and may be essential for satisfactory tree growth. Heavy track-type tractors of 100 horsepower or more are needed to pull deep rippers, but when farmers decide upon extensive shelterbelt planting where ripping is necessary it may be possible to get the work done on contract from a local authority or road contractor. Deep ripping, that is to 18-24 inches, to break up a pan not only allows roots access to a greater volume of soil but permits better recharge of soil moisture from rain which would otherwise be lost as run-off.

THE ADDITION OF FERTILISERS

With the exception of skeletal soil types most of the soils of the arid and semi-arid zones are at least moderately fertile by standards for tree requirements. In the latter zone most of the cleared land represents the better land in any district, and may not need the addition of more than small quantities of fertilisers in the early years of tree establishment. Compared with this most of the land available for commercial tree plantations in the sub-humid and humid zones is comparatively poor and may be deficient in both major and minor elements; of the former, phosphorus and calcium are the most likely, and of the latter, zinc, boron and magnesium. Most farmers will have knowledge of conditions on their own farms, especially if they engage in crop production, but if not, information should be sought from the local agricultural officer. Where there is no detailed knowledge of specific deficiencies on a particular soil type it is wiser to either apply the manures which are found satisfactory for crops on the same soil type or to use a balanced mixed fertiliser, possibly one containing nitrogen in a slowly available form, for example John Innes Base. Where nitrogen is not included then superphosphate, with minor elements appropriate to the district, is a good choice. The application of large doses of artificial fertilisers in dry country may prove harmful to trees, since it can lead to higher ion concentrations near the roots than in an area of high rainfall, for example thirty inches or more. Allowance should also be made for soil salinity.

Fertiliser should not contact roots of newly-planted trees; if applied at the time of planting it should be mixed with a large quantity of soil, put some inches below the roots or placed in a slit some inches away from the roots. The broadcast application of fertilisers on the surface is not recommended since it becomes available to other vegetation which may compete with the young tree; where grass is a problem, broadcasting of fertilisers is particularly undesirable, although it may produce quite good results when the trees are several years old and canopy is almost established in the case of plantations. When applying fertilisers to young trees at planting, only small quantities should be used, for example as little as ¼ oz. of John Innes Base or 1-2 oz of superphosphate. The best way to go about the work is to carefully weigh out dosage and relate this to a teaspoon or other suitable container for subsequent visual measurement. Reference should be made to Chapter 16 of *Growing Trees on Australian Farms* for further information on soil deficiencies and to Table 16.2 for composition and cost of important fertilisers likely to be used for trees.

WATERING

The watering of newly-planted shrubs and trees is usually essential to ensure successful establishment in arid zones and is often desirable in semi-arid areas. Once established, the continuation of watering may be desirable if better than normal early growth is desired, as in some street and park planting where moderate size is wanted as soon as possible either for the sake of appearance or to ensure that the plants quickly pass the stage at which damage by vandalism is most likely. In the context of this book watering will be considered mainly with regard to ornamental planting and the establishment of trees for shade or in shelterbelts for stock and crops. The techniques and economics of watering with timber production as the main objective, a condition which is usually only justified when the wood has a very high economic or social value and where water can be obtained cheaply in the large quantities needed for such purposes, will not be considered. Techniques discussed will be those where either measured quantities of water are applied to individual trees or where watering is by seepage from channels or pipes on gentle sloping ground. Methods of application involving the use of overhead sprays and other mainly horticultural and agricultural techniques are less likely to be used and readers are referred to the appropriate handbooks for information.

SELECTION OF SPECIES AND PLANTING STOCK

In the planning of a specific project it will be assumed that thought has been given to the selection of the tree or shrub species to be planted. In general it is important that the species should be able to grow at least moderately well with minimum attention, preferably with watering only at the time of planting. There are some exceptions, of course, such as along the major rivers which flow through the dry country, where water can be readily and economically pumped from such sources. Attention has been drawn elsewhere, and is emphasised again here, that even the Murray and the Murrumbidgee Rivers have not been truly perennial during some of the worst droughts of the past; though these cessations of flow are less likely in the future with the development of numerous weirs and major dams on them and their tributaries. Still, it is highly undesirable to place reliance on unlimited availability of water from such sources; there could be times when withdrawal may be restricted by statutory authority and, in such a case, watering may necessarily have to be restricted to selected trees or parks.

Many authorities are strongly in favour of using maximum size planting stock, for example 6-10 feet high for most evergreens intended for ornamental purposes. From the viewpoint of growing as quickly as practicable to the stage when the necessity for protection from public damage is reduced to a low level, large stock size is desirable, but to the writer there is another aspect which merits serious consideration. It is that since most plants for use in the dry country, and especially for ornamental planting, are grown in containers which are relatively small compared with the size of the plant above soil level, the roots in the container are so intertwined that the condition of being 'root bound' occurs. This is a serious condition and one from which a tree may never fully recover; it nearly always leads to poor development of the principal lateral roots, sometimes to self strangulation and eventual death and, almost certainly, to greater liability of windthrow. The smallest size plant compatible with other factors is highly desirable, apart from being more economic in cost of nursery production, transport, planting and a reduced demand for water. The size of trees and shrubs used should be considered for each project.

TOPOGRAPHY AND SOILS

Other factors being equal, for most species gentle slopes are preferable to completely flat sites. Gentle slopes permit watering by infiltration, increase the chances of salt already present in the soil or introduced in watering being washed out when there are rains of sufficient intensity to result in run-off, and reduces the possibility of poor soil aeration for those species which cannot tolerate such conditions.

The nature of the soil and the size of the holes prepared for new plants are important factors. Where the soil is a deep sand or a friable loam, the size and depth of the hole, provided that it is adequate for spread of the roots of the young plant, is usually not critical. On the other hand, where the sub-soil is a heavy clay special conditions apply. It is often the practice to remove the clay from a hole larger than is necessary for the young plant and refill the hole with a sandy loam or other suitable material, whilst the impermeable clay is used to form the edges of a basin to hold water around the site. This practice is excellent for the first few years but a time soon comes, especially if water has been used rather liberally, when the roots have extended to the perimeter of the sides of heavy clay. The alternatives for the rooting system are that the roots remain in the better-textured, well-watered small space, or penetrate extensively into the outer, heavier soil; in general there will be a concentration of roots in the confined lighter-textured soil—a condition which will not only make the tree dependent upon continued regular watering but also increase the liability to windthrow.

WATER SUPPLY AND QUALITY

In southern Australia winter planting is the general practice, whilst in the north summer is the season. In the wetter areas of the semi-arid zone in the south the advent of cooler weather and the beginning of the period of maximum rainfall coincide with the beginning of the planting season. Whilst some watering for large, evergreen stock may be desirable at the time of planting it should not be necessary for other plants except during unusually dry periods in late winter or early spring. Small one and two-year-old tubed or bare rooted stock for shelterbelts or small woodlots should, in general, not need watering in areas of over 18 inches annual rainfall. Deciduous stock, for example large poplar and willow cuttings, may benefit from some watering if rainfall is below average. In the lower rainfall parts of the semi-arid zone and throughout the arid zone more watering will be required.

Planting in northern Australia usually commences as soon as the ground is suitable after the break of the summer monsoon; some watering will be necessary, especially with large plants, during the dry season.

When planning ornamental planting, especially in the arid zone, attention should be given both to the amount and cost of water available and to its quality. Inland bore water and even that from the lower reaches of the larger rivers may have a salt content which is too high for many types of vegetation. When the only water available has an appreciable salt content consideration should be given to selecting trees and shrubs which are tolerant of moderate levels of salt in the soil. If the water comes from a township sewerage and drainage system the presence of detergents and other agents inimical to plant growth may limit its use; also if the water is piped back to trees through the settlement, safeguards should be taken against accidental human consumption, unless the water has been re-chlorinated.

Transport of the water from the source, for example river, lagoon, bore or sewerage plant, may be effectively done on a small scale by tankers which may have a dual role for fire control. For distances over about one mile pumping of the water to suitable elevated tanks or installation of a pressurised reticulation system may be more economic. In the case of long shelterbelts it may be desirable to install pipes along each belt, with taps suitably placed for the use of long hoses by men who water each tree individually. In the case of small avenues on farms and stations, further labour can be saved by feeding water from a tank into either metal or plastic pipes which have small holes drilled in them opposite each tree; the size of the holes can be adjusted to allow only a small trickle to each tree. If each branch of the distribution system is provided with a control valve, then labour in watering is reduced to a minimum.

METHODS OF APPLICATION OF WATER

The principal methods are:

1. *Direct application to each plant*

It is desirable to create a small basin around each tree to hold the water applied; this can be readily done by slightly dishing the soil near the stem and building a wall of a retentive material, such as clay, around the perimeter. In addition, especially in arid areas, it has been found desirable to insert into the soil to a depth of 15-30 inches, a vertical pipe 3-6 inches in diameter. The pipes can be of any material, but earthenware or asbestos cement are suitable, and should have holes about ½ inch diameter spaced at 4-6 inches along their length. The purpose of the pipe is to encourage percolation in the deeper soil layers. Illustrations of this method at Woomera, South Australia, are shown in Plate 42.

The amount of water needed in this and other methods will vary with the climate, size of the plant and permeability of the sub-soil. Watering should be reduced during the season of least demand by the plant. Under the more severe conditions in the arid zone, watering may be needed weekly, and more often under heat wave conditions combined with strong, hot winds. In the milder parts of the semi-arid zone watering may only be needed during the period of greatest water stress, which may vary from the annual dry season to lengthy periods when prolonged droughts occur. Even in Melbourne and Canberra, during the 1967-68 drought watering of park and street trees became necessary. In each case water was not drawn from the public supply; Melbourne used industrial waste water in part and Canberra drew from the artificial lake.

It is undesirable to apply more water than the plant needs; not only does excessive watering inhibit root extension and so create a condition demanding more water but may add excessive salt to the soil.

2. *Infiltration from channels*

This method can be used on soils of moderate or low permeability on sloping sites. The layout of the channels is almost along the contours, but with just sufficient slope to allow water to flow slowly. The location should be secured by precision levelling prior to planting; individual channels may be 4-6 inches deep and branches 100-200 yards long. Trees are planted on the lower side but close to the channels. Soil between trees should be kept free of weeds. The satisfactory operation of such a scheme involves water storage and control devices which come within the scope of water and irrigation engineers. It is not considered that this method is likely to be used to any extent in Australia, other than in existing irrigation areas where shelterbelts are needed for the protection of fruit crops. Infiltration from pipe lines has been discussed elsewhere.

3. *Artificial inundation*

The basis of this method are small areas of completely level land, often only a few acres in extent, surrounded by control banks, and a generous supply of water. Such basins can also be constructed in step structures on gentle slopes. The ground is flooded for periods of from a few hours to one or two days at a time, on two or several occasions during the growing season. It is a method of irrigation which has been used for thousands of years, especially for crops such as rice, but is not considered to have much application in Australia, although some stands of trees such as river red gum (*E. camaldulensis*) have grown under conditions of natural inundation and certain northern Australian species which grow on flats exist because of the annual soaking such areas get during the monsoons.

ESTABLISHMENT

This section will discuss spacing, selection of type of nursery stock to be planted, handling of the plants, time of planting, planting method, initial watering, refilling and a short note on direct seeding.

SPACING

A decision on spacing is necessary in the early stages of planning the establishment of ornamental trees or shelterbelts. If the property owner is intending to raise his own nursery stock the period could be as short as one year for seedling stock, but is more likely to be two to three years; even longer may be necessary for advanced stock to be used in streets and parks. In the case of intended purchase of stock it is usually desirable to make advance arrangements with nurseries, especially if rare species or large numbers are envisaged. In the case of the commonly-used species in the sub-humid and humid zones, for example radiata pine, it is usually possible to buy thousands without prior arrangements, but the position is quite different with the immense variety of species which can be used in the dry country.

TYPE OF PLANTS

There is an optimum size of tree stock to be used if a balance is to be obtained between economy in establishment and successful establishment. In the case of most evergreen plants, for example eucalypts and pines, a seedling 10-15 inches high with a good root system is large enough on the one hand to withstand competition from low ground vegetation or from harmful agencies other than grazing and, on the other hand, is small enough not to experience too great a shock in transfer from the nursery to the field and this normally leads to a high percentage 'take'. In addition, the smaller the tree the less is its water demand in the first year, but the root system must be deep enough in the soil not to be affected by drying of the top 2-3 inches of the ground. In the case of deciduous species plants 2-3 years old and up to 4-5 feet in height may be preferable when only used in small numbers and there are no problems in suitable preparation of the planting site. For public planting in streets and parks still larger plants may be used; evergreens up to 10 feet in height, and deciduous species, such as poplars, up to 20 or 30 feet can be used if suitable site preparation and subsequent watering is practical. Here the question is not economy of initial establishment but is the total cost to obtain a desired effect.

In the case of evergreens a decision may be necessary between open-rooted and tubed or potted stock. For the scale of operations of most farmers or local authorities in rural towns dealing only in a few hundred plants a year, tubed or potted plants, using a soil mixture which will not readily fall apart at planting, usually gives better and more reliable results. The subject is dealt with more fully in Chapters 18 and 19 of this book and Chapter 11 of *Growing Trees on Australian Farms* and the accompanying plates.

HANDLING OF PLANTING STOCK

As a general principle, the shorter the time from lifting in the nursery to planting in the permanent position the better. Even in the case of open-rooted evergreen stock, however, correct lifting, packing and storage will enable the plants to survive from several days to weeks if the bundles are opened and 'heeled-in'. Plants in pots or tubes are not usually seriously affected unless placed in exposed positions and left unwatered for long periods. Deciduous plants are even more tolerant of transfer to a new site, but even with these it is inadvisable to leave the roots exposed to wind and sun or to frost. On receipt at a farm, unless time and conditions are suitable for immediate planting, bundles of open-rooted stock should be treated as mentioned above. Light watering may be desirable under very dry conditions but care should be taken that the soil on the fine root hairs is not removed by excessive application of water. Plants in pots or tubes can either be left standing on the ground in a sheltered position or placed in a shade house or farm building but, in any case, where there is no direct sun on them or exposure to frost. In these positions they should be lightly watered at intervals from one to several days, depending upon the time of year.

The transfer of plants from the farm nursery or where they are temporarily stored is usually most conveniently done by a trailer attached to car or tractor. In the case of open-rooted stock they should be tied in bundles, and groups of bundles placed in boxes or tins covered loosely with either plastic sheets or dampened sacking. Especially with plastic sheeting care needs to be taken that the plants are not allowed to overheat by standing in the sun for long periods. During planting open-rooted stock should be carried in cut-down 4-5 gallon tins, canvas bags or even boxes with handles, which will hold 100 or more plants. The fine root hairs of all plants are very sensitive to desiccation and it is particularly important with evergreens that they should be protected from drying out at all times; on a windy day exposure only long enough for an operator to smoke a cigarette may be sufficient to seriously affect survival. The care of deciduous tree stock is less demanding but roots should be protected from exposure at all times.

SEASON OF PLANTING

Some aspects of this have been discussed in Chapter 18. With open-rooted stock the principal criteria are dormancy in growth, suitable soil moisture, and an atmosphere which is less drying than at other seasons; tubed or potted plants will tolerate more adverse conditions and in some areas, for example in the summer-rainfall area of northern Australia, it may be desirable to plant in early summer at the commencement of the wet season. This is particularly the case if some watering is possible until the summer rains have wetted the soils to the optimum level; such early planting provides the longest possible period for the plants to become established before the end of the wet season. In the case of a small number of plants, for example a dozen or so for ornamental use around a homestead, planting can be done at almost any time of the year provided that adequate watering can be done until the plants are established, but periods of hot and drying winds should be avoided.

REFILLING

This is the term given to the replacement of dead plants in newly-planted areas. The aim in all planting is, of course, to have 100 per cent success but, inevitably, some losses are likely in large projects—usually due to drought or severe frost if the original planting stock was satisfactory. Factors in refilling are:

(a) If necessary it should be done as soon as possible, preferably before the end of the same planting season and, in any case, within twelve months.

(b) It is desirable that any tree stock used in the subsequent planting season should be at least as large as that originally used; larger if there is reasonable chance of successful establishment.

(c) The necessity for refilling will vary with the type of planting, that is with a single row shelterbelt or an ornamental avenue 100 per cent establishment is desirable, but with a small woodlot of trees at a close spacing of, say 8 × 8 feet, a 85-90 per cent take is usually considered quite acceptable.

INITIAL WATERING

In the arid zone, watering for the first six months or so is essential and is usually desirable in the semi-arid zone. Pits which have been prepared in advance can be soaked a few days before planting and it may be desirable to form a basin around the edge of the pit to hold water. After planting a good soaking of the pit at suitable intervals is more satisfactory than many light waterings, as the latter encourages the development of surface rooting. Reference to watering as a routine practice is discussed elsewhere.

PLANTING METHODS

On most soils a spade is a good planting tool. If the site has not been prepared as recommended earlier in this chapter, it is desirable to chip away the grass from an area at least 18 inches square, but greater if there is a dense sward. The soil should be dug in the centre, making sure that neither grass sod nor heavy clay subsoil is positioned at the level where the roots of the young plant will be placed. For small planting stock, that is up to about 18 inches high, a slit is made in the area dug and the plant inserted to the depth it was in the nursery, or slightly deeper, and the soil firmed. This can be done either by ramming the soil with the heel in a direction towards the tree but taking care to avoid damaging the stem, or by placing the spade in the soil about 3-4 inches away from the young tree and levering towards it; this secondary slit should be closed with the heel or spade. For larger plants, especially those with a large root system, it is necessary to dig a pit and spread out the root system evenly before returning soil to the pit and tamping lightly. Plants over 6 feet in height, and particularly evergreens, may require staking for a year or so until the root system has developed enough to hold the trees firm.

Illustrations of how to restrain trees in position are given in Plate 90 and also Fig. 5.2 of *Growing Trees on Australian Farms.*

Alternative tools to the spade are mattocks for use on stony sites or where roots remain in the soil from trees which have been felled, and steel dibbles made from old crow-bars or pieces of piping for use when planting tubed stock. Trees for shelterbelts or woodlots should be planted in rows at regular spacing; one method for obtaining this is given in Chapter 13 and illustrated in Fig. 13.1 of *Growing Trees on Australian Farms.*

The greatest possible care should be taken to ensure that roots are not twisted when placed in slits or pits; avoid giving a young plant a twisting action when planting, as this will inevitably lead to a twist in the root system. The bad results of this practice may not be apparent for several years but then the trees may die for no obvious above-ground reason. If they are dug up, however, it will be found that the twisted lateral roots have caused strangulation of the main roots of the trees; where the strangulation effect does not cause death it may seriously reduce wind stability.

To recapitulate the more important criteria for successful tree planting:

1. The roots of young trees, especially those of open-rooted stock, should not be unnecessarily exposed to wind and sun, or left where frosts can settle on them.
2. Care should be taken that the roots of plants are not allowed to become dry; this is particularly important for the root hairs of evergreen stock.
3. The planting hole must be of adequate size so that roots are not cramped.
4. Roots must not be bent or twisted; very long roots can be trimmed with clippers (sheep shears are excellent for this purpose).
5. The soil should be firmed around the roots and if the plant is several feet high it may be necessary to support it with stakes.
6. Care should be taken that there are no air cavities in the planting hole.

PROTECTION

Nearly all young trees in their first year or two are susceptible to damage by grazing animals; hares, for instance, appear to delight in nipping off the tops of young pines for sheer devilry, since they do not eat them. One hare in a night has been known to go along rows of radiata pine (*Pinus radiata*) in the first six months after being planted, and nip off the tops of dozens of young pines, dropping the tops alongside the plant attacked. Fencing, including wire netting as a barrier against domesticated animals as well as some native species, is usually essential until the plants have grown beyond the stage at which damage is likely; this period may vary from a couple of years for damage by rabbits and hares to several years for cows and horses. If shelter is wanted down to near ground level, then it may be necessary to retain fencing for many years or even indefinitely. In the case of ornamental trees or public plantings the use of large stock and wire guards are usually necessary. When land holders are planning long avenues or screens of widely spaced ornamental species it may be desirable to plant in stages and move the wire guards from the older trees to the most recently planted.

The use of substances applied to young trees to act as repellents appears attractive, especially for protection against rabbits and hares, but in general the efficiency of them has not proved high enough to justify treatment. The use of a mixture of bitter aloes and soft soap applied to the stem of young pines has reduced damage by rabbits and hares, as well as the application of tarry materials; but a substance which is economic, efficient and readily applied has yet to be developed.

CHAPTER 21

Tree maintenance and silvicultural treatment

UNDER most conditions once trees have been satisfactorily established they need comparatively little attention from early age to maturity; this is particularly the case if they have been carefully selected with due regard to the environment in which they have been planted. Maintenance of the tree and its site will be considered with regard to fencing, repair of wind damage, control of competition from other plants, insect control, disease control, fencing and soil fertility. The cultural measures of pruning, thinning and natural regeneration will be treated only briefly since they have been dealt with more comprehensively in Chapters 14, 15 and 17 of *Growing Trees on Australian Farms*. The more specialised aspects of treating valuable and old specimens, often referred to as 'tree-doctoring', is covered in Chapter 22 of this book.

FENCING

Until trees have reached the stage at which further protection from grazing animals is considered unnecessary, fencing should be maintained in an effective condition. Wire netting, in particular, should be inspected regularly for holes which may permit entry of hares and rabbits. Fortunately wombats are not as plentiful in the dry country as in the humid zones where they bulldoze their way through fencing and netting, allowing entry of plant-destroying animals.

WIND DAMAGE

There are two aspects of wind damage; the first is that of trees being blown over, most common with young plants in the first few years of life but liable to occur at any age, and the second is damage to the main stem or crowns of older trees. Wind blow is most likely to occur on silty clay soils when gale force winds follow soaking rains; very occasionally on unstable soils and under particularly unfavourable conditions of rain and wind large areas of mature trees may be affected. This latter state of affairs may not be as serious as appears at first consideration, if there is an active utilisation operation under way; it simply means that felling is transferred to the affected area and replanting is done in due course.

With trees only 1-2 inches in diameter at ground level, leans may be corrected by standing the stems erect and either placing large clods of earth or a small log against the lee side from the prevailing wind direction for the smaller trees, or using props for the larger specimens. If the climatic conditions which have caused the damage only occur at infrequent intervals, that is once in 10-20 years, the trees will resume their erect growth and develop a rooting system which will withstand the next storm. The technique of supporting small trees is shown in Plate 38 of *Growing Trees on Australian Farms*, whilst one method of staking is given in Fig. 5.3 of the same book. In the case of small numbers of trees, for example those used for shade and shelter around a homestead, it may be desirable to give the trees support at the time of planting; examples are given in Plate 90.

The occurrence of lean in trees and the opening of the top of the soil around the base of the stem due to a rotating movement of the tree is more likely to occur when:

1. Trees are planted on unstable, poorly-drained soils. When such sites cannot be avoided it is usually beneficial to make small mounds either by hand or line ploughing and plant the trees on these; this practice, however, is more suitable for humid climates than the dry country.
2. Young plants are not placed sufficiently deep in the ground and the soil is not tamped around the stem.
3. Large stock grown in small containers are used or the root system has been too heavily pruned on leaving the nursery. However attractive a 5-6 foot plant in a small container may appear, it will not have a suitable root system.
4. Plantations have not been thinned at the desirable stage of development. On unfavourable sites frequent light thinnings are much better than heavier thinnings at longer intervals.

When tops are blown out of trees or large branches broken by wind, suitable pruning may remedy the position in some cases; further aspects of this are discussed in Chapter 20. Strongly developed bows in the lower trunks of trees being grown for timber are very undesirable, since abnormal wood structure results (compression wood in conifers and tension wood in hardwoods), which causes difficulties in sawing and seasoning, and is generally unacceptable to the pulp and paper industry.

COMPETITION FROM OTHER PLANTS

During the first few years of life it is particularly important that young trees should be protected against competing vegetation such as grass, bracken fern, shrubs, vines or even other tree growth, for example coppice from eucalypt stumps; older and larger trees have usually grown beyond the influence of competing vegetation. The better the early growth of young trees the sooner they will reach the stage at which they provide the shelter, shade or other benefits for which they were planted. Competition may not only be for above-ground growing space and light, but also for moisture and nutriment from the soil. Not all trees have the same reaction to restriction of light; most of the dry country species, including acacias, eucalypts and cypress pines, are 'intolerant' and consequently need comparatively wide spacing and freedom from shading by other vegetation. Trees which will stand a high degree of shade are called 'tolerant'; the best examples of these are many of the species of the rainforest.

Weeds, that is any unwanted plants, can be controlled either by mechanical or chemical means. Shallow soil cultivation, wherever topography, freedom from large stones or timber on the ground permits, is effective both in destroying weeds and providing a fine soil tilth which is useful in minimising evaporation. Most equipment used on farms or in orchards is suitable for this work, including tractors with light blades attached, but some forms of rotary cultivators should not be used. Careful operation is necessary, of course, so that machines do not damage young plants and in some cases where there is particularly vigorous weed growth some hand work immediately adjacent to the young trees may be necessary. In areas inaccessible to machines all weed control will need to be done by hand, using spades, dutch hoes, mattocks or other suitable tools, including secateurs for the cutting of vines.

A useful alternative to cultivation is mowing. Depending upon the scale of operations and the firmness of the soil, equipment can range from the small domestic rotary mower, to the larger self-propelled machines which are particularly useful for the smaller scale of public plantings, or the heavier tractor-drawn implements. When the soil is light and wind erosion is feared, mowing is a useful means of weed control and can be repeated until the young trees are above competition. The smaller machines have the advantage that growth can be cut to within a few inches of tree stems with little risk of damage. In the case of fast-growing annual vegetation, for instance bracken fern fronds or soft shrubs, control on steep or rocky sites will usually have to be done by hand, using slashers or heavy hoes.

Chemical control through the use of weedicides offers still another means of control, but care needs to be taken both in the dosage used and the conditions of application. Particular attention must be given to the use of 'selective' weedicides since a solution stronger than that recommended by the manufacturers may either kill the young trees or have a serious stunting effect. Unless the property owner has considerable experience in the use of weedicides he is recommended to seek advice from the local agricultural officer, forester or technical expert of a chemical company; even then he should make small-scale trials before treating all his trees. In preference to using selective weedicides, for small-scale operations it is possible to use a hand-operated spray with a shield attached which allows coverage of the weeds near a plant but protects the young tree; this equipment is illustrated in Plate 36 of *Growing Trees on Australian Farms.*

Care needs to be taken with the use of weedicides, not only in ensuring that the correct dilutions of sprays are used but that spraying is done under suitable weather conditions (calm weather being essential for the control of spray drift in small-scale operations) and that the operator uses suitable protective clothing, particularly goggles and gloves. The operator should wear long trousers and have his shirt sleeves rolled down; *a casual attitude to the use of dangerous chemicals may result in serious long term effects which may not be apparent immediately after contamination.* Appendix II gives information on the properties of weedicides important in forestry and other tree-growing activities. Care in the use of insecticides, most of which have considerable similarity with some of the weedicides, is discussed in some detail in Chapter 18 of *Growing Trees on Australian Farms.*

CONTROL OF INSECTS

For the property owner with only a few shelterbelts and small woodlots, possibly only representing 1-5 per cent of his holding, insect damage rarely attains epidemic proportions. The trees most likely to be affected are those in the nursery, during the first two or three years after planting, or those in old age. Vigorous saplings, poles and those trees which have not become overmature are less likely to be attacked. In the nursery, white grubs, wireworms (beetle larvae), cutworms (moth larvae), and eelworms (nematodes) may damage the roots, whilst aphids and other sap-sucking insects attack the aerial parts. Woolly aphis on coniferous trees 2-5 years old may be a sign of an unsuitable site; if the site is favourable then the growth of the tree will suppress the aphis without chemical assistance. Nearly all trees are subject to attack by defoliators, but eucalypts are particularly susceptible at all stages of growth; if the trees are not too large, orchard-type equipment can be used for control, though there is always the serious danger that the use of chemicals will also destroy insects or birds which prey on the leaf eaters. In trees approaching over-maturity, wood borers may be a major pest, whilst a beetle (*Hylastes* spp.) which breeds under the bark of *Pinus* logs kills young pines. Any very sick old trees or any that are dying should be felled, the branches and foliage burnt completely and the logs either removed, possibly for sale, or the bark stripped and burnt. These dying trees and material on the ground can act as foci for the infestation of the remaining trees and to maintain the latter as healthy as possible good standards of forest sanitation should be maintained at all times.

A final recommendation; do not wait until it is obvious from a distance that trees are seriously affected by insects—periodically, say two or three times a year, walk or ride horseback along the shelterbelts, etc., and look for signs of defoliation, dead tips of branchlets or unusual insect activity.

CONTROL OF TREE DISEASES

Tree diseases can be classified into non-parasitic and parasitic. Common causes of the former are unsuitable soils, including deficiency of minor elements, and an otherwise unsuitable environment. Nutritional problems are discussed elsewhere in this book and in Chapter 16 of *Growing Trees on Australian Farms*, whilst the second cause emphasises the desirability, rather the necessity, of selecting species which are suitable for the environment of the farm upon which they are to be planted.

Most parasitic diseases present complex problems which often require specialist pathologists to identify; treatment may not be direct and, in many cases, is outside the scope of the private individual. Simple fungal diseases in nurseries, however, can often be controlled by spraying with Bordeaux mixture or other nursery or horticultural formulation. The following notes are quoted from Chapter 19 by Dr W. A. Heather in *Growing Trees on Australian Farms*:

CULTURAL PRACTICES AND DISEASE CONTROL

1. In general, a vigorous tree will remain a healthy tree. Many weak pathogens become seriously parasitic on trees on unfavourable sites or on those which are suffering serious competition from other tree undergrowth. The farmer

The natural regeneration of individual trees and shrubs or of a forest requires, amongst other things, a combination of seed supply, sufficient rainfall at the right time and protection from birds and animals. Good natural regeneration is considered a sign that the environment is suitable. In the case of an introduced species it is said to be *acclimatised* when it commences to spread naturally. In the dry country the establishment of natural regeneration usually only occurs at long intervals when there has been above average rainfall for a year or more.

Top left: White cypress pine (*Callitris glauca*) in a State Forest between Forbes and Grenfell on the Western Slopes of New South Wales. In the background is a thinned mature forest and a regeneration area shows in the centre. In the foreground a giant rotary slasher has been used to thin the original very thick seedling crop; the seed trees are still present whilst the unwanted eucalypts (mainly *Eucalyptus albens*) have been frilled and poisoned with 5 per cent 2, 4, 5, T.

Centre left: This shows river red gum (*E. camaldulensis*) in a slight depression on the roadside where the young plants have not been damaged by stock.

Bottom left: Cooba (*Acacia salicina*) showing good natural regeneration at Broken Hill where the average annual rainfall is only 9 inches. The site is not watered but is protected from stock.

Bottom right: The athel tree (*Tamarix articulata*) at Broken Hill. This introduced tree is regenerating on a very adverse site, with salt showing on the surface. Not watered.

Plate 89 SOME ASPECTS OF NATURAL REGENERATION IN THE DRY COUNTRY

should endeavour by sound choice of species for the site, by cultivation before establishment, by controlling undergrowth and by fertilising and thinning, to maintain his stand in a vigorous and healthy condition.

2. Hygiene is as important in tending forest stands as in all forms of crop and stock production. Pathogens build up populations in stands containing weakened, dying or dead standing trees. High pathogen populations will frequently result in parasitism by organisms which are usually of little significance.

3. The selection of species for establishment is all important. The farmer should use species whose environmental requirement are suited to the particular area. The advice of specialists should be sought if the species is not already established in the area.

4. In tending and thinning operations damage to roots, stems and crowns should be avoided so far as possible. Wounds and fire scars provide the most significant points of entry for stem-rotting pathogens. Stems with serious butt fire scars should be removed.

5. Coppice stumps of stands managed on a coppice system should be cut as low as possible and should be free of heart rot. Penetration of the heartwood of standards from coppiced stumps is the most important single source of coppice heartwood infection by disease organisms. Infected stumps should be disposed of by grubbing or burning at the time of clearing; they are a source of potential root rot and stem decay in the established crop.

6. Seedlings for establishment should be vigorous and healthy. An undersized chlorotic seedling is very likely to be infected by soil-inhabiting fungi. Seedlings should not be taken out of local quarantine areas. These are likely to be diseased even if superficially they appear healthy. They are likely to extend the areas of important forest diseases.

7. The established area should be examined critically at 6-monthly intervals and necessary thinning, pruning and hygiene operations carried out. Any unusual symptoms of ill health should be reported to forest authorities. There have been examples where important introduced pathogens have been present in stands for more than 2 years before notification to the authorities. Such irresponsibility poses a threat to the national forest economy.

8. Spraying for the control of fungal diseases is generally only practicable with isolated ornamentals. Sprays should be used in accordance with the manufacturer's instructions. According to the size of the tree a knapsack or motorised fogging spray can be used. A useful general reference on fungicide sprays (Horfall 1956) is given at the end of the chapter.

9. Viruses have not been extensively studied in forest trees. They have been recorded in poplar, sandalwood, elms and even conifers. In establishment of trees by vegetative propagation it is wise to follow the usual procedures of selection of material, sterilisation of tools, etc., adopted for pome and stone fruit tree propagation.

10. The plant quarantine section of the Commonwealth Department of Health should be consulted about any proposed introduction of plant propagating material from overseas.

SOIL FERTILITY

Most of the soils of arid and semi-arid zones are sufficiently fertile to support throughout a rotation the adequate growth of trees in shelterbelts and those used for ornament. Decline in fertility in tree growth is, however, much more difficult to measure than with field crops. Where better than average growth is wanted for any site the addition of artificial fertilisers at intervals of several years will usually give the desired result, provided that the rainfall is sufficient, preferably over 20 inches a year, or enough water is applied. The extensive use of artificial fertilisers in the dry country is not likely to be worthwhile, especially where the rainfall is below 15 inches in southern Australia and about 25 inches in the north. Most shelterbelts will benefit, of course, from superphosphate and other manures applied to wheat crops and pastures. The decline of production in the second rotation of coniferous plantations on rather poor podzolic soils as well as on better soils in the humid zones is a complicated problem which cannot be completely assessed by measurement of the chemical elements in the soil; it is not thought likely that this will become a serious factor in the dry country. Further aspects of the maintenance of forest soil fertility is discussed in Chapter 14 of *Growing Trees on Australian Farms*.

SILVICULTURAL OPERATIONS

There are three principal aspects of pruning: one is pruning to improve the shape of trees used for amenity purposes; another is the removal of branches, especially in conifers, to improve log form and produce knot-free timber in commercial plantations; and the third is the specialised pruning and repair ('tree-doctoring') of valuable and old trees, which is the subject of Chapter 22.

PRUNING

Some tree species, particularly hardwoods, may tend to develop a number of major branches at the expense of a single main stem; where this is so the position can usually be remedied by pruning the unwanted branches at an early stage, possibly when the branches are only ½-1 inch in diameter. Particularly in the case of street and park trees it may be desirable to develop a clear stem for the lower 6-15 feet, above which it may not be important whether the main stem divides or not. Conifers normally have a well-defined main stem but sometimes two equal shoots may develop, giving what is known as 'double-leaders'. Particularly if these commence below mid-height and both develop to large stems, there is a considerable chance of wind damage—one leader is broken off and, inevitably, the other follows, so that the tree is a total loss. When the double leaders first arise is the time to remove one of them, after which the other will develop normally. A good quality orchard-type secateur is satisfactory for cutting material up to about one inch in diameter; for branches 1-2¼ inches in size large pruners are needed, and for still bigger branches pruning saws are preferable. All cuts should be made in as near a vertical position as practicable in order to shed rain and, in the case of conifers, to encourage the flow of resin over the wound; the dressing of large stubs is included in Chapter 22.

When coniferous species are wanted for shade trees, for example Aleppo pine (*Pinus halepensis*) or Canary Island pine (*P. canariensis*) it may be desirable to deliberately encourage side branch development by cutting out the leader at a height of 8-15 feet when it is only 1¼-2 inches in diameter. Sometimes when this is done too early in life one of the laterals will replace the lost leader, but this can be treated in the same way if it appears that there will not be the desired development of a bushy crown. Aleppo pine when grown as individual trees in the open tends to develop a bushy crown without assistance but Canary Island pine has a main stem which shows strong dominance. Stone pine (*P. pinea*), on the other hand, naturally develops an umbrella-shaped crown ideal for shade, without any cultural stimulus.

There is a considerable difference between tree groups as to their tolerance to pruning. Most hardwoods, in general, will stand regular and heavy pruning and new shoots will appear from just below the cut ends of the branches; conifers, on the other hand, do not show similar growth and are less suitable for regular pruning unless it is done at the tips of the branches, as in training a hedge.

The optimum season of the year varies with species. With the exception of ornamentals grown for their bloom most deciduous trees should be pruned in the dormant season; other hardwoods are, in general, best pruned in late winter or early in the growing season so that callus growth quickly commences to heal the stub. Conifers pruned in spring and early summer are likely to have a copious flow of resin which covers the wound and protects it from fungal spores, but where the wood wasp, *Sirex*, occurs this flow of resin is undesirable and pruning should be restricted to winter.

PRUNING FOR TIMBER QUALITY

In farm woodlots it is usually good forest practice to prune branches on the lower 6-8 feet of the stem as soon as the tree will tolerate their removal; this is as soon as the foliage commences to die and, in any case, so that branches are not removed for more than 40 per cent of the height of the tree. Even if subsequent high pruning is not done this 'low' pruning is useful as a fire control measure and facilitates movement in the forest for thinning operations.

Pruning for only 6-8 feet may be of benefit to the form of posts and strainers taken from early thinnings but greater heights are needed for saw logs and poles, whilst peeler logs are usually in multiples of 6 feet 6 inches, so that when timber production is the objective pruning operations should be done to at least 13 feet, but preferably to as near 20 feet as possible. Above that height, special equipment and techniques are necessary; costs also increase considerably and are only justified when a particularly valuable product becomes available as a result of the operation. In order to ensure that the central knotty core remains as small as possible, that is five to seven inches diameter, it is necessary to prune in progressive stages and three would be the minimum in pruning to 20 feet.

The principal tools for pruning to about 8-9 feet are pruning shears; above that curved pruning saws on light handles are the most useful general implement, having one length of handle for pruning to 8-14 feet and another for 14-20 feet.

Illustrations of pruning equipment and operations are given in Plates 39-41 of *Growing Trees on Australian Farms*, whilst Chapter 14 of the same book discusses commercial pruning operations.

THINNING

Thinning operations in a woodlot or forest consists in the progressive removal of selected trees from a developing stand, with the objective of improving growth, form and size of those which remain. First priority for thinning are any diseased, dead or dying specimens, followed by suppressed trees and any dominant trees with very defective stems or crowns, especially those which are affecting the development of better trees; if the thinning level is still not sufficiently intensive, further trees are selected from the sub-dominant or dominated class (these are trees which are not in the upper canopy but the leaders of which still have free access to light). Commercial thinnings usually commence when the trees are approaching one-third to half mature height and continue at intervals so that there may be three to six or more operations during a rotation. Further information on thinning is given in Chapter 15 of *Growing Trees on Australian Farms.*

Thinning should not be done in shelterbelts or amongst ornamental trees except for the removal of badly diseased or damaged specimens.

NATURAL REGENERATION

The establishment of woodlots from direct sowing of seed or by planting is known as 'artificial regeneration'; in contrast to this is 'natural regeneration', which is the renewal of a forest crop by seed from trees already existing on an area, or by coppice or root suckers. Natural regeneration has limited value for the renewal of ornamental trees or narrow shelterbelts but has possibilities for cypress pine and open eucalypt forest in the semi-arid zone and, combined with appropriate conservation practices, for the regeneration of trees and shrub growth of rangelands both in the arid and semi-arid zones.

For natural regeneration to be satisfactory there are a number of conditions which must be suitable:

1. The trees or shrubs already existing on the area must be of acceptable species. If there are undesirable species, for example those of very poor form or growth or with poisonous foliage or fruit, it may be necessary to fell them some years in advance of the regeneration operation so that seed will become non-viable, or the regeneration can be destroyed either by grazing or, if poisonous, by cutting and burning. In the case of natural forests or plantation woodlots which have been under management for a large part of a rotation, opportunities will have arisen in the normal sequence of thinning operations for the removal of trees showing poor form or growth, whilst with mixed woods it is possible to change the balance of species.

The position of very open rangelands of low economic value, for example with a carrying capacity of one sheep to 15-20 acres, is that the cost of felling and burning undesirable species is usually not justified.

2. There must be sufficient 'seed trees' left to provide for seeding of the area when a good seed year arrives. What constitutes a sufficient number varies both with the height of the specimens and the type of seed, that is whether it is readily carried by wind, and the class of forest or woodland which it is desired to regenerate. In the case of a moderately dense cypress pine forest which it is planned to manage for timber production it may be necessary to have 3-10 seed trees per acre but, on the other hand, if it is a mixed grazing and timber proposition possibly 1-2

trees per acre would be adequate. For open rangelands where only scattered clumps of shrubs or small trees are desired one seed-bearing plant to several acres may be enough to maintain the species concerned.

3. For coppice growth of eucalypts there needs to be an adequate stocking of stumps from which the coppice shoots can develop and for a fully-stocked forest this may be as high as 200-300 per acre. After two to four coppice regeneration cycles, stumps lose their power to produce vigorous shoots and it is therefore desirable to replace say 25-50 per cent of the older coppice stumps with younger stumps each rotation.

4. Where natural regeneration from seed is intended it is necessary to wait for a 'seed year'. Most tree species do not produce heavy crops of seed each year. The periods between seed years vary widely and for many Australian trees in the higher rainfall zones are from 2 to 4 years; during the intervening years the seed crop may be light or even absent. The seeding of shrubs in the dry country show a close correlation with the rainfall pattern in the preceding twelve months.

It is usually possible to forecast the advent of a good seed year by the development of flower buds; in the case of the smaller shrubs this may be 3-6 months before the production of ripe seed, for eucalypts up to one year and with conifers 1½-2 years. In the arid zone, shrubs which get additional moisture, for example from a drainage channel or run-off from a road, may flower and seed more freely than those not so advantageously placed. It is therefore important to examine a good cross-section of topography when making an estimate of future seed production.

5. Having obtained a good seed year, it is also necessary to have better than average rainfall. The interval between the coincidence of good seed years and adequate rainfall is greater in arid zones than in humid zones, and may range from 5-20 years.

6. Protection from animals.

The period of protection necessary will vary both with the type of animal, including both rabbits and hares as well as domesticated animals, and the rate of growth and palatability of the young plants. In the case of woodlots protection can be afforded by fencing the whole area under regeneration, but for extensive operations on rangelands it may only be possible to protect scattered small areas, say of an acre or two.

With coppice regeneration where there are more than 2-3 evenly-spaced shoots on a stump it is usually desirable to remove any additional shoots when they are 2-4 feet high. A larger number can be left when it is desired to produce the maximum quantity of small-sized material, for example stakes and poles 1-2 inches in diameter.

In the dry country, especially in the higher rainfall parts of the semi-arid zone, by means of natural regeneration over a long period landowners can convert an overmature, unproductive eucalypt forest into a stand producing useful quantities of fencing material, light poles and readily handled firewood. Illustrations of this type of conversion is given in Plate 47 of *Growing Trees on Australian Farms* and further information on the treatment of natural forests and silvicultural systems is given in Chapter 17 of the same book.

FURTHER READING

Horfall, J. G. (1956). *Principles of Fungicidal Action.* Chronica Botanica, Waltham, Mass.

Jacobs, M. R. (1955). *Growth Habits of the Eucalypts.* Forestry and Timber Bureau, Canberra.

Martin, Hubert (ed.) (1969). *Insecticides and Fungicide Handbook for Crop Protection.* 3rd ed. Blackwell Scientific Publications.

CHAPTER 22

The care of valuable trees

by V. S. Russell

THE purposes for which trees are grown or retained around farm homesteads, suburban homes and in streets, parks and gardens are primarily aesthetic appeal, the amelioration of climatic factors such as wind and temperature affecting man, and the provision of shelter for birds and animals.

The aesthetic appeal of trees and the shade and shelter they provide combine to improve the environment in which we live, extending some of the advantages of the indoor environment to the outdoors. Man appears to have an instinctive need for overhead shelter reflected in a sense of unease when completely exposed that is not easily explained in terms of cold logic. This instinct is demonstrated by the natural tendency of man to move into the proximity of trees, if not actually beneath the canopy, while relaxing out of doors, even during the most pleasant weather.

For these reasons, trees around the home have a value not easily assessed in commercial terms, although their presence markedly increases the total real estate value of any development of man. The care and maintenance of such trees therefore assumes a commercial as well as an aesthetic value.

The age at which trees reach maturity varies from 30 to 100 years or more, and from this time on the growth of new tissue simply keeps pace, for a period, with the death of older tissue until with increasing age, growth renewal becomes inadequate, more and more dead twigs and branches appear, and gradual decline leads to death. This picture of the normal tree can be changed by a variety of factors such as drought, fire, inundation and severe insect or fungal attack, but particularly is it affected by man's activities.

The usual aim of growing trees for timber production is to promote the most rapid growth desirable and fell them when growth rate declines to the point at which its value fails to exceed the interest rate expected on capital investment. This practice is in marked contrast to the growing of trees for purposes of ornament, shade and shelter, when their greatest value is usually attained after they reach maturity. The aims of all cultural practices from this point onwards are directed to maintaining health of the trees and postponing the period of decline and death as long as possible.

The value of trees is never more quickly realised than when lost by death or destruction and this chapter is concerned with their preservation and maintenance

STRUCTURE OF A TREE

In order to care for trees adequately an understanding of the structure and functions of the various parts of a tree is essential.

THE ROOTS

These perform four primary functions, namely, anchorage, absorption of water and mineral nutrients in solution, conduction of these nutrients to the stem, and storage of elaborated food materials manufactured by the leaves.

The anchorage which roots provide is dependent upon the development, lateral extent and depth of the primary root system. The root system varies with species, and is affected by soil type and condition and the degree of competition and shelter provided by neighbouring trees. Trees exposed to wind all their lives develop a stronger primary root system under the stimulus of wind sway of the stem and branches than trees in shelter. Removal of this shelter, for example by felling too many neighbouring trees, frequently results in subsequent loss by windthrow of many of those remaining. Similarly, the prolonged retention of stays used to support a newly-transplanted tree removes the stimulus of the tree to provide an adequate anchoring root system; when the stays are eventually removed the tree is liable to windthrow in the first storm.

At the point of attachment to the trunk, roots are relatively large and few in number but they divide repeatedly, becoming progressively smaller in diameter until at the tips they are very fine and numerous. Although the total volume of the root system may be from one tenth to a half of the total volume of the tree, the main bulk of the roots is involved only in anchorage, conduction and storage of food material. Absorption of mineral nutrients from the soil occurs through the very fine root hairs which consist of elongated single cells borne in very large numbers near the ends of the fine rootlets. The root hairs are very delicate and short-lived, but new ones are continually being formed. The tips of the rootlets are protected from damage in the soil by root caps, but suffer damage to varying degrees in any transplanting operation by breakage and desiccation. The extent of the setback a tree suffers in transplanting is largely dependent upon the care with which these fine rootlets are preserved.

The absorption of mineral nutrients from the soil can only occur by solution in soil water and absorption through the cell walls of the root hairs. Absorption of water occurs partly by osmosis, that is the process by which water moves from the dilute solution of the soil water through the root cell walls to the more concentrated solution in the cells themselves, and partly by the 'transpiration pull' which develops by evaporation from the leaves. Mineral nutrients in solution have a natural tendency to diffuse from concentrated solutions to more dilute solutions to equalise the concentration. For mineral nutrients to be absorbed from the soil into the root cells, work must be done to oppose the natural diffusion gradient and the energy for this work is provided by the process of respiration, that is the oxidation of food materials manufactured in the leaves and transported to the roots. In this process oxygen is used and carbon dioxide produced, both of these gases being absorbed or excreted in solution in the soil water. Since oxygen has limited solubility in water, soils must be well aerated to ensure a continuing supply of oxygen. The conducting tissues of mature roots are similar in form and function to those in the stem and so will be described in that section.

The health and vigour of a tree is dependent to a very great extent on the health of the root system but because the latter is out of sight it is often out of mind. Usually, the first obvious signs of a diseased or unhealthy root system become apparent in the foliage. It must be remembered that root systems are just as subject to attack by insect pests, fungal and bacterial diseases as the crown, but the control of such attacks is much more difficult and frequently impossible. The maintenance of a healthy root system is best achieved by avoiding mechanical damage and soil compaction and maintaining adequate but not excessive levels of mosture, mineral nutrients and aeration.

THE STEM AND BRANCHES

The functions of the stem and branches of a tree are structural support of the foliage, flowers and fruits; conduction of water, mineral nutrients and elaborated food materials; and storage of food materials.

The greatest part of the stem consists of wood tissue, the xylem, which is composed of conducting cells and fibres which give tensile strength to the tissue. It is the xylem which gives a tree its strength and stability. In young stems the xylem is generally pale in colour, but as the tree grows older the inner xylem tissues die and darken in colour with the deposition on the cell walls of various organic substances. These tissues no longer take part in the conduction of food materials but are still of importance in providing structural support. This dark wood is the heartwood and it is surrounded by a cylinder of pale-coloured xylem tissue, the sapwood, which is still alive. It is through the conducting cells of the sapwood that the mineral nutrients in solution pass from the roots to the leaves.

Outside the xylem is a thin layer of cells, the wood cambium, which by cell division produces from its inner surface a new ring of xylem tissue each growing season; these are the annual rings. From its outer surface the cambium produces phloem cells which conduct elaborated food materials downwards from the leaves to the stem and roots. It is the phloem which makes up the live, moist inner bark of the trunk and branches and it is the cambium layer which is torn apart when live bark is stripped from a tree, leaving some cambial cells on the wood and some on the bark.

Because of the expansion in diameter of the trunk of a growing tree arising by growth from the cambium, the outer cylinder of bark is stretched and develops numbers of longitudinal fissures which, if they extended through the phloem to the cambium, would allow drying out and open the way to attack by disease or insects. This is prevented by a second layer of cambium tissue, the cork cambium or phellogen, which lies in the outer edge of the phloem and continually produces cork cells forming the outer bark or 'periderm'. As the periderm is stretched by outward growth of the stem it cracks and a new phellogen is formed deeper in the phloem, the cells outside this layer then dying and forming part of the outer bark. The periderm is impervious to the passage of water except through pores called lenticels on young stems and branchlets of some species, and it provides insulation from extremes of temperature, and protection from attack by disease and insects. Some trees retain their outer bark, which may be comparatively soft, fibrous and inflammable as in the stringybarks or very hard and resistant to fire as in the ironbarks but, in all cases, the bark on old trees is deeply fissured. Other trees shed their bark more or less annually either in small plates as in spotted gum or long ribbons, as in manna gum, and deep fissures cannot develop.

In addition to movement of water and mineral nutrients and elaborated food material through the xylem and phloem, some lateral movement of food materials occurs through the medullary rays. These appear in a cross section of a stem as numerous radial lines running from the outer edge of the xylem towards the centre of the stem. The medullary rays facilitate the entry of disease organisms into the sapwood when the bark is damaged.

An understanding of the way in which food materials move in a stem is of importance in pruning and treatment of wounds. It explains, too, why a tree dies comparatively slowly if a ring of bark only is removed from round the stem; the roots are starved of food materials from the leaves while the leaves still receive, for a time, the water and mineral nutrients passing upwards through the sapwood. If the sapwood is severed over the whole of the circumference the tree dies much more rapidly, but the stored food material in the roots and lower stem is frequently sufficient to promote the growth of vigorous coppice shoots or suckers.

Longitudinal extension of both stems and roots occurs only over a short distance behind the growing tips and all growth in length originates from the meristematic tissue in these growing tips. The term meristematic refers to tissues which are capable of active cell division and differentiation into specialised tissues. The wood cambium and cork cambium are also meristematic but growth from these gives rise to lateral expansion only.

It is only from the cambium that new tissue called callus growth can be generated to heal wounds of the stem, branches or roots caused by mechanical damage, insect attack or fire. An appreciation of this fact is of the utmost importance in the treatment of such injuries.

THE LEAVES

The function of leaves is to manufacture the food materials required for growth. This process, called photosynthesis, is highly complex and need not be discussed in detail here, but essentially the process results in the combination of carbon dioxide from the atmosphere and water, absorbed either through the roots or from the atmosphere, to form sugars and other carbohydrates under the stimulus of sunlight on the green colouring matter (chlorophyll) in the leaves. These carbohydrates may be utilised directly or after combination with other elements in various cells of the tree for the growth of tissues in the stem, bark and roots, for the formation of fruits and seeds, and storage. Only plants containing chlorophyll have the power to manufacture their own food materials and all other forms of life are dependent on them either directly or indirectly for food.

The surface of leaves is pitted by numerous tiny pores, the stomates, which allow the entry of carbon dioxide for photosynthesis and the emission of oxygen, a product of the same process, and water vapour which is lost in transpiration. It was once believed that the passage of water into or out of the leaves occurred only through the stomates, because the thin waxy cuticle covering the leaf surface resists the passage of water. It is now known that a material in the cuticle permits direct absorption of water and mineral nutrients in solution into the leaves. This fact is of importance in 'foliar feeding' which will be discussed later in this chapter, and explains the rapid recovery of wilting leaves following light spray irrigation.

Leaves of trees have a normal life of only a few months to a few years, even when not damaged by insects or other causes. Since the tree is entirely dependent on the leaves for the manufacture of its food materials, defoliation results in a set-back to growth and the tree must rely on its stored food materials to replace foliage before growth in length or diameter can continue. Hence the maintenance of a healthy crown is essential to promote maximum growth.

The amount of water transpired by the leaves of a large tree on a hot summer day is enormous (as much as 80-150 gallons where the water supply is not limited) and this is very much greater than would be lost by evaporation from the area of bare ground the tree covers. Of the total quantity of water absorbed by the roots as much as 98 per cent is lost in transpiration, the balance being retained by the plant cells for maintenance of turgidity in the plant tissues and for chemical combination in food materials. Tree species vary enormously in the efficiency with which they utilise available water supplies. In the process of evolution they have developed a variety of mechanisms to reduce losses by transpiration during periods of drought stress but *the growing of trees away from their normal environment requires careful attention to their water requirements.*

THE CARE AND MAINTENANCE OF TREES

The most important aspects of the care and maintenance of trees are transplanting, pruning, the treatment of injuries, pest and disease control and the maintenance of suitable soil conditions.

TRANSPLANTING

Nursery practice and the raising of seedlings has been discussed in Chapter 19. The process of transplanting trees from the nursery to their final site is a most important factor in their satisfactory survival and growth to maturity. Many failures in the establishment of expensive trees in a new site are due to incorrect transplanting practices.

The natural desire for large trees in the shortest possible time too often results in poor survival of large transplants but sometimes it is many years before the ill effects of certain practices become apparent. Probably the first and most serious fault in the handling of open-rooted stock is exposure of the root system to desiccation by sun or wind or damage by frost. Most deciduous species can be transplanted successfully during the dormant season as large open-rooted stock but even though devoid of leaves, which would show a quick response to drying of the root system, every effort should be made to avoid desiccation at the time of lifting, during transportation and at the planting site. A simple and most effective protective measure is the puddling of the roots immediately after lifting in a slurry of water and clay-loam soil mixed to the consistency of thick soup. During transport the crowns as well as the roots must be protected from wind and sun by a cover such as polythene or canvas. Hessian alone is quite inadequate and if polythene is used care must be taken to prevent sweating. At the planting site, if any delay in planting is likely to occur the plants should be 'heeled-in' in wet sand or, preferably, sawdust. The use of heavy clay soil for this purpose results in poor contact between soil and roots and further damage to the roots when the plant is removed.

A reputable nursery will provide open-rooted stock with the crowns pruned to compensate for the damage to the root system that is inevitable in the lifting operation. Buyers should refuse to accept stock which has a poor balance of crown and root system. Large evergreens are frequently sold with a soil ball wrapped in hessian and bound securely to prevent breakage. The purpose of the soil ball is to minimise damage to the roots, to keep them moist, and maintain intimate contact with the soil. The root ball must never be allowed to break or dry out and the crown must also be protected from sun and wind, particularly during transportation. Conifers frequently fail to show a quick response to drying as the foliage remains green and does not wilt, so it is only by inspection of the root ball that dryness becomes apparent. Although the balling of stock reduces root damage some loss of roots always occurs and compensation for this by pruning of the crown is essential. Again the customer should only accept plants with a good root to crown ratio.

Open-rooted and balled stock are best planted in winter, or early spring, but balled stock may also be planted in autumn. Container stock by comparison can be planted at any time of the year if sufficient care is used to avoid root damage and maintain adequate soil moisture levels.

Intending buyers of large plants in containers should be wary of stock in cans even as large as 5-gallon drums. The restriction to growth causes circling of the roots around the sides and bottom of the container and this is a condition from which a plant never recovers. When planted out the young tree often fails to make adequate growth and is highly susceptible to wind-throw. Even when some roots do grow out, one or more of the encircling roots can cause strangulation of the main feeder roots and subsequent death of the tree, possibly many years after planting. All bruised or broken roots should be pruned cleanly immediately before planting and such pruning of the crown as may be necessary to compensate for root loss should be completed at this time.

There are four serious mistakes commonly made in transplanting and the first of these is lack of forethought in regard to drainage. A tree that is planted in a hole in tight clay soil, particularly when a more porous back fill is used, may as well have been planted in a concrete sump; it is bound to suffer from waterlogging at some stage of its life. Clay soils should be deeply ripped and the trees planted to a depth not greater than the ripping. A more expensive alternative is the provision of agricultural tile or rubble drains which permit drainage from the bottom of the hole. Drainage operations must be completed in advance of planting.

The second mistake is planting at an incorrect depth, usually too deeply. This may be due to the digging of too deep and large a hole, and subsequent subsidence, or more commonly to the belief that trees should be planted deeper than they grew in the nursery in order to improve stability. In either case the plant suffers from poor soil aeration and from ponding of water at the base of the stem. These factors are more serious in clay soils than in sandy soils, but trees should always be planted at the same depth in the field as they grew in the nursery.

The third mistake is the deformation of the roots of an open-rooted tree by twisting or wedging them in a planting hole that is too small. The hole should be large enough to accommodate the bulk of the roots without bending them. Extra long roots can be reduced in length by cutting cleanly with sharp secateurs. If a post-hole digger has been used the sides of the hole tend to be smooth and may appear polished. This surface should be roughened with a spade or bar before the tree is planted, to promote root penetration and discourage circling of the roots.

The fourth mistake is faulty tamping and leaving of air pockets when back filling. The soil should be free of clods and undecomposed organic matter and should be replaced without transposing sub-soil and top-soil. If the natural soil is so poor that other soil must be used for satisfactory growth, the hole should be large enough to provide growing space for the roots of the maturing tree and attention given to possible drainage problems. Open-rooted trees should be shaken gently during backfilling to work the soil down into the root mass and the soil firmly tamped, to eliminate air pockets. With container and balled stock the holes should be checked individually for correct depth by measurement, or by trial if the trees are not too heavy, before removing the plants from their wrappings or containers. To minimise damage to the roots of a balled tree it should be lowered into the hole before cutting the bindings. As much of the hessian is then cut away from the sides as possible, leaving the single layer at the base to rot away. With containers, the base should be cut away before lowering the tree into position. The container is then cut down one side and removed. Any encircling roots should be pruned before backfilling.

A saucer-shaped depression should be made around each tree immediately after planting and filled with water. It must be remembered that container stock in the nursery probably received almost daily watering and will suffer severely if watered only at long intervals in the field. The use of a layer of mulch, 3-4 inches deep, will markedly reduce loss of water by evaporation, weed competition and the effects of wide temperature fluctuations, but mulch must be kept away from the base of the stem to avoid the risk of collar rot.

Large trees should be staked or guyed promptly after planting but the use of fertiliser is best postponed for about six months until the root system has regenerated and is capable of utilising added nutrients. An exception to this rule can be made if safe, slow-release fertilisers are available at a reasonable price.

Grass and weed growth provide severe competition for young trees and should be kept in check by cultivation or judicious use of weedicides known to be harmless to the trees grown, or from which the trees can be shielded.

PRUNING

The pruning of ornamental trees, as distinct from the highly specialised methods applied to orchard species, is primarily concerned with training to produce a satisfactory stem and well-formed crown, the removal of dead or dying branches, and branches which may be a hazard to life or property.

Training should be commenced early in the life of the tree since if delayed it results in large wounds and greater difficulty in encouraging the even development and form usually desired. Trees that are open-grown tend to develop more massive branches close to the ground and less height growth than trees in forest formation. To develop a clean trunk and promote growth in the leader the lower limbs should be progressively removed, unless required for shelter from wind, always bearing in mind that extensive removal of branches carrying healthy foliage results in a check to growth. It is better, therefore, to remove a few of the lowest branches each year until the required height of clear stem is obtained, than to do all the work in a single operation. Most eucalypts are self-pruning and the genus has developed a highly efficient method of shedding the lower branches up to about three quarters of an inch in diameter. Only branches over this diameter likely to develop into large limbs should be removed from the lower part of the stem. Live branches may also be removed if competing with the leader, overlapping other limbs and causing abrasion, or because of undue competition for sunlight.

The development of double leaders should be prevented, not only to improve the form of the tree but because the acute angle between them is a potential weakness likely to cause splitting and breakage of one or both leaders of the maturing tree during high winds. Wounds caused by such breakages are unsightly and difficult to heal and render the tree highly susceptible to attack by insects and fungal diseases. The removal of one leader early in the life of the tree leaves only a small wound and the plant's food material is directed towards the better development of the remaining leader. Where an ornamental tree has been allowed to develop two or more leaders so large that removal of one would seriously affect the form of the tree it should be braced to avoid storm breakage. This practice will be described in detail later in this chapter.

When branches must be shortened back or entirely removed they should be cut to a side branch or flush with the trunk of the tree. A long branch stub does not heal but tends to die back to the stem, branch or bud from which the food material must come to promote the callus tissue which heals the wounds.

Callus tissue does not readily grow upwards to cover a branch stub and as the dying back of the bark further interferes with healing, the dead stub remains a possible site of insect and disease attack. When cutting back to a side branch or bud these should be chosen to provide future growth in the direction required to best shape the crown. The cut should start just above the branch or bud and slope back at an angle of about 45 degrees. A right-angled cut does not heal as readily and a cut at too sharp an angle weakens the branch.

When cutting back to the main stem, the cut should be nearly flush and as parallel with the stem as practicable; and the edges trimmed smooth with a sharp blade. The pointing of the cut above and below the wound with knife or chisel will facilitate healing still further. The painting of all pruning wounds with a suitable preservative will prevent insect and disease attack, but in practice this treatment is generally restricted to wounds over an inch or two in diameter.

When removing a large branch it should first be severed a foot or two from the trunk by first making an undercut until the blade of the saw begins to bind and then cutting the upper side of the branch a little beyond the under cut. This avoids the splitting and tearing of the wood and bark down the stem, and leaves only a stub to be supported while the final cut is made. If the falling of a heavy limb is likely to cause damage to property it must be supported by ropes slung over branches above it or, preferably, by ropes through pulley blocks with a sound anchorage in the upper limbs and trunk. The cutting of branches in this situation should be left to a skilled tree surgeon since the severed branch can easily swing back and seriously injure the pruner. There is always an element of danger involved in the cutting of large limbs in the crown of a tree and every precaution should be taken to avoid accident.

Dead, damaged or diseased branches should always be removed as soon as practicable after they are noticed, as they are focal points from which disease and insect attack can spread to healthy tissues. They must be cut back to healthy growing tissue in the same manner as that described for live branches, but an added precaution is needed; pruning tools must be sterilised before being used on another tree, and preferably before use on a healthy limb of the same tree if disease is apparent. A cheap, effective sterilant is methylated spirits which does not cause rusting of tools and as it evaporates quickly it does not harm healthy tissue, but precautions must be taken against fire.

POLLARDING

This is the practice of severely heading back a tree by cutting off the tops of all branches above a certain length to utilise the prunings, to prevent interference with overhead wires or buildings, or to stimulate greater density of foliage in the crown. Pollarding is necessary for commercial reasons with such species as willows grown for canes, the growth of which is stimulated by severe pruning, and trees utilised for stock fodder in times of drought.

With few exceptions the pollarding of ornamental trees, like the surgical amputation of human limbs, is a practice to be avoided as far as possible; not only are such operations expensive but the results are frequently unsightly and the subject suffers in the process. The exceptions are unhealthy and stagheaded trees which may be revitalised by heavy pruning and species such as sugar gum (*Eucalyptus cladocalyx*) which, when allowed to grow freely, develop large open crowns somewhat sparsely foliaged. The pollarding of such species stimulates development of a more compact, densely-foliaged crown capable of casting heavy shade.

The pollarding of street trees frequently demonstrates nothing but the lack of forethought of the planners in choosing species which would interfere with overhead wires or neighbouring buildings. In choosing trees for any particular site thought must be given to the space which they will occupy and the species chosen in accordance with a knowledge of their natural dimensions of crown and root system when mature. This forethought not only results in the planting of trees of greater aesthetic appeal but saves the high cost of a repetitive and time-consuming operation.

GUYING, STAKING AND BRACING

GUYING AND STAKING

These are the principal means used to provide mechanical support to a tree in the first year or two after transplanting. The aim is simply to provide support against wind to supplement the anchorage provided by the roots until the soil has consolidated and the roots have grown sufficiently to provide all that is required.

Large trees require three guy wires or cables attached to a loop encased in garden hose or other protective covering, and encircling the stem 8-10 feet from the ground. The ends of the guys are attached to stakes driven into the ground at evenly spaced intervals about 8-10 feet from the tree. Guy wires of this kind are a danger to pedestrians and in public areas they should be made clearly visible by attachment of white-painted slats. An alternative in such areas is the use of three wooden props with either natural forks at the ends or a Y-shaped metal prong inserted into the end of each pole. The projecting arms of the prongs should be padded and propped against the base of a lateral branch. The advantage of guy wires is that they can be slackened in the second year to allow some wind sway and thus stimulate the development of a satisfactory anchoring root system. Guy wires should be removed altogether by the end of the second growing season. A danger with the use of wooden props is that they are rigid and their abrupt removal leaves the tree susceptible to wind-throw.

Stakes provide sufficient support for smaller trees. If the tree is open-rooted, one stout stake is satisfactory and it should be driven into place before the tree is planted. During planting the roots are so arranged that the stem is within 3-4 inches of the stake. After replacement of the soil a wire, run through a piece of old garden hose, is fixed near the top of the stake passing in figure-8 fashion around both tree and stake.

If the tree is planted with a ball of soil around the roots, a single stake close enough to the stem to provide the necessary support is liable to cause root damage when driven into place. In this situation two stakes must be used set far enough from the stem to avoid damage to the root ball and the stem is then supported between them with two loops of wire run through protective hose.

If stock guards are required they must be large enough and strong enough to protect the growing tree from damage by any stock expected to be present in the area. Four stout stakes or posts should be used, with wire mesh fixed around them, and the tops should be braced with struts. Two opposite posts can be used to support the trees. The stem of the tree must be free enough to allow for expansion in diameter and neither wire nor rope should be bound around it. Again, the ties should be slackened in the second season after transplanting and the trees allowed to develop wind firmness. During this period they should be examined periodically to ensure that the loops are so fixed that no serious abrasion occurs.

Trees which have developed a lean due to strong winds and sodden ground conditions can be straightened and supported with guy wires or props as described for newly transplanted trees, provided the damage is treated promptly while the soil is still wet.

BRACING AND CABLING

Bracing and cabling are terms used to describe the structural support given to adjoining branches or stems of trees with deep V-shaped crotches. Such junctions are weak and liable to split in high winds. A steel rod, threaded at both ends, is passed through holes drilled in line through two adjoining limbs or stems and nuts are fitted over diamond- or oval-shaped washers set in line with the surface of the stems. The bark should be cut away neatly to allow the washers to sit flush against

Top line: These four illustrations show various methods for staking and guying trees.*

Centre left: Large tree prepared for transplanting; butt padded to avoid abrasion during shifting. The upper branches were bound with hessian in the previous year to prevent sunscald following preparatory root pruning.†

Bottom left: Large balled transplants require support during transport to avoid damage to the root ball and branches.†

Bottom right: Bracing with steel bolts prevents splitting of deep V-crotches.*

(* *Photos V. Russell.* † *Photos P. Hahn.*)

Plate 90 THE CARE OF VALUABLE TREES—I

Top line: These three illustrations are of the same tree.

Left: Shows the damage done by a car.

Centre: The first stage in repair aims at trimming the edges of the wound to a line where the live bark and cambium are undamaged. The shape of the final cut should be elliptical or pointed in line with the stem.

Right: Wound dressing is applied to the whole of the exposed surface.

Centre line: Removal of a large branch.

Left: The pruning of a large branch requires three cuts. The undercut is made first, several inches from the trunk. The second cut is made on the upper side of the branch a short distance from the first cut.

Right: The final cut is made close to the stem. The cavity revealed in this example requires cleaning before application of a wound dressing.

Bottom left: A tree ring-barked by vandals shows the development across the wound of callus arising from traces of undamaged cambial tissue.

Bottom right: The practice of bridge grafting which may prove successful in saving a valuable tree ring-barked by vandals.

(*Photos V. Russell.*)

Plate 91 THE CARE OF VALUABLE TREES—II

the sapwood and the surplus ends of the rod cut off. In time, callus growth will occlude the whole of the nut and bolt head. *Under no circumstances should wire be wound around the limbs to bind them together since this will cut through the phloem and cambial tissue and seriously interfere with translocation of plant food materials.*

Some species are particularly prone to the development of a number of massive, acute-angled branches, and where a danger of splitting is apparent they should be braced together by bolts or by steel cables attached to eye bolts passing through the limbs. If cables are used care must be taken to set the eye bolts in line with the cables to avoid twisting of the limbs under stress of wind sway.

When logs from trees in which steel rods or other metallic material has been used for support in early life are disposed of, any sawmillers who may wish to purchase them should be warned of the fact. There can be grave danger to life when saws strike such objects.

THE TREATMENT OF WOUNDS

Wounds on trees may be caused by a variety of agents such as pruning, motor cars, small boys, lovers and lightning, but the principles of treatment are the same; it should be carried out as soon as possible after the damage is noted, the growth of healing callus promoted and exposed tissue coated with a protective dressing.

Large wounds caused by impact require smoothing over to remove splinters or stubs of wood. The edges of the wound should then be trimmed smoothly with a sharp blade to a line where the live bark and cambium is undamaged. The shape of the final cut should be elliptical or pointed in line with the stem or branch. Cuts made in the bark straight across the stem do not heal so readily, as the bark has a tendency to lift away from the wood and die back. A most useful tool for trimming a large wound is a tomahawk with a very keen edge. It is not to be used with a chopping action, except for the first rough trimming, but with a rolling action, pressure being applied to the back of the head with one hand while the other guides the handle. The edges of the wound are then coated with a dressing that is not toxic to live tissue, and shellac is excellent for this purpose. When this has dried the whole wound should be coated with a suitable proprietary brand of bituminous paint.

Damage by motor vehicles varies from slight to complete destruction but a severe impact may be enough to shatter the bark for three quarters of the circumference of the tree without causing breakage of the stem. Such trees may still be saved if the wound is treated promptly but they should be inspected frequently for signs of stress. Greater attention must be given to their water and mineral nutrient requirements and they may respond well to soluble complete fertilisers applied periodically either by foliar spray or soil injection.

Branches which have been severely damaged or broken by storms must be cut back either to a side branch or the main stem, in the manner previously described in the section on pruning. When large branches are removed the cut surfaces are liable to checking and the cracks so formed allow water, insects and disease organisms to penetrate deeply. This can be prevented if treatment is carried out promptly after removal of the branch. After the surface has been painted with bituminous paint, a layer of heavy canvas is tacked over the wound, the edges being cut smaller than the wound to allow for callus growth. The surface of the canvas is then thoroughly coated with the same paint. As with all large wounds the tree should be inspected regularly and a fresh coating applied when necessary.

Where splitting of a deep V-crotch has commenced, a mercuric fungicide should be sprayed into it and allowed to run down the split to cover as much of the exposed surfaces as possible. The limbs must then be braced with a steel rod or cable, as previously described, and the split completely closed. Grafting wax is of value for sealing the edges of the split before application of bituminous paint. Steel braces used should be left permanently in place because, although healing may be complete, the site of the split remains a potential weakness.

The carving of names on trees is a common pastime of children and people with little else to advertise and the application of bituminous paint is usually sufficient to promote healing and avoid disease or insect attack.

RINGBARKING

Ringbarking by vandals results in wounds much more difficult to treat than wounds caused by simple impact but, provided the sapwood is not seriously damaged and treatment is prompt, the position is not always hopeless. Frequently it will be found that small strips of inner bark and cambial tissue have been left intact and these should be carefully preserved. If the ring is narrow, simple packing with wet sterile clay, wrapping with moist hessian and covering the whole area with black polythene bound above and below, will prove sufficient. This area should then be shaded to prevent heating by direct sunlight. The tree should be inspected frequently and the hessian moistened when necessary until the callus tissue has grown over the ring.

Wider rings are more difficult to treat and may require bridge grafting. This involves the insertion at approximately 3 to 4-inch intervals around the stem of several segments of a branch, about half an inch in diameter, between the upper and lower cut edges. The principles of grafting apply here, the aim being to obtain the closest possible contact between the cambium of the scions or 'bridges' and the cambium of the tree stem. Each end of the scion is cut to a wedge shape and the ends inserted beneath the bark above and below the cut. The scions must be small enough in diameter to allow them to be bent without breakage and to permit insertion without further damage to the bark. The scions are most readily obtained by pruning part of the crown of the damaged tree. This avoids any possibility of incompatibility between scion and stock and ensures the use of fresh material. After application of grafting wax to seal the junctions of the scions and stem, the whole area should be painted with a wound dressing and shaded from the sun.

An alternative to the use of bridging scions is the use of strips of live bark cut from branches or stems of similar diameter, and neatly fitted across the gap for its full circumference. These are tacked in place and bound firmly with polythene tape after the joins have been sealed with grafting wax. The elements most affecting success in both methods are neatness of fit and speed of operation, and care must be taken to avoid inadvertently reversing the direction of sap flow in the scions.

Ringbarking may stimulate the growth of coppice shoots or suckers below the ring and these should not be removed until healing is complete because they make up for some of the food material the root system no longer receives from the crown of the tree.

CAVITY TREATMENT

Callus tissue does not form readily over a gaping hole and if the removal of a large branch reveals a cavity this should be filled or covered to prevent water and vermin entering the cavity. Concrete and cement mortar have been frequently used to fill cavities but such materials have many disadvantages. They are completely rigid and do not allow movement with wind sway. They swell or shrink differentially from wood under the influence of changes in temperature and this prevents formation of a good seal against water between the wood and the filling. One of the worst features is that if or when callus tissue forms over the concrete the filling tends to be forgotten by the time the tree has to be felled because of old age or disease and the effects on a chain saw can be quite dramatic and very dangerous to the operator. Asphaltic mixtures of bitumen with sand or, preferably, sawdust are more satisfactory but have a tendency to soften and run or slump in hot weather.

The objects of treating cavities are to give structural support to the tree and prevent accumulation of water, to destroy harbours for vermin and to retard the growth of wood-rotting fungi. It must be realised that the filling of a cavity, unless it is small and completely accessible, does not prevent the spread of wood-rotting fungi already present, since these will have extended from the area of obvious wood rot into apparently sound timber and removal of this may seriously reduce structural stability. The filling of a large cavity, if done properly, is the most expensive treatment a tree is likely to receive and consideration should be given to the age, vitality and value of the tree before the decision to fill a cavity is made.

Whether a cavity is to be filled, covered or left open it must be cleaned, and all exposed surfaces of the heartwood painted or sprayed with at least two coats of creosote. If sapwood has been exposed during cleaning it should be coated with a bituminous paint, because creosote is toxic to live tissues. Water must be drained out, if necessary by boring an upward sloping hole through bark and sapwood into the base of the cavity and inserting a drainpipe.

Covering a large cavity with sheet metal is cheaper than filling it, and satisfies the same objectives except provision of structural support. A smooth ledge is cut with hammer and chisel into sound wood around the edges of the cavity, and exposed tissue painted immediately with shellac. Heavy paper or cardboard can be used to make a template so that metal sheeting (copper or galvanised iron) can be cut to give a neat fit. The metal sheet is then nailed in place and painted with bituminous paint to completely seal the edges.

If a decision is made to fill an extensive cavity in preference to leaving it open or covering it, the work should only be undertaken by someone experienced in the requirements of such work. If non-flexible materials such as cement mortar or concrete are used, allowance must be made for drainage of water which would otherwise collect between the wood and the filling material. The filling must be placed in layers about 6 inches deep with polythene sheeting or roofing paper between each layer so that some movement by wind sway is possible. Cement fills are generally more satisfactory in small cavities or cavities at the base of the trunk since wind sway tends to cause the filler in a large cavity to crumble or damages the wood still further by friction and grinding. Finally, the filling should not extend outwards beyond the cambium around the edge of the cavity, because this inhibits development of callus tissue.

WOUND DRESSINGS

The ideal dressing for a wound disinfects the area treated, remains toxic to disease and insects, but non-toxic to living plant tissues; it prevents the development of fissures due to drying of exposed wood; it is readily adhesive, waterproof, flexible and does not crack on drying. Unfortunately no such material exists at present and materials currently available have, at best, only a few of these requirements.

Shellac is one of the most useful and least harmful dressings for live bark and cambial tissue, but it is not durable and after drying should be coated with a more durable material. Household paints for exterior use may injure cambial tissues and although more durable than shellac, are not as long-lasting as bituminous paints. Creosote paints are highly effective for destroying wood-decay fungi and insects, but they are also toxic to live plant tissues and their use should be restricted to treatment of heartwood in cavities. Bordeaux pastes are antiseptic, but harmful to growing tissue and non-durable.

The most suitable general-purpose tree paints are those with a bitumen base and there are a number of these on the market. An example is the proprietary brand 'Stayseal' which is durable, flexible, waterproof and does not crack on drying, but contains no fungicide. It consists of 65 per cent bitumen and 35 per cent solvents. Such mixtures are more harmful to cambial tissues than shellac and may suppress callus growth to some extent, but ease of application and durability make them very useful as general purpose wound dressings.

TREATMENT OF OTHER INJURIES

SUNSCALD

Sunscald may result from severe pruning, defoliation due to drought, or removal of neighbouring trees with consequent loss of shade in hot weather. Large trees recently transplanted are particularly susceptible to sunscald and the stem and larger branches may be protected by spiral wrapping and overlapping of strips of hessian or bitumenised paper. This material should be removed after the first summer as it provides a harbour for insect pests and the accumulation of moisture fosters development of fungal cankers. Attention should be given to water and nutrient requirements to stimulate growth of foliage.

Sunscald is most severe on thin-barked species and causes death by desiccation of phloem and cambial tissues inadequately protected by the bark. The dead bark may adhere firmly to the sapwood and recovery by growth of callus from the outer edges of the scald is slow and leaves a permanent fault in the wood. Treatment involves stripping away all dead bark tissue, trimming the edges of the scald, painting with a preservative paint and shading.

LIGHTNING

Lightning causes damage more frequently than is generally realised since it is usually only the rare and spectacular damage that is recognised. Lightning damage is extremely variable and may leave little or no evidence or merely cause a few strips of bark to be blasted off. The electric current tends to follow the moist tissues of the inner bark and cambium, and the heat generated may be sufficient to explosively vaporise the sap and blast off strips of bark or shatter the wood without leaving any sign of scorching. The damage can appear to be mechanical, such as that due to scraping by a machine, but this would damage branches in addition. Lightning frequently strips pieces of bark off the stem close to branches without the latter being affected. Apparently undamaged trees adjacent to one that has obviously been struck will often die, but this may take a period of three years or more and there is often no visible external sign of damage. It may be that these trees were connected to the first by natural grafting of the roots or at least in very close contact with the roots of the tree struck and if the soil is dry, the electric current would tend to follow the sapstream of the roots until it was dissipated. Presumably the trees die because of damage to phloem and cambium tissues and subsequent inability of the root system to provide the requirements of the tree.

When lightning has shattered an appreciable part of the main stem, treatment is usually a waste of time, as the tree would probably die in spite of wound treatment. If only a few strips of bark have been removed, injuries may be treated as for other mechanical wounds, but attempts to trace the passage of the current between these strips is difficult and likely to cause unnecessary damage. The tree should be inspected frequently for signs of lifting bark and treated if necessary.

Particularly valuable trees exposed to the risk of lightning can be protected with conductors as used for buildings, the only difference being that for trees allowance must be made for wind sway by using flexible materials and the conductor must be extended periodically as the tree grows.

SOIL GRADING OPERATIONS

Grading operations, by changing the natural soil level, are a frequent cause of death of trees in newly built-up areas. Species vary markedly in their response to changes in the grade level of the soil in which they grow and many of the eucalypts are particularly sensitive. Unless a species is known to be little affected by the kind of grade change made necessary, for example by building construction, every precaution should be taken to minimise the harmful effects. Removal of even a few inches of topsoil can destroy a large portion of the feeding root system and result in desiccation of part of that remaining. If roots are buried deeply by a fill they suffer from lack of oxygen due to poor aeration. If only a few inches of fill is to be placed this must be of highly porous material. If more is required tile drains should be laid radially in the manner of spokes of a wheel, sloping away from a well, the 'hub' built around the trunk of the tree. The well allows room for growth and prevents soil from contacting the trunk. The outer ends of the 'spokes' are then connected with a circular rim of drain tiles just beyond the branch spread and provision made for free drainage from the lowest point. The whole area is then covered with coarse gravel, or crushed rock to within 6-8 inches of the finished grade and porous top-soil used to grade level.

BITUMEN SEALING

Bitumen sealing close to the trunks of established trees in a newly-developed parking area is a frequent cause of slow death of the trees for a number of reasons. The first is that the seal prevents penetration of water to the root system, the second that grade changes and compaction are usually necessary in preparing the base for sealing and third, that normal aeration is prevented. There is little that can be done as a curative measure without great expense, but the problem can be avoided by laying drainage tiles and crushed rock, as described in the section above on soil grading operations, before preparing a base for sealing. The drainage tiles not only allow for aeration but also for irrigation. The cost of such work must be weighed against the value and expected life of the trees.

SOIL COMPACTION

Compaction of the soil under trees occurs due to trampling by humans and domestic stock and parking of cars. Natural leaf litter is pulverised and blown away, water penetration is severely restricted and aeration seriously retarded. The combination of these factors frequently results in loss of vigour and slow deterioration of the trees. Cultivation to improve the situation may do more harm than good by destroying a large portion of the feeding root system. The most suitable treatment is by water and fertiliser injection with a high pressure spray nozzle. This method will be described in the section on the use of fertilisers.

DISEASE AND INSECT PEST CONTROL

This is a most important aspect of the care of trees and correct identification of the disease or insect pest, or at least the group to which the disease or pest belongs, is essential for effective treatment. To deal with such a subject adequately requires a volume by itself and in portion of a chapter the description of a few aspects and examples would be of little value for an appreciation of such a complex subject. The reader is therefore referred to other books on the subject and, if sound local knowledge is not available, he should seek advice on specific problems from departments of forestry, agriculture or other organisations carrying out research in pest and disease control. Advice is freely available from such organisations in all Australian States.

For people seeking such advice it is important for them to appreciate that the best description is not as valuable as a good sample. If the technical name of the tree affected is unknown, a pressed and dried specimen of the foliage, and fruits, flowers or buds should be submitted. If an insect is soft-bodied, one or preferably several, should be preserved in methylated spirits before despatch; if hard-bodied it can be killed with an insecticide and either pinned or otherwise supported in a package so that it arrives in an undamaged condition. Samples of the damage caused are an advantage in identification of insect pests, and essential for diagnosis of diseases. The use of photographs which clearly show the damage inflicted may prove a substitute for samples but it must be realised that no entomologist or plant pathologist will make an identification nor prescribe treatment without adequate information, and the better the information provided, the more promptly an answer can be supplied.

At times there may be several different species of insects present on an unhealthy tree and there is a danger that the wrong one may be blamed for the major damage. To avoid such mistaken identity samples of all the insects present should be supplied and it is then essential to number them and keep duplicate numbered specimens for reference.

IRRIGATION

Ornamental trees may require irrigation for three years or more after establishment depending on the species chosen and the rainfall, and even mature trees may require supplementary water during droughts. Frequent light watering should be avoided as it encourages shallow rooting. Sufficient water should be applied on each occasion to ensure deep penetration. The usual method is to build a circular mound around the tree to form a basin of sufficient diameter to hold the water required. The mound built for watering a mature tree during drought should have at least the same diameter as the crown. A mound is preferred to a trench since the latter may cause damage to the root system.

Evaporation may cause considerable loss of the water applied but this can be reduced by mulching. In arid areas where conservation of water assumes greatest importance water use can be made more effective by other means. When the tree is first planted a post-hole auger may be used to dig two holes at the limit of root extension on opposite sides of the tree to a depth of at least two feet. The holes are then filled with coarse gravel or crushed rock and water is poured into these when required. As the tree matures additional watering points may be made as required near the limit of crown spread. An alternative to the use of gravel is the insertion of two six inch diameter drainage tiles placed vertically one above the other in each hole and capping with a suitable cover to prevent damage and gradual filling with leaf litter and soil.

THE USE OF FERTILISERS

The major constituents of the carbohydrates that form the basis of the food material and much of the tissues manufactured by plants, are carbon, hydrogen and oxygen which are derived from water and the atmosphere, but all other essential elements, with the partial exception of nitrogen, must be obtained from the soil. Nitrogen may be extracted from the atmosphere and fixed in the soil by certain bacteria, particularly the various species of *Rhizobium* which form nodules on the roots of legumes. This nitrogen eventually becomes available to other plants but frequently trees have no source of nitrogen except that derived rrom organic or inorganic materials in the soil and this element, together with phosphorus and potassium, is frequently deficient for satisfactory growth of trees. The other major essential elements for plant nutrition are sulphur, calcium and magnesium, but these are usually present in soils in adequate quantities.

There are also a number of other essential elements required in very small amounts and these include iron, boron, manganese, copper, molybdenum and zinc. Diagnosis of deficiencies of these elements can be made only by specialists in plant nutrition. In alkaline soils iron may be present in abundance but not in a form available to plants and if the problem is rightly diagnosed, the addition of iron chelate will correct the apparent deficiency.

The major constituents of commercial fertilisers are nitrogen (N), phosphorus (P) and potassium (K) and the concentration of each is usually given as a ratio of percentages, for example 10:8:6 indicates 10 per cent nitrogen, 8 per cent phosphorus and 6 per cent potassium. Such a fertiliser provides a good balance of the nutrients most commonly in demand by trees.

RATES AND FREQUENCY OF APPLICATION

Rates and frequency of application of fertilisers vary with the species, age and vigour of the trees and where response to fertiliser is not known it is better to use too little than too much. More fertiliser can always be added if necessary but an excess may prove toxic. Generally a rate of 2-3 oz of a 10:8:6 or 10:9:8 fertiliser per square yard of ground treated can be regarded as safe and adequate for most Australian soils. It is only the nitrogen in such a fertiliser which is readily leached from the soil and may require replenishment annually. Phosphorus and potassium are relatively immobile in the soil so the use of a complete fertiliser is not normally required more frequently than every 3-5 years.

METHODS OF APPLICATION

Methods of application of fertiliser vary from spreading by hand or machine to soil injection. Fertiliser is most easily applied by spreading over the surface of the ground, but regular surface application of a complete fertiliser has several disadvantages. Firstly, it encourages weeds, and removal of these by chipping or hoeing removes a considerable proportion of the fertiliser applied. Secondly, the practice encourages shallow rooting of the tree since phosphorus and, to a lesser extent, potassium are not leached deeply into the soil by water but remain in the top inch or two. Shallow-rooted trees suffer most during dry spells and require more frequent watering. Thirdly, if a tree is growing in a lawned area, the grass benefits from the fertiliser more than the tree.

Punch bars, crowbars and drills operated by compressed air have all been used for making holes into which fertiliser can be poured but all of these implements can cause serious root damage and they all tend to aggravate compaction problems. Soil compaction can be reduced, fertiliser placed where it can most readily be utilised, and root damage avoided by using hydraulic pressure of a water jet from a high pressure spray unit. The pressure required is about 200-250 lb per square inch for reasonable speed in operation, and the lance should be about 3 feet long with a ⅛-inch nozzle, and a trigger-type valve is essential. The use of an inverted metal dish about 8-10 inches in diameter, and with a hole in the centre through which the lance slides freely, prevents splashing of the operator. The water jet is directed more or less vertically into the ground and in other than stony soil the lance can be pushed in quite rapidly to a depth of 2 feet or more.

The roots of mature trees usually extend a little beyond the branch spread and there are few feeding roots within the inner third of the radius of this circle. The hole should therefore be made in a band in the outer two thirds of this circle, at a spacing of 12-18 inches and to a depth of 18-24 inches. When the ground has dried a pelleted free-flowing fertiliser of 10:8:6 to 10:6:4 composition can be poured into the holes at a rate of 2-3 oz per square yard of ground surface treated. A more rapid response can be obtained if soluble fertiliser such as 'Aquasol', 'Thrive' or 'Zest' is added to the water in the soil injection unit at a rate of about 1 lb per 100 gallons. This method is recommended to stimulate growth of trees which are lacking in vigour and to aid recovery from injury.

Foliage sprays of soluble fertilisers are of value in assisting trees to recover from damage to the root system caused by transplanting or by excavation. The rates used should be the same as those given above for soil injection, and the foliage should be sprayed to the point of heavy run-off. The amount of nutrient absorbed depends on the type of leaf and the total area of leaf surface; the method is of least value with trees which have been partially defoliated. The response to this method is not lasting and the spraying should be repeated at intervals of 7-10 days for several weeks.

FURTHER READING

Anderson, R. F. (1960). *Forest and Shade Tree Entomology.* John Wiley.

Conover, H. S. (1958). *Grounds Maintenance Handbook.* McGraw-Hill.

Hadlington, P. W. (1966). *Common Insect Pests of Trees.* Forestry Commission of N.S.W.

Peace, T. R. (1962). *Pathology of Trees and Shrubs.* Clarendon Press, Oxford.

Pirone, P. P. (1966). *Tree Maintenance.* Oxford University Press.

CHAPTER 23

The preservation of timber in service

by F. A. Dale

TIMBER is the most versatile building material in existence. It is light, strong, easy to work and to handle. Because it is an organic material it can be damaged by the elements and by living organisms, but given the proper protection it can be made to last indefinitely.

This chapter deals with the preservation of timber in the broadest sense, that is its protection against the weather, biological attack and fire. It can, of course, be damaged by overloading and physical wear, but these are outside the scope of this chapter. The main hazards affecting timber in service are:

Shrinkage and distortion
Weathering
Decay
Insects
Fire

In order to understand how timber is affected by shrinkage and distortion and how to protect it, some understanding of its physical properties is needed. It is a cellulosic material having considerable affinity for water. Green timber contains 50-150 per cent of its dry weight of water which keeps its cells and fibres swollen. As the wood dries these elements contract across the grain, but usually to only a negligible extent in a lengthwise direction. Because shrinkage in the tangential direction is roughly twice that occurring in the radial direction, that is from the centre to the surface of the tree, and because the surface of a piece dries more rapidly than the inside, stresses are set up which can cause distortion, splitting and surface checking. These may not affect its use for many purposes but are serious in others.

Most timbers do not start to shrink until the moisture content has fallen to 'fibre saturation point', at about 25-30 per cent moisture content, when all the free water has left the cells, like a sponge which has been wrung but is still damp. It then shrinks progressively until 'equilibrium moisture content' (e.m.c.) is reached, when it is in balance with the moisture content of the air around it. The e.m.c. in inland Australia varies from 6-15 per cent, depending on the locality and time of year. Hence it is important that when dimensional stability is critical, timber should be dried to the moisture content appropriate to its final location. This applies particularly to furniture, joinery and flooring where any movement in service is conspicuous and can cause binding or lack of fit. On the other hand, timber can be used green or partly dried for many structural purposes, particularly out of doors. Building frames and fences fall in this class. However, in much of inland Australia, where the e.m.c. is low, green timber used for building purposes dries out and distorts so rapidly that it may be unusable. In this case properly dried timber is much to be preferred.

Some Australian hardwoods suffer from an additional and abnormal form of shrinkage called 'collapse', in which the cells collapse and distort as the timber loses its free water. This causes severe distortion and makes the timber harder and more difficult to work. It can be overcome by 'reconditioning' of kiln or air-dried timber, but if this is not practical for building frames these must be erected before collapse becomes serious. Even then fixing of sheathing and lining can be affected.

Softwoods, such as radiata pine, generally shrink less than hardwoods but often contain knots and cross grain which cause distortion in drying. Growth stresses in fast-grown or immature timber can accentuate distortion in drying.

The effects of shrinkage can be overcome in most cases. Selection of the most suitable type and grade of timber for a particular use is very important. Timbers that dry with minimum degrade should be used where possible. Timber can be restrained as it dries by weighting drying stacks and these can be covered to prevent too rapid drying. End splits can be minimised by painting or greasing the end grain to reduce the much faster drying which takes place from these surfaces. Complete control of degrade may only be practical, however, for the saw-miller or timber merchant who can afford drying kilns and proper air-drying facilities.

The effects of shrinkage and distortion can also be minimised by proper design,

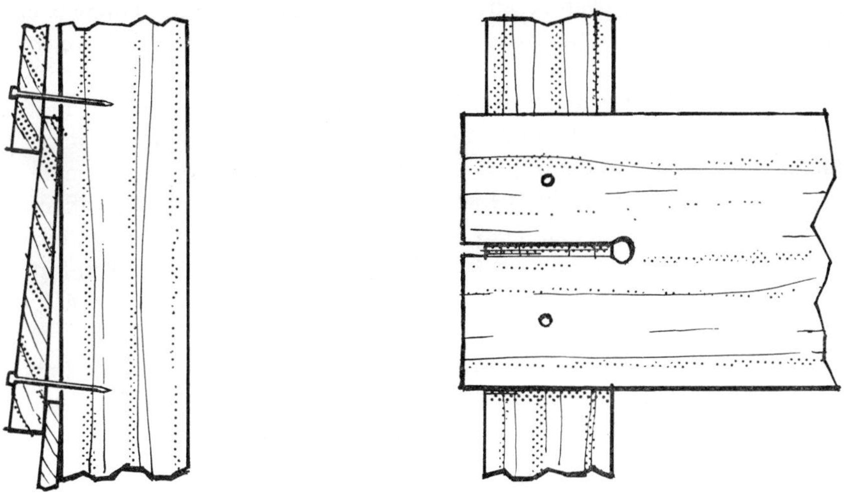

(a) A method of fastening weather-board to allow movement.

(b) Slotting the ends of boards or structural members reduces the possibility of splitting.

FIG. 23.1

The correct design of timber can minimise the effects of shrinkage.

particularly in exposed structures. Wide boards are more likely to split than narrow ones, particularly at the ends, and if they are used the fastenings should allow for this (see Figure 23.1). These precautions are much more important with timbers erected green or partly dried and with hardwoods. Dry softwoods are much less likely to cause trouble. Additional protection can be obtained by painting or oiling and this will be discussed under weathering.

Round timbers used as transmission poles or fence posts will always split as they dry out in service to an extent depending on their size, species and preservative treatment. These splits may affect the appearance, but only in rare cases do they affect strength or serviceability. Where appearance is important naturally durable timbers, for example the heavy hardwoods and softwoods such as cypress pine are preferred, but light hardwoods or radiata pine can be treated with preservative oils to reduce splitting.

WEATHERING

Timber exposed to the weather will suffer from moisture changes and sunlight. The surface will bleach and the fibres may separate. In many timbers, for example jarrah (*Eucalyptus marginata*) the effects may be confined to the outer 1/32 inch or so but in others, such as karri (*E. diversicolor*), checking may continue into the wood until it finally disintegrates. Horizontal surfaces which collect rainwater, for example the tops of rail sleepers, will suffer worst. As timber gets older surface breakdown may be accelerated by decay and in this condition it will catch fire more readily.

Painting is the best known and often the most effective protection, but it is expensive. If not properly maintained a neglected surface looks worse than unpainted timber and may even accentuate decay by trapping water. Oiling is cheaper and easier and is just as attractive for many purposes, but it will not give the same protection as painting unless repeated every year or two, particularly with hardwoods. The use of varnish for exterior timbers is not recommended as it needs more work than either painting or oiling to maintain its original appearance. Many manufacturers now make natural oil finishes based on the Madison formula developed by the U.S. Forest Products Laboratory, which include a preservative to prevent surface decay and mould. They are often pigmented so that as the oil vehicle weathers away the pigment remains to maintain the appearance, and will last longer on rough sawn surfaces than smooth dressed ones. A little wax is often added to make them water-repellent.

Cheaper finishes such as creosote oil, pentachlorphenol in mineral oil and even sump oil, used alone or in combination, will give a uniform color and water-repellent surface for a time, depending on the oil used. Generally speaking heavy, dark oil finishes will last longer than light or clear finishes. They are suitable for fences, rails, palings, sawn weatherboards or other rough timbers. The preservative component again prevents mould or decay.

Even naturally weather resistant timbers such as jarrah, cypress pine (*Callitris* spp.) and brush box (*Tristania conferta*) will benefit from the use of preservative oil finishes and their regular use will allow virtually any timber to be exposed to the weather. Any timber used must be free of sapwood, which is much more liable to decay and attack than heartwood.

Another approach is to impregnate suitable timbers with a preservative to make them resistant to decay and insects and preferably with the inclusion of a water repellent. Radiata pine, which is mostly sapwood, is ideal, but the sapwood of nearly all timbers can be impregnated. Timber so treated needs no further maintenance, but oil stains may be used to mask surface bleaching.

DECAY

Rot or decay is caused by fungi, living organisms which are always present in damp soil and whose microscopic spores are carried by wind and water. Mushrooms and toadstools are typical fruiting bodies of fungi. Wood destroying fungi feed on wood under suitable conditions of moisture and temperature in the presence of oxygen. Timber at less than 20 per cent moisture content is too dry and timber submerged in water is too wet for them to survive. Below 40°F fungi cannot destroy wood and they die very quickly at temperatures above 150°F.

Timber in the ground is readily attacked but timber above ground may be attacked if it is repeatedly wetted and does not dry out quickly. Timber in fresh water containing dissolved oxygen may be attacked by soft rot caused by another type of fungus, but this is much less common than ordinary decay. Fungal attack

TABLE 23.1

THE DURABILITY OF SOME AUSTRALIAN TIMBERS

Class 1(*a*)	wandoo	*Eucalyptus wandoo*
	grey box	*Eucalyptus* spp.
	ironbark	*Eucalyptus* spp.
	tallowwood	*E. microcorys*
	cypress pine	*Callitris* spp.
Class 2	jarrah	*E. marginata*
	red gum	*Eucalyptus* spp.
	blackbutt	*E. pilularis*
	turpentine	*Syncarpia glomulifera*
Class 3	messmate (changed name)	*E. obliqua*
	karri	*E. diversicolor*
	brush box	*Tristania conferta*
	rose gum	*E. grandis*
Class 4	mountain ash	*E. regnans*
	radiata pine	*Pinus radiata*
	hoop pine	*Araucaria cunninghamii*
	silky oak	*Grevillea robusta*

(*a*) These ratings apply to service in the ground and are reasonably valid for termite resistance as well as decay.

Class 1—Timbers of the highest natural durability which may be expected to resist both decay and termite attack for at least 25 years and sometimes 50 years.

Class 2—Timbers of high natural durability which may be expected to have a life of about 15 to 25 years.

Class 3—Timbers of only moderate durability which may be expected to have a life of about 8 to 15 years.

Class 4—Timbers of low durability which may last from about 1 to 8 years. These timbers have about the same durability as untreated sapwood, which is generally regarded as Class 4, irrespective of species.

softens and weakens wood and makes it much more susceptible to fire when it dries out. Visible evidence of attack are strands or threads through or over the wood and in advanced decay the presence of fruiting bodies such as mushrooms, toadstools or 'conks' on the wood or nearby.

The sapwood of nearly all timbers will decay easily, but the heartwood can vary enormously in its resistance. In general, light, soft timbers are much less durable than heavy, hard timbers, but there are notable exceptions.

Obviously where timber is likely to be attacked it is best to use the most durable timber available. Nowadays these are often scarce or too expensive. Sometimes a durable timber may be too heavy, too hard or too difficult to work for a particular purpose, but less durable timbers can be protected either by keeping them dry or by preservative treatment with chemicals poisonous to fungi.

Roofing is often the cheapest and most effective way of keeping timber dry above ground but painting is effective if properly maintained so that water cannot lodge in the joints, particularly in the end grain. With the wide variety of timbers now being used considerable extra protection can be obtained by using a water-repellent preservative before painting, as this helps to exclude water and prevent decay.

INSECTS

In inland Australia there are large areas of low rainfall where the risk of decay in wood above ground is negligible or very slight. Insects of many types attack timber of widely differing moisture content, but Australia has been lucky in having many native timbers such as jarrah, red gum, ironbark and cypress pine which are very resistant. By far the most destructive insects are subterranean termites ('white ants') which travel underground and attack live and dead trees as well as unprotected susceptible timber in buildings, poles and fence posts throughout mainland Australia. Because they avoid the light they can do great damage before they are discovered; like other insects they must have moisture and so require contact with the ground to survive. Termites live in colonies of many thousands of individuals and may forage for hundreds of yards for food, so that killing a few or removing the affected timber rarely affects the colony, which must be eradicated by more subtle means, such as dusting with arsenic. Quite simple measures, however, can be used to protect buildings, poles or fence posts.

Buildings can be protected by chemically treating the soil with insecticides, by mechanical barriers, for example ant caps, or by using only resistant or preservative-treated timber in their construction. The choice of method depends on the hazard, type of construction and materials available. Chemically treated soil barriers are best suited to slab-on-ground construction, while mechanical barriers are satisfactory for buildings well off the ground so that they can be regularly inspected. Treated or durable timber can be used as an added precaution where the hazard is high or where the other methods are impractical or inconvenient.

Poles and posts will not be attacked if they are of resistant or treated timber, but chemical soil treatment may be needed where the hazard is very high, for example in the Northern Territory or the north-west of Western Australia.

Of the timber borers the most important is the *Lyctus* or powder-post beetle, the larva of which attacks the sapwood of some hardwoods containing starch. In the southern States this is confined to a few timbers such as messmate with narrow sapwood the loss of which, if destroyed by insects, has negligible effect on structural strength. It can be a nuisance, however, if it attacks hardwood trim or flooring or if it emerges through plaster sheet from infested framing. The simplest protection in these cases is to exclude susceptible sapwood from mouldings and limit the amount in framing timbers.

In Queensland and New South Wales there are a number of timbers with very wide susceptible sapwood, such as brown tulip oak (*Argyrodendron trifoliolatum*), which must be immunised against attack beforehand and this is controlled by State legislation. Other timbers with susceptible sapwood, such as spotted gum (*Eucalyptus maculata*), must be treated before use in the round for posts or poles, otherwise the sapwood will be rapidly destroyed. Recovery or yield of sawn timber from the log is increased if the sapwood is treated.

PRESERVATIVES

Timber can be protected against decay and insects if properly treated with preservatives. These can be divided into two classes, preservative oils and water-borne preservatives.

Preservative oils include creosote oil and pentachlorphenol in heavy mineral oil, which are used for preserving fence posts, poles, rail sleepers, cross arms and construction timbers. Pentachlorphenol, copper naphthenate, tributyl-tin oxide and other compounds dissolved in lighter oils are used for protection of external timbers, usually followed by painting to prevent loss due to weathering.

The most important water-borne preservatives are the copper-chrome-arsenic or CCA preservatives which fix in the wood and become virtually insoluble after treatment. The arsenic in CCA preservatives is particularly effective in preventing termite attack. A recent development is the incorporation of a very small amount of arsenic in creosote oil for the same purpose, so that there is resistance to termites even at low retentions.

The preservative oils are water repellent and help to prevent splitting and checking. The water-borne preservatives give clean, paintable surfaces but are not water repellent, although a recently-developed additive for one proprietary preservative substantially reduces checking of exposed timber.

Where timber is only required to be immunised against borer attack in buildings this can be done with a low loading of simple water soluble chemicals such as boric acid, borax and sodium fluoride. These are mainly used for the prevention of *Lyctus* attack in susceptible hardwoods.

THE PROPER USE OF PRESERVATIVES

Surface applications of any preservative by brushing, spray or dip are of limited value. On timber above ground the surface application of preservative oils, if properly maintained, will protect against weathering and superficial decay. It will not protect susceptible sapwood against insects such as *Lyctus*, which often lay their eggs before the preservative can be applied and which will emerge through it. Nor will it protect against termites if they can bypass the surface treatment to reach the wood beneath. These surface films are readily oxidised, evaporated or washed away by rain.

If the timber can be impregnated with preservative so that the latter is protected from weathering or leaching (washing out) much better protection is obtained, and if enough insoluble preservative is introduced this protection can last for over 70 years. It is safe to assume that moderate loadings of preservative oils or fixed water-borne salts will protect timber in the ground for 30-40 years.

The heartwood of most timbers is very hard to impregnate but the sapwood of practically any timber can be treated fairly easily. This means that round timber can usually be protected by treating the sapwood band alone and this is the basis of the commercial treatment of posts and poles now well established throughout Australia. Some timbers, particularly softwoods like radiata pine, have a very wide sapwood and sometimes a permeable heartwood. These can be treated after sawing, when they are most useful for all types of construction.

Most commercial treatment is done under a pressure of 200 psi, but the heartwood of light and medium density eucalypts can be impregnated if higher pressures, up to 1,000 psi, are employed. Two such plants using preservative oils and one using CCA salts are now established in Australia and can treat sawn hardwoods for structural purposes.

PREPARING FOR TREATMENT

Timber for treatment with preservative oil must be dried below fibre saturation point (25 per cent) for satisfactory impregnation and, in general, the heavier the timber the lower the moisture content required. Timber for treatment with fixed water-borne preservatives can be treated at a higher figure, 30 per cent for hardwoods and up to 40 per cent for softwoods.

Round timber should have all inner and outer bark removed when it is prepared. It should be stacked to air dry on proper foundations at least 12 inches above weed-free soil. Each layer of posts or poles should be separated by several sawn strips or rounds to allow maximum air movement (Fig. 23.2). If the timbers tend to split at the ends extra length should be allowed for docking unuseable material. Alternatively, posts may be dried in long lengths which are cut to post size before treatment. End coating with grease or wax emulsion will reduce end splitting as also will drying under cover in hot, dry weather.

Ideally the moisture content of drying timber should be checked with a moisture meter; failing this smaller timbers may be weighed at intervals until the weight is reasonably constant. Pine may safely be treated if it has dried to half its green weight.

Adequate drying is essential for good treatment. Under the best conditions round posts will dry in 2-3 months, but drying slows down and may cease in cold, wet weather. Rain wetting will raise the surface moisture content, and treatment after prolonged heavy rain must be delayed until the wood has re-dried. Pine is easy to air-dry, but prolonged wet or very humid weather can slow or even stop drying. Under these conditions 'blue stain' and sometimes decay can occur before

Top left: POOR STACKING. These stacks are too close to the ground, too high and too close together. Posts in these stacks will vary in moisture content and treatment will give variable results.

Top right: GOOD STACKING. Drying stacks for posts or poles should be raised above ground with plenty of room for air circulation. The ground must be well drained and free of weeds and rubbish. Posts dried in bundles take 1-2 months longer to dry but are easier to handle. Even in these stacks the bottom posts are too close to the ground, 12-18″ is the minimum. Covers on top of the stacks will prevent re-wetting, mould or decay and excessive splitting. The ideal is an open drying shed.

Below left: PORTABLE LOW-PRESSURE TREATMENT PLANT. This plant will treat hardwoods with CCA solution at 50-100 psi in 2-3 hours. One engine drives both vacuum and pressure pumps. All valves are quick acting ball cocks. Note flanged end for connecting long extension to treat single poles. Towing weight (empty) about 2 tons. Price about $(A)3250 in 1970.

Below right: Farmers can assemble their own plant like this, although they usually buy the treating cylinder. Pumps, storage and mix tanks can be connected by hoses. Having the plant in a shed allows wet weather treatment. A fence to keep out stock is desirable.

it is ready to treat. Blue stain affects the absorption of preservative and decay affects the strength of the timber. In inland Australia drying conditions are good for most of the year, but very cold weather or monsoon rains will upset them.

METHODS OF PRESERVATION

Dry round or permeable sawn timber can be impregnated with preservative by one of the following methods:

Pressure treatment
Hot and cold bath
Cold soaking

Pressure treatment is the most versatile, reliable and widely-used method nowadays. In its simplest form, dry timber is put in a vessel which is then sealed and filled with liquid preservative. Pressure is applied to the liquid and more preservative is pumped in until the wood will accept no more. The liquid is then drawn off and the timber removed. This treatment is suitable for both oil and water-borne preservatives, but the former usually require heating to improve penetration and give a reasonably clean surface.

(a) *Full-cell treatment.* In this method an initial vacuum is drawn to remove air from the wood and maintained while filling the cylinder to ensure maximum absorption of liquid. A short final vacuum is used to remove 'drip' or surplus preservative. This is the method most widely used with water-borne preservatives.

(b) *Empty-cell treatments.* In the simplest of these, the Lowry process, the cylinder is filled at atmospheric pressure. Some surplus preservative is removed as air escapes from the wood during the final vacuum. If the same penetration is required with a minimum final retention of preservative the cylinder is first filled with compressed air and the liquid pumped in afterwards. This ensures maximum 'kickback' of preservative during the final vacuum. This is the Rueping method. Both forms of empty-cell treatment are used with hardwood and softwood when treating with preservative oils.

(c) *Low-pressure treatment.* Hardwoods can often be treated with CCA preservatives at pressures of 50-100 psi instead of the usual 200 psi. This means that the equipment can be of lighter construction and can be made portable, so that it can be taken into the forest or plantation (see Plate 92). These plants are particularly suitable for remote areas and have been used for treating teatree (*Melaleuca* spp.), brigalow (*Acacia harpophylla*), mallee (*Eucalyptus* spp.) and other timbers not usually considered for treatment. Full-cell treatment is mostly used but even the Lowry treatment is effective.

Where smaller quantities are to be treated there are several simple methods which need less equipment and know-how.

The most effective of these is the *hot and cold bath treatment* in which the dry timber is heated in hot creosote or oil until all the air is expelled from the sapwood, at about 180-210°F. The bath is then allowed to cool and the oil is forced in by atmospheric pressure to replace the expelled air. Alternatively the timber may be heated in boiling water or steam and then quenched in cold preservative. The process is simple but takes longer (6-12 hours) than pressure treatment and it is difficult to prevent over-treatment in light, very permeable timbers such as radiata pine. If the preservative oil is heated over an open fire there is also some risk of fire if it becomes overheated. As a do-it-yourself method it is attractive because the butts of posts or poles can be heavily treated for maximum service in the ground.

Cold soaking with preservative oils is the simplest of all dry timber treatments but is only effective in lighter, more permeable timbers and must be continued until the wood will absorb no more preservative. Control of retention is virtually impossible and treatment may take a week or even longer. However, this method also allows for heavy butt treatment followed by a light top treatment for the upper parts of posts and poles. This can reduce the cost without risk. The top of a post or pole requires much less preservative than the butt for the same length of service.

Green round timber can be treated with fixed water-borne preservative by the simplest of all methods, the *sap-replacement process*. This must be done only with freshly-cut timber from which all the bark has been removed. One end of the post is submerged in preservative solution and as the sap evaporates from the exposed surface the solution is drawn up to replace it. If done carefully in good drying weather this treatment can give reasonable groundline protection for posts and poles but it is not always successful and careful control, including examination of the penetration obtained, is essential. The whole process takes about a week, including top treatment which is obtained by subsequently submerging the top end for a shorter period.

Although the process is attractive because of its simplicity and low cost it has two disadvantages. One is that only enough posts or poles can be cut as can be treated at a time and any delay can spoil the treatment. The other is that in one State at least, stringent regulations are now in force covering the home use of preservatives containing arsenic and these must be complied with before treatment.

Another process of this type, applicable to green poles, is the Boucherie or *sap-displacement process*, in which water-borne solution is applied under pressure to one end so that the sap is displaced and flows out the other. When the solution flowing from the other end is without dilution the treatment is complete. This usually takes only an hour or two but more sophisticated equipment is needed than for sap replacement and its use is limited.

Another type of treatment altogether is *diffusion* in which chemicals in aqueous solution are induced to move through wet wood. Variations of this process are used to immunise large quantities of sawn timber and veneer but it is not suitable for fixed preservatives with which this chapter is mainly concerned.

Another treatment which should be mentioned is that of *Boultonising* or boiling under vacuum, now being used for the simultaneous drying and impregnation of round timber with creosote oil. Its use is limited to commercial treatment plants where the extra cost of equipment and treatment is justified by the saving obtained by the elimination of air drying stock.

TABLE 23.2

SUMMARY OF METHODS OF TIMBER PRESERVATION

Preservative	*Process*	*Advantages*	*Disadvantages*	*Uses*
Creosote and pentachlorphenol in heavy oil	Hot and cold Bath	Simple, very effective	Fire risk, hard to control	Posts, poles
	Cold Soak	Simple, cheap	Hard to control, suits light timbers only	Posts, timber above ground
	Pressure Methods	Most reliable, quick good control	More costly equipment, heat needed	All purposes
'Penta' in light oil and 'natural' oil finish preservatives	Brush	Cheap, quick	Superficial, must be repeated	Rough boards, fences
	Soak	Simple, better penetration	Variable penetration	Joinery
'Fixed' CCA waterborne preservatives	Brush	Quick	Very superficial	Colour only
	Soak	Simple	Very variable penetration	Palings, rough boards
	Pressure Methods	Quick, versatile, reliable	More costly equipment	All purposes
	Low Pressure	Simple, relatively cheap	Only suitable for hardwoods	Posts, small poles
	Sap replacement, displacement	Very simple, cheap; use on green posts	Variable penetration, limited output	Posts, small poles

CHOOSING THE BEST TREATMENT

Table 23.2 is a guide to readers who may find it difficult to assess the relative merits of each preservative and process in relation to their particular timber protection problem.

SAFETY IN HANDLING PRESERVATIVES

The common preservatives, creosote, pentachlorphenol in oil and the CCA compounds are as safe to handle as the insecticides now freely available, if commonsense precautions are observed.

The preservative oils because of their pungent smell and irritating properties are rarely, if ever, ingested accidentally, but they can irritate sensitive skin, the eyes and mucous membranes. Plastic gloves and aprons should be worn when handling treated timber as repeated contact can cause dermatitis. Immediate washing with soap or mild detergent and water will remove spots or splashes on the skin.

The sale and use of the CCA preservatives is controlled by poisons legislation. For obvious reasons the dry salts and solutions must be kept right away from children and animals. Gloves and aprons must be worn when treating timber and a respirator is also essential when mixing dry powder.

Washing with water will remove CCA solution from the skin although a stain may remain. *Once dry, CCA-treated timber is safe to handle and may be used in coolrooms or other foodstores and in aqueducts, but it must not be burnt except in special incinerators.* Fires made with offcuts can give off arsenical fumes and the ash is poisonous.

PROTECTION AGAINST FIRE

It is impossible to cover this aspect of wood protection adequately in a brief note, but the following principles and recommendations apply.

In general, heavy timbers are harder to ignite and take longer to burn than light timbers. Timber that has decayed, particularly the sapwood, will ignite and burn very easily when dry. For this reason alone, full-length treatment of posts and poles is desirable. Timbers of large cross-section are desirable for maximum fire resistance as they are harder to ignite and take longer to burn.

Preservative treatment enhances fire resistance by preventing decay and creosote treatment makes round posts and poles very fire-resistant. Although creosote-treated timber will ignite readily the dense smoke so produced generally quenches the flames as soon as the source of ignition is removed. CCA-treated timber is not easy to ignite but once it does ignite often continues to burn because of 'afterglow' which is peculiar to these preservatives. Recent development of a new fire-retardant variation of the CCA formula by the C.S.I.R.O. is designed to make posts treated with it reasonably resistant to grass fires without being any less resistant to termites and decay.

Radiata pine and other permeable timbers can be impregnated with fire-retardant salts and at high loadings very good fire resistance can be obtained. These salts, however, are water soluble, so that their use is confined to timber used out of the ground and under cover. Timber so treated can be used outside only if kept properly painted.

Fire-retardant paints have limited use inside buildings and are not suitable for outdoor exposure, but heavily-pigmented alkyd-based 'mat' finishes are reasonably fire retardant and may be used outside.

To sum up, heavy timber construction and creosote-treated round timbers are very fire resistant, while lighter structures can be given reasonable protection by simple techniques. The chances of a building being destroyed by fire, however, in general depend far more on the quantity and flammability of the contents than the materials and type of construction.

More detailed information on timber preservation can be obtained by a study of the publications listed below, or by inquiry to State forest services, the Timber Preservers' Association of Australia, Division of Forest Products of the Commonwealth Scientific and Industrial Research Organisation, or the timber promotion organisations in each State.

FURTHER READING

C.S.I.R.O. Leaflet No. 12 *Round Fence Posts; Their Preservative Treatment.*

C.S.I.R.O. Forest Products Newsletter reprints:
No. 223—*Recognition of the Commoner Types of Borer.*
No. 260—*Decay in Timber.*
No. 316-7—*Preservative Treatments for Timber Construction.*
No. 328—*Fence Posts and Fire.*
No. 335—*Good Housekeeping and Timber Preservation.*
No. 360—*Sap Displacement Simplified.*
No. 365—*Natural Finishes on Wood.*

C.S.I.R.O. Division of Forest Products. *Equipment for Low Pressure Fence Post Treatment. Eradicating Termites from Buildings.*

CHAPTER 24

Fire control in the arid and semi-arid lands of Australia

by A. G. McArthur

THE arid lands of Australia for the purpose of this paper are almost synonymous with the area of 'rangelands' as described by Perry (1967). Rangelands are defined as 'those areas of the continent where rainfall on a given piece of land is inadequate for economic crop production or pasture improvement'.

For fire control purposes the definition of arid and humid land goes a little further than the simplified but very meaningful definition adopted by Perry. Humid lands are taken to be those which produce sufficient fuel to carry fire in most years except those of severe drought. Arid and semi-arid rangelands are defined as those areas which only produce sufficient fuel to carry a fire in years of very much above annual rainfall. This may be on a frequency of once every 10 to 15 years, or even longer. A further break-down is then made within the arid region and an area of desert or semi-desert country is delineated where fires would seldom, if ever, burn large areas due to lack of fuel continuity in all seasons. Small spinifex fires may occur in the depressions between sand-ridges but would have a very small size potential due to the frequency of natural fire barriers in this type of desert or semi-desert country. Even in this country, however, the spinifex may develop to a stage where large-scale fires may occur on a frequency of once every 50 years or so, again especially in a year of exceptionally heavy rainfall.

Within the region where fires may occur in most years, an area of predominantly forest fuel-types occurs in the higher rainfall zone close to the coast, with an area which is predominantly open grassland, both natural and improved, with some savannah woodland formations, occurring towards the zone of lower rainfall. This area coincides with the main wheat-growing areas south of the Tropic of Capricorn and the main beef cattle areas in the sub-tropics. North of the Tropic there is no distinction made between forest and grassland areas. In this region the forest is more open and grass is the predominant fuel beneath the trees.

The boundaries of the arid lands tend to fluctuate somewhat during cycles of above average rainfall and more arid periods. The boundaries of these lands have been defined more clearly in this paper by relating them to the pasture maps produced by Perry (1960) and Collins (1967). The south-eastern and south-western boundary of the semi-arid land generally includes most of the mallee scrub associations. This then grades into saltbush and acacia associations.

North of the Tropic the boundary between humid and arid lands is harder to define. In the Northern Territory it is taken to be the boundary between open forests and savannah woodlands and arid woodland and scrub after Williams (1955). In Queensland it is approximately the western edge of the brigalow lands.

The boundaries of the various defined fuel-type regions in Australia are shown in Figure 24.1. These boundaries depart substantially from those given by McArthur (1968) in the dealing with 'Rural Fire Protection' in humid areas and by Luke (1961). The relative fire frequency to be expected over the Australian continent has become much more clearly defined over recent years with the occurrence of large-scale fires in the arid interior following a succession of above average rainfall years and by the improved fire reporting network now available.

Region C is semi-arid to arid land supporting insufficient stands of grass to support a running fire except in years of very much above average rainfall. The

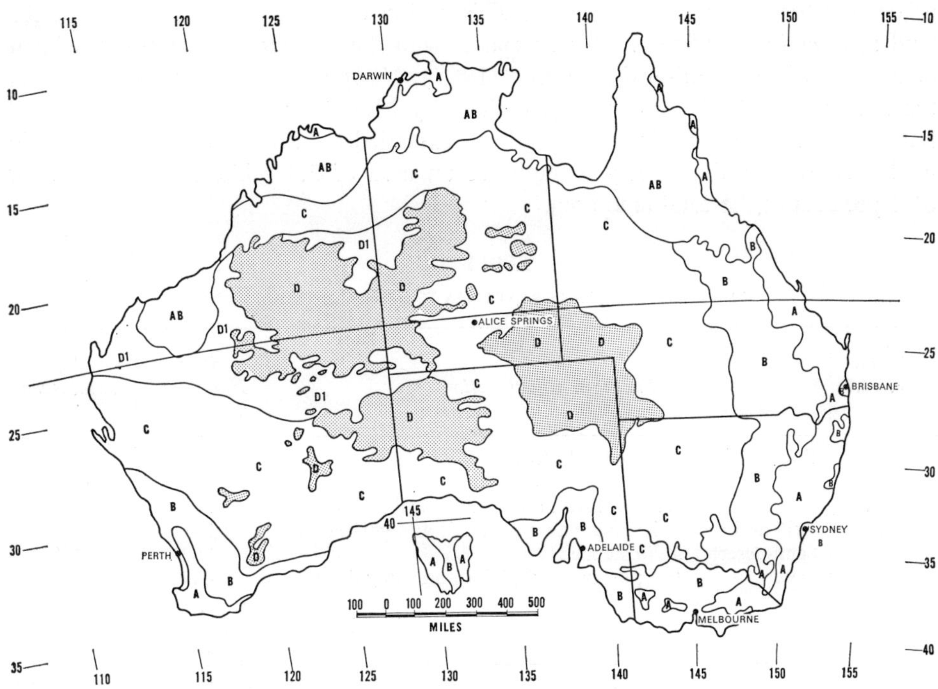

REGION

A Predominantly high closed forest with some areas of open grasslands. Sufficient fuel available for fires every season.

B Predominantly open grasslands and savannah woodlands. Sufficient fuel available for fires in most seasons.

AB Sub-tropical open low forest and tall tropical grasses. Sufficient fuel available for fire in most seasons.

C Arid and semi-arid savannah woodlands and grasslands. Sufficient grass fuel available to support widespread fires in occasional seasons following above average rainfall.

D1 Semi-desert Triodia Steppe.

D Desert sand-dunes; desert sand-plains and stony desert.

FIG. 24.1

Main fueltype regions in Australia.

fire frequency may vary from 1 in 10 to 1 in 15 or 20 years bordering on the semi-desert region. This region subject to occasional fire seasons covers an area of 1,341,000 square miles or 45.2% of the continent.

Region D1 is the *Triodia* steppe region defined by Gardner (1944) and may support a running fire in years of exceptional rainfall. It covers an area of 202,000 square miles or 6.8 per cent of the continent.

Region D includes the semi-desert and desert sand-dunes, sand-plains and stony desert country. It covers an area of 447,000 square miles or 15.1 per cent of the continent. Even in this region of 'semi-desert' it is likely that the spinifex vegetation may build up to a level where an exceptional rainfall may give sufficient growth of inter-tussock short grasses and forbs to carry a running fire under strong wind conditions. The fire frequency of this region may be once in 50 years or so.

Regions C, D1 and D of Figure 24.1 together form the arid lands of Australia as defined for fire control purposes. This region comprises 67.1 per cent of the land area of the continent. This compares with the 74 per cent of the continent defined as arid or semi-arid by Perry (1967), Newman and Condon (1969) and the 55.2 per cent of the continent defined by McArthur (1968) for fire control purposes. Other authors in this book have extended the area of arid and semi-arid land even further than Perry to embrace an area of 2.4 million square miles or 81.1 per cent of the total land area.

TABLE 24.1

LAND CLASSIFICATION IN AUSTRALIA FOR FIRE CONTROL PURPOSES

Region (Fig. 24.1)	*Fuel-type Classification*	*Area Thousands Sq Miles*	*% Total Area*
A	Predominantly high closed forest	204	6.8
B	Predominantly open grasslands	402	13.6
AB	Sub-tropical open low forest and tall tropical grasses	372	12.5
C	Arid and semi-arid savannah woodlands and grasslands	1,341	45.2
D1	Semi-desert-*Triodia* steppe	202	6.8
D	Desert sand-dunes; desert sand-plains and stony desert	447	15.1
	Total	2,968	100.0

Regions A, B and AB in Figure 24.1 and Table 24.1 are the humid and sub-humid lands which carry sufficient fuel to support a fire in most years. This fire-prone area covers 978,000 square miles or 32.9 per cent of the continent.

FIRE OCCURRENCE IN THE ARID LANDS

Although a considerable amount of literature can be found relating to fire control in humid and sub-humid lands, scarcely any references can be found relating to fire behaviour, fire occurrence or fire effects in the arid lands of Australia. Whatever literature exists is almost entirely confined to observations made by the early explorers or by present-day botanists and ecologists.

One could almost assume that fire is not an important environmental factor of these lands, yet the ecological and botanical literature makes frequent reference to the effect of fire and the fact that it can have a profound influence on plant succession, soil erosion and grazing potential. It is certain that the occurrence of fires in the arid and semi-arid regions of Australia is much more frequent than most fire control organisations had previously assumed. An examination of recently-produced photomaps covering these arid lands reveals a complex mosaic of fire patterns dating back in time from when the aerial photographs were taken. One such example of this burnt mosaic is shown in Plate 92.

Perhaps the absence of literature relating to fire control in the arid lands is a reflection of the fact that a fire-suppression organisation as we know it in the humid lands is virtually non-existent. This is due mainly to the fact that these areas are very sparsely populated and still relatively undeveloped after some hundred years or so of pastoral occupation. Over the past two or three decades, some of the largest fires ever to have occurred in Australian fire control history have originated in the semi-arid area of some 1.341 million square miles designated as Region C in Figure 24.1.

One of these was the Balladonia fire in Western Australia reported by Rodger (1961) in the Royal Commission *Report* on the 1960-61 fire season in that State.

> The largest fire of the 1960-61 fire season was the Balladonia fire which burnt in the Eucla Land Division. It was probably one of the largest single fires that has occurred in the history of fire control in Australia.
>
> This region generally carries a sparse stand of low shrubs and scattered patches of light timber which will not carry a fire any distance. During the summer and autumn of 1960, however, it received good rains and the growth of grass which followed was apparently sufficient to carry a fire during the following summer.
>
> The cause of this fire was not ascertained, but it was possibly lit by travellers along the Eyre Highway.
>
> It commenced in the vicinity of the 300-mile peg on the Highway about the middle of October.
>
> It burnt slowly westward during October and November on an ever-widening face and, by the first week in December, was approaching pastoral country some 150 miles east of Norseman. In the 50-day period the fire had travelled 150 miles and had developed a front of approximately 50 miles.
>
> On 6 December the owners of Noondoonia Station requested that action be taken to arrest the western spread of the fire as it was threatening their own and other pastoral holdings. Firebreaks were cut in front of the fire with the earth-moving machinery from the Dundas Shire Council and, by December 11, the western front was brought under control from about the 146-mile peg east of Norseman on the Eyre Highway.
>
> No accurate estimate of the area burnt has been made, but it is possibly in the vicinity of 3,750,000 acres. The country over which the fire burnt is mainly uninhabited Crown land owing to lack of fresh water and no damage to stock or property was reported.

In the same fire season huge areas of land were burnt in the Meekatharra-Mt Magnet area of Western Australia. These fires were mostly caused by lightning strikes in prolific grasslands resulting from the same cyclonic depression which produced the abundant pastures which sustained the Balladonia fire described by the Royal Commission *Report*.

Extensive fires have been reported in the Northern Territory since the very early days of exploration over 100 years ago. On 23 April 1860 John McDouall Stuart planted a flag on Central Mount Stuart, almost exactly in the geographical centre of the Australian continent. Soon after this date the party continued to move northwards . . . 'At Tennant Creek named for John Tennant of Port Lincoln, where the horses were nearly all lame and the country burnt black with bushfires'.

After several unsuccessful attempts to reach the northern coast of the continent, Stuart set out on his fifth attempt on 20 December 1861 and finally reached the sea on Van Diemen's Gulf on 24 July 1862. The most fearful part of Stuart's journey 'was the road back through flame and smoke for the first thousand miles, waterholes drying, skeleton horses dying'.

In more recent times, two serious fire seasons occurred in the arid lands of the Northern Territory during 1968-69 and 1969-70.

In 1968 the largest fire complex was the Killarney-Top Springs fire which burnt from 29 August to 6 September.

This fire (or fires) had two sources. The first fire started on undeveloped country on Gorrie Station some 50 miles south-south-east of Katherine and burnt in a generally south-westerly direction. The second fire apparently started on the Stuart Highway near Dunmarra a hundred miles to the south-east of the Gorrie fire. This fire was thought to be burning for a fortnight before it was reported. It burnt generally in a westerly direction.

Both fires were burning in inaccessible dissected peneplain country which is largely unoccupied and of little commercial value at the present time.

On 30 August, the Gorrie fire made a run of 15 miles towards Top Springs Station on a front of 20-30 miles.

Two days later the Dunmarra fire, still moving westwards, moved down the jump-up on to the grazing country of Birrimba Station on a 15 to 20-mile front. Over 75 per cent of Birrimba Station was burnt out and the main fire-fighting effort was to protect the homestead and home paddocks. This fire also moved across to Top Springs Station and joined with the Gorrie fire on a broad front.

During the same period there was another large fire burning on a 45-mile front on Wave Hill Station which moved in a south-easterly direction into semi-desert country. At the same time another major fire was burning in the Newcastle Waters area.

The Gorrie, Dunmarra and Wave Hill fires formed a continuous burnt area which covered some 13½ million acres. Even this estimate of the area burnt may be conservative as the eventual extent of the fire in the semi-desert country was not determined.

Further widespread fires occurred in the Northern Territory during 1969, and were largely concentrated in the Alice Springs region and in the Dry River-Victoria River region. In all, some 76 fires were reported to the Northern Territory Bush Fire Council, but the Council only recorded burnt area estimates on 27 of these 76 fires, covering some 18,000 square miles.

Fire control people in the Territory conservatively estimate that 50,000 square miles of country was burnt in 1968 and at least a further 50,000 square miles in

1969. These are probably quite conservative estimates as little concern is shown for fires burning in the desert areas. A hundred million acres could well have burnt in the two seasons.

At least some fire control progress is now being made in the Northern Territory. With the formation of the Bush Fire Council in 1966, determined efforts were made in 1968 and 1969 to control the spread of some of these huge fires with the use of mechanical equipment such as graders. Yet the effort is still pitifully small due to the low population density of this land. On the Gorrie and Dunmarra fires of August 1968 it was only possible to muster 4 graders and 20 men out of a possible 25 in the entire area.

Another region where very large fires have occurred is that part of Region C extending from Charleville in Queensland down to Cobar, Ivanhoe and Mildura on the Murray River in Victoria. In the 1931-32 fire season huge fires occurred around Menindee and Ivanhoe in October 1931 and around Tilpa and Cobar in mid-December, extending to around Deniliquin by the end of December 1931.

In the 1939-40 fire season huge fires occurred between Roto and Hillston on 5-6 January 1940.

Serious fires again occurred in the Cobar-Roto-Hillston area in the 1956-57 fire season and somewhat less serious fires occurred in the region during the 1957-58 fire season.

Very large areas were again burnt in the Cobar-Mt Hope region of New South Wales in the 1969-70 fire season and further west towards Ivanhoe. It is conservatively estimated that around 2½ million acres burnt in the semi-arid Western Division of New South Wales during November and December 1969.

Thus, it would appear that a very serious fire season is likely to develop on the average once every 10-15 years. This region of New South Wales has large areas of mallee scrub and in an average to below average rainfall year there is not sufficient fuel continuity to carry a running fire. Yet most of the mallee eucalypts have decorticating bark hanging from the limbs and litter accumulates at the base of the mallee clumps over the years if not removed by burning. The fuel continuity link is provided by an exceptionally high rainfall year, when grass will grow between the mallee clumps. The decorticating bark hanging from the trees then aggravates the fire suppression problem and reasonably long-distance spotting of 2-3 miles can occur in these mixed grass and mallee scrub areas.

Generally the fire suppression organisation can handle fires which occur in open grassland areas but find difficulty of access and spotting too much to cope with fires in the mallee scrubs during these occasional years.

Prolific cured grass is the key to the occurrence of widespread fires in Region C of the continent and to explain the occurrence of bad fire seasons one has therefore to examine rainfall records for evidence indicating a prolific growth of grass.

Luke (1970) has made a recent study of cyclonic fire occurrence in this region as it relates to above average rainfall. Good rains occurred in the Ivanhoe area in the spring of 1930 and again in the winter of 1931. It is assumed that the 1930 rains resulted in a prolific growth of grass but this may not have cured in time to bring about a bad summer fire season early in 1931, largely due to the fact that there was a December rainfall of 480 points which undoubtedly prevented the grassland curing and would bring on a new flush of green grass. The heavy grass growth resulting from this above-average winter, spring and summer rainfall would have cured in the autumn months and would then be augmented by new growth resulting from considerably above-average rainfall in the May-June period of 1931. Rainfall from July onwards was very much below average and the grasslands would have been fully cured by late September or early October, giving the conditions for the severe fires which occurred early in October.

In December 1931, as mentioned previously, extensive bush fires raged between Cobar and Tilpa on a 50-mile front. Rainfall at Cobar in 1930 was well above average and during the period from March to June 1931 rainfall was about three times a long-term average. This undoubtedly produced prolific grass throughout the mallee country, which cured in spring and early summer months giving the very severe fire season in December. Again, two good seasons appear to have made their contribution to the total fuel mass as was the case previously mentioned at Ivanhoe.

The same pattern was repeated in January 1940 when very widespread fires occurred near Roto and Hillston. Both Hillston and Ivanhoe had very little rain in 1938, but rainfall was almost double normal in 1939. There was no rain at all in December 1939 and one can visualise the grass state in January 1940 as being dense and continuous, resulting from the good growing conditions from February to November and in a fully cured state due to the hot and dry December.

The same pattern occurred in the 1957-58 fire season in the semi-arid areas of western New South Wales when large fires occurred round Mt Hope and Hillston. Rainfall during 1956 was exceptionally high, being as much as twice to 2½ times normal for some stations. However, the spring of 1956 and the summer of 1956-57 were dry. Undoubtedly the grasslands were fully cured and carried a heavy volume of fuel, explaining the very large areas burnt by fires in this year.

During 1957 rainfall was well below average but it appears that sufficient grass carried over from the lush year of 1956 to allow a large fire to develop near Mt Hope during the summer of 1957-58. Again, this fire burnt in mixed mallee-grassland areas and was described as being a very difficult fire to handle and extremely severe with many clumps of mallee being burnt to the ground.

The pattern is again similar in the 1969-70 fire season when very large areas were burnt in the Cobar, Mt Hope and Menindee region of New South Wales. The very heavy grass growth appears to have resulted from exceptionally heavy rainfalls during February and March 1969, followed by average to slightly above average rainfalls during the winter months. The prolific grass growth resulting from the good growing conditions then cured in a dry spring and the stage was again set for extensive grass fires in both the mallee country and in the saltbush-bluebush plains. The rainfall at Hay was 10 inches above average in 1969 and 50 per cent above average at Cobar.

This pattern of association between very much above-average rainfall and associated heavy grass growth appears to be repeated right throughout the semi-arid region of Australia and could well even extend into the semi-desert country. Rainfall variability is a characteristic feature of the arid lands and it appears that even a single rain influence can produce prolific grass growth sufficient to carry extensive fires. Winkworth (1967) has recorded over 10 inches of rain falling at Central Mt Wedge Station west of Alice Springs in a single day in an area where the average annual rainfall is 10 inches. The same type of pattern resulted in the huge Balladonia fire, previously described, in late 1960 where prolific grass growth resulted from a single cyclonic depression which moved across the southern part of Western Australia towards the Nullarbor Plain.

The pattern of widespread fire occurrence in this semi-arid region is also related to the stocking rate in the area to a very large extent. Frequently these above-average rain periods follow on from a drought period during which stocking rates are drastically reduced. When the heavy rains occur and prolific grass growth results, there is insufficient stock to make any significant impression on the pastures and so the fuel remains in a burnable condition for upwards of two years. This has certainly occurred in the Western Division of New South Wales several times and obviously also occurred in the Northern Territory during 1968 and 1969. In this region the drought-breaking rains followed a virtual drought period which extended over a period of five to seven years.

THE AUSTRALIAN FIRE CLIMATE

The fire season in Australia progresses fairly steadily from the north to the south, coinciding with the seasonal change in pressure systems. North of the Tropic serious grass fires generally occur in the winter and spring months. The fire season then extends southwards through Queensland and New South Wales into Victoria and

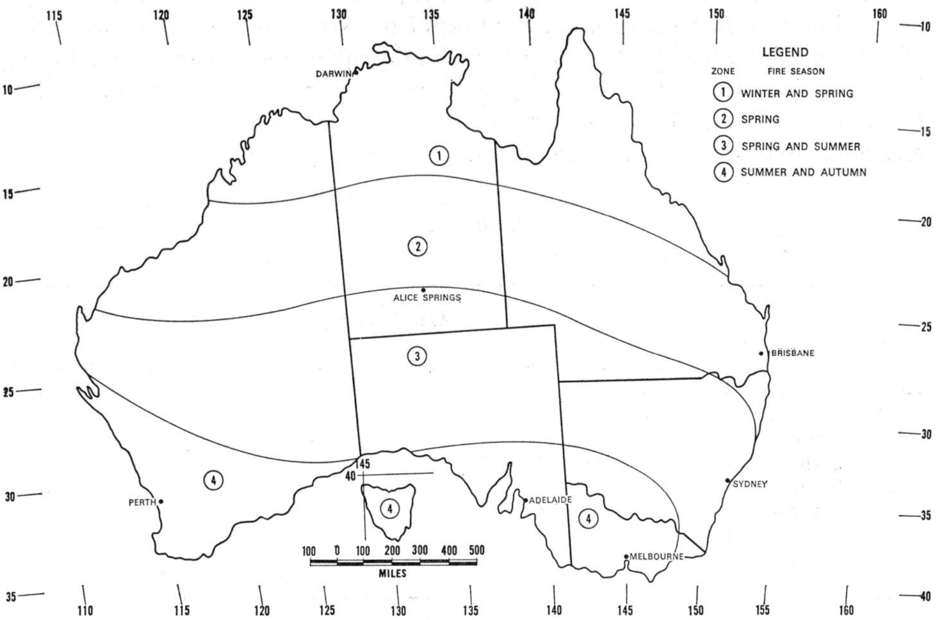

FIG. 24.2
Zones of seasonal fire occurrence in Australia.

Tasmania. A similar pattern also occurs in Western Australia, where south of latitude 32°S. major fire occurrence is generally confined to the summer months but occasionally extends into the autumn.

This seasonal change or 'march of the fire season' is illustrated in Figure 24.2, which gives the seasonal fire occurrence pattern throughout Australia. The major and minor months of the fire season are given in Table 24.2 below.

Fires require adequate fuel and suitable meteorological conditions for sustained spread. Even in the higher rainfall and more closely-settled areas of Australia there are many years when there may be insufficient fuel to support widespread fires due to prevailing drought conditions. Even if pastoral conditions are favourable, summer rains may prevent the grasslands from reaching a fully cured state, thus preventing damaging fires from occurring. This may happen during 50 per cent of years in the uniform rainfall zone, but not in the well-defined winter or summer rainfall zones.

If grass growth occurs in the arid lands, the grass will certainly become fully cured in the dry hot months and thus a serious fire season will result any time rainfall has been sufficient to produce abundant grass growth.

Although capital values per acre may be no more than a few cents per acre in the arid and semi-arid lands, it must be remembered that in the absence of a highly organised fire suppression system, which requires a man-power availability considerably in excess of what will ever be available in these regions, fires, although generally spreading at much slower rates than in the more settled areas, can burn extremely large areas, as instanced in the examples given for the Northern Territory where in 1968 a complex of three fires burned something in excess of 13½ million acres. Even at a low value of perhaps 5 cents per acre, the total monetary damage can add up to a formidable figure for fires of this size. Apart from monetary loss of grazing values, the effect of bush fires in arid environments can have extremely long-lasting effects on the vegetation and soil erosion processes and in the long term, damage can be just as serious as large fires which occur in the more humid and closely-settled lands of the continent.

TABLE 24.2

SEASONAL ZONES OF FIRE OCCURRENCE IN AUSTRALIA

Zone	*Fire season*	*Months of the fire season*											
		July	*Aug.*	*Sept.*	*Oct.*	*Nov.*	*Dec.*	*Jan.*	*Feb.*	*Mar.*	*Apr.*	*May*	*June*
(1)	Winter and spring	O	N	N	N	N	O	O	—	—	—	O	O
(2)	Spring	—	O	N	N	N	N	O	O	—	—	—	—
(3)	Spring and summer	—	—	O	N	N	N	N	O	O	—	—	—
(4)	Summer and autumn	—	—	—	—	O	N	N	N	N	O	—	—

N = normal months when serious fires are likely to occur.
O = occasional months when serious fires may occur.
— = months where fire occurrence is very unlikely.

FACTORS AFFECTING FIRE BEHAVIOUR

The factors affecting fire behaviour in the arid rangelands of Australia are exactly the same as in any other region of the world. However, the relative importance of the various weather and fuel variables can be quite different. In the humid regions weather factors are the dominating variable affecting fire behaviour, but in the arid lands weather is a much more constant factor of the environment than it is in the higher rainfall regions. For many months of the fire season the maximum daily temperature will be in the order of 95-105°F or greater, humidities invariably between 10 and 30 per cent and wind velocities ranging from 5-15 miles per hour.

It is perhaps fortunate that the very strong wind velocities which periodically occur south of latitude 35°S. associated with the passage of a cold front, are not a frequent occurrence in the tropics and sub-tropics. Here, constant trade winds averaging between 10 and 15 miles an hour are more the rule than the exception. If these extremely strong wind velocities, which can reach an average of 35 mph over a period of 6 hours or more, occurred frequently in the tropics, most of these regions would be burnt in any year when there was sufficient fuel present to carry a fire.

Weather has both a cumulative and immediate effect on fire behaviour. Weather factors therefore must be evaluated on a time scale varying from a year, season, month, week, day, hour or even periods of only a few minutes. Such evaluations require consideration of many factors such as air circulation, cloud coverage, precipitation, temperature, relative humidity, wind velocity and direction, solar radiation and other atmospheric variables. Weather however, is only one of the factors controlling fire behaviour. The interaction of weather with fuel characteristics, topography and the dynamics of combustion processes must also be considered before it is possible to obtain an accurate estimate of fire behaviour.

The more common factors affecting fire behaviour include:

1. *Fuel Characteristics*

As far as the arid rangelands of Australia are concerned, fuel quantity and fuel continuity becomes perhaps the dominating factor affecting fire behaviour in this type of environment. The amount of dry pasture available in an area has a very significant effect on rate of spread, flame heights, fire intensity and so on. All other factors being equal, a fire will spread twice as fast in a two tons per acre grassland as in a one ton per acre pasture. In other words, rate of spread varies directly with the quantity of fuel available for combustion.

One of the constant characteristics of grass fuels in arid areas is the presence of spinifex and other tussock-forming grasses. These grass species are not only very characteristic of the desert or semi-desert areas, but are also a very common component of fuels in region C, the region where periodic or occasional fires occur. A basic understanding of fire behaviour in these lands therefore depends on some knowledge of the growth characteristics and the process of fuel accumulation in these tussock rangelands.

Spinifex is generally a coarse tussock-forming grass with strongly folded pungent leaves. Tussocks can be up to two to three feet in diameter and 12-18 inches high and regular in shape. Tall flowering culms rise to five feet, giving an impression of denseness. Frequently the tussocks can become ring- or crescent-shaped.

The tussocks are generally separated by bare ground two or three times the area occupied by the plants. Bare areas lighter than normal are frequently caused by termitaria or ant-beds. Frequently a sparse layer of small trees and shrubs occurs throughout but contributes little to the total volume of plant material.

In an area studied by Winkworth (1967) in central Australia total plant cover ranged from 28 to 44 per cent and a proportion of the spinifex was dead on all sites. Death was due to drought, with the addition in southern sites of a proportion of dead material normally found in the centre of the ring-shaped tussocks.

Of greater importance from a fire control viewpoint, was the mean standing crop rates. These, in Winkworth's study area, ranged from 2,800 to 5,000 pounds per acre, of which 96 to 98 per cent was invariably spinifex.

One interesting aspect of Winkworth's study was the fact that perhaps 80 per cent of the spinifex in the sand plain pasture land under study was in regenerative stages owing to fire or a degenerative stage owing to drought.

Spinifex communities regenerating from fire were very extensive and one area alone, known to be burnt by a single fire, covered about 4,000 square miles. This suggests that all spinifex communities have a cyclic development from fire to fire.

Studies of burnt plots near Alice Springs for seven years by Winkworth (1967) indicate clearly that the vegetation is regenerating towards a replica of its original state. The time taken to achieve maturity is yet unknown. The dry weight production of spinifex in seven years was 734 pounds per acre, which is 27 per cent of the 'mature' stands of the central area. This would imply that the spinifex areas studied by Winkworth would only burn perhaps every 10-15 years. This was 'hard' spinifex (*Triodia basedowii*).

Perry (1960) in describing the characteristics of the 'soft' spinifex (*Triodia pungens*) plains states 'that when this country is used for grazing, it is burnt as often as it will carry a fire, which is about once every 4 or 5 years. During the first year or so after burning there is a greater proportion of the more palatable species and often the young growth of spinifex is softer and more palatable. Within 3 to 5 years competition from spinifex reduces the other species to a sparse cover and the tussock becomes large enough and close enough to support another fire.'

Spinifex almost exclusively occupies the sand plains of central Australia and this suggests a true adaption to the soil environment. The vegetation is an edaphic climax adapted to repeated burning and no evidence could be found to show that it is seral proceeding a mulga woodland as suggested by Chippendale (1958). Increases in shrub density occur after fire followed by decreases. Shrub invasion of undistributed communities has not been observed by Winkworth.

A study of Burbidge (1943) on the ecological succession observed during regeneration of *Triodia pungens* after burning throw some further light on the burning characteristics of spinifex grass. The study area was about 60 miles east-south-east of Port Hedland in the region designated AB on the fuel-type map, Figure 24.1, and classified as open savannah woodland by Gardner (1944) and semi-desert savannah by Teakle (1938). In the study area the spinifex formed a pyramidal tussock varying in size up to six feet in diameter and about the same in height. *Triodia pungens* is described by Burbidge as a coarse tussock-forming resinous grass.

One important observation made by Burbidge in this region is the fact that spinifex does not necessarily regenerate after fire from a central butt. Over a wide area of this region, fire causes the complete destruction of the plants and regeneration takes place by means of seedling establishment. It appeared that it was only along the coastal strip from Port Hedland to the northern end of the 80-mile beach that burning left a tussock with a viable butt whose re-growth was available for grazing animals within a few months. Through the inland country along the De Grey, Shaw and Congan Rivers, up to Marble Bar, the tussocks are completely destroyed and it is at least a year before re-growth is available to the sheep.

Burbidge goes on to say that fortunately the *Triodia* tussocks are widely spaced and it is difficult to keep a running fire going. Thus only small patches from two or three tussocks up to about one acre are burnt in each place. The spacing of the ground cover has probably played an important role in protecting the vegetation from the consequences of biotic activities during the 70 years of pastoral occupation.

The general practice in this region is to burn during the mustering period in April and June when the men, riding the paddocks, drop lighted matches at random. Apart from the more or less concentrated effort, sporadic burning is carried out at any time during the year when suitable conditions exist for the spread of fire.

It is evident that if the tussocks are completely destroyed in May or June, the ground will lie bare until the following January or March when the summer rains bring about seedling regeneration.

Burbidge also found that burning followed by prolonged and heavy stocking can destroy regenerating spinifex and convert the country into an *Erogrostis eriopoda*-dominant association. This in turn can also be removed by heavy grazing and replaced by annuals. Eventually this could lead to severe erosion—and substantially reduce the fire-carrying capacity of these semi-desert grasslands.

Chippendale (1958) made a botanical survey of a 50-square-mile area on Hamilton Downs in September 1955. Spinifex in this area was mainly *Triodia basedowii.* The area is about 120 to 130 miles west-north-west of Alice Springs and is in the sand plain desert country described by Winkworth (1967). The block studied by Chippendale forms an example of what is regarded as 'desert' by pastoralists, but is actually 'semi-desert'. In this area fires had been over the area twice in the last 10 years, the latest time being about two years before the survey.

Another interesting observation of Chippendale's was the fact that in the *Triodia basedowii* association small grasses are not plentiful and occur mainly following late summer or early winter rains in small colonies between the large clumps of the dominant spinifex species. This gives some lead to what happens when an exceptionally heavy rainfall occurs at a time favourable for plant growth. Obviously this inter-tussock growth of short grasses and forbs would then be sufficient to provide the necessary fuel continuity for the rapid spread of fire.

The significant fact which emerges from these observations of botanists and ecologists is that, in a normal condition, spinifex forms a widely-spaced fuel mass carrying a reasonable volume per acre, but it would be difficult for fire to travel from tussock to tussock unless a reasonably strong wind was blowing. This question of fuel continuity becomes a dominant variable controlling fire behaviour in arid and semi-arid regions, and it is for this reason that extensive fires in these lands only occur at widely-spaced intervals when abnormal rainfall is received.

Not only is fuel quantity important from a fire control viewpoint but the fineness of the fuel also vitally affects fire behaviour characteristics such as rate of spread. The finer the fuel type, the more rapidly a fire will spread through the pasture. Extremely fine pastures found in the more humid areas may not carry a particularly heavy quantity of fuel, but a fire will burn through them extremely rapidly. Coarse grasses generally tend to burn more slowly, but have a much greater heat output and are frequently more difficult to control because of this.

The burning characteristics of spinifex and associated grasses and herbs are not known, but the comment by Burbidge (1943) that it is a resinous grass may give some clue to its burning characteristics. This might imply that it has a high heat yield and could be a very hot burning fire. A common height of 5 or 6 feet as suggested by some of the references quoted would also produce very high flame heights and this is perhaps the key to the whole fire behaviour characteristic of this type of rangeland. With high flame heights it would not need a particularly strong wind velocity to bend the flames over and produce flame contact from tussock to tussock and so initiate the chain reaction required for a fire front to proceed unhindered through this type of grass.

Another important fuel characteristic relating to fire behaviour is the fact that freshly-cured pastures contain various compounds which make them highly flammable through the liberation of volatile gases. The advent of rain on these pastures will quickly leach the volatile compounds, and the pasture immediately becomes less flammable and the whole level of fire danger decreases.

Thus, it can generally be assumed that if there is a large carry-over of grass from one good season to the next, that the grasslands in the following season would be less flammable than those in the season immediately following the curing of the grasses.

2. *Fuel Moisture Content*

The fuel moisture content of fine grasses responds very rapidly to changes in temperature and relative humidity and the resulting change fairly quickly affects the behaviour of the fire. As the moisture content increases so does the rate of spread decrease. All other factors remaining constant, the fire spreads at a slower rate and is much more readily controlled.

Temperature exerts a relatively small influence on fuel moisture content and the main controlling factor is the relative humidity. In arid lands these factors of fuel moisture content are, however, relatively unimportant because of the fairly constant high temperatures and low humidities which prevail in these regions during the months of the fire season.

Provided the relative humidity is low, fires can burn with high intensity at low temperatures, in the range of 50 to 60°F. The significance of this fact is, providing there is abundant pastures present, and a reasonably strong wind blowing, fires in the arid and semi-arid regions can burn at reasonably fast rates of spread throughout the night.

A typical example of this type of fire behaviour was the Trangie fire in 1952. Trangie is just on the edge of the semi-arid region in north-western New South Wales. The Trangie fire started one afternoon under conditions of high temperature, low humidity and moderate winds and the suppression forces found difficulty in bringing it under control before nightfall. A cool change passed through the area with associated strong wind velocities and the fire continued to burn throughout the night at an average rate of spread of around 2-3 miles per hour. The temperature was in the low 50s and the fire-fighters were wearing coats to keep warm. In total, this fire burnt something in the order of 125,000 acres and the major part of this area was burnt during the night-time hours under low temperatures, but also under conditions of low humidity and fairly constant winds of 15 to 20 mph.

Associated with high temperatures and low humidities in the tropics are generally very unstable air masses with instability for at least 6 to 8 thousand feet. Evidence of this is seen in frequent willy-willys or whirleys indicating the movement of convection cells across the countryside, and the presence of this instability, combined with low fuel moisture content, adds considerably to fire suppression difficulty.

3. *Wind Velocity*

Grasslands are usually exposed to the full force of the wind and this is frequently the straw that breaks the camel's back. The rate of spread of a fire varies roughly with the square of wind velocity. For example, a fire burning under a 20 mph wind will spread at a rate four times faster than a fire burning under a 10 mph wind. This relationship does not hold at very low wind velocities since the curve tends towards a linear relationship at wind velocities below 6 to 7 mph. It also does not hold at wind velocities greater than 25 to 28 mph, where the front of a fire tends to be blown out or splits into a number of narrow tongues and the spread process tends to become ineffective and the rate of spread of the fire decreases (McArthur, 1968A).

In the tropics and sub-tropics strong wind velocities such as are experienced in latitudes south of 35°S. are not a common feature of the environment and a more usual wind velocity is in the order of 15 to 20 mph. However, much stronger winds of short duration can sometimes be associated with the down-draft from a dry thunderstorm. Frequently these down-draft winds in the sub-tropics can reach 25 to 30 mph for half an hour or so. As many fires in the semi-arid lands start from lightning strikes associated with these dry thunderstorms, the initial spread of the fire can be quite rapid whilst remaining under the influence of these down-draft winds.

A fire burning under low wind velocities presents little suppression difficulty to fire control units and will spread quite slowly even when burning in a heavy continuous pasture. The fact that most pastures in the semi-arid rangelands are somewhat sparse or tussocky with a discontinuous fuel layer would tend to upset the rate of spread-wind velocity relationship, which has been determined fairly precisely for fires in the higher rainfall regions of the continent. In discontinuous tussock pastures the spread process might be quite spasmodic at wind velocities less than 10 mph and, overall, would be relatively slow. It would appear that wind velocities would have to exceed 10 to 15 mph to maintain a constant moving flame front through the discontinuous type of fuel characteristic of these lands.

Even when the short grasses and forbs fill up the inter-tussock spaces, the relatively low fuel quantity of this inter-tussock area would tend to give relatively low rates of spread even under reasonably strong wind velocities.

The maximum rate of spread of grass fires in heavy pastures under the worst possible conditions in the temperate regions is around 8 mph. It would seem that in the arid and semi-arid regions, rates of spread would be much less than this because of the presence of natural fuel barriers such as sand-ridges, bare inter-tussock spaces and so on, and in these areas a maximum spread of 4 mph would appear to be about the worst possible. The Gorrie fire in the Northern Territory on 30 August 1968 made a run of 15 miles in one day, which would give an average rate of spread of 1½ to 2 miles per hour. The flame heights on this fire were generally 10 to 15 feet on broad fronts described as fast moving.

The rate of spread, of course, would depend largely on grass quantity and wind speed. For this particular fire, field estimates gave the wind velocity at 20 to 25 mph with a temperature ranging between 90 and 105°F and low humidities in the order of 10 to 15 per cent. The wind velocities are possibly over-estimated and are probably referring to gust velocities rather than an average wind speed.

4. *The Curing Process*

This is an important variable affecting fire behaviour in the higher rainfall regions of the continent but is of minor significance in the arid and semi-arid rangelands. Certainly, grass would begin to grow shortly after a significant rainfall, but in the absence of follow-up rains and in the presence of continued high temperatures and moisture stress pastures in these regions would cure very rapidly every year. Frequently during extended drought years, no seasonal grass growth will occur at all and the arid areas may remain in a grassless condition for 2 or 3 years.

In the temperate regions it takes six to eight weeks for grasslands to fully cure from the time seed heads are fully developed, in the absence of significant rain. In the arid regions the period to reach a fully-cured state would possibly be half this period. Some drought-resistant spinifex species retain some greenness throughout the year, but a high proportion of the tussock contains dead material from previous years.

The pattern of seasonal rainfall will determine the amount of pasture growth and the length of the growing season in the arid lands. However, it is not the curing process which is the significant variable, but whether there has been sufficient rain to bring on a substantial growth of pasture so as to provide the necessary amount of fuel to sustain the combustion process, especially in the inter-tussock areas.

5. *Topography*

The general characteristic of the arid lands is their monotonous flat topography and thus the effect of topography is not nearly as significant in the arid lands as it is in the more humid hilly lands. However, it is worthwhile mentioning that fires travel much more quickly up-slope than on level ground, and compared to a fire burning on level ground a grassfire will travel twice as fast up a slope of 10° and four times as fast up a 20° slope. There is of course, a corresponding reduction in rate of spread on a down-slope run.

In the 1968 fires towards Top Springs in the Northern Territory, there was one recorded instance of a head fire travelling one mile in five minutes. This gives an average rate of spread of 12 mph and it must be presumed that the fire was travelling up-slope during this period because a rate of spread of 12 mph on level ground is virtually impossible.

6. *Atmospheric Instability*

As mentioned previously this is virtually a constant factor in the arid zone. Instability cannot be measured simply but there are many warning signs available which indicate its existence, such as the presence of dust devils or whirleys. Clear visibility is another good indication that instability exists. If a fire is burning under conditions of atmospheric instability it will develop a convection column much more rapidly and spread at much faster rates than one burning under stable conditions.

Severe instability causes considerable suppression difficulty through the formation of massive fire whirlwinds, especially under heavy pasture conditions. In the arid lands on and north of the Tropic, sometimes the main spread process is by means of these whirlwinds which tend to develop at the head and move the fire along with them.

7. *General*

The level of fire danger, which is an expression of the chances of a fire starting, its rate of spread, damage potential and its difficulty of suppression, existing in any particular area is determined largely by climatic factors in the immediate past growing season. The characteristics of this season, the amount of rain which has fallen and so on, will determine whether pastures are abundant or virtually non-existent. If seasonal conditions have been such that good grass growth has occurred, a high level of fire danger is certainly going to exist throughout the following hot and dry months of the year. On this level of fuel availability, day to day weather factors are superimposed.

In this day to day, week by week weather pattern there is a constant change in combinations of temperature, humidity and wind velocity and wind direction. Fires start easily when temperatures are high and humidities are low and they will spread rapidly on days of high wind velocity.

In the more settled regions of the continent where initial attack systems are highly organised, a very close watch must be kept on the day to day variation in these weather conditions. In the sparsely-settled arid regions with very rudimentary fire suppression organisations, the development of initial attack systems depending on these weather patterns is quite unrealistic. Fire occurrence becomes more a matter of random distribution as they are generally started by lightning strikes and by Aboriginal activity, and a fire can virtually be in existence for some days or even weeks before it is detected. Whether or not suppression action is then taken will depend on whether the fire is threatening improved country or the more highly-productive pasture types on individual station properties.

TABLE 24.3

RELATIONSHIP BETWEEN WIND VELOCITY AND RATE OF FORWARD PROGRESS OF GRASSFIRES BURNING IN VARYING FUEL QUANTITIES AIR TEMPERATURES 100°F RELATIVE HUMIDITY 15% FUEL MOISTURE CONTENT 3%

Average Wind Velocity	*Maximum rate of forward spread in miles per hour for given fuel quantities (in tons per acre)*						*Fire Danger Index*
mph	½ *Sparse*	*1 Light*	*2 Average*	*3 Heavy*	*4 Heavy*	*5 Very Heavy*	
5	0.08	0.12	0.34	0.50	0.68	0.82	6
10	0.22	0.45	0.90	1.50	1.75	2.10	16
15	0.42	0.87	1.70	2.40	3.3	4.0	29
20	0.75	1.5	2.9	4.3	5.8	7.0	53
25	1.1	2.25	4.5	6.5	9.0	10.8	80
30	1.1	2.25	4.5	6.5	9.0	10.8	100

The use of fire danger meters, which are simple slide rule devices which integrate the effect of the various meteorological perimeters of rate of spread could not be recommended for use in the arid lands because of the very extensive and almost non-existent fire suppression organisation existing in these areas. It may be of some interest, however, to tabulate the rate of spread of grass fires under various wind velocities in sparse, moderate and abundant pasture lands. Here the emphasis is given to fuel quantity because we have already indicated that this is really the controlling factor in determining the rate of spread of fires in the arid lands. These spread rates are shown in Table 24.3 and are typical of a temperature about 100°F and a relative humidity of around 15 per cent.

One important factor vital to all fire control operations is the effect of time on the area spread of fire. This factor will be of tremendous importance in the arid and semi-arid regions because of the necessarily long periods involved between the start of a fire and the time that suppression action can be taken on it.

The area burnt by a fire spreading at a constant rate varies as the square of the burning time. In other words, if the area burnt in the first hour is, say, 1,000 acres, the area burnt in 2 hours will be 4,000 acres, 16,000 acres in 4 hours and 64,000 acres in 8 hours. This type of relationship gives some idea why fires in these arid lands, which might be burning for days on end, cover such tremendous areas. The area which may be burnt by a grassfire at various times from origin is shown in Table 24.4.

Because of the great distances involved, the sparseness of the rural population and in many cases, difficulty of access, it is hard to imagine that these lands can ever maintain an efficient fire suppression organisation which will have a chance of keeping the area burnt by a fire to a small acreage. Time is a virtually unlimited factor in these lands and as a consequence, fires will almost burn an unlimited area. The extent of this area will invariably be determined by prevailing weather conditions and not so much by the efforts of man to extinguish them. In arid tussock grasslands, the spread of a fire will generally cease when either the wind velocity drops to a near calm or a shower of rain occurs from an isolated thunderstorm.

TABLE 24.4

THE MAXIMUM AREA OF A GRASS FIRE AT VARIOUS FIRE DANGER INDICES AND TIME FROM COMMENCEMENT

Fire Danger Index	*Max. area in acres at various times from the start (hours)(a)*					
	½	1	2	4	8	16
5	16	100	400	1,600	6,400	25,600
10	36	225	900	3,600	14,400	57,600
20	85	520	2,100	8,500	34,000	136,000
40	200	1,200	4,800	19,000	76,000	304,000
50	260	1,600	6,200	25,000	100,000	400,000
70	425	2,550	10,000	40,000	160,000	640,000
100	700	4,200	17,000	68,000	272,000	1,088,000

(*a*) These figures assume that the head fire burns unchecked. Suppression action which may be only partially successful will reduce these areas.

However, towards the fringe of the arid lands, population density is much higher and there are many areas where an effective rural fire brigade organisation exists. In these regions, bushfire brigades have a fair chance of containing fires to a reasonable area provided certain precautions are taken, such as the cutting or burning of fire breaks and also the development and maintenance of efficient fire suppression equipment.

ORGANISING TO FIGHT FIRES

In the more closely settled areas of Australia rural fire control is very much a community effort and a highly organised volunteer rural bushfire brigade system operates in most parts of Region B (Fig. 24.1). In the arid lands, fire control depends basically on the efforts of each individual station owner and his employees, and whether or not they can protect the more valuable grazing pastures of the station will depend entirely on the equipment they have and how they are organised to fight the outbreak once it starts.

The average size of properties in the arid land of Australia is extremely large. Table 24.5 is derived from data given by Waring (1969) and shows the average acreage for both sheep and cattle stations in the arid lands of each State.

TABLE 24.5

AVERAGE SIZE OF PROPERTIES CARRYING SHEEP OR CATTLE IN THE ARID AND SEMI-ARID REGIONS OF AUSTRALIA 1962-1965 (AFTER WARING, 1969)

State	*Predominantly Sheep*		*Predominantly Beef Cattle*	
	No. Properties	*Av. Acreage*	*No. Properties*	*Av. Acreage*
N.S.W.	3,761	29,628	1,318	12,825
Qld	2,992	32,675	786	214,731
S.A.	774	58,171	—	—
W.A.	353	426,008	45	922,719
N.T.	—	—	134	1,134,435

This gives some idea of the critically-low population density existing throughout the lands of Western Australia, Queensland and the Northern Territory. By contrast, the arid lands of New South Wales and South Australia are relatively closely settled and at least in some of these areas there would be sufficient manpower and equipment available to maintain a reasonably efficient fire control organisation.

The conventional tanker units of the higher rainfall areas must, of necessity, be replaced by alternative means of fire suppression. This will generally take the form of mechanical equipment suitable for cutting fire breaks through the pasture lands. This equipment largely consists of rubber-tyred graders which can move at high speed and rubber-tyred tractors capable of pulling a disc plough or what might be called a fire plough.

There are many areas in the semi-arid lands of central Queensland where station owners have developed remarkably efficient single-man fire-fighting units. These consist of rubber-tyred tractors pulling a single or double disc plough which cuts a bare earth trail at fast speeds. Attached to the plough is an automatic lighter set at the level of the grass on the inside of the ploughed line. This lighter is generally fed from a liquid petroleum gas cylinder and so the process of trail building and back-burning is done simultaneously and a back burn can be put in at the rate of 15 to 20 miles per hour. Even if the back burn jumps the narrow fire trail all that the operator has to do is to wheel around and cut another line around the outbreak. Generally the hop-over can be quickly brought under control and the main task of building line and back-burning continued.

The inference here, is of course, that most fire-fighting operations in the arid and semi-arid rangelands will take the form of indirect methods of fire fighting as distinct from the more common direct method of attack practised in the more closely settled areas of the continent. In direct attack, water or a water-chemical mixture is applied directly to the fire edge from tanker units and the edge is extinguished as the units move progressively around the fire perimeter.

When using indirect methods of attack, the fire trail or bare earth trail is kept some distance away from the edge of the fire and the intervening area is burnt out. Naturally it is preferable to construct the fire trail and carry out the back-burning operation as close to the fire edge as possible. In these circumstances it might be termed a 'parallel' method of attack. This close parallel method of indirect attack is easy to apply in open grassland areas but frequently encounters difficulty in mallee and acacia scrub associations or in broken topography.

One basic problem encountered when carrying out indirect attack on a fire is the fact that fires spread very slowly against the wind and this is especially so when there is any lack of fuel continuity. Because of this, it is frequently necessary to widen or speed up the spread of the back fire into the main fire front so as to produce a sufficient width of burnt ground before the main fire meets the back-burning line.

In western areas of New South Wales, if manpower is available, lines are frequently run at right angles from the back-burning line 5 or 6 chains in towards the main fire, so as to give adequate width of line before the faster moving head fire meets the back-burning line. Essentially, if the line is narrow and only a few

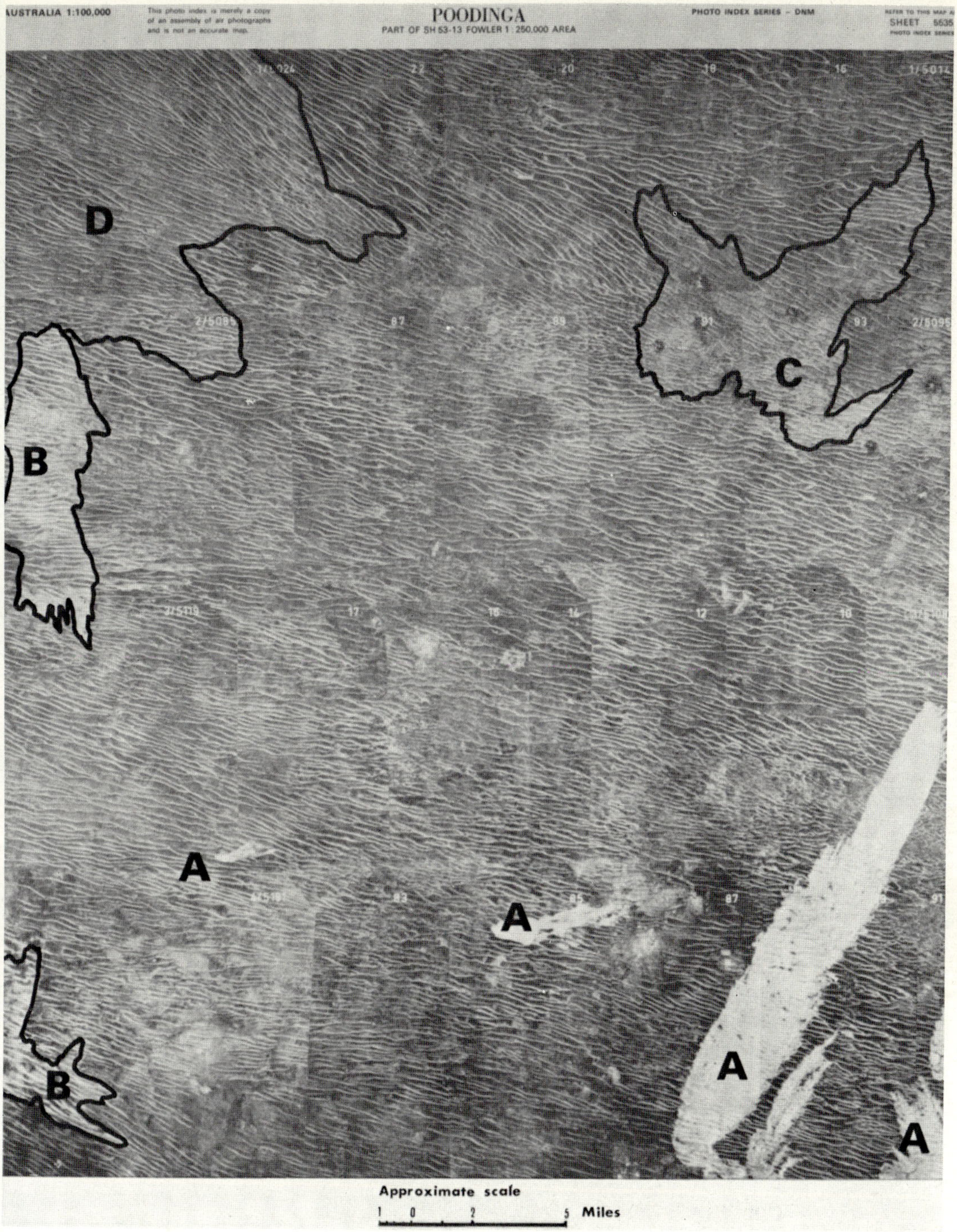

A Fires occurring within the past year

B Fires occurring within the past 5-10 years

C Fires occurring within the past 10-20 years

D An ancient fire pattern where the vegetation has almost recovered to pre-fire density.

VEGETATION Mallee sand-dune formation, vicinity Great Australian Bight. Lat. 31°, long. 133° 30′.

(*Date of photography—April, 1963.*)

Plate 93 MOSAIC OF FIRE PATTERNS FREQUENTLY OBSERVED IN THE ARID REGIONS OF AUSTRALIA

Top: Heavy grass growth in semi-arid mallee eucalypt associations following a season of above average rainfall. This allows rapid fire-spread through the continuous fuel-complex. Roto-Mt. Hope region of New South Wales 1970.

Bottom: The heavy accumulation of fuel which forms around the base of mallee species in a long unburnt condition. The heat-flux resulting from the burning of this material in a bushfire generally causes the death of the above-ground portion of the tree but not the massive below-ground lignotuber.

(*Photos W. Hurditch.*)

Plate 94 HEAVY GROWTH IN SEMI-ARID MALLEE EUCALYPT ASSOCIATIONS AND THE ACCUMULATION OF FUEL

Top: Decorticating bark hanging in strips from a species of mallee eucalypt. This provides the material for long-distance spotting, and creates extreme suppression difficulty.

Bottom: Spinifex grass forming a discontinuous fuel complex in mallee eucalypt formation. Fire will carry through this fuel under high wind conditions. Roto area of New South Wales, 1970.

(*Photo W. Hurditch.*)

Plate 95 DECORTICATING BARK FROM MALLEE EUCALYPTS AND SPINIFEX GRASS AS FUEL TYPES IN THE DRY COUNTRY

Top: More open stands of mallee eucalypts with a sparse growth of grass between tree clumps. Fire tankers and mechanical equipment can move through this class of country.

Bottom: Epicormic shoots growing from a mallee species killed by bushfires five years previously.

(*Photo W. Hurditch.*)

Plate 96 MORE OPEN STANDS OF MALLEE EUCALYPTS AND EPICORMIC GROWTH AFTER FIRE DAMAGE

feet wide, there is a distinct possibility that the main fire will burn over it, especially if it is burning with high flame heights. Thus, one might visualise this fast mobile back-burning team as a two-man effort where the trail is cut and lit by one machine and another vehicle perhaps moves parallel to it towards the main fire making sure that the width of the burnt ground is adequate to hold the line. The success of most back-burning operations generally depends on a double-lighting technique. An ideal team would have a ruggedly-constructed tanker unit available to bring any hop-overs under control.

This problem of adequate width of the back burn is generally only a problem at the front of the fire and does not create any particular problem on the flanks of the fire. The flanks of a fire are generally burning quite slowly and a narrow width of 10 or 12 feet is generally quite adequate to halt the slow side-fire spread. Often the back-burning line is constructed too far away from the main fire edge and frequently the two fires do not meet. This can cause some trouble if the main fires make a major run on subsequent days under 'blow-up' conditions. If there are any patches of timber, spot fires may be thrown across the back burn line.

The basic principle of tactical fire fighting is to stop the forward progress of the fire. Thus fire suppression action should commence at or in front of the head and cut a fire trail and back burn into the approaching fire front. Once the forward spread is stopped, parallel firelines can be then cut around both flanks, back-burning as the trail building proceeds. This is, in effect, a pincer movement commencing from the front.

If the fire is spreading very rapidly and has high flame heights the reverse technique is often used. An anchor point is established at the rear of the fire and back-burning units move in a pincer movement around both sides of the fire until they eventually catch up with the head fire. How far the units have to go before they catch up with the head will depend entirely on the weather conditions and how rapidly the fire is spreading.

Efficient fire suppression is generally a matter of systematic organisation making the most effective use of the resources available to the fire-fighting forces. There is generally a tendency to waste time and effort on frontal attack on fast-moving high intensity head fires when the logical approach is to work from the rear of the fire progressively extinguishing both flanks. It is a matter of fact that if a fire can be kept relatively narrow it can run for a very long distance without burning large areas. Big fires are generally associated with broad-spreading fronts which have been allowed to develop because of lack of attention being given to bringing the flanks under control. Uncontrolled flanks are very sensitive to changes in wind direction and any sustained spread on a broad front will result in a rapid increase in area burnt.

South of latitude 30°S. the presence of mallee scrub becomes a major problem in fire suppression operations in New South Wales, Victoria, South Australia and Western Australia. As mentioned previously, large amounts of fuel can accumulate in these mallee scrubs over a period of 10 to 15 years and long strips of the decorticating bark hang from the limbs. This material provides a source of fire-brand material which can be carried fairly long distances ahead of the main fire in a mass spotting process. In addition to this problem, fire-fighting units and even mechanical equipment such as bulldozers and graders find difficulty in travelling through these fairly dense mallee scrubs. In these circumstances, there is a tendency to establish back-burning lines along a convenient road or railway line some considerable distance ahead of the main fire or to the side of it and back burn from these pre-existing barriers. These extensive back burns frequently result in excessively large areas of country being burnt out. There have been many instances where the area burnt by a back burn has been larger than the area which would have been burnt by the main fire if allowed to burn unhindered. Frequently the back burn and the main fire never link up.

It is difficult to envisage fire control in the arid and semi-arid rangelands of Australia ever being any more than an extensive form of fire management due to the existing low levels of productivity and shortages of manpower and equipment in these regions. Seldom will a fire be attacked as an entity and almost invariably the fire suppression organisation will come into being when the fire is of some considerable size and as a consequence the fire-fighting operation, of necessity, becomes a series of rearguard actions. Frequently the men and equipment on one side of a large fire have very little knowledge of operations on the other side of the fire, especially when we think in terms of fires of some million acres or more. Aerial reconnaissance and the use of radio equipment will tend to overcome this lack of co-ordination inherent in many fire suppression operations in the arid lands.

In general, the major effort in the very sparsely populated regions will be directed towards protecting the homestead areas and any area of pasture which has a high carrying capacity. In these cases, it might well be possible for the station owners to give more thought to selective grazing around certain areas and even to the design of a firebreak system to protect these more vulnerable areas.

Firebreaks, certainly in western New South Wales, used to consist of two parallel ploughed lines from two to five chains apart and as soon as the grass became burnable it was a regular operation to burn out the intervening strips of grass. This was generally done in the late evening or even night time during late spring or early summer. It appears that this fire prevention measure is not practised nearly as much now as it was 30 or 40 years ago.

At least property owners in these arid and semi-arid lands have one big advantage over fire control organisations in the more settled areas of the continent. As large quantities of fuel are the determining factor in creating a bad fire season, property owners have a visual representation that they have very much above average pastures in a particular year following abundant rainfalls and at least they have some months to prepare themselves for the almost certain occurrence of extensive fires. In such a situation, station owners should certainly give greater attention to the establishment of firebreaks and the protection of their vulnerable homestead blocks by systematic rotational grazing and even patch burning in the areas of lower productivity, such as semi-desert spinifex pastures.

One important aspect which is frequently overlooked, especially in the region of central Queensland where brigalow lands are being cleared for pasture, even extending into the drier acacia land, is the fact that wherever possible strips of the natural vegetation should be left as windbreaks in a systematic pattern. Not only do these strips of the natural vegetation act as windbreaks, but in most cases they act as an effective firebreak because fuel does not generally accumulate under these drier type scrubs to the extent where grassfires will burn through them.

Another point to remember is the fact that tussock grasses such as spinifex frequently carry enough dead material to enable patch burning to be carried out, while the more productive short grass-forb areas are still in a green and non-flammable condition. More attention to the control burning of the low carrying-capacity lands in a good season could provide an effective firebreak system which might at least prevent some of the fires attaining the huge areas which they have done in recent years of above average rainfall.

In the mallee areas of New South Wales, South Australia and Western Australia, more attention should be given to the control burning of these associations more frequently than has been done in the past. There is ample evidence from the accounts of early explorers that a lot of this mallee country was burnt fairly frequently by the Aborigines, generally in winter time. There is no doubt that greater attention should be paid to fuel reduction in these areas, which in a good season and after a long period of fuel accumulation, give tremendous difficulty to fire-fighters in containing fires within these timbered areas.

DAMAGE CAUSED BY FIRES IN ARID LANDS

Fires in arid lands have two immediate effects. Firstly, there is the obvious damage caused to capital improvements including buildings, fencing, stock and loss of pasture. These losses can be readily evaluated and expressed in monetary terms. Losses will vary according to the intensity of agriculture practised and may range from a few cents per acre in the arid regions to quite high values towards the more humid boundaries of the semi-arid pastoral lands.

Secondly, there is the effect of fire on site productivity and erosion processes and even in some parts on aesthetic values. This form of damage is hard to estimate and little work has been done so far in this field. The damage which is caused by a particular fire depends on a complex of factors including the intensity of the fire, but most importantly, the time of the year that the fire occurs. A fire which occurs when the soil moisture is still at a reasonably high level will generally permit any tussock-forming grasses to put out fresh shoots immediately following the fire and perhaps allow some species to re-vegetate from seedling germination. The worst situation which can exist is for a fire to occur just after the grasslands have become fully cured and, in this case, there will be no re-vegetation of the pasture until significant rainfalls occur. In the sub-tropical areas this might mean a period of 8 to 10 months, but in the winter rainfall zone there can of course be a complete failure of the next season's rain and the pastures can remain in a deteriorating condition for upwards of 18 months. Under these circumstances, tussock grasses are subject to severe wind erosion and quite a high proportion of the perennial tussocks may be killed and extensive wind erosion can result. The Mitchell grass country of Queensland and the Northern Territory appears to be prone to extensive damage from this cause. It is for this reason that fire control systems are probably more highly developed in this region, certainly in Queensland, than in most other semi-arid lands of Australia.

In general, the effect of fire in the arid lands is closely associated with grazing pressure on the burnt pastures. If there is no grazing pressure present, a natural ecological succession will generally follow the fire and little damage should occur from it. In the presence of heavy grazing quite drastic disturbances often result and considerable deterioration of the pastures can occur.

Mention has already been made of the study by Burbidge (1943) in the Port Hedland area of Western Australia. Burbidge found that the effect of man and his grazing animals on the *Triodia* association appears so far to have been very slight in comparison with the effect on native vegetation in other parts of the State. Nevertheless, there were places where major changes in vegetation can be recognised. The *Triodia* on these areas has been destroyed, and replaced with *Eragrostis eriopoda*-dominant association. This disturbance and the normal sequence is not merely a regressive change in the succession and it seems best to define it as a disclimax. That the disclimax is a 'permanent' change, was indicated by the fact that in one place the coarse woolly bases so characteristic of *Eragrostis eriopoda* association was being removed in its turn by the grazing animal. The result was a growth of annuals. This represents a return to the condition seen in the early stages of the sere on a bare area.

This is a classic example of the combined effect of burning and intensive grazing resulting in a major change of the dominant species in the area.

Stocker (1966), dealing with fire in the Northern Territory, makes the observation that in the arid southern section of the Territory, there is ample evidence of extensive fires. Observations indicate that most of the trees and shrubs are quite sensitive to fire and usually die after they are burnt. He goes on to comment that burning the vegetation of the low rainfall zone in the Territory seems undesirable. In the past when the only grazing animals were kangaroos, an occasional fire would have had little lasting effect on the vegetation, because the regeneration would not be heavily grazed. Now, with the introduction of cattle, horses, donkeys, camels and rabbits, the regeneration resulting from fires is likely to be heavily over-grazed, unless exceptional rains follow immediately after burning. Burning under present conditions would predispose the burnt areas to erosion.

Gardner (1956) comments in detail on the fire factor in relation to the vegetation of Western Australia. He states:

> . . . bush fires appear to be most prevalent or to have the most marked periodicity in the heathlands and the mallee thickets in which conflagration of extensive occurrences have been observed, some of them travelling on a broad front for many miles, leaving in their trail bare soil or landscapes in which only dead stems remain. Such fires do not occur to the same extent in the sclerophyllous woodland country, nor do they appear to be a feature of the mesophytic forest areas under natural conditions.
>
> Under natural conditions, fires appear to originate mainly by lightning, especially in lateritic or haematitic country.
>
> The effect of the bush fire varies with the nature of the formation. In shrub and mallee thickets, if of sufficient density, the effect is the total destruction of the aerial parts of the woody plant.
>
> This is particularly true of those forms dominated by *Eucalyptus* and *Melaleuca*—both genera-rich plants yielding essential oils. All that remains after the fire are the woody bulb-like bases of the mallee *Eucalyptus* and certain species of

> *Melaleuca* with similar structure, for example *Melaleuca uncinata*. In the savannah formations of the tropics the grass and other herbaceous growth is destroyed, but many of the trees, especially the bloodwoods, ironbarks and smoothbarks (gum trees) are unaffected. Many species of *Acacia* and the *Proteaceae* are liable to be killed outright, but some with thick suberose bark such as *Acacia pyrifolia* and a number of cork bark species of *Hakea* are unaffected. The dry grass carries a fire which is rapid in action so that its heating effect on the trees is relatively transient.
>
> The open nature of the mulga formation and the presence of so much plant debris on the surface of the soil would indicate that fires are not an occurrence of frequent periodicity in this formation. And here again, the highly inflammable nature of the dry *Compositae* would result in a rapidly moving fire unable to burn the rough and hard-barked mulga shrubs and trees.

Gardner also comments:

> . . . that the mallee structure is particularly adapted to resist burning even when severe and there is generally an immediate regeneration by vegetative shoots from adventitious buds in the bulb-like stock.

In conclusion, Gardner states:

> . . . it is postulated that there is sufficient evidence to show that fire as a factor affecting plants is one that has been in existence from remote times, and that the flora of Australia, in part, is pyrophilous. The growth form of the mallee *Eucalyptus*, its power of repeated regeneration almost indefinitely through its woody stock by means of adventitious shoots, and its habitat, all favour this theory. The structure of the fruit in certain primitive families well illustrates the adaptation of the plant to fire, and will allow no other explanation. The general response of the flora following burning of a periodic nature, including the existence of several species of the so-called 'fire bushes' still further supports this theory. Evidence has been advanced showing that absolute protection from fire in the thicket formations has a detrimental effect on the vegetation.

Hooper (1969) commenting on the damage resulting from the Gorrie fire, previously described, which burnt in August 1968, states '. . . over 2 thousand square miles of pastoral country were burnt in a total area of 21,000 square miles. Following the 1969 wet season these areas have deteriorated from a pastoral viewpoint, due to heavy weed growth. Various estimates range up to 5 years for the country to return to full grazing productivity. Birrimba Station was in the early stages of development and the socio-economic effect of the fires could be serious'.

The value of this grazing land between Katherine and Wave Hill is quite considerable and Hooper states '. . . an offer of nearly $2,000,000 was made for a nearby improved station of comparable size in 1968 and the offer was refused'.

It is basic to all consideration of fire damage whether in the arid and semi-arid lands or in the higher rainfall areas of the continent, that the effect of any particular fire is primarily determined by the intensity of the fire. In the case of grass fires, intensities can range from a low of perhaps 50-100 b.t.u's/sec/ft of fire front to a high of 20-30,000 btu/sec/ft of fire front in the case of a fast-moving grass fire in heavy abundant pastures. The higher the intensity of a fire, the greater is the flame temperature and the greater the heat flux extending down into the soil.

In the case of very high intensity fires, such as frequently occur in years of abundant continuous pastures, the root system of the tussock grasses can be killed

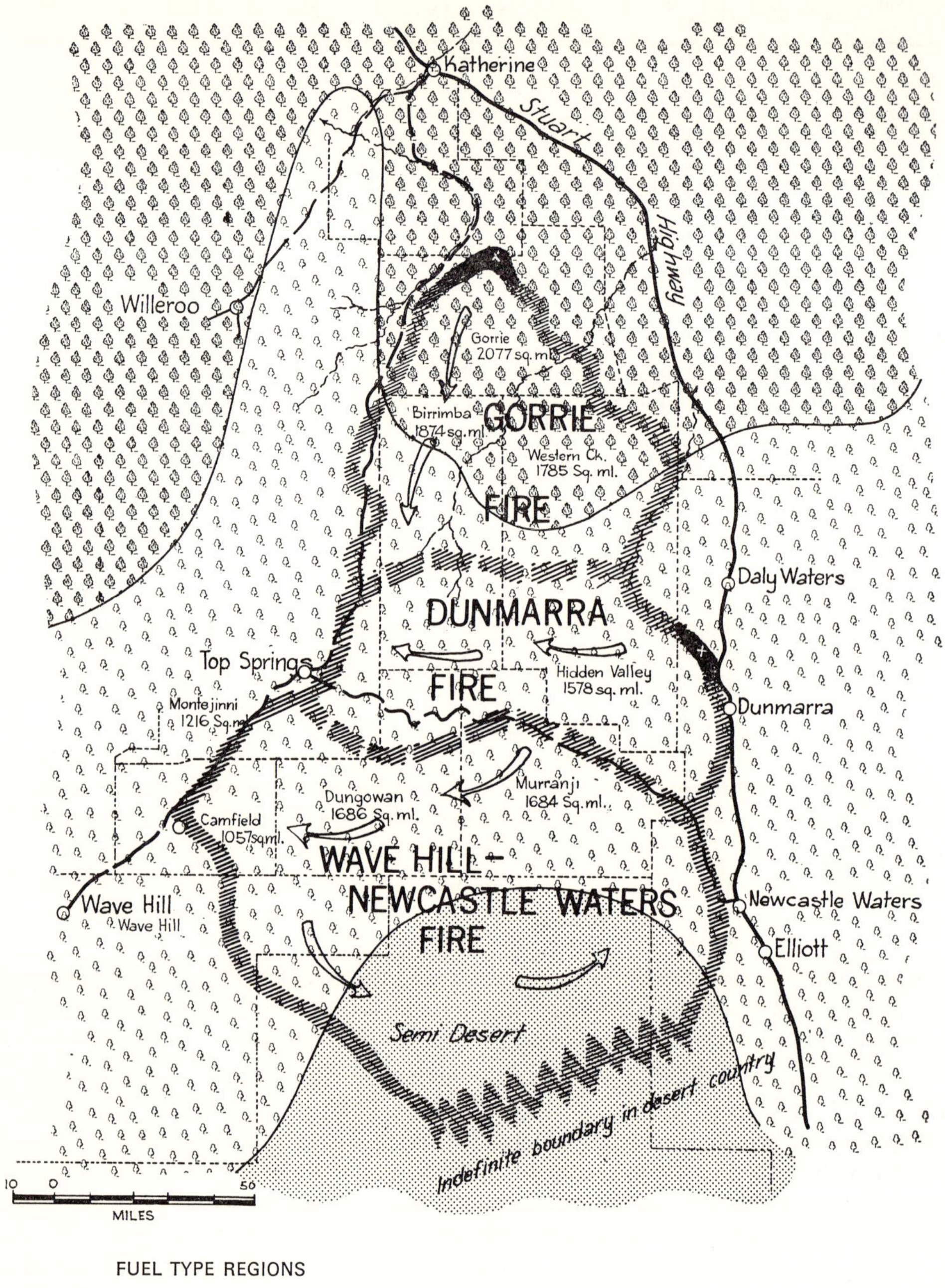

FUEL TYPE REGIONS

Sub-tropical open forest and tall tropical grasses

Arid and semi-arid savannah woodlands and grasslands

Semi-desert and desert country

FIG. 24.3

Plan showing approximate area burnt by three major bushfires in the semi-arid and arid zones of the Northern Territory during August-September, 1968.

outright and recovery of the country will then depend on seedling regeneration. Obviously one of the keys to the question of relative fire damage is the association of grazing with the effect of any fire. In ungrazed areas it is quite apparent that fire damage will not be severe and in general, the vegetation will return to its original form. With associated grazing pressures, drastic disturbances obviously can occur.

The other very significant factor is the weather conditions prevailing after the fire. If the fire is quickly followed by significant rain the effect may be transient. If followed by an extended drought period, drastic deterioration of the pasture lands can occur and severe erosion processes can be initiated.

One interesting effect of bush fires in the semi-arid areas was mentioned in the Royal Commission *Report* (1901), enquiring into the conditions of the Crown tenants in the Western Division of New South Wales. An invasion of woody shrubs has periodically occurred in the Cobar district of New South Wales. The first such invasion was in the 1890s following a period of above-average rainfall and a similar situation occurred in the early 1950s following another prolonged period of above average rainfall. As early as 1892 Dickson (1892) said that settlement in the Cobar district had increased the cypress pine to such an extent that less stock could be carried there. The Royal Commission mentioned that the lack of bush fires following pastoral occupation had been an important factor in allowing scrub and timber regeneration to grow unhindered and reach maturity.

An investigation of this scrub invasion problem is being carried out at the present time by the C.S.I.R.O., Division of Plant Industry, and it would appear that the only economic way of reducing the scrub competition is by burning. Whether or not the use of fire as a management tool can be practised in these lands is doubtful and it may even be doubted whether such land should be used for pastoral pursuits. In the long-term it may well be proved that the land utilisation in this region should be for timber production and not pastoral pursuits. This country is an important buffer between the more arid lands to the west and the more productive lands to the east and could well be a natural erosion barrier to the prevailing dry continental winds in this part of the continent.

It is important to realise that the use of fire is a long-practised management tool in the sub-tropics and especially so in Region AB in Figure 24.1, where the use of fire as a management tool to reduce rank unpalatable grass growth is practised almost annually. In an average year probably well over 25 per cent of this huge area of some 372,000 square miles is burnt to promote better grazing values either just before or just after the wet summer period.

The same situation appears to be practised over a large proportion of the spinifex country and the more arid lands towards the centre of the continent and fire could well be a natural phenomenon of this country. Certainly the accounts of the early explorers such as Giles, Stuart and Gregory indicate that this country was fired by the Aborigines or by lightning whenever it would burn.

With increasing attention being given to better pasture species in many of these lands, there will be a continuing need for greater fire protection of the more highly-productive pastures in the arid lands and it could well be that fire protection is in a transitional stage at the present time. There could well be conflict between the fairly regular burning practices in the harder-type pastures and the necessary protection which must be given to the more palatable and generally more vulnerable short grass species.

Whatever changes take place in pastoral activity in these arid lands, it is difficult to imagine that fire control will ever be anything more than an extensively-practised measure and it would appear that the occurrence of very large fires will be a constant factor of this environment for many more decades.

A high proportion of fires which occur in these arid lands will result from lightning and their occurrence is fairly random and unpredictable. If a large number of lightning fires start following a good season which has produced abundant continuous grasslands over huge areas, the result will always be extensive conflagrations burning many millions of acres. There are significant advances being made, however, in fire control of these lands, especially in the protection of the more highly-productive pastures and it could be confidently expected that there will be a continuing reduction in the area burnt annually. As far as the more humid tropic lands are concerned, fire will continue to play a dominant role in their management for many years and a reduction in the area burnt annually in this region is difficult to imagine.

In the future, the deliberate use of fire may be given a sounder scientific basis and the majority of burning should be carried out at a time of the year which is more favourable to pasture production and result in a lessening of the damage which has resulted from many of these past fires. If the deliberate use of fire is placed on a more systematic and scientific basis, the extent of wildfire damage should be correspondingly reduced.

The area of land burnt each year in the various fuel-type regions shown in Figure 24.1 have been calculated from fire statistics available for the more

TABLE 24.6

ESTIMATED AREA OF LAND BURNT ANNUALLY IN THE VARIOUS FUEL-TYPE REGIONS OF AUSTRALIA

Region	*Area* (*thousand sq miles*)	*Area* (*million acres*)	*Estimated Area Burnt in last 10 years* (*million acres*)	*Area Burnt Annually* (*million acres*)	*Per cent of Total Area Burnt Annually*
A	204	131	20	2(*a*)	1.53
B	402	257	10	1(*b*)	0.39
AB	372	238	595	59(*c*)	25.00
C	1,341	858	100	10(*d*)	1.16
D1	202	129	129	13(*e*)	10.00
D	447	286	143	14.3(*f*)	5.00
Continent	2,968	1,899	997	99.7	5.25

(a) Wildfires only. Approximately 2 million acres per annum in addition would be control-burnt.
(b) Wildfires only. Possibly an additional 100,000 acres would be control-burnt annually.
(c) An estimate that 25 per cent of this sub-tropical open forest area is burnt by cattlemen to remove unpalatable grass resulting from the summer rains and also to promote and prolong green grass into the dry period.
(d) Largely wildfires resulting from lightning, Aboriginal burning and some deliberate burning by cattlemen to improve rough grazing values.
(e) Estimated that 10 per cent of the *Triodia* steppe country is burnt by property owners to improve grazing values, together with some occasional large wildfires.
(f) Estimate of area burnt by wildfires caused by lightning, Aboriginal activity and some deliberate burning for grazing purposes.

settled areas of the continent, and estimated for those regions which do not have an organised system of fire control. These estimates of annual burnt area are given in Table 24.6.

The average annual area burnt in the humid lands of the continent is 62 million acres, whilst the area burnt annually in the arid and semi-arid regions of Australia is 37 million acres.

This appears to be a huge area, but is considered to be on the conservative side. No more accurate figure can be determined until a more accurate reporting network is established. There is little doubt that this will only occur when the remote-sensing earth resources satellites come into operation within the next two or three years. Only then will we have some accurate measure of the huge areas burnt periodically in the sub-tropical humid lands and in the sparsely settled arid regions of the continent.

FURTHER READING

Anon. (1901). *Royal Commission to inquire into the conditions of the Crown Tenants, Western Division of New South Wales.* Govt. Printer, Sydney.

Burbidge, N. T. (1943). 'Ecological succession observed during regeneration of *Triodia pungens* R.Br. after burning'. *J. Roy. Soc. W.A.*, 28 : 149-156.

Chippendale, G. M. (1958). 'Notes on the vegetation of a desert area in Central Australia'. *Trans. Roy. Soc. S.Aust.*, 81 : 31-41.

Collins, H. G. (1967). *Rural Land Utilization.* Comm. Inst. of Valuers, John Andrew & Co., Sydney.

Dickson, S. (1892). 'The effects of settlement and pastoral occupation in Australia upon the indigenous vegetation'. *Trans. Roy. Soc. S.Aust.*, 15 : 195-206.

Gardner, C. A. (1944). 'The Vegetation of Western Australia'. *J. Roy. Soc. W.A.*, 28 : 11-87.

Gardner, C. A. (1956). 'The fire factor in relation to the Vegetation of Western Australia'. *W.A. Naturalist*, 5 : 166-173.

Hooper, R. (1968). Notes on fires in the Northern Territory. (Pers. communication.)

Luke, R. H. (1961). *Bush fire control in Australia.* Hodder and Stoughton, Sydney.

Luke, R. H. (1970). Fire Occurrence in the Western Division of New South Wales. (Departmental Report.) Chief Secretary's Dept. Sydney.

McArthur, A. G. (1968). 'Rural Fire Protection'. Chapter 20, in *Growing Trees on Australian Farms.* Govt. Printer, Canberra.

McArthur, A. G. (1968A). 'The Tasmanian Bushfires of February 1967 and associated fire behaviour characteristics'. *Aust. Fire Prot. Assoc. 2nd Aust. Conf. on Fire*, Sydney.

Newman, J. C. & Condon, R. W. (1969). 'Land use and present conditions'. In *Arid Lands of Australia.* Aust. Nat. Univ. Press, Canberra.

Perry, R. A. (1960). *Pasture lands of the Northern Territory, Australia.* (C.S.I.R.O. Aust. Land Res. Set No. 5.)

Perry, R. A. (1967). 'The need for rangelands research in Australia'. *Proc. ecol. Soc. Aust.*, 2 : 1-14.

Rodger, G. J. (1961). *Report of the Royal Commission on the bushfires of December, 1960 and January, February and March, 1961 in Western Australia.* Govt. Printer, Perth.

Stocker, G. C. (1966). 'Effects of fires on vegetation in the Northern Territory'. *Aust. For. J.*, 30 : 223-30.

Teakle, L. J. H. (1938). 'A regional classification of the soils of Western Australia'. *J. Roy. Soc. W.A.*, 24 : 123-195.

Waring, E. J. (1969). 'Some economic aspects of the pastoral industry in Australia'. In *Arid Lands of Australia.* Aust. Nat. Univ. Press, Canberra.

Williams, R. J. (1955). 'Vegetation Regions' (map and notes). In *Atlas of Australian Resources.* Dept. Nat. Dev., Canberra.

Winkworth, R. E. (1967). 'The composition of several arid spinifex grasslands of Central Australia in relation to rainfall, soil water relations and nutrients'. *Aust. J. Bot.*, 15 : 107-30.

Appendix I

TABLE I.1

ANNUAL RAINFALL—PERCENTILE VALUES AND LENGTH OF RECORDS FOR SELECTED STATIONS IN THE ARID AND SEMI-ARID ZONES OF AUSTRALIA(*a*)

	Climatic Zone	*Rainfall (inches)*					
		Lowest on record	*Percentile values* 10	50	90	*Highest on record*	*Years of record*
WESTERN AUSTRALIA							
North Kimberley—							
Wyndham	S-A	14.38	17.46	25.68	38.24	56.34	79
West Kimberley—							
Fitzroy Crossing	S-A	7.41	11.53	20.48	28.67	35.19	72
East Kimberley							
Hall's Creek	S-A	8.44	11.17	17.82	28.81	42.02	75
De Grey—							
Marble Bar	A	2.80	7.13	12.19	20.41	29.20	71
Port Hedland	A	1.25	4.82	10.73	22.74	40.13	68
Fortescue—							
Onslow	A	0.57	2.35	9.18	21.94	39.27	80
Roy Hill	A	0.88	4.86	9.02	17.49	25.15	66
West Gascoyne—							
Carnarvon	A	2.66	4.30	8.42	14.69	25.36	83
East Gascoyne—							
Mt Magnet	A	2.89	3.98	8.21	14.39	25.27	71
Meekatharra	A	2.59	3.28	7.02	14.06	16.52	22
Mundiwindi	A	1.03	4.22	9.16	17.86	32.11	50
North Eastern—							
Giles	A	1.49	2.52	8.60	13.20	13.46	13
South Eastern—							
Kalgoorlie (*b*)	S-A	4.75	5.74	9.05	14.76	19.11	69
Laverton	A	2.58	4.09	7.66	13.98	17.82	67
Norseman	S-A	5.43	7.03	10.45	14.77	23.99	69
Salmon Gums	S-A	6.28	8.24	13.08	19.42	22.36	46
Southern Cross	S-A	4.64	6.47	10.75	15.47	22.64	79
Eucla—							
Eucla	S-A	4.42	6.69	9.79	13.59	17.05	75
Rawlinna	S-A	3.10	3.36	6.55	12.32	16.95	52
North Coastal—							
Geraldton	S-A	10.32	12.18	18.20	23.85	33.65	81
Mullewa	S-A	6.85	8.10	13.01	18.85	23.64	70
Central North—							
Merredin	S-A	5.12	9.29	12.34	16.92	22.19	61
Central South—							
Corrigin	S-A	8.44	10.52	14.96	20.16	23.96	56
Narrogin(*c*)	S-A	10.56	14.95	20.11	26.01	29.17	74
Lake Grace	S-A	8.37	9.35	14.27	18.18	23.48	53

(*a*) Data taken from publication of the Bureau of Meteorology, *Review of Australia's Water Resources; Monthly Rainfall and Evaporation* (1968), published for the Australian Water Resources Council.
(*b*) Kalgoorlie is on the border of the arid and semi-arid zones.
(*c*) Narrogin is on the border of the semi-arid and sub-humid zones.

TABLE I.1—*continued*

	Climatic Zone	*Rainfall (inches)* Lowest on record	Percentile values 10	50	90	Highest on record	Years of record
SOUTH AUSTRALIA							
North West Interior—							
Coober Pedy	A	1.20	2.12	4.81	10.26	12.95	54
Tarcoola	A	1.99	3.24	6.38	11.12	13.76	62
Far North Interior—							
Marree	A	2.13	3.30	5.39	9.41	16.08	80
Oodnadatta	A	1.16	1.87	4.22	8.26	11.63	56
Western—							
Ceduna	S-A	5.73	7.64	10.77	15.29	19.13	59
Cook	A	2.42	3.33	6.08	10.00	12.75	51
Port Lincoln	S-A	11.40	14.69	18.48	24.25	29.99	100
Upper North—							
Port Augusta(*d*)	S-A	2.21	5.93	9.26	13.42	20.90	106
Upper Murray Valley—							
Renmark	S-A	4.12	6.70	10.08	14.62	17.10	76
VICTORIA							
Mildura	S-A	4.99	7.37	10.72	15.91	17.21	12
Ouyen	S-A	5.78	7.74	13.50	17.36	21.29	53
NEW SOUTH WALES							
Western—							
Bourke	S-A	4.02	6.44	12.43	22.99	33.62	88
Broken Hill	A	2.24	5.15	9.12	14.47	17.61	77
Cobar	S-A	4.56	7.04	13.55	21.60	31.48	84
Ivanhoe	A	2.82	5.98	10.60	18.36	24.80	80
Tibooburra	A	1.88	3.59	6.84	14.82	25.81	74
Riverina—							
Deniliquin	S-A	5.54	10.34	15.16	22.99	28.72	106
Griffith	S-A	7.34	9.19	15.58	23.44	27.61	54
Wagga Wagga(*e*)	S-A	8.86	14.00	22.26	28.73	38.91	27
Central Plains—							
Condobolin	S-A	8.29	11.16	15.73	24.26	35.54	85
Coonabarabran	S-A	12.64	17.70	27.45	39.71	62.85	86
North West Plains—							
Moree	S-A	7.97	14.94	21.76	31.33	43.60	86
Walgett	S-A	6.74	10.25	17.87	28.30	36.30	87
QUEENSLAND							
Carpentaria—							
Burketown	S-A	7.38	14.07	27.54	45.77	75.99	79
Cloncurry	S-A	5.29	8.97	16.90	28.22	41.23	82
Hughenden	S-A	5.91	8.29	19.23	30.02	42.71	82
Central Coast—							
Charters Towers	S-A	4.29	13.23	24.97	39.68	48.13	84
Western—							
Birdsville	A	1.30	2.04	4.51	12.43	21.33	68
Boulia	A	0.95	3.86	8.00	20.92	31.43	79
Camooweal	S-A	5.91	8.04	14.14	24.13	32.27	74
Windorah	A	3.31	4.20	10.14	20.28	38.98	82

(*d*) Port Augusta is on the border of the arid and semi-arid zones.
(*e*) Wagga is on the border of the semi-arid and sub-humid zones.

TABLE I.1—*continued*

	Climatic Zone	Rainfall (inches)					
		Lowest on record	Percentile values			Highest on record	Years of record
			10	50	90		
QUEENSLAND—*continued*							
Central—							
Barcaldine	S-A	5.74	9.68	18.58	35.08	46.39	79
Clermont	S-A	5.45	15.66	24.63	42.17	51.00	95
South West—							
Charleville	S-A	7.97	10.45	18.71	31.59	47.32	85
Cunnamulla	S-A	5.30	6.66	13.67	22.88	36.40	87
Thargomindah	A	1.98	4.10	10.40	18.28	28.45	87
Maranoa—							
Roma	S-A	6.44	14.43	21.88	34.88	60.07	92
St. George	S-A	4.93	11.81	18.92	30.45	39.52	85
NORTHERN TERRITORY							
Northern Rivers—							
Daly Waters	S-A	8.99	16.57	24.22	35.60	45.95	93
Katherine	S-A	17.30	24.31	39.93	51.09	58.91	93
Tennant Creek	A	3.72	6.80	13.37	22.46	34.02	92
Northern Plateau—							
Alice Springs	A	2.37	5.41	10.09	16.98	28.57	92
Barrow Creek	A	2.76	4.84	10.44	18.15	38.86	95
Finke	A	0.83	1.69	4.99	9.67	12.31	89

TABLE I.2

RAINFALL—MEAN MONTHLY AND MEAN ANNUAL; PERIOD OF YEARS OVER WHICH RECORDS HAVE BEEN MADE FOR SELECTED STATIONS IN THE ARID AND SEMI-ARID ZONES OF AUSTRALIA(*a*)

	Period	*Monthly average rainfall(a) (points of .01 inch)*												*Mean annual*
		Jan.	*Feb.*	*Mar.*	*Apr.*	*May*	*June*	*July*	*Aug.*	*Sept.*	*Oct.*	*Nov.*	*Dec.*	
WESTERN AUSTRALIA														
North Kimberley—														
Wyndham	1886–1965	770	645	476	104	30	20	17	2	7	44	181	434	2,730
West Kimberley—														
Fitzroy Crossing	1893–1964	586	523	279	72	41	37	31	5	5	17	96	346	2,043
East Kimberley—														
Hall's Creek	1890–1960	519	452	258	68	35	24	27	10	5	52	128	298	1,886
De Grey—														
Marble Bar	1895–1964	289	292	203	81	91	93	50	20	4	18	35	136	1,317
Port Hedland	1897–1965	210	260	290	87	123	111	41	31	5	6	11	53	1,228
Fortescue—														
Onslow	1886–1964	97	182	203	78	171	169	76	36	4	2	7	14	1,039
Roy Hill	1900–1964	165	210	185	88	75	66	46	24	7	17	29	98	1,010
West Gascoyne—														
Carnarvon	1883–1964	44	80	70	56	156	227	160	69	21	13	5	12	913
East Gascoyne—														
Mt Magnet	1894–1964	95	110	102	61	102	118	104	81	34	22	27	42	898
Meekatharra	1908–1964	141	102	132	76	102	110	65	49	16	15	31	45	884
Mundiwindi	1915–1964	174	181	185	85	83	73	32	30	13	31	45	104	1,036
North Eastern—														
Giles	1956–1968	88	121	48	55	101	87	52	61	10	23	58	110	814
South Eastern—														
Kalgoorlie(*b*)	1896–1964	69	91	102	84	110	109	94	88	50	59	56	59	971
Laverton	1899–1964	93	87	125	89	93	93	58	50	30	29	57	62	866
Norseman	1897–1964	62	97	88	92	118	115	109	104	75	82	72	73	1,087
Salmon Gums	1922–1964	83	95	110	104	131	155	146	131	125	109	77	76	1,342
Southern Cross	1889–1964	50	75	94	80	134	148	154	118	72	59	51	46	1,081
Eucla—														
Eucla	1876–1964	60	70	79	105	122	106	92	90	71	75	64	58	987
Rawlinna	1915–1964	57	63	71	74	70	71	56	66	44	52	48	50	722
North Coastal—														
Geraldton	1877–1964	23	29	52	93	274	478	380	267	122	71	26	15	1,830
Mullewa	1896–1964	50	59	80	68	185	265	241	174	81	53	30	28	1,314
Central North—														
Merredin	1903–1964	40	51	87	87	165	206	217	157	100	77	50	53	1,290
Central South—														
Corrigin	1910–1964	42	61	90	88	200	260	259	196	116	92	55	54	1,513
Narrogin(*c*)	1891–1964	38	63	86	119	257	363	367	278	188	133	53	50	1,998
Lake Grace	1914–1964	52	73	97	92	189	224	206	166	119	92	54	51	1,415
SOUTH AUSTRALIA														
North West Interior—														
Coober Pedy	1931–1966	47	105	49	20	49	54	19	29	22	52	35	52	533
Tarcoola	1904–1956	37	85	44	35	58	73	46	62	46	69	55	55	665
Far North Interior—														
Marree	1885–1964	51	64	60	42	56	62	35	33	40	48	46	69	607
Oodnadatta	1892–1965	61	78	49	23	39	53	27	25	29	43	33	46	506
Western—														
Ceduna	1906–1964	29	63	52	74	145	161	151	140	90	93	76	61	1,135
Cook	1918–1964	39	46	58	55	50	73	49	53	34	62	52	53	624
Port Lincoln	1866–1964	53	59	73	140	227	298	305	258	191	141	92	69	1,906
Upper North—														
Port Augusta(*d*)	1860–1962	59	66	66	75	102	107	77	88	88	91	71	63	953
Upper Murray Valley—														
Renmark	1899–1964	60	76	60	69	105	103	87	100	111	104	84	74	1,033
VICTORIA														
Mildura	1890–1965	75	76	69	63	104	116	93	104	94	104	81	79	1,058
Ouyen	1913–1963	77	94	81	81	128	127	122	128	125	146	100	98	1,307
NEW SOUTH WALES														
Western—														
Bourke	1871–1964	142	162	135	112	103	118	89	77	74	109	117	136	1,374
Broken Hill	1889–1964	74	92	71	64	87	93	69	72	69	88	77	76	933
Cobar	1882–1964	129	138	120	104	110	126	92	116	91	118	117	142	1,403
Ivanhoe	1885–1964	87	105	98	70	98	110	82	88	83	102	93	105	1,121
Tibooburra	1886–1965	81	111	99	51	58	68	57	44	41	70	58	87	825
Riverina—														
Deniliquin	1859–1964	104	111	127	123	160	172	137	141	147	161	113	107	1,608
Griffith	1914–1964	109	107	138	128	140	163	128	155	123	170	115	127	1,603
Wagga Wagga(*e*)	1872–1964	149	151	169	166	184	249	193	195	182	213	160	147	2,158

Notes: (*a*) Data taken from publication of the Bureau of Meteorology, *Rainfall Statistics Australia—long period averages (all years), standard period normals of rainfall and raindays (1931-1960) (1960).*

(*b*) Kalgoorlie is on the border of the arid and semi-arid zones.

(*c*) Narrogin is on the border of the semi-arid and sub-humid zones.

(*d*) Port Augusta is on the border of the arid and semi-arid zones.

(*e*) Wagga is on the border of the semi-arid and sub-humid zones.

TABLE I.2—*continued*

	Period	Monthly average rainfall(*a*) (*points of* .01 *inch*)												Mean annual
		Jan.	*Feb.*	*Mar.*	*Apr.*	*May*	*June*	*July*	*Aug.*	*Sept.*	*Oct.*	*Nov.*	*Dec.*	
N.S.WALES—*continued*														
Central Plains—														
Condobolin	1881–1964	160	151	155	132	138	153	127	140	114	152	131	160	1,713
Coonabarabran	1880–1964	294	329	250	220	201	238	210	205	186	213	223	243	2,812
North West Plains—														
Moree	1880–1964	267	271	235	130	157	179	155	127	135	193	211	220	2,280
Walgett	1878–1964	230	229	167	131	148	158	128	107	109	145	153	159	1,864
QUEENSLAND														
Carpentaria—														
Burketown	1886–1964	837	703	570	93	21	26	6	4	7	50	138	431	2,886
Cloncurry	1884–1964	431	457	246	75	58	53	33	15	28	54	122	269	1,841
Hughenden	1884–1964	452	388	235	98	68	79	48	29	40	82	124	271	1,914
Central Coast—														
Charters Towers	1882–1964	536	515	417	149	76	115	64	48	65	77	152	321	2,535
Western—														
Birdsville	1892–1964	79	96	69	38	41	45	39	19	27	45	50	57	605
Boulia	1886–1964	171	196	157	56	49	48	35	26	28	62	72	118	1,018
Camooweal	1891–1964	352	343	219	51	42	52	23	12	22	52	115	224	1,507
Mt Isa	1926–1964	331	361	208	66	77	73	24	9	26	84	95	223	1,577
Windorah	1887–1964	149	191	176	77	63	71	56	39	45	75	79	124	1,145
Central—														
Barcaldine	1886–1964	328	314	275	148	121	111	96	56	60	121	137	226	1,993
Clermont	1870–1964	492	455	310	154	129	151	108	68	87	139	208	354	2,655
South West—														
Charleville	1874–1964	273	270	257	138	121	122	116	75	80	140	162	222	1,976
Cunnamulla	1879–1964	157	218	156	100	107	112	87	64	77	99	106	149	1,432
Thargomindah	1879–1964	146	144	108	71	81	77	52	51	46	77	94	122	1,069
Maranoa—														
Roma	1870–1964	326	306	267	131	137	142	146	94	130	194	215	254	2,342
St. George	1881–1964	276	261	219	125	139	143	130	95	109	147	176	199	2,019
NORTHERN TERRITORY														
Northern Rivers—														
Daly Waters	1873–1965	606	577	452	92	24	21	4	8	17	75	209	394	2,479
Katherine	1873–1963	914	793	615	134	22	8	3	2	23	117	330	785	3,746
Tennant Creek	1874–1963	356	336	181	59	44	29	24	12	23	55	102	196	1,417
Northern Plateau—														
Alice Springs	1873–1963	155	167	112	67	64	58	41	37	33	80	94	143	1,051
Barrow Creek	1874–1963	211	212	149	52	58	48	34	18	26	69	106	157	1,140
Finke	1938–1963	69	144	38	56	50	42	49	28	12	49	29	51	617

TABLE I.3

LOCATION, TEMPERATURE AND FROST INFORMATION FOR SELECTED STATIONS IN THE ARID AND SEMI-ARID ZONES OF AUSTRALIA(a)

State and locality	*Climatic Zones*	*Alt. (feet)*	*Lat. °S*	*Long. °E*	*Mean temperature (°F)*					*Frosts based on 32°F screen*		*Frost free average period (days) to 36°F in the screen*
					January(a)		*July*		*Year*	*Average number (year)*	*Lowest on record (°F)*	
					Min.	*Max.*	*Min.*	*Max.*				
WESTERN AUSTRALIA												
North Kimberley—												
Wyndham	S-A	23	15.4	128.1	81.4	98.5 Nov.	66.2	85.0	84.4	...	50.0	365
West Kimberley—												
Fitzroy Crossing	S-A	450	18.2	125.6	77.1	106.2 Nov.	52.3	84.4	81.9	...	34.8	365
East Kimberley—												
Hall's Creek	S-A	1,225	18.2	127.8	74.2	100.5 Nov.	47.6	80.1	78.1	N.A.	30.0	359
De Grey—												
Marble Bar	A	595	21.2	119.7	78.1	107.5 Dec.	52.4	80.6	81.9	...	34.0	365
Port Hedland	A	25	20.3	118.4	77.6	95.3 Mar.	55.6	79.3	76.9	...	42.0	365
Fortescue—												
Onslow	A	14	21.7	114.9	74.2	96.4	51.5	77.3	76.1	...	37.5	365
West Gascoyne—												
Carnarvon	A	15	24.9	113.6	72.4	88.1 Feb.	51.6	71.7	71.5	...	37.0	365
East Gascoyne—												
Mt Magnet	A	1,401	24.3	116.9	73.5	101.5	42.9	63.9	71.0	N.A.	29.3	N.A.
Meekatharra	A	1,675	26.6	118.5	73.1	100.4	44.0	67.5	71.9	0.1	31.2	341
Mundiwindi	A	1,840	23.9	120.2	73.6	100.5	41.4	70.0	72.7	5.5	22.4	300
North Eastern—												
Giles	A	1,900	25.0	128.3	75.1	99.3	44.8	67.2	73.1	N.A.	30.0	N.A.

South Eastern—													
Kalgoorlie(*b*)	A	1,247	30.7	121.5	64.2	93.2	42.9	62.5	66.2	0.6	30.0	297
Laverton	A	1,530	28.7	122.4	68.9	96.7	41.4	64.5	68.7	3.6	24.5	289
Salmon Gums	S-A	816	33.0	121.6	55.8	84.7	39.3	60.6	60.6	11.6	24.0	184
Southern Cross	S-A	1,170	31.2	119.3	62.2	94.3	39.1	61.7	64.5	10.8	23.0	228
Eucla—													
Eucla	S-A	15	31.7	129.0	62.5	78.3	44.2	64.6	63.2	0.6	28.0	335
Rawlinna	S-A	607	31.0	125.2	58.9	90.0	39.3	64.2	64.2	3.1	27.8	279
North Coastal—													
Geraldton	S-A	13	28.7	114.6	66.3	84.5	51.7	67.7	67.7	...	33.4	365
Mullewa	S-A	925	28.5	115.5	66.7	97.4	43.8	65.4	68.7	0.2	29.4	310
Central North—													
Merredin	S-A	1,046	31.5	118.3	61.8	92.3	40.2	60.1	63.5	6.2	25.0	242
Central South—													
Corrigin	S-A	968	32.3	117.9	60.5	91.6	40.0	61.8	62.0	15	24.2	N.A.
Narrogin(*c*)	S-A	1,114	32.9	117.1	56.3	87.3	41.3	57.9	60.2	3.5	27.2	225
Lake Grace	S-A	946	33.1	118.5	57.9	90.1	40.8	58.5	62.1	N.A.	30.0	N.A.
SOUTH AUSTRALIA													
North West Interior—													
Coober Pedy	A	700	29.0	134.7	Long term average not available						32.0(*d*)	
Tarcoola	A	395	30.7	134.6	63.3	93.9	38.5	64.5	65.6	10.1	23.5	254
Far North Interior—													
Marree	A	160	29.6	138.1	67.8	97.7	40.0	65.6	68.3	4	27.0	N.A.
Oodnadatta	A	384	27.5	135.5	71.4	98.0	41.9	66.8	70.2	N.A.	27.4	N.A.
Western—													
Ceduna	S-A	43	32.1	133.7	58.8	81.5	43.8	62.6	61.9			
Cook	A	404	30.6	130.4	58.4	89.7	39.4	64.3	63.8	8.1	25.8	242
Port Lincoln	S-A	13	34.7	135.9	58.5	77.7 Feb.	46.4	60.2	60.7	...	32.0	365
Upper North—													
Port Augusta(*e*)	S-A	18	32.5	137.7	65.3	89.5	43.9	62.8	66.2	0.1	31.4	349
Upper Murray Valley—													
Renmark	S-A	80	34.1	140.7	62.6	91.1	41.2	60.4	63.4	N.A.	25.9	N.A.
VICTORIA													
Mildura	S-A	177	34.2	142.2	61.7	90.0 Feb.	40.5	59.5	63.3	1.1	23.0	272
Ouyen	S-A	165	35.1	142.3	58.7	89.4	39.8	58.7	61.3	9.8	27.0	229

For footnotes see p. 325.

TABLE I.3—*continued*

State and locality	*Climatic Zones*	*Alt. (feet)*	*Lat. °S*	*Long. °E*	*Mean temperature (°F)*					*Frosts based on 32°F screen*		*Frost free average period (days) to 36°F in the screen*
					January(a)		*July*		*Year*	*Average number (year)*	*Lowest on record (°F)*	
					Min.	*Max.*	*Min.*	*Max.*				
NEW SOUTH WALES												
Western—												
Bourke	S-A	361	30.2	146.0	69.3	98.0	40.8	63.8	68.7	1.6	25.0	299
Broken Hill	A	1,000	31.9	141.5	64.5	90.5	41.2	59.5	64.4	2.3	27.0	298
Cobar	S-A	822	31.5	145.8	66.7	94.2	39.3	60.0	65.7	12.4	24.4	240
Ivanhoe	A	288	32.9	144.3	63.1	96.5	38.0	62.0	64.6	17	22.0	N.A.
Tibooburra	A	700	29.4	142.0	70.0	96.7	41.2	63.8	68.7	1.3	23.0	304
Riverina—												
Deniliquin	S-A	311	35.2	145.0	60.5	88.0	39.9	57.5	61.8	3.8	26.0	252
Griffith	S-A	420	34.3	146.0	61.8	89.2	38.4	57.8	62.0	16.3	23.5	212
Wagga Wagga(*f*)	S-A	612	35.1	147.4	61.5	89.8	37.8	57.1	61.6	13.1	22.0	226
Central Plains—												
Condobolin	S-A	653	33.1	147.2	65.5	94.4	38.2	60.3	65.0	11.7	20.0	240
Coonabarabran	S-A	1,673	31.3	149.3	58.9	88.3	33.3	57.7	59.8	52.1	11.4	186
North West Plains—												
Moree	S-A	686	29.5	149.8	67.4	96.0	39.0	64.8	67.6	7.0	19.0	255
Walgett	S-A	436	36.2	148.2	68.5	95.2	40.6	63.1	67.5	7.6	23.0	255
QUEENSLAND												
Carpentaria—												
Burketown	S-A	30	17.8	139.3	76.6	95.6 Dec.	55.0	81.2	78.6	...	38.0	365
Cloncurry	S-A	633	20.7	140.5	76.2	100.4 Dec.	51.5	76.4	77.9	...	34.5	365
Hughenden	S-A	1,074	20.8	144.2	70.7	99.4 Dec.	46.5	77.1	75.1	N.A.	27.0	358
Central Coast—												
Charters Towers	S-A	1,019	20.0	146.1	69.7	94.7	51.6	76.0	74.4	...	35.5	365

Western—											
Boulia	A	478	22.9	139.6	76.0	101.8	45.8	72.8	75.4		26.4	365
Camooweal	S-A	738	19.9	138.3	74.5	100.2 Dec.	47.7	77.2	76.9		29.0	348
Windorah	A	390	25.4	142.6	74.5	100.8	43.0	69.7	73.2	0.1	28.0	328
Central—												
Barcaldine	S-A	869	23.6	145.2	73.4	96.6	45.1	72.0	73.1	0.4	24.0	330
Clermont	S-A	870	22.8	147.6	69.1	95.2 Dec.	42.8	72.8	71.7	2.1	21.1	319
South West—												
Charleville	S-A	965	26.4	146.2	70.8	97.6	40.1	68.3	70.4	7.2	23.0	273
Cunnamulla	S-A	629	28.1	145.6	71.3	97.4	42.3	65.1	69.9	0.8	26.0	309
Thargomindah	A	402	28.0	143.7	74.1	98.1	42.4	65.3	71.0	0.5	23.7	328
Maranoa—												
Roma	S-A	1,010	26.5	148.7	68.3	94.4	39.3	67.4	68.5	18.4	21.4	244
St George	S-A	650	28.0	148.6	70.5	96.5	40.8	65.5	68.9	4.9	24.0	274
NORTHERN TERRITORY												
Northern Rivers—												
Daly Waters	S-A	691	16.3	133.4	74.8	102.1 Nov.	54.0	83.9	80.4	N.A.	30.2	365
Katherine	S-A	373	14.1	132.7	76.5	100.6 Nov.	55.8	86.4	80.9	...	37.0	365
Tennant Creek	S-A	1,075	19.6	134.2	75.5	99.8 Dec.	51.1	75.4	77.4	...	36.4	365
Wave Hill	S-A	700	17.5	130.9	75.0	102.3 Nov.	55.0	81.8 June	80.1		38.0	365
Northern Plateau—												
Alice Springs	A	1,901	22.6	133.6	69.8	95.3	38.9	66.9	69.1	12.0	22.0	263
Barrow Creek	A	1,679	21.5	133.9	75.8	98.1	47.3	72.6	75.0	1.0	30.0	364
Finke	A	883	25.6	134.6	72.8	99.3	42.4	67.1	71.0	N.A.	26.0	N.A.

(*a*) Data taken from Bureau of Meteorology publications, *Climatic Averages* (1956) for temperature, *Frost in the Australian Region* (1945) by J. C. Foley or specially supplied by the Bureau.
(*b*) Kalgoorlie is on the border of the arid and semi-arid zones.
(*c*) Narrogin is on the border of the semi-arid and sub-humid zones.
(*d*) Only 8/7/64-8/7/67.
(*e*) Port Augusta is on the border of the arid and semi-arid zones.
f) Wagga is on the border of the semi-arid and sub-humid zones.

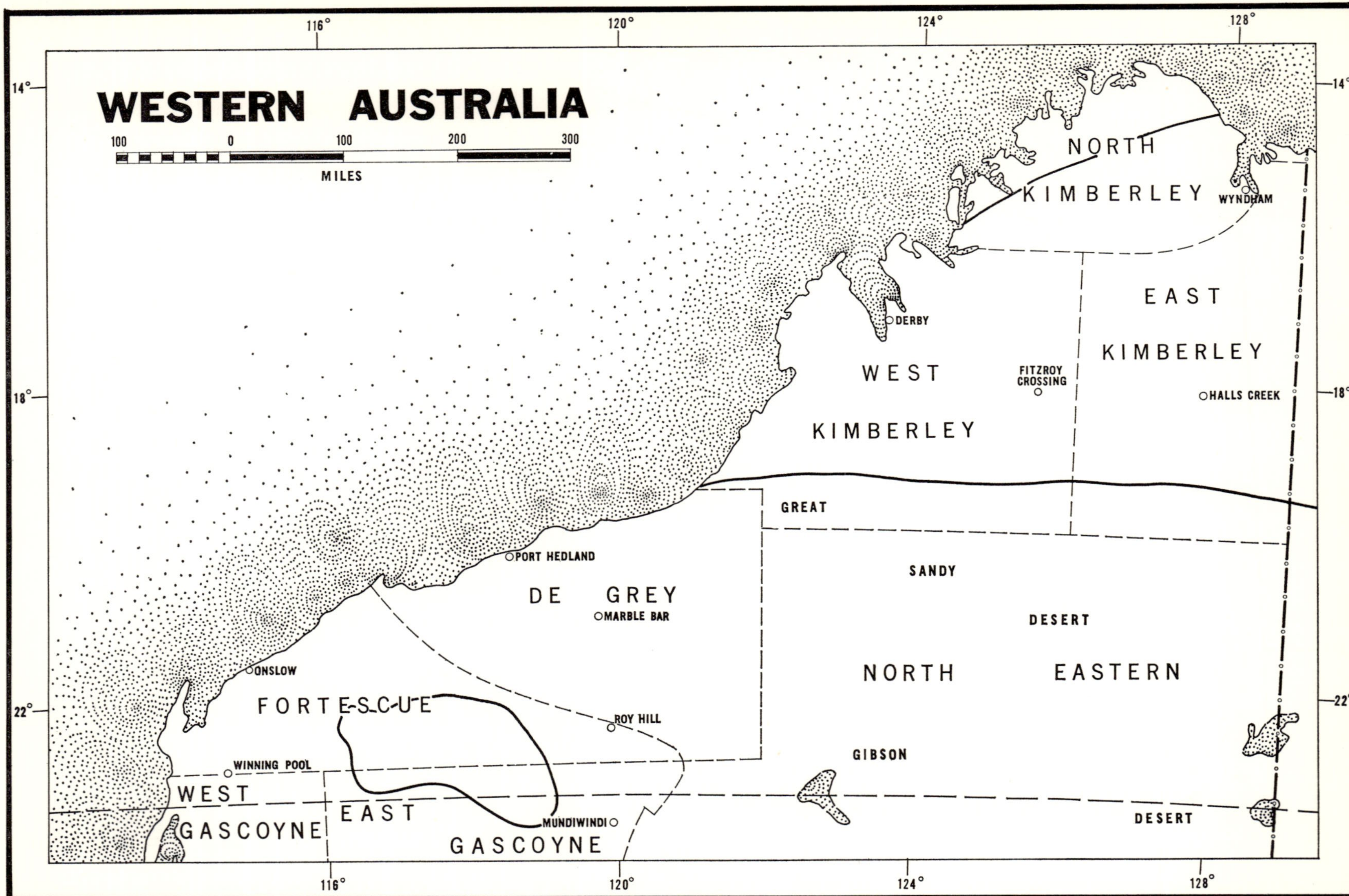

WESTERN AUSTRALIA
100
0
100
200
300
MILES
116°
120°
124°
128°
14°
18°
22°
NORTH
KIMBERLEY
WYNDHAM
EAST
KIMBERLEY
DERBY
WEST
KIMBERLEY
FITZROY CROSSING
HALLS CREEK
GREAT
SANDY
DESERT
NORTH
EASTERN
PORT HEDLAND
DE GREY
MARBLE BAR
ONSLOW
FORTESCUE
ROY HILL
GIBSON
DESERT
WINNING POOL
WEST
GASCOYNE
EAST
GASCOYNE
MUNDIWINDI

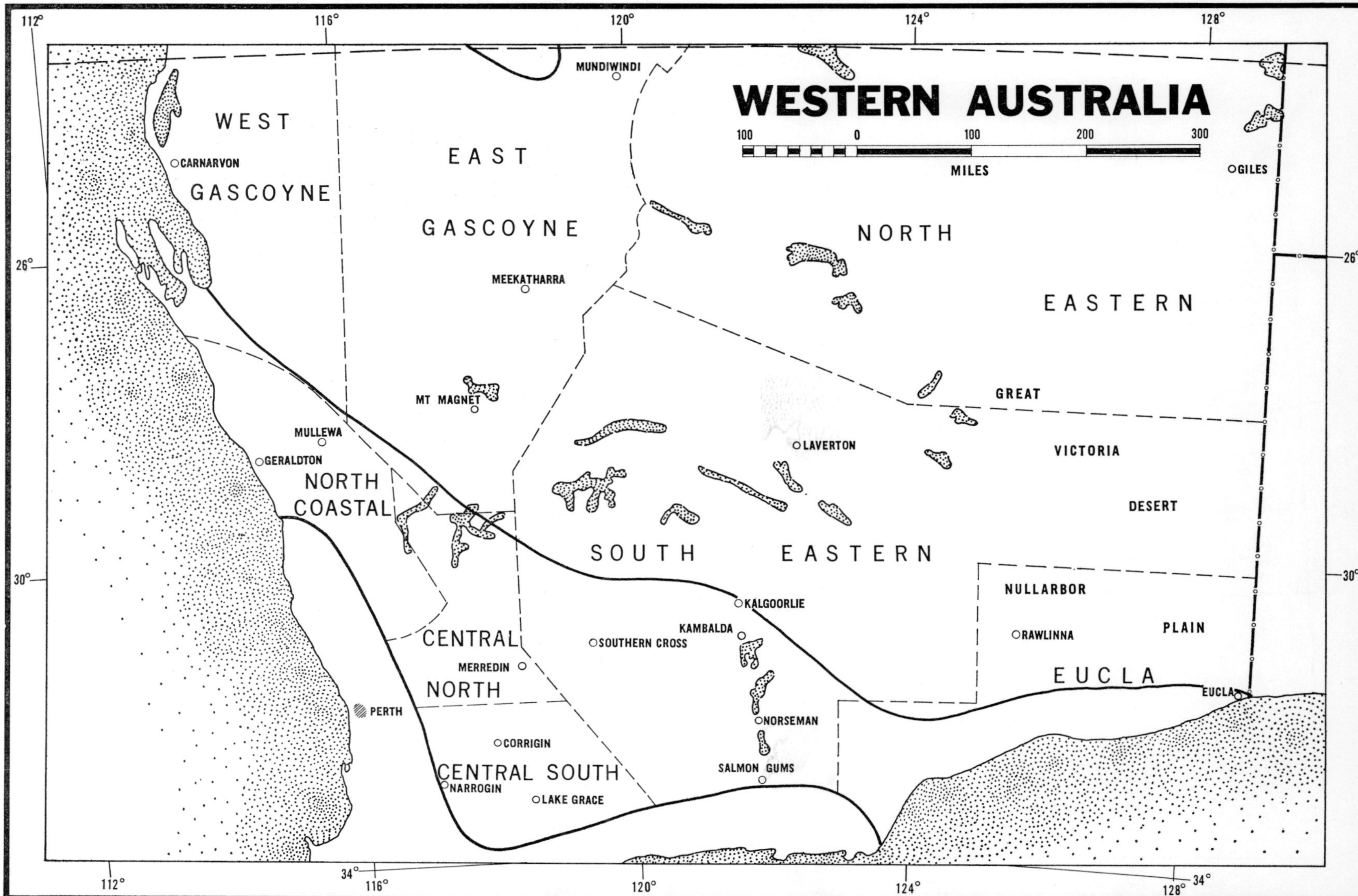
WESTERN AUSTRALIA
100
0
100
200
300
MILES
WEST
GASCOYNE
EAST
GASCOYNE
NORTH
EASTERN
GREAT
VICTORIA
DESERT
NORTH
COASTAL
SOUTH
EASTERN
NULLARBOR
PLAIN
EUCLA
CENTRAL
NORTH
CENTRAL SOUTH
CARNARVON
MUNDIWINDI
GILES
MEEKATHARRA
MT MAGNET
MULLEWA
GERALDTON
LAVERTON
KALGOORLIE
KAMBALDA
SOUTHERN CROSS
MERREDIN
PERTH
CORRIGIN
NARROGIN
LAKE GRACE
NORSEMAN
SALMON GUMS
RAWLINNA
EUCLA
112°
116°
120°
124°
128°
26°
30°
34°

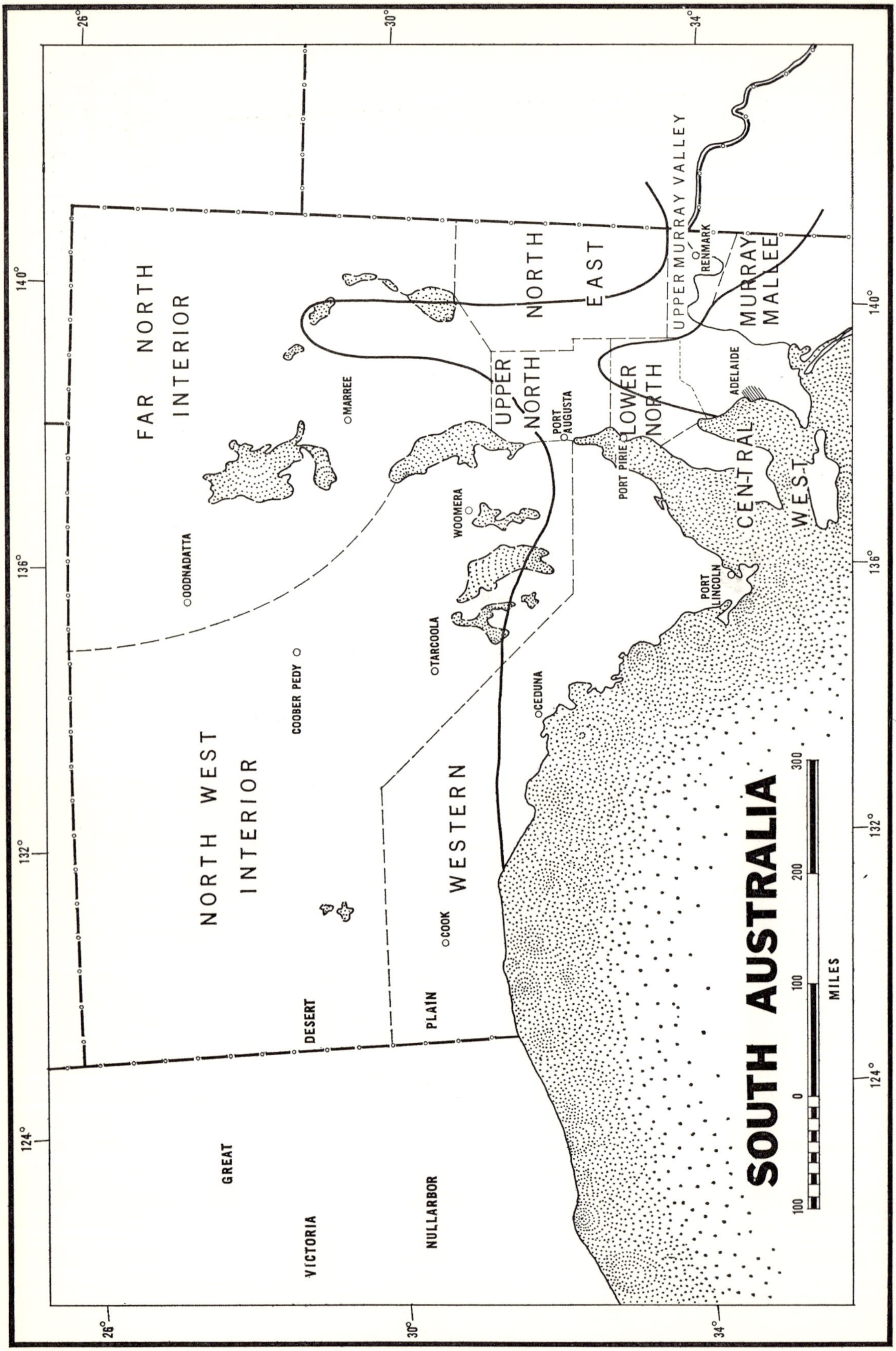
SOUTH AUSTRALIA
MILES
100
0
100
200
300
FAR NORTH INTERIOR
NORTH WEST INTERIOR
UPPER NORTH
NORTH EAST
LOWER NORTH
UPPER MURRAY VALLEY
MURRAY MALLEE
CENTRAL
WEST
WESTERN
GREAT VICTORIA DESERT
NULLARBOR PLAIN
OODNADATTA
MARREE
COOBER PEDY
TARCOOLA
WOOMERA
COOK
CEDUNA
PORT AUGUSTA
PORT PIRIE
PORT LINCOLN
ADELAIDE
RENMARK
26°
30°
34°
124°
132°
136°
140°

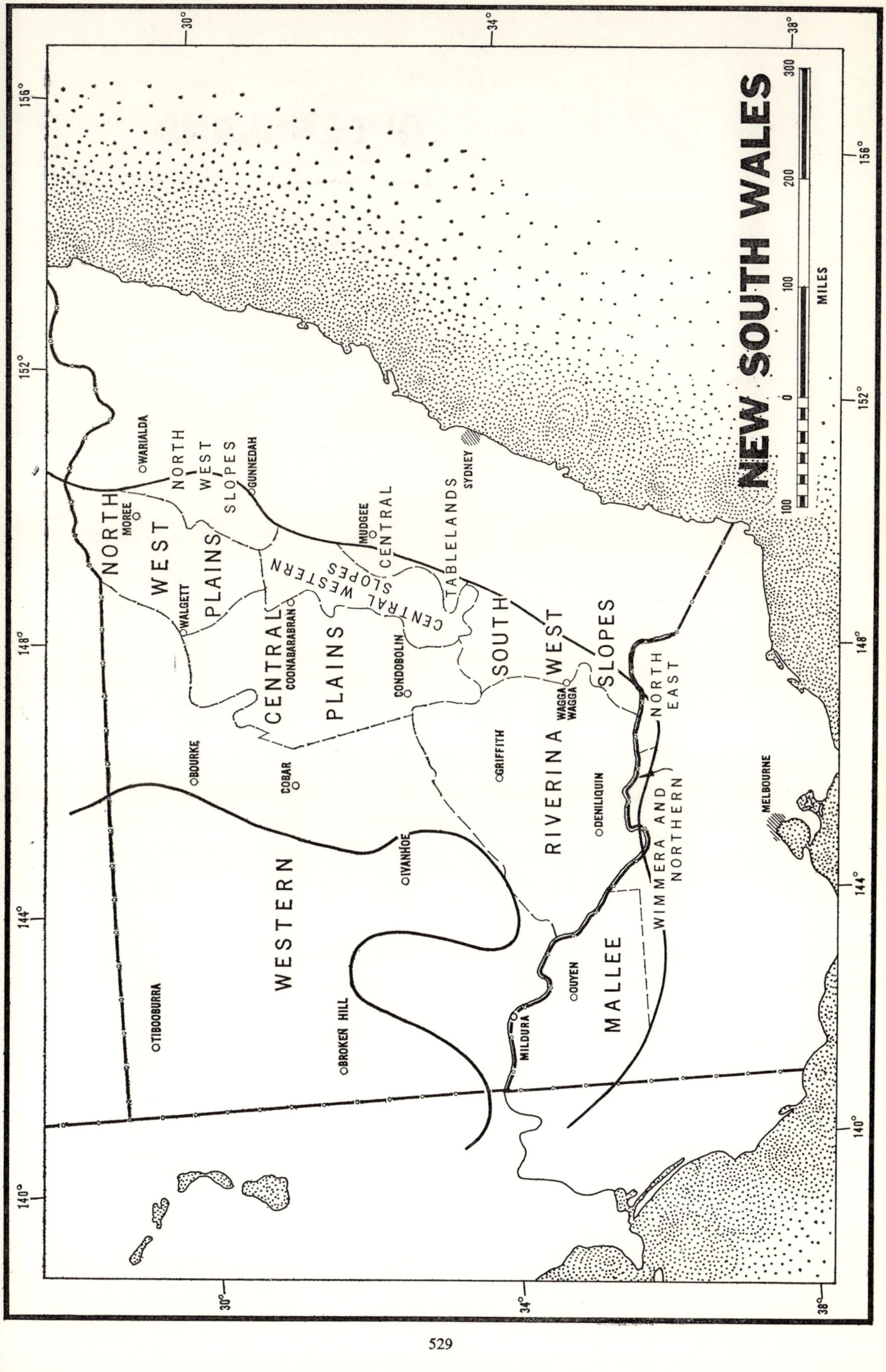
NEW SOUTH WALES
MILES
100 0 100 200 300
NORTH WEST
NORTH WEST SLOPES
PLAINS
CENTRAL TABLELANDS
CENTRAL WESTERN SLOPES
CENTRAL PLAINS
SOUTH WEST SLOPES
RIVERINA
NORTH EAST
WIMMERA AND NORTHERN
MALLEE
WESTERN
WARIALDA
GUNNEDAH
MOREE
SYDNEY
MUDGEE
WALGETT
COONABARABRAN
CONDOBOLIN
WAGGA WAGGA
GRIFFITH
BOURKE
COBAR
DENILIQUIN
MELBOURNE
IVANHOE
OUYEN
MILDURA
TIBOOBURRA
BROKEN HILL
30°
34°
38°
140°
144°
148°
152°
156°

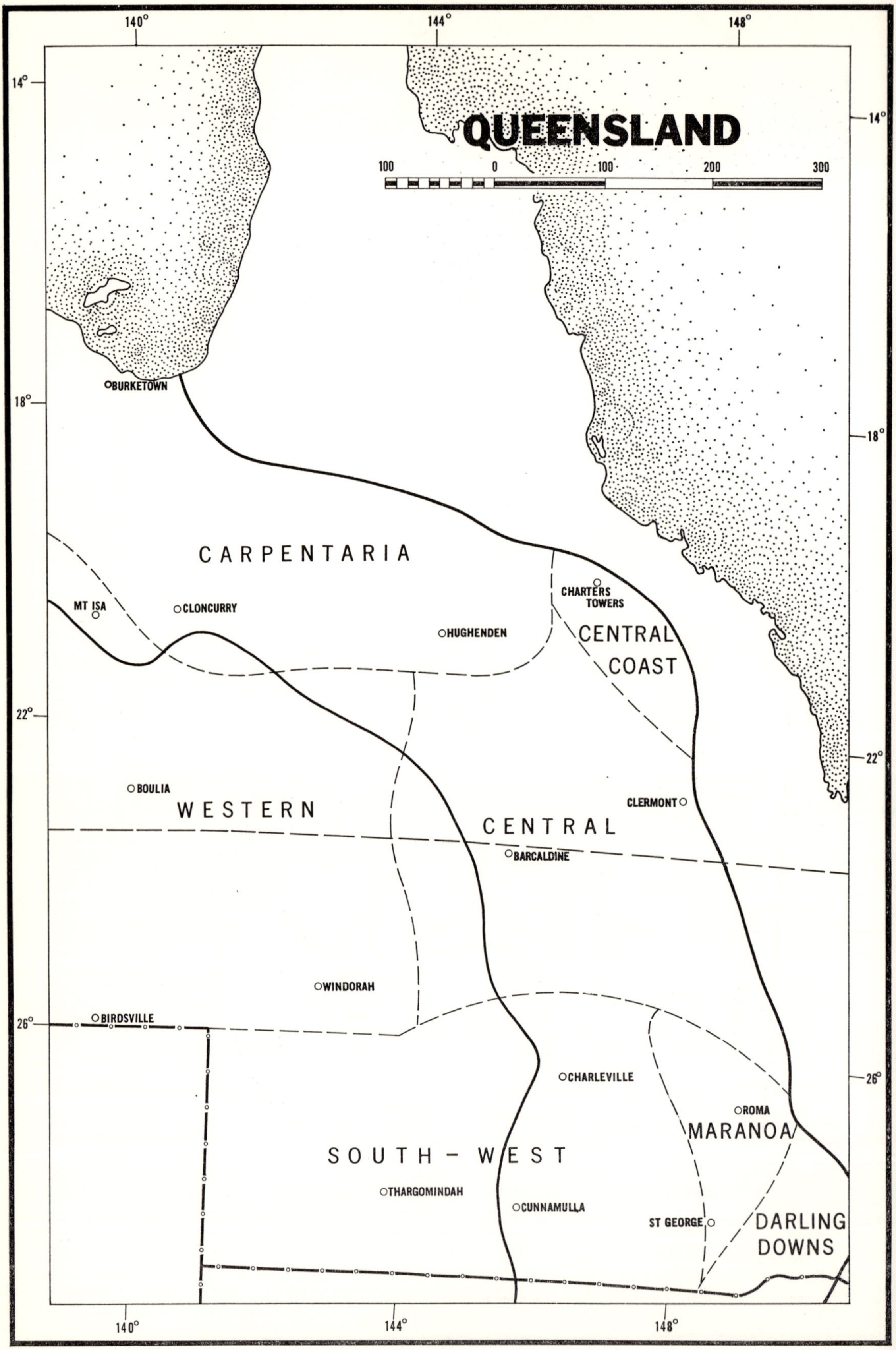
QUEENSLAND
100
0
100
200
300
140°
144°
148°
14°
18°
22°
26°
CARPENTARIA
WESTERN
CENTRAL
CENTRAL COAST
SOUTH - WEST
MARANOA
DARLING DOWNS
BURKETOWN
MT ISA
CLONCURRY
HUGHENDEN
CHARTERS TOWERS
BOULIA
CLERMONT
BARCALDINE
WINDORAH
BIRDSVILLE
CHARLEVILLE
ROMA
THARGOMINDAH
CUNNAMULLA
ST GEORGE

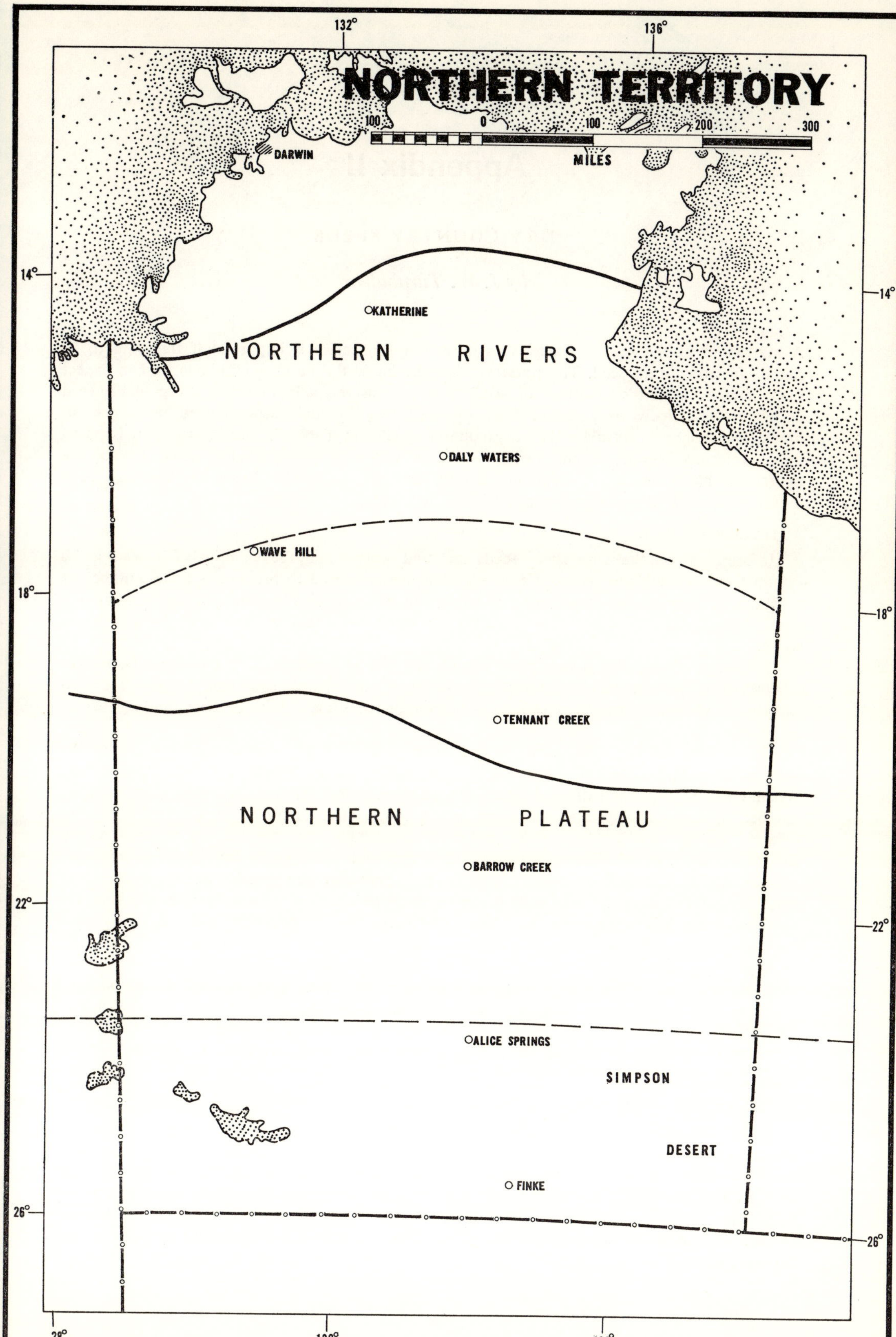
NORTHERN TERRITORY
100
0
100
200
300
MILES
132°
136°
DARWIN
14°
KATHERINE
NORTHERN RIVERS
DALY WATERS
WAVE HILL
18°
TENNANT CREEK
NORTHERN PLATEAU
BARROW CREEK
22°
ALICE SPRINGS
SIMPSON
DESERT
FINKE
26°
28°
132°
136°

36

Appendix II

DRY COUNTRY SEEDS

by J. W. Turnbull

A knowledge of the seed characteristics of individual species is important if trees are to be raised efficiently from seed. The nurseryman must know the number of plants he can expect from a given weight or number of seeds and he must be able to make every viable seed germinate as quickly as possible. Data on seed viability and recommended treatments to break dormancy are given for 117 dry country species in Table 1. The information is based on seed testing records of the Forest Research Institute, Canberra. For some species the figures are based on only a small number of tests but, used with discretion, they can be a useful guide until more comprehensive data are available.

NUMBER OF VIABLE SEEDS PER UNIT WEIGHT

Information on the number of seeds per unit weight and the viability of a seed lot is essential to calculate the approximate weight of seed required to produce a given number of seedlings. It also enables the density of sowing to be controlled so that the seedlings do not come up too thickly and compete with each other, or come up too sparsely and waste space.

The range of seed size is very great and for the species listed in Table 1 it varies from over 20,000 seeds per ounce for *Agonis*, *Callistemon*, *Melaleuca* and many *Casuarina* and *Eucalyptus* species, to less than 100 seeds per ounce for some species of *Eremophila*, *Owenia* and *Santalum*.

For larger-seeded species, such as acacias, it is a relatively simple matter to count the seeds in a given sample and carry out a germination test to determine viability percentage. When eucalypt seed is extracted from the capsule, however, it is a mixture of a small number of fully-developed viable seeds and a much larger number of unfertilised ovules referred to as 'chaff'. The viable seed and the chaff may be indistinguishable, as in *Eucalyptus gillii,* or of different colour but of the same size, as in *Eucalyptus stricklandii,* and not readily separated by sieving. It is, therefore, not common practice to remove the chaff from commercial eucalypt seed and the percentage germination is usually low although the viability of fertile seed may be very high. The percentage germination figure is rarely used for eucalypts and the viability is expressed as the number of viable seeds per unit weight of seeds plus chaff. The number of viable seeds per unit weight is also a more appropriate measure for other species with a very small seed, for example, *Melaleuca* and *Casuarina* spp.

When calculating the quantity of seed required for sowing, it must be remembered that the effective germination in the nursery will be less than that indicated by laboratory germination tests. As a general rule, it is expected that from one-third to one-half of the viable seeds will produce usable seedlings.

DORMANCY AND PRE-TREATMENT

Seeds will normally germinate when the embryo receives an adequate supply of water and oxygen at a suitable temperature. If viable seeds fail to germinate even when external conditions of water, temperature and oxygen are favourable they are said to be 'dormant' or in a state of dormancy. Dormant seeds will not germinate until they undergo some physiological or physical change. In nature these changes take place over a longer or shorter period, depending on the species, but the changes can be induced relatively quickly by various treatments.

Dormancy acts to ensure that germination takes place when conditions are such that the seedling will have the maximum probability of survival, or to permit only a percentage of seeds to germinate at any one time so that the probability that some seed will germinate under favourable conditions is increased. As seeds are often the only means of ensuring the survival of a species it is not surprising that in the course of evolution many species have developed dormancy mechanisms. Dormancy can be due to the immaturity of the embryo, impermeability of the seed coat to water and gases, mechanical prevention of embryo development for example by a hard seed coat, special requirements for temperature or light, or the presence of inhibitory substances.

Seed dormancy can present problems to the nurseryman as affected seeds often germinate irregularly over two or three years, resulting in sparse seed beds containing stock of different ages and sizes and tying up nursery area for long periods with consequent increased costs. To ensure quick and uniform germination a number of dormancy-breaking seed treatments have been developed. The treatments are usually either chemical or physical, depending on the type of dormancy.

Freshly-picked seeds of many species will often not germinate, germinate only partially, or germinate within a very restricted range of conditions. Normally this dormancy persists only a few weeks but it can be much longer and the seeds are then said to require a period of 'after-ripening'. After-ripening often occurs in dry storage but many species need moist conditions. The process of after-ripening can be hastened by a technique known as cold stratification. Cold stratification is carried out by placing the seeds between layers of moist peat, sand, vermiculite or other water-retaining media and keeping them at a temperature from 0-5°C for 21 to 90 days, or occasionally longer. This technique is successful in breaking dormancy of a wide range of eucalypts and a number of introduced conifers such as *Pinus canariensis*. Very small seeds may become lost in the stratifying medium and they are best treated by placing them on moist filter paper over vermiculite; or by soaking in water for 24 hours and allowing them to surface dry before sealing in a polyethylene bag.

The presence or absence of light controls the germination of seeds of some species. Species such as *Eucalyptus camaldulensis* and *Eucalyptus microtheca* will germinate fully only if exposed to light during germination. The germination appears to be promoted by the red end of the visible spectrum. On the other hand some seeds, such as *Banksia marginata*, are partially inhibited by light and should be germinated in absolute darkness. This factor is important in sowing seeds in the nursery as the deep sowing of light-requiring species may result in poor germination and conversely species inhibited by light should be well covered.

Failure of seeds to germinate may be the result of a hard or impermeable seed coat which prevents the entrance of water or oxygen. This characteristic is widespread in leguminous genera such as *Acacia, Cassia* or *Robinia*. To overcome seed-coat dormancy the seed coat must be made more permeable.

Mechanical Seed Treatment. When only a few seeds have to be treated a simple method is to nick the seed coat with a razor blade, or to file a small hole in it. It is sometimes recommended that the whole seed coat be carefully removed after soaking in water to soften it, for example, in some species of *Grevillea* and *Calytrix*. Larger quantities of seed can be scarified by tumbling the seed with sharp dry sand or in a drum with an abrasive lining.

Chemical Seed Treatment. Soaking seed in concentrated (i.e. undiluted, commercial grade) sulphuric acid is the most commonly employed method of chemical scarification of hard seed coats. The acid treatment has the basic disadvantage that precautions must be taken to protect the operator. Care must also be taken to remove all traces of acid from the seed by a thorough washing in running water. Chemicals such as calcium hypochlorite, sodium hydroxide, hydrochloric acid and ethyl alcohol have been used successfully to break dormancy in some species but as a general rule sulphuric acid is more effective on hard-coated species.

Hot Water Seed Treatment. A simpler and safer method of breaking seed-coat dormancy, which requires no special equipment, is to use hot water. A volume of water, equivalent to three or four times the volume of the seeds to be treated, is heated until the temperature is from 90-100°C. The hot water is then removed from the heat source and the seeds are allowed to soak in the water as it cools. It is advisable to run a small-scale test to determine the relationship between the volume of water and the weight of seed before treating a large quantity of seed of any one species. An excess of water results in slower cooling, which may result in damage to the most sensitive species. The seeds may be left in the water for 24-48 hours to imbibe; the ones that have been effectively treated will swell and those which do not swell can be re-treated. For some species, for example *Acacia botrycephala*, it has been found that a longer exposure to boiling water is advantageous. Boiling the seeds for one hour may substantially increase the germination. After the hot water treatment the seeds should be sown immediately as they cannot be dried and returned to storage. Hot water treatment is prescribed for most species of *Acacia*, *Albizia*, *Bauhinia*, *Cassia* and *Robinia*.

SEED STORAGE

The longevity of seeds is determined by morphological and physiological features of the seed and the storage environment. A seed contains an embryonic plant in which respiration and transpiration continue at a greater or lesser rate depending mainly on the temperature and humidity of the environment. Seeds have no means of replenishing the reserve food material consumed in metabolism and the loss of viability is largely governed by the rate of destruction of reserves. The most important factors which affect the metabolic rate and longevity are the species of seed, pre-storage treatment, stage of maturity, viability and moisture content when stored, the temperature regime and atmospheric composition, including humidity, of the storage chamber and the extent of infection by fungi and insects.

Only fully-ripened seeds with high viability are likely to remain viable for long periods. Seed longevity may be related to the characteristics of the seed coat; seeds with hard and impermeable coats frequently have prolonged viability.

The essence of good storage is to reduce the level of seed respiration by maintaining the critical factors, temperature and humidity, at the minimum safe and practicable level. The precise conditions must be varied to suit the species but it is rarely sufficient to control one factor at an optimum level without controlling the other.

The choice of storage method depends largely on the seed characteristics and the anticipated length of storage.

Dry Storage. All seeds that tolerate desiccation are best stored by firstly ensuring that the moisture content is reduced to an adequate level and then storing in sealed containers at low temperature. Alternative dry storage methods in common use are:

(a) *Temperature and humidity uncontrolled.* This is the simplest possible method; the seeds are stored under cover in bags with provision for ventilation and protection. It can be employed for *Acacia* spp. which are relatively insensitive to external conditions, but it is not recommended for valuable seeds.

(b) *In sealed containers.* The containers must be air-tight. Cans are frequently used although for small amounts of seed glass jars or plastic containers are suitable. The containers should be kept in an unheated room or cellar to reduce temperature fluctuations. This is a good method for relatively short term storage of most dry country species, for example *Casuarina*, *Eucalyptus* and *Melaleuca* spp.

(c) *In sealed containers at low temperature.* Controlled low levels of seed moisture and temperature provide satisfactory conditions for prolonged storage of most species. The moisture content of the seed should normally be between 7 per cent and 10 per cent and the storage temperature between 1°C and 5°C. For some species for example, *Eucalyptus microtheca*, these conditions are essential to prevent rapid loss of viability.

Moist Storage. Some seeds cannot tolerate drying and cannot normally be preserved for more than a few months unless storage temperature is controlled. This is not a common characteristic in dry country species but may be encountered with the acorns of *Quercus* spp. The best method of storing this seed is to maintain the moisture content at a steady high level by mixing the seed with moist sand, peat or vermiculite and maintaining the temperature just above freezing point. Aeration of the seed is essential because of the relatively high metabolic rate.

TABLE II.1

SUMMARY OF SEED VIABILITY, WEIGHT AND PRETREATMENT

Species	*Number of tests*	*Approx. No. of Viable Seeds per oz*	*Range of Viable Seeds per oz*	*Average No. of Seeds per oz*	*Average Germination %*	*Pretreatment(a)*	*Remarks*
Acacia acuminata	3	650	500–850	1,700	30	1	
A. aneura	9	1,300	1,000–2,000	1,950	70	1	
A. brachystachya	1	600	—	1,500	40	1	
A. cambagei	1	1,050	—	1,400	75	2	Boiling water is lethal
A. cyanophylla	4	975	650–1,300	1,400	70	1	
A. excelsa	1	225	—	625	35	1	
A. homalophylla	2	1,850	1,800–1,900	2,600	70	1	
A. ligulata	1	500	—	550	90	1	
A. microbotrya	1	825		2,400	35	1	
A. notabilis	1				75	1	
A. oswaldii	2	290	210–360	330	85	1	
A. pendula	2	975	750–1,200	1,500	65	3	Soak for 1 hour in H_2SO_4
A. peuce	1	210		250	85	1	
A. pycnantha	5	1,100	600–1,850	1,500	75	1	
A. rigens	1	4,000		5,700	70	1	
A. salicina	3	410	310–480	650	65	1	
A. spectabalis	1	850		1,000	85	1	
A. stenophylla	2	340	260–430	600	65	1	
A. tetragonophylla	1	1,150		1,900	60	1	
A. victoriae	6	775	550–950	1,050	75	1	
Agonis flexuosa	1	25,000				0	
Angophora melanoxylon	2	3,500	2,200–4,800	5,000	70	0	
Atalaya hemiglauca	1	250		330	75	0	
Atriplex nummularia	2	775	750–800	3,900	20	2, 3 or 4	Removal of seed coat plus stratification effective
Brachychiton rupestre	1	320		430	75	0	
Callistemon viminalis	1	90,000				0	
Callitris endlicheri	2	350	200–480	4,300	8	0	Dormancy above 20°C
C. glauca	2	575	380–750	1,900	30	0	Dormancy above 20°C
C. preissii ssp. *preissii*	2	175	160–190	2,200	8	0	Dormancy above 15°C
Cassia artemesiodes	2	1,200	500–1,900	1,900	65	1 or 2	
C. sturtii	1	170		1,700	10	1 or 2	
Casuarina cristata	3	5,200	900–11,000			0	
C. decaisneana	9	1,000	600–1,700	2,000	50	0	
C. dielsiana	1	5,700		7,100	80	0	
C. glauca	7	8,600	1,200–19,500			0	
C. huegeliana	4	4,700	3,000–8,200			0	
C. luehmannii	2	3,750	3,400–4,100			0	
C. stricta	4	6,100	4,700–8,700			0	
C. suberosa	1	7,300				0	
Dodonea attenuata	1	1,700		2,800	60	1	
Eremophila latrobii	1			225		1	
E. longifolia	1	50		100	50	1	
Eucalyptus argophloia	2	42,000	19,000–66,000			0	
E. alba (Northern Territory)	3	2,100	1,300–3,300			0	
E. alba (Queensland)	3	9,700	6,400–16,000			0	
E. angulosa	5	1,400	700–2,400			0	Photosensitive
E. astringens	16	2,900	1,000–4,600			0	
E. behriana	8	11,000	3,400–14,500			0	
E. brockwayi	7	10,000	6,500–15,000			0	
E. camaldulensis	132	19,000	1,200–62,000			0	Photosensitive
E. campaspe	3	6,600	2,300–9,200			0	
E. cladocalyx	14	2,600	400–4,400			0	
E. clelandii	4	8,400	4,500–15,000			2	
E. conica	1	17,000				0	
E. corrugata	4	3,100	1,700–4,400			0	
E. crebra	18	18,000	7,400–34,000			0	
E. dichromophloia	15	1,300	400–3,000			0	
E. dumosa	4	12,000	10,400–13,900			0	
E. dundasii	5	11,000	8,800–15,100			0	
E. eremophila	6	7,800	4,500–10,000			0	
E. exserta	5	1,900	2,300–19,000			0	Sensitive to excess moisture
E. flocktoniae	5	4,600	3,400–5,700			2 or 4	
E. forrestiana	3	1,100	1,000–1,300			0	
E. gamophylla	5	4,600	1,800–8,300			0	
E. gardneri	4	4,000	1,900–5,900			2	
E. gillii	1	1,200				0	
E. gonglyocarpa	2	1,900	1,600–2,200			0	
E. gracilis	5	16,000	3,600–35,700			2	
E. incrassata	4	1,100	150–3,000			0	
E. intertexta	8	4,700	3,500–8,300			0	
E. kondininensis	3	10,000	6,200–16,700			2	

(*a*) Key to pre-treatment is as follows:
0 = No pretreatment
1 = Hot water
2 = Scarify
3 = Concentrated sulphuric acid (H_2SO_4)
4 = Cold stratification

TABLE II.1—*continued*

Species	*Number of tests*	*Approx. No. of Viable Seeds per oz*	*Range of Viable Seeds per oz*	*Average No. of Seeds per oz*	*Average Germination %*	*Pre-treatment(a)*	*Remarks*
E. largiflorens	8	11,000	1,800–22,000			0	
E. lesouefii	4	4,700	1,400–8,900			0	
E. lehmannii	3	2,100	1,400–3,500			0	
E. leucoxylon	27	6,100	1,100–13,000			0	
E. longicornis	5	13,000	7,900–22,000			0	Sensitive to excess moisture
E. loxophleba	11	15,000	8,200–42,000			0	
E. melanophloia	7	3,500	600–8,100			0	
E. melanoxylon	3	8,600	4,000–16,700			0	
E. melliodora	16	8,800	1,300–32,000			0	
E. microtheca	6	16,000	10,200–28,000			0	Photosensitive; store at 5°C
E. occidentalis	13	5,100	1,400–8,500			0	
E. ochrophloia	7	2,500	1,200–4,100			0	
E. odorata	4	12,000	6,100–15,600			0	
E. oleosa	14	3,100	600–6,400			0	
E. orgadophila	3	3,300	1,600–6,200			0	
E. papuana	5	1,800	150–4,700			0	
E. pilligaensis	4	16,500	7,600–25,000			0	
E. platypus var. *platypus*	2	4,700	3,600–5,800			0	
E. platypus var. *heterophylla*	2	6,200	6,100–6,300			0	
E. polycarpa	7	1,000	300–1,900			0	
E. populnea	11	58,000	15,000–150,000			0	
E. redunca	1	4,800				0	
E. salmonophloia	13	15,000	6,500–26,000			0	
E. salubris	6	7,500	4,800–10,000			0	
E. sargentii	6	6,000	3,000–8,200			0	
E. sideroxylon ssp. *sideroxylon*	20	6,800	2,100–21,000			0	
E. spathulata	2	12,000	9,600–14,000			0	
E. stricklandii	11	3,200	1,400–4,400			0	
E. tessellaris	7	4,500	1,800–7,500			0	
E. thozetiana	3	9,800	3,400–21,000			0	
E. torquata	27	2,600	1,300–4,700			0	
E. trachyphloia	3	2,500	1,900–3,600			0	
E. viridis	6	16,000	11,000–21,000			0	
E. wandoo	6	7,200	4,300–10,200			0	
E. woollsiana	8	33,000	7,700–75,000			0	
Grevillea striata	2	600	110–1,100	2,800	20	0	
Hakea leucoptera	1	750		1,200	65	0	
Heterodendron oleifolium	1	170		950	15		
Lagunaria patersonii	1	610		875	70	0	Slow germination
Melaleuca alternifolia	3	80,000	30,000–127,000			0	
M. pauciflora	1	36,000				0	
Owenia acidula	1			75		2	
O. reticulata	1	2		2	100	2	
Petalostigma quadriloculare	1	640		750	85		
Pittosporum phylliraeoides	2	500	450–550	1,300	40		Slow germination
Santalum lanceolatum	1	6		13	50	2	
Ventilago viminalis	1	47		950	5		Cold stratification will not break dormancy

FURTHER READING

Biasutti Owen, E. (1956). *The Storage of Seeds for the Maintenance of Viability*. Bull. 43. Bur. Pastures Field Crops, Hurley.

Food and Agriculture Organisation (1955). *Tree Seed Notes: Arid Areas and Humid Tropics*. For. Development Paper 10. Rome.

Goor, A. Y., & Barney, C. W. (1968). *Forest Tree Planting in Arid Zones*. Ronald Press, New York.

Larsen, E. (1965). *Germination of Eucalyptus Seed*. Aust. For. Timb. Bur. Leafl. 94. For. Timb. Bur., Canberra.

Magini, E. (1962). 'Forest seed handling, equipment and procedures. II: Seed treatments, storage, testing and transport'. *Unasylva*, 16 : 20-35.

United States Department of Agriculture (1948). *Woody Plant Seed Manual*. Misc. Pub. 654. Washington D.C.

United States Department of Agriculture (1961). *Seeds. Year Book of Agriculture 1961*. Washington D.C.

Appendix III

CONVERSION FACTORS AND METRIC EQUIVALENTS OF ENGLISH UNITS OF MEASUREMENTS

The metric equivalents have been taken from British Standard 350 Part 1 (1944) and Part 2 (1962) and all figures have been checked on a modern calculator at the Forest Research Institute.

LINEAR

Inches to millimetres	×	25.40
Feet to metres	×	0.3048
Miles to kilometres	×	1.609 344

Inches to millimetres

0.1 in	2.540 mm
0.2	5.080
0.3	7.620
0.4	10.160
0.5	12.700
0.6	15.240
0.7	17.780
0.8	20.320
0.9	22.860
1.0	25.400
1.25	31.750
1.50	38.100
1.75	44.450

Inches to centimetres

2.0 in	5.080 cm
2.5	6.350
3.0	7.620
3.5	8.890
4.0	10.160
4.5	11.430

Inches to centimetres (Cont.)

5.0	12.700
5.5	13.970
6.0	15.240
7.0	17.780
8.0	20.320
9.0	22.860
10.0	25.400
11.0	27.940
12.0	30.480

Feet to metres

1.0 ft	0.3048 m
2.0	0.6096
3.0	0.9144
4.0	1.2192
5.0	1.5240
6.0	1.8288
7.0	2.1336
8.0	2.4384
9.0	2.7432
10.0	3.0480

Feet to metres (Cont.)

15.0	4.5720
20.0	6.0960
25.0	7.6200
30.0	9.1440
35.0	10.6680
40.0	12.1920
45.0	13.7160
50.0	15.2400
55.0	16.7640
60.0	18.2880
65.0	19.8120
70.0	21.3360
75.0	22.8600
80.0	24.3840
85.0	25.9080
90.0	27.4320
95.0	28.9560
100.0	30.4800
110	33.5280
120	36.5760
130	39.6240
140	42.6720

Feet to metres (Cont.)

150	45.7200
160	48.7680
170	51.8160
180	54.8640
190	57.9120
200	60.9600
250	76.2000
300	91.4400

Miles to kilometres

1.0 mile	1.6093 km
2.0	3.2187
3.0	4.8280
4.0	6.4374
5.0	8.0467
6.0	9.6561
7.0	11.2654
8.0	12.8748
9.0	14.4841
10.0	16.0934

Continued on next page

AREA

Square feet to square metres × 0.092 903 04
Acres to hectares × 0.404 686

Square feet to square metres

1.0 ft^2	0.09290 m^2
2.0	0.18581
3.0	0.27871
4.0	0.37161
5.0	0.46452
6.0	0.55742
7.0	0.65032
8.0	0.74322
9.0	0.83613
10.0	0.92903

Square miles to square kilometres taken from the basis of square yard to square metres × 0.836 127 36

1.0 sq mls	2.5900 sq km
2.0	5.1800
3.0	7.7700
4.0	10.3600
5.0	12.9499
6.0	15.5399
7.0	18.1299
8.0	20.7199
9.0	23.3099
10.0	25.8999

Acres to hectares

1.0	0.40469
2.0	0.80937
3.0	1.21406
4.0	1.61874
5.0	2.02343
6.0	2.42811
7.0	2.83280
8.0	3.23749
9.0	3.64217
10.0	4.04686

Square feet per acre to square metres per hectare based on one lineal foot = 0.3048 lineal metre or square foot to square metres × 0.09290304; and hectares to acres × 2.471052; sq ft/acre to sq m/hectare × 0.2296358

1 sq ft/acre	0.22964 sq m/hectare
2	0.45927
3	0.68891
4	0.91854
5	1.14818
6	1.37781
7	1.60745
8	1.83709
9	2.06672
10	2.29636

VOLUME

Cubic inches to cubic centimetres × 16.387 064
Cubic feet to cubic metres based on linear factor of 1 foot = 0.3048 metre

1.0 ft^3	0.02832 m^3
2.0	0.05663
3.0	0.08495
4.0	0.11327
5.0	0.14158
6.0	0.16990
7.0	0.19822
8.0	0.22653
9.0	0.25485
10.0	0.28317

Cubic feet per acre to cubic metres per hectare based on cubic feet to cubic metres × 0.02832; hectares to acres × 2.471052; ft^3/acre to m^3/hectare × .06997239

1.0 ft^3/ac.	0.06998 m^3/hectare
2.0	0.13996
3.0	0.20994
4.0	0.27993
5.0	0.34991
6.0	0.41989
7.0	0.48987
8.0	0.55985
9.0	0.62983
10.0	0.69982

TEMPERATURE

Degrees fahrenheit to degrees centigrade

One degree F to degree C × 0.556
One degree C to degree F × 1.800
(Adjusting for 32°F = 0°C)

°F	°C	°F	°C
32°F	0.0°C	55 °F	12.8 °C
33	0.6	60	15.6
34	1.1	65	18.3
35	1.7	70	21.1
36	2.2	75	23.9
37	2.8	80	26.7
38	3.3	85	29.4
39	3.9	90	32.2
40	4.4	95	35.0
45	7.2	100	37.8
50	10.0		

MASS

16 ounces (avoirdupois) to grammes = × 453.592 37
Pounds to kilogrammes = × 0.453 592 37

oz	gm	lb	kg
0.5 oz	14.175 gm	1.0 lb	0.4536 kg
1.0	28.350	2.0	0.9072
1.5	42.524	3.0	1.3608
2.0	56.699	4.0	1.8144
2.5	70.874	5.0	2.2680
3.0	85.049	6.0	2.7215
3.5	99.223	7.0	3.1751
4.0	113.398	8.0	3.6287
4.5	127.573	9.0	4.0823
5.0	141.748	10.0	4.5359
5.5	155.922		
6.0	170.097		
6.5	184.272	20.0	9.0718
7.0	198.447	30.0	13.6078
7.5	212.621	40.0	18.1437
8.0	226.796	50.0	22.6796
		60.0	27.2155
9.0	255.146	70.0	31.7515
10.0	283.495	80.0	36.2874
11.0	311.845	90.0	40.8233
12.0	340.194	100.0	45 3592
13.0	368.544		
14.0	396.893		
15.0	425.243		
16.0	453.592		

Index

Bold page number indicates main reference

This book is in BS Format.
Text set in 10 on 12 pt. and 8 on 9 pt. Times Roman.
Printed on Burnie English Finish, 85 gsm.
Plates printed on Gemini White Art, 98 gsm.
Bound in Winterbottom Coverex.

Designed by Kim Young, Design Section,
Australian Government Publishing Service.

R70/10

CLIMATIC REGIONS

PORTIONS OF THE STATES WHICH ARE COVERED BY SEPARATE MAPS ARE SHOWN IN OUTLINE

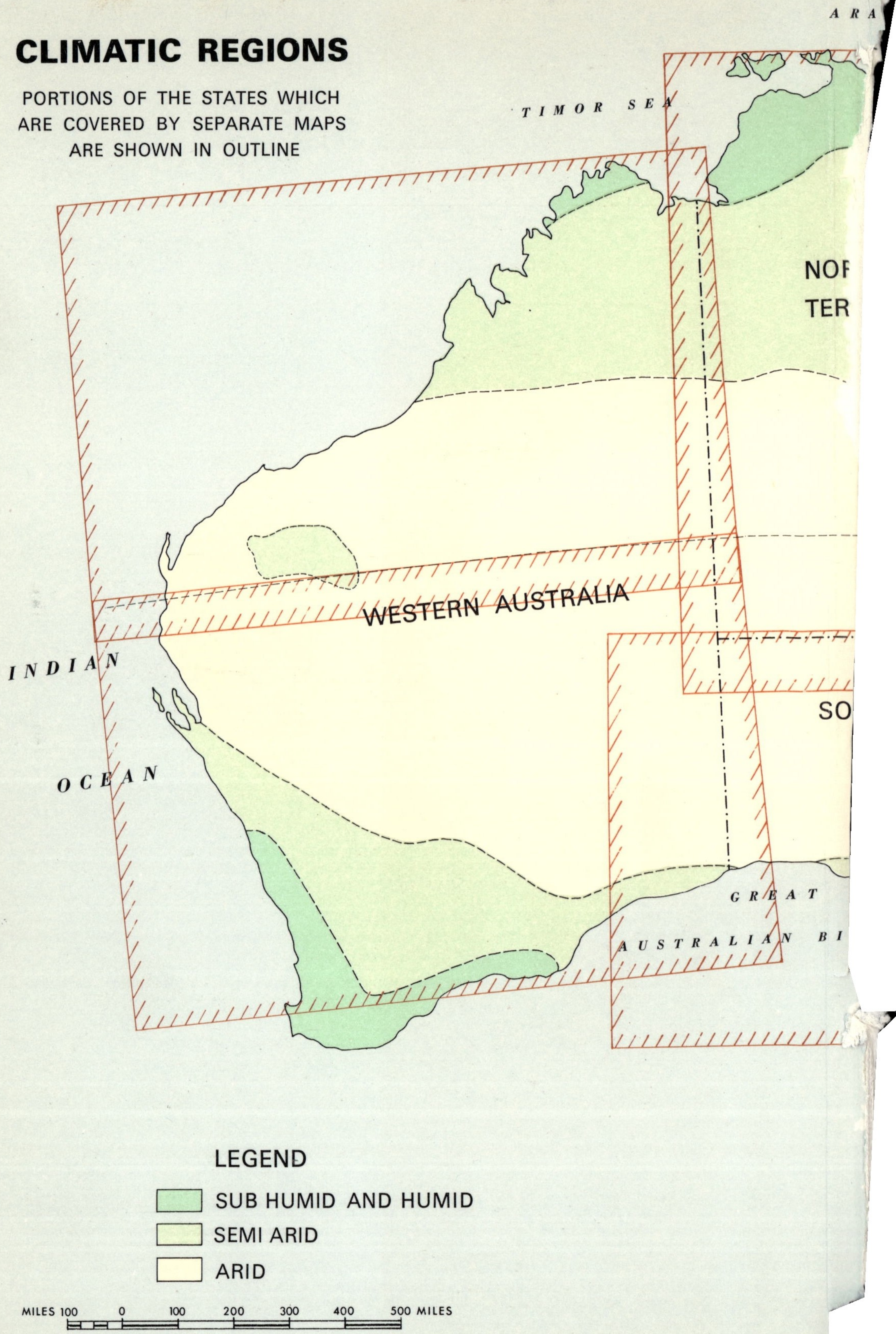